ADVANCED MATERIALS, MECHANICAL AND STRUCTURAL ENGINEERING

PROCEEDINGS OF THE 2ND INTERNATIONAL CONFERENCE OF ADVANCED MATERIALS, MECHANICAL AND STRUCTURAL ENGINEERING (AMMSE 2015), JE-JU ISLAND, SOUTH KOREA, 18–20 SEPTEMBER, 2015

Advanced Materials, Mechanical and Structural Engineering

Editors

Seung Ho Hong
Department of Civil and Environmental Engineering, West Virginia University, Morgantown, WV, USA

Junwon Seo
South Dakota State University, Brookings, SD, USA

Kihoon Moon
Korea Expressway Corporation, Gimcheon City, Republic of Korea

CRC Press is an imprint of the
Taylor & Francis Group, an **informa** business
A BALKEMA BOOK

CRC Press/Balkema is an imprint of the Taylor & Francis Group, an informa business

Typeset by V Publishing Solutions Pvt Ltd., Chennai, India

Published by: CRC Press/Balkema
P.O. Box 11320, 2301 EH Leiden, The Netherlands
e-mail: Pub.NL@taylorandfrancis.com
www.crcpress.com – www.taylorandfrancis.com

ISBN: 978-1-138-02908-8 (Hbk)
ISBN: 978-1-315-64468-4 (eBook PDF)

Advanced Materials, Mechanical and Structural Engineering – Hong, Seo & Moon (Eds)
© 2016 Taylor & Francis Group, London, ISBN: 978-1-138-02908-8

Table of contents

Advanced Materials, Mechanical and Structural Engineering – Hong, Seo & Moon (Eds)
© 2016 Taylor & Francis Group, London, ISBN: 978-1-138-02908-8

Preface

AMMSE 2015 is an annual international conference on Advanced Material, Mechanics and Structural Engineering (AMMSE 2015) which took place in Jeju Island, South-Korea, on September 18–20, 2015. The purpose of the conference was to establish platforms for collaborative research projects in the field of advanced material, mechanics and structural engineering. In the conference, engineers from different industries and researchers from various academic institutions exchanged and shared their experiences, presented their research results, explored collaborations, and sparked new ideas with the aim of developing new projects and exploiting new technologies in the field.

This book is a collection of accepted papers. The papers presented in the proceedings were peer-reviewed by 2–3 expert referees. This volume contains four main subject areas: 1. Advanced material and application studies, 2. Management systems and civil engineering application, 3. Mechanical and structures engineering applications, and 4. Sensors, hydraulic and electric power engineering. The committee of AMMSE 2015 would like to express their sincere thanks to all authors for their high-quality research papers and careful presentations. Also, we would ike to thank the reviewers for their careful comments and advices. Finally, thanks are expressed to CRC Press/Balkema as well for producing this volume.

The Organizing Committee of AMMSE 2015
Committee Chair Prof. Seung Ho Hong,
West Virginia University

Advanced Materials, Mechanical and Structural Engineering – Hong, Seo & Moon (Eds)
© 2016 Taylor & Francis Group, London, ISBN: 978-1-138-02908-8

Organization

CIVIL AND ENVIRONMENTAL ENGINEERING-WEST VIRGINIA UNIVERSITY

West Virginia University (WVU) is a public land-grant research university in Morgantown, West Virginia, United State. The total number of students and academic faculty members is 30,000 and 2,000 respectively. Based on the highest level of research activity, the Carnegie Foundation designated WVU as the Highest Research University.

Among the professional degree programs in 15 different colleges at WVU, the Civil and Environmental Engineering department is planning to be a global leader in designing, constructing, managing, and renewing built and natural environments, hereby contributing to a sustainable world, shaping public policy, and enhancing health, safety, and the quality of life. Through cross-disciplinary research, we are developing technological innovation that defines and solves complex problems at the interface of built, natural, and social systems, and engaging in service to the level of state, nation, and the world.

Advanced Materials, Mechanical and Structural Engineering – Hong, Seo & Moon (Eds)
 ISBN: 978-1-138-02908-8

Inhibitory effect of *Salvia officinalis* L. oil on Candida biofilm

B. Thaweboon & S. Thaweboon
Faculty of Dentistry, Mahidol University, Bangkok, Thailand

ABSTRACT: This study aims to determine the effect of *Salvia officinalis* L. (sage) essential oil on Candida biofilm formation in vitro. Sage oil (Hong-Huat Company, Thailand) was prepared in Tween-80 and diluted to concentrations of 0.5–100 mg/mL. Biofilms of *Candida albicans* (ATCC10231 and clinical strain) were grown in 96-well plates with Yeast Nitrogen Base medium (YNB) supplemented with 100 mM glucose in shaking incubator at 37°C for 24 h. After washing, each concentration of sage oil with YNB medium was added and further incubated for 24 h at 37°C. Evaluation of biofilm was assessed through the XTT reduction assay. A 0.2% chlorhexidine solution was used as positive control. It was found that 90% Candida biofilm reduction was demonstrated at concentrations of >5 mg/mL oil whereas chlorhexidine exhibited 89% biofilm reduction. In conclusion, our results indicate that sage oil may be a good alternative to current treatment for oral Candida infection.

1 INTRODUCTION

Fungal infection is one of the major health problems that frequently occur in the oral cavity. The consequences are exacerbated by a concomitant increase in resistance to traditional antifungal agents (Oberoi et al. 2012) due to the rise in the immunodeficient and immunocompromised populations globally. Among all causative agents, *Candida albicans* is the principal species and considered the most virulent. In this context, the identification of effective alternative therapies to the current antifungal agents is important.

The use of medicinal plants to improve oral health has been observed with increasing interest by researchers in the study of their biological properties and active ingredients responsible for their therapeutic effects. Salvia officinalis L. or sage is a member of the Lamiaceae or mint family which is widely used both in culinary and medicinal preparations. Sage leaves and its essential oil possess carminative, antispasmodic, antiseptic, anti-inflammatory, and mucolytic properties (Martins et al. 2015, Marchev et al. 2014). In dentistry, essential oil is applied for the treatment of inflammation and infections of the mucous membranes of the throat and mouth (stomatitis, gingivitis, and pharyngitis (Taheri et al. 2011). Previous studies have reported antimicrobial properties of sage oil against gram negative and gram positive bacteria, dermatophytes, and fungi (Abu-Darwish et al. 2013).

The aim of the present study was to assess the antifungal effect of sage oil against the biofilm of *C. albicans.*

2 MATERIALS AND METHODS

2.1 *Oil preparing*

Salvia officinalis L. or sage oil (Hong-Huat Company, Thailand) was prepared in Tween-80 and diluted to concentrations of 0.5–50 mg/mL.

2.2 *Candida*

Candida albicans ATCC 10231 and strain isolated from the oral lesion of patient were obtained from the culture collection of Oral Microbiology Department, Faculty of Dentistry, Mahidol University, Thailand. They were maintained on Yeast Nitrogen Base (YNB; Difco, USA) agar slant at 4°C. Few colonies from the agar were inoculated in YNB medium with 50 mM glucose and incubated at 30°C in shaking incubator, at 100 rpm for 24 h. Cells from this culture were harvested by centrifugation (3200 g, 5 min). The Cell pellet was washed twice with Phosphate Buffer Saline (PBS, pH 7.4). Candida suspensions were prepared in YNB supplemented with 100 mM glucose to yield a concentration of approximately 107 CFU/mL using McFarland standard.

2.3 *Biofilm formation*

For initial adhesion to a solid phase, 100 μL of Candida suspension was added to each well of 96-well tissue culture plate and incubated in shaking incubator (75 rpm) at 37°C for 90 min. After that, the non-adhered cells were washed 3 times with PBS and further incubated in 200 μL of YNB supplemented with 100 mM glucose for 24 h to allow biofilm formation.

2.4 *Effect of oil on biofilm*

After Candida biofilm was developed on the bottom of well, YNB medium supplemented with 100 mM glucose along with various concentrations of sage oil (0.5–100 mg/mL) was added to each well. The plate was incubated in shaking incubator for another 24 h at 37°C. Planktonic cells were removed by washing and biofilms were observed under an inverted microscope. Evaluation of biofilm was assessed through the XTT [2, 3-bis (2-methoxy-4-nitro-sulfophenyl)-2H-tetrazolium-5-carboxanilide] (Sigma-Aldrich, UK) reduction assay. XTT solution was prepared by mixing 1 mg/mL XTT salt in PBS and stored at −20°C. Prior use, menadione solution prepared in acetone was added to XTT to a final concentration of 4 μM. 100 μL of XTT-menadione solution was added to biofilm and incubated in shaking incubator in the dark at 37°C for 5 h. The color developed from the water soluble formazan product was measured at 492 nm. A 0.2% chlorhexidine gluconate solution and Tween-80 were used as positive and negative controls respectively.

All experiments were performed triplicate in three separate occasions.

3 RESULTS

The biofilm formation of standard and reference strains of *C. albicans* biofilm was demonstrated in Figure 1. It was clearly shown that all concentrations of sage oil used in the present study could affect *C. albicans* biofilm formation. 90% biofilm reduction was observed at concentration of >5 mg/mL sage oil compared to negative control using Tween-80 whereas 62% reduction was observed at 2.5 mg/mL (Table 1). For the positive control, chlorhexidine exhibited 89% biofilm reduction. There were no significant differences between the inhibitory effects on standard and clinical strain of C. albicans.

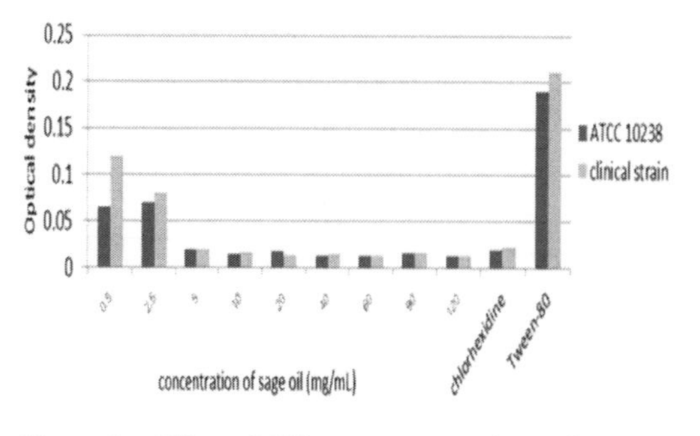

Figure 1. Effect of different concentrations of sage oil on *C. albicans* biofilm.

Table 1. Percentage of Candida biofilm reduction effect of *S. officinalis*.

S. officinalis (mg/mL)	Candida biofilm reduction (%) ATCC 10238	Clinical strain
0.5	65.26	42.86
2.5	63.16	61.90
5.0	90.00	90.95
10.0	92.10	91.90
20.0	90.53	93.33
40.0	93.16	92.86
60.0	92.63	93.81
80.0	91.58	92.38
100.0	91.05	91.90
120.0	92.63	93.81
chlorhexidine	89.47	89.05

* Values are percentage of reduction compared to negative control, Tween–80.

4 DISCUSSION

Salvia officinalis L. or sage is one of the oldest medicinal plants used in traditional medicine by different cultures to treat oral inflammation and digestive disorders (Raal et al. 2007). Essential oil extracted from the leaves has been demonstrated to have antimicrobial activities against many types of fungi including Candida (Abu-Darwish et al. 2013, Sookto et al. 2013, Pinto et al. 2007).

Candida species, especially *C. albicans*, are regarded as the most common opportunistic pathogens in oral cavity. One of the major factors contributing to the virulence of these fungi is their flexibility in adapting to a variety of different environments and attachment to the surface and growing in microbial communities known as biofilm. From the perspective of disease pathogenesis, the most important feature of biofilm development is the high resistance to antimicrobial agents that can be up to 1000-fold greater than that of planktonic cells. In the present study, sage oil exhibited inhibitory effects against biofilm of both strains of *C. albicans*. (90% biofilm reduction was demonstrated at the concentration of 5–120 mg/mL of the oil whereas 62% reduction was observed at 2.5 mg/mL (the concentration reported previously as MIC and MLC values of planktonic cells) (Sookto et al. 2013). However, the inhibition was not in a dose-dependent manner. The higher concentration of oil did not result in greater inhibition.

This may be due to the penetration limitation of oil into biofilm matrix to reach the target cells or the occurrence of some resistant microbial cells.

Many investigations have confirmed the pharmaceutical importance of sage due to its diverse bioactivities. The pharmaceutical important compounds of sage oil are sterols, phenolics (caffeic acid, rosmarinic acid, luteolin), and terpenoids (1,8-cineole, cis- and trans-thujone, camphor, oleanolic acid, ursolic acid, carnosic acid) (Martins et al. 2015, Abu-Darwish et al. 2013). However, environmental conditions such as growing conditions (soil, climate, light, altitude), harvesting, and processing are factors that can directly influence quantitative compositions of the oil (Farhat et al. 2009).

Since essential oils are complex mixtures of several compounds, it is difficult to attribute their biological activity to a particular constituent. Usually, major compounds are the ones responsible for the antifungal activity of the essential oils. However, some studies show that minor components may have a crucial role in the biological activity of the oils (Koroch et al. 2007). In the case of sage oil, cisthujone, 1,8-cineole, and camphor are suggested to be the main components responsible for its antifungal activity against Candida and other filamentous fungi (Pinto et al. 2007). However, there are very limited numbers of papers dealing with the mechanisms of antifungal property of sage oil. It has been proposed that the interaction of essential oil with lipid components in the cellular structure of yeast cells, thereby increasing membrane permeability and electrolyte imbalance impairs initial adhesion of yeast cells and then affects biofilm formation (Braga et al. 2008).

Considering the toxicity of sage oil, it is Generally Recognized As Safe (GRAS) by the Food and Drug Administration in USA (2012) and drugs and tinctures prepared from these plants are listed in several official Pharmacopoeias (Marchev et al. 2014). Previous study in cell cultures stated that bioactive concentrations of sage oil did not affect mammalian macrophages and keratinocytes viability, making them suitable to be incorporated in skin care formulations for cosmetic and pharmaceutical purposes (Abu-Darwish et al. 2013).

5 CONCLUSION

In this part results from this study demonstrated that *S. officinalis* oil at concentrations of 5–120 mg/mL could inhibit (90% of *C. albicans* biofilm formation. The effect was similar on both standard strain and strain isolated from the patient's lesion. It may be a good alternative to current treatment for oral Candida infection. However, further studies are required to evaluate the effect on other yeast strains or cytotoxicity before considering for use in patients.

ACKNOWLEDGEMENT

This study was supported by grant from Mahidol University Faculty of Dentistry, Thailand (DTID 451).

REFERENCES

Abu-Darwish, M.S., Cabral, C. & Ferreira, I.V. et al. 2013. Essential oil of common sage (Salvia officinalis L.) from Jordan: assessment of safety in mammalian cells and its antifungal and anti-inflammatory potential. *BioMed Research International,* 2013: 1–9.

Braga, P.C., Culici, M., Alfieri, M. & Dal, S.M. 2008. Thymol inhibits Candida albicans biofilm formation and mature biofilm. *International Journal of Antimicrobial Agents,* 31: 472–477.

Farhat, M.B., Jordan, M.J., Chaouch-Hamada, R., Landoulsi, A. & Sotomayor, J.A. 2009. Variations in essential oil, phenolic compounds, and antioxidant activity of Tunisian cultivated Salvia officinalis L. *Journal of Agricultural and Food Chemistry,* 57: 10349–10356.

Koroch, A.R., Zyglado, J.A. & Juliani, H.R. 2007. *Bioactivity of essential oils and their components.* In R.G. Berger (ed.), Flavours and Fragrances. Chemistry, Bioprocessing and Sustainability: 87–115. Berlin: Springer-Verlagpp.

Marchev, A., Haas, C. & Schulz, S. et al. 2014. Sage in vitro cultures: a promising tool for the production of bioactive terpenes and phenolic substances. *Biotechnology Letter,* 36: 211–221.

Martins, N., Barros, L., Santos-Buelga, C., Henriques, M., Silva, S. & Ferreira, I.C. 2015. Evaluation of bioactive properties and phenolic compounds in different extracts prepared from Salvia officinalis L. *Food Chemistry,* 1: 378–385.

Oberoi, J.K., Wattal, C., Goel, N., Raveendran, R. Datta, S. & Prasad, K. 2012. Non-albicans Candida species in blood stream infections in a tertiary care hospital at New Delhi, India. *Indian Journal Medical Research,* 136: 997–1003.

Pinto, E., Salgueiro, L.R., Cavaleiro, C., Palmeira, A. & Goncalves, M.J. 2007. In vitro susceptibility of some species of yeasts and filamentous fungi to essential oils of Salvia officinalis. *Industrial Crops and Products,* 26: 135–141.

Raal, A., Orav, A. & Arak, E. 2007. Composition of the essen-tial oil of Salvia officinalis L. from various European countries. *Natural Product Research,* 21: 406–411.

Sookto, T., Srithavaj, T., Thaweboon, S., Thaweboon, B. & Shrestha, B. 2013. In vitro effects of Salvia officinalis L. essential oil on Candida albicans. *Asian Pacific Journal of Tropical Biomedicine,* 3: 376–380.

Taheri, J.B., Azimi, S., Rafieian, N. & Zanjani, H.A. 2011. Herbs in dentistry. *International Dental Journal,* 61: 287–296.

Advanced Materials, Mechanical and Structural Engineering – Hong, Seo & Moon (Eds)
© 2016 Taylor & Francis Group, London, ISBN: 978-1-138-02908-8

Adhesion and biofilm formation of *Streptococcus mutans* on dental sealant incorporated with silver-nanoparticle

S. Thaweboon & B. Thaweboon
Faculty of Dentistry, Mahidol University, Bangkok, Thailand

ABSTRACT: This study aims to investigate the adhesion and biofilm formation of *Streptococcus mutans* on dental sealant incorporated with silver-nanoparticle. Dental sealants containing 0.5%, 1.0% and 1.5% $AgZrPO_4$ nanoparticle were prepared. Mixture of bacterial suspension (2×10^6 cells/mL) and pooled saliva was added onto dental sealant. For adhesion assay, the plate was incubated in shaking incubator at 37°C for 3 h. Biofilm formation was done as the adhesion assay and further incubated for 24 h. The amount of adhering bacteria and biofilm formation was determined by crystal violet technique. Significant reduction of *S. mutans* biofilm formation was observed in the groups of 0.5%, 1% and 1.5% silver-nanoparticle compared with control even though no inhibitory effect was observed on initial adhesion. The incorporation of low concentrations of silver-nanoparticle to dental sealant clearly exhibited the suppressive effect on *S. mutans* biofilm formation. The use of these sealants could have a potential for caries prevention.

1 INTRODUCTION

Dental caries is one of the serious public health problems affecting children and adults worldwide. The prevalence of the disease is high in children. It is the most important oral disease and is of medical, social and economic importance. The disease is the result of a dental decay process in which mainly acidogenic and aciduric bacteria residing in a complex biofilm degrade tooth structure, leading to demineralization and cavitation. Unlike classical infectious diseases, which are caused by microbial pathogens, dental caries is caused by resident oral microflora. *Streptococcus mutans* appears to play an important role in the initiation of dental caries since its activities lead to colonization of the tooth surface, dental plaque or oral biofilm formation and demineralization of tooth enamel by acids (Loesche. 1986).

Dental sealant which is a material introduced into the occlusal pits and fissures of the teeth as a protective layer has been widely used to prevent dental caries. According to American Dental Association expert panel's conclusion, the reduction of dental caries incidence in children and adolescents after placement of pit and fissure sealant was reported from 60–86% (Beauchamp et al. 2008). In addition, application of dental sealant could delay the progression of carious lesions in adolescents and adults (Beauchamp et al. 2008). On the other hand, the miroleakage and microbial effect on food residue on sealant is also reported as the major cause of secondary caries on these tooth surfaces (Simonsen & Neal. 2011). The incorporation of the antimicrobial agent to sealant materials could have the potential benefit of additional caries protection. Today, the silver-nanoparticle is going favoured as an antimicrobial agent due to its broad spectrum, low toxicity and lack of bacterial resistance (Peng et al. 2012). Therefore, the aim of the present study was to investigate the adhesion and biofilm formation of *S. mutans* on dental sealant incorporated with silver-nanoparticle.

2 MATERIALS AND METHODS

2.1 *Microorganisms*

Streptococcus mutans KPSK2 was obtained from a culture collection of Oral Microbiology Department, Faculty of Dentistry, Mahidol University, Thailand. The bacteria were grown in 50 mL of brain heart infusion broth at 37°C in 5% CO_2 atmosphere for 24 h. After that, they were centrifuged at 5000 rpm for 15 min and the pellet was resuspended in phosphate buffer saline solution to 2×10^6 cells/mL.

2.2 *Saliva*

Paraffin stimulated whole saliva was collected from four healthy volunteers who had good oral hygiene, no history of systemic disease and not taken any antibiotics or other drugs during and two weeks prior to the experiment. The saliva samples were pooled, centrifuged at 10,000 rpm for 20 min to remove cell debris and then filtered through the Millipore membrane (0.22 μm).

2.3 *Dental sealant*

Dental sealant used in this study was Helioseal fissure sealant (Ivoclar Vivadent, USA). The sealant composition was: bisphenol A glycidyl methacrylate (Bis-GMA), triethyleneglycol dimethacrylate (TEGDMA) and titanium dioxide. Silver zirconium phosphate ($AgZrPO_4$) nanoparticles (National Direct Network Company, Thailand) were added to dental sealant at the concentrations of 0.5%, 1.0% and 1.5% w/w. Dental sealant samples were prepared on flat bottoms of 96-well plate.

2.4 *Adhesion assay*

One hundred μL of bacterial suspension and 100 μL of pooled saliva were mixed together and added onto the prepared dental sealant with different concentrations of silver nanoparticle on 96-well plates. The plates were incubated in shaking incubator at 37°C, 100 rpm for 3 h. After washing to remove the non-adhered bacterial cells, the adhered cells were fixed by 99% methanol and stained with 0.5% crystal violet for 20 min. Two hundred μL of 33% acetic acid was added and the optical density was determined at 575 nm.

2.5 *Biofilm assay*

The experiment was done as in the adhesion assay. Two hundred μL of brain heart infusion broth supplemented with 50 mM glucose was added to the adhered bacterial cells on dental sealant and further incubated in shaking incubator at 37°C, 100 rpm for 24 h. For biofilm quantification, the plates were immersed in deionized water and slowly shaken to remove remaining planktonic or loosely bound bacterial cells. The amount of biofilm formation was evaluated by staining with crystal violet and measuring the optical density at 575 nm.

Dental sealant without a silver nanoparticle was used as a control.

2.6 *Statistical analysis*

Experiments were done triplicate in three separate occasions. Comparison of S. mutans adhesion on dental sealants was analyzed by one-way analysis of variance (ANOVA) and Dunnett's test to determine significant differences. In the evaluation of biofilm formation, Kruskal Wallis rank analysis of variance was applied to assess any differences among groups. Pairwise comparisons were performed by Mann-Whitney U tests. Data were considered statistically significant at p-value <0.05M.

Table 1. *S. mutans* adhesion on dental sealants containing various concentrations of silver-nanoparticle.

Dental sealant with $AgZrPO_4$	Optical density (mean ± SD)	p-value
0.5%	0.622 ± 0.003	0.54
1.0%	0.607 ± 0.004	0.85
1.5%	0.621 ± 0.008	0.61
Control	0.612 ± 0.005	

Table 2. *S. mutans* biofilm formation on dental sealants containing various concentrations of silver-nanoparticle.

Dental sealant with $AgZrPO_4$	Optical density (mean rank)	p-value
0.5%	5.33	0.001
1.0%	5.00	0.000
1.5%	5.00	0.000
Control	13.67	

3 RESULT

No significant difference of *S. mutans* adhesion was observed among any groups of dental sealant incorporated with silver-nanoparticle or control group, silver-nanoparticle free sealant (Table 1). However, significant inhibition of *S. mutans* biofilm development was demonstrated in the groups of sealant containing 0.5%, 1% and 1.5% silver-nanoparticle compared with control (Table 2). The reduction was not different among each concentration of silver nanoparticle.

4 DISCUSSION

The adhesion of cariogenic bacteria such as *S. mutans* to tooth surfaces is the primary and essential prerequisite for the development of biofilm related to dental caries. It was found that secondary caries under and around restorative filling materials (Sarrett. 2007, Friedl et al. 1995). Therefore, materials with a low susceptibility to bacterial adhering or biofilm developing are preferable.

The antimicrobial properties of silver have long been recognized and used extensively in applications for disinfecting medical devices and home appliances for water treatment (Bosetti et al. 2002, Li et al. 2008). Several kinds of inorganic materials containing silver using different carriers such as zeolite, activated carbon and phosphate were developed to extend lifetime of silver. The prolonged constant dissociation of silver ions into the surrounding environment provides a higher clinical efficacy of these silver containing materials. In addition, these materials do not cause toxic effect on humans or disturb the color of the products

(Sekhon & Kambojet. 2010 a,b). In dentistry, silver particles have been developed and investigated for a range of possible applications, for example, incorporated into denture base materials (Bajracharya et al. 2014), dental implants (Almaguer-Flores et al. 2010) and filling materials (Burgers et al. 2009). In our study, the incorporation of 0.5–1.5% silver-nanoparticles to dental sealant did not show any inhibitory effect on bacterial initial adhesion. This was contrary to the result of previous study (Burgers et al. 2009) which demonstrated that the addition of microparticulate silver to resin composite could reduce the number of adhering bacteria. The difference may be from the artificial saliva employed in their initial adhesion assay and whole saliva used in the present study since multiple proteins in whole saliva mediate the adhesion of bacteria to the surfaces (Rudney. 1999).

Considering the prolonged outcome of bacterial biofilm development, dental sealant containing silver-nanoparticles clearly exhibited the suppressive effect compared with control without silver particle (Table 2). However, no significant difference was observed among each concentration of silver. Our data seem consistent with other previous studies that have reported antibacterial efficacy of dental resin incorporating silver-nanoparticles on cariogenic biofilm (Cheng et al. 2012, Akhavan et al. 2013). In addition, the activity has been shown to be time-dependent, such that extending the length of silver-nanoparticles exposure to 24 h enhanced the inhibitory effects on microorganisms (Bahador et al. 2013). Several mechanisms have been proposed for the antimicrobial property of silver-nanoparticles: (1) adhesion of particles to the surface altering membrane properties. Silver-nanoparticles have been reported to degrade lipopolysaccharide molecules, accumulate inside the membrane forming pits and increase membrane permeability (Sondi & Salopek-Sondi. 2004); (2) particles penetrating inside the bacterial cell, resulting in DNA damage; (3) interaction of particles with thiol and carbonyl groups in proteins, resulting in inactivation of respiratory chains of bacteria, ATP production and growth inhibition (Chen et al. 2011, Li et al. 2011).

Regarding adverse effect on the materials, it has been shown that the low amount of silver-nanoparticles (0.5%) incorporated to dental resin did not change color or mechanical properties of the polymer (Bajracharya et al. 2014, Monteiro et al. 2012).

Results from our investigation suggest potential anti-biofilm property of dental sealant containing low concentrations of silver-nanoparticle (0.5–1.5%). However, it should be noted that the activities of materials were influenced by silver particles dispersing as well as the homogeneous incorporation into the resin.

ACKNOWLEDGEMENT

The authors would like to thank the Faculty of Dentistry, Mahidol University for financial support (DTID 451 Research fund).

REFERENCES

Akhavan, A., Sodagar, A., Mojtahedzadeh, F. & Sodagar, K. 2013. Investigating the effect of incorporating nanosilver/nanohydroxyapatite particles on the shear bond strength of orthodontic adhesives. *Acta Odontologica Scandinavica*, 71: 1038–1042.

Almaguer-Flores, A., Ximenez-Fyvie, L.A. & Rodil, S.E. 2010. Oral bacterial adhesion on amorphous carbon and titanium films: effect of surface roughness and culture media. *Journal of Biomedical Material Research*, 92: 196–204.

Bahador, A., Sodagar, A., Azizy, B., Kassaee, M.Z., Pourakbari, B. & Arab, S. 2013. Anti-cariogenic effect of polymethylmethacrylate with in situ generated silver nanoparticles on planktonic and biofilm bacteria. *Annals of Biologica Research*, 4: 211–219.

Bajracharya, S., Thaweboon, S., Thaweboon, B., Wonglamsam, A. & Srithavaj, T. 2014. In vitro evaluation of Candida albicans biofilm formation on denture base PMMA resin incorporated with silver nanoparticles and its effect on flexural strength. *Advanced Materials Research*, 905: 51–55.

Beauchamp, J., Caufield, P.W., Crall, J.J., Donly, K., Feigal, R. & Gooch, B. et al. 2008. Evidence-based clinical recommendations for the use of pit and fissure sealant: a report of the American Dental Association Council on Scientific Affairs. *Journal of the American Dental Association*, 139: 257–268.

Bosetti, M., Masse, A., Tobin, E. & Cannas, M. 2002. Silver coated materials for external fixation devices: in vitro biocompatibility and genotoxicity. *Biomaterials*, 23: 887–892.

Bürgers, R., Eidt, A., Frankenberger, R., Rosentritt, M., Schweikl, H. & Handel, G. et al. 2009. The anti-adherence activity and bactericidal effect of microparticulate silver additives in composite resin materials. *Archives of Oral Biology*, 54: 595–601.

Chen, M., Yang, Z., Wu, H., Pan, X., Xie, X. & Wu, C. 2011. Antimicrobial activity and the mechanism of silver nanoparticle thermosensitive gel. *International Journal of Nanomedicine*, 6: 2873–2877.

Cheng, L., Weir, M.D., Xu, H.H., Kraigsley, A.M., Lin, N.J. & Lin-Gibson, S. et al. 2012. Antibacterial and physical properties of calcium-phosphate and calcium-fluoride nanocomposites with chlorhexidine. *Dental Materials*, 28: 573–583.

Friedl, K.H., Hiller, K.A. & Schmalz, G. 1995. Placement and replacement of composite restorations in Germany. *Operative Dentistry*, 20: 34–88.

Li, D., Diao, J., Zhang, J. & Liu, J. 2011. Fabrication of new chitosan-based composite sponge containing silver nanoparticles and its antibacterial properties for wound dressing. *Journal of Nanoscience and Nanotechnology*, 11: 4733–4738.

Li, Q., Mahendra, S., Lyon, D.Y., Brunet, L., Liga, M.V., Li, D. & Alvarez, P.J. 2008. Antimicrobial nanomaterials for water disinfection and microbial control: potential applications and implications. *Water Research*, 42: 459–602.

Loesche, W.J. 1986. Role of *Streptococcus mutans* in human dental decay. *Microbiology Reviews*, 50: 353–380.

Monteiro, D.R., Gorup, L.F., Takamiya, A.S., de Camargo, E.R., Filho, A.C. & Barbosa, D.B. 2012. Silver distribution and release from an antimicrobial denture base resin containing silver colloidal nanoparticles. *Journal of Prosthodontics*, 21: 7–15.

Peng, J.J. Botelho, M.G. & Matinlinna, J.P. 2012. Silver compounds used in dentistry for caries management: a review. *Journal of Dentistry*, 40: 531–541.

Rudney, J.D., Hickey, K.L. & Ji, Z. 1999. Cumulative correlations of lysozyme, lactoferrin, peroxidase, S-IgA, amylase, and total protein concentrations with adherence of oral viridans streptococci to micropiates coated with human saliva. *Journal of Dental Research*, 78: 759–768.

Sarrett, D.C. 2007. Prediction of clinical outcomes of restorations based on in vivo marginal quality evaluation. *The Journal of Adhesive Dentistry*, 9: 117–120.

Sekhon, B.S. & Kamboj, S.R. 2010. Inorganic nanomedicine-part 1. *Nanomedicine*, 6: 516–522.

Sekhon, B.S. & Kamboj, S.R. 2010. Inorganic nanomedicine-part 2. *Nanomedicine*, 6: 612–618.

Simonsen, R.J. & Neal, R.C. 2011. A review of the clinical application and performance of pit and fissure sealants. *Australian Dental Journal*, 56: 45–58.

Sondi, I. & Salopek-Sondi, B. 2004. Silver nanoparticles as antimicrobial agent: a case study on *E. coli* as a model for Gram-negative bacteria. *Journal of Colloid and Interface Science*, 275: 177–182.

Advanced Materials, Mechanical and Structural Engineering – Hong, Seo & Moon (Eds)
 ISBN: 978-1-138-02908-8

Research on turning Ni-based super alloys with green cooling and lubricating technology

H. Wang, J.J. Liu & F.S. Ni
Engineering Research Center of Dredging Technology of Ministry of Education, Hohai University, Changzhou, China

ABSTRACT: The machinability of nickel-based super alloy GH4169 is poor. Traditional machining methods involve the application of cutting fluids with the active additives, which causes environmental pollution and health problems. In this paper, the MQL (Minimum Quantity Lubrication) using organic alcohols was applied on turning instead of cutting fluids for green cutting GH4169. The effects of the MQL, wet and dry cutting applications on cutting force, chip deformation and tool wear were examined using the coated carbide tool YBG202. The results of the experiments indicated that application of the MQL could reduce tool wear and produced lower cutting force, and the values were reduced by about 20 percent compared with dry cutting, and it is has an advantage in chip curl. The results indicate that clean production was achieved in metal cutting associated with MQL using alcohols as cooling and lubricating fluids.

1 INTRODUCTION

Nickel-based super alloy GH4169 is widely employed in the aerospace industry, particularly in the hot sections of gas turbine engines, due to their high-temperature strength and high corrosion resistance. They are known to be one of the most difficult-to-cut materials and their machinability is poor, especially the cutting force, high cutting temperature and short tool life (Dudzinski et al. 2004, Ezugwu et al. 2003).

Traditional machining methods for nickel-based super alloys involve the application of cutting fluids with active additives, which can reduce the negative effects of heat and friction on both the tool and the workpiece and also decrease the cutting force and cutting temperature (Trent 1991). However, they lead to environmental and health problems. In addition, the cost of dealing with disposal is expensive and increases the production costs. With the development of societal environmental awareness and governmental regulation, the negative effects of conventional cutting fluids on the environment and human health should be considered in metal machining.

In the 21st century, the machining technology is expected to be clean, ecological and low energy cost. In recent decades, machining without the application of any cutting fluids (dry cutting or green cutting) is becoming more and more popular. But there are still some problems about the dry cutting of nickel-based super alloys; one such problem is the high temperature in the cutting zone that negatively affects the tool life. Therefore, the new green cutting technique, which is ecological and economical and has superior cooling and lubricating parameters, has become the focus of attention in metal machining, such as minimum quantity of lubricant (Kamata & Obikawa 2007), spray cooling (Ezugwu & Bonney 2004), cryogenic cooling (Wang & Rajurkar 2000) and water vapor (Han et al. 2008). All of them have positive effects of machining nickel-based super alloys, can decrease the cutting force and cutting temperature, reduce the tool wear, and increase the production efficiency.

In this study, the application of the MQL (Minimum Quantity Lubrication) using organic alcohols as a coolant and lubricant was investigated. The effects of the MQL on cutting force F_c and chip deformation were examined, compared with cutting fluids and dry cutting conditions in turning of GH4169 with the coated carbide tool YBG202.

2 CUTTING TEST

2.1 *Machine, tool and workpiece*

The experiment was set up on a universal turning machine CA6140 with a vertical milling head, powered by a 7.5 Kw electric motor, with a speed range of 10-1400 rev/min and a feed range of 0.014-3.16 mm/rev.

The insert is a carbide tool YBG202 coated with MT-TiCN, Al_2O_3 and TiN. The geometric parameters of the tool are listed in Table 1.

Ni-based super alloy GH4169 was selected as the workpiece in this experiment. The chemical composition of GH4169 is given in Table 2, and its physical and mechanical properties are provided in Table 3.

2.2 *Experimental system*

The experimental system is shown in Figure 1. The cutting test was investigated by a single-factor method. The cutting force was measured by a piezoelectric dynamometer and charge amplifier YE5850 manufactured by the Dalian University of Technology, China. The tool wear VB was measured by a CCD microscope shown in Figure 2.

Table 1. Tool geometry parameters.

Parameter	Value
Rake angle (γ_0)	14°
Clearance angle (α_0)	6°
Major cutting edge angle (κ_r)	60°
Minor cutting edge angle (κ_r')	30°
Inclination angle (λ_s)	–6°
Cutting edge radius (r_ε)	0.5 mm

Table 2. Chemical components of the workpiece (W_t)%.

Ni	Cr	Nb	Mo	Ti	C	Al	Si	Mn
51.75	17.00	5.11	2.93	1.04	0.042	0.41	0.21	0.03

Table 3. Physical and mechanical properties of the workpiece.

Property	Value
Yield strength σ_s (MPa)	1260
Tensile strength σ_s (MPa)	1430
Elongation δ(%)	24
Cross-section contraction ratio ϕ (%)	40
Hardness HBS	390

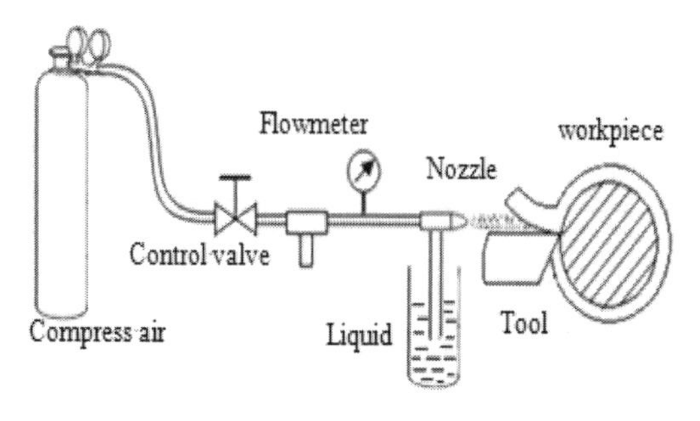

Figure 1. Schematic of the experimental system.

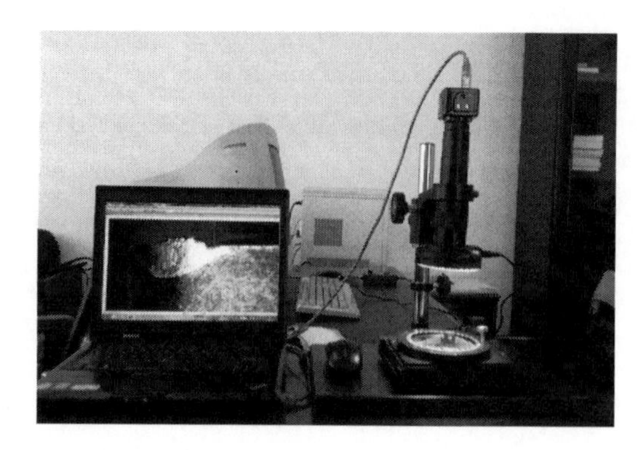

Figure 2. The tool wear measuring device.

2.3 *Cooling and lubricating condition*

The experiments during the MQL, wet and dry cutting were carried out using the same cutting parameters. The cutting fluids were prepared as water-based emulsion that was applied as flood cooling and lubrication. The convention was 5%, the temperature was 2°C, the flux of emulsion was 350 ml/min, and the diameter of the pipeline was 5 mm. The cutting fluids in the MQL were processed by 30% ethylene glycol, 20% ethanol, 20% glycerol and water, and all of the composition of substances were green and healthy. The fluids flow in the MQL was 2.0 ml/s. The pressure of the MQL was 0.25 MPa, During the application of the MQL, the distance between the nozzle and the cutting zone was 20 mm.

2.4 *Methods and cutting parameters*

In this study, the experiments were investigated by the single-factor method. The cutting force was measured by a piezoelectric dynamometer. The process was as follows: First, fastening the cutting velocity $v_c = 44$ m/min and the feed $f = 0.1$ mm/z, varying the depth of cut $a_p = 0.4,0.8,1.2,1.6$ mm; then fastening $v_c = 44$ m/min and $a_p = 0.8$ mm, varying $f = 0.05,0.1,0.15,0.2$ mm/r; next, fastening $a_p = 0.8$ mm and $f = 0.1$ mm/z, varying $v_c = 35, 44, 55, 70$ m/min; lastly, fastening $a_p = 1$ mm and $f = 0.1$ mm/z, $v_c = 66$ m/min, measured the tool wear under different lubrication conditions.

3 RESULTS AND DISCUSSION

3.1 *Cutting force results*

The variety of cutting force with cutting parameters is showed in Figures 3 and 5.

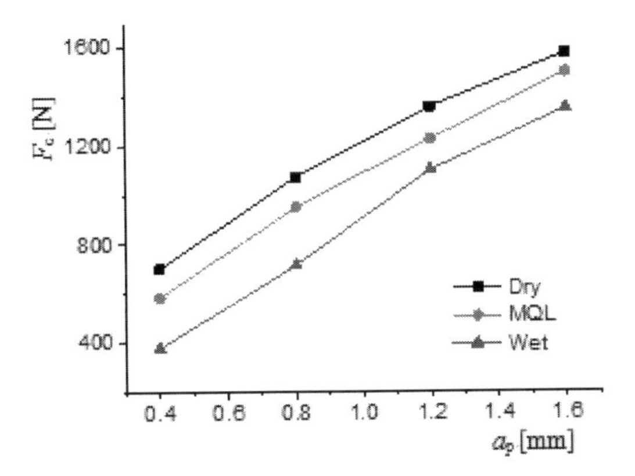

Figure 3. The curve of Fc-a_p.

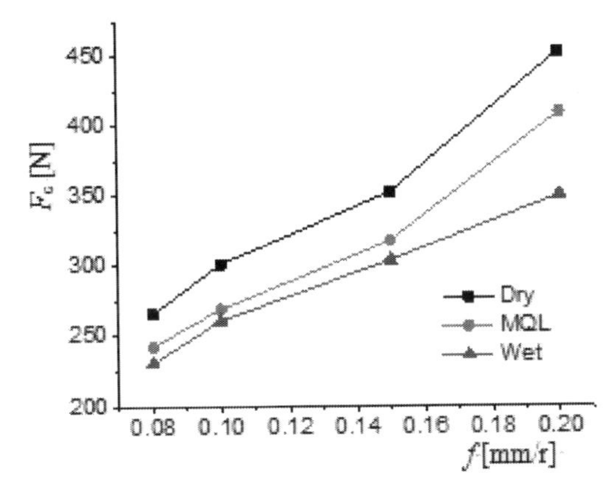

Figure 4. The curve of F_f-f.

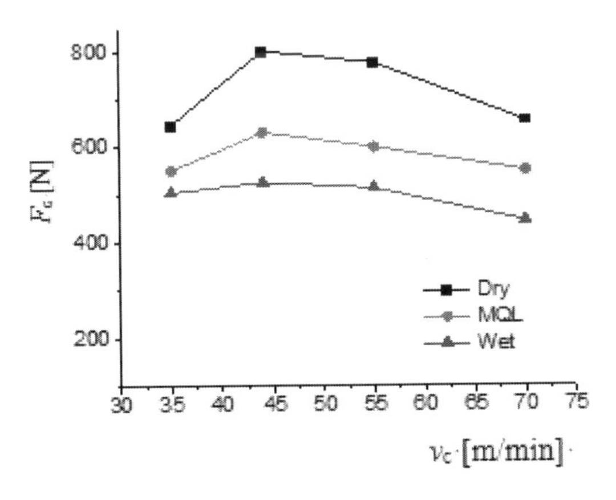

Figure 5. The curve of v_c on Fc.

As shown in Figures 3 and 5, the change trend in the cutting force during application of the MQL and wet cutting was the same as during dry cutting while changing the cutting parameters. As shown in Figures 3 and 4 the cutting force was increased with increasing a_p and f. Figure 5 show that the cutting force Fc was increased within the cutting speed scope of 35 m/min to 45 m/min under the MQL and wet cutting, but with the increasing cutting speed, the cutting force was decreased. The cutting force Fc was reduced by about 10–25% compared with dry cutting, respectively. And this effect was similar to the wet cutting.

3.2 *Chip formation*

The variety of chip formation with v_c is presented in Table 4.

Table 4 presents a comparison of chips obtained during the dry, wet and MQL applications at v_c = 44–70 m/min, f = 0.1 mm/r and a_p = 0.8 mm. It was observed that the chip style was C-shaped during dry cutting, and the high temperature resulted in the chip's frangible poor scalability, which in turn led to process vibration and poor surface quality. During the application of the MQL and wet cutting, chips were broken off like a short helical tubular coiled, which was mainly due to strong cooling to reduce the cutting temperature that tends to curl the chips.

3.3 *Tool wear*

The effect of lubricating on tool flank wear is provided in Table 5. The cutting condition was 1.0 mm depth of cut at a 66 m/min cutting speed

Table 4. The variety of chip formation with v_c (a_p = 0.8 mm, f = 0.1 mm/r).

V_c m/min	Dry	Wet	MQL
35			
44m/min			
55m/min			
70m/min			

Table 5. The effect of lubrication on flank wear.

T(s)	Dry	MQL	Wet
10	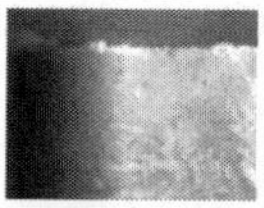	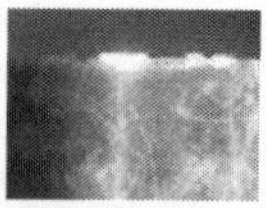	
30	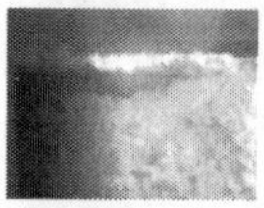	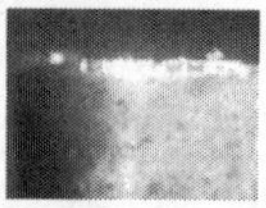	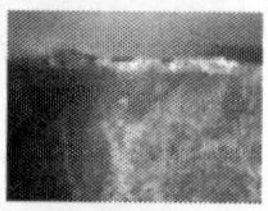
60	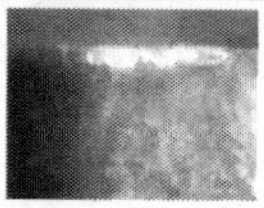	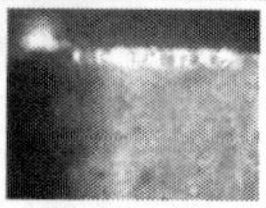	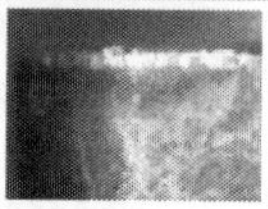
90	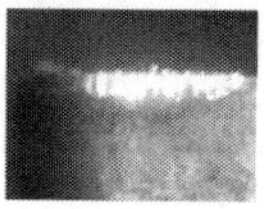		

and a 0.1 mm/rev feed speed. There were obviously adhesive wear and abrasive wear.

It showed that the wear of the insert increased as the cutting time increased. Under the dry cutting condition, the insert wear was higher than under the MQL and wet conditions, and the wear was least in the wet condition.

4 CONCLUSIONS

The experimental investigation on the application of MQL, wet and dry cutting in the turning of GH4169 with the coated carbide tool YBC251 was carried out. The aim of this study was to examine the effects of the MQL on the metal machining process. The investigation drew some conclusions as follows:

1. Alcohols can be used as cutting fluids to achieve green cutting, which can reduce energy consumption and environmental pollution relative to the dry cutting and traditional emulsion in the machining of nickel-based super alloys.
2. Under the MQL condition, the cutting force Fc was reduced by about 10–25% than under the dry cutting condition. The chip curled smoothly. Further, the optimal cutting speed for machining the nickel-base super alloy was 55 m/min, with the nozzle diameter of 20 mm and the air pressure of 0.25 Mpa.
3. The results of the tool wear test indicated that the flank wear of the insert increased as the cutting time increased. Under the dry cutting condition, the insert wear was higher than the MQL and wet conditions, and the wear was least in the wet condition.

ACKNOWLEDGMENT

This research was financially supported by the Natural Science Foundation of Jiangsu Province, China (BK20140247) and the Jiangsu Province University Students' Innovation and Entrepreneurship Training Program 201410294041X.

REFERENCES

Dudzinski, D., Devillez, A. & Moufki, A. 2004. A review of developments towards dry and high speed machining of Inconel 718 alloy, *International Journal of Machine Tools & Manufacture,* 44(4): 439–456.

Ezugwu, E.O. & Bonney, J. 2004. Effect of high-pressure coolant supply when machining nickel-base, Inconel 718, alloy with coated carbide tools, *Journal of Materials Processing Technology*, 153–154(1–3): 1045-1050.

Ezugwu, E.O., Bonney, J. & Tamane, Y. 2003. An overview of the machinability of aeroengine alloys, *Journal of Materials Processing Technology*, 134(2): 233–253.

Han, R.D., Zhang, Y. & Wang, Y. 2008. The Effect of Superheated Water Vapor as Coolant and Lubricant on Chip Formation of Difficult-to-Cut Materials in Green Cutting, *Key Engineering Materials*, 375–376: 172–176.

Kamata, Y. & Obikawa, T. 2007. High speed MQL finish-turning of Inconel 718 with different coated tools, *Journal of Materials Processing Technology*, 192–193: 281–286.

Trent, E.M. 1991. *Metal Cutting* (3rd Ed), Butterworth, London, UK.

Wang, Z.Y. & Rajurkar, K.P. 2000. Cryogenic machining of hard-to-cut materials, *Wear*, 239(2): 168–175.

Advanced Materials, Mechanical and Structural Engineering – Hong, Seo & Moon (Eds)
 ISBN: 978-1-138-02908-8

Fabrication of polycrystalline silicon elements for micro- and nanomechanical accelerometers

O.A. Ageev, E.Yu. Gusev, J.Yu. Jityaeva, A.S. Kolomiytsev & A.V. Bykov
Institute of Nanotechnology, Electronics and Electronics Equipment Engineering, Southern Federal University, Taganrog, Russia

ABSTRACT: This paper presents results of the deposition polycrystalline silicon films by plasma-enhanced chemical vapor deposition for the formation inertial mass of MEMS/NEMS sensors. The effect of power, chamber pressure and temperature on the structural properties of silicon films have been investigated. Amorphous, nano- and polycrystalline structure of layers have been shown using reflection high-energy electron diffraction and ellipsometry. It has been proved that amorphous phase turned into nanocrystalline at 540°C, and into polycrystalline at 610°C. Atomic force microscopy data analysis clearly showed that the grain size and Root-Mean Square roughness (RMS) increased with the temperature and pressure. However, the RMS roughness has reached maximum at 600°C with following decreasing. Thus, deposition conditions of nano- and polycrystalline silicon films with the grain size of 40–250 nm, and RMS roughness of 1.1–3.5 nm have been obtained. The microhardness and the Young modulus were equaled 15–20 GPa and 150–250 GPa, respectively. The polysilicon films were doping up concentration of charge carriers of 1.9–2.4 · 10^{20} cm^{-3}. Perforated inertial mass structures have been fabricated by optical lithography and focused ion beam lithography.

1 INTRODUCTION

A wide variety of MEMS sensors and actuators based on polysilicon as structural material owing to its likely mechanical properties and fabricability (Bhushman 2010, French 2002, Kaher 1995). Polysilicon surface micromachining offers the possibility of producing micromechanical accelerometers and gyroscopes using standard IC processing techniques (Berman & Krim 2013, Bhushman 2010, French 2002).

The Plasma-Enhanced Chemical Vapor Deposition (PECVD) method from silane plasma is widely used to deposit the amorphous and nanocrystalline and polycrystalline silicon (Berman & Krim 2013, French 2002, Perrin et al. 1996, Sniegowski & Boer 2000). The high technological sensitivity of silicon phases allows to vary of material properties but makes it difficult to obtain it reproducibility (French 2002, Velichko 2012). One of the main problems is a precision control of the electrical and physical properties of the films, which largely depends on the structure, the type and size of the grains (French 2002). For micro- and nanoaccelerometers applications, the carrier concentration level of structural films is expected to be more than 10^{19} cm^{-3} and its mechanical properties should be similar to theoretical values. This is particularly important when design and modeling steps carried out previously.

In the present paper, the polysilicon film for MEMS elements were prepared from (SiH_4 + Ar) plasma in a conventional capacitive coupled rf (13.56 MHz) PECVD system (French 2002, Gusev 2015). The low thermal sensitivity of the deposition rate allowed to obtain high (±5%) homogeneity film properties. The structural evolution of the samples was investigated by means of reflection high-energy electron diffraction and ellipsometry, X-ray diffraction and Raman scattering measurements.

In the present paper, the polysilicon film for MEMS elements were prepared from (SiH_4 + Ar) plasma in a conventional capacitive coupled rf (13.56 MHz) PECVD system (French 2002, Gusev 2015, Velichko 2012). The low thermal sensitivity of the deposition rate allowed to obtain high (±5%) homogeneity film properties. The structural evolution of the samples was investigated by means of reflection high-energy electron diffraction and ellipsometry, X-ray diffraction and Raman scattering measurements.

The electrical parameters of these films were measured by contactless and Hall/van der Pauw techniques. The purpose of this work is to investigate the mechanical, structural and electrical properties of the polysilicon films to be able to apply them for the micro- and nano-mechanical accelerometers and MEMS applications.

2 EXPERIMENTAL

Various polysilicon films were prepared by plasma enhanced chemical vapor deposition (PlasmaLab 100 Oxford Instruments) (Gusev 2015, Velichko 2012). The substrate was 100 mm n-type silicon wafer Si (100). Silicon dioxide films were grown on the substrates to the desired thickness in the range of 0.3–2.0 μm using inductively coupled plasma CVD (SemiTEq ICPd81). PECVD is used the reduction reaction of silane under Argon (Ar). An overview about the possible reactions in silane plasmas is given in (Perrin et al. 1996). The reaction for silane is given by $SiH_4 \rightarrow SiH_2 + H_2\uparrow$, $SiH_2 \rightarrow Si + H_2\uparrow$. The flow gases Ar: SiH_4 = 9:1 was constant at 500 cm^3/min.

The influence of deposition parameters, such as power, working pressure and temperature on deposition rate and properties of polysilicon films were investigated in our experiments. The crystalline structure of deposited films was examined by reflection high-energy electron diffraction (RHEED, in Pulsed Laser Deposition module of Nanotechnological complex with advanced analytical facilities (Inv. № P3011370305537/4)) (Ageev 2011).

The surface morphology, mechanical properties and film thickness were observed with an atomic force microscopy (NTEGRA Vita Probe Nanolaboratory) and a scanning electron microscopy with focused ion beam (Nova Nanolab 600), respectively (Ageev 2011, 2014, Gusev 2015, Velichko 2012). The index of refraction was determined by ellipsometer (LEF-3M). AFM images were analyzed by Image Analysis 3.5 program.

A diamond three sided Berkovich pyramid with an apex angle of θ = 70° between the edge and height was used as the indenter. The probe was indented into the sample surface at a constant speed; the process was accompanied by recording the values of the load and the appropriate depth of indenter penetration in the material, on the basis of which the resulting dependence (load curve) was plotted. Scratch force was 700–900 μN.

The polycrystalline silicon films with thickness of 300 nm and 2.0 μm were doped using of diffusion oven (SD.OM-3M) with the liquid PCl_3 diffusant. The prediffusion and drive-in steps temperatures were 850°C and 1150°C, respectively. Concentration, mobility of charge carriers and resistivity after doping were measured using Hall/van-der-Pauw and contactless methods (Ecopia HMS-3000/1T and LEI 1510M40).

Polysilicon inertial masses were fabricated by optical lithography (SUSS MJB4) and dry etching under aluminum mask (SemiTEq ICPe68), as well as focused ion lithography (Nova Nanolab 600).

3 RESULTS AND DISCUSSION

The silicon films with the crystal size from 40 nm to 250 nm were fabricated using plasma-enhanced chemical vapor deposition. Reflection high-energy electron diffraction images analysis revealed the existence of amorphous, nanocrystalline and polycrystalline phase in the structure of the films (Figure 1). The amorphous film was formed under temperature from 500°C to 540–550°C, the nano- and polycrystalline films from 550°C to 600°C and from 610°C to 700°C, respectively. The existence of the above-mentioned phases was proved by ellipsometry. The refractive index values were estimated from 4.5 to 5.0 for amorphous phase and from 3.5 to 4.0 and 4.5 to 5.5 for nano- and polycrystalline, respectively.

The influences of temperature on grain size and Root-Mean-Square (RMS) roughness of silicon films are shown in Figure 2. The working pressure and the power were fixed as 1000 mTorr and 40 W. RMS roughness reached maximum of 4 nm at 600°C with following decreasing. Grain size increased from 70 nm to 210 nm under the temperature from 550°C to 700°C (Figure 3). The nanocrystalline phase changed to polycrystalline at 610°C, that is in a good agreement with (French 2002, Kaher 1995).

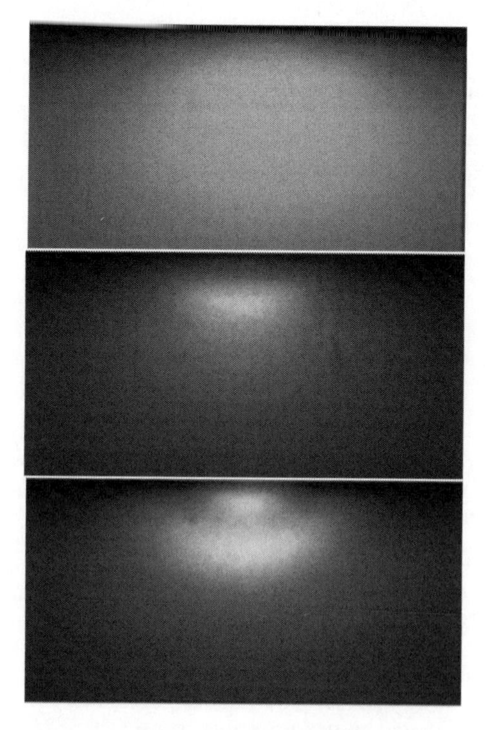

Figure 1. RHEED image of amorphous, nano- and polycrystalline silicon films (from top to bottom).

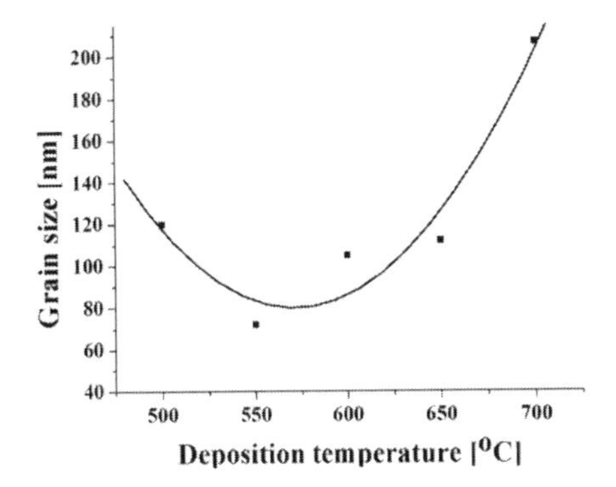

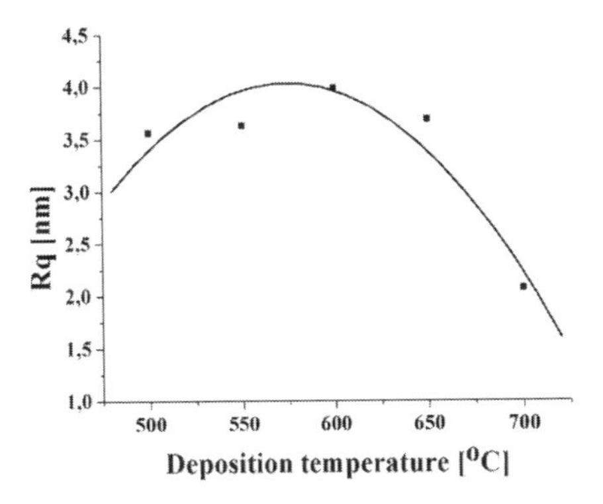

Figure 2. Variation of grain size (top) and RMS roughness (bottom) plotted as a function of deposition temperature.

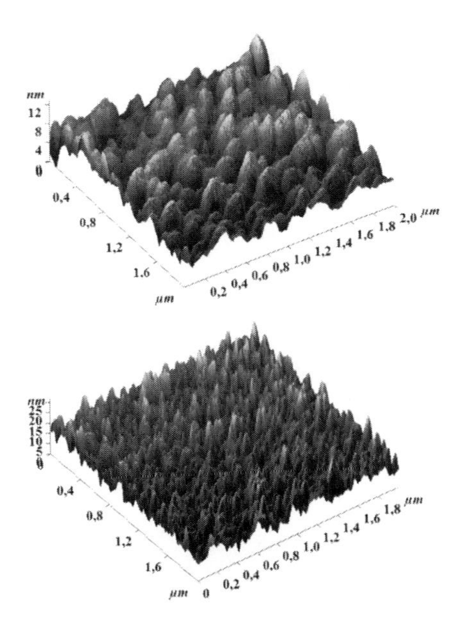

Figure 3. AFM images of the Si films deposited at 550°C (top) and 700°C (bottom).

The analysis of atomic force microscopy data showed the grain size and root-mean square roughness increased with the temperature (see Figure 3) and pressure. The crystal diameter and the RMS roughness increased by 6–8% and 75–80%, respectively with working pressure from 1 to 2 Torr. This behavior is in agreement with data reported by other authors (French 2002, Kaher 1995).

Experimental data of polysilicon film deposition at 700°C, gas flow SiH_4:Ar = 50:450 cm^3/min and 1 Torr and indentation are listed in Table 1.

The samples # 9 and # 11 can be considered as optimal in respect to theoretical and experimental data correlations. The Young modulus spread grows with increment of film thickness (deposition

Table 1. Mechanical properties of polysilicon films.

#	Power, W	Time, min	Micro-hardness, GPa	Young modulus, GPa
1	40	40	16.71 ± 4.76	253.61 ± 55.99
8	40	5	14.66 ± 2.58	217.18 ± 37.36
9	20	5	19.07 ± 2.99	172.18 ± 4.88
10	40	10	18.96 ± 4.34	147.03 ± 58.02
11	20	40	19.33 ± 1.35	170.52 ± 47.31

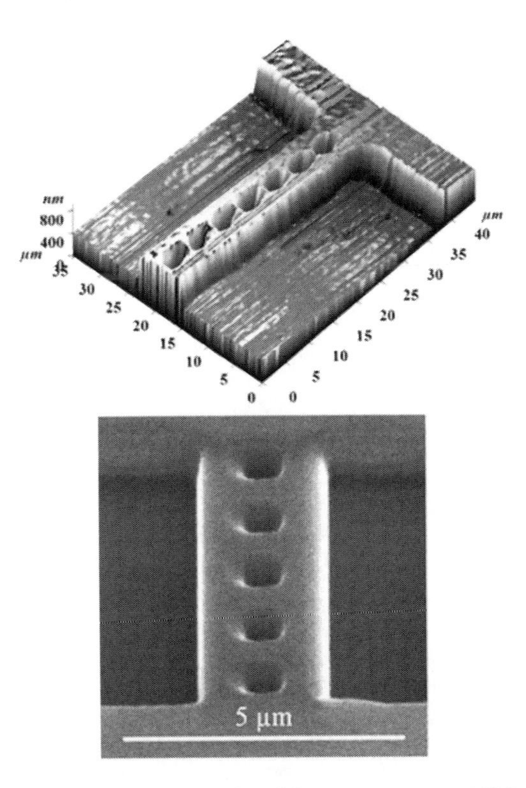

Figure 4. Polysilicon inertial mass structure: AFM (top) and SEM (bottom) images.

time and power), the grain size (crystallite) and depends on the ratio of grain boundaries, interface boundaries and triple lines parts. The obtained results on microhardness require further clarification for Hall-Petch relation.

Concentration, mobility of charge carriers and resistivity of doped films were measured using Hall/van-der-Pauw techniques and contactless method. Values were amounted of $1.9–2.4 \cdot 10^{20}$ cm^{-3}, 27.5–34.1 $cm^2/(V \cdot s)$ and $9.35–9.7 \cdot 10^{-4}$ $\Omega \cdot cm$. Sheet resistance for 200 nm and 2.0 μm film thickness were 2.0 and 8.9 Ω/sq, respectively.

At the final stage, polysilicon inertial masses were fabricated by optical lithography (Figure 4, top) and focused ion beam lithography (Figure 4, bottom) for further investigation. Underlying sacrificial layer was lithography patterned and removed by wet etching (Gusev 2015). Fabricated polysilicon films, structures and elements with mechanical parameters closed to theoretical values and carrier concentration of $\sim 10^{20}$ cm^{-3} are desired and suitable to formation of inertial masses for micro- and nanomechanical accelerometer applications.

4 CONCLUSIONS

Uniform nano- and polycrystalline silicon films were deposited on to SiO_2/n-Si(100) substrate by PECVD. RHEED analysis, ellipsometry and AFM revealed the crystalline phases. The morphology, structures, mechanical properties and refractive index of the films were adjusted by the selection of rf power, chamber pressure and temperature. The dependences of the grain size and RMS roughness on deposition parameters were observed: the grain size and RMS roughness were in the range of 40–250 nm and 1.1–3.5 nm, respectively. According to AFM indentations the microhardness and Young module of obtained films were 14.66–19.33 GPa and 147.03–253.61 GPa, respectively. Concentration, mobility of charge carriers and resistivity of doped polysilicon films were $1.9–2.4 \cdot 10^{20}$ cm^{-3}, 27.5–34.1 $cm^2/V \cdot s$ and $9.35–9.7 \cdot 10^{-4}$ $\Omega \cdot cm$, respectively. Sheet resistance for 200 nm and 2000 nm film thickness were 2.0 and 8.9 Ω/sq, respectively.

Polysilicon inertial masses based on the doped films were fabricated by optical lithography and focused ion beam lithography.

Fabricated polysilicon films, structures and elements with mechanical parameters close to theoretical values and carrier concentration of $\sim 10^{20}$ cm^{-3} are desired and suitable to formation of inertial masses for micro- and nanomechanical accelerometer applications.

The results can be used to develop manufacturing processes for fabrication of micro- and nanomechanical accelerometers and gyroscopes by surface micromachining as well as in studies of micro-and nanosystem technology.

ACKNOWLEDGEMENT

This research was financially supported by the Ministry of Education and Science of Russian Federation within the contract 14.575.21.0045 (unique identifier RFMEFI57514X0045).

The equipment of the Centre of Collective Use of Equipment "Nanotechnologies" of Southern Federal University was used for this study.

REFERENCES

Ageev, O.A., Kolomiytsev, A.S., Mikhaylichenko, A.V., Smirnov, V.A., Ptashnik, V.V., Solodovnik, M.S., Fedotov, A.A., Zamburg, E.G., Klimin, V.S., Ilin, O.I., Gromov, A.L. & Rukomoykin, A.V. 2011. Nanoscale structures' production based on modular nanotechnologycal platform nanofab. *Izvestiya SFedU. Engineering Sciences,* 1: 109–116.

Ageev, O.A., Alekseev, A.M., Vnukova, A.V., Gromov, A.L., Kolomiytsev, A.S., Konoplev, B.G. & Lisitsyn, S.A. 2014. Studying the resolving power of nanosized profiling using focused ion beams. *Nanotechnologies in Russia,* 9: 26–30.

Berman, D. & Krim, J. 2013. Surface science, MEMS and NEMS: Progress and opportunities for surface science research performed on, or by, microdevices. *Progress in Surface Science*, 88: 171–211.

Bhushan, B. 2010. *Springer Handbook of Nanotechnology*, third ed. Heidelberg, Dordrecht, London, New York: Springer.

French, P.J. 2002. Polysilicon: a versatile material for Microsystems. *Sensors and Actuators A Physical,* 99: 3–12.

Gusev, E.Yu., Jityaeva, J.Y., Kolomiytsev, A.S., Gamaleev, V.A., Kots, I.N. & Bykov, A.V. 2015. Wet etching of silicon dioxide sacrificial layer for mems structures forming. *Proceeding of the International Conference "Physics and Mechanics of New Materials and Their Applications"*: 100–101. Azov.

Gusev, E.Yu., Jityaeva, J.Y., Gamaleev, V.A., Kolomiytsev, A.S., Kots, I.N. & Bykov, A.V. 2015. Research of wet SiO_2 sacrificial layer etching for MEMS structures forming based on poly-Si/SiO_2/Si. *Izvestiya SFedU. Engineering Sciences,* 2: 236–245.

Kareh, El.B. 1995. *Fundamentals of Semiconductor Processing Technology,* Boston: Springer.

Perrin, J., Leroy, O. & Bordage, M.C. 1996. Cross-sections, rate constants and transport coefficients in silane plasma. *Contributions to Plasma Physics,* 36: 3–49.

Sniegowski, J.J. & Boer, M.P. 2000. IC-Compatible polysilicon surface micromachining. *Annual Review of Materials Research*, 30: 299–333.

Velichko, R.V., Gusev, E.Yu., Gamaleev, V.A., Mikhno, A.S. & Bychkova, A.S. 2012. PECVD analysis of nano- and polycrystalline silicon films. *Fundamental Research*, 11–5: 1176–1179.

Advanced Materials, Mechanical and Structural Engineering – Hong, Seo & Moon (Eds)
© 2016 Taylor & Francis Group, London, ISBN: 978-1-138-02908-8

Corrosion fatigue crack propagation behavior of LY12CZ aluminum alloy

X.G. Huang
College of Pipeline and Civil Engineering, China University of Petroleum, Qingdao, China

Z.Y. Han
School of Petroleum Engineering, China University of Petroleum, Qingdao, China

G.T. Feng
Drilling Technology Research Institute, Shengli Petroleum Engineering Co. Ltd., Sinopec, Dongying, China

ABSTRACT: This study was undertaken to identify the synergistic interactions between mechanical and environmental factors on the crack propagation behavior of LY12CZ (Al-Cu-Mg) aluminum alloy in artificial seawater (3.5wt.% NaCl solution). Clamping device of Single-edge Notched Bending (SENB) specimen was specially designed for MTS809 axially loading test system. Corrosion Fatigue Crack Propagation (CFCP) tests had been carried out at different frequencies and stress ratios, in the air and artificial seawater with different pH values. Crack propagation rates were found to increase with the decline of loading frequency and pH value of the solution. Stress ratio was found only significantly to affect the crack propagation near the threshold region. The mechanics of corrosive environment accelerating crack propagation at the neutral solution was mainly contributed to the anodic dissolution at the crack tip while the influence of hydrogen-induced damage gradually revealed with the increase of acidity.

1 INTRODUCTION

Corrosion fatigue was an environmental degradation phenomenon caused by the synergetic effect of cyclic stress and corrosive environments. Corrosion fatigue failure of aluminum alloys had become the major issue in the reliability and durability of metal components and structures (Schijve 1994, Bellinger & Komorowski 1997). The development of effective CFCP rate prediction technologies was essential for the reliability analysis of structures, as well as the successful implementation of life extension strategies (Menan & Henaff 2010, Ruiz & Elices 1996, Fontea et al. 2003).

In general, corrosion fatigue crack propagation has a strong dependence on material and corrosive environment. Various factors, such as the microstructural feature, the magnitude of mean stress, the amplitude of cyclic stress and the frequency of stress cycle, may govern the fatigue crack propagation behavior of metal materials. For example, Kermanidis et al. (Kermanidis et al. 2005) and Sadananda et al. (Sadananda & Vasudevan 2004) reported that CFCP rate increased with the increasing stress Ratio (R) in a 2024 T731 aluminum and a structural steel respectively. But there existed a contrary opinion that the CFCP rate increased with increasing stress ratio (Murakami et al. 1991). The lower frequency was considered to cause the increase of crack propagation rates in corrosive environments (Wang 2008). However, the effect was thought to saturate at very low frequency (Ruiz & Elices 1996, Gangloff 1989). It was accepted that the essence of crack propagation acceleration in corrosive environment was local material damage caused by an electrochemical process at the crack tip, especially reflected as hydrogen embrittlement and anode dissolution (Robertson 2001, Wang et al. 1995). Menon et al. (Menan & Henaff 2009) held that fatigue crack propagation enhancement in corrosive solution was mainly induced by the hydrogen embrittlement at the crack-tip. Zhao et al. (Zhao et al. 2012) investigated the effects of strain on the corrosion behavior of X80 steel in 3.5wt.% NaCl solution and analyzed the contribution of anodic dissolution to the CFCP. Shen et al. (Shen & Lv 2001) carried out experimental research for the relation between strain and corrosion current and calculated the corrosive damage of anode dissolution on CFCP of 7475-T761 aluminum alloy in 3.5 wt.% NaCl solution. But at high frequency, the advection effect on the rate of corrosion fatigue for the cases of diffusion and mixed kinetic control had to be considered (Engelhardt & Macdonald 2010). All those studies indicated that in corrosion fatigue crack propagation domains, the detrimental synergistic effects between fatigue damage and corrosion taking place at the crack tip still need to be identified.

The study presented here was undertaken to improve the understanding of environmentally-assisted mechanisms and identify synergistic interactions between fatigue and corrosion in the crack propagation of aluminum alloys. In this aim, three-point bending CFCP test samples of LY12CZ aluminum alloys were designed. The tests had been carried out in the air and artificial seawater respectively to study the influence of frequency, stress ratio, and pH value on the CFCP rate. The probable mechanisms that governed the CFCP behavior of LY12CZ aluminum alloy were discussed in terms of anodic dissolution, hydrogen-induced damage and so on.

2 EXPERIMENTAL STUDY

2.1 *Material and specimen*

The chemical composition and mechanical properties of LY12CZ aluminum alloy were shown in Tables 1 and 2, respectively.

The CFCP tests were performed on the pre-cracked SENB specimens. Standard specimens S = 120 mm, W = 30 mm, B = 15 mm, with a notch of 11 mm, from the as-received material to the shape were shown in Figure 1. The crack plane was located at mid-span of the plate for a higher sensitivity to corrosion fatigue crack propagation (ASTM. E647-00. 1998).

Table 1. Chemical composition of LY12CZ aluminum alloy. (wt.%).

Cu	Mg	Mn	Zn	Si	Fe	Al
4.3	1.5	0.6	0.3	0.5	0.5	balance

Table 2. Mechanical properties of the material.

Young's modulus	Yield strength	Tensile strength	Elongation
72 Gpa	320 Mpa	460 Mpa	17.4%

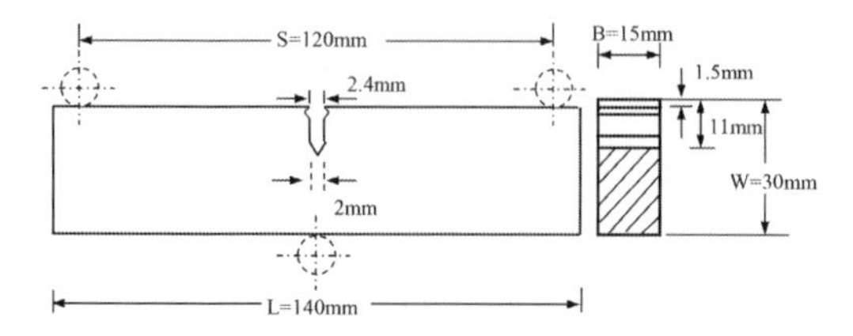

Figure 1. Corrosion fatigue specimen details.

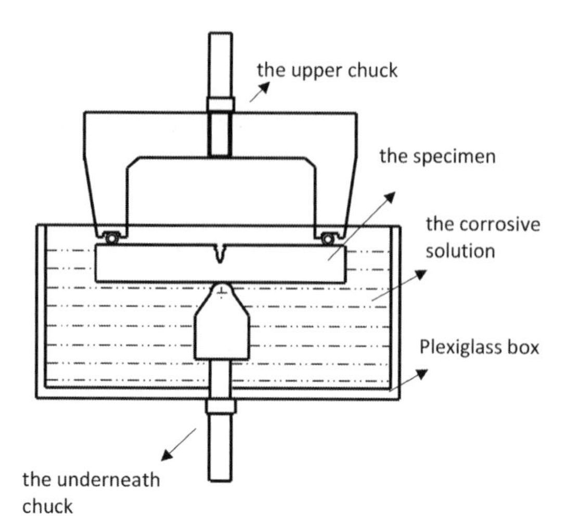

Figure 2. Loading equipment of SENB specimen for corrosion fatigue.

A special specimen-clamping device and a container for corrosion solution were fabricated to ensure that the test could be carried out on the MTS809 axial-loading fatigue testing machine.

2.2 *Experimental procedure*

Fatigue crack propagation tests were performed under load control using constant amplitude cyclic loading, i.e. under increasing ΔK. Pre cracking initiated in the air at 5 Hz to a crack propagation from the notch of about 1.5 mm, by a sinusoidal waveform at loading ratio 0.1. To study the influences of frequency, stress ratio, and pH value on fatigue crack propagation, the specimens were tested at eight different loading-environment combinations (4 specimens per combination), shown in Table 3. For specimens in the seawater, the test stopped when a final crack propagation increment of about 1 mm (GB/T 6398-2000. 2000). The post-fracture analysis was performed on the fractured surfaces of the specimens by a Scanning Electron Microscope (SEM) to study corrosion fatigue striations on the surface and, at the same time, to analyze the mechanism of corrosion environment accelerating fatigue crack propagation with respect to the surface area examined.

3 EXPERIMENTAL RESULTS AND DISCUSSION

3.1 *Experimental results*

The corrosion fatigue crack propagation rates da/dN measured at f = 5 Hz in the seawater were presented in Figure 3 as a function of ΔK and compared to data obtained in the air for selected

Table 3. Loading-environment combinations for corrosion fatigue tests.

No.	Stress ratio	Frequency (Hz)	Environment
1	0.1	5	in air
2	0.1	5	3.5%NaCl (pH = 7)
3	0.1	1	3.5%NaCl (pH = 7)
4	0.1	10	3.5%NaCl (pH = 7)
5	0.3	5	3.5%NaCl (pH = 7)
6	0.5	5	3.5%NaCl (pH = 7)
7	0.1	5	3.5%NaCl (pH = 5)
8	0.1	5	3.5%NaCl (pH = 3)

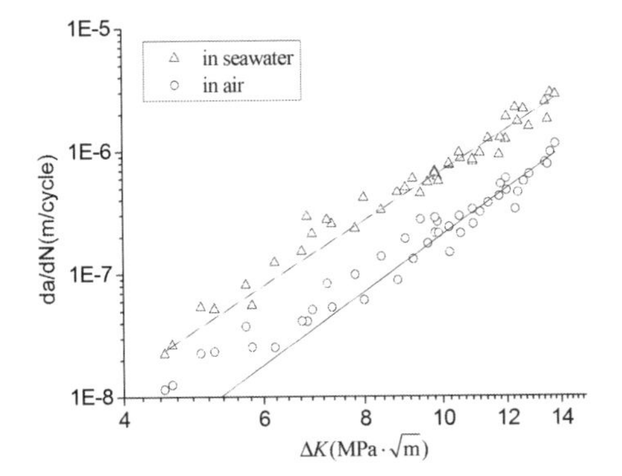

Figure 3. $\mathrm{d}a/\mathrm{d}N$ at f = 5 Hz in the air and seawater.

values of ΔK. The crack propagation rates in the seawater are much higher than those measured in the air, especially at low ΔK values, and the accelerating effect of the corrosive environment on the crack propagation was very significant. The crack propagation rate in the air and seawater increased with the increase of ΔK, and ΔK might be the main factor that controls corrosion fatigue crack propagation of LY12CZ aluminum alloy.

Loading frequency had a little effect on the fatigue performance in the inert medium, so the frequency effect was usually negligible on fatigue problems. However, in the corrosive medium, the frequency was an important factor affecting fatigue characteristics of the material. The corrosion fatigue crack propagation rates at different frequencies were shown in Figure 4. The cyclic loading frequency had a significant influence on corrosion fatigue crack propagation. The lower of frequency was the rise of corrosion fatigue crack propagation rate appeared. The fresh material at the crack tip had more time for full interactions with corrosive environment at low load frequency, exacerbating the corrosion damage of the material at crack tip, and this process consequently accelerated the CFCP.

The influence of stress ratio on corrosion fatigue crack propagation rate was displayed in Figure 5.

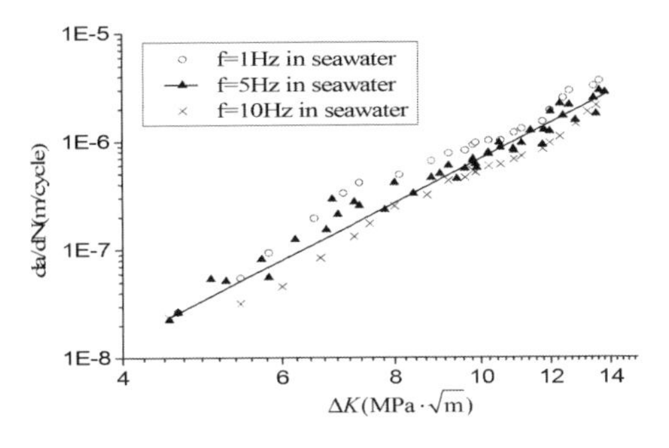

Figure 4. $\mathrm{d}a/\mathrm{d}N$ in the seawater at different frequencies.

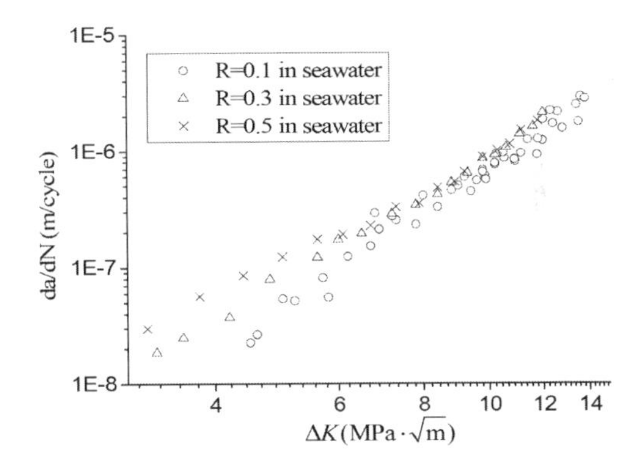

Figure 5. $\mathrm{d}a/\mathrm{d}N$ in the seawater at different stress ratios.

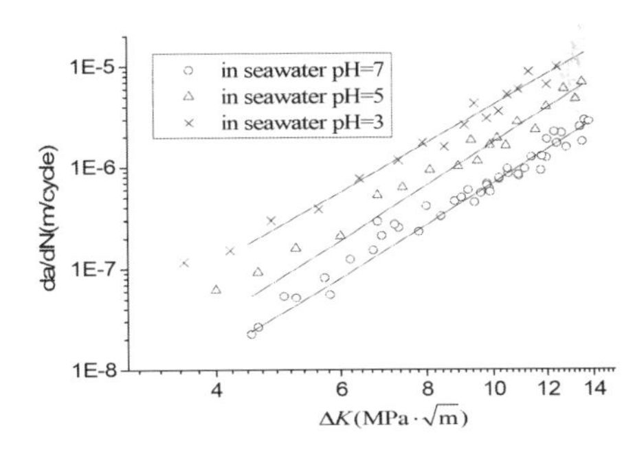

Figure 6. $\mathrm{d}a/\mathrm{d}N$ in the seawater with different pH value.

The crack propagation appeared an evident threshold effect during the descent of ΔK. In the near-threshold region, stress ratio had a significantly influence on crack propagation rate, and the crack propagated much rapidly at the high level of stress ratio. However, this threshold effect tapered off with the increase of ΔK. So it was believed that stress ratio only significantly affected the CFCP in the near-threshold region (Vasudevan & Suresh 1982).

Figure 6 displayed the $\mathrm{d}a/\mathrm{d}N$ varying with ΔK in the seawater of different pH values. As was seen that the acidity of the seawater significantly affected the crack propagation. The stronger the acidity was, the higher the crack propagation rate was. The strong acidity of corrosion solution greatly promoted the anodic dissolution of freshmen metal at the crack tip. Moreover, the hydrogen, generated from hydrogen ions in the solution by the replacement reaction, adhered to the surface at the crack tip and then diffused into the plastic zone. The mechanical performance deterioration caused by hydrogen-induced damage greatly accelerated the crack propagation of corrosion fatigue.

3.2 *Discussion*

Fracture surfaces of the specimen were observed by SEM, to study the mechanism that NaCl solution promoted fatigue crack propagation of the LY12CZ aluminum alloy. The SEM fractography in neutral seawater (pH = 7.0, f = 5 Hz) was shown in Figure 7. As seen from the picture, there obviously existed numerous visible pits produced by anodic dissolution. It fully showed that anodic dissolution at crack tip played a major role in accelerating corrosion fatigue crack propagation. The hydrolysis of Al^{3+} produced by anodic dissolution increased the solution acidity at the crack tip and accelerated the rupture of surface oxide film. The repeated dissolution of newborn metal at crack tip promoted the crack propagation.

The SEM fractography in weakly acidic seawater (pH = 5.0) was displayed in Figure 8. There were many clear fatigue striations and flocculent reaction products, but no apparent corrosion pits in the picture. The superfluous hydrogen ions in acidic solution produced the hydrogen by replacement reaction. The hydrogen atoms then entered the plastic zone and led to the hydrogen-induced embrittlement of the material at the crack tip.

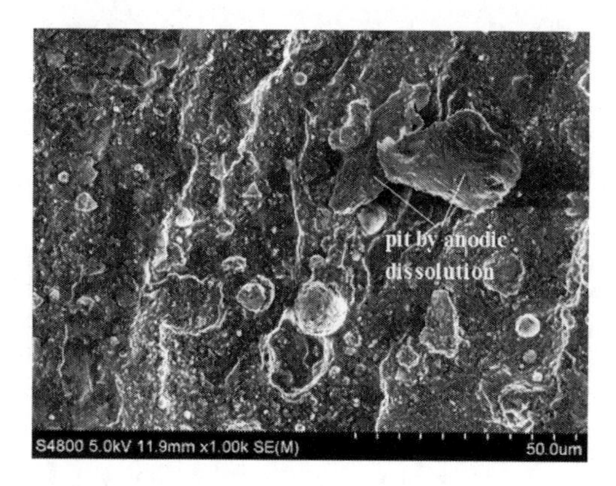

Figure 7. SEM fractography in neutral seawater.

Figure 8. SEM fractography in the weakly anodic seawater.

Therefore, the crack propagation rate in the acidic solution was much higher than that in the neutral solution for the hydrogen-induced damage was not obvious in the neutral solution.

4 CONCLUSIONS

Corrosion fatigue crack propagation tests had been carried out to analyze the mechanism that the corrosion environment accelerated fatigue crack propagation of LY12CZ aluminum alloy. The typical conclusions were:

1. The corrosive environment had an accelerating effect on fatigue crack propagation, and fatigue crack propagation rate in corrosive environment was larger than that in the air.
2. The load frequency, stress ratio, and pH value of solution affected the crack growing behavior of LY12CZ aluminum alloy. Lower frequency caused an increase of crack propagation rate for the more serious of corrosive damage of materials at the crack tip. The stress ratio only significantly affected corrosion fatigue crack propagation near the threshold region, and the crack grows rapidly at a high stress ratio. The accelerating effect of pH value on crack propagation is very significant, attributing to the anodic dissolution and hydrogen-induced damage in the low pH solution.

ACKNOWLEDGEMENT

This work was supported by the National Nature Science Foundation of China (No. 51404286) and Nature Science Foundation of Shandong Province (No. ZR 2014EEQ018).

REFERENCES

ASTM. E647-00. 1998. In Annual Book of ASTM Standards.

Bellinger, N.C. & Komorowski, J.P. 1997. Corrosion pillowing stresses in fuselage lap joints. *AIAA Journal*, 35(2): 317–320.

Engelhardt, G.R. & Macdonald, D.D. 2010. Modelling the crack propagation rate for corrosion fatigue at high frequency of applied stress. *Corrosion Science*, 52: 1115–1122.

Fontea, M., da Romeirob, F., Freitasc, M., de Stanzl-Tscheggd, S.E., Tschegge, E.K. & VasudÉvanf, A.K. 2003. The effect of microstructure and environment on fatigue crack propagation in 7049 aluminium alloy at negative stress ratios. *International Journal of Fatigue*, 25(9–11): 1209–1216.

Gangloff, R.P. 1989. *Corrosion Fatigue crack propagation in metals*. USA: National Association of Corrosion Engineers.

GB/T 6398-2000. 2000. *Standard Test Method for Fatigue Crack Growth Rates of Metallic Materials in China*.

Kermanidis, A.T. Petroyianis, P.V. & Pantelakis, S.G. 2005. Fatigue and damage tolerance behaviour of corroded 2024 T351 aircraft aluminum alloy. *Theoretical and Applied Fracture Mechanics*, 43(1): 121–132.

Menan, F. & Henaff, G. 2009. Influence of frequency and exposure to a saline solution on the corrosion fatigue crack propagation behavior of the aluminum alloy 2024. *International Journal of Fatigue*, 31(11–12): 1684–1695.

Menan, F. & Henaff, G. 2010. Synergistic action of fatigue and corrosion during crack growth in the 2024 aluminium alloy. *Procediaedia Engineering*, 2(1): 1441–1450.

Murakami, R.I., Kim, Y.H. & Ferguson, W.G. 1991. The effects of microstructure and fracture surface roughness on near threshold fatigue crack propagation characteristics of a two-phase cast stainless steel. *Fatigue & Fracture of Engineering Materials & Structures*, 14(7): 741–748.

Robertson, I.M. 2001. The effect of hydrogen on dislocation dynamics. *Engineering Fracture Mechanics*, 68(6): 671–692.

Ruiz, J. & Elices, M. 1996. Environmental fatigue in a 7000 series aluminum alloy. *Corrosion Science*, 38(10): 1815–1837.

Sadananda, K. & Vasudevan, A.K. 2004. Crack tip driving forces and crack growth representation under fatigue. *International Journal of Fatigue*, 26(1): 39–47.

Schijve, J. 1994. Fatigue of aircraft materials and structures. *International Journal of Fatigue*, 16(1): 21–32.

Shen, H.J. & Lv, G.Z. 2001. Effect of anode dissolution on corrosive fatigue crack propagation. *Journal of Northwestern Polytechnical University*, 19(2): 225–228.

Vasudevan, A.K. & Suresh, S. 1982. Influence of corrosion deposits on near-threshold fatigue crack growth behavior in 2XXX and 7XXX series aluminum alloys. *Metallurgical Transactions A*, 13: 2271–2280.

Wang, R. 2008. A fracture model of corrosion fatigue crack propagation of aluminum alloys based on the material elements fracture ahead of a crack tip. *Engineering Fracture Mechanics*, 30: 1376–1386.

Wang, Z.F., Wei, X.J., Li, J. & Ke, W. 1995. Effect of anodic dissolution and hydrogen entrance on mechanical properties of material at crack tip. *Corrosion Science and Protection Technology*, 7(2): 157–161.

Zhao, W.M., Xin, R.F., He, Z.R. & Wang, Y. 2012. Contribution of anodic dissolution to the corrosion fatigue crack propagation of X80 steel in 3.5 wt.% NaCl solution. *Corrosion Science*, 63: 387–392.

Advanced Materials, Mechanical and Structural Engineering – Hong, Seo & Moon (Eds)
© 2016 Taylor & Francis Group, London, ISBN: 978-1-138-02908-8

Formation and photocatalysis of novel ternary Fe-based composites

Y. Teng
Department of Materials Science and Engineering, CAS Key Laboratory of Materials for Energy Conversion, University of Science and Technology of China, Hefei, China

L.X. Song
Department of Materials Science and Engineering, CAS Key Laboratory of Materials for Energy Conversion, University of Science and Technology of China, Hefei, China
Department of Chemistry, University of Science and Technology of China, Hefei, China

W. Liu
Department of Materials Science and Engineering, CAS Key Laboratory of Materials for Energy Conversion, University of Science and Technology of China, Hefei, China

R.R. Xu
Department of Chemistry, University of Science and Technology of China, Hefei, China

ABSTRACT: Herein we report the first example of the creation of ternary $Fe/Fe_3O_4/Fe_2O_3$ nanocomposites with nanostructures by a sintering process using Fc/β-CD/$FeCl_3$ complex as a precursor. The nanorod composite exhibited excellent photodegradation efficiency for rhodamine B and very high photocatalytic selectivity for the organic dyes. We believe that this study represents an important advance regarding Fe-based nanocomposites.

1 INTRODUCTION

In recent years, a great deal of research has been focused on the synthesis of iron-containing nanostructured composites, because these materials are provided with encouraging magnetic and catalytic properties of zero-valent iron nanoparticles and protect metallic iron(0) from oxidation (Zeng et al. 2004, Noradoun et al. 2003). Special methods are usually required to control the reduction process of Fe^{3+} and protect the high active iron(0) nanoparticles in order to gain nanocomposites with controllable components. (Nurmi et al. 2004, Dumitrache et al. 2005) Furthermore, tuning the morphology and composition of the composites is one of the most challenging issues. Ai and coworkers utilized the reduction of ferric ions with sodium borohydride to obtain core-shell Fe@Fe_2O_3 nanowires, and successfully tuned the shell thickness through changing water-aging time (Ai et al. 2013). However, to the best of our knowledge, ternary $Fe/Fe_3O_4/Fe_2O_3$ nanocomposites have not been reported so far.

In a previous work, we constructed a series of α-Fe_2O_3 crystal materials with fascinating polyhedral morphologies: octahedral, cuboctahedral and truncated cubic structures through a novel solid-phase sintering process using adducts of β-cyclodextrin (β-CD) and $FeCl_3$ as precursors (Wang et al. 2012). The size and shape of the polyhedra were controlled by sintering temperatures and compositions of the adducts. This result shed a light of the application of the organic hosts in the controllable synthesis of the nanoparticles. Thus, Ferrocene (Fc) was introduced into an adduct system of β-CD and $FeCl_3$ to see whether or not a coordination compound containing Fe could affect the synthesis process.

Herein, we report a simple and controllable fabrication of ternary $Fe/Fe_3O_4/Fe_2O_3$ nanocomposites for the first time. The synthesis was realized simply through sintering a series adducts included an inorganic salt, an organic host and a coordination compound. Different from these studies on iron-containing nanoparticles obtained by reducing metal salts using borohydride derivatives in organic solvents (Tsai & Dye 1991, Bonnemann et al. 1990), our preparation takes the advantage of the self-reduction of the precursors.

2 EXPERIMENTAL SECTION

2.1 *Materials*

Ferric chloride ($FeCl_3 \cdot 6H_2O$), β-CD and Fc were obtained from Shanghai Chemical Reagent Company. Methylene Blue (MB) and Rhodamine B (RhB) were from Aladdin Chemistry Co. Ltd. All other chemicals were of general-purpose reagent grade, unless otherwise stated.

2.2 *Preparation of solid iron-based materials*

$FeCl_3 \cdot 6H_2O$ (1 mmol, 0.270 mg), Fc (0.5 mmol, 0.093 mg) were mixed to a round-bottom flask with 50 mL deionized distilled water, followed by vigorous stirring at 333 K for 30 min. Then 1134 mg β-CD (1 mmol) was added to the round-bottom flask and stirred for 4 h. Water was drawn off by rotary evaporation under vacuum at 333 K. A greenish blue residue was obtained, washed by cold anhydrous ethanol and dried in vacuum. Then, the composite was sintered in tube furnace (Nabertherm, M7/11, with a program controller) at 773 K for 3 h in nitrogen. After cooling down to room temperature, the black power was obtained, washed by anhydrous ethanol several times, dried in vacuum and marked as TF-1. Besides, the other product, marked as TF-2, was obtained by sintering the mixture of $FeCl_3 \cdot 6H_2O$/Fc/β-CD (1.1.1, molar ratio).

2.3 *Preparation of photocatalysts*

First, 5.0 mg of TF photocatalyst was placed in the aqueous solution of MB and RhB (20 mL, 3.0 mg · L^{-1}), respectively. Then, the suspensions were stirred for 30 min in dark to reach adsorption equilibrium. Subsequently, the suspensions were transferred into three-necked flasks and irradiated under visible light for 2 h. Finally, the solutions were collected for measurements of UV-Vis spectroscopy to determine the content of the dyes in the solutions with a wavelength range from 190 to 900 nm.

2.4 *Instruments and methods*

X-Ray Diffraction patterns (XRD) of the prepared samples were recorded on a Philips X'Pert Pro X-ray diffractometer equipped with Cu Kα radiation (40 kV, 40 mA) and analyzed in the range of $10° \leq 2\theta \leq 70°$. The surface morphology of the samples was observed by a JSM-6700F Field-Emission Scanning Electron Microscope (FE-SEM). X-ray Photoelectron Spectrum (XPS) was obtained by an ESCALAB250 spectrometer, with Al excitation radiation (1486.6 eV) in ultra-high vacuum conditions (2.00×10^{-9} Torr). All the values of binding energy were referenced to C1 s neutral carbon peak at 284.6 eV with an energy resolution of 0.16 eV to compensate for the surface charging effect. Absorption spectra were recorded on a Shimadzu UV 2401-(PC) spectrometer at room temperature over the wavelength range from 190 to 800 nm, using quartz cells with a 1 cm optical path.

3 RESULTS AND DISCUSSION

3.1 *Formation and structure of the TF-1 and TF-2 composites*

Figure 1 displays the XRD patterns of the TF-1 and TF-2 composites. For TF-1, the two characteristic diffraction peaks at 44.7 and 65.0° (indicated by the red asterisks) are attributed to the (110) and (200) phases of crystal structure of metallic Fe (cubic system, JCPDS 87-0722) (Guo et al. 2011). At the same time, the characteristic diffraction peaks at 30.1 (220), 43.1 (400), 53.4 (422), 56.9 (333) and 62.5° (440) (cubic system, JCPDS 89-4319, indicated by the green asterisks) (Shi et al. 2011) are due to Fe_3O_4. The other four peaks are indexed to α-Fe_2O_3 at 33.2 (104), 40.9 (113), 43.5 (202) and 57.5° (122) (hexagonal system, JCPDS 89-0599, indicated by the blue asterisks) (Genty et al. 2012). It is the first example to construct the mixed-valence (0, +2, +3) iron-based ternary composites by a chemical route. For TF-2, the same peaks of Fe, Fe_3O_4, and α-Fe_2O_3 were observed.

The XPS spectra of the TF-1 in Figure 2 shows three peaks corresponding to Fe $2p_{1/2}$ and $2p_{3/2}$. As the binding energies of Fe $2p_{3/2}$ and $2p_{1/2}$ in metallic Fe are around 707 and 720 eV, respectively (Lu et al. 2007). The peak centered at 719.5 is attributable to Fe $2p_{1/2}$ in metallic Fe. In addition, the binding energies of Fe $2p_{3/2}$ and $2p_{1/2}$ in Fe_2O_3 and Fe_3O_4 are around 709 and 723 eV, respectively

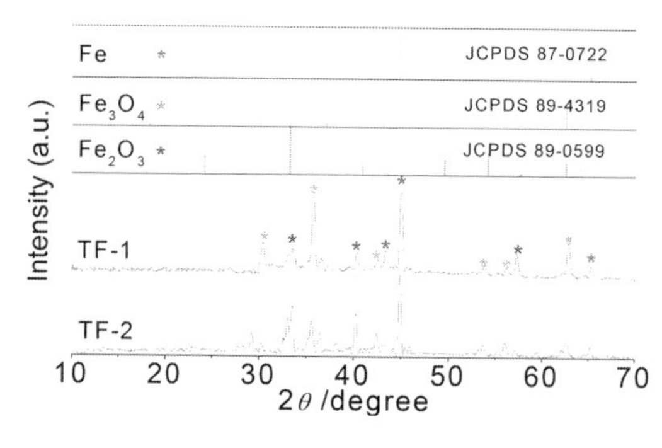

Figure 1. XRD patterns of the TF-1 and TF-2 composites.

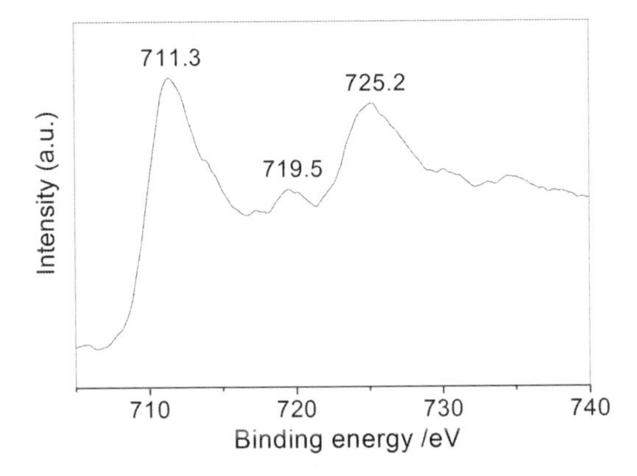

Figure 2. Fe 2p XPS spectra of TF-1.

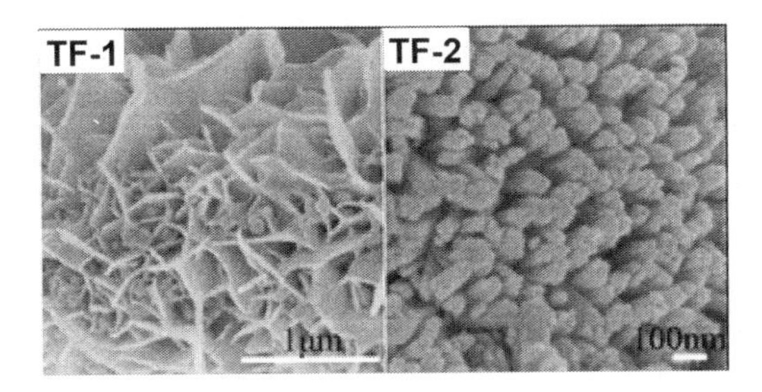

Figure 3. FE-SEM images of the TF-1 and TF-2 nanocomposites.

(Lu et al. 2007). The binding energies centered at 711.3 and 725.2 eV are the general reflection of Fe $2p_{3/2}$ and $2p_{1/2}$ in the TF-1, respectively.

These results show that the TF-1 and TF-2 are both ternary Fe-based composites in-situ originated during the sintering process, and the initial ratio of the precursors has caused a change in stacking properties of the materials.

The FE-SEM images of the two composites in Figure 3 exhibit different kinds of morphologies. TF-1 shows a self-assembled structure with numerous nanosheets (thickness 20~50 nm, width, 200~500 nm). However, the TF-2 demonstrates a nanorod like structure (diameter 80~100 nm).

3.2 *Proposed mechanism of the TF-1 and TF-2 composites*

Three Fe-based materials were formed using binary precursors ($FeCl_3 \cdot 6H_2O$/β-CD, 1:1; $FeCl_3 \cdot 6H_2O$/Fc, 1:1; and Fc/β-CD, 1:1, marked as BF-1, BF-2 and BF-3, respectively) in order to figure out the formation process of the two ternary nanocomposites. The SEM images in Figures 4 A, B and C show that,

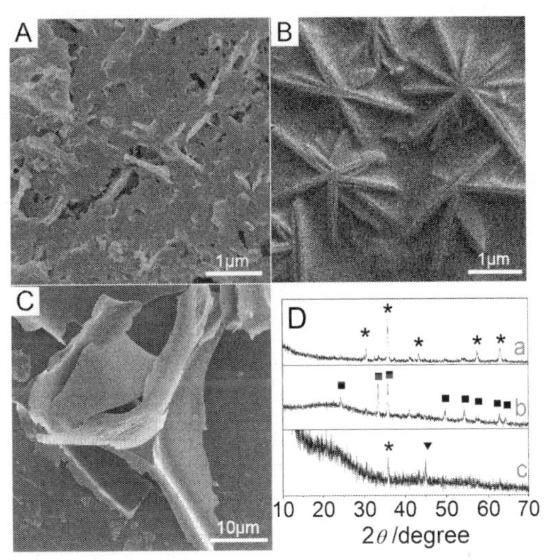

Figure 4. FE-SEM images and XRD patterns of the BF-1 (A), BF-2 (B) and BF-3 (C). (*, ▲ and ■ stand for Fe_3O_4, Fe and α-Fe_2O_3, respectively.)

the BF-1, 2, and 3 exhibit a membrane like structure, a flower like structure assembled by nanorods, and a lamellar structure, respectively, which are totally different from TF-1 or -2. More importantly, the components of the materials are distinguished from each other (Figure 4D). For example, the BF-1 and BF-2 have structures indexed to the pure Fe_3O_4 (JCPDS 89-4319) (Shi et al. 2011) and pure α-Fe_2O_3 (JCPDS 89-0599) (Genty et al. 2012), respectively. However, the XRD peaks of BF-3 can be attributed to a composite of Fe and Fe_3O_4.

These results suggested the important role of β-CD in the generation of metallic Fe and Fe(II)-compound in this construction process. On the one hand, the inclusion interaction between β-CD and Fc might prevent Fc from sublimation at higher temperatures (Liu et al. 1999). On the other hand, carbon produced by β-CD provides a reductive environment for the formation of the mixed valence composites (Teng et al. 2014). The other interesting finding is that metallic iron only occurs when Fc was introduced, indicating that Fc might be the origin of zero-valent iron. Thus, more Fc in the precursor of TF-2 might lead to more metallic Fe in the final product, resulting in a change in the structure and morphology of the two composites.

$$\text{Fc} + \beta\text{-CD} \rightarrow \beta\text{-CD-Fc} \quad (1)$$

$$FeCl_3 + 3H_2O \rightarrow Fe(OH)_3 + 3HCl \quad (2)$$

$$2Fe(OH)_3 \rightarrow Fe_2O_3 + 3H_2O \quad (3)$$

$$\beta\text{-CD} \rightarrow C + H_2O \quad (4)$$

$$Fc \rightarrow Fe \tag{5}$$

$$6Fe_2O_3 \rightarrow 4Fe_3O_4 + CO_2 \tag{6}$$

Taken together with our previous findings, a possible formation and transformation mechanism of the Fe-based composites was proposed as follows (Equations 1–6). First, the inclusion complex β-CD-Fc formed in the stirring process of the precursor, and the mixture of $FeCl_3$ and β-CD-Fc was sintered at 773 K thereafter (Equation 1) (Liu et al. 1999). Second, the α-Fe_2O_3 crystal generated (Equations 2 and 3) from the hydrolysis of $FeCl_3$ (Wang et al. 2012). Third, some of β-CD molecules were carbonized to produce carbon particles (Equation 4) (Giordano et al. 2001). Then, Fc was decomposed to metallic Fe with the protection of nitrogen (Equation 5) (Wang et al. 2012). Finally, some of the as-obtained α-Fe_2O_3 was reduced to Fe_3O_4 by the in-situ carbon particles produced during synthesis (Equation 6).

3.3 *Magnetic properties of the TF-1 and TF-2 composites*

We further investigated the magnetic properties of the TF-1 and TF-2 composites using Superconducting Quantum Interference Device (SQUID). Figure 5 illustrates the field dependences of magnetizations of the two nanomaterials at 298 K. Two interesting phenomena are developed from this figure. First, the two Fe-based nanocomposites exhibit superparamagnetism. Second, the nanosheets had a saturation magnetization (*M*s) of 12.5 emu/g, while the nanorods possessed a *M*s value of 7.5 emu/g. The smaller crystal sizes of TF-2 nanorods might be the main reason for the low *M*s. The smooth change of magnetization

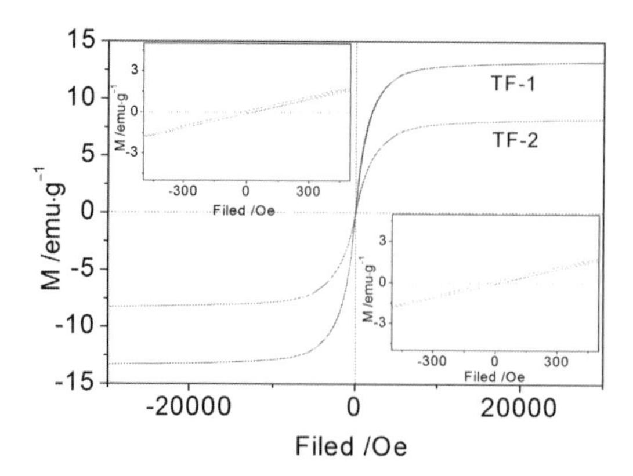

Figure 5. Field dependence magnetic hysteresis loops of the TF-1 and TF-2 composites at 298 K.

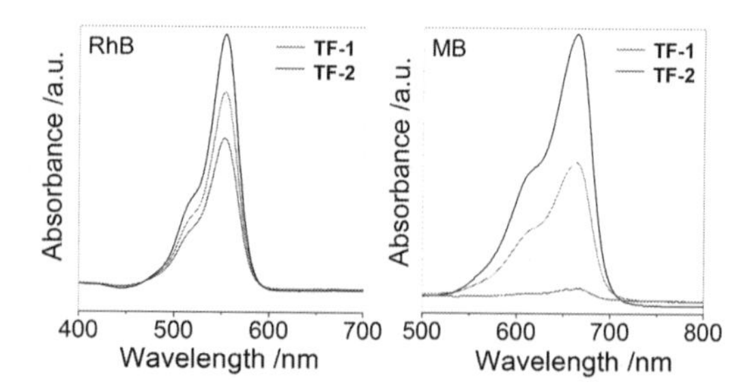

Figure 6. UV-Vis spectra of RhB and MB solutions in the absence and presence of the TF-1 and TF-2 kept for 2 h under visible light irradiation.

with given field implied that the three components (Fe, Fe_2O_3 and Fe_3O_4) were in intimate contact and exchange coupled (Zeng et al. 2002).

3.4 *Photodegradation of dyes catalyzed by the TF-1 and TF-2 composites*

Figure 6 displays the absorption spectra of the solutions photodegraded by the as-synthesized TF-1 and TF-2 nanomaterials. It can be seen that the strong absorption peaks of the RhB solutions at 554 nm and the MB solutions at 664 nm were decreased after irradiation in the presence of the catalysts (Li et al. 2014). Photodegradation degrees (α) can be calculated by Equation 7 (Song et al. 2012).

$$\alpha = (A_0 - A)/A_0 \times 100\% \tag{7}$$

In the equation, A_0 and A are the maximum absorption values of a dye solution after light irradiation with and without the two materials, respectively. Almost all MB and 40.4% RhB was degraded by the TF-2, whereas the TF-1 only degraded 48.6% and 22.5% of the RhB and MB dyes, respectively. Two interesting phenomena were observed. I). For the same dye, the TF-2 shows a better catalytic performance than the TF-1. This might due to the smaller particle size and lager surface area of the rod-like structure of the TF-1 compared with the TF-2. II) RhB and MB exhibited quite different photodegradation properties for the same catalyst. We consider that this might be attributed to the structural distinction (eg. MB contains a sulfonate functional group and an azo structure, while RhB has a xanthen group and N,N-diethylamino structures) of the two dyes (Song et al. 2012, Khedkar et al. 2012), because different structure may result in different adsorption interactions and photocatalytic mechanism (Wang et al. 2008). Therefore, we concluded that the catalytic selectivity of the TF catalysts is dependent on chemical structure of the organic dyes.

4 CONCLUSIONS

Two ternary $Fe/Fe_3O_4/Fe_2O_3$ nanocomposites were constructed through a sintering process using $Fc/\beta\text{-}CD/FeCl_3$ complex as a precursor. Our results reveal that the structural and physical properties of the ternary nanocomposites can be mediated by the initial amount of Fc in the precursor. Importantly, β-CD plays a key role in protecting metallic Fe and reducing Fe_2O_3 during the sintering process by generating carbon. Further, the TF-2 nanorods exhibit an excellent photocatalytic activity for MB and very high photocatalytic selectivity for the organic dyes: MB > RhB. Overall, this work represents an important step toward the controllable synthesis of Fe-based nanocomposites.

ACKNOWLEDGEMENT

The authors are grateful to the Fundamental Research Funds for the Central Universities (No. WK 2060190052) and the Natural Science Foundation of Anhui Province (No. 090416228) for financial support of this work.

REFERENCES

Ai, Z.H., Gao, Z.T., Zhang, L.Z., He, W.W. & Yin, J.J. 2013. Core-shell structure dependent reactivity of $Fe@Fe_2O_3$ nanowires on aerobic degradation of 4–chlorophenol. *Environmental Science & Technology*, 47(10): 5344–5352.

Bonnemann, H., Brijoux, W. & Joussen, T. 1990. The preparation of finely divided metal and alloy powders. *Angewandte Chemie International Edition*, 29(3): 273–275.

Dumitrache, F., Morjan, I., Alexandrescu, R., Ciupina, V., Prodan, G., Voicu, I., Fleaca, C., Albu, L., Savoiu, M., Sandu, I., Popovici, E. & Soare, I. 2005. Iron-iron oxide core-shell nanoparticles synthesized by laser pyrolysis followed by superficial oxidation. *Applied Surface Science*, 247(1–4): 25–31.

Genty, E., Cousin, R., Capelle, S., Gennequin, C. & Siffert, S. 2012. Catalytic Oxidation of Toluene and CO over Nanocatalysts Derived from Hydrotalcite-Like Compounds (X62+Al23+): Effect of the Bivalent Cation. *European Journal of Inorganic Chemistry*, 2012(16): 2802–2811.

Giordano, F., Novak, C. & Moyano, J.R. 2001. Thermal analysis of cyclodextrins and their inclusion compounds. *Thermochimica Acta*, 380(2): 123–151.

Guo, X.W., Fang, X.P., Mao, Y., Wang, Z.X., Wu, F. & Chen, L.Q. 2011. Capacitive Energy Storage on Fe/Li3PO4 Grain Boundaries. *The Journal of Physical Chemistry C*, 115(9): 3803–3808.

Khedkar, J.K., Jagtap, K.K., Pinjari, R.V., Ray, A.K. & Gejji, S.P. 2012. Binding of rhodamine B and kiton red S to cu-curbit[7]uril: density functional investigations. *Journal of Molecular Modeling*, 18(8): 3743–3750.

Li, H.P., Liu, J.Y., Liang, X.F., Hou, W.G. & Tao, X.T. 2014. Enhanced visible light photocatalytic activity of bismuth ox-ybromide lamellas with decreasing lamella thicknesses. *Journal of Materials Chemistry A*, 2(23): 8926–8932.

Liu, J., Mendoza, S., Roman, E., Lynn, M.J., Xu, R.L. & Kaifer, A.E. 1999. Cyclodextrin-modified gold nanospheres. Host-guest interactions at work to control colloidal properties. *Journal of the American Chemical Society*. 121(17): 4304–4305.

Lu, L.R., Ai, Z.H., Li, J.P., Zheng, Z., Li, Q. & Zhang, L.Z. 2007. Synthesis and characterization of Fe-Fe2O3 core-shell nanowires and nanonecklaces. *Crystal Growth & Design*, 7(2): 459–464.

Noradoun, C., Engelmann, M.D., McLaughlin, M., Hutcheson, R., Breen, K., Paszczynski, A. & Cheng, I.F. 2003. Destruction of chlorinated phenols by dioxygen activation under aqueous room temperature and pressure conditions. *Industrial & Engineering Chemistry Research*, 42(21): 5024–5030.

Nurmi, J.T., Tratnyek, P.G., Sarathy, V., Baer, D.R., Amonette, J.E., Pecher, K., Wang, C., Linehan, J.C., Matson, D.W., Penn, R.L. & Driessen, M.D. Characterization and properties of metallic iron nanoparticles: Spectroscopy, electro chemistry, and kinetics. *Environmental Science & Technology*, 39(5): 1221–1230.

Shi, W.H., Zhu, J.X., Sim, D.H., Tay, Y.Y., Lu, Z.Y., Zhang, X.J., Sharma, Y., Srinivasan, M., Zhang, H., Hong, H.H. & Yan, Q.Y. 2011. Achieving high specific charge capacitances in Fe3O4/reduced graphene oxide nanocomposites. *Journal of Materials Chemistry*, 21(10): 3422–3427.

Song, L.X., Teng, Y. & Chen, J. 2012. Structure, Property, and Function of Gallium/Urea and Gallium/Polyethylene Glycol Composites and Their Sintering Products: beta- and gamma-Gallium Oxide Nanocrystals. *The Journal of Physical Chemistry C*, 116(43): 22859–22866.

Teng, Y., Song, L.X., Ponchel, A., Monflier, E., Shao, Z.C., Xia, J. & Yang, Z.K. 2014. Temperature-dependent formation of Ru-based nanocomposites: structures and properties. *RSC Advances*. 4(51): 26847–26854.

Tsai, K.L. & Dye, J.L. 1991. Nanoscale metal particles by ho-mogeneous reduction with alkalides or electrides. *Journal of the American Chemical Society*, 113(5): 1650–1652.

Wang, H., Yan, N., Li, Y., Zhou, X.H., Chen, J., Yu, B.X., Gong, M. & Chen, Q.W. 2012. Fe nanoparticle-functionalized multi-walled carbon nanotubes: one-pot synthesis and their applications in magnetic removal of heavy metal ions. *Journal of Materials Chemistry*, 22(18): 9230–9236.

Wang, L.B., Song, L.X., Dang, Z., Chen, J., Yang, J. & Zeng, J. 2012. Controlled growth and magnetic properties of alpha-Fe_2O_3 nanocrystals: Octahedra,

cuboctahedra and truncated cubes. *Cryst Eng Comm,* 14(10): 3355–3358.
Wang, Q., Chen, C.C., Zhao, D., Ma, W.H. & Zhao, J.C. 2008. Change of adsorption modes of dyes on fluorinated TiO_2 and its effect on photocatalytic degradation of dyes under visible irradiation. *Langmuir,* 24(14): 7338–7345.
Zeng, H., Li, J., Liu, J.P., Wang, Z.L. & Sun, S.H. 2002. Exchange-coupled nanocomposite magnets by nanoparticle self-assembly. *Nature,* 420(6914): 395–398.
Zeng, H., Li, J., Wang, Z.L., Liu, J.P. & Sun, S.H. 2004. Bimagnetic core/shell FePt/Fe_3O_4 nanoparticles. *Nano Letters*, 4 (1): 187–190.

Advanced Materials, Mechanical and Structural Engineering – Hong, Seo & Moon (Eds)
© 2016 Taylor & Francis Group, London, ISBN: 978-1-138-02908-8

Determination of creep parameters of materials by indentation load relaxation approach

W.Z. Yan & Z.F. Yue
Advanced Materials Test Center, Department of Engineering Mechanics, Northwestern Polytechnical University, Xi'an, P.R. China

ABSTRACT: The present work explored the possibility of retrieving tensile creep parameters from indentation load relaxation test. The relationship between the indentation relaxation and uni-axial tensile relaxation was built. Assuming the creep of material obey the Norton creep law, it is deduced that the indentation load relaxation rate exhibits a power law dependence on the current indentation load, and the exponent is identical with that from the Norton creep law. Two new procedures were then proposed to obtain tensile creep parameters directly from the indentation load relaxation curves.

1 INTRODUCTION

The growing application of high temperature materials, such as gas turbines, chemical reactors and nuclear materials, challenges the researchers to assess the mechanical properties of materials at high temperature (Webster & Ainsworth 1994). In such applications, creep becomes important due to its thermal activated mechanism (Mavoori et al. 1997).

Engineering approaches often neglect short primary and tertiary creep regimes and describe the creep process by the stress and temperature dependence of the secondary creep rate, which can be written in power law form for most engineering alloy:

$$\dot{\varepsilon} = B(T)\cdot\sigma^n = B'\cdot\exp\left(-Q_{app}/RT\right)\cdot\sigma^n = A\sigma^n \quad (1)$$

where, $B(T)$ is a temperature-dependent constant, σ is the applied stress, n is the stress exponent, B' is the material and microstructure dependent creep constant, Q_{app} is the apparent activation energy of creep, and R and T are gas constant and absolute temperature. When we are not explicitly interested in the temperature dependence of creep, we integrate the temperature dependence into the constant A, which then accounts for micro-structural effects and for the temperature dependence.

There are many approaches reported in literature to evaluate material creep properties. Traditional method is the uni-axial tensile creep test under a constant tensile load (Carroll et al. 1989). However, this method is not practicable because the design of specimen gripping, precision extensometer and stable temperature control become increasingly sophisticated with the growth of temperature (Woodford 1996). Meanwhile, the preparation of specimens is expensive and time-consuming. These shortages of traditional creep test motivated the researchers to explore new methods to obtain material creep properties. Another novel method to determine material creep properties is the 'indentation creep' test using a flat-ended punch (Yu & Li 1977). Recently, the effectiveness of indentation creep test has been validated by the determination of creep parameters of nickel based single crystal super alloy (Yan et al. 2010).

Most of previous researches on creep test were carried out employing a constant load aims to generate a constant creep stress. In engineering, however, creep usually occurs in the form of stress relaxation. It was first shown by Hart (1976) that a carefully conducted stress relaxation test was capable of establishing the stress vs. strain rate relationship over a wide range of strain rates with relatively small amounts of plastic strain. The 'Bending Stress Relaxation (BSR)' method (Morscher & Dicarlo 1992, Katoh & Snead 2005, Ivankovic et al. 2006) has been exploited to qualitatively evaluate creep properties of materials serve at high temperature. However, no procedure that enables uni-axial tensile creep parameters to be directly derived from creep stress relaxation curves can be found in literature.

The present work aims at providing some simple and effective procedures to retrieve material creep parameters by indentation load relaxation approach. The indentation load relaxation data was converted into tensile creep data by making use of conversion factors.

2 ANALYTICAL ASSESSMENT OF INDENTATION LOAD RELAXATION USING A FLAT CYLINDRICAL INDENTER

For the tensile stress relaxation process, the stress re-laxation ratio is defined as (Morscher & Dicarlo 1992):

$$m = \sigma(t)/\sigma(0) \tag{2}$$

where, $\sigma(t)$ is the current stress during stress relaxation process, $\sigma(0)$ is the initial stress before relaxation. The creep strain can be expressed as (Morscher & Dicarlo 1992):

$$\begin{aligned}\varepsilon c(t) &= \varepsilon(t)[(1/m)-1] \\ &= [\sigma(t)/E][(1/m)-1]\end{aligned} \tag{3}$$

where, $\varepsilon(t)$ is the current elastic strain, $\varepsilon_c(t)$ is the current creep strain and E is the Young's modulus.

It is a common practice to correlate the indentation data with tensile data using conversion factors (Sastry 2005). The current elastic strain of tensile test can be predicted from indentation data by:

$$\varepsilon(t) = C_1 \frac{h(t)}{D} = C_1 \frac{F(t)}{DS} \tag{4}$$

where, $F(t)$ is the current indentation load, $h(t)$ is the pseudo indentation depth corresponding to the cur-rent indentation load, D is the diameter of the indenter, C_1 is the conversion factor used with $C_1 = 1$, which is an acceptable simplification for most materials (Sastry 2005), S is the indentation stiffness:

$$S = \frac{Ed}{1-\nu^2} = \frac{F}{h} \tag{5}$$

where, ν is the Possion's ratio.

The creep stress can be converted between the indentation stress and the tensile stress by:

$$\sigma(t)_{-I} = C_0 \sigma(t)_{-T} \tag{6}$$

where, C_0 is the stress conversion factor within the range of 2.77~3.85 (Liu et al. 2007), $\sigma(t)_{-T}$ is the tensile stress, $\sigma(t)_{-I}$ is the indentation stress defined as the indentation load F divided by the cylindrical indenter area.

Submitting Equations 4 and 6 into Equation 3, we have:

$$\begin{aligned}\varepsilon_c(t) &= C_1 \frac{F(t)}{DS}\left[(1/m)-1\right] \\ &= C_1 \frac{C_0 \sigma(t)_{-T} \cdot \pi (D/2)^2}{DS}\left[(1/m)-1\right] \\ &= \frac{C_1 C_0 \pi D}{4S}\sigma(0)_{-T} - \frac{C_1 C_0 \pi D}{4S}\sigma(t)_{-T}\end{aligned} \tag{7}$$

Differentiating Equation 7, we obtain:

$$\dot{\varepsilon}_c(t) = -\frac{C_1 C_0 \pi D}{4S}\dot{\sigma}(t)_{-T} = A\left[\sigma(t)_{-T}\right]^n \tag{8}$$

For simplicity, Equation 8 can be transformed into:

$$\dot{\sigma}(t)_{-T} = B\left[\sigma(t)_{-T}\right]^n \tag{9}$$

where,

$$B = -\frac{4AS}{C_1 C_0 \pi D} = -\frac{4AE}{C_1 C_0 (1-\nu^2)\pi} \tag{10}$$

Inserting Equation 6 to Equation 9, we obtain:

$$\dot{\sigma}(t)_{-I} = \frac{B}{C_0^{n-1}}\left[\sigma(t)_{-I}\right]^n = Y\left[\sigma(t)_{-I}\right]^n \tag{11}$$

where,

$$Y = -\frac{4AE}{C_1 C_0{}^n (1-\nu^2)\pi} \tag{12}$$

Equation 11 can be substituted by:

$$\frac{\dot{F}(t)}{\pi R^2} = Y\left[\frac{F(t)}{\pi R^2}\right]^n \tag{13}$$

where, R is the radius of the indenter. Equation 13 can be further written by:

$$\dot{F}(t) = Z\left[F(t)\right]^n \tag{14}$$

where,

$$Z = \frac{4AE}{C_1 C_0{}^n (1-\nu^2)\pi^n R^{2n-2}} \tag{15}$$

From Equation 14, it can be seen that the indentation load relaxation rate shows a power law dependence on the current indentation load, and the exponent is identical with the creep stress exponent from Norton creep formula.

3 MODEL DEVELOPMENT

Two finite element models are developed in the present study, they are indentation relaxation model and uni-axial tensile relaxation model, respectively (see Figure 1).

A flat cylindrical indenter with a diameter of $D = 0.5$ mm was employed in the indentation relaxation model. Axisymmetric model was used for the simulation of indentation load relaxation. The size of specimen was chosen as $10D \times 10D \times 10D$ to eliminate the boundary effect. The mesh of the indented zone was refined. The model was divided into 900 linear quadrilateral elements of type CGAX4R. The nodes in the indented zone were constrained using Multiple Point Constraint (MPC) to guarantee the displacements in y direction of each node are the same. The indentation load relaxation analysis was realized by two steps:

Step 1: Static, general analysis step in which the initial indentation load $F|_{t=0}$ was applied on the indented zone.

Step 2: Visco analysis step in which the nodes in the indented zone was constrained in y direction and their reaction forces were output.

The simulation of tensile stress relaxation employed a cylindrical specimen with a length of 100 mm and a diameter of 20 mm. The fix end was constrained in x-, y- and z-directions while the loading end was constrained in x- and y-directions. The tensile load P was applied on the loading end. The mechanical parameters that input the FE model were listed in Table 1 (Yan et al. 2009).

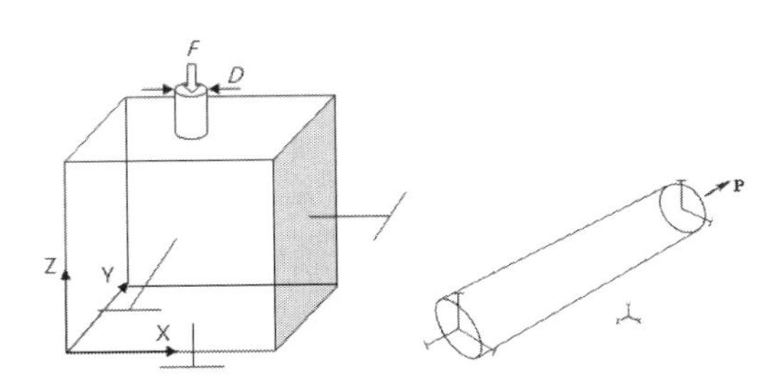

Figure 1. Schematic view and finite element model of the (a) indentation and (b) tensile relaxation.

Table 1. Mechanical properties used in FE models.

Properties	Value
Young's modulus (MPa)	138000
Possion's ratio	0.33
Creep pre-factor (s^{-1} MPa^{n})	1.78×10^{-17}
Creep exponent	5.0

4 RESULTS AND DISCUSSION

4.1 *The indentation relaxation curve*

The normalized indentation load relaxation curves with different initial indentation loads were shown in Figure 2a. It can be seen that a higher initial indentation load results in a faster decrease of normalized indentation load for the same relaxation time, which is similar with the tensile stress relaxation curve (Taub & Luborsky 1981). The relaxation kinetics of the normalized indentation relaxation curve can be remarkably well fitted by the so-called Kohlrausch-Williams-Watt (KWW) relaxation equation (Cumbrera et al. 1993):

$$m = F(t)/F|_{t=0} = \exp\left[-\left(\frac{t}{\tau}\right)^{b}\right] \tag{16}$$

where, τ is the characteristic relaxation time which can be regarded as the waiting time of the activation process. The parameter b is called the correlation factor and can be related tentatively to the mobility of a structural unit at the atomic or molecular scale. The value of b ranges from 0 to 1 (Ngai et al. 1988, Perez et al. 1988). The best fit parameters for the normalized indentation relaxation curves were listed in Table 2. It is seen that the characteristic relaxation time decreases with the increase of initial indentation load, which indicates that deformation is easier to be thermally activated under a higher indentation load.

Figure 2b was obtained by differentiating the indentation load relaxation curves. It can be seen that the indentation load relaxation rate experienced two stages: the transient decrease stage and the steady stage. In the steady state, the indentation load relaxation rate converges at a constant value with the lapse of time, which is also in consistence with the tensile stress relaxation (Mavoori et al. 1997).

Table 2. Load relaxation characteristics under different initial indentation loads.

Initial indentation load $F|_{t=0}$ (N)	Characteristic relaxation time τ(s)	Correlation factor b
118	5820	0.81
157	4419	0.68
196	3271	0.58
235	2403	0.52

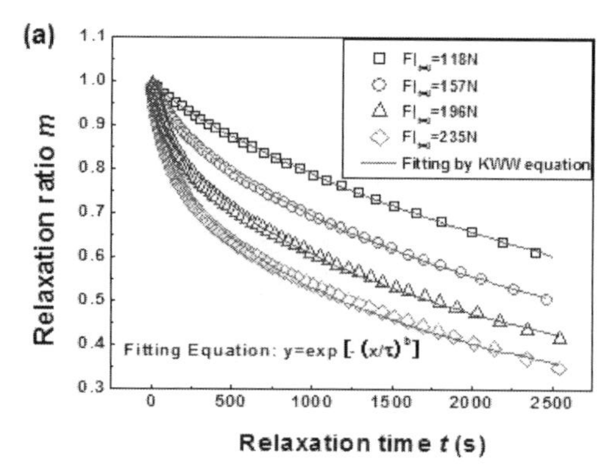

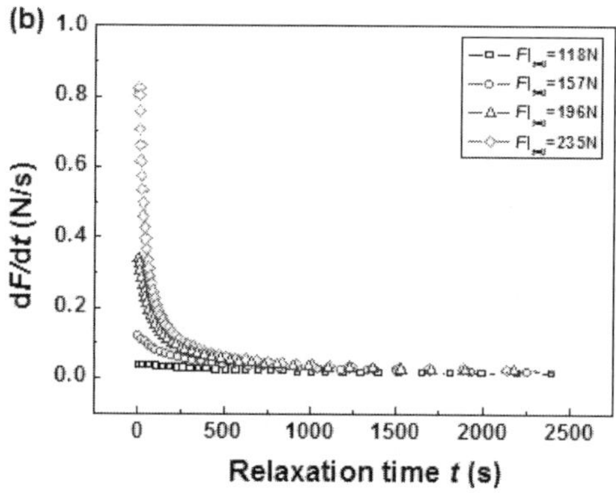

Figure 2. The evolution of (a) normalized indentation load and (b) indentation load relaxation rate as a function of relaxation time.

4.2 *Determination of creep stress exponent from a single indentation load relaxation curve (procedure 1)*

From Equation 14, we know that the creep stress exponent can be determined from the following log-log relation:

$$\lg \dot{F} = \lg Z + n \lg F \quad (17)$$

and the creep stress exponent can be expressed by:

$$n = \frac{d\left(\lg \dot{F}\right)}{d\left(\lg F\right)} \quad (18)$$

The indentation load relaxation rate as a function of current indentation load was plotted in the double logarithmic grid (see Figure 3), yielding the slope of the curve as the creep stress exponent (refer to Equation 18). For comparison purpose we consider the indentation load relaxation curves with two different initial indentation loads.

Several interesting features can be seen in Figure 3. There is a well transition of indentation load relaxation rate between the low load regime and high load regime. The linear fitting of data in high load regime yields a reasonable creep stress exponent, although it exhibits a slight dependence on the initial indentation load $F|_{t=0}$ (see Figure 4). The data in low load regime can not be used to derive the creep stress exponent since it yields a much smaller value than the accurate value. This phenomenon has been reported by other researchers. For example, in the four-point bending stress relaxation test on polycrystalline ceramic, creep stress exponent of 1.17 and 4.12 have been observed in low stress regime and high stress regime, respectively (Shetty & Gordon, 1979). So far, the dependence of the derived stress exponent on the magnitude of creep stress can not be rationalized yet. Further studies need to be carried out to clarify this phenomenon.

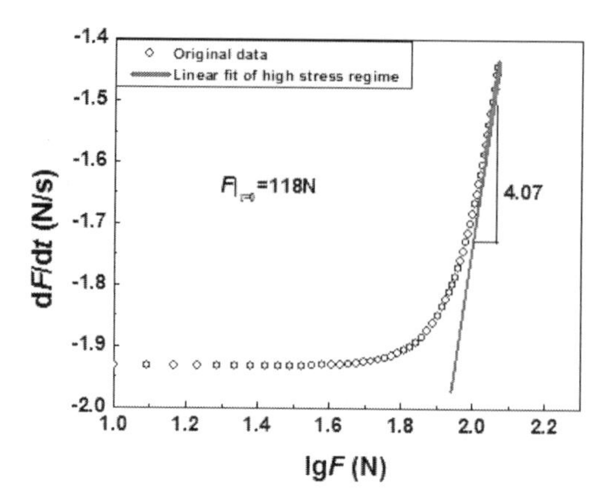

Figure 3. Determination of creep stress exponent using a single indentation load relaxation curve.

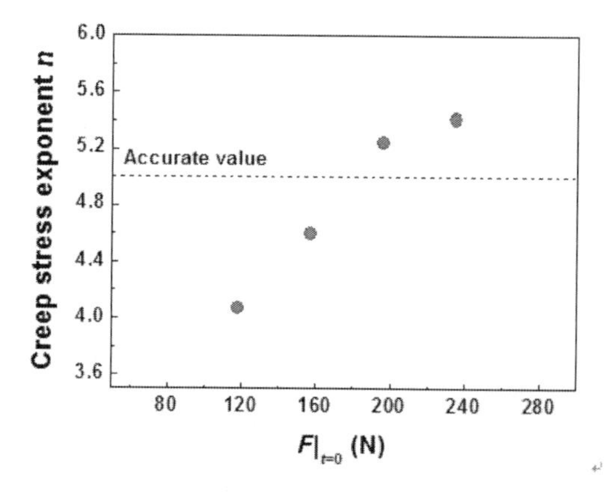

Figure 4. Effect of initial indentation load on the creep stress exponent derived by procedure 1.

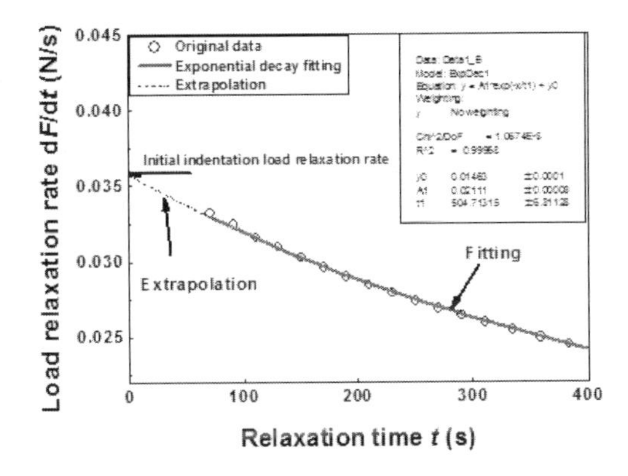

Figure 5. Determination of initial indentation load relaxation rate by extrapolation method.

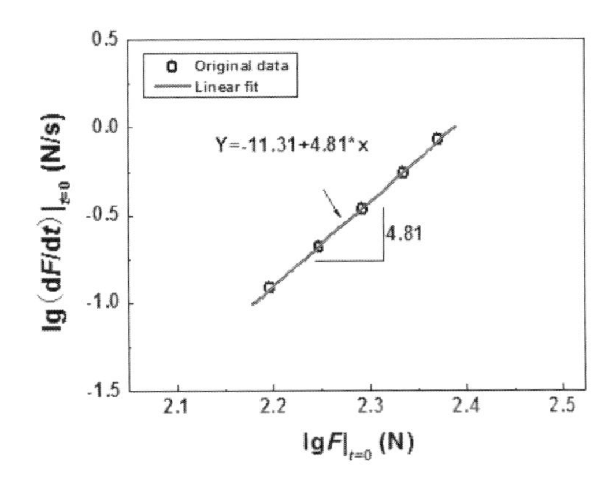

Figure 6. Determination of creep stress exponent using multiple indentation load relaxation curve.

4.3 *Determination of creep stress exponent n and creep constant A using multiple indentation load relaxation curves (procedure 2)*

Here we propose a new procedure concerning the initial indentation load and the initial indentation load relaxation rate. The key problem of this procedure is the extracting of the initial indentation load relaxation rate form the relaxation curve. Here, we put forward the extrapolation approach to obtain the initial indentation load relaxation rate. The indentation load relaxation rate was plotted as a function of relaxation time. The data (in high load regime) was then fitted using an exponential decay function and extrapolated to zero relaxation time ($t = 0$), and the corresponding value can be considered as the initial indentation load relaxation rate.

The initial indentation load relaxation rates corresponding to initial indentation load of $F|_{t=0}$ = 118, 157, 196 and 235 N were extracted using the extrapolation approach and plotted as a function of initial indentation load in log-log form as shown in Figure 6.

Synthesizing the fitting results of Figure 6 and Equation 17, we have:

$$\lg Z = \lg\left[\frac{4AE}{C_1 C_0^n \left(1-\nu^2\right)\pi^n R^{2n-2}}\right]$$

$$= -11.31 \tag{19}$$

from which the creep constant A can be solved.

The derived creep stress exponent and creep constant were compared with the accurate values (see Table 3). It can be seen that the creep parameters derived from procedure 2 agree reasonably well with the accurate values.

Table 3. Creep parameters obtained from procedure 2 and the accurate values.

Creep parameter	Procedure 2	Accurate value
Creep pre-factor A (s^{-1}/MPa^n)	1.01×10^{-17}	1.78×10^{-17}
Stress exponent n	4.81	5.0

5 SUMMARY AND CONCLUSIONS

The present study started from an analytical assessment of the indentation load relaxation test. The relationship between indentation load relaxation data and uni-axial tensile relaxation data was built.

Based on theoretical analysis, two procedures were proposed to derive tensile creep parameters directly from indentation load relaxation curve. The first procedure is capable of deriving tensile creep parameters from a single indentation load relaxation curve. In the second procedure, the extrapolation approach was put forward to extract the initial indentation load relaxation rate. Although multiple relaxation curves are required, the second procedure shows a higher reliability and accuracy.

ACKNOWLEDGEMENT

The authors appreciate the financial supports from National High Technology Research and Development Program of China (2007 AA04Z404), the Introduction of High-level Personnel Research and Start-up Funded Project of Northwestern Polytechnical University (G2015 KY0309).

REFERENCES

Carroll, D.F., Wiederhom, S.M. & Roberts, D.E. 1989. Technique for Tensile Creep Testing of Ceramics. *Journal of the American Ceramic Society*, 72(9): 1610–1614.

Cumbrera, F.L., Sanchez-Bajo, F., Guiberteau, F., Solier, J.D. & Muñoz, A. 1993. The Williams-Watts dependence as a common phenomenological approach to relaxation processes in condensed matter. *Journal of Materials Science,* 28(19): 5387–5396.

Hart, E.W. 1976. Constitutive Relations for the Nonelastic Deformation of Metals. *Journal of Engineering Materials and Technology,* 98(3): 193–202.

Ivankovic, H., Tkalcec, E. & Rein, R. 2006. Microstructure and high temperature 4-point bending creep of sol–gel derived mullite ceramics, *Journal of the European Ceramic Society,* 26: 1637–1646.

Katoh, Y. & Snead, L.L. 2005. Bend stress relaxation creep of CVD silicon carbide. *Ceramic Engineering & Science Proceedings,* 26(2): 265–272.

Liu, Y.J., Zhao, B., Xu, B.X. & Yue, Z.F. 2007. Experimental and numerical study of the method to determine the creep parameters from the indentation creep testing. *Materials Science and Engineering A*, 456: 103–108.

Mavoori, H., Chin, J. & Vaynman, S. 1997. Creep, Stress Relaxation, and Plastic Deformation in Sn-Ag and Sn-Zn Eutectic Solders. *Journal of Electronic Materials*, 26(7): 783–790.

Morscher, G.N. & Dicarlo, J.A. 1992. A simple test for thermomechanical evaluation of ceramic fibers. *Journal of the American Ceramic Society.* 75(1): 136–140.

Ngai, K.L., Rajagopal, A.K. & Teitler, S. 1988. Slowing down of relaxation in a complex system by constraint dynamics. *The Journal of Chemical Physics,* 88(8): 5086–5094.

Perez, J., Cavaille, J.Y. & Etienne, S. 1988. Physical interpretation of the theological behavior of amorphous polymers through the glass transition. *Applied Physics Reviews*, 23: 125–135.

Sastry, D.H. 2005. Impression creep technique-An overview. *Materials Science and Engineering: A,* 409: 67–75.

Shetty, D.K. & Gordon, R.S. 1979. Stress-relaxation technique for deformation studies in four-point bend tests: application to polycrystalline ceramics at elevated temperatures. *Journal of Materials Science,* 14: 2163–2171.

Taub, A.I. & Luborsky, F.E. 1981. Creep, stress relaxation and structural change of amorphous alloys. *Acta Metallurgica*, 29(12): 1939–1948.

Webster, G.A. & Ainsworth, R.A. 1994. *High temperature component life assessment:* 103–114. London: Chapman & Hall.

Woodford, D.A. 1996. Creep design analysis of silicon nitride using stress relaxation data. *Materials & Design*, 11(3): 127–132.

Yan, W.Z., Wen, S.F., Liu, J. & Yue, Z.F. 2010. Comparison between impression creep and uni-axial tensile creep performed on nickel-based single crystal superalloys. *Materials Science and Engineering: A,* 527: 1850–1855.

Yan, W.Z., Zhao, B. & Yue, Z.F. 2009. *Effect of Surface Roughness on Determination of Creep Parameters using Impression Creep Technique.* Materials research society symposium proceeding 1224, Boston, MA, United states, 1224-GG05-12.

Yu, H.Y. & Li, J.C.M. 1977. Computer simulation of impression creep by the finite element method. *Journal of Materials Science*, 12: 2214–2222.

Advanced Materials, Mechanical and Structural Engineering – Hong, Seo & Moon (Eds)
© 2016 Taylor & Francis Group, London, ISBN: 978-1-138-02908-8

Dielectric relaxation and molecular mobility in the fullerene containing polymer nanocomposites

N.A. Nikonorova
Institute of High-Molecular Compounds, Russian Academy of Sciences, Saint-Petersburg, Russia

A.A. Kononov, R.A. Castro & N.I. Anisimova
Herzen State Pedagogical University, Saint-Petersburg, Russia

ABSTRACT: The paper views the results of research of molecular mobility and dielectric relaxation in polymer nanocomposites PPO + C_{60} in the range of frequencies 10^{-2}–10^{7} Hz and temperatures +20°C–+250°C. The existence of the relaxation β-process with the temperature shifting to the area of higher frequencies that corresponds to the mechanism of dipolar relaxation is discovered. The introduction of fullerene in the polymer matrix increases the activation energy of the β-process.

1 INTRODUCTION

Polymers are ideal matrixes for creation of new materials with the given properties and, in particular, for receiving the Pervaporative Membranes (PM). Modification of PM by introduction of carbon nanoparticles to a polymeric matrix leads to the receiving of membranes with the increased level of selectivity and productivity. The properties of nanocomposites depend not only on the chemical constitution of a matrix, concentration and the nature of nanoparticles, but also on the molecular mobility determined by the inside- and intermolecular interactions (Castro & Lushin 2014, Afanasjev et al. 2012). In this research the molecular mobility of nanocomposites based on Polyphenylene Oxide (PPO) with 0, 1, 2, 4 И 8% of fullerene (C_{60}) is investigated by the method of the dielectric spectroscopy.

2 METHODS AND MATERIALS

The composite PPO + C_{60} was obtained by mixing the solutions of PPO in chloroform (concentration of 2 masses.%) and C_{60} in toluene (concentration of 0.14 masses.%) in amounts providing the demanded fullerene content in the composite. The resulting solution was left for 3–4 days to conduct the interaction between the polymer and molecules of C_{60}. Then, the solution of the composite was treated by ultrasound for 40 min and filtered by using the Shott filter to remove dust impurities. The films were obtained by evaporation of toluene from the mixture of C_{60} and PS solutions on a cellophane surface at a temperature of 40°C.

The dielectric spectra (temperature-frequency dependence of dielectric parameters) were obtained with the broadband spectrometer "Novocontrol 81" in the range of frequencies 10^{-2}–10^{7} Hz and temperatures +20°C–+250°C.

The investigation of the surface structure of the samples was performed under low vacuum (up to 350 PA) with the use of the scanning supermicroscope (SEM) of Carl Zeiss EVO 40. The scanning supermicroscope (SEM) Carl Zeiss EVO 40 is designed for obtaining images of objects in "direct" electrons and electrons of a backscattering. The maximum resolution of the microscope is 3 nanometers.

When considering the surface of phenylene-oxide similar inhomogeneities were not detected.

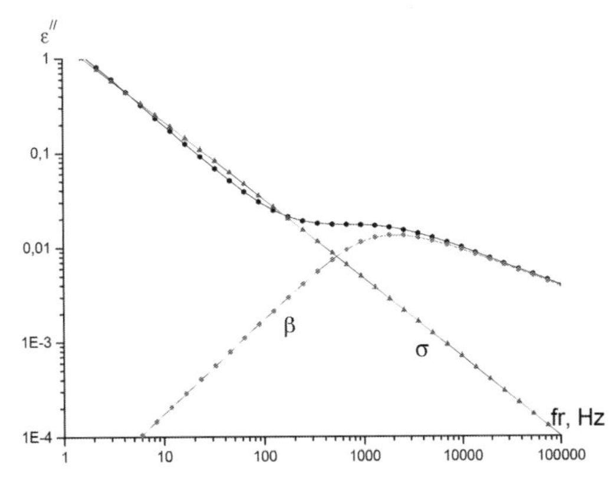

Figure 1. Frequency dependence of ε// at temperature 250°C.

The cut is rather uniform in zone 1, zones 2 and 3 correspond to the defects of an sample which appeared as a result of deformation when cutting (Figure 4). In Figure 2 the scan of a cut of sample PPO+4%C_{60} at resolution of 20 microns is submitted. The surface is inhomogeneous, covered with black streaks, which are better seen in Figure 3.

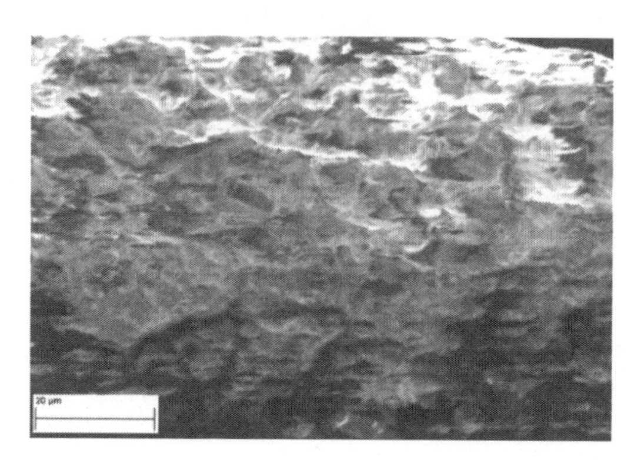

Figure 2. The scan of a cut of exemplar PPO+4%C60 at resolution of 20 microns.

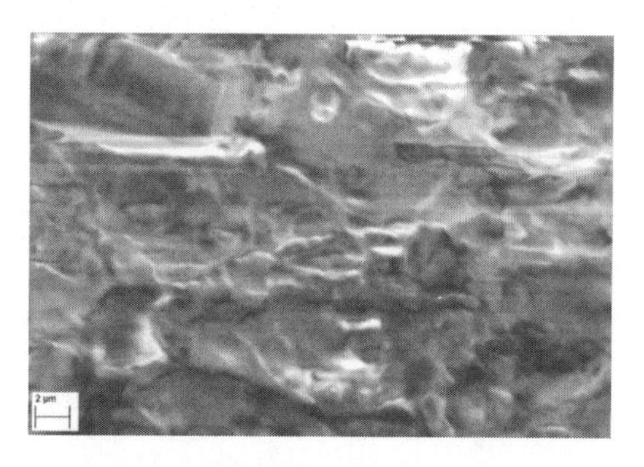

Figure 3. The scan of a cut of exemplar PPO+4%C_{60} at resolution of 2 microns.

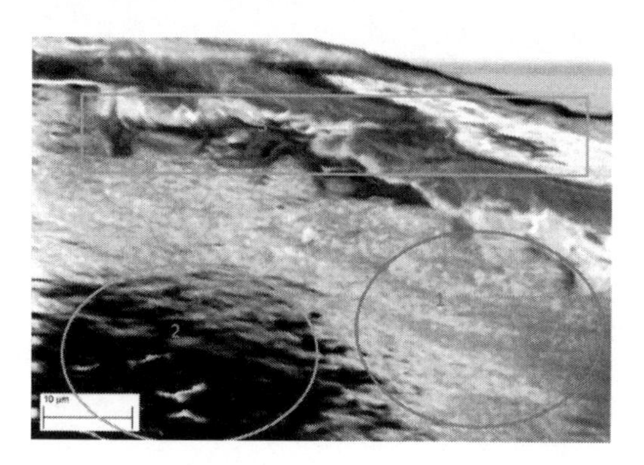

Figure 4. The scan of a cut of an exemplar of PPO at resolution of 10 microns.

When considering the surface of a clear exemplar of phenyleneoxide similar inhomogeneities were not detected. The cut is rather uniform in zone 1, zones 2 and 3 correspond to the defects of an exemplar which appeared as a result of deformation when cutting (Figure 4).

3 RESULTS AND DISCUSSION

The relaxation β-process of the dipole polarization was revealed. Figure 1 shows division of the experimental frequency dependence of $\varepsilon^{//}$ into the β-process and contribution due to conduction for the composite PPO+8%C_{60}.

The comparison of molecular mobility parameters for the relaxation process of dipole polarization for nanocomposites with different concentrations of C_{60} and model systems gives the possibility to link the β-process with the local mobility of polar groups. The molecular mobility parameters for the β-process are defined by the empirical equation by Havriliak-Negami (Kremer & Schonhals 2003):

$$\varepsilon^*(\omega) = \varepsilon_\infty + \frac{\Delta\varepsilon}{\left[1+(i\omega\tau)^{\alpha_{HN}}\right]^{\beta_{HN}}} \tag{1}$$

where, $\varepsilon\ \infty$—high frequency limit of dielectric capacitance's real part, $\Delta\varepsilon$—dielectric increment (difference between low frequency and high frequency limits), $\omega = 2\pi f$, αHN and βHN—form parameters, describing symmetrical ($\beta = 1$ – Cole-Cole model) and asymmetrical ($\alpha = 1$ – Cole-Davidson model) dispersion of relaxation function. The most probable relaxation time (τ_{max}) of this process is defined as:

$$\tau_{MAKC} = \tau_{HN}\left[\frac{\sin\left(\dfrac{\pi(\alpha_{HN})\beta_{HN}}{2(\beta_{HN}+1)}\right)}{\sin\left(\dfrac{\pi(\alpha_{HN})}{2(\beta_{HN}+1)}\right)}\right]^{1/(\alpha_{HN})} \tag{2}$$

The PPO macromolecule contains a single polar—O-group in the main chain. The polarization of this group proceeds in two stages. In the glassy state the molecular mobility of phenylene links and adjacent polar groups is the cause of the β-process. Frequency of a maximum of a loss factor increases with growth of temperature. This is characteristic for the relaxation process of dipole polarization. Figure 5 shows the frequency dependence of $\varepsilon^{//}$ at different temperatures.

The change of fullerene concentration in the composite leads to displacement of relaxation

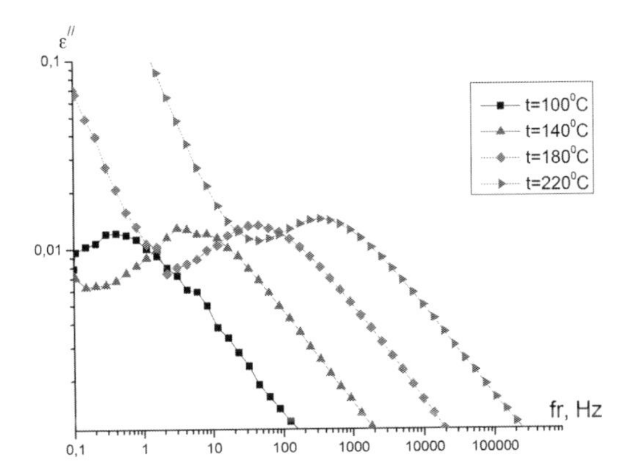

Figure 5. The frequency dependence of ε'' at different temperatures for samples with 8% of fullerene.

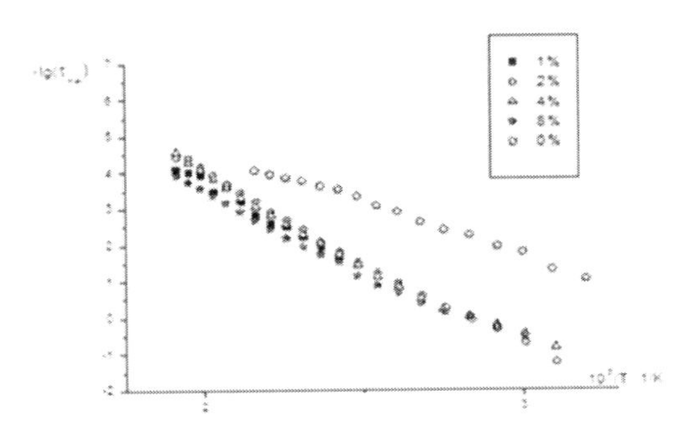

Figure 6. The temperature dependence of relaxation times τ_{max} for different percentages of C_{60} in the PPO.

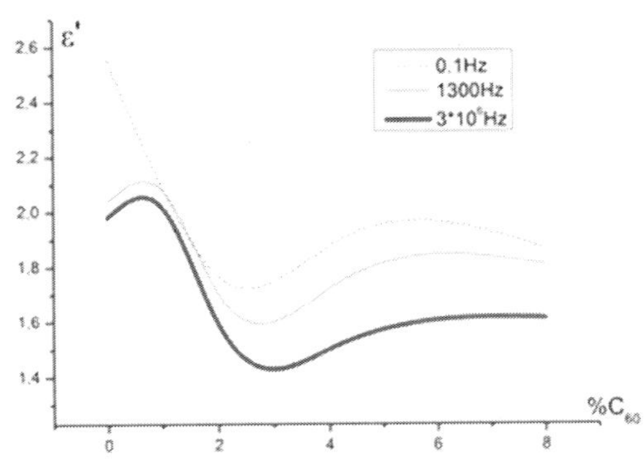

Figure 7. Dependence ε' from percentage filler content in PPO.

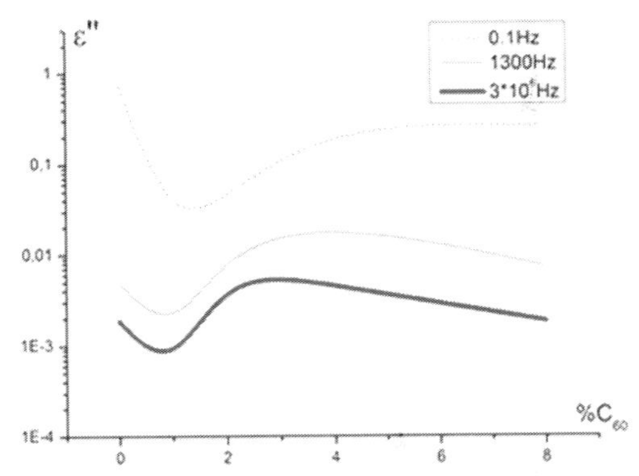

Figure 8. Dependence ε'' from percentage filler content in PPO.

process to the high temperatures region, i.e. to the reduction of molecular mobility. For the PPO-0% the parameters of the equation of Arrhenius $-\log\tau_0$ and E_a are 10.6 and 13.6 kcal/mol, for the PPO-1%, 2%, and 4% are 13 and 22 kcal/mol respectively. For the PPO-8% relaxation time are close to the fullerene containing samples, but E_a drops to 19 kcal/mol. It's possible to suppose that in the nanocomposites the fullerene introduces between the methyl groups, this decreases the free volume required for the reorientation of the phenyl rings. With the 8% concentration fullerene can form aggregates (clusters) that are embedded between the macromolecules, which leads to decrease of internal rotation barriers.

From Figures 7 and 8 it is visible that the dielectric permittivity and the loss factor depend on the percentage of the excipient C_{60} non-linearly, the minimum of losses is observed in Figure 8. The maximum losses correspond to the lowest frequencies which suggests that lowering the frequency results in the growth of polarization of a dielectric in a variation field, and, therefore, in relaxational losses. The minimum of losses is observed in the case when the excipient is in the molecular-dispersed form in the matrix of polymer and between the molecules of the polymer and the excipient the chemical bond complicating the process of polarization and reducing the relaxation losses is detected. The percentage increase in the content of C_{60} leads to higher losses, which may be attributed to relaxation processes connected with the movement of fullerene clusters, appearing already in the composite with 2% of the excipient, and to the conductivity increase. The further monotonic reduction of losses may be connected with the increase of the clusters' volume and the decrease of their quantity with the growing excipient content. The maximum value of losses is observed in the pure polymer. As the molecules and side groups of chains of the PPO are not connected by particles of fulleren their polarization is most facilitated.

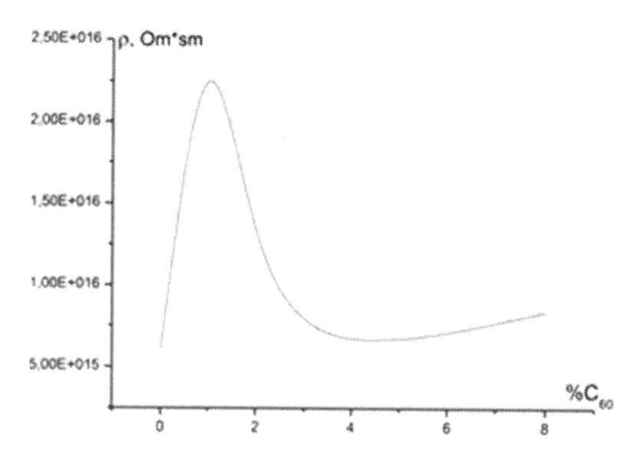

Figure 9. Dependence ρ from percentage of an excipient in PPO at $f = 0.1$ Hz.

Figure 9 shows the dependence of specific resistance on the percentage of an excipient in the polymer. The greatest resistance corresponds to 1% of the content of the excipient, which may be caused by physical cross-linking of the polymer chains to the fullerene molecules that impedes the movement of charge carriers and consequently reduces the conductivity. The further decrease in resistance may be attributed to the activation of a new type of conductivity, which is bound to formation of the continuous clusters of molecules of fulleren.

4 CONCLUSIONS

The relaxation β-process connected with the mobility of the phenylene links and adjacent polar groups is found in the PPO + C_{60}. The increase in the filler concentration leads to the increase of the β-process activation energy.

ACKNOWLEDGEMENT

The reported study was partially supported by the Ministry of Education and Science of the Russian Federation (project № 2015/376).

REFERENCES

Afanasjev, V.P., Vendik, I.B., Vendig, O.G., Medevedeva, N.Yu., Odit M.A., Sitnikova, M.F., Petrov, A.A., Sokolova, I.M., Chigirev, D.A. & Castro, R.A. 2012. Analysis of the dielectric spectra of composite sealing coatings in a wide frequency range, *Glass Physics and Chemistry*, 38(1): 63–70.

Castro, R.A. & Lushin, E.N. 2014. Dielectric spectroscopy of polymer nanocomposites based on tetrazol and KNO_3/*Journal of Physics: Conference Series,* 012096. DOI: 10.1088/1742–6596/541/1/012096., 541.

Kremer, K. & Schonhals, A. (Eds.) 2003. *Broadband dielectric spectroscopy* Springer Berlin Heidelberg, 729.

Advanced Materials, Mechanical and Structural Engineering – Hong, Seo & Moon (Eds)
 ISBN: 978-1-138-02908-8

Optimum material gradient for a Functionally Graded Endodontic Prefabricated Post: FEA

W.M.K. Helal
College of Mechanical and Electrical Engineering, Harbin Engineering University, Harbin, P.R. China
Department of Mechanical Engineering, Faculty of Engineering, Kafrelsheikh University, Kafrelsheikh, Egypt

D.Y. Shi
College of Mechanical and Electrical Engineering, Harbin Engineering University, Harbin, P.R. China

ABSTRACT: In the treatment process of our teeth, an Endodontic Prefabricated Post (EPP) plays a vital role. Scientists are always trying to improve its performance. A comparison between the performances of an EPP made of homogeneous material and an EPP made of Functionally Graded Material (FGM) was made in the present investigation. The Finite Element Analysis (FEA) was adopted in order to investigate this comparison. Then, the optimization technique was employed to carry out the optimum material gradient of a Functionally Graded Endodontic Prefabricated Post (FGEPP). The results indicated that the performance of an EPP made of FGM is better than that of an EPP made of homogeneous material.

1 INTRODUCTION

In the treatment process of our teeth, an EPP in the Endodontically Treaded Tooth (ETT) plays a vital role. Scientists are always trying to improve its performance.

In the market, two general types of post can be found: parallel and tapered. Post retention can be defined as: 'the possibility range of a post for resisting vertical dislodging forces' (Richard & James 2004). A number of scientists have recommended that a parallel post is more retentive than a tapered post (Standlee et al. 1978, Felton et al. 1991, Johnson et al. 1978, Qualtrough et al. 2003). Thus, a parallel post is considered in this work.

In our engineering applications, pure materials are of little use because of the demand of conflicting property requirement. An application may need a material that is hard as well as ductile. However, in nature, no such material exists (Rasheedat & Esther 2012).

Scientists have found a new material to achieve the requirements in our engineering applications. This new material called the functionally graded material or Functionally Gradient Material (FGM) can be defined as: 'a material whose properties vary from one surface to another'.

A comparison between the performances of an EPP made of homogeneous material and an EPP made of FGM is the primary purpose of the present investigation. And the main purpose of the present work is to determine the optimum material gradient of a FGEPP.

2 MATERIALS AND METHODS

2.1 *Material gradient of a FGEPP*

In the FGM simulation case, a comparison among the distribution functions for the constituents' volume fraction has been carried out by Helal and Shi (2014). Based on the results, the optimum material gradient for FGMs can be described by using a modified sigmoid function. Thus, in the present work, the modulus of Elasticity (E) of FGEPP will be considered to vary continuously throughout the height of the EPP, according to the volume fraction of the constituent materials based on the sigmoid function.

Delale and Erdoogan (1983) indicated that the effect of the Poisson ratio (υ) on the deformation is much less than that of E. Thus, in the present investigation, the value of υ is constant and set to 0.3 (Helal et al. 2015).

Now, let us consider the case of a FGEPP of width W, total height h with a coordinate system, as shown in Figure 1. Figure 1 shows the FGEPP graded vertically from bottom to top. It should be noted that the values of E vary over the EPP from $E_2 = 10$ GPa to $E_1 = 210$ GPa.

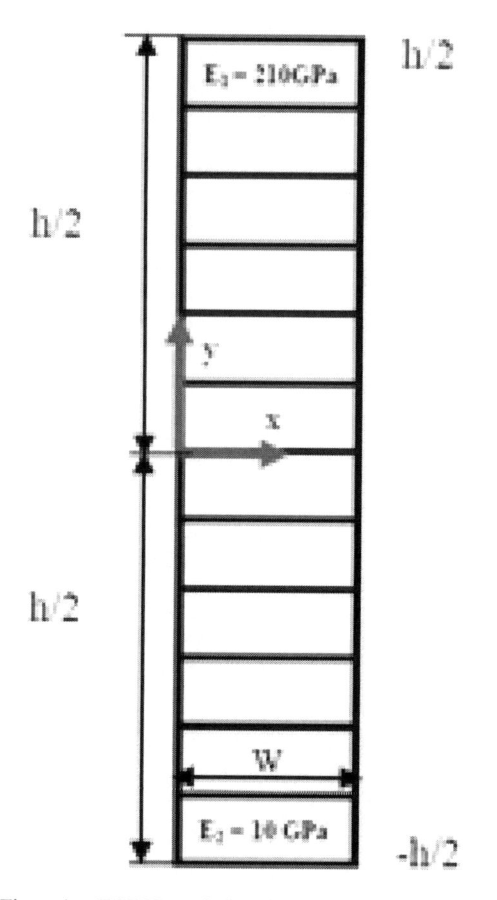

Figure 1. FGEPP graded vertically from bottom to top.

2.2 *Characteristics of the sigmoid—functionally graded endodontic prefabricated posts (S-FGEPPs)*

The behavior of a sigmoid function is defined by two power-law functions. These functions are defined by

$$\Psi_1(y) = 1 - \frac{1}{2}\left(\frac{h/2 - y}{h/2}\right)^w \text{ for } 0 \le y \le h/2 \tag{1}$$

$$\Psi_2(y) = \frac{1}{2}\left(\frac{h/2 + y}{h/2}\right)^w \text{ for } -h/2 \le y \le 0 \tag{2}$$

where w is the material gradient index.

By using the rule of mixture, E of the sigmoid (S-FGEPPs) can be evaluated as follows:

$$E(y) = \Psi_1(y)E_1 + [1 - \Psi_1(y)]E_2 \quad \text{for } 0 \le y \le h/2 \tag{3}$$

$$E(y) = \Psi_2(y)E_1 + [1 - \Psi_2(y)]E_2 \quad \text{for } -h/2 \le y \le 0 \tag{4}$$

Figure 2 shows the variation of E, which is given by Equations (3) and (4), that represents the sigmoid distribution.

2.3 *Finite element analysis*

The FEA was adopted in order to investigate a comparison between the performances of an EPP made of homogeneous material and an EPP made of a FGM.

The structure of a tested problem FE model is adopted from the literature, which includes: endodontic post, dentin, gutta-percha, Periodontal Ligament (PDL), and cortical bone, as plotted in Figure 3.

Table 1 presents the mechanical properties, E (GPa) and v (non-dimensional) of the materials used in the present investigation. It should be noted that, when comparing between the performances of an EPP made of homogeneous material and an EPP made of a FGM, the values of E of the homogeneous EPP will be considered as 10, 110 and 210 GPa.

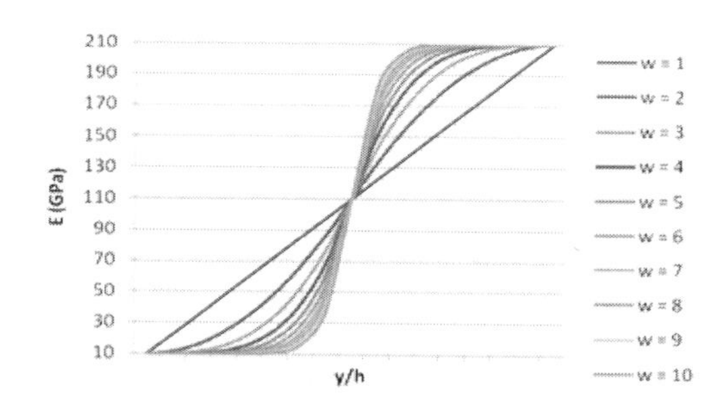

Figure 2. The variation of E in S-FGEPPs.

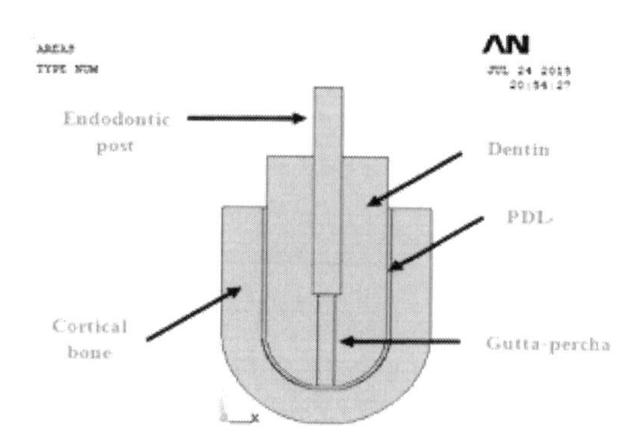

Figure 3. Structure of the FE model.

In the present study, the ANSYS package was employed to perform the tested problem model. The model was built in ANSYS Rel.12.1 by means of the Parametric Design Language (APDL).

In order to generate the tested problem model in the x-y plane, plane elements (PLANE 82) were used. The element (PLANE 82) was used as a plane stress.

It should be noted that, all components of the FE model were considered as complete bonding. All exterior nodes on the cortical bone surface were considered as the fixed boundary condition. A compressive load of magnitude 200 N was applied on the center node at the top of the EPP surface, as shown in Figure 4.

2.4 *Optimization technique*

In order to solve the present problem, the optimization technique was developed for the current problem. The model consists of objective function, design variable and state variable. The design objective is to minimize the maximum von Mises stress in the FGEPP.

Equations (1) and (2) are modified and become (Helal and Shi 2014):

Table 1. Mechanical properties of the materials used in this analysis.

Material	E(GPa)	υ	References
Gutta-percha	0.96E-3	0.3	(Joshi et al. 2001)
Dentin	18.6	0.4	(Joshi et al. 2001)
PDL	0.0689	0.45	(Hong et al. 2001)
Cortical bone	13.7	0.3	(Hong et al. 2001)

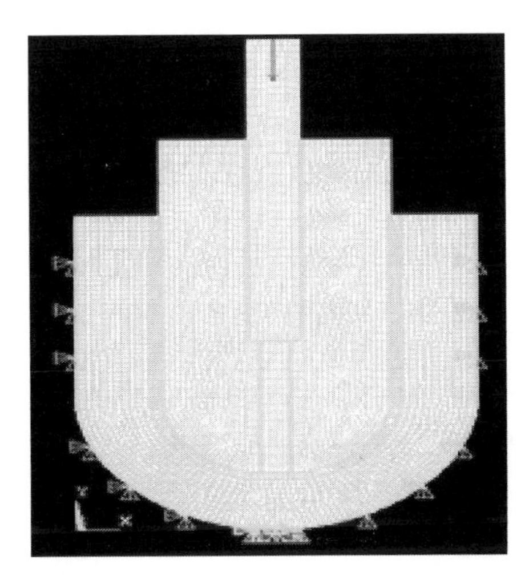

Figure 4. Boundary condition and external loads.

$$\Psi_1(y) = 1 - \frac{1}{2}\left(\frac{h/2 - y}{h/2}\right)^w \quad \text{for } 0 \le y \le h/2 \tag{5}$$

$$\Psi_2(y) = \frac{1}{2}\left(\frac{h/2 + y}{h/2}\right)^D \quad \text{for } -h/2 \le y \le 0 \tag{6}$$

where w and D are material gradient indices.

In this problem, the design variable is the volume fraction Ψ(y) of the S-FGEPPP, which is described by the material gradient indices w and D. Furthermore, the objective function is to minimize the maximum von Mises stress in a FGEPP. In the present investigation, the constraints used are as follows:

1. To retain the values of the design variables w and D within the permissible limits used in the literature (Fuchiyama et al. 1993, Fuchiyama et al. 2001) as $0 \le w \le 10$, and also $0 \le D \le 10$.
2. To retain the maximum shear stress in the FGEPP model (τ_{FGEPP}) less than or equal to the value obtained for the initial case (τ_i).

The optimum design procedure flow chart is shown in Figure 5.

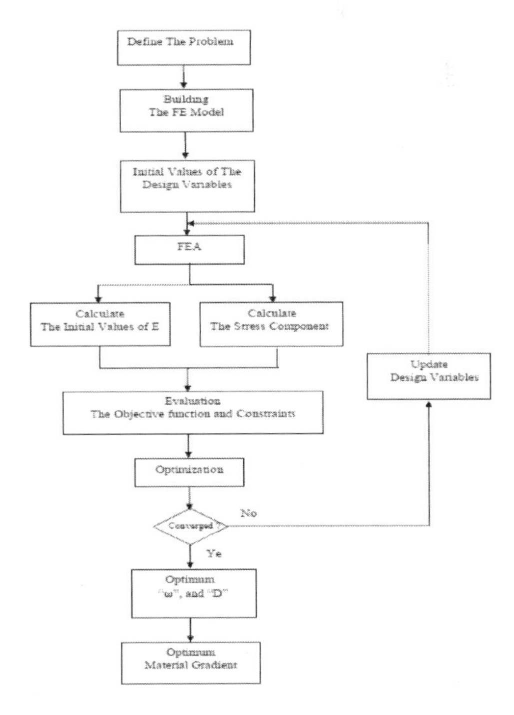

Figure 5. Optimum design procedure flow chart.

3 RESULTS AND DISCUSSION

When comparing von Mises stress, shear stress and deformation for the homogeneous case (E = 10, 110, and 210 GPa), the results obtained indicate that the performance of an ETT improved when the E of an EPP was equal to 210 GPa.

When comparing von Mises stress, shear stress and deformation for the FGM case ($w = D = 1$, and $w = D = 10$), the results obtained indicate that the performance of an ETT improved when the material gradient indices $w = D = 1$ were used.

To minimize the maximum von Mises stress in an EPP made of FGM, the FEA and optimization technique were developed. After the optimization process was finished, the optimum values of the material gradient indices of EPP made of FGM, w and D, were obtained, being equal to 8.9873 and 0.653, respectively. Figure 6 shows the optimum material gradient of a FGEPP.

When comparing von Mises stress, shear stress and deformation for the homogeneous case (E = 210 GPa) and the FGEPP case (w = 8.9873, and D = 0.653), the results obtained indicate that the performance of a FGPP with the optimum material gradient was better than that of an EPP made of homogeneous material.

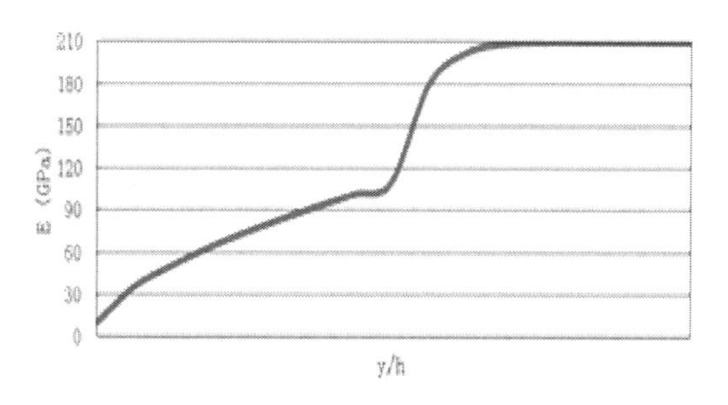

Figure 6. Optimum material gradient of a FGEPP.

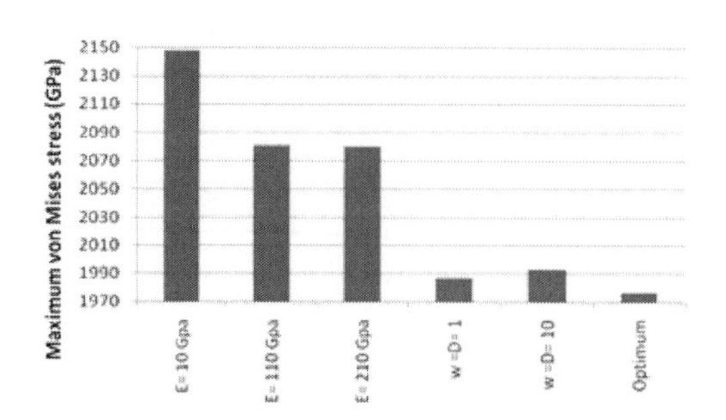

Figure 7. Simulation cases, von Mises stress, of the homogeneous and FGM cases.

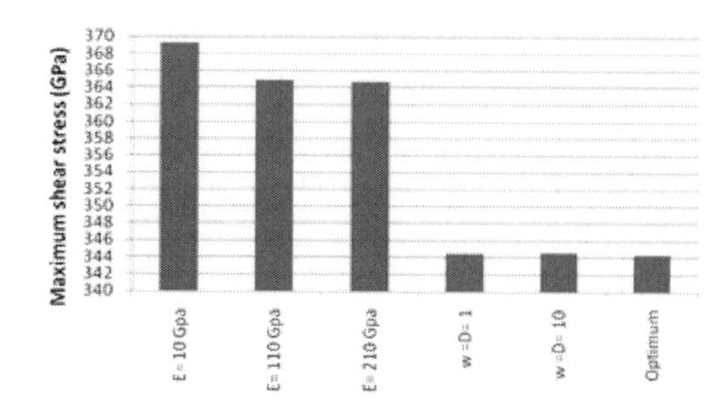

Figure 8. Simulation cases, shear stress, of the homogeneous and FGM cases.

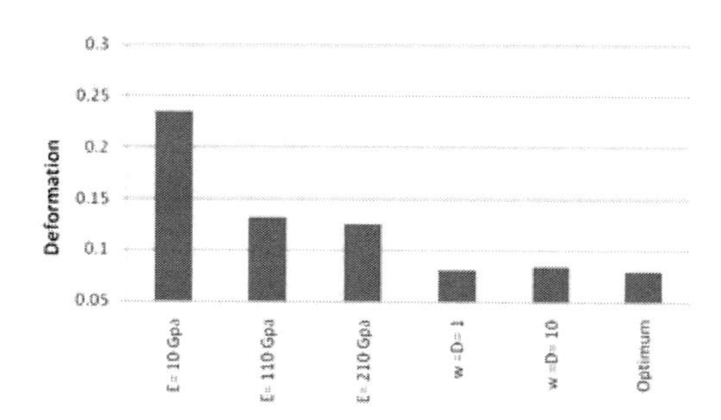

Figure 9. Simulation cases, deformation, of the homogeneous and FGM cases.

4 CONCLUSIONS

When comparing von Mises stress, shear stress and deformation for the homogeneous case (E = 10, 110, and 210 GPa), the performance of an ETT improved when the E of an EPP was equal to 210 GPa.

The optimum material gradient indices w and D were investigated, being equal to 8.9873, and 0.653, respectively. The performance of an EPP made of FGM was better than that of an EPP made of homogeneous material.

ACKNOWLEDGMENT

This paper was funded by the International Exchange Program of Harbin Engineering University for Innovation-oriented Talents Cultivation. The authors would like to thank Kafr El-Sheikh University, Egypt, for providing the facilities.

REFERENCES

Delale, F. & Erdogan, F. 1983. The Crack Problem for a Nonhomogeneous Plane, *ASME Journal of Applied Mechanics*, 50(3): 609–614.

Felton, D.A., Webb, E.L., Kanoy, B.E. & Dugoni, J. 1991. Threaded endodontic dowels: Effect of post

design on incidence of root fracture, *Journal of Prosthetic Dentistry*, 65(6): 179–187.
Fuchiyama, T., Noda, N., Tsuji, T. & Obata, Y. 1993. *Analysis of thermal stress and stress intensity factor of functionally gradient materials*, in Ceramic Transactions: Functionally Gradient Materials, I. B. Holt, Ed., American Ceramic Society, Westerville, Ohio, USA. 34: 425–432.
Fuchiyama, T. & Noda, N. 2001. *Multiple crack growths in the functionally graded plate under thermal shock*, in Proceedings of the 4th International Congress on Thermal Stresses, Osaka, Japan, 121–124.
Hong, S.Y., Lisa, A.L.. Anthony, M. & David, A.F. 2001. The effects of dowel design and load direction on dowel-and-core restorations, *Journal of Prosthetic Dentistry*, 85: 558–567.
Johnson, J.K. & Sakamura, J.S. 1978. Dowel form and tensile force, *Journal of Prosthetic Dentistry*, 40(6): 645–649.
Joshi, S., Mukherjee, A., Kheur, M. & Mehta, A. 2001. Mechanical performance of endodontically treated teeth, *Finite elements in analysis and design,* 37(8): 587–601.
Qualtrough, A.J., Chandler, N.P. & Purton, D.G. 2003. A comparison of the retention of tooth-colored posts, *Quintessence International*, 34(3): 199–201.
Rasheedat, M., Mahamood, & Akinlabi, E.T. 2012. *Functionally Graded Material: An Overview,* Proceeding of the WCE, 2012, 3: 5.
Richard, S.S., DDS & James, W.R. 2004. Post Placement and Restoration of Endodontically Treated Teeth: A Literature Review. *Journal of Endodontics*, 30(5): 289–301.
Standlee, J.P., Caputo, A.A. & Hanson, E.C. 1978. Retention of endodontic dowels: Effects of cement, dowel length, diameter, and design, *Journal of Prosthetic Dentistry*, 39(4): 401–405.
Wasim, M.K., Helal & Shi, D.Y. 2014. Optimum Material Gradient for Functionally Graded Rectangular Plate with the Finite Element Method, *Indian Journal of Materials Science*, 2014: 501935.
Wasim, M.K., Helal & Shi, D.Y. 2015. Optimum Gradation Direction for a Functionally Graded Endodontic Prefabricated Parallel Post: A Finite Element Method, *Journal of Biomimetics, Biomaterials and Biomedical Engineering,* 24: 56–69.

Advanced Materials, Mechanical and Structural Engineering – Hong, Seo & Moon (Eds)
 ISBN: 978-1-138-02908-8

Membrane conformity assessment for Zinc-Bromine flow Batteries

M. Kim & J.H. Jeon
Department of Energy and Advanced Material Engineering, Dongguk University, Seoul, Korea

ABSTRACT: The aim of this paper is to evaluate membrane conformity assessment in determining a membrane employed in the Zinc-Bromine flow Battery (ZBB). For this purpose, two different types of membranes are used in 25 cm^2 ZBB cell operations: one is SF600, which is a porous membrane based on a hydrocarbon, and the other is Nafion 115, which is a cation-ion exchange membrane based on a sulfonated tetrafluoroethylene. Performance comparisons are carried out using a 2.0 M $ZnBr_2$ electrolyte solution with or without a bromine-complexing agent (Methyl-Ethyl-Pyrrolidinium bromide, MEP). The experimental results indicate that in the case without MEP, Nafion 115 allows higher performance of ZBB cells than SF600, while in the case with MEP, SF600 results in higher stability and durability than Nafion 115. In conclusion, the intrinsic chemical properties of the bromine-complexing agent as well as membrane material have a significant effect on the membrane performance.

1 INTRODUCTION

A Zinc Bromine flow Battery (ZBB) is an attractive and useful technology for the application of utility energy storage (Joseph & Shahidehpour 2006), but the distinctive electrochemical characteristics of the electrolyte during the cell operation cause a lot of problems. One of the problems is bromine, which has a strong corrosive and fast vapor ability. This induces bromine crossover that causes the self-discharge of ZBB. And the adsorption of bromine on the membrane surface also increases membrane resistivity. Especially, toxic bromine reduces the stability and durability of ZBB performance. $ZnBr_2$ electrolyte solution typically employs a bromine-complexing agent (QBr(aq)) as a supporting additive in order to reduce bromine crossover and avoid bromine gasification (Daniel 1980). This attempt is limited by the performance of ZBB cell materials including electrolytes. This is because during charging, the complexed polybromide (i.e., $n(Br_2) \rightarrow QBr(aq)$ ($Q(Br_2)nBr$) forms a heavy black second phase (precipitate) in the catholyte and has a bad effect on the stability and durability of the membrane used in the ZBB cell (Pell 1994).

In the ZBB, a noble membrane with ionic conductivity, physical strength and chemical stability is required. More importantly, high stability and durability must also be guaranteed during charging and discharging. This paper deals with membrane conformity assessment by the charge and discharge operations of the ZBB cell employing the $ZnBr_2$ electrolyte solution with or without the bromine-complexing agent. Two different types of membranes are compared in this work: one is SF600 (Asahi, thickness: 0.6 mm) (Phillip 1999), an oligomer porous membrane based on a hydrocarbon, and the other is Nafion 115 (Dupont, thickness 0.127 mm) (Kim et al. 2010), a cation exchange membrane based on a sulfonated tetrafluoroethylene. Using the ZBB cell experiments of membranes, this paper provides a new insight into the development of the separator for ZBBs.

2 EXPERIMENTAL METHODS

2.1 *Experimental preparation*

For membrane performance comparison, as shown in Figure 1, a 25 cm^2 ZBB cell was used in this work, consisting of a carbon felt (Toyobo, 4 mm), bipolar plate (SGL, 0.6 mm) and membrane (SF600 or Nafion 115). The carbon felt was not used in a zinc-half cell where metallic zinc is plated on the electrode surface during charging (Dennis et al. 2012, Shin et al. 2012). For the experimental preparation, SF600 or Nafion 115 membrane was treated for 5 hours in the distilled water and the aqueous zinc-bromide solution (2.0 M $ZnBr_2$ electrolyte with 0.5 M $ZnCl_2$) was used with or without 0.7 M Methyl Ethyl Pyrrolidinium (MEP) as a bromine-complexing-agent. The MEP plays a role in preventing the bromine cross over and bromine gasification (Weber et al. 2011). The cell operation was carried out for 30 cycles and the current density was kept constant at 20 mA/cm^2. The charge-discharge cycling operation was performed by WBCS3000 (Won-A tech. Korea) with a constant charge time (1 hr) at room temperature (Li et al. 2015).

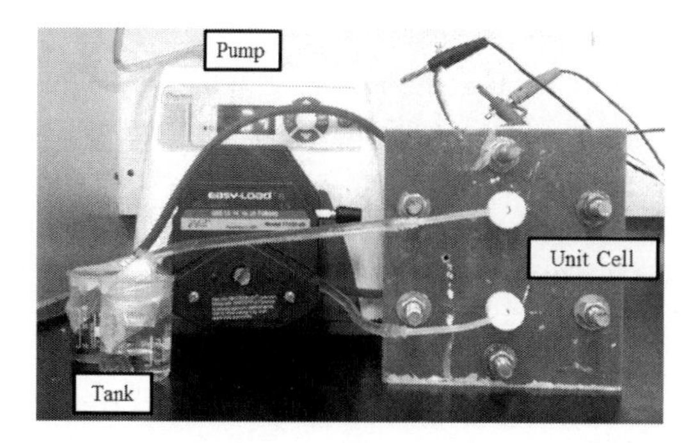

Figure 1. Cyclic operation test-bed (active area 25 cm^2 and current density 20 mA/cm^2).

2.2 *Weight-change measurement of membranes before and after the cell operation*

Membrane durability for long-time cell operation (i.e. membrane damage against bromine during the charge operation) was evaluated. This is due to the fact that toxic bromine can damage the separation layer. For these reasons, in this paper, the weight change of the membrane was measured before and after the cell operation. Weight measurement was performed by EL403 (Mettler Toledo, Switzerland).

2.3 *Observation with SEM*

In order to investigate the zinc particle and signs of damage from bromine, the SEM (JEOL Ltd., Japan) focused on the surface morphology of the membrane obtained after washing the membranes with distilled water.

3 RESULTS AND DISCUSSION

3.1 *Evaluation through ZBB efficiency*

Figures 2 (without MEP) and 3 (with MEP) show the charge-discharge curves of the cells using the SF600 and Nafion 115, respectively, in order to compare the operating performance. As shown in Figure 2, Nafion 115 has longer charge and discharge curves (without MEP) than SF600, but the charge-discharge curve (with MEP) is not complete due to the fact that anomalies occur, as shown in Figure 3(b).

However, the operating performance of SF600 becomes better than Nafion 115 in the case without MEP. Figures 4 and 5 show the electro-chemical performance of the ZnBr2 electrolyte solution with and without MEP, where CE and VE denote the coulombic efficiency and the voltaic efficiency, respectively. When not using the MEP (Figure 4), the VE of SF600 and Nafion 115 is similar but the

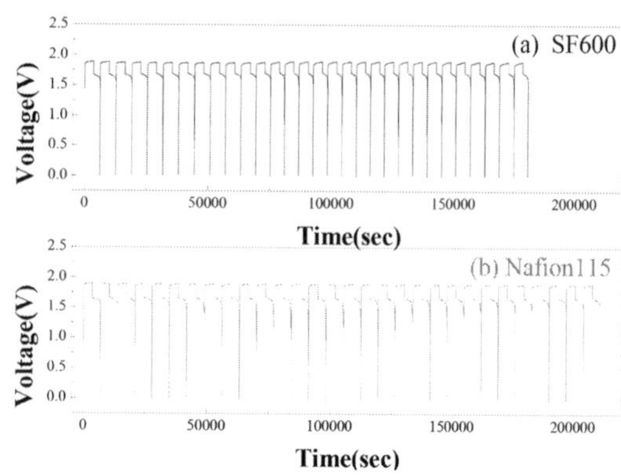

Figure 2. Cyclic operations due to the application of (a) SF600 and (b) Nafion 115 without MEP (current density 20 mA/cm^2).

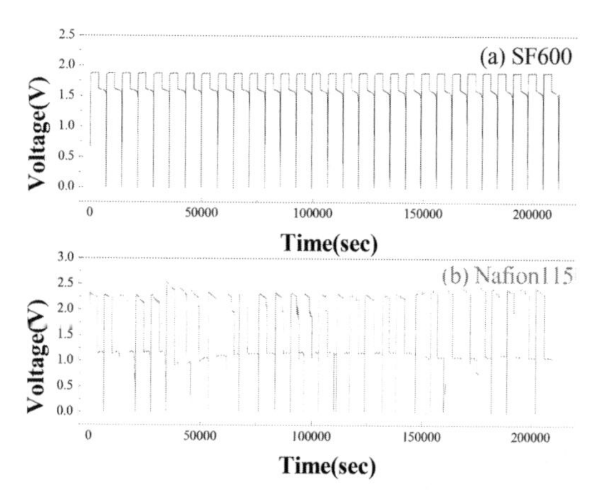

Figure 3. Cyclic operations due to the application of (a) SF600 and (b) Nafion 115 with MEP (current density 20 mA/cm^2).

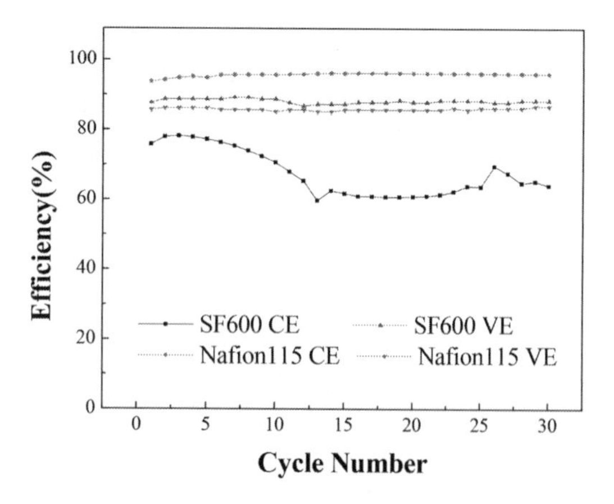

Figure 4. Coulombic efficiency and voltaic efficiency curves due to the cyclic operation of the ZBB cell employing the electrolyte without MEP.

average Energy Efficiency (EE) of SF600 (59.29%) is higher than that of Nafion 115 (81.82%). It is due to the fact that coulombic stability of the cell using Nafion 115 is higher than using SF600. As shown in Figure 5, the VE of the SF600 is significantly higher than that of Nafion 115, and the average EE of SF600 (78.24%) is higher than that of Nafion 115 (46.33%). Besides, coulombic stability of the cell using SF600 is better than that using Nafion 115. It can be assumed that SF600 is more suitable than Nafion 115 when using MEP. Table 1 presents the best efficiencies using Nafion 115 and SF600 without MEP.

The result indicates that the CE is similar but the VE of the cell using SF600 is lower than that using Nafion 115 due to the influence of MEP. It can be assumed that MEP increases the resistance of the cell because the bromine complex can be adsorbed on the SF600 surface.

3.2 *Changes in membrane weight due to the experimental step*

The membrane was washed with distilled water before the experiment to prevent the adsorption of the electrolyte on the membrane during the cell operation test (Bellows et al. 1999). Then, the weight of each membrane was increased due to absorption of distilled water (Goswami et al. 2008). The weight rate of SF600 was greater than that of Nafion115, as given in Tables 2 and 3. It can be assumed that SF600 absorbs more water than Nafion 115 because the former is a porous membrane. After the cell operation test, the weight of each membrane was measured when the electrolyte was dried. And the weight of SF600 was found to be heavier than that when it absorbed the water before the experiment. However, in contrast to the case of SF600, the weight of Nafion 115 was reduced.

It can be assumed that the influence of the weight loss due to the corrosive bromine is larger than that of the adsorption of the precipitate. This is because SF600 has a larger ion exchange rate than Nafion 115. This is because SF600 is a porous membrane

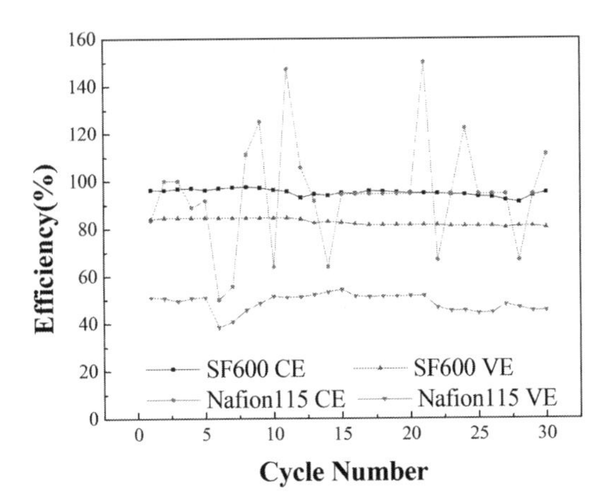

Figure 5. Coulombic efficiency and voltaic efficiency curves due to the cyclic operation of the ZBB cell employing the electrolyte with MEP.

Table 1. Efficiencies of each membrane in best performance.

	Coulombic (%)	Voltaic (%)	Energy (%)
SF600*	94.90	82.43	78.24
Nafion 115**	95.42	85.75	81.82

*Efficiencies of the ZBB cell with SF600 when using MEP.
**Efficiencies of the ZBB cell with Nafion 115 when not using MEP.

Table 2. Weight of the membranes in each condition. It was shown that the effects of the precipitate were generated during the charge cycle.

	2.0 M $ZnBr_2$ + 0.5 M $ZnCl_2$	
	SF600(g)	Nafion 115(g)
Before washing	2.152	1.388
After washing*	4.241	1.560
After the test**	6.431	1.539
After the test and washing***	5.202	1.425

*Before the experiment, it was a process of washing the membrane in distilled water.
**The condition of drying without washing after the experiment.
***Dried membrane that was washed with distilled water after the experiment.

Table 3. Weight of the membranes in each condition.

	2.0 M $ZnBr_2$ + 0.5 M $ZnCl_2$ + 0.7 M MEP	
	SF600(g)	Nafion 115(g)
Before washing	2.336	1.541
After washing*	4.438	1.772
After the test**	5.797	1.690
After the test and washing***	4.741	1.622

*Before the experiment, it was a process of washing the membrane in distilled water.
**The condition of drying without washing after the experiment.
***Dried membrane that was washed with distilled water after the experiment.

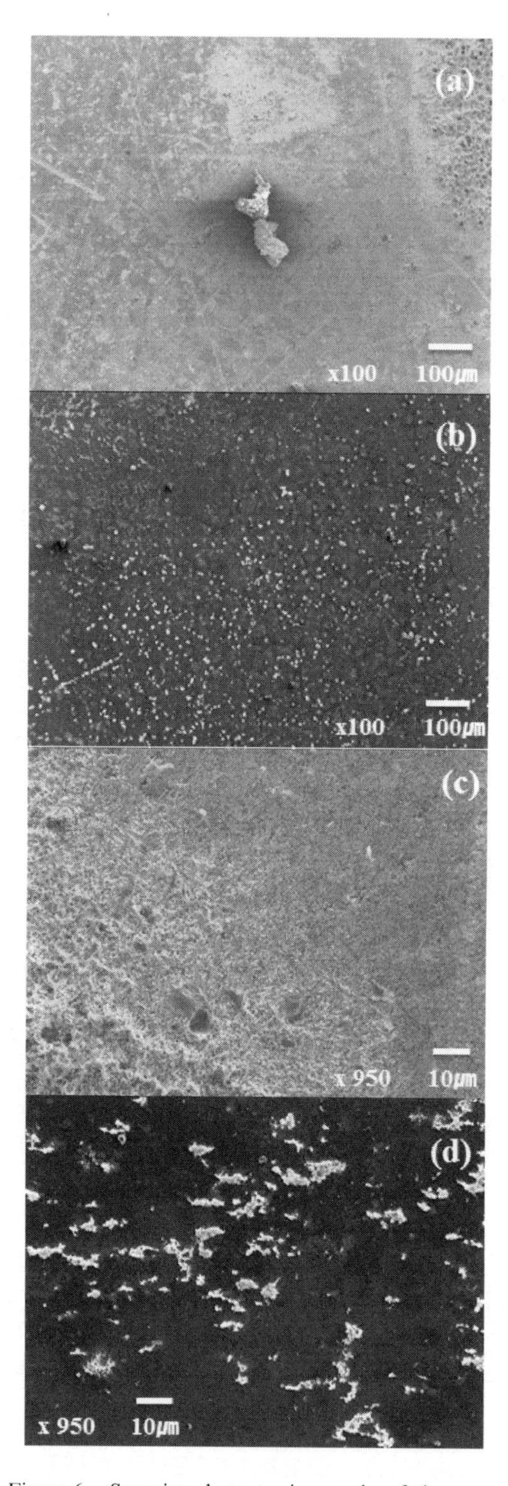

Figure 6. Scanning electron micrographs of the membrane surface after the charge-discharge cycling test with different membranes when using MEP: (a), (c) with SF600, (b), (d) with Nafion 115.

whose ions can move in either direction. But Nafion 115 is a cation membrane whose anion cannot move toward the anode (Yeager & Steck 1981). Therefore, using Nafion 115, only half of the anions are involved in the ion reaction rate. So SF600 has inevitably more precipitate than Nafion 115.

3.3 *Observation of the membrane surface*

The SEM images of the membrane surface are shown in Figure 6. The images are used for observing the surface of SF600 (Figure 6 (a) & (c)) and Nafion 115 (Figure 6 (b) & (d)). The magnification of Figure 6 (a) & (b) is 100 times, and that of (c) and (d) is 950 times. As it can be seen, the zinc particle on the surface of SF600 (Figure 6(a)) is not more well-distributed than that on the surface of Nafion 115 (Figure 6(b)) In addition, the particles are attached on the surface of the membrane, as shown in Figure 6 (c), but it is can be seen from Figure 6 (d) that the particle is well dispersed. This means that the diffused zinc deposition formed during the charging process has an advantage of prolonging the charge-discharge life of the batteries (Kan et al. 1998).

4 CONCLUSIONS

This paper describes the performance comparison of SF600 (hydrocarbon-based membrane) and Nafion 115 (sulfonated tetrafluoroethylene-based membrane) through the ZBB cell operation, in order to find a separator for high stability and durability of ZBBs. The experimental results indicate that the intrinsic chemical properties of a bromine-complexing agent as well as membrane material have a significant effect on the membrane performance and ZBB cell durability. In addition to the bromine crossover (by a porous property) and high membrane thickness (0.6 mm), in the case of hydrocarbon-based SF600, there is a significant disadvantage such that bromine complexes are adsorbed onto the separator surface in the bromine-half cell during charging and the cell-charge voltage is increased due to the resistance increment of the SF600 surface. On the other hand, the sulfonated tetrafluoroethylene-based Nafion 115 membrane (thickness 0.127 mm) has a fatal weakness against the bromine-complexing agent, while there are no problems such as bromine crossover or bromine adsorption during charging. Consequently, a hydrocarbon-based membrane is much more suitable to the separator for the ZBB than a sulfonated tetrafluoroethylene-based membrane if a bromine-complexing agent is used for the protection of bromine crossover and gasification.

ACKNOWLEDGMENT

This work was supported by the Technology Innovation Program (10043787, Development of Ion Selective Membranes for Redox Flow Battery) funded by the Ministry of Trade, Industry & Energy (MI, Korea).

REFERENCES

Bellows, R.J. et al. 1999. Neutron Imaging Technique for In Situ Measurement of Water Transport Gradients within Nafion in Polymer Electrolyte Fuel Cells. *Journal of the Electrochemical Society*, 146(3): 1099–1103.

Daniel, J. 1980. Bromine Complexation in Zinc-Bromine Circulating Batteries. *Journal* of *the Electro-chemical Society,* 127(3): 528–532.

Dennis, K. et al. 2012. *Reversible polarity operation and switching method for ZnBr flow battery when connected to common DC bus*. United States: US0326672.

Goswami, S. et al. 2008. Wetting and Absorption of Water Drops on Nafion Films. *American Chemical Society*, 24: 8627–8633.

Joseph, A & Shahidehpour, M. 2006. *Battery storage systems in electric power systems*. Power Engineering Society General Meeting.

Kan, J. et al. 1998. Effect of inhibitors on Zn-dendrite formation for zinc-polyniline secondary battery. *Journal of Power Sources*, 74: 113–116.

Kim, T. et al. 2010. Characteristics of Membrane Electrode Assemblies for PEMFC. *Journal of Advanced Engineering and Technology,* 3(2): 199–203.

Li, B. et al. 2015. Ambipolar zinc-polyiodide electrolyte for a high-energy density aqueous redox flow battery. *Nature Communications,* 6(6303), doi:10.1038/ncomms7303.

Pell, W. 1994. *Zinc/Bromine battery electrolytes: Electrochemical, physicochemical and spectroscopic studies*. Ottawa: University of Ottawa.

Phillip, E. 1999. *Development of Zinc/Bromine Batteries for Load-Leveling Application: Phase 1 Final Report*. Milwaukee: Sandia National Laboratories.

Shin, K. et al. 2012. Effects of *Electrolyte* Concentration on Growth of Dendritic Zinc in Aqueous Solutions, *The Korean Hydrogen & New Energy Society*, 23(4): 390–396.

Weber, A.Z. et al. 2011. Redox flow batteries: a review. *Journal of Applied Electrochemistry,* 41(10): 1137–1164.

Yeager, H.L. & Steck, A. 1981. Cation and Water Diffusion in Nafion Ion Exchange Membranes: Influence of Polymer Structure. *Journal of the Electrochemical Society*, 128(9): 1880–1884.

Advanced Materials, Mechanical and Structural Engineering – Hong, Seo & Moon (Eds)
© 2016 Taylor & Francis Group, London, ISBN: 978-1-138-02908-8

Effects of ligating methods and materials on IFD properties of NiTi wire

J.H. Paek & Y.J. Ahn
Department of Medical Engineering, Kyung Hee University, Seoul, Korea

Y.G. Park
Department of Orthodontics, College of Dentistry, Kyung Hee University, Seoul, Korea

H.K. Park & S. Choi
Department of Medical Engineering, Kyung Hee University, Seoul, Korea
Department of Biomedical Engineering, College of Medicine, Kyung Hee University, Seoul, Korea

ABSTRACT: This study examined the effect of different ligating systems on the Initial Force Delivery (IFD) characteristics of superelastic NiTi orthodontic wire using a modified three-point bending test. It comprised a combination of two types of ligature methods and six types of brackets. For the three-point bending test, each wire was ligated on each side of the bracket with an interbracket distance of 14.23 mm. The wire was deflected lingually from the horizontal line by putting the weight in the container of 10–200 g at 10 g intervals. The results indicated that superelastic NiTi wires were deflected more in the stainless steel ligature groups than in the elastomeric ligature groups. Wire deflection was higher in the self-ligating bracket groups than in the conventional ligating bracket groups. This study developed a method to predict the IFD of wire by observing the wire geometric configuration using a low-cost device based on the three-point bending.

1 INTRODUCTION

The purpose of orthodontic treatment is to relocate malpositioned teeth through the application of mechanical forces. It is generally accepted that optimal orthodontic tooth movement occurs under a low and continuous force (Ren et al. 2003). A force of 1–2 N is generally recommended as a suitable force for orthodontic tooth movement. The force to exceed the capillary blood pressure reduces the cellularity of the periodontal ligament that can result in a slowing down or cessation of tooth movement (Choi et al. 2015).

In the initial alignment stage (Figure 1) of fixed appliance treatment, superelastic NiTi archwires have been widely used by orthodontists because of the advantage of effective tooth movement by a light continuous force and their shape memory effect (Warita et al. 1996). NiTi wires exhibit a clinically applicable character because they show a typical nonlinear recoverable Load-Deflection (LD) behavior (Figure 2 A). The phase transition between the austenite and martensite phases is manifested by a flat or nearly flat plateau in the LD curve. To study those LD characteristics

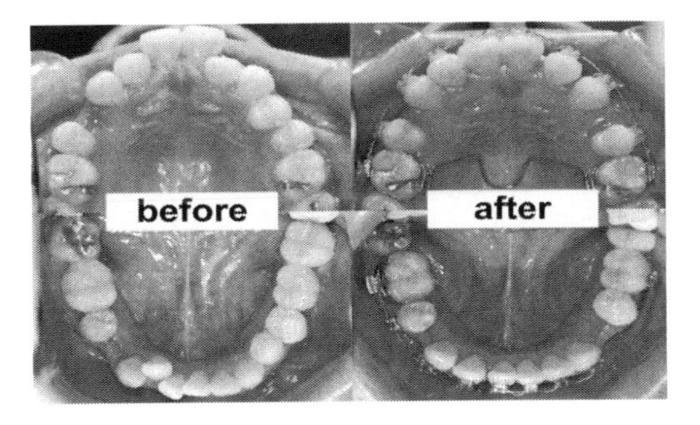

Figure 1. Intraoral photos before and after the initial alignment stage.

of NiTi wires, many experiments have been performed using three-point bending tests on expensive load cells (Nakano et al. 1999). Since the unloading behavior of a wire represents the force delivery characteristic of a wire in an orthodontic appliance, most studies have focused more on the unloading curves (Bartzela et al. 2007). However, there is no report on the starting point of the unloading curve, which represents the magnitude of force deflecting the high resilient wire to the dis-

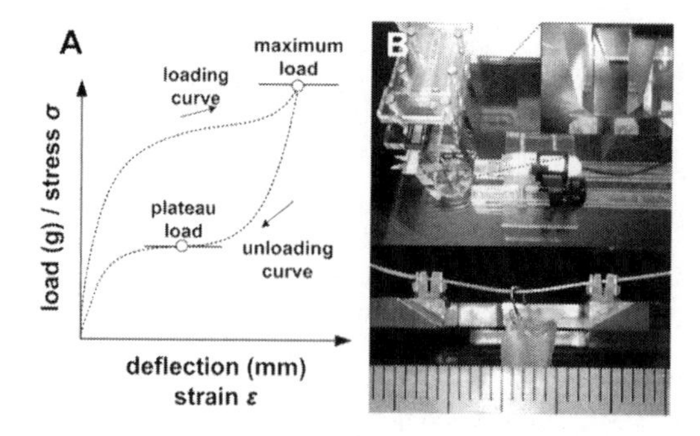

Figure 2. (A) LD characteristics of the superelastic NiTi orthodontic wire in the three-point bending test and (B) low-cost device for evaluating the geometric configuration of orthodontic wires using the modified three-point bending test.

placed tooth for a full bracket engagement. Engaging the wire into the brackets activates the wire and exerts internal forces on the tooth. Finally, it is equal to the initial force that the tooth will receive. In this study, this force is defined as the Initial Force Delivery (IFD). Therefore, this study assessed the effect of different ligating systems on the IFD characteristics of superelastic NiTi orthodontic wires using a modified three-point bending.

2 EXPERIMENTAL SETUPS

2.1 *Sample preparation*

The 0.014-inch Sentalloy M® (GAC, Central Islip, NY, USA) was used as superelastic NiTi wires. Since these wires were mainly available as preformed arches, the two posterior sections from each archwire were cut and used. The prepared wires were assigned to eight groups, according to a combination of six types of orthodontic brackets (Table 1). All ligations were carried out by the same orthodontist who was well practiced in the procedure to eliminate the interpersonal variations in the ligation procedure.

2.2 *Three-point bending test*

To examine the relationship between the load and deflection in bending wires, a modified three-point bending test using the self-made device was performed. This low-cost device (Figure 2B) consists of digital calipers, **USB** microscope and acrylic structure. For the three-point bending design, the wire samples were inserted and ligated on each side of the bracket slots and displaced at the midpoint. Two brackets of each group were bonded to both horizontal surfaces of the digital caliper using an instant adhesive. The mesial plane of the each bracket base was lined up with the mesial plane of the caliper. As the bracket base faced the floor, the three-point bending test in the bucco-lingual plane, which simulated the first-order wire deflection to the lingual side, was performed. The interbracket distance was modulated by digital calipers (14.23 mm). This interbracket distance was derived from five patients by CT scans that were obtained using a high-speed CT system before starting the orthodontic treatment. The DICOM data were acquired. The reformatted 2D images of the jaw and teeth were produced using Vworks 5.0.

The engaged superelastic NiTi wire was connected to the acrylic container at its half-way point using a metal hook without a bracket. The weight was placed in a container from 10 g to 200 g at 10 g intervals. The wire was deflected lingually from the horizontal line. The wire was reconverted before increasing the weight each time. Eight specimens of each wire were tested. All measurements were taken at a constant room temperature. To estimate the wire geometric configuration, pictures were taken every moment using a microscope with a metal ruler for data compensation. Five clinicians measured the amount of wire deflection vertically using Motic Images Plus version 2.0.

Table 1. Eight ligating system-based experimental groups.*

Group	Bracket	Ligating system	Manufacturer
A	Mini-Taurus®	Metal CLB+SSL method	Rocky Mountain Orthodontics (Denver, CO, USA)
B	Mini-Taurus®	Metal CLB+EL method	Rocky Mountain Orthodontics
C	Clarity®	Ceramic CLB+SSL method	3M Unitek (St Paul, MN, USA)
D	Clarity®	Ceramic CLB+EL method	3M Unitek
E	Damon II®	Passive type SLB	Ormco SDS (Glendora, CA, USA)
F	Speed®	Active type SLB	Strite Industries (Strite Industries, Cambridge, Ontario, Canada)
G	Clarity-SL®	Passive type SLB	3M Unitek (Monrovia, CA, USA)
H	Clippy-C®	Active type SLB	Tomy (Tokyo, Japan)

* CLB, conventional ligating bracket, SSL, stainless steel ligature (0.009 inch; Ortho Organizers, Carlsbad, CA, USA), EL, elastomeric ligature (TP Orthodontics, La Porte, IN, USA), SLB, self-ligating bracket.

2.3 *Statistics*

Quantitative data are expressed as mean ± SD. Statistical analyses were performed to compare the mean values obtained from each group using a two-tailed Student's t-test. One-way ANOVA was performed to compare the differences in the mean deflection values with respect to the loads in each group. Additional *post hoc* comparisons were performed using a Student-Newman-Keuls test where appropriate. P-values <0.05 were considered significant.

3 RESULTS

In all the experimental groups, linear graphs were achieved except for group C and all points of the graphs represent the IFD of the Sentalloy® NiTi wire. Those graphs are distinct from the loading part of the LD curve typical for shape memory alloys, as the weight was not loaded continuously. The load in these graphs means the IFD to deflect the Sentalloy® NiTi wire from a flat condition at every moment.

Figure 3 shows the mean deflection values with respect to the loads in the CLB groups. The amount of wire deflection was greater in group A (SSL method) than in group B (EL method). The amount of wire deflection was greater in group C (SSL method) than in group D (EL method). The maximum wire deflection in groups A to D was 2.50 ± 0.04 mm, 1.73 ± 0.04 mm, 2.41 ± 0.06 mm, and 2.15 ± 0.05 mm, respectively. ANOVA revealed significant differences ($P < 0.005$) in the deflection values with respect to the loads in each CLB group except for the 10 g and 20 g loads ($P = 0.002$ and 0.073), whereas the *post hoc* comparisons revealed a significant difference ($P < 0.05$) between each group.

Figure 4 shows the mean deflection values with respect to the loads in the SLB groups. The system listed in descending order of deflection magnitude until 170 g was groups E to H ($P < 0.001$). After a 180 g loading, the descending order of the deflection magnitude was changed to groups F, E, G and H ($P < 0.001$). The maximum deflection of the wire in groups E to H was 4.26 ± 0.05 mm, 4.31 ± 0.02 mm, 4.22 ± 0.04 mm, and 3.93 ± 0.03 mm, respectively. ANOVA revealed significant differences ($P < 0.001$) in the deflection values with respect to the loads in each SLB group except for a load of 180 g ($P = 0.006$), whereas *post hoc* comparisons revealed a significant difference ($P < 0.05$) between each group.

Figure 5 shows the results for the SSL groups (the average of groups A and C) and EL groups (the average of groups B and D). The wire deflection magnitude was significantly different between the SSL and EL groups on the 10, 20, 30, 70, 140, 160, and 170 g load points ($P < 0.05$). The wire was deflected more in the SSL groups than in the EL groups.

Figure 6 shows the results of the passive SLB groups (the average of groups E and G) and active SLB groups (the average of groups F and H). There were no significant differences between these groups.

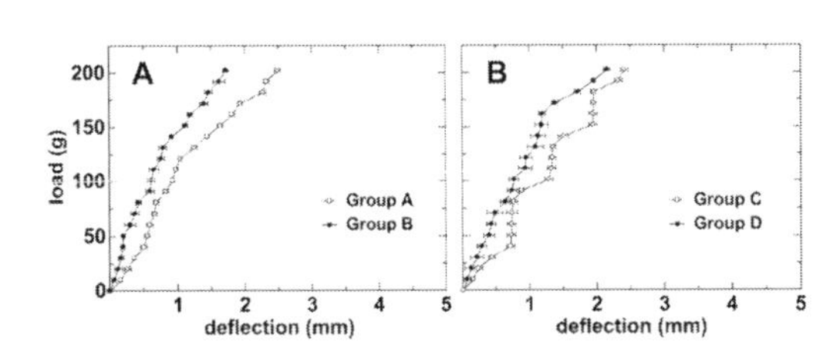

Figure 3. Initial force LD characteristic curves of Sentalloy® NiTi wire in the CLB groups.

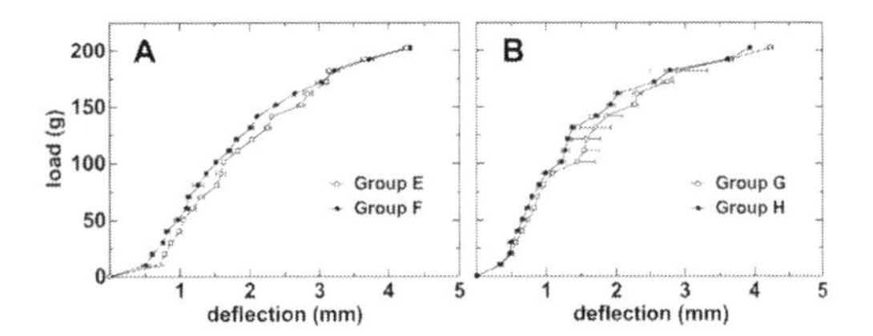

Figure 4. Initial force LD characteristic curves of Sentalloy® NiTi wire in SLB groups.

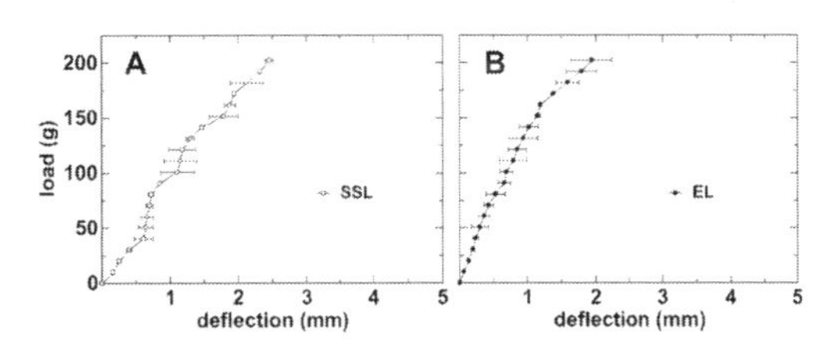

Figure 5. Initial force LD characteristic curves of Sentalloy® NiTi wire in the (A) SSL and (B) EL groups.

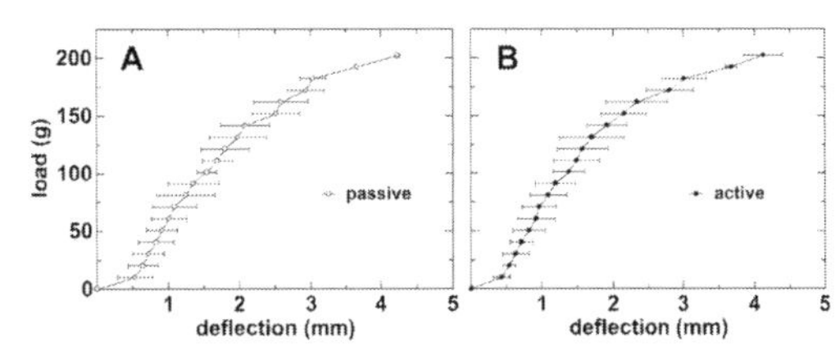

Figure 6. Initial force LD characteristic curves of Sentalloy® NiTi wire in the (A) passive and (B) active SLB groups.

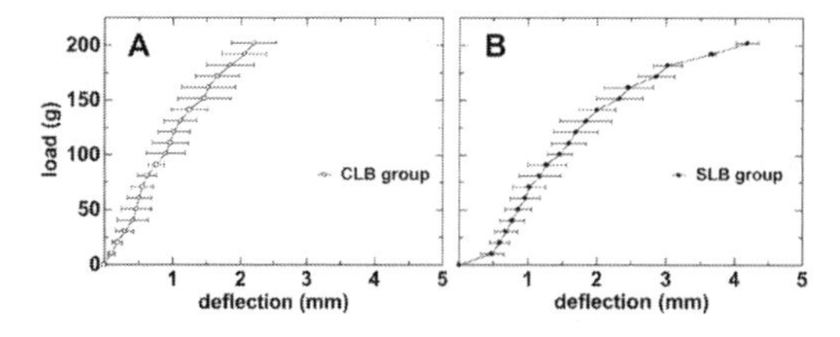

Figure 7. Initial force LD characteristic curves of Sentalloy® NiTi wire in the (A) CLB and (B) SLB groups.

Figure 7 shows the results of the CLB groups (the average of groups A to D) and SLB groups (the average of groups E to H). The wire deflection magnitude was greater in the SLB groups than in the CLB groups ($P < 0.05$).

4 DISCUSSION

This study presents a modified three-point bending based on a simple and inexpensive device. The three-point bending method is not directly transferable to the clinical setting and resembles the bucco-lingual movements (Oltjen et al. 1997). This method focuses more on the physical and biomechanical properties of the wires, offers reproducibility, and is useful for purely theoretical evaluations. It is a standardized testing method that allows a comparison with other studies (Tonner & Waters 1994, Choi et al. 2012).

The most appropriate wire tests may be those that reproduce the condition encountered clinically, with the wire constrained as a part of a fully fixed appliance. On the other hand, the present study developed a method for predicting the IFD of an orthodontic wire by observing the wire geometric configuration using a simple device. The load in each group equals the sum of the wire shear force and frictional resistance to sliding at both bracket points, which is comprised of ligation- and wire-derived components. In the experimental setup, the wire was deflected in the lingual direction. At both sides of the brackets, the wire contacted not only the mesio-lingual side of the slot base, but also the disto-buccal aspects of the bracket slot, which were different in all the investigated bracket types: the SSL and EL methods in the CLB groups, a stainless steel rigid face in Damon II®, NiTi spring clip in Speed®, NiTi pointed clips in Clarity-SL®, and a rhodium-coated sliding clip in Clippy-C®. Therefore, the wire locking mechanism between the deflected wire and the bracket's buccal aspect might affect the results. All experimental brackets have a metal composition slot base except for the Clippy-C®, which is made up of a highly temperature-resistant alumina-coated ceramic. The metal slot and ceramic body are combined in the Clarity® and Clarity-SL® esthetic brackets to decrease the friction resistance.

In the CLB groups at 200 g, the Clarity® group with SSL (group C) showed the greatest deflection (2.50 ± 0.04 mm), whereas the Mini-Taurus® group with EL (group B) showed the smallest deflection (1.73 ± 0.04 mm). This suggests that the EL method increases the load to deflect the NiTi wire (Figure 4). Clinically, the increased load corresponds to the force required in the EL method to engage the wire in the slot, which means patients may feel more discomfort with the EL method and less with the SSL method. The elasticity of EL is expected to give a continuously tightening pressure to the wire in the bracket slot, whereas this is not possible in the case of SSL. This pressure, which can be regarded as the frictional force, might restrain the intrinsic mechanical behavior of the NiTi wire. This assumption is consistent with that of Ireland et al. (1991), who reported that the EL method produced significantly more friction than the SSL method. As against general expectation, the Clarity® group showed a lower IFD than the Mini-Taurus® group. The Clarity® bracket consisting of a metal slot base would not make much of a difference to the metal brackets with respect to friction.

In the SLB groups at a 200 g load, the Speed® (group F) showed the greatest deflection (4.31 ± 0.02 mm), followed by the Damon II® (group E, 4.26 ± 0.04 mm), Clarity-SL® (group G, 4.22 ± 0.04 mm), and Clippy-C® (group H, 3.93 ± 0.03 mm). On the other hand, from 10 g to 170 g, the order of wire deflection magnitude was Damon II® >Speed®> Clarity-SL® >Clippy-C®. The graph showed an apparent reversal of deflection order between Damon II® and Speed® during loading from 180 g to 200 g (Figure 5). The major difference between the passive and active SLB groups was the function and structure of the bracket facial wall. In the active SLB groups, the margin of the active clip invades the bracket slot that makes the deflected wire contact the facial wall more easily than the passive type. Therefore, the friction resistance might increase in the active SLB groups. This assumption is reasonable to explain the phenomenon from 10 g to 170 g. From 180 g to 200 g, it is believed that both the flexible active clips in the Speed® and bracket slot width differences were due to these results. A comparison of the differences between the passive and active SLB groups deserves little consideration because all SLB groups in this study have various facial wall structures, mesio-distal slot widths and compositions of the slot base, which are factors affecting the frictional force. There were no significant differences between these groups.

The orthodontic NiTi wire was deflected in the lingual direction at the middle point and an IFD of a 0.014-inch orthodontic NiTi wire could be predicted in this study. In the Damon® bracket system (passive SLB), a force of 79 g was needed to deflect the wire to a 1.5 mm lingually malaligned tooth. This magnitude is suitable for the tooth movement in this *in vitro* experiment. On the other hand, in the Mini-Taurus® metal bracket system (CLB+EL), a force of 184 g was needed to deflect the wire 1.5 mm lingually. This force is recommended in the existing literature for incisor tooth movements that are exceeded substantially. The IFD of the Mini-Taurus® metal CLB with EL was the greatest among the groups examined, approximately two times that of the Damon®. Most of the SLB groups showed a lower IFD than the CLB groups. Of course, this predicting algorithm is not identical to that *in vivo*, as the biological resistance, mastication force, wear of materials, and wet environment affect the aforementioned calculations. According to general expectations, the wire deflection magnitude was greater in the SLB groups than in the CLB groups. This is consistent with previously reported studies in which the SLB groups showed a lower frictional force than CLB groups (Cordasco et al. 2009, Franchi et al., 2008, Matarese et al. 2008).

This study will serve as a fundamental reference for further studies aiming to produce a predicting algorithm at the force level and consistency of force exertion over a range of activations displayed from orthodontic wires.

5 CONCLUSIONS

This study examined the effect of different ligating systems on the IFD characteristics of superelastic NiTi wires using a modified three-point bending test. As a result, the following findings were obtained:

1. This study developed a method for predicting the IFD of orthodontic wires through observations of the wire geometric configuration with a low-cost device.
2. The expression of the IFD characteristics of orthodontic NiTi wires is affected by various ligating systems.
3. The higher friction resistance due to the EL module pressure in the EL groups showed a higher IFD of the orthodontic NiTi wire than that in the SSL groups.
4. There were no significant differences in the IFD between the passive SLB groups and active SLB groups.
5. The SLB groups showed a lower friction resistance, and a lower IFD of the orthodontic NiTi wire than the CLB groups.

ACKNOWLEDGMENTS

This study was supported by a grant from the Korean Health Technology Research & Development Project, by the Ministry of Health & Welfare, Republic of Korea (HI14C2241).

APPENDIX

The prepared wires were assigned to eight groups (Table 1) according to a combination of six types of orthodontic brackets including the Mini-Taurus® (CLB, Rocky Mountain Orthodontics, Denver, CO, USA), the Clarity® (CLB, 3M Unitek, St Paul, MN, USA), the Damon II® (Passive SLB, Ormco SDS, Glendora, CA, USA), the Speed® (Active SLB, Strite Industries, Cambridge, Ontario, Canada), the Clarity-SL® (Passive SLB, 3M Unitek, Monrovia, CA, USA) and the Clippy-C® (Active SLB, Tomy, Tokyo, Japan), and two types of ligature methods including the 0.009-inch stainless steel ligature (SSL, Ortho Organizers, Carlsbad, CA, USA) and the elastomeric ligature (EL, TP Orthodontics, La Porte, IN, USA).

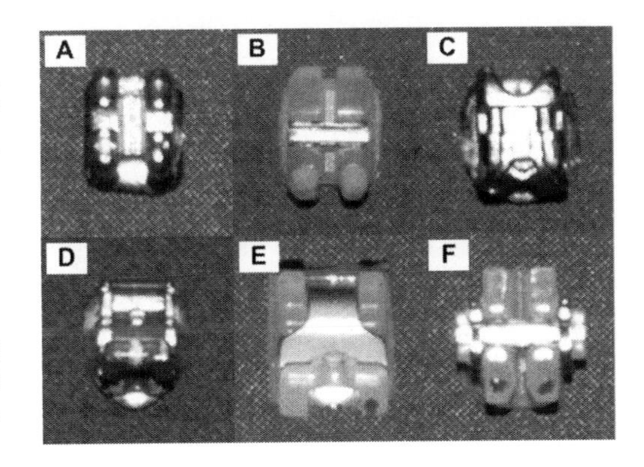

Figure A1. Six types of orthodontic brackets used in this study. (A) Mini-Taurus® (Rocky Mountain Orthodontics), (B) Clari-ty® (3M Unitek), (C) Damon II® (Ormco SDS), (D) Speed® (Strite Industries), (E) Clarity-SL® (3M Unitek) and (F) Clippy-C® (Tomy)

REFERENCES

Bartzela, T.N., Senn, C. & Wichelhaus, A. 2007. Load-deflection characteristics of superelastic nickel-titanium wires. *The Angle Orthodontist*, 77(6): 991–998.

Choi, S., Joo, H.J., Cheong, Y., Park, Y.G. & Park, H.K. 2012. Effects of self-ligating brackets on the surfaces of stainless steel wires following clinical use: AFM investigation. *Journal of Microscopy*, 246(1): 53–59.

Choi, S., Park, D.J., Kim, K.A., Park, K.H., Park, H.K. & Park, Y.G. 2015. In vitro sliding-driven morphological changes in representative esthetic NiTi

archwire surfaces. *Microscopy Research and Technique*, 78(10): 926–934.
Cordasco, G., Farronato, G., Festa, F., Nucera, R., Parazzoli, E. & Grossi, G.B. 2009. In vitro evaluation of the frictional forces between brackets and archwire with three passive self-ligating brackets. *European Journal of Orthodontics*, 31(6): 643–646.
Franchi, L., Baccetti, T., Camporesi, M. & Barbato, E. 2008. Forces released during sliding mechanics with passive self-ligating brackets or nonconventional elastomeric ligatures. *American Journal of Orthodontics and Dentofacial Orthopedics*, 133(1): 87–90.
Ireland, A.J., Sherriff, M. & McDonald, F. 1991. Effect of bracket and wire composition on frictional forces. *European Journal of Orthodontics*, 13(4): 322–328.
Matarese, G., Nucera, R., Militi, A., Mazza, M., Portelli, M., Festa, F. & Cordasco, G. 2008. Evaluation of frictional forces during dental alignment: an experimental model with 3 nonleveled brackets. *American Journal of Orthodontics and Dentofacial Orthopedics*, 133(5): 708–715.
Nakano, H., Satoh, S., Norris, R., Jin, T., Kamegai, T. & Ishi-kawa, F. 1999. Mechanical properties of several nickel-titanium alloy wires in three-point bending tests. *American Journal of Orthodontics and Dentofacial Orthopedics*, 115(4): 390–395.
Oltjen, J.M., Duncanson Jr, M.G., Ghosh, J., Nanda, R.S. & Currier, G.F. 1997. Stiffness-deflection behavior of selected orthodontic wires. *The Angle Orthodontist*, 67(3): 209–218.
Reitan, K. 1967. Clinical and histologic observations on tooth movement during and after orthodontic treatment. *American Journal of Orthodontics and Dentofacial Orthopedics*, 53(10): 721–745.
Ren, Y., Maltha, J.C. & Kuijpers-Jagtman, A.M. 2003. Optimum force magnitude for orthodontic tooth movement: a system atic literature review. *The Angle Orthodontist*, 73(1): 86–92.
Tonner, R.I.M. & Waters, N.E. 1994. The characteristics of super-elastic NiTi wires in three-point bending. Part 2: Intra-batch variation. *European Journal of Orthodontics*, 16(5): 421–425.
Warita, H., Iida, J. & Yamaguchi, S. 1996. A study on experimental tooth movement with Ni-Ti alloy orthodontic wires: comparison between light continuous and light dissipating force. *The Journal of Japan Orthodontic Society*, 55(6): 515–527.

Advanced Materials, Mechanical and Structural Engineering – Hong, Seo & Moon (Eds)
© 2016 Taylor & Francis Group, London, ISBN: 978-1-138-02908-8

Arsenic flux dependence of GaAs (001) homoepitaxial growth

O.A. Ageev, M.S. Solodovnik & S.V. Balakirev
Southern Federal University, Taganrog, Russian Federation

ABSTRACT: We study arsenic flux effect on the homoepitaxial growth of GaAs (001) using the kinetic Monte Carlo method. Gallium atoms and different arsenic species (As_2 and As_4) are involved in the simulation. We show that the surface concentration of physisorbed arsenic tetramers rises with increase of arsenic flux whereas the concentration of physisorbed dimers has a peak at arsenic flux of $1 \cdot 10^{15}$ cm^{-2} s^{-1} at growth rate of 0.1 nm/s. The calculated surface density of GaAs islands yields $2 \cdot 10^{12}$ cm^{-2}, in agreement with experiments. The island density rises with arsenic flux increase, and this dependence remains the same at a range of growth rates from 0.03 to 0.3 nm/s.

1 INTRODUCTION

In recent years, much attention has been focused on kinetic processes during Molecular Beam Epitaxy (MBE) of III-V compound semiconductors, such as GaAs. This is fundamentally important not only for crystal growth but also for fabricating of integrated circuits, optoelectronic and photonic devices. Understanding the microscopic phenomena is a critical step toward optimizing the characteristics of epitaxial structures by altering the growth conditions.

By now, diffusion and nucleation processes on the technologically important GaAs (001) surface have been particularly investigated using various techniques such as reflection high-energy electron diffration (Hata et al. 1990), Scanning Tunneling Microscopy (STM) (LaBella et al. 2000), ab initio calculations (Kratzer et al. 1999) and Kinetic Monte Carlo (KMC) simulations (Amrani et al. 2011). Theoretical models have been widely used to investigate surface reconstructions, activation barriers, surface diffusion of adatoms, and the influence of substrate temperature on them. However, almost nothing is known about the arsenic flux dependence of GaAs growth. Furthermore, modeling is mostly performed with considering epitaxy from As_2 flux (Kratzer & Scheffler 2002) or even one-component case (Smilauer et al. 1993), whereas widely used MBE growth from As_4 flux is more complex and requires involving many species and microscopic processes in the simulation (Avery et al. 1997).

In this paper, we report KMC simulation of arsenic flux effect on the growth of GaAs on the GaAs (001) – (2 × 4) reconstructed surface during MBE from Ga and As_4 vapor fluxes. The KMC technique enables simulating in a wide range of vapor fluxes and at the large length and time scales that cannot be reached by quantum ab initio calculations.

2 DESCRIPTION OF THE MODEL

The crystal is represented as a two-dimensional lattice having the physically adequate zinc blende structure of GaAs. The starting surface is prepared in the $\beta 2$ (2 × 4) reconstruction of GaAs (001) which implies that arsenic dimer rows and missing trenches alternate across the lattice.

A basic parameter of the model is a simple event which is performed with probability specified by an activation barrier. Activation energy Ei of an event i depends on type of microscopic process, type of a particle, its location and adatom environment. The frequency fi of occurrence of an event i is evaluated by the Arrhenius law:

$$f_i = \nu_0 \exp\left(E_i / kT\right) \quad (1)$$

where, ν_0 is the attempt frequency set at 10^{13} s^{-1} (Kley et al. 1997); k is the Boltzmann constant; and T is the substrate temperature.

Since the simulation suggests the growth on the reconstructed surface from the fluxes of components with substantially different behavior, a large number of events are necessary to be taken into account. In this simulation, we consider more than 40 events. Activation energies of basic events were obtained from both experimental and previous theoretical studies. Intermediate values were derived by adding the binding energy of a conglomerate, as obtained in density functional theory (Kratzer, P. et al. 1999), to the barrier for the corresponding basic event. In the following the processes included in the present simulation are described in detail.

Gallium adsorption: Ga atoms are deposited on the surface and occupy random sites in the gallium sublattice. The flux of atoms is determined by the growth rate v:

$$J_{Ga} = 0.5v/\sigma \qquad (2)$$

where, σ is area occupied by one atom.

Ga atoms adsorb with unit sticking probability and don't desorp that is typical for GaAs growth temperatures (Tok et al. 1997b).

Gallium surface diffusion and incorporation: Ga adatom migration is implemented by consecutive hops to adjacent sublattice sites. The diffusion is anisotropic so that activation barrier for hopping is 1.3 eV along the [110] direction and 1.5 eV along the [110] direction (LaBella et al. 2000). A Ga adatom can diffuse in each state, but if it reaches a strongly bound configuration (Kratzer & Scheffler 2002), it is not likely to move anymore and can be considered as incorporated into the crystal.

Arsenic physisorption and surface diffusion: Arsenic is deposited from As_4 flux and occupy random double sites in the arsenic sublattice. While each Ga atom stays on the surface, arsenic molecules find sufficiently strong binding sites only after gallium has been deposited (Tok et al. 1997b). Calculations based on density-functional theory found a chemical adsorption (chemisorption) pathway for the arsenic molecules close to Ga adatoms (Morgan et al. 1999), whereas arsenic interact with the surface through weak molecular adsorption (physisorption). The presence of this weakly-bound state results in increase of As_4 diffusion rate. As_4 molecule migrates on the surface with low activation barrier of 0.2 eV in their innate form (Tok et al. 1997a). As it meets a pair of Ga atoms, it can dissociate to two dimers, as the modulated-beam study suggests (Foxon & Joyce 1975). Thereby, Ga adatoms, As_2 and As_4 molecules coexist on the surface. The contribution of atomic As is neglected in the simulation since As_2 dimers don't dissociate under typical growth conditions (Kratzer & Scheffler 2002).

Arsenic desorption: Only physisorped molecules can re-evaporate from the surface. The activation barriers for desorption of As_2 and As_4 molecules without nearest neighbors is 0.48 and 0.37 eV, respectively (Garcia et al. 1989).

Arsenic chemisorption: The chemisorption of As_4 occurs along with the pairwise dissociation on two Ga adatoms. One of As_2 dimers is incorporated into the crystal with activation energy of 0.16 eV or less (Avery et al. 1997). The second dimer is chemisorped too, if there is another pair of Ga atoms next to the dimer. Otherwise, it starts the surface diffusion.

A reverse transition to the physisorped state is possible. The activation barrier for this process also depends on the location and environment of a dimer and varies from 1.7 to 2.6 eV (Morgan et al. 1999).

3 RESULTS AND DISCUSSION

The KMC simulations show that islands preferentially form in trench sites and favor elongation along the [110] direction (Figure 1). The growth then proceeds by lateral attachment of more gallium and arsenic atoms with (2×4) reconstruction maintaining. The morphology of islands formed at simular GaAs growth parameters is in consistence with STM images (Avery et al. 1997) and theoretical studies (Kratzer & Scheffler 2002).

For quantitative analysis of arsenic flux effect on the growth we study the arsenic flux dependence of the island surface density. An important feature of the growth in submonolayer regime is the saturation of the island density after a certain amount of material has been deposited. Our simulations show that the saturation is reached after deposition of ~0.1 Monolayer (ML) GaAs at $T = 580$ C and $v = 0.03$ nm/s (Figure 2), in agreement with experiments (Itoh, M. et al. 2000). It is observed

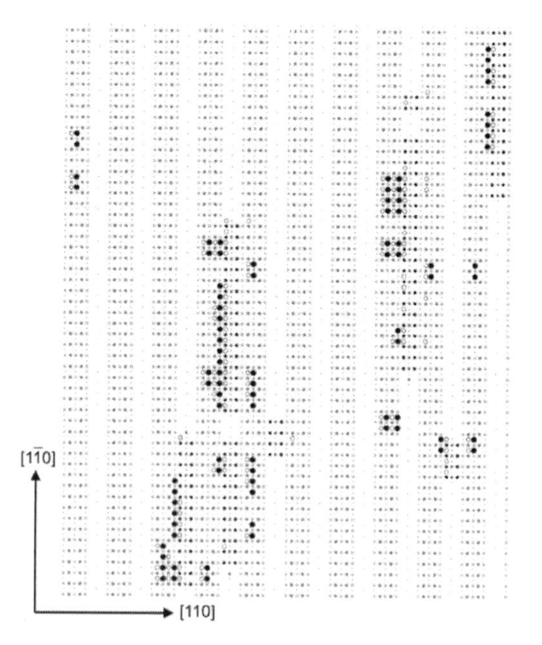

Figure 1. Island morphology in a simulation area of 160 Å × 200 Å after deposition of 0.1 ML GaAs at $T = 580°C$, $v = 0.03$ nm/s, $J_{AS4} = 1 \cdot 10^{14}$ cm^{-2} s^{-1}. The open circles represent Ga atoms and the filled circles As atoms. The starting surface is marked in grey and freshly deposited material in black.

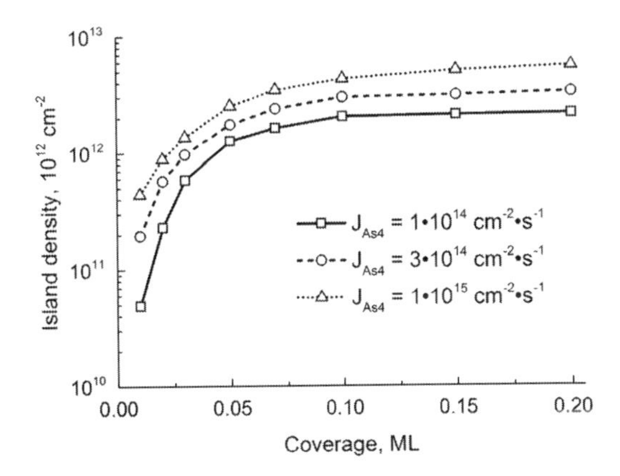

Figure 2. Time evolution of the island densities at various arsenic fluxes ($T = 580°C$, $v = 0.03$ nm/s).

Figure 3. Saturation island density as a function of V/III flux ratio ($\theta = 0.1$ ML, $T = 580°C$).

for each arsenic flux J_{AS4} from $1 \cdot 10^{14}$ to $1 \cdot 10^{15}$ cm^{-2} s^{-1}, but the saturation island density N is different. The simulation yields $N = 2 \cdot 10^{12}$ cm^{-2} at $J_{AS4} = 1 \cdot 10^{14}$ cm^{-2} s^{-1}, in consistence with experimental data (Avery, A.R. et al. 1997).

Figure 2 also shows an increase of the island density nearly three times when the arsenic flux increases to $1.3 \cdot 10^{15}$ cm^{-2} s^{-1}. A large amount of arsenic on the surface leads to the gallium diffusion suppression and intensification of nucleation processes. Pairs of Ga atoms covered with As_2 molecules act as main nucleation centers (Balakirev et al. 2014).

The saturation island density is calculated for different growth rates (Figure 3). The arsenic flux varies over a wide range for each of them, so it is reasonable to plot the island density against V/III flux ratio. It is known that the island density rises with the growth rate increase (Dubrovskii & Cirlin 2005), and our simulations confirm that, as evidenced in Figure 3.

We also observe a significant influence of V/III flux ratio on the island density. It is the strongest at $v = 0.03$ nm/s when the island density increases about twofold with the increase of V/III flux ratio from 3 to 40. The further increase of both growth rate and V/III flux ratio brings the island density to the limit value. The enhanced nucleation results in the formation of a large number of little clusters (less than 10 atoms) which consume the deposited material and do not let larger islands grow. Hence, for a certain value of coverage θ there is a peak value of the island density which is about $6 \cdot 10^{12}$ cm^{-2} at $\theta = 0.1$ ML GaAs.

In order to estimate which arsenic species make the largest contribution to the island formation, we calculate its surface concentration (Figure 4).

As discussed above, arsenic is deposited on the surface in the form of As_4 tetramers and can either desorb or dissociate to two As_2 dimers with chemisorption of one or both of them. Free As_2 molecules can also migrate on the surface in physisorbed state until they desorb or chemisorb. Figure 4 shows that the concentration of physisorbed As_4 molecules is much more than that of physisorbed dimers. This is attributed to greater chemical activity of As_2 molecules having dangling bonds (Morgan, C.G. et al. 1999). During one measurement step they typically have time to find the most favorable site in the lattice or desorb.

The important factor affecting physisorbed As_2 concentration is relative amounts of deposited Ga and As. While physisorbed As_4 concentration increases with the arsenic flux increase and can even increase the concentration of chemisorbed arsenic (Figure 4c), physisorbed As_2 concentration has a peak at $J_{AS4} = 1 \cdot 10^{15}$ cm^{-2} s^{-1} (Figure 4b) that corresponds to V/III flux ratio equal 10 at $v = 0.1$ nm/s. Larger arsenic fluxes lead to the shortage of Ga atoms to be chemisorbed on, and smaller arsenic fluxes causes arsenic shortage.

The concentration of chemisorbed dimers is the same at each arsenic flux, since it is determined by an amount of Ga on the surface and represents material used for the island formation. Making a comparison between Figure 3 and Figure 4, we can see that $3 \cdot 10^{13}$ chemisorbed dimers yield $3.3 \cdot 10^{12}$, $4.5 \cdot 10^{12}$, $5.5 \cdot 10^{12}$ islands at V/III flux ratio equal 3, 10 and 40, respectively. It means that an average island contains about 36, 27 and 22 atoms at respective technological parameters.

4 CONCLUSIONS

The KMC simulations show that the arsenic flux alteration enables to control GaAs epitaxial film characteristics over a wide range. An amount of gallium on the surface defines an amount of arsenic that is chemisorbed and incorporated into the crystal. However, the island density strongly depends on the arsenic flux. Its saturation value increases with the arsenic flux increase and approaches to a limit at larger growth rates. Arsenic on the surface is mainly represented in the form of physisorbed tetramers, whereas physisorbed dimers are quickly chemisorbed and cease to move. The largest concentration of physisorbed As_2 molecules is observed at V/III flux ratio of 10, as it is enough Ga atoms to prevent As desorption but not so many to chemisorb a lot of dimers.

ACKNOWLEDGEMENT

This work was supported by the Russian Science Foundation Grant No. 15-19-10006. The results were obtained using the equipment of Common Use Center and Education and Research Center "Nanotechnologies" of Southern Federal University.

REFERENCES

Amrani, A. et al. 2011. A Monte Carlo investigation of Gallium and Arsenic migration on GaAs (100) surface. *Applied Nanoscience,* 1: 59–65.

Avery, A.R. et al. 1997. Nucleation and Growth of Islands on GaAs Surfaces. *Physical Review Letters,* 79(20): 3938–3941.

Balakirev, S.V. et al. 2014. Model of the initial stage of GaAs homoepitaxial growth by MBE considering growth components flux ratio. *Proceedings of the SFU. Engineering,* 9: 94–105.

Dubrovskii, V.G. & Cirlin, G.E. 2005. Growth kinetics of thin films formed by nucleation mechanisms of the layer formation. *Semiconductors,* 39(11): 1312–1319.

Foxon, C.T. & Joyce, B.A. 1975. Interaction kinetics of As4 and Ga on {100} GaAs surfaces using a modulated molecular beam technique. *Surface Science,* 50: 434.

Hata, M. et al. 1990. Distributions of growth rates on patterned surfaces measured by scanning microprobe reflection high-energy electron diffraction. *Journal of Vacuum Science & Technology B,* 8: 692.

Itoh, M. et al. 2000. Transformation kinetics of homoepitaxial islands on GaAs (001). *Surface Science,* 464: 200.

Kley, A. et al. 1997. Novel Diffusion Mechanism on the GaAs (001) Surface: The Role of Adatom-Dimer Interaction. *Physical Review Letters,* 79(26): 5278.

Kratzer, P. et al. 1999. Model for nucleation in GaAs homoepitaxy derived from first principles. *Physical Review B,* 59(23): 15246.

Kratzer, P. & Scheffler, M. 2002. Reaction-Limited Island Nucleation in Molecular Beam Epitaxy of Compound Semi-conductors. *Physical Review Letters,* 88(3): 036102.

LaBella, V.P. et al. 2000. Monte Carlo derived diffusion parameters for Ga on the GaAs (001) – (2 × 4) surface: A molecular beam epitaxy–scanning tunneling microscopy study. *Journal of Vacuum Science & Technology A,* 18(4): 1526–1531.

Morgan, C.G. et al. 1999. Arsenic Dimer Dynamics during MBE Growth: Theoretical Evidence for a Novel Chemisorption State of As2 Molecules on GaAs Surfaces. *Physical Review Letters,* 82(24): 4886.

Smilauer, P. et al. 1993. Reentrant layer-by-layer growth: A numerical study. *Physical Review B,* 47: 4119.

Tok, E.S. et al. 1997a. Incorporation kinetics of As2 and As4 on GaAs (110). *Surface Science,* 371: 277.

Tok, E.S. et al. 1997b. Arsenic incorporation kinetics in GaAs (001) homoepitaxy revisited. *Surface Science,* 374: 397.

Advanced Materials, Mechanical and Structural Engineering – Hong, Seo & Moon (Eds)
 ISBN: 978-1-138-02908-8

Influence of fly ash on the properties of Recycled Aggregate Concrete

Y.L. Zhang & W.X. Tang
School of Civil Engineering and Architecture, Zhejiang University of Science and Technology, Hangzhou, China

ABSTRACT: This paper discussed the influence of the Recycled concrete Aggregate (RA), water/binder ratio and fly ash on the workability, compressive strength and SEM morphology of Recycled Aggregate Concrete (RAC). The first batch specimens were prepared with the same water/cement ratio of 0.55 but with different replacements of RA, i.e. 0, 20%, 30%, 40%, 50%, 60%, 70%, 80% and 100%, respectively. It presents the fluidity and compressive strength of RAC decreases with the increasing RA, with the exception of the group with 30% RA, whose compressive strength is even higher than the control concrete. Another three groups were prepared with 100% RA and with different water/binder ratios of 0.50, 0.55 and 0.6, each using 0%, 10%, 20% and 30% fly ash, respectively, as cement replacement. It shows the compressive strength of RAC decreases after the addition of fly ash, but the group with 20% fly ash keeps the level very close to that with 10% fly ash. The SEM tests imply that the fly ash can improve the interface between the cement stone and the aggregate, while at 28 days, the compressive strength of RAC with fly ash is still lower than the control specimens due to its slow hydration speed.

1 INTRODUCTION

In the process of maintenance and demolition of old buildings, roads and bridges, a large amount of waste concrete will be produced, which can be reprocessed as Recycled concrete Aggregate (RA) in fresh concrete. Recycled coarse aggregate has high porosity and water absorption, and low strength.

Meanwhile, the complicated interfacial transition zones in RA and between RA and new mortar, old mortar and new mortar, can influence the performance of Recycled Aggregate Concrete (RAC) (Cui et al. 2011, Geng et al. 2012). There should be an optimal dosage of RA in RAC and different methods to improve the integral performance of RAC (Kwan et al. 2012). The addition of mineral admixtures is generally regarded as one of the efficient ways to improve the interface zones in RAC (Li et al. 2012, Xiao 2008). Pozzolanic materials are highly effective in improving the durability of RAC (Ann et al. 2008, Somna et al. 2012). In this paper, the slump, compressive strength and micro-morphology of RAC with and without fly ash are investigated, aiming to analyze the influence of fly ash on RAC.

2 EXPERIMENTAL PROCEDURE

2.1 *Raw materials*

1. Cement: Portland ordinary cement 42.5 was used.

Table 1. Properties of aggregates.

Items	RA	Natural aggregate
Apparent density	2600 kg/m^3	2720 kg/m^3
Water absorption	6.08%	1.80%
Crush index	12.5%	7.5%
Gradation	5.0–31.5	5.0–31.5

2. Fine aggregate: natural river sand was used with a water content of 2.04%, a fineness modulus of 2.56 and with good gradation.
3. Fly ash I: local fly ash was obtained from Lanxi city in Zhejiang Province.
4. Coarse aggregate: recycled concrete aggregates (RA) were crushed from some small beams of reinforced concrete kept in the laboratory, with a similar gradation of 5.0~31.5 to the natural gravels used. Other properties are listed in Table 1.

2.2 *Mix proportions of concrete*

The basic mix proportion of ordinary concrete was designed as summarized in Table 2. According to the basic mix proportion, the first group D was used to compare the influence of Recycled concrete Aggregates (RA) on Recycled Aggregate Concrete (RAC) with a W/C ratio of 0.55 and a RA of 0%, 20%, 30%, 40%, 50%, 60%, 70%, 80% and 100%, respectively, as the replacement of natural gravels, with each group correspondingly

Table 2. Basic mix proportion of concrete.

Raw material	Cement/ kg	Water/ kg	W/C	Sand/ kg	Gravel/ kg	Sand ratio
Mass	380	190	0.5	674	1252	35%

Table 3. Slump of RAC with different replacements of RA.

Group	D0	D2	D3	D4	D5	D6	D7	D8	D10
RA/%	0	20	30	40	50	60	70	80	100
Slump/mm	53	50	49	48	44	43	40	37	35

named as D0, D2, D3, D4, D5, D6, D7, D8 and D10 (Table 3). Another three groups, A, B and C, respectively, with different water/binder (*W/B*) ratios of 0.5, 0.55 and 0.6, were used to analyze the fly ash function on RAC. Each group includes four dosages (weight ratio) of fly ash as replacement of cement, i.e. 0%, 10%, 20% and 30%, respectively.

3 SPECIMEN MAKING AND TESTS

The recycled aggregates were immersed in water for 24 hours and then exposed to air for 2 hours. For each group, the workability of concrete was first tested and then the cubic specimens of concrete were made. The engineering plastic molds with a dimension of 100 mm × 100 mm × 100 mm were used for making concrete specimens. A total of 180 cubic specimens were completed to be tested at 7 d, 14 d and 28 d. The specimens were moved to the standard curing room for first 24 hours and kept cured in standard curing room conditions after being removed from the molds. The micro-morphology of each group was tested at same ages using SEM.

4 EXERIMENTAL RESULTS

4.1 *Workability of RAC without or with fly ash*

The results of the slump tests of RAC without fly ash are summarized in Table 3. With the increasing RA, the slump of RAC with the same W/C ratio decreased, while both the viscosity and water retention were good. The slump value of RAC with 100% RA decreased to 35 mm, from 53 mm of that with 0% RA, while the viscosity and water retention appeared much better than the latter one, which may be due to the bigger friction between the coarse surface of RA and the cement paste.

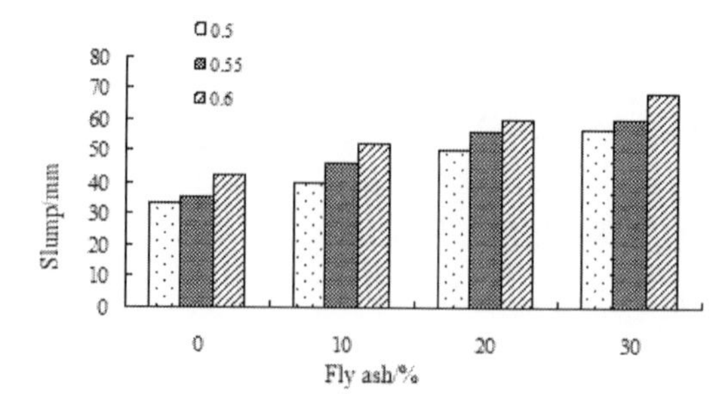

Figure 1. Slump of RAC with fly ash.

The slump results of RAC with fly ash are shown in Figure 1. When the fresh RAC has the same *W/B* ratio, it indicates that the slump increases with the increasing fly ash dosage. Meanwhile, when the fly ash dosage is kept the same, the slump of the fresh RAC with 100% RA increases with the increasing *W/B* ratio.

4.2 *Compressive strength*

4.2.1 *Influence of RA on the compressive strength of RAC without fly ash*

The results of the compressive strength of recycled concrete with different dosages of RA are shown in Figure 2. In general, with the increasing replacement amounts of RA, the compressive strength of RAC decreases, except D3 with 30% RA. Group D3 has much higher compressive strength than the other groups including the control one with no RA. Between 30% and 50% replacement amounts of RA, the compressive strength of the specimens decreases sharply, while eliminating in the case with more than 50% RA.

Normally, it is explained that the RA includes the old mortar part that results in the lower elastic modulus and strength than that of natural aggregates. Furthermore, more tiny cracks are developed inside the RA during the period of crushing, sieving, cleaning and drying. And the poor binding of the interface between the RA and cement mortar also contributes to the less compressive strength of the recycled concrete than that of the ordinary concrete. The phenomenon of higher compressive strength of recycled concrete with a certain dosage of RA reflects the difference from the above finding.

4.2.2 *Influence of fly ash on the compressive strength of RAC with 100% RA*

The results of the compressive strength of RAC with 100% RA instead of natural gravel are presented in Table 4. Three groups of RAC specimens with a *W/B* ratio of 0.5, 0.55 and 0.60 have the

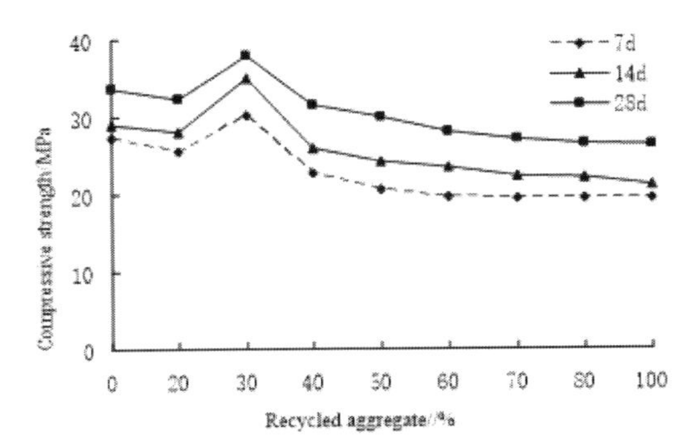

Figure 2. Compressive strength of recycled aggregate concrete.

Table 4. Compressive strength of RAC with fly ash.

W/B	Group	Fly ash/%	Compressive strength/ MPa 7 d	14 d	28 d
0.50	A0	0	23.23	27.23	32.68
	A1	10	20.72	23.07	29.94
	A2	20	21.51	24.02	28.73
	A3	30	18.60	21.27	27.79
0.55	B0	0	19.40	21.09	26.42
	B1	10	16.30	19.93	25.81
	B2	20	17.91	20.09	26.02
	B3	30	14.51	16.78	24.79
0.60	C0	0	16.97	20.11	23.90
	C1	10	14.93	16.90	21.43
	C2	20	15.03	16.57	21.75
	C3	30	12.73	15.03	19.43

dosages of fly ash of 0%, 10%, 20% and 30% to replace cement, respectively.

For each group with different ratios of *W/B*, the compressive strength at 7 d, 14 d and 28 d decreases apparently after the addition of fly ash. When the dosage of fly ash is varied from 10% to 20%, the specimens show a very close compressive strength. After the fly ash dosage of more than 20%, the compressive strength decreases slowly, which may be due to both roles of corresponding reduction of cement part in the system with the increasing fly ash and the weak hydration of fly ash before 28 d. Some other experiments showed the similar results (Xiao & Huang 2006). Consequently, we can know that there exists the limited dosage of fly ash to be used in recycled concrete. To improve its integral pozzolanic function, the optimum dosage of fly ash should be verified. Considering the same group here, it can be still found that the compressive strength increases much faster after 14 d than before 14 d. Meanwhile, all the compressive strengths of the specimens with fly ash have a higher increasing amplitude from 14 d to 28 d than that of the control ones without fly ash; while in earlier curing ages, from 7 d to 14 d, the difference among each other is very similar, which shows that fly ash hydrates slowly in earlier curing ages but contributes much to recycled concrete with 100% RA to improve its strength.

As for groups with a W/*B* ratio of 0.5, the compressive strength of A0, A1, A2 and A3 at 14 d, respectively, increases by 17.22%, 11.34%, 11.67% and 21.89% compared with that at 7d, while correspondingly at 28 d, 20.01%, 29.78%, 19.61% and 30.65% compared with that at 14 d. Both groups with *W*/B ratios of 0.55 and 0.60 have the similar change.

In addition, it is quite obvious that the compressive strength of RAC increases with the decreasing W/B ratio, which indicates that the W/B ratio is a key factor to the concrete strength, which is the same as in the ordinary concrete.

4.3 *Influence of fly ash on the micro-morphology of RAC*

The micro-morphology images of RAC without fly ash for groups D3 and D5 at 7 d, 14 d and 28 d, respectively, are shown in Figure 3. The images of the micro-morphology of groups A1 and A2 at 7d, 14 d and 28 d, respectively, are shown in Figure 4. It obviously shows the loose interface between the aggregates and cement mortar in RAC, leading to a weak micro-structure. Some interfaces are filled with a few hydrating products, while some others do not have the same appearance, which may imply the cracks developed when loading. From Figure 4, it can be seen that the interface is modified with more ball grains of fly ash as the filler. However, ball grains of fly ash are very clear in pictures no matter at 7 d, 14 d or 28 d, though much fewer at 28d, which proves the slow hydration speed of fly ash and the lower strength of RAC with fly ash at earlier days.

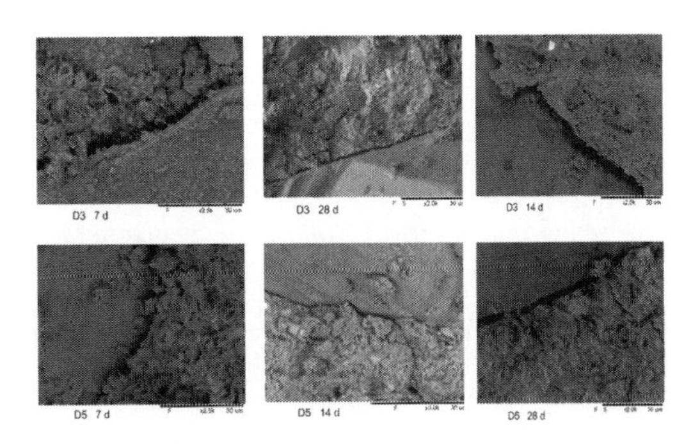

Figure 3. SEM images of RAC without fly ash at different curing ages (groups D3 and D5).

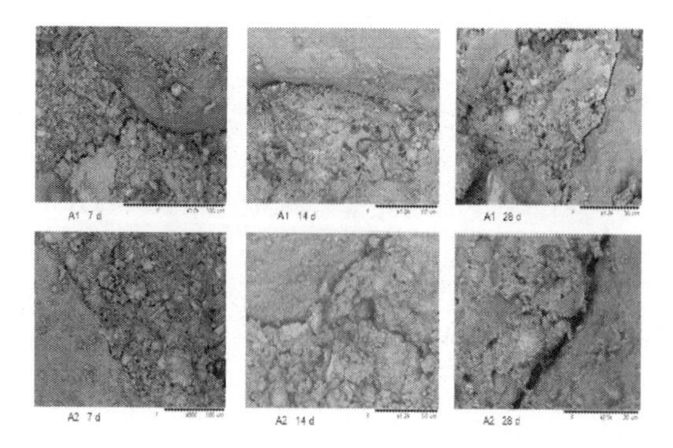

Figure 4. SEM images of RAC with fly ash at different curing ages (groups A1 and A2).

Figure 5. SEM images of RAC with fly ash at 14 d (groups A1, A2, B1 and B2).

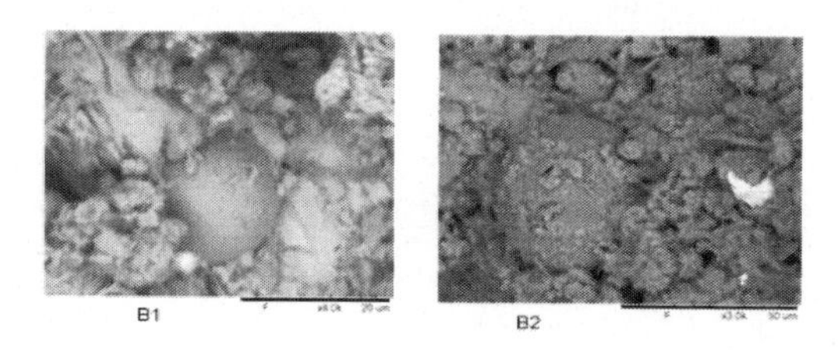

Figure 6. SEMS images of RAC with fly ash at 28 d (groups B1 and B2).

From Figure 5, it can be seen that there are hydrating products with the gel shapes attached on the surface of fly ash balls, which means that with the age increasing, fly ash starts the secondary hydration and evokes the development of gels such as hydrated calcium silicates and calcium aluminates. It also indicates few apparent $Ca(OH)_2$ crystals in the system, which is usually due to the partial $Ca(OH)_2$ consumption by the secondary hydration of fly ash and leads to finer grains and pores, correspondingly to dense cement stone. But even at 28 d (Figure 6), the hydration progress of fly ash is not completed, and the ball grains can be seen with slightly more or much more gels on the surface, which is the reason why the compressive strength of all specimens with fly ash is less than the control specimens at the same ages, even at 28 days.

5 CONCLUSIONS

1. With the increasing RA replacement amounts of natural aggregate, the slump of RAC decreases. While with the increasing dosage of fly ash, the slump of RAC with 100% RA increases.
2. The compressive strength of RAC with 30% RA is much higher than that of the other groups without or with other replacement amounts of RA.
3. The compressive strength of RAC with 100% RA decreases with the increasing fly ash, but the group with 20% fly ash keeps a very close level to that with 10% fly ash. The optimized dosage of fly ash in RAC with 100% RA could be no more than 20%.
4. Fly ash can modify the interface between the cement stone and the aggregate, while at 28 days, the compressive strength of RAC with fly ash is still lower than the control specimens due to its slow hydration speed.

REFERENCES

Ann, K.Y., Moon, H.Y., Kim Y.B. & Ryou, J. 2008. Durability of recycled aggregate concrete using pozzolanic materials. *Waste Management*, 28(6): 993–999.

Cui, Z.L., Lu, S.S. & Wang, Z.S. 2011. Influence of Mortar Transition Zone with Different Strength Class on Recycled Aggregate Concrete, *Bulletin of the Chinese Ceramic Society*, 30(3): 545−549.

Geng, O., Chen, C., Gu, R.J., Zheng, J.J. & Chai, B.S. 2012. Development law of interfacial microscopic structure in recycled coarse aggregate concrete, *Journal of Building Materials*, 12(3): 340–344.

Kwan, W.H., Ramli, M., Kam, K.J. & Sulieman, M.Z. 2012. Influence of the amount of recycled coarse aggregate in concrete design and durability properties. *Construction and Building Materials*, 26(1): 565–573.

Li, H., Sun, W. & Zuo, X.B. 2012. Effect of Mineral Admixtures on Sulfate Attack Resistance of Cement-Based Materials, *Journal of the Chinese Ceramic Society*, 40(8): 1119–1126.

Somna, R., Jaturapitakkul C. & Amde, A. 2012. Effect of ground fly ash and ground bagasse ash on the durability of recycled aggregate concrete, *Cement & Concrete Composites,* 34(7): 848–854.

Xiao, J.Z. 2008. *Recycled Concrete,* China Architecture & Building Press, Beijing.

Xiao, J.Z. & Huang, Y.B. 2006. Residual compressive strength of recycled concrete after high temperature, *Journal of Building Materials,* 9(3): 255–259.

Advanced Materials, Mechanical and Structural Engineering – Hong, Seo & Moon (Eds)
© 2016 Taylor & Francis Group, London, ISBN: 978-1-138-02908-8

The propagation of coupling plastic waves in a rate-sensitive material under compression and torsion

B. Wang, Q.Z. Song & Z.P. Tang
University of Science and Technology of China, Hefei, China

ABSTRACT: Plastic waves under combined stress are different from simply longitudinal waves and transverse waves since the longitudinal and transverse waves are coupled in that situation. Since there are only a few of experimental results of this phenomenon, in this paper, we build a system to apply a longitudinal impact to a pre-torqued tube based on the Split Hopkinson Pressure Bar (SHPB), and conduct an experimental research on 304 stainless steel that is rate-sensitive. The results show that the plastic wave structures of this material under compression and torsion are Coupling Fast Waves (CFW) and Coupling Slow Waves (CSW) with continuous changing speeds, and there is no constant region between the CFW and CSW. By using the numerical simulation, we obtain the stress data in the experimental procedure, which has not been discussed so far. The results indicate that the stress path structure of the rate-sensitive material is basically similar to the path structure of the rate non-sensitive material, and the longitudinal and shear components decrease with time.

1 INTRODUCTION

In most cases, materials and structures are under complex stress, and the response behavior is different with one-dimensional stress or strain (Tang & Aidun 2009). Ting (1969, 1985) conducted a series of theoretical research on plastic waves under complex stress in a rate non-sensitive material, and put forward the theoretical resolution. Different from the usual longitudinal waves or transverse waves that propagate independently, longitudinal parts and transverse parts are coupled and propagate with the same speed under complex stress. The waves can be divided into two groups according to the speed range and different pattern of coupling: the waves whose speed is always greater than the elastic transverse waves are called Coupling Fast Waves (CFW), while the waves whose speed is always less than the elastic transverse waves are called Coupling Slow Waves (CSW). These two groups of waves will be divided by a constant region since the wave speed is not continuous in most cases.

Clifton et al. (1970a, 1970b) conducted experiments in which rate non-sensitive 3003 aluminum tubes were subjected to a static plastic torque followed by a longitudinal compressive impact, and the observed strain time profiles corresponded with the theoretical resolution very well, and proved the validation of the theoretical research on plastic waves under complex stress. Then, they (Hsu & Clifton 1974a, 1974b) conducted experiments of the same kind in the rate-sensitive material α-Ti, and the results indicated that there was no constant region between the CFW and CSW, which was different from the rate-independent theory.

Since the report on the coupling wave structure of the rate-sensitive material α-Ti observed by Clifton et al., none of the studies have reported the same kind of experiment for other rate-sensitive materials so far. Whether the wave structure is widely existed in rate-sensitive materials and the path and evolution pattern in the stress space is not clearly understood. Therefore, we choose the rate-sensitive material 304 stainless steel and conduct the experiment of coupling plastic waves. By using the numerical simulation, we explore the path and evolution pattern in the stress space.

2 EXPERIMENTAL METHOD AND DEVICE

The dynamical axial force and moment will propagate along the X axis in a thin-walled tube, as shown in Figure 1. For the rate non-sensitive material, based on the theoretical resolution (Ting 1969, 1985), plastic waves can be divided into the CFW and the CSW. The paths of the CFW are along a series of ellipses bigger than the initial yield surface, while the paths of the CSW are perpendicular to the paths of CFW, as shown in Figure 2. The speed of these two kinds of waves decreases along

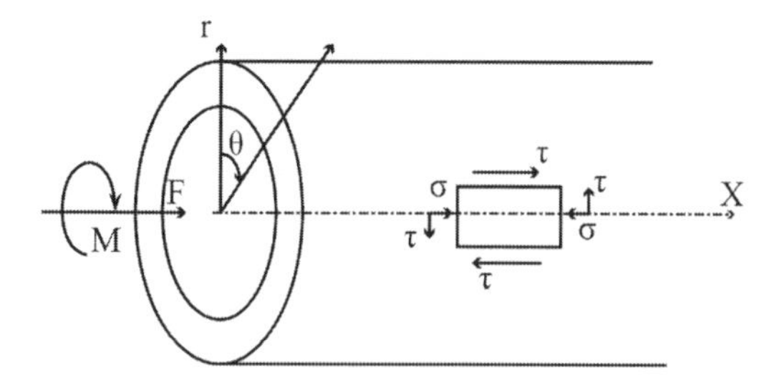

Figure 1. A thin-walled tube under combined compression-torsion loading.

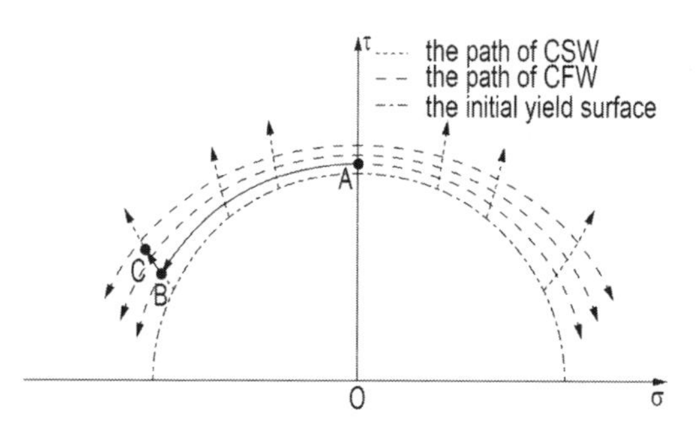

Figure 2. Coupling plastic wave paths of the rate non-sensitive material.

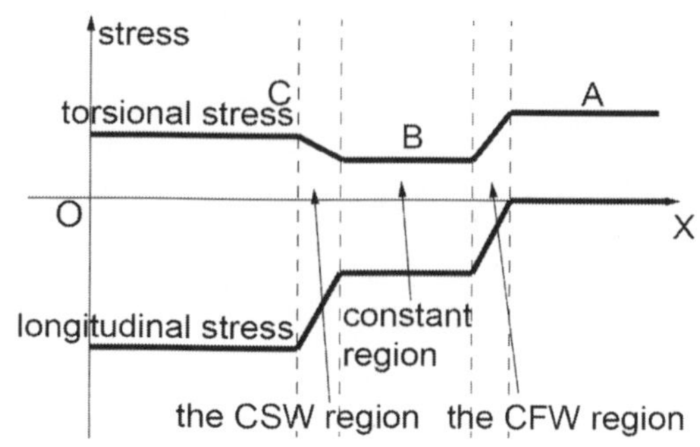

Figure 3. The wave profile of the rate non-sensitive material.

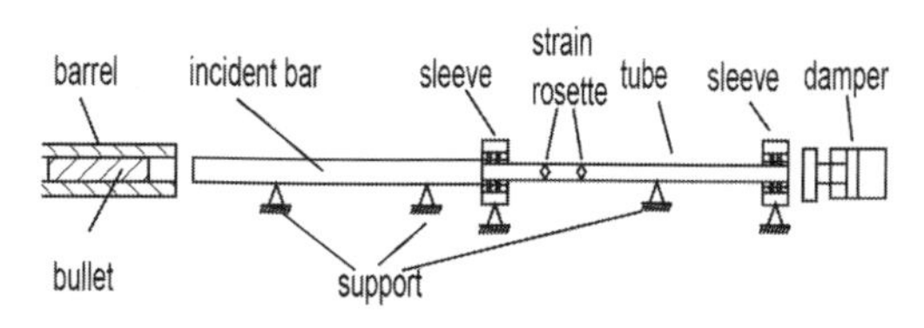

Figure 4. The schematic diagram of the impact of a pre-torqued tube.

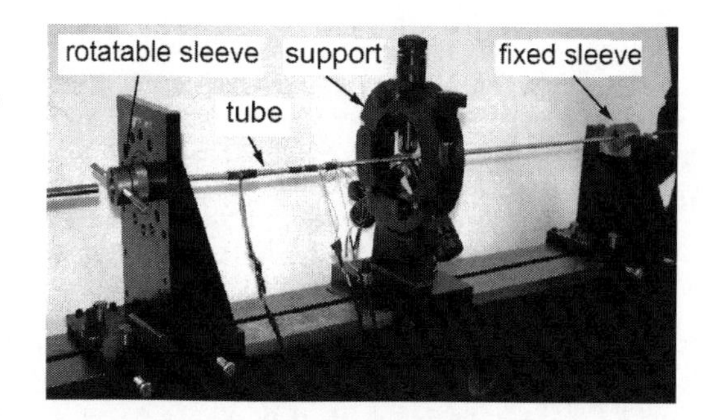

Figure 5. The twisting part of the system.

the arrow direction. In this paper, we define tension to be positive and compression to be negative.

In order to observe the coupling plastic waves, the specimen should be under complex stress loading. Owing to the difficulty of applying axial force and moment impact at the same time, a simple way is to restrict the rotational angle and apply the axial impact, as the path A-B-C shown in Figure 2. The specimen is twisted in the plastic region at the τ axis (let this point be A) and the angle is held at first, and then the axial impact is applied to a point in the second quadrant (let this point be C). Since the plastic waves should be along the paths of CFW and CSW, shown in Figure 2, the wave structures should be the CFW A-B, the constant region B, and the CSW B-C, as shown in Figure 3.

To apply the longitudinal impact to a pre-torqued tube, we build an experimental system based on the Split Hopkinson Pressure Bar (SHPB), as shown in Figure 4. Pins are implanted at the both ends of the tube, and both ends are put into two sleeves mounted at supports; one sleeve is rotatable while another is fixed. There are slide ways in both sleeves so that the rotation of pins at the tube ends is restricted while the longitudinal movement is allowed. With the sleeves and pins, we can maintain the tube torqued and then launch the bullet to produce a compressive impact. The twisting part is shown in Figure 5.

3 EXPERIMENTAL RESULTS

The positions that are 100 mm and 200 mm away from the impact end of the tube are denoted as #1 and #2, respectively. Rectangular strain rosettes are pasted at these positions and the longitudinal and torsional strain time profiles are recorded. After a series of experiments, we obtain the complete waveforms in four experiments, the results

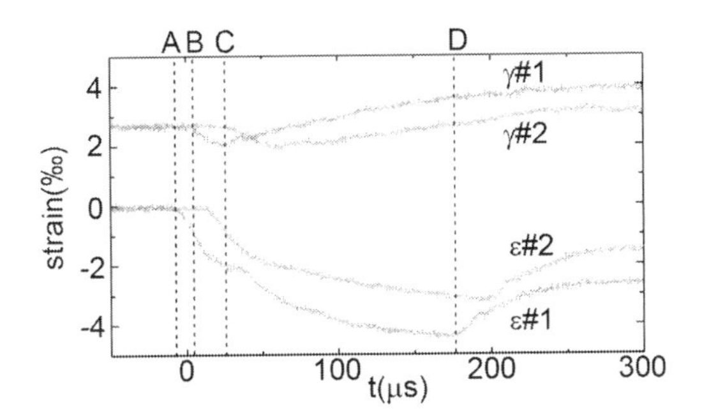

Figure 6. The typical strain time profile.

exhibit well repeatability and the typical results are shown in Figure 6.

We choose the strain time profiles at position #1 to analyze in detail, as shown in Figure 6. Before the impact, the torsional strain of the tube maintains 0.27% while the longitudinal strain is zero, and this state of strain is denoted as A. The longitudinal compressive strain increases to 0.11% while the torsional strain remains 0.27% when the longitudinal impact arrived at first, and this state of strain is denoted as B. Then, the longitudinal compressive strain keeps increasing to 0.2% while the torsional strain decreases to 0.2% gradually, and this state of strain is denoted as C. The longitudinal compressive strain keeps increasing to 0.44% while the torsional strain keeps increasing to 0.36% before the unloading wave arrives, and this state of strain is denoted as D. The wave structures of position #2 are almost the same with the wave structure at position #1.

The quasi-static torsional yield strain is about 0.21%, which is less than the pre-torqued strain 0.27%. According to the rate-independent theory, there should be no elastic longitudinal waves. But for the rate-sensitive material, as segment AB exhibits, there should be elastic longitudinal waves due to the strengthen effect. The segment BC should be the coupling fast waves while the segment CD should be the coupling slow waves of this material, respectively, different from the rate-independent theory. There is no constant region between the CFW and the CSW, which means that the speed of these waves is continuous.

The experimental results of the rate-sensitive 304 stainless steel exhibit some difference with the rate-independent theory. First, there should be elastic waves due to the strengthen effect under combined strain; second, there is no constant region between the CFW and the CSW. These results of 304 stainless steel agree with those of α-Ti reported by Clifton et al. (Hsu & Clifton 1974b), which means that these characteristics of the rate-sensitive material exist widely.

4 NUMERICAL SIMULATION

The characteristics of coupling plastic waves in the strain space are explored by experiments, but the path and evolution pattern in the stress space is not clearly understood yet. To understand the coupling plastic waves better, we performed a numerical simulation of the experimental process.

The kinetic equations are

$$\sigma_X = \rho_0 u_t \tag{1}$$

$$\tau_X = \rho_0 v_t \tag{2}$$

$$u_X = \varepsilon_t \tag{3}$$

$$v_X = \gamma_t \tag{4}$$

where ρ_0 is the density; σ and τ are the longitudinal and shear stresses; u and v are the longitudinal and circumferential particle velocities; and ε and γ are the longitudinal and shear strains, respectively. Using the Johnson-Cook constitutive (Johnson & Cook 1983) and ignoring the thermal part, we obtain

$$\bar{\sigma} = [\mathrm{A}+\mathrm{B}(\bar{\varepsilon}^p)^n]\left(1+C\ln\frac{\dot{\bar{\varepsilon}}^p}{\dot{\varepsilon}_0}\right) \tag{5}$$

where $\bar{\sigma}$ is the equivalent stress; $\bar{\varepsilon}^p$ is the equivalent plastic strain; $\dot{\bar{\varepsilon}}^p$ is the equivalent plastic strain rate; $\dot{\varepsilon}_0$ is the referential strain rate; and A, B, C, n are parameters.

We use the lax scheme (Lax 1954) to finite difference Equations (1)–(4), and extend the constitutive Equation (5) to combined stress by the radial return algorithm (Lu et al. 2003). Let $\dot{\varepsilon}_0$ be 0.01 s^{-1}, and the parameters A = 200 Mpa, B = 50000 Mpa, C = 0.12, n = 0.8. The comparison of the simulated strain time profiles and experimental results at positions #1 and #2 of the typical experiment is shown in Figure 7.

We can see that the simulation fits well with the experimental results. By using the simulation, we can obtain the stress of different times and positions in the experimental process. The stress of the rate-sensitive material is related to time and position, which is given by

$$\sigma = \sigma(X,t) \tag{6}$$

$$\tau = \tau(X,t) \tag{7}$$

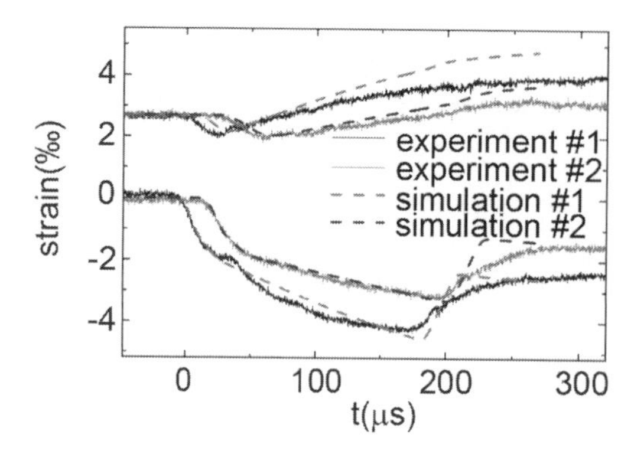

Figure 7. The comparison of the simulation and the experiment.

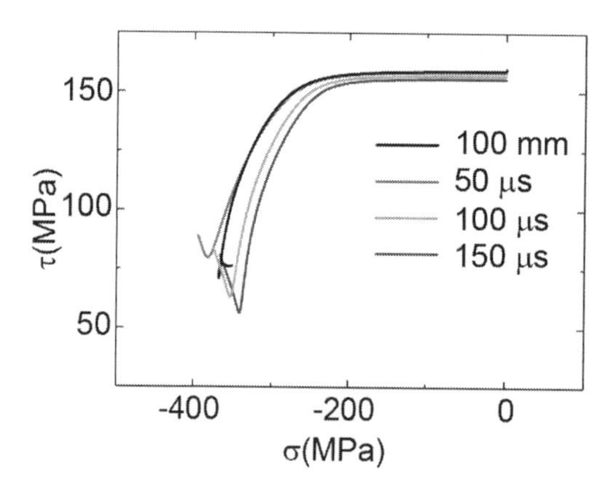

Figure 8. The stress paths of simulation in the σ-τ plane.

so, there are two types of stress paths, with a fixed time or position, as shown in Figures 8 and 9:

$$\tau = f(\sigma, t) \tag{8}$$

$$\tau = g(\sigma, X) \tag{9}$$

The four profiles shown in Figure 8 are $\tau = f(\sigma, 50\ \mu s)$, $\tau = f(\sigma, 100\ \mu s)$, $\tau = f(\sigma, 150\ \mu s)$, and $\tau = g(\sigma, 100\ mm)$ in the σ–τ plane, respectively. The compressive stress σ increases while τ remains unchanged at first on the four profiles, which reflects the elastic longitudinal wave due to the strengthen effect and corresponds with the experimental observation. After the segment of elastic wave, the three profiles of $\tau = f(\sigma, 50\ \mu s)$, $\tau = f(\sigma, 100\ \mu s)$, $\tau = f(\sigma, 150\ \mu s)$ exhibit the same kind of path structure with the rate-independent theory, shown in Figure 2, but the amplitudes of σ and

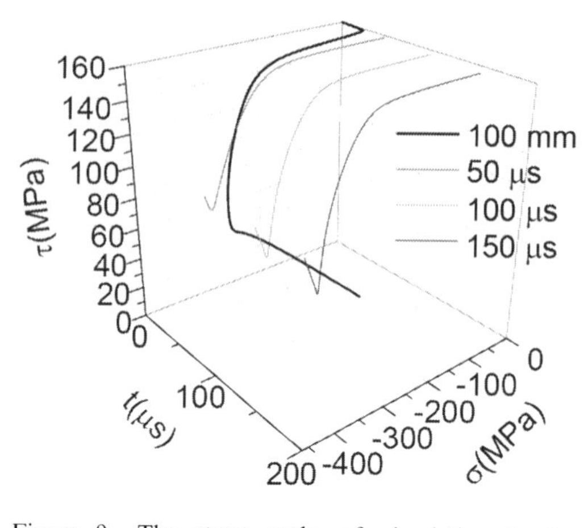

Figure 9. The stress paths of simulation in the $\sigma-\tau-t$ space.

τ decrease with time, which reflects the influence of stress relaxation. The path structure of $\tau = g(\sigma, 100mm)$ is not obvious since the structure and relaxation are coupled. The four profiles in the $\sigma-\tau-t$ space are shown in Figure 9. It can be clearly seen that $\tau = f(\sigma, t)$ with different times can form a surface in this space, and $\tau = g(\sigma, x)$ will evolve on this surface.

The simulation results indicate that the path structure of rate-sensitive coupling plastic waves is basically similar to the rate-independent theory, but the longitudinal and shear components decrease with time due to the relaxation effect.

5 SUMMARY

In this paper, we build a system to apply the longitudinal impact to a pre-torqued tube based on the SHPB, and conduct an experimental research of 304 stainless steel that is rate-sensitive. The experimental results indicate that there should be an elastic wave at first even pre-torqued to plasticity, the coupling fast waves and coupling slow waves are observed and there is no constant region between them. These results of 304 stainless steel agree with those for α-Ti reported by Clifton et al, which means that these characteristics of the rate-sensitive material exist widely.

By using the numerical simulation, we obtain the stress of different times and positions in the experimental process. The results indicate that the stress path structure of the rate-sensitive material is basically similar to the path structure of the rate non-sensitive material, but the longitudinal and shear components decrease with time due to the relaxation effect.

Since there is a lack of enough experimental results for coupling plastic waves in rate-sensitive materials, our experimental results of 304 stainless steel are good complements. And we achieve a better understanding of the coupling plastic waves in the rate-sensitive material by performing an experiment and simulation. A potential application of this phenomenon is to examine the validation of different constitutive equations for the plastic flow under complex and dynamic stress.

ACKNOWLEDGMENT

This work was supported by the National Natural Science Foundation of China (Grant No. 11072240, 11272311).

REFERENCES

Hsu, J.C.C. & Clifton, R.J. 1974a. Plastic waves in a rate sensitive material—I. Waves of uniaxial stress. *Journal of the Mechanics and Physics of Solids,* 22: 233–253.

Hsu, J.C.C. & Clifton, R.J. 1974b. Plastic waves in a rate sensitive material—II. Waves of combined stress. *Journal of the Mechanics and Physics of Solids,* 22: 255–266.

Johnson, G.R. & Cook, W.H. 1983. A constitutive model and data for metals subjected to large strains, high strain rates and high temperatures. In: *Proceedings of the 7th International Symposium on Ballistics.* Hague, Netherlands.

Lax, P.D. 1954. Weak solutions of nonlinear hyperbolic equations and their numerical computation. *Communications on Pure and Applied Mathematics,* 7: 159–193.

Lipkin, J. & Clifton, R.J. 1970a. Plastic waves of combined stresses due to longitudinal impact of a pretorqued tube—Part I: Experimental results. *Journal of Applied Mechanics,* 37: 1107–1112.

Lipkin, J. & Clifton, R.J. 1970b. Plastic waves of combined stresses due to longitudinal impact of a pretorqued tube—Part 2: Comparison of theory with experiment. *Journal of Applied Mechanics,* 37: 1113–1120.

Lu, J. & Zhuang, Z. et al. 2003. Application of numerical simulation to SHPB test to investigate the dynamic compressive behavior of material with failure. *Key Engineering Materials,* 243: 433–438.

Tang, Z.P. & Aidun, J.B. 2009. Combined compression and shear waves in solids. In Horie, Y. (eds). *Shock Wave Science and Technology Reference Library*, 3: 109–167. Berlin: Springer.

Ting, T.C.T. 1969. Plane waves due to combined compressive and shear stresses in a half space. *Journal of Applied Mechanics,* 36: 189–197.

Ting, T.C.T. 1985. *The Nonlinear Stress Waves in Solids.* Beijing: The friendship press of china.

Advanced Materials, Mechanical and Structural Engineering – Hong, Seo & Moon (Eds)
© 2016 Taylor & Francis Group, London, ISBN: 978-1-138-02908-8

Synthesis and characterization of CdS nanoparticles using a chemical precipitation method

Ch.V. Reddy & J.S. Shim
School of Mechanical Engineering, Yeungnam University, Gyeongsan, South Korea

ABSTRACT: In this paper, we report a chemical synthesis route for the preparation of CdS nanoparticles. The nanoparticles were characterized using X-ray Diffraction (XRD), Scanning Electron Microscopy (SEM), Transmission Electron Microscopy (TEM) and X-ray Photoelectron Spectroscopy (XPS) measurements. From the XRD data, it was confirmed that the prepared nanoparticles showed a hexagonal crystal structure. The surface morphology of the prepared sample was examined using SEM. TEM observation showed that the CdS nanoparticles synthesized by chemical synthesis was well dispersed and the average crystallite size was found to be ~18 nm. X-ray photoelectron spectroscopy analysis showed the presence of the binding energies of Cd and S bonds.

1 INTRODUCTION

Semiconductor Nanoparticles (NPs) displays interesting electronic and optical properties that are attributed to the quantum confinement effect and their large surface to volume ratio of atoms, both of which are fundamental and of technological interest. Currently, cadmium sulfide (CdS) nanoparticles have attracted considerable interest among the researchers because of its discrete energy levels, tunable band gap, size-dependent optical properties. CdS with a band gap of 2.42 eV at room temperature is a typical wide direct band-gap II–VI semiconductor that is extensively used in solar cells, light emitting diodes, biological imaging and labeling devices, catalysts and other optoelectronic devices (Singh et al. 2004, Zhang et al. 2007, Yao et al. 2006).

It is well known that the properties of CdS nanomaterials significantly depend on their sizes and spatial architectures. Therefore, in recent years, many efforts have been devoted to the fabrication of CdS nanomaterials with the desired size and structure by different methods. Xie et al. reported a kind of branch-like CdS micro patterns using thiosemicarbazide both as a sulfur source and as a capping ligand in a methanol/water system (Wang 2000). Gao and co-workers synthesized three-dimensional CdS nanocrystals using hexamethylenetetramine [$(CH_2)4\ N_4$, HMT] as a capping reagent (Gao et al. 2008).

Based on the composition of the nanoparticles, active sites may vary and generate different amounts of reactive species. The reactivity of nanoparticles is a function of their physico-chemical properties, such as size, surface characteristics, crystal phase and concentrations, agglomeration behavior and suspension stability (Jiang et al. 2008, Almquist & Biswas 2002, Reddy et al. 2012). Various studies have reported that particle size is an important parameter that affects the catalytic activity of the materials; the size and mobility of the nanoparticles will influence the microbial inactivation. By considering all these enormous applications of CdS in various industrial applications, a large-scale and inexpensive preparation method is required. Co-precipitation is one of the most promising techniques owing to the large-scale preparation ability without using a special apparatus.

The dendrite-like CdS hierarchical nanostructures obtained by a facile and effective hydrothermal route at a mild temperature have been reported (Yu et al. 2013). In addition, CdS nanochains (Ge et al. 2008), mesoporous tubular structures (Zhang et al. 2011) and bicrystalline CdS nanoribbons (Fan et al. 2009) have also been obtained. Herein, in the present investigation, CdS nanoparticles were successfully prepared by the simple chemical precipitation method without using any surfactants. A possible growth mechanism of the as-prepared CdS product is proposed. The as-synthesized nanoparticles were characterized using XRD, SEM, TEM and XPS measurements.

2 EXPERIMENTAL PROCEDURE

To synthesize CdS nanoparticles, the following materials were used. All the chemical reagents were of analytical grade without further purification. Cadmium acetate and sodium sulfide were used as precursors. Deionized water was used for all dilution and sample preparations. All the chemicals were above 99% purity. All the glassware used in this experimental work was acid washed.

3 SYNTHESIS OF CDS NANOPARTICLES

In a typical procedure, 0.2 mol of cadmium acetate [$Cd(CH_3COO)_2.4H_2O$] in 50 mL of deionized water–ethanol matrix (equal volumes) and an equal molar amount of sodium sulfide [$Na_2S.xH_2O$] 0.1 mol in another deionized water–ethanol matrix were added to the above solution dropwise with continuous stirring. The mixture was stirred for 4 h magnetically at 80°C until a homogeneous white solution was obtained. The obtained dispersions were washed with deionized water and ethanol several times to remove impurities. After washing, the solution was centrifuged at 10,000 rpm for about 30 min. The settled powder was collected and dried in a hot air oven at 120°C for 2 h.

The CdS nanoparticles were characterized using the powder X-ray Diffraction (XRD) pattern with Cu Ka radiation on a PANalytical X'Pert PRO diffractometer. The scanning electron microscopy (SEM) images were recorded using a FE-SEM/EDS (1) S 4100 instrument. The Transmission Electron Microscope (TEM) images were recorded using a TEM HITACHI H-7600 instrument. X-ray Photoelectron Spectroscopy (XPS) was performed using a Thermo Scientific (K Alpha surface analysis) instrument.

4 RESULTS AND DISCUSSION

The XRD pattern of CdS nanoparticles is shown in Figure 1. All the diffraction peaks can be indexed to the hexagonal CdS structure, which is well consistence with the standard JCPDS No: 41-1049. The peaks obtained corresponded to the (100), (002), (101), (102), (110) (103), (201), (004), (203), (211) and (114) planes of CdS, with lattice cell parameters a = 0.1410 and c = 0.6719 nm. No diffraction peaks of other crystalline forms were detected, which demonstrated that the sample had high purity and good crystallinity. These results are consistent with those reported previously (Gajanan et al. 2015, Yu et al. 2014). The average crystallite size was calculated using Debye-Scherrer's formula: $D = 0.9\lambda / \beta \cos \theta$, where λ is the wavelength (Cu Kα); β is the Full Width at Half Maximum (FWHM); and θ is the diffraction angle. The average crystallite size of the sample was found to be 18 nm.

Scanning electron microscopy is widely used to obtain data about the surface morphology of as-synthesized nanoparticles. The SEM image of CdS nanoparticles is shown in Figure 2 with different magnifications. From the SEM micrographs, the formation of non-uniformly distributed spherical and flake-like structures with a size below 100 nm was clearly observed. The microstructures of the prepared sample were further studied using TEM. The TEM images of CdS nanoparticles are shown in Figure 3. A spherical-like shape of some CdS particles was observed. The majority of the CdS nanoparticles present in this powder had sizes ranging between 18 and 20 nm, which was in good agreement with the particle size deduced from the CdS XRD spectrum.

XPS is a surface analytical technique that can provide useful information on the complete chemical composition of the sample surface. Figure 4 shows the XPS spectra for Cd3d and S 2p core levels recorded for CdS nanoparticles. Figure 4 (a) shows the XPS spectrum of the Cd 3d level of CdS nanoparticles. The Cd 3d features consist of the main $3d_{5/2}$ and $3d_{3/2}$ spin-orbit components. The binding energies of 404.8 and 411.5 eV, respec-

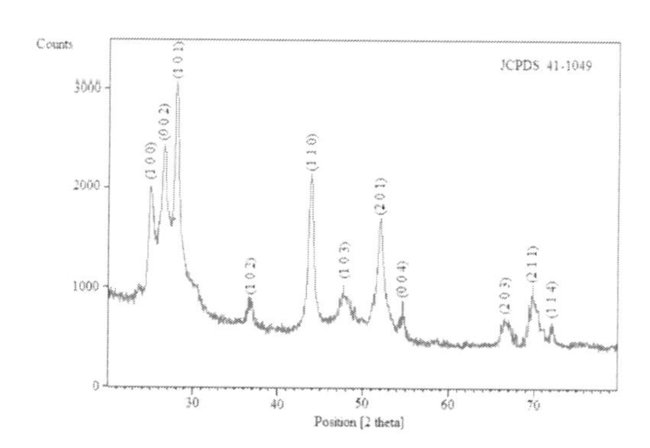

Figure 1. XRD patterns of CdS nanoparticles.

Figure 2. SEM images of CdS nanoparticles.

tively, were found for the $3d_{5/2}$ and $3d_{3/2}$ components. The binding energy of Cd $3d_{5/2}$ is attributed to the Cd^{2+} bonding state, which agrees well with the previous report (Maliki et al. 2003).

Figure 4(b) shows the XPS spectra of S 2p levels of CdS nanoparticles. The S 2p spectra were deconvoluted into single states of S $2p_{3/2}$

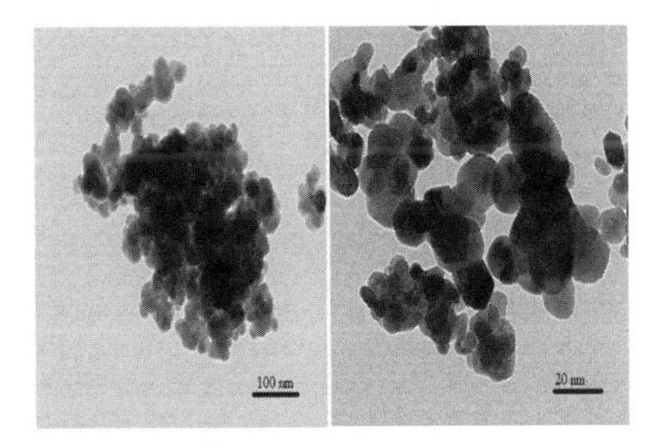

Figure 3. TEM images of CdS nanoparticles with different magnifications.

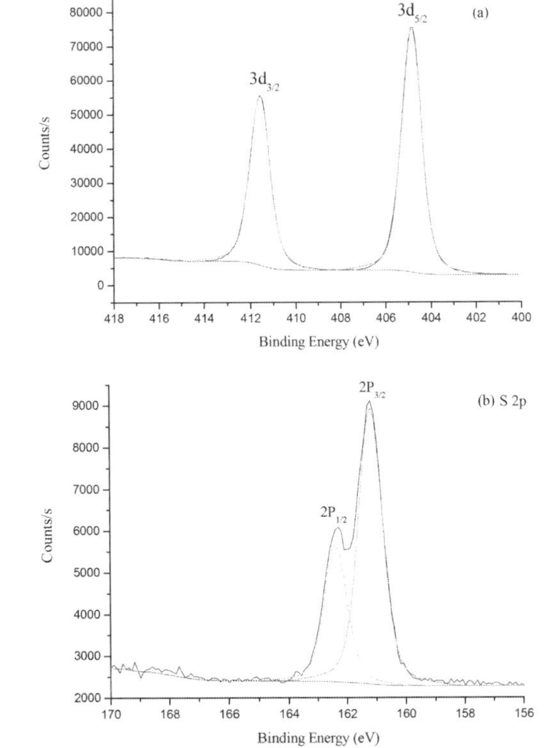

Figure 4. XPS spectra of as-synthesized CdS nanoparticles: (a) Cd 3d and (b) S 2p.

and S $2p_{1/2}$ at 161.9 and 163.1 eV, respectively. S 2p peaks observed at 161.9 eV and 163.1 eV are attributed to the presence of metal sulfide. The binding energy of S 2p at 161.9 eV also agreed well with the blue shift of the binding energy compared with the observed S 2p levels in bulk CdS (Winkler et al. 1999). This more intense peak is assigned to S^{2-} present in the solid solution (Dutkova et al. 2012). The appearance of a low-intensity peak at 163.1 eV is due to the oxidation of sulfur in nanoparticles (Wagner et al. 1978).

The Binding Energy (B.E) values are characteristics of sulfides (Chavan et al. 2006). The XPS results confirm the formation of sulfides on the surface, indicating effective passivation. Experimentally, binding energy in XPS measurement is defined as the energy difference between the Fermi level and the orbital electron energy level (Weightman 1982). Considering that the fixed energy difference between the valence band top and the orbital electron energy level, it is natural to conclude that the sample with a larger binding energy had a higher carrier density.

5 CONCLUSIONS

In summary, CdS nanoparticles were prepared by using the chemical precipitation method. The as-prepared nanoparticles were characterized using various measurements. The structural and surface morphological investigations from XRD and TEM measurements are in agreement. From the XRD analysis, the diffraction peaks pertaining to 24.80°, 26.50°, and 28.18° are attributed to the (100), (002) and (101) planes of the hexagonal structure. The average crystallite size of CdS nanoparticles was 18 nm, which was also confirmed by the TEM measurement. The surface morphology of the prepared nanoparticles exhibited a spherical and flake-like structure. From the XPS spectra, the binding energies of Cd $3d_{5/2}$ and $3d_{3/2}$ were found to be 404.8 and 411.5 eV, respectively. The binding energy of $Cd3d_{5/2}$ was attributed to the Cd^{2}+ bonding state.

REFERENCES

Almquist, C.B. & Biswas, P. 2002. Role of Synthesis Method and Particle Size of Nanostructured TiO_2 on Its Photoactivity, *Journal of Catalysis,* 212: 145–156.

Chavan, A. et al. 2006. Surface passivation and capping of GaSb photodiode by chemical bath deposition of CdS, *Journal of Applied Physics,* 100(6): 064512.

Dutková, E., Baláž, P., Pourghahramani, P., Balek, V., Nguyen, A.V., Šatka, A., Kováč, J. & Ficeriová, J.

2012. Mechanochemically Synthesised ZnxCd1-xS Nanoparticles for Solar Energy Applications, *Journal of Nano Research,* 18–19: 247–256.
Fan, X., Zhang, M.L., Shafiq, I., Zhang, W.J., Lee, C.S. & Lee, S.T. 2009. Bicrystalline CdS nanoribbons, *Crystal Growth & Design,* 9: 1375–1377.
Gajanan, P., Supriya, D. & Shrivastava, A.K. 2015. Effect of Gd3+doping and reaction temperature on structural and optical properties of CdS nanoparticles, *Materials Science and Engineering B,* 200: 59–66.
Gao, F., Lu, Q.Y., Meng, X.K. & Komarneni, S. 2008. CdS nanorod-based structures: from two and three-dimensional leaves to flowers, *The Journal of Physical Chemistry C,* 112: 13359–13365.
Ge, C., Xu, M., Fang, J., Lei, J. & Ju, H. 2008. Luminescent cadmium sulfide nanochains templated on unfixed deoxyribonucleic acid and their fractal alignment by droplet dewetting, *The Journal of Physical Chemistry C,* 112: 10602–10608.
Jiang, J., Oberdorster, G., Elder, A., Gelein, R., Mercer, P. & Biswas, P. 2008. Does Nanoparticle Activity Depend upon Size and Crystal Phase? *Nanotoxicology,* 2: 33–42.
Maliki, H.E., Bernède, J.C., Marsillac, S., Pinel, J., Castel, X. & Pouzet, J. 2003. Study of the influence of annealing on the properties of CBD-CdS thin films, *Applied Surface Science,* 205: 65–79.
Reddy, V., Krishna, R., Raghavendra Rao T., Udayachandran, T.U.S., Reddy, Y.P., Rao, P.S. & Ravikumar, R.V.S.S.N. 2012. Synthesis and spectral characterizations of Fe3+ doped b-BaB2O4 nano crystallite powder. *Journal of Molecular Structure,* 1012: 17–21.
Singh, R.S., Rangari, V.K., Sanagapalli, S., Jayaraman, V., Mahendra, S. & Singh, V.P. 2004. Nanostructured CdTe, CdS and TiO2 for thin film solar cell applications, *Solar Energy Materials and Solar Cells,* 82: 315–330.
Wagner, C.D., Riggs, W.M., Davis, L.E. & Moulder, J.F. 1978. *Handbook of X-ray photoelectron spectroscopy,* in: G.E. Muilenberg (Ed.), Perkin Elmer Corporation.
Wang, Z.L. 2000. Characterizing the structure and properties of individual wire-like nanoentities, *Advanced Materials,* 12: 1295–1298.
Weightman, P. 1982. X-ray-excited Auger and photoelectron spectroscopy, *Reports on Progress in Physics,* 45: 753–814.
Winkler, U., Eich, D., Chen, Z.H., Fink, R., Kulkarni, S.K. & Umbach, E. 1999. Detailed investigation of CdS nanoparticle surfaces by high resolution photoelectron spectroscopy, *Chemical Physics Letters,* 306: 95–102.
Yao, W.T., Yu, S.H., Liu, S.J., Chen, J.P., Liu, X.M. & Li, F.Q. 2006. Architectural control syntheses of CdS and CdSe nanoflowers, branched nanowires, and nanotrees via a solvothermal approach in a mixed solution and their photocatalytic property, *The Journal of Physical Chemistry B,* 110: 11704–11710.
Yu, Z., Wu, X., Wang, J., Jia, W.N., Zhu, G.S. & Qu, F.Y. 2013. Facile template-free synthesis and visible-light driven photocatalytic performances of dendritic CdS hierarchical structures, *Dalton Transactions,* 42: 4633–4638.
Yu, Z., Yin, B., Qu, F.Y. & Wu, X. 2014. Synthesis of self-assembled CdS nanospheres and their photocatalytic activities by photodegradation of organic dye molecules, *Chemical Engineering Journal,* 258: 203–209.
Zhang, J., Di, X.W., Liu, Z.L., Xu, G., Xu, S.M. & Zhou, X.P. 2007. Multicolored luminescent CdS nanocrystals, *Transactions of Nonferrous Metals Society of China,* 17: 1367–1372.
Zhang, W.M., Li, N., Tang, D.H. & Wang, Y.Y. 2011. Tubular CdS with mesoporous structure self-Templated by cadmium complexes of oleate, *Microporous and Mesoporous Materials,* 143: 249–251.

Advanced Materials, Mechanical and Structural Engineering – Hong, Seo & Moon (Eds)
 ISBN: 978-1-138-02908-8

Making concrete using alkali-activated Ladle Slag and Electric Arc Furnace Slag aggregates

W.H. Zhong & W.H. Huang
Department of Civil Engineering, National Central University, Zhongli, Taiwan, R.O.C.

T.H. Lu
Re-source Technology Co. Ltd., Houlong Township, Miaoli County, Taiwan, R.O.C.

ABSTRACT: Ladle Slag (LS) and Electric Arc Furnace (EAF) slag are the by-products of EAF steel-making. This study aimed at producing a concrete with alkali-activated slag-only binder (non-Portland cement), and aggregate consisting of large proportions of EAF slag. A laboratory investigation was conducted on the compressive strength, volume stability and durability of the concrete mixed using alkali-activated slags, EAF slag fine aggregate and natural coarse aggregate. The results indicate that a blend of 50% Blast Furnace Slag (BFS) and 50% ladle slag can be activated successively with alkali as a cementitious material in making concrete. The compressive strength of the concrete specimens prepared using such cementitious material and EAF slag aggregate was found to be close to that made with natural aggregate.

1 INTRODUCTION

Electric Arc Furnace Slag (EAFS) and Ladle Slag (LS) are produced by the oxidation and reduction processes in EAF steel-making. In the past, a majority of LS has been placed in landfills. Due to the limited availability of land and possible environmental concerns, there is an urgent need to find a means of utilizing LS as engineering materials. In the meantime, supplies of natural fine aggregate in Taiwan are dwindling. Therefore, it is proposed that EAFS be processed by magnetic-screening, crushing and sieving procedures, to produce fine aggregates to replace natural fine aggregates for making concrete.

Alkali activation, a relatively new technology for producing concrete binder, uses an alkali activator to activate materials such as silicon aluminum materials to form binders. This kind of binder can be used to replace Portland cement for making concrete. Shi and Day (1995) observed the reaction between activator and cementing components, and divided the reaction into two phases: (1) the use of the high pH alkaline activator to destroy the crystal morphology of the silicon and aluminum oxide, and to convert into silicon and aluminum ion in order to facilitate the reaction; (2) the dissociation of the alkaline activator into an anion or an anionic group, where the cementing components absorb Ca^{2+} to generate hydration products. The threshold pH for the activating capability of an alkaline activator is 11.5. The initial pH of an alkaline activator has an important role in dissolving the slag for the early formation of Ca compounds that result from the reaction between the anion and the anionic group of the alkaline activator and Ca^{2+} dissolving from the surface of slag grains. Further hydration of alkali-activated cement is dominated by Ca compounds rather than by the initial pH of the alkaline activator. Zhong et al. (2013) pointed out that the use of workable mixture of Alkali-Activated LS (AALS) exhibited AALS pastes similar to Portland cement pastes, and AALS can be used to produce concrete.

Previous studies indicated that EAFS is suitable to replace natural aggregate for making concrete. However, appropriate pretreatment and storage are prerequisites for its successful use. Special attention must be paid to the crushing process to produce a suitable grade, as coarse EAFS has been reported to cause expansion of the concrete upon hydration. Manso et al. (2006) and Pellegrion et al. (2009) pointed out that EAFS achieves the chemical and physical stabilization and can be made as a recycled material for aggregate. It has also been found that the strength of concrete made using EAFS as aggregate is higher than that using natural aggregate, while the two concretes exhibited similar durability in terms of resistance to sulfate attack.

The objective of this study is to activate LS as a cementitious material and to use EAFS as a fine aggregate along with alkali-activated LS to make concrete. More specifically, this study is undertaken to understand the interaction between alkali-activated LS and EAFS fine aggregate, and to suggest a recycling method of AALS and EAFS based on the research findings. The primary research questions to be addressed in this study are as follows: (a) compressive strength, potential for alkali-aggregate reaction and drying shrinkage of alkali-activated LS mortar with EAFS fine aggregate; (b) mix proportion of concrete made with EAFS fine aggregate and AALS.

2 MATERIALS AND METHODS

2.1 *Materials and mix proportion of mortars and concretes*

The materials used to produce the binder are LS and Blast Furnace Slag (BFS) at 50:50 weight percent. The LS and BFS were produced from EAF and blast furnace steel-making plants in Taiwan. The chemical composition and physical properties of the LS and BFS are given in Table 1.

The alkaline activator used in this study was a mixture of sodium silicate solution and sodium hydroxide powder of industrial grade. The sodium silicate solution, with a SiO_2/Na_2O ratio (i.e. modulus of silicate, Ms) of 3.13, is composed of 28.35% SiO_2, 9.06% Na_2O and 62.09% water by mass.

The coarse aggregate consisted of limestone with a maximum size of 19 mm, and the fine aggregates consisted of locally available river sand and EAFS. The properties of the aggregates used in the study are summarized in Table 2. Figure 1 shows the grain size distribution of EAFS. In this study, EAFS was used to replace 25–100% natural fine aggregate for proportioning mortar and concrete mixes.

The activators investigated are liquid sodium silicate and sodium hydroxide powder. Sodium silicate and sodium hydroxide were blended to provide an Ms of 0.75. The activator was then mixed with the binder at a w/b ratio of 0.5 and a total Na content of 4% (wt of the binder). The resulting paste composition is given in Table 3.

The mortar samples were prepared based on the data from Table 4. Mortar cube specimens were prepared for testing the compressive strength, while the

Table 1. The chemical composition and physical properties of the LS and BFS.

Composition (%)	LS	BFS
SiO_2	25.25	33.6
CaO	56.80	40.8
Al_2O_3	3.30	15.0
Fe_2O_3	0.88	0.3
MgO	7.82	6.1
SO_3	2.87	1.8
Physical property		
Blaine fineness (m^2/kg)	180	400
Specific gravity	2.88	2.83

Table 2. Property of aggregates.

Property	Fine aggregate		Coarse aggregate
	River sand	EAFS	
Specific gravity	2.62	3.06	2.6
Absorption (%)	0.4	2.1	0.9
Fineness modulus	2.6	3.1	–
Dry unit weight(kg/m^3)	–	–	1571

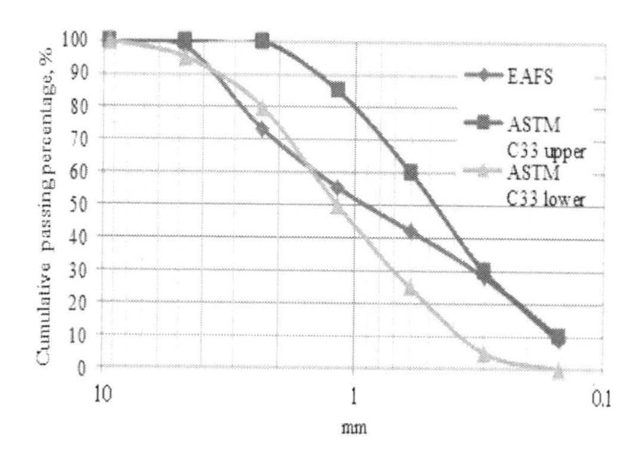

Figure 1. Grain size distribution of EAFS.

Table 3. The mix composition of AALS paste.

w/b*	Na (%)	Ms	LS/BFS (wt.%)
0.5	4	0.75	50/50

*Water/binder ratio.

Table 4. The mix composition of mortar (wt.%).

Component	Cube	AAR bars	Drying shrinkage bars
Paste/Aggregate [weight ratio]	1/2.75	1/2.25	1/2
River sand/EAFS [wt.%]	100/0 75/50 50/50 25/75 0/100	100/0 75/25 50/50 0/100 Reactive aggr.	100/0 75/25 50/50 0/100

Table 5. The mix composition of concrete [kg/m³].

Component	Control	EAFS
LS	214	214
BFS	214	214
Coarse aggregate	1075	1075
Natural fine aggregate	665	0
EAFS fine aggregate	0	777
Sodium silicate	43	43
Sodium hydroxide	19	19
Water	176	176

mortar bars were used for testing the potential alkali reactivity of cement-aggregate (Alkali-Aggregate Reaction, AAR) combinations and drying shrinkage. Mortar was mixed in a desktop mixer. Activators were blended with water first. Fine aggregates were then added to the mixer and mixed at a low speed for 30 seconds, followed by medium-speed mixing for 30 seconds. Then, the mixer was kept at rest for 1.5 minutes, followed by a remixing of 1 minute.

Two concrete mixes were prepared using alkali-activated LS+BFS as the binder. The control mix used natural aggregates while the EAFS mix used EAFS as the fine aggregate. The concrete mixture proportions are summarized in Table 5.

2.2 *Testing methods*

Cubes (50 × 50 × 50 mm) of mortar for compressive strength testing were in triplicate at 3, 7, 28, 90 days after casting. The samples were exposed to curing conditions of 23°C and 100% RH after demolding.

A test on the potential of alkali-aggregate reaction was performed using mortar prisms of 25 × 25 × 285 mm. Four mortar prisms were used for each mix. The prisms were demolded 24 h after casting and placed in plastic containers with sealed lids and a certain amount of water at the bottom. The containers were held at 38 ± 2°C, and the prisms were kept in a vertical position without touching the water. The change in the length of the mortar prisms was determined at ages of 14 days and 1, 2, 3, 4 and 6 months.

Dry shrinkage measurements were performed using mortar dry shrinkage prisms of 25 × 25 × 285 mm. The first reading was taken immediately after demolding (24 hr after casting) and curing (48 hr in water) with a total time of 72 hr. The prisms were stored in a drying cabinet with the temperature and relative humidity maintained at 23 ± 2°C and 50 ± 4%, respectively.

Workability of concrete was assessed by slump loss vs. elapsed time relationship. Compressive strength testing for concrete was conducted on cylindrical specimens with a height of 100 mm. And a total of three cylinders were tested for the determination of each strength.

3 RESULTS AND DISCUSSION

3.1 *Compressive strength*

Figure 2 shows the development of compressive strength with time for mortars with varying amounts of EAFS. The control mix (0%EAFS) showed the highest 3-day strength. The EAFS mixes showed tardy early strength development, followed by identical strength with the control mix at later ages. Mortar mix with 50% EAFS showed higher strength than others at ages of 7–90 days. There was a non-significant difference in strength between samples made with natural aggregate and EAFS aggregate.

3.2 *Alkali-Aggregate Reaction (AAR)*

Figure 3 shows the expansion measured for reactive aggregate, control and EAFS mortars at 6 months of storage. The expansion of mortar with reactive aggregate was high from the start of the experiment, and reached 0.05% after first 30 days. After 12 months, it was approximately 0.11%. Thus, the

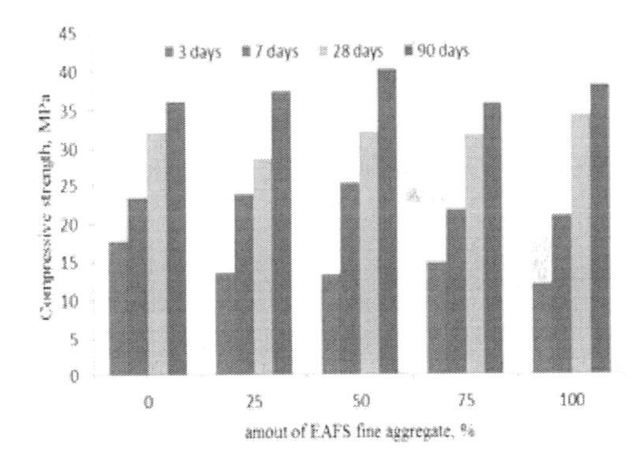

Figure 2. Compressive strength of AALS with EAFS fine aggregate mortar cubes.

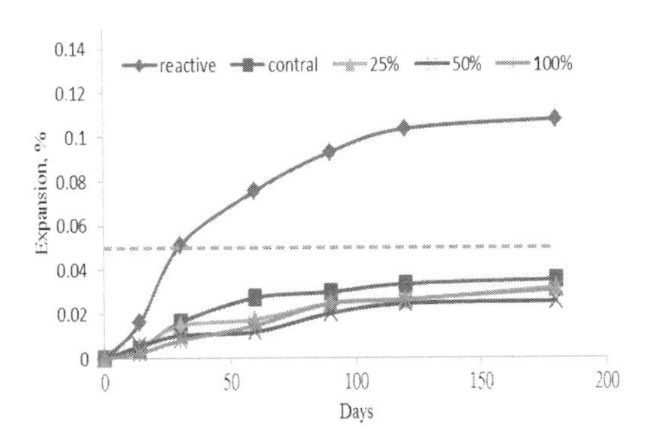

Figure 3. Expansion of mortar prisms subjected to the alkali-aggregate reaction.

expansion of reactive aggregate was above the limit of 0.05% at 6 months set by the standard ASTM C 227. On the contrary, expansion of the control and EAFS mixes was slow and reached 0.03%–0.04% after 6 months, which was within the limit set by the standard. This indicates that alkali-activated LS does not cause the cement-aggregate reaction with EAFS. Puertas et al. (2009) pointed out that the use of siliceous aggregate exhibited a significantly higher alkali-aggregate reaction expansion than the non-reactive and reactive calcareous aggregates with alkali-activated slag using sodium silicate. They compared different curing conditions and aggregates for testing and concluded that the calcareous aggregates in alkali-activated slag proved to be resistant to the alkali-aggregate reaction. EAFS fine aggregate was similar to the non-reactive calcareous aggregate and hardly caused the alkali-aggregate reaction.

3.3 *Drying shrinkage*

The results of drying shrinkage measurements are shown in Figure 4. All mortar prisms showed the similar shrinkage at the testing ages. Drying shrinkage of control mix (100%) was similar to the 100% EAFS mix at all testing ages. It was worth noting that the high absorption of EAFS aggregate on drying shrinkage was not observed. This indicates that the use of EAFS does not result in unfavorable effects on drying shrinkage of AALS mortars.

3.4 *Testing on concrete*

Slump loss versus time relationship is illustrated in Figure 5. The control mix demonstrated better workability than the EAFS mix. At the first 30 min, the control mix showed slump similar to the initial slump, while EAFS mix started to exhibit slump loss relative to the initial slump. After 30 min, the slump loss of the control mix increased and the total slump loss at 90 min was similar to that of the EAFS mix.

Figure 6 shows the strength development with time following curing. Two concrete mixes showed almost identical 3-day strength. After 7 days, the EAFS concrete mix showed a slightly higher compressive strength than the control mix. This can be attributed to the rough surface texture of EAFS that improves the bound between the cementing paste and the aggregate.

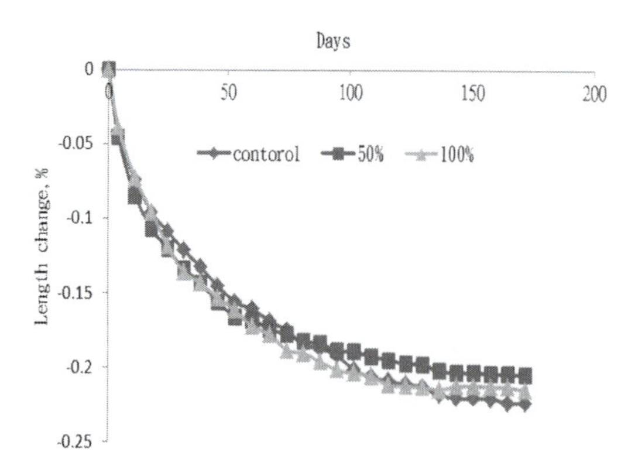

Figure 4. Drying shrinkage of mortar prisms made with AALS and EAFS fine aggregate.

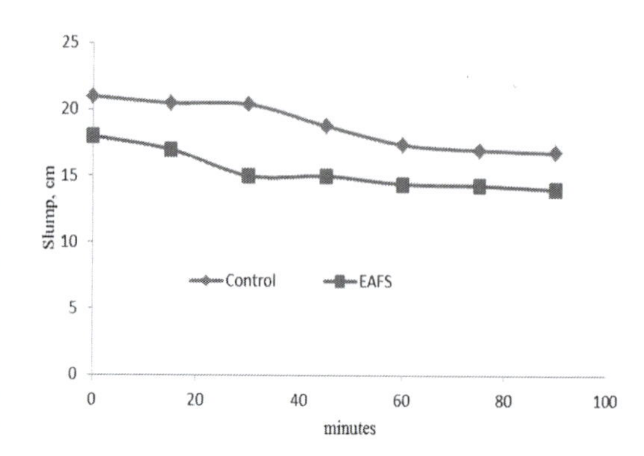

Figure 5. Slump loss versus time for concrete made with AALS and EAFS fine aggregate.

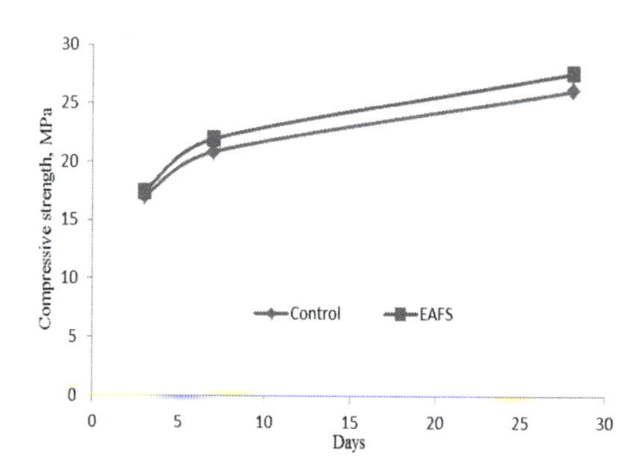

Figure 6. Compressive strength of concrete made with AALS and EAFS fine aggregate.

4 CONCLUSIONS

Based on the laboratory investigations conducted on AALS binder and EAFS aggregate, the following conclusions are drawn:

1. Crushed EAFS fine aggregate has a grain size distribution similar to coarse sand and exhibits high water absorption and specific gravity.

2. The use of EAFS fine aggregate in AALS mortar showed lower early (3-day) strength than that of natural aggregate. At later ages, mortars using EAFS aggregate showed similar or higher strength than that with natural aggregate, regardless of the amounts of EAFS used.
3. EAFS fine aggregate was demonstrated to be non-reactive to the alkali-aggregate reaction in the AALS cementing system. The drying shrinkage of EAFS fine aggregate was similar to natural aggregate.
4. The higher absorption of EAFS fine aggregate caused a decline in the slump of AALS concrete. Concrete made with EAFS fine aggregate and AALS showed a slightly higher strength than that with natural fine aggregate.
5. EAFS proved to be a suitable fine aggregate and can be used along with the AALS binder for making concrete.

REFERENCES

Manso, J.M., Polanco, J.A., Losanez, M. & Gonzalez, J.J. 2006. Durability of concrete made with EAF slag as aggregate, *Cement and Concrete Composites*, 28: 528–534.

Pellegrion, C. & Gaddo, V. 2009. Mechanical and durability characteristics of concrete containing EAF slag as aggregate, *Cement and Concrete Composites*, 31: 663–671.

Puertas, F., Palacios, M., Gil-Maroto, A. & Vázquez, T. 2009. Alkali-aggregate behaviour of alkali-activated slag mortars: Effect of aggregate type, *Cement and Concrete Composites*, 31: 277–284.

Shi, C. & Day, R.L. 1995. A calorimetric study of early hydration of alkali-slag cements, *Cement and Concrete Research*, 25: 1333–1346.

Zhong, W.H., Lu, T.H. & Huang, W.H. 2013. Alkali-Activated EAF Reducing Slag as Binder for Concrete, *Advanced Materials Research*, 723: 580–587.

Advanced Materials, Mechanical and Structural Engineering – Hong, Seo & Moon (Eds)
© 2016 Taylor & Francis Group, London, ISBN: 978-1-138-02908-8

Study of morphological and chemical characteristics of corrosion scales in steel pipes

A. Andrianov, V. Chukhin & V. Orlov
Department of Water Supply, Moscow State University of Civil Engineering (National Research University), Moscow, Russia

ABSTRACT: Morphological and chemical characteristics of steel corrosion scales were systematically investigated in this work. Special attention was paid to tubercle structure and its composition. The corrosion scales were characterized by Scanning Electron Microscopy (SEM) and Energy Dispersive X-ray Spectrometry (EDS). The study of four scale samples harvested from old steel drinking water pipes showed four characteristic regions that differ significantly in structure and composition: the surface layer, the hard shell layer, the core and the base layer. The last one is almost never mentioned in the literature. The obtained results allowed us to supposed that iron-reducing and iron-oxidization bacteria have a significant impact on corrosion processes, especially on tubercles formation and their microstructure.

1 INTRODUCTION

Corrosion of the steel and iron pipes has a great impact on water quality and pipelines lifetime in water distribution systems. Uniform corrosion increases iron content in drinking water and tuberculation dramatically increases hydraulic flow resistance. Also pitting corrosion affects pipe durability and is regarded as the main reason of steel pipe failures.

The majority of publications are devoted to iron pipe corrosion and steel pipe corrosion is described to a less extent. Meanwhile a majority of pipelines in drinking and industrial water distribution systems in Russia (more than 60%) are conducted using uncoated steel pipes. Mostly their age have already exceeded the useful lifespan and their performance always risks to provide a failure. This situation emphasizes the significance of corrosion processes that occur in water distribution systems. The ability to predict corrosion conditions that lead to a pipe failure can help water system operators to implement efficient pipeline rehabilitation. Thus corrosion and tuberculation can be regarded as two main reasons of water quality deterioration that provide certain risks for human health and leads to loss of consumers' confidence (Husband & Boxall 2011).

Sontheimer et al. (1981) found out that once iron pipe is put into operation, corrosion occurs quickly during the first few years and then slows down due to formation of corrosion deposits on the internal pipe surface that prevents its contact with dissolved oxygen. However, despite the fact that deposits prevent oxygen flow to cathode areas on the pipe surface, corrosion process continues at a lower rate.

All corrosion models can be divided into three categories: electrochemical, chemical and microbiological. Gerke et al. (2008) studied the physicochemical properties of the five iron pipe samples covered with tubercles. All pipes tested in this study were operated for the same time and under the same water conditions, but tubercles structure and their composition differed significantly. The internal morphology studies revealed that all tubercles included three areas: a porous core, a hard layer (shell) and a loose surface layer. The internal morphology of some tubercles was more complex and included core regions with marbled structure due to thin veinlets of Fe_3O_4 (Gerke et al. 2008).

Sander et al. (1997) supported the chemical theory and proposed that corrosion rate depends on the nature and concentration of carbonate surface complexes. The initial corrosion rate at the iron surface depends on oxide dissolution and inhomogeneous oxide film formation. Surface complexes dissolve oxide film and provide free surface for further corrosion. The authors suggested that corrosion situation in pipe networks is determined by inhomogeneous oxide layers formation and surface complexation. The main distinguishing feature of this model is the predominance of surface interactions over electrostatic effects.

Electrochemical model gives a good description of the corrosion process, but does not explain some aspects of tubercles formation and morphology. Many research works are devoted to investigation of microorganisms' role in corrosion acceleration (microbiologically influenced corrosion).

Wang et al. (2012) showed that biofilms are always formed on the pipe surface despite the applied disinfection measures, and biofilm formation significantly affects iron corrosion and change the structure, composition and morphology of corrosion scales. Bacteria that participated in corrosion scale build-up were identified as *Sediminibacterium sp.* (iron-oxidizing bacteria), *Shewanella sp.* (iron-reducing bacteria) and *Limnobacter thioxidans strain* (sulfur-oxidizing bacteria). The authors determined that during the primary period these corrosion-inducing bacteria promoted iron corrosion due to synergistic interactions between the metal surface, abiotic corrosion and bacterial cells together with their metabolites. Later anaerobic iron-oxidizing bacteria become predominant corrosion bacteria that prevent further corrosion due to formation of protective layers.

The results of earlier studies conducted at Moscow water distribution system demonstrated that biofilm was always present inside the steel pipes. Microbiological analysis identified a small amount of filamentous iron bacteria *Leptothrix* and unicellular *Gallionella*. These microorganisms can oxidize iron (II) into iron (III) and produce iron hydroxide in the form of tuberous (ocher) rust flakes.

The major goal of our present work was to investigate mechanisms of formation, morphology and elemental composition of typical corrosion scales in Moscow water distribution system and find the traces of microbiological corrosion processes.

2 MATERIALS AND METHODS

2.1 *Moscow water supply system*

Moscow water supply system produces by 5 WTPs up to 3.54 million cubic meters of drinking water every day. Raw water is taken from Volga River and Moscow River using cascade of reservoirs. The water is distributed via 12,724 km of mains and distribution network pipelines with inner diameter ranging from 100 to 1400 mm. Pipelines were made of steel (61%), cast and ductile iron (36%), plastic and other materials (3%). Most of steel pipes (about 80%) need urgent repair or replacement due to lack of protective covering, low quality of steel and age. The mean age for steel pipelines is 24 years, but a lot of pipes are older than 50 years and are still in operation. The most frequent failures (up to 68%) are attributed to emerging of fistulas in steel pipes.

Over the past 20 years water consumption in Moscow decreased by 44%. This had a negative impact on flow velocities in the pipelines, especially in distribution network. More than 30% of existing pipelines are operated under flow velocities less than 0.2 m/s, in another 30% flow velocities range from 0.2 to 0.5 m/s. The residence time of water reaches 24 hours. These factors lead to water quality deterioration in the distribution network as well as at the consumers' taps. The initial water quality monitoring in distribution network demonstrated short-term increases in turbidity, color, iron and total microbial count. But most of the time these water quality characteristics conform to drinking water standards.

2.2 *Drinking water quality*

The main commonly controlled water quality characteristics in Moscow distribution network during the period of 2001–2012 were: turbidity (0.2–0.45 NTU), colour (7 grad), total iron (0.05–0.1 mg/L), aluminum (0.07–0.14 mg/L), TDS (120–250 ppm) and temperature (0.5–24°C). It was also detected that turbidity and total iron concentrations slightly increased downstream the distribution system.

Stability is an important indicator of water quality that characterizes corrosion processes occurring in water supply network. The calculated values of Langelier Saturation Index in Moscow tap water for the period 2001–2012 varied from −1.82 to +0.06 and Ryzner Stability Index values varied from 7.5 to 10.5. Fluctuations in water stability indexes values corresponded to variations of the surface water quality characteristics and reagent doses. Negative LSI values were observed during the spring flood and rarely during autumn rains. Thus, the Moscow drinking water can be recognized as low-corrosive type.

2.3 *Corrosion deposits samples*

Four samples were cut out from old steel pipes from different sites of water distribution system in the south and south-east districts of Moscow city. All pipes were replaced during the emergency repair works on mains or distribution pipelines. Identification of each pipe segment, service life and pipe diameter are presented in the Table 1.

All pipes segments were intensively corroded and contained numerous tubercles. All samples

Table 1. Pipe segment, pipe diameter and service life.

Pipe segment	Pipe diameter (mm)	Service life (years)	Cathodic protection
Pipe A	300	36	–
Pipe B	300	37	Partially
Pipe C	900	43	–
Pipe D	1400	36	+

were dried indoors in the air during 10–14 days. Pipe segments A, B and C had a solid-core, dense but fragile corrosion scales covered with tubercles. The typical size of tubercles was 5 × 5 mm (in the base) for pipe B and 25 × 50 mm for pipes A and C. The typical height was 3–15 mm (pipe A), 3–10 mm (pipe B), 3–20 mm (pipe C). Below the scale layer the pipe surface showed traces of pitting corrosion with a size of 1–3 mm in depth and 2–5 mm in diameter for pipe A, 0.5–1.5 mm in depth and 2–20 mm in diameter for pipe B, 1–3 mm in depth and 2–5 mm in diameter for pipe C.

Corrosion scales on the inner surface of pipe segment D significantly differed from the other samples—the deposits were solid, very dense and strong, with a layered structure and flat tubercles with height about 1–3 mm. Steel surface under scale layer was also corroded, but corrosion traces were flat with area diameter from 5 to 20 mm and depth about 0.5 mm, rarely –1 mm.

2.4 *Samples analysis*

Digital images of outward, inward and cross-section views of corrosion scales were made by using SEM (Quanta 250 FEI). The elemental composition of targeted areas was determined using EDX (APOLLO X SDD EDAX).

3 RESULTS AND DISCUSSION

3.1 *Corrosion scales morphology and composition*

A state of the art review (Sontheimer et al. 2002, Sarin et al. 2004, Gerke et al. 2008) shows that all tubercles discovered and described by different authors demonstrate similar structures and morphology regularities irrespectively of their environment, age and water quality. Figure 1 shows a simple schematic layer structure and basic chemical

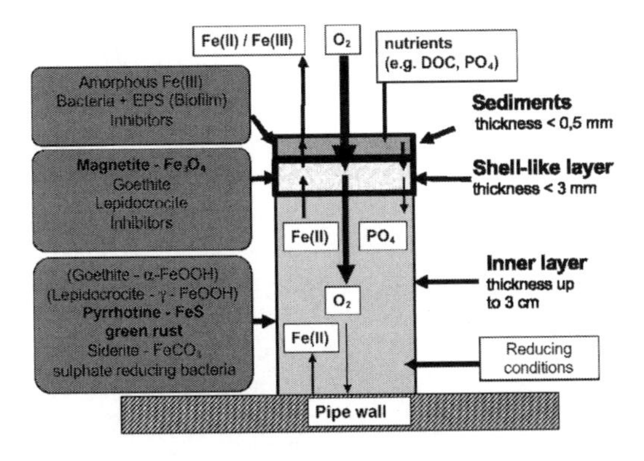

Figure 1. Schematic layer structure in ages unprotected cast iron and steel pipes (Wricke B. et al. 2007).

reactions in corrosion deposits. The authors identify three characteristic structures: sediments, shell-like layer and inner layer (core).

SEM-EDX study of samples harvested from pipe A, B, C and D showed following characteristic regions that differ significantly in structure and composition.

The thin surface layer composed of steel corrosion products with a high content of impurities deposited from water: silica, aluminum and calcium oxides, sulfates, phosphates, organic matter etc. The mineral phase in the surface layer has an amorphous structure (Figures 2a, b).

Hard layer (shell) mainly comprises FeOOH and Fe_3O_4 and has a metallic luster (Figures 2c, d). Almost all samples have a hard layer short-circuited to the pipe wall. The thickness of shell does not exceed 3 mm. The criterion of constant thickness of the hard layer of a growing tubercle should be continuous build-up on the water side and a simultaneous dissolution on the pipe side. Otherwise the cracking of the hard layer should occur. However, no evidence of this theory is shown in the literature. The experiments described in (McEnaney & Smith 1980) show that the hard layer can be formed by cathodic reduction of γ-FeOOH loose deposits to dense crystalline magnetite Fe_3O_4 when corrosion deposits come into contact with deoxygenated water. The transport of iron ions in the solution occurs through the pores in the crystal structure formed within the tubercle, and magnetite provides a path for electrons.

The core (Figures 2e, f, g) preferably comprises steel corrosion products and minor amounts of other impurities. The deeper layers are composed of different crystalline structures such as FeOOH (goethite and lepidocrocite), Fe_2O_3, Fe_3O_4, FeO, $FeCO_3$. Near the pipe surface the iron content in the corrosion scales increases.

The bottom layer (base) is formed in place where metal corrosion occurs and comprises the steel corrosion products with trace amounts of steel additives (carbon, manganese, sulfur) and a small amount of other impurities (silica, chlorine). This layer has a homogeneous crystal structure (Figures 2h, i, k). The absence of aluminum, calcium, magnesium and other elements typical for the surface layer indicates that the core layer is formed without direct contact with interior liquid of the tubercles or flowing water in the pipe. Dense structure of the bottom layer protects the pipe from rapid corrosion.

3.2 *Microbiological corrosion*

The tubercles formation on the pipe surface can be explained by the biofilm contribution, which is formed on pipe inner surface with time, even when

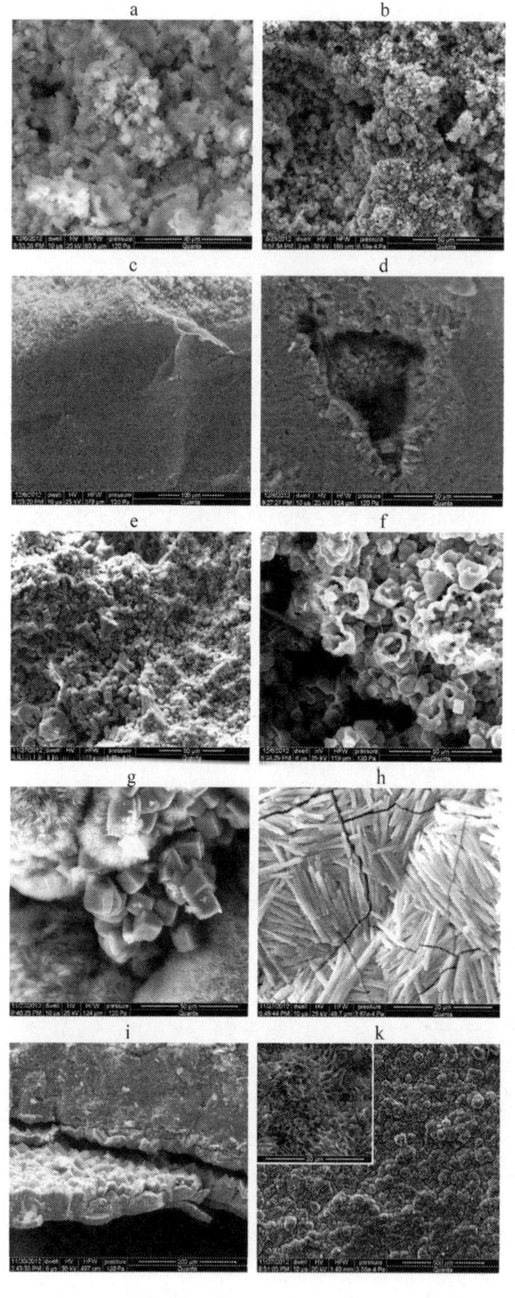

Figure 2. The SEM images of pipe tubercles: surface layer, pipe C (a); surface layer, pipe A (b); hard shell layer, pipe C (c, d); core, pipe C (e, f); core, pipe A (g); base layer, pipe D, bottom view (h) and cross-section (i), base layer, pipe B (k).

the disinfection reagents are used. The samples obtained from the excavation sites were already dry; therefore it was not possible to determine the presence of living biofilm on the scale surfaces.

Table 2. EDS analysis of primary elemental composition of corrosion scale samples [at%].

Sample	C	O	Fe	Al	Si
Thin surface layer					
Pipe A	9–12	43–47	9–26	12–22	6–12
Pipe B	8–11	23–37	32–56	8–14	3–12
Pipe C	0–10	32–40	31–54	6–15	2–5
Pipe D	0	35–42	24–40	9–12	14–26
Hard layer (shell)					
Pipe A	6–10	20–31	58–78	0–1	0–1.4
Pipe C	3.6	26.8	66.4	1.6	1.2
Core region					
Pipe A	5–7	21–34	57–75	0–1.5	0–1.5
Pipe B	5–10	33–38	51–60	0	0
Pipe C	4	30	63	0	0
Pipe D	0	28–40	59–70	0–0.5	0
Bottom layer					
Pipe A	<4	16	80	0	0
Pipe B	0	24–28	68–72	0–0.5	0–0.8
Pipe C	3–7	22–25	57–66	0	0
Pipe D	0	36	60–62	0	0.6
(Figure 2h)	4	15	78	0	0

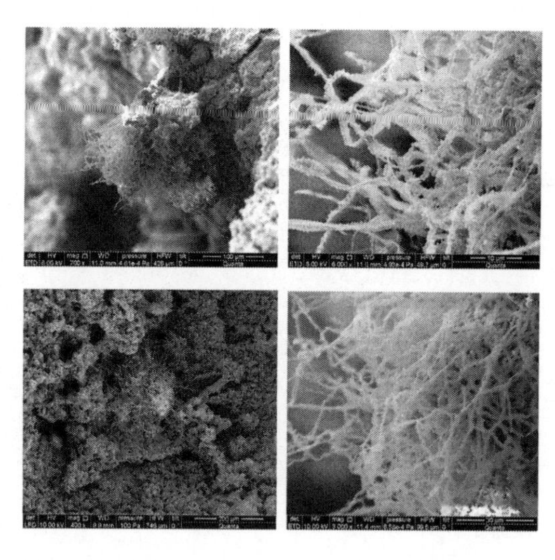

Figure 3. Micrographs of bacterial colonies from core region (pipe C).

However, a careful examination by optical microscopy of internal structure of tubercles extracted from the pipe C showed a very small clusters of microorganisms which looked like small spheres about 0.1 mm in diameter and surrounded by filaments.

The micrographs of these biological objects were obtained using SEM (Figure 3). Based on the literature data, it was concluded that this is an anaerobic filamentous form of bacteria, most likely *Leptothrix sp.* The micrographs show that this bacteria form colonies inside the tubercle cavities and are firmly attached to iron oxides walls.

The participation of bacteria in tubercle formation can be described by the following considerations. The initial corrosion leads to formation of thin loose deposits consisting of iron hydroxide. As water flows through the pipe, colonies of aerobic iron-oxidizing bacteria and probably facultative anaerobic iron-reducing bacteria (*Leptothrix sp.* and *Shewanella sp.*) are formed in this layer on the pipe surface. These bacteria metabolizm substantially contributes to the corrosion process. Polysaccharides produced by bacteria form a film on the pipe surface. This film has a good adhesion property and prevents oxygen diffusion to the pipe wall. During the biofilm formation its thickness increases and it becomes impermeable to oxygen. Thus anaerobic conditions are created inside the tubercle and aerobic iron-oxidizing bacteria species become replaced by anaerobic iron-reducing species. Part of the iron (II) is transported towards deposits following diffusion mechanism where it interacts with dissolved oxygen and forms again a dense layer due to the cathodic reduction of FeOOH and its conversion into Fe_3O_4. Thus, a constant thickness of the dense layer is controlled by interaction aerobic and anaerobic bacteria. Another amount of iron (II) participates in the tubercles buildup. This process is initiated around nucleus, such as: crystalline iron structures from dead cells or carbon dioxide bubbles formed during the microbial oxidation of organic matter.

The EDX analysis of tubercles showed that all layers except the bottom layer contain significant amount of carbon. This can be explained by the presence of organic and inorganic compounds. However, there is no carbon in the bottom layer. Furthermore, it was found that when tubercle is breaking away from the pipe surface the fracture plane coincides with this layer. That fact explains a binder role of carbon material involved in corrosion scales build-up, and the weakness of bottom layer due to the absence of organic carbon. Probably, most of the detected carbon is obtained from polysaccharides that adhere the pipe surface and corrosion scales.

As organic matter content is limited in the tubercle core, the iron bacteria lack nutrition, combine into the colony and corrosion process within this tubercle stops. However, during the further operation of the pipeline new portions of organic and inorganic substances deposit on tubercle surface and it starts to grow again. This can explain the presence of vein-like (marmoreal) features, which are discovered by Gerke et al. (2008). Apparently the formation of a dense layer of corrosion scales reduces steel or cast-iron corrosion rate but does not completely prevent corrosion. This can be confirmed by a constant slight iron increase in the tap water.

Among the list of research works devoted to microbiological corrosion, some studies of Sulfate-Reducing Bacteria (SRB) should be mentioned. Seth & Edyvean (2006) often detected the SRB presence in the drinking water distribution system and their ability to colonize in a new iron and steel pipes.

Seth & Edyvean (2006) found that SRB formed colonies on the pipes surface prior to tubercles and black slime formation and appearance of hydrogen sulfide odor. The SRB activity resulted in increased sulfur content in the corrosion deposits. It should be mentioned that the sulfur content in all samples (taken from pipes A–D) does not exceed 1.5%, which can be, apparently, attributed to low sulphate content in the feed water and the absence of SRB.

4 CONCLUSIONS

The examination of corrosion scales obtained from four old steel pipes operated in Moscow water distribution system showed that the discovered tubercles have similar morphological and mineralogical characteristics. This conforms to conclusions presented in state of the art publications. The authors emphasize the presence of bottom layer which plays an important role in the tubercle growth mechanism. Based on literature analysis and study of deposits morphology it was concluded that iron-oxidizing and iron-reducing bacteria can make a substantial contribution to corrosion processes and be responsible for tubercle growth and structure formation. The main distinguishing features of these corrosion scales can be defined as: large volume, low density and high carbon content.

REFERENCES

Gerke, T.L., Maynard, J.B., Schock, M.R. & Lytle, D.L. 2008. Physiochemical characterization of five iron tubercles from a single drinking water distribution system: possible new insights on their formation and growth. *Corrosion Science,* 50(7): 2030–2039.

Husband, P.S. & Boxall, J.B. 2011. Asset deterioration and discoloration in water distribution systems. *Water Research,* 45(1): 113–124.

Lin, J., Ellaway, M. & Adrien, R. 2001. Study of corrosion material accumulated on the inner wall of steel water pipe. *Corrosion Science,* 43(11): 2065–2081.

McEnaney, B. & Smith, D.C. 1980. The reductive dissolution of γ-FeOOH in corrosion scales formed on cast iron in near-neutral waters. *Corrosion Science,* 20(7): 873–886.

Sander, A., Berghult, B., Ahlberg, E., Elfström Broo, A., Lind Johansson, E. & Hedberg, T. 1997. Iron

corrosion in drinking water distribution systems—surface complexation aspects. *Corrosion Science,* 39(1): 77–93.
Sarin, P., Snoeyink, V.L., Bebee, J., Jim, K.K., Beckett, M.A., Kriven, W.M. & Clement, J.A. 2004. Iron release from corroded iron pipes in drinking water distribution systems: effect of dissolved oxygen. *Water Research,* 38(5): 1259–1269.
Seth, A.D. & Edyvean, R.G.J. 2006. The function of sulfate-reducing bacteria in corrosion of potable water mains. *International Biodeterioration & Biodegradation,* 58(3/4): 108–111.
Sontheimer, H., Kolle, W. & Snoeyink, V.L. 1981. The siderite model of the formation of corrosion-resistant scales. *J. AWWA,* 73(11): 572–579.
Vreeburg, J.H.G., Schippers, D., Verberk, J.Q.J.C. & van Dijk, J.C. 2008. Impact of particles on sediment accumulation in a drinking water distribution system. *Water Research,* 42(16): 4233–4242.
Wang, H., Hu, C., Hu, X., Yang, M. & Qu, J. 2012. Effects of disinfectant and biofilm on the corrosion of cast iron pipes in a reclaimed water distribution system. *Water Research,* 46(4): 1070–1078.
Wricke, B. et al. 2007. *Particles in relation to water quality deterioration and problems in the network: State-of-the-art review.* Techneau, 28 June 2007, Report D 5.5.1 + D 5.5.2.

Advanced Materials, Mechanical and Structural Engineering – Hong, Seo & Moon (Eds)
 ISBN: 978-1-138-02908-8

Principles of optimal structure formation of ceramic semi-dry pressed brick

A.Yu. Stolboushkin & A.I. Ivanov
Institute of Thermophysics Named After S.S. Kutateladze Siberian Brunch of the Russian Academy of Science, Novosibirsk, Russia

O.A. Fomina & A.S. Fomin
Siberian State Industrial University, Novokuznetsk, Russia

G.I. Storozhenko
Ltd. "Baskey Keramik", Chelyabinsk, Russia

ABSTRACT: The principles for the creation of optimal structures of ceramic semi-dry pressed brick are formulated. It is found out that for low- and moderate-plasticity clay raw material, its refinement to a class of −0.3+0 mm is required. Best grain packaging of the grinded fine raw material during compaction is achieved due to its preliminary aggregation. It is established that the rational granulometric composition of a press powder is ensured by material granulation in intensive mixers. Experimentally and in the industrial conditions, it is confirmed that the bricks produced from fine-grained granulated material have a uniform, defect-free texture of a ceramic crock, providing an increase (up to 1.5 times) in physical and mechanical properties of products. An effective wall ceramics with a uniformly distributed system of freeze-resistant macropores is developed due to the introduction of a granulated foam glass into the batch.

1 INTRODUCTION

The main objective of any production is to obtain high-quality products, which in the ceramic industry is achieved by studying the formation processes of optimal structures of adobe products as well as ceramic crock, and their effective control.

The method of semi-dry pressing of a ceramic brick, being actively introduced at some point of time in the past due to the seeming simplicity of technology, has in recent years been subjected to a technological and instrumental reconstruction (Gurov et al. 2013, Tatski et al. 2007, Gurov et al. 2012). "Reset" of the method is carried out by addressing the root causes of poor performance properties of the products: low quality of raw material grinding, granulometric heterogeneity and moisture of press powders, which result in the well-known structural defects in semi-dry pressed bricks (Figure 1).

Samples of ceramic bricks, manufactured at brick factories in West Siberia by the traditional method of semi-dry pressing, were selected in order to investigate their structure. To examine the internal defects, the products were cut into fragments (Figure 2). In the bricks produced at Verkh-Koensk and Berdsk brick factories, cracks and cavities caused by the improper stock preparation and adobe molding were discovered (Figure 3).

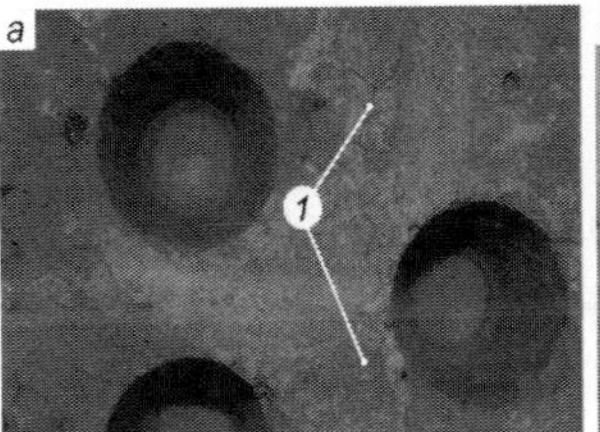

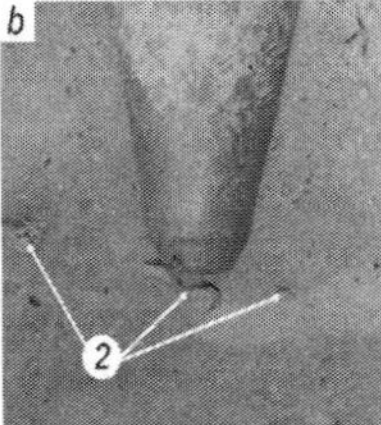

Figure 1. Top view (a) and cross-section (b) of ceramic semi-dry pressed brick with semi-closed voids: 1—drying cracks, 2—caverns.

Practice has shown that the satisfactory (according to the Russian State Standards) quality of ceramic semi-dry pressed brick can be achieved by:

- granulation of a clay mass with the subsequent drying of granules in the dryer drum up to the mixing moisture content of the molding powder (Kondratenko et al. 2001);
- fine grinding and mechanical activation of a clay raw material (Storozhenko et al. 2001);
- use of new technological methods and equipment for processing of raw clay and pressing of products (Kondratenko et al. 2003, Shlegel et al. 2009).

Figure 2. Cutting scheme of a ceramic semi-dry pressed brick into fragments.

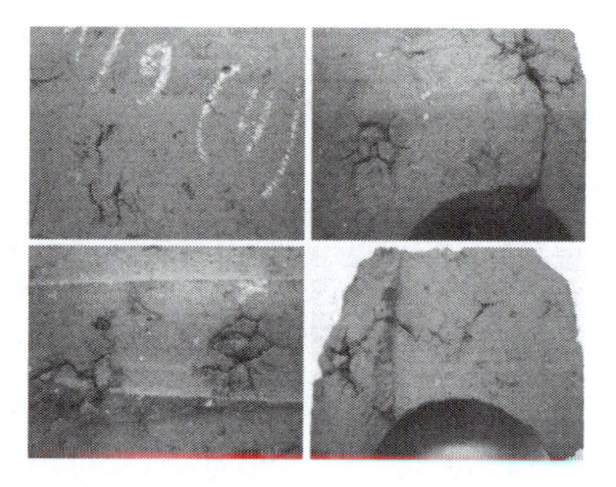

Figure 3. Structural defects in the internal areas of a ceramic semi-dry pressed brick.

However, taken separately, the technological advances do not yet allow to solve the main problem—the same quality of ceramic products produced by soft-mud molding and semi-dry pressing.

2 TECHNOLOGICAL PRINCIPLES OF OPTIMAL STRUCTURE FORMATION

The authors offer an integrated solution of the problem, which is based on the formation of optimal structures of ceramic bricks via modeling processes of fine grinding, granulation, pressing, drying and burning of products.

To create optimal structures of ceramic semi-dry pressed bricks in the laboratory and during industrial tests, the following technological principles were implemented.

First, low-plasticity clay material was subjected to drying and mechanical activation (grinding up to a class of −0.3 + 0 mm) in the jet rotary mill-dryer or oscillating mill. Such mixture preparation before obtaining the press powder is especially effective and necessary for carbonized raw materials to eliminate the harmful effects of carbonate inclusions on the crock quality. It should be noted that during the soft-mud process of brick molding, the fine grinding of clays is carried out on rollers with a gap clearance of 0.7–0.8 mm (Grubacic 2009).

To create optimal conditions for products molding from fine activated raw material, new technological approaches to its aggregation are needed. It is impossible to achieve the required level of moisture and homogenization of press powder by using the traditional equipment. Rational granulometric composition for the best packaging of press powders during pressing is provided by intensive mixers (Stolboushkin et al. 2012). Granulated masses have a greater mobility and lower values of elastic deformation and internal energy compared with dispersed powders that can significantly reduce the pressing pressure of products.

Change in the conditions of raw materials preparation and granulation of press powders contributes to obtaining a homogeneous structure of capillary-porous body and improves the adobe hydraulic conductivity. It was established that under optimum pressing pressures during the process of adobe compaction up to plastic deformation of granules at their boundaries, there is a concentration of the liquid phase due to the moisture squeezing from the granules' surface. Then, while burning in the boundary layer, an intensive generation of a pyroplastic phase takes place due to the predominance of low-melt components of a batch that migrate to the surface of the granules during pressing. It results in the formation of a sintered crock in these zones and, finally, increases the strength of contacts between granules (Stolboushkin 2009).

3 CHARACTERISTICS OF RAW MATERIAL

Implementation of the described principles was carried out for low- and moderate-plasticity raw clay material traditionally used in semi-dry pressing, as shown in the below example of loam from Berdsk deposit in Novosibirsk Oblast (Russia). Raw material is low-disperse, non-caking, highly sensitive to drying, with a high content of coloring oxides. The chemical composition is given in Table 1.

4 EXPERIMENTAL PROCEDURE

Pilot industrial tests of technological approaches to the formation of an optimal structure of ceramic brick were carried out at LLC "Berdsk Brick Factory" (Russia). A pilot batch of bricks was produced using a new method of mixture preparation including drying and grinding in a jet rotary mill-dryer up to a class of −0.3+0 mm, and granulation of powder in the intensive mixer up to the formation of granules of diameter 1–3 mm

Table 1. The chemical composition.

Mass fraction of components, % (on absolutely dry substance)							
SiO_2	TiO_2	Al_2O_3	Fe_2O_3	MgO	CaO	R_2O	LOI
60.5	0.86	13.3	5.35	1.62	5.18	3.60	8.63

simultaneously moisturizing up to a molding moisture of 10.5–11%. Brick molding from the granulated and factory press powder was carried out on hydraulic press. Drying and burning of the products were carried out at a temperature of 1000°C in the tunnel kiln in accordance with the process requirements for brick production at a factory.

Based on the technological principles of forming the ceramic composites with matrix structure (Stolboushkin 2011), we developed an effective manufacturing technology of semi-dry pressed wall ceramics from Berdsk loam and granulated foam glass.

5 STUDY OF STRUCTURE

Complex investigations of the structure and phase composition of ceramic brick were carried out with the use of petrographic, electron microscopy and X-ray analysis methods.

The investigations of the ceramic crock structure (Figure 4) showed that the texture of the factory brick is large-brecciated with fragments of initial raw minerals (Figure 4a), along with their perimeter macropores, indicating the defects caused by adobe pressing. On the contrary, in the brick made from finely granulated material, the texture is eutectophyric (Figure 4b). Fine-grained mineral neoformations are imbedded into the holocrystalline bulk (Figure 5c, d; Figure 6c, d). Homogeneous, defect-free texture of the crock enhances the quality of semi-dry pressed ceramic wall materials produced by the new method of mixture preparation.

It should be noted that the developed principles of formation of ceramic products' structures significantly extend the capabilities of the semi-dry pressing technology. Fine grinding allows to mix efficiently the activated powder with any kind of additives (bulk coloring, burning out, structure-forming ones) and granulation—to receive a wide range of press powders for the production of not only wall but also construction ceramics.

The results obtained are determined by the specific physical and chemical processes in ceramic crock formation from the granulated press powder, in which the hollow foam glass granules are evenly distributed. At a temperature of 850–900°C, the granule shell softens and the generated liquid phase penetrates into the body of the ceramic brick. Due to the surface tension forces, the approaching of solid phase particles takes place. Besides, the particles of clay minerals and unbound quartz dissolve in it and new crystalline phases are generated from the melt. Thus, in the ceramic brick, an evenly distributed pore system is formed, the walls of which are waterproof and durable as a result of the solid phase flow along the border "granule-crock" (Figure 7).

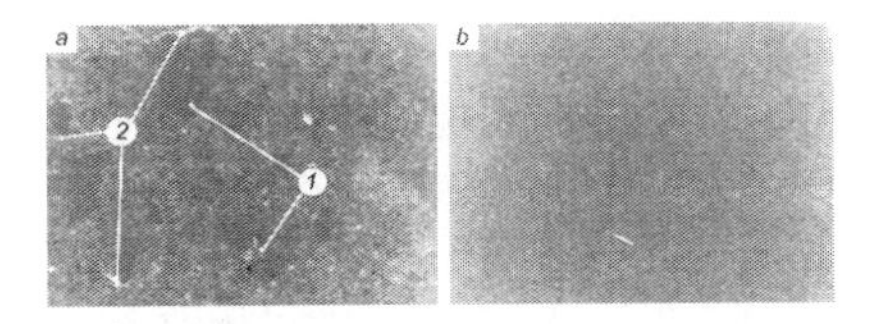

Figure 4. Macrostructure of the ceramic brick produced from Berdsk loam with a factory mixture preparation (a) and produced according to the developed method (b). Shooting conditions: polished section, reflected light, magnification × 50, nicols II: 1—fragments of initial minerals; 2—macropores.

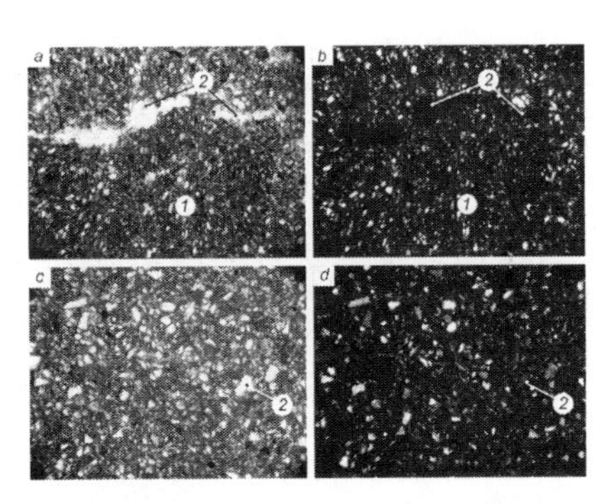

Figure 5. Micrographs of the ceramic brick produced from Berdsk loam with factory mixture preparation (a, b) and produced according to the developed method (c, d). Shooting conditions: thin section, transmitted light, magnification × 50, nicols II (a); nicols X (b); magnification × 100, nicols II (c); nicols X (d): 1—fragments of initial minerals; 2—pores.

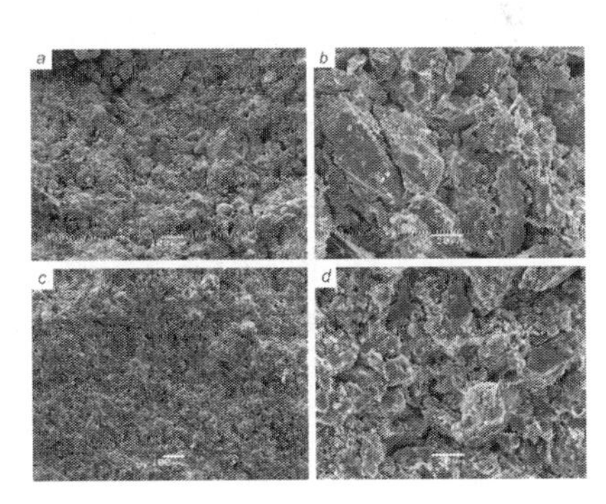

Figure 6. SEM micrographs of the ceramic brick structure produced from Berdsk loam with factory mixture preparation (a, b) and produced according to the developed method (c, d).

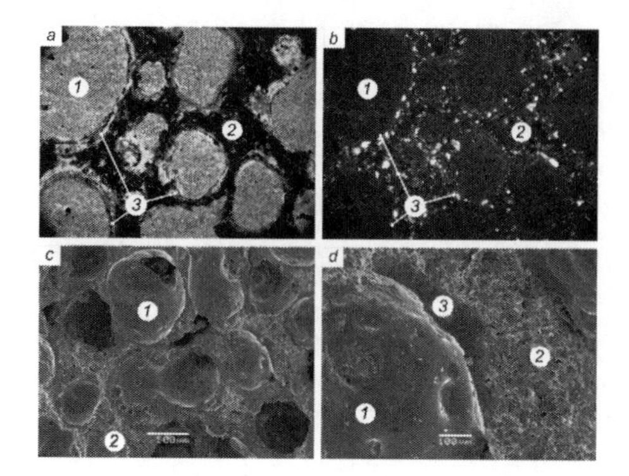

Figure 7. Micrographs of the brick structure with filler from granulated foam glass crystalline material. Shooting conditions: thin section, transmitted light, magnification × 50, nicols II (a); nicols X (b); SEM (c, d): 1—vitrified pore; 2—ceramic frame; 3—glass crystalline smelting at the border of the pore with a ceramic frame.

Table 2. Physical and mechanical properties of the samples depending on the method of press-powder preparation.

Method of press powder preparation	Ultimate strength, MPa		Average density, kg/m³	Water absorption, %	Frost resistance, cycles
	Compression	Bending			
Traditional (drying, grinding)	12.1	2.2	1770	13.6	35
Proposed (mechanical activation, granulation)	19.3	2.5	1675	11.6	>50

6 RESULTS AND DISCUSSION

The results of the physico-mechanical tests, performed in the factory's laboratory indicate that the ceramic brick made from the granulated Berdsk loam, produced in accordance with the proposed method, essentially surpasses in quality its analogue produced by the traditional method (Table 2).

In the industrial conditions, ceramic bricks produced from Berdsk loam and granulated foam glass showed a compressive strength of more than 16 MPa, an average density below 1000 kg/m³ and a water absorption capacity of 6–7%.

7 CONCLUSIONS

Laboratory studies and pilot industrial tests of methods for the creation of optimal structures of ceramic products have demonstrated the potential for progressive development of semi-dry pressing technology. The proposed method increases the physical and mechanical properties of ceramic bricks (compression strength up to 1.5 times). It can be demanded by brick manufacturers and compete with the soft-mud process of brick molding in the conditions of inevitable transition of the sub-sector producing ceramic wall materials to the usage of lower-quality raw materials.

ACKNOWLEDGMENT

This work was supported by the Ministry of Education and Science of the Russian Federation (project ID RFMEF160714X0106).

REFERENCES

Grubacic, V. 2009. Company BEDESCHI: second century in the lead of machine manufacturing for the ceramic industry. *Construction Materials,* 4: 30–31.

Gurov, N.G., Gurova, O.E. & Storozhenko, G.I. 2013. Innovative ways of technological and equipment reconstruction of semidry pressing factories. *Construction Materials,* 12: 52–55.

Gurov, N.G., Naumov, A.A. & Ivanov, N.N. 2012. Ways of increase frost resistance of semidry pressing brick. *Construction Materials,* 3: 40–42.

Kondratenko, V.A. & Peshkov, V.N. 2001. New technological line for production semidry pressing face ceramic brick. *Construction Materials,* 5: 41–42.

Kondratenko, V.A. Peshkov, V.N. & Slednev, D.V. 2003. Modern technology and equipment for production of semidry pressing ceramic brick. *Construction Materials,* 2: 18–19.

Shlegel, I.F., Shaevich, G., Ya. Mikhailets, S.N., Andrianov, A.V., Bakhta, A.O., Ivanov, V.G., Makarov, S.G., Miroshnikov, V.E., Noskov, A.V. & Titov, G.V. 2009. The new complex ShL 400 for church brick production. *Construction Materials,* 4: 32–36.

Stolboushkin, A.Yu. 2011. Theoretical Grounds of Formation of Ceramic Matrix Composites on the Basis of Anthropogenic and Natural Raw Materials. *Construction Materials,* 2: 10–15.

Stolboushkin, A.Yu., Ivanov, A.I., Zorya, V.N., Storozhenko, G.I. & Druzhinin S.V. 2012. Features of granulation of anthropogenic and natural raw materials for wall ceramic. *Construction Materials,* 5: 85–89.

Stolboushkin, A.Yu., Stolboushkina, O.A. & Berdov, G.I. 2009. Optimization of parameters of pressing of granulated anthropogenic and natural raw materials for ceramic brick production. *Construction Materials,* 4: 30–31.

Storozhenko, G.I., Pak, Yu. A., Boldyrev, G.V., Yaroshuk, V.G., Yaroshuk, A.G. & Sobyanin, N.V. 2001. Production of ceramic brick from activated loamy raw at medium power factories. *Construction Materials,* 12: 72–73.

Tatski, L.N., Mashkina, E.V. & Storozhenko, G.I. 2007. Two steps activation of raw materials in technology of wall ceramic. *Construction Materials,* 9: 11–13.

Advanced Materials, Mechanical and Structural Engineering – Hong, Seo & Moon (Eds)
© 2016 Taylor & Francis Group, London, ISBN: 978-1-138-02908-8

Tension behavior of filled silicone rubbers at different temperatures

Y.N. Lv, L.M. Guo & Y. Wang
Department of Modern Mechanics, University of Science and Technology of China, Hefei, Anhui, P.R. China

Z.F. Deng
Institute of Structural Mechanics, China Academy of Engineering Physics, Mianyang, Sichuan, P.R. China

ABSTRACT: The effect of temperature on the uniaxial tension stress-strain behavior of filled silicon rubbers was investigated. Silicone rubber samples of two hardness (Shore A 50 and Shore A 80) were tested at a strain rate of 0.001 s^{-1} and temperatures of −20, 20 and 50°C, respectively. The non-contact optical measuring technique, automated grid method for strain measurement, was used to accurately capture the full-field deformation information of the tension sample. Experimental results indicate that the tension responses of filled silicon rubbers exhibit obvious nonlinear elastic characteristics within a moderate strain range and the values of stiffness and nominal stress at a given elongation are dependent on the temperature. The filled silicone rubber can be regarded approximately as an incompressible material. Based on the incompressibility hypothesis, a modified hyperelastic constitutive model was proposed to describe the temperature-dependent tension behavior of silicon rubbers.

1 INTRODUCTION

Due to their good mechanical properties, excellent resistant to extreme environments and temperatures and good bio-compatibility, the filled silicone rubbers are increasingly used as engineering materials in automotive and bio-medical applications. Extensive investigations on the mechanical responses at room temperature have been reported for the unfilled and the filled silicone rubbers and their composites. For example, Podnos et al. (2006) performed uniaxial tension, biaxial tension and pure shear tests to investigate the mechanical behavior of filled silicone rubbers. Meunier et al. (2008) analyzed the mechanical behavior of an unfilled silicone rubber using tension, pure shear, compression, plane strain compression and bulge tests and also performed finite element simulations to evaluate the choice of the hyperelastic models. Korochkina et al. (2008) combined experimental and numerical investigations to evaluate the applicability of various constitutive models for predicting the nonlinear behavior of three silicone rubber compounds that are encountered in the pad printing process. Machado et al. (2012) experimentally characterized the anisotropy induced by the Mullins effect in a particle-reinforced silicone rubber. Bailly et al. (2014) investigated the tension behavior of silicone rubber membranes reinforced with architectured fiber networks and found that the mechanical behavior of the membranes is hyperelastic. Benevides & Nunes (2015) studied the mechanical behavior of the alumina-filled silicone rubber subjected to pure shear loadings at finite strains. It can be found that most of aforementioned investigations dealt with the mechanical responses of silicone rubbers at large strain deformation with nominal strains more than 100% and up to 300%. It is necessary to achieve an accurate measurement and reliable characterization of the tension behavior of silicone rubbers at moderate strains for engineering applications. However, the investigations on the accurate measurement at moderate strains for rubber-like materials and the temperature effect on the tension behavior of silicone rubbers are reported little.

The purpose of the present paper is to investigate the tension responses of the filled silicone rubbers over a wide range of temperatures. Non-contact optical measurement technique was utilized to accurately obtain the deformation field information at moderate strains. The effect of temperature on the stress-strain behavior of silicone rubbers was examined and the phenomenologically based constitutive model was proposed to describe the tension behavior within the moderate strain range.

2 EXPERIMENTAL

2.1 *Material*

The materials used in the present investigation were filled silicone rubbers supplied by Shin-Etsu Chemical Co., Ltd. in Japan. Two kinds of silicone rubbers were studied in the hardness levels of Shore A 50 and 80.

2.2 *Tension tests*

Quasi-static uniaxial tension tests were carried out on the INSTRON-E3000 testing machines operated in the displacement control mode. The flat samples with 14 mm in gage width, 170 mm in gage length and 3 mm in thickness were used. The loading was performed at a constant nominal strain rate of 0.001 s^{-1} over a temperature range of −20 to 50°C.

For tension testing of rubber-like materials, the extensometer was conventionally used to measure the large deformation of the samples. However, due to the contact between the extensometer and the soft sample, the strain measurement results are not accurate enough at small strains. Another simple method to measure the deformation of the sample is to evaluate the strain by the crosshead displacements of testing machine. However, this method is not accurate because the end effects are included and the measured strain does not correspond to the actual strain in the uniform deformation zone in the sample. In order to accurately capture the full-field deformation information at moderate strain level with a strain less than 100%, the non-contact measurement technique, automated grid technique, was employed to conduct the strain measurement in the present paper. Compared with other optical methods such as the moiré interferometry, holography interferometry and digital image correlation (Rastogi 2000, Nunes 2010), the optical system and sample preparation of the automated grid technique are simpler and no special illumination is required (Buisson & Ravi-chandar 1990, Sirkis & Lim 1991, Nazarenko et al. 1994). The measurement system consists of a CCD video camera and an image-processing system. The implementation process of the automated grid method includes preparation of grid pattern, recognition of the reference and deformed images and displacement calculation of the grids. Normally, the painting method is adopted to prepare grid pattern. The template with piercing grid is made and is aligned closely with the specimen, and then the white paint is applied to the template. In order to get more detailed results, the grid should be made small and dense enough. In the present paper, the mixture of the silicone grease and the magnesium power was used instead of the white paint to avoid obscure boundary of the grids when taking away the template. Here, the grid pattern was applied to each sample in a diameter of 1 mm. The images (570 pixels × 570 pixels) were recorded at a frequency of 1 Hz. The strain measurement precision can be reached at the 10^{-3} strain level. It is assumed that sample material deforms isotropically in the two lateral directions. Therefore, only one CCD was placed in front of the sample to record the in-plane deformation information. Figure 1 shows the front view images of a tension sample. It can be seen that the edge of the grid is clear enough and the painted grid is firmly adhered to the sample in the course of loading, which ensures the accuracy of the motion analysis of the grids and the strain calculation of the sample.

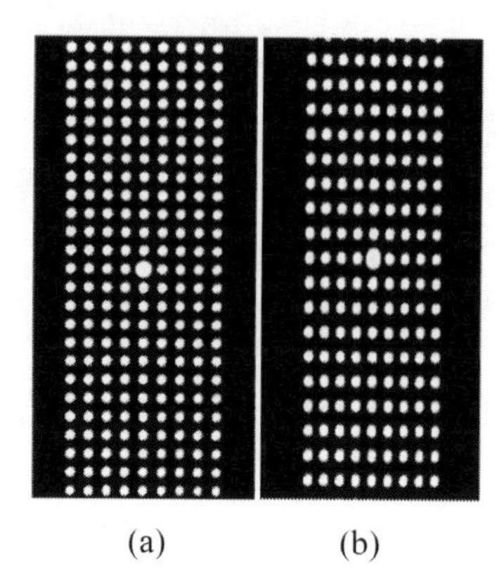

(a) (b)

Figure 1. Front view images of a tension sample: (a) reference image, and (b) deformed image.

3 RESULT AND DISCUSSION

In order to investigate the specimen geometry on the stress-strain response of the silicone rubber, the full-field strain of the tension sample was measured. Experimental results show that the uniform deformation field at the middle section of the specimen can be obtained using the present specimen geometry, namely the requirement of the uniaxial tension experimental condition can be satisfied. The nominal stress versus nominal axial-strain curves obtained under uniaxial tension at quasi-static strain rate are shown in Figure 2. It is found that the curves for two silicone rubbers exhibit an obvious nonlinear elastic behavior within the tested strain range at all testing temperatures. Especially, the nonlinear behavior can be observed at small strain less than 10%. Also, it can be seen that the tension responses of the silicone rubbers are sensitive to the temperature. For the filled silicone rubber with hardness of 50, the values of stiffness increase with increasing testing temperature, which exhibits a temperature hardening phenomenon. However, the stiffness of the filled silicone rubber with hardness of 80 decreases with the increase of temperature, which is comparable to the existing results on silicone rubbers (Rey et al. 2013). At room temperature, the hard silicone rubber (Shore A 80)

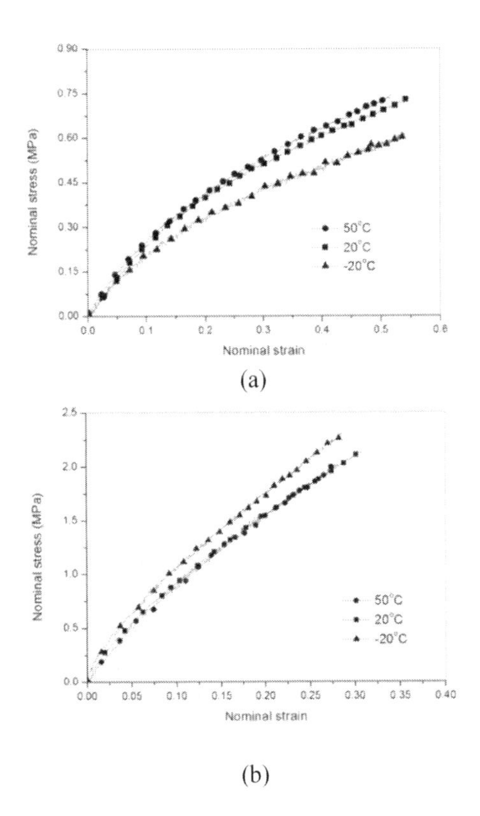

Figure 2. Tension responses at loading rate of 0.001 s^{-1} as a function of temperature for samples with (a) hardness of 50 and (b) hardness of 80.

exhibits about 0.8 MPa increase in nominal stress in comparison with the soft one (Shore A 50) at a strain of 10%. Furthermore, the difference of the values of the nominal stress increases with increasing strain. Similar trend was also found at testing temperatures of −20 and 50°C. However, such difference of nominal stress at a given strain between these two silicone rubbers is sensitive to the testing temperature. For example, the hard rubber exhibits about 0.9 MPa and 0.3 MPa increases in nominal stress at a strain of 10% compared with the soft one at −20 and 50°C, respectively.

Figure 3 shows the relationship of axial and transverse strains in uniaxial extension at different temperatures for the sample with hardness of 50. The compressibility of the tested rubber can be observed. According to the incompressible assumption, the relationship between the nominal transverse strain, ε_t, and the nominal longitudinal strain, ε_l, should be,

$$\varepsilon_t = -1 + \frac{1}{\sqrt{1+\varepsilon_l}} \tag{1}$$

The solid line in Figure 3 represents the incompressible curve. It can be seen that the measured curve of longitudinal strain—transverse strain at room temperature is close to the incompressible curve, which indicates that the tested rubber can be approximately regarded as an incompressible material. The measured Poisson ratio of the silicone rubber with hardness of 50 is 0.49 within the moderate strain range tested in the present paper. Moreover, the Poisson ratio changes little with temperature. Similar results can be observed for the filled silicone rubber with hardness of 80.

A considerable number of constitutive models based on continuum mechanics have been proposed to describe the hyperelastic behavior of isotropic rubber-like materials (Mooney 1940, Rivlin 1948, Yeoh 1990, Arruda & Boyce 1993, Lion 1997, Boyce & Arruda 2000). According to the description of the strain energy density function, these constitutive models can be divided into two categories. One is the phenomenologically based model which the strain energy density is represented in terms of Cauchy-Green deformation tensor invariants (I_1, I_2 and I_3) or principal stretches (λ_1, λ_2 and λ_3). The other is the physically-based model based on kinetic theory and statistics of molecular-chain network.

Due to mathematical simplicity, the phenomenologically based constitutive models are widely used for engineering purposes. According to the conventional assumption of isotropic incompressibility for rubbery materials, the strain energy density function can be represented by the first and second tensor invariants ($I_3 = 1$). Following the analysis of Rivlin, the strain energy density is expressed by a polynomial series involving ($I_1 - 3$) and ($I_2 - 3$) as follows.

$$W = \sum_{i+j=1}^{N} C_{ij}(I_1-3)^i(I_2-3)^j \tag{2}$$

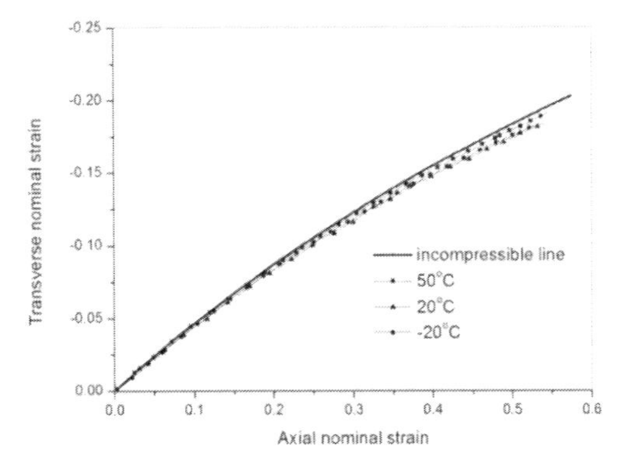

Figure 3. Relationship of axial and transverse nominal strains in uniaxial extension at different temperatures for the sample with hardness of 50.

where, $I_1 = \lambda_1^2+\lambda_2^2+\lambda_3^2$ and $I_1 = \lambda_1^2\ \lambda_2^2+\lambda_2^2\ \lambda_3^2+\lambda_3^2\ \lambda_1^2$. Based on this expression, several well-known phenomenologically based models were derived. For example, when N is taken as 1, the Mooney-Rivlin model is obtained as

$$W_{MR} = C_{10}(I_1 - 3) + C_{01}(I_2 - 3) \tag{3}$$

If the contribution of second invariant to the strain energy is omitted and N = 3 is taken, the Yeoh model can be obtained as

$$W_{Yeoh} = C_1(I_1 - 3) + C_2(I_1 - 3)^2 + C_3(I_1 - 3)^3 \tag{4}$$

where, C_1, C_2 and C_3 = model parameters.

The eight-chain model proposed by Arruda and Boyce is a well-used physically-based model which uses the following strain energy function.

$$W_{eight-chain} = nkTN\left(\beta\sqrt{\frac{I_1}{3N}} + \ln\frac{\beta}{\sinh\beta}\right) \tag{5}$$

where, n = chain density; k = Boltzmann's constant; N = the number of rigid links of an equal length; and T = temperature. β is the inverse Langevin function, $\beta = L^{-1}[\sqrt{I_1/(3N)}$, for the Langevin function defined as $L(\beta) = coth\beta - 1/\beta$.

For the uniaxial tension loading condition, the principal values of the nominal stress are $\sigma_1 = \sigma$ and $\sigma_2 = \sigma_3 = 0$. The principal values of the stretch ratio are $\lambda_1 = \lambda$ and $\lambda_2 = \lambda_3 = \lambda^{-1/2}$. Therefore, the relation of the nominal stress and the longitudinal stretch ratio can be simplified as follows.

$$\sigma = 2\lambda\left(1 - \frac{1}{\lambda^3}\right)\left[\frac{\partial W}{\partial I_1} + \frac{1}{\lambda}\frac{\partial W}{\partial I_2}\right] \tag{6}$$

The comparison among the uniaxial tension experimental data and the aforementioned constitutive models is presented in Figure 4. It is seen that both phenomenologically based models and the molecular network model have the ability to describe the non-linear behavior of the silicone rubber. However, the model results are not consistent well with the experimental data within the small strain range. It can be extrapolated that the better model results can be given if more material parameters are included in the phenomenologically based model. For example, as shown in Equation 2, five and nine model parameters are introduced when N = 2 and N = 3 are taken. However, it is not easy to accurately determine the model parameters in actual engineering applications especially when the temperature dependence of the model parameters is taken into account. In this paper, a modified strain energy density function was proposed based on the Rivlin model as follows.

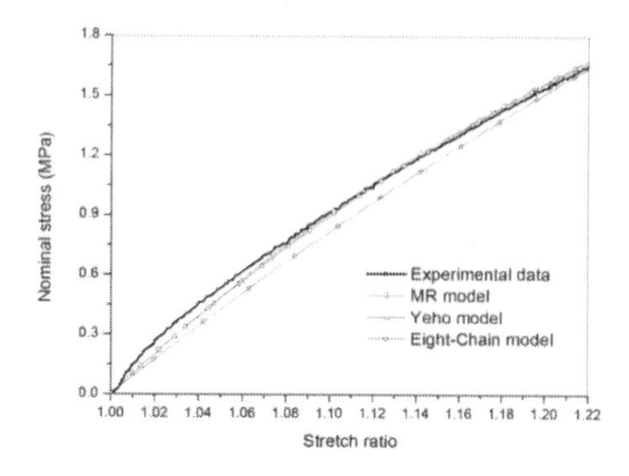

Figure 4. Comparison of the model results and the uniaxial ex-tension data in small strains.

$$W_{MR-modified} = C(I_1 - 3) + \frac{a}{b+1}(I_2 - 3)^{b+1} \tag{7}$$

where, C, a and b = model parameters.

The values of model parameters were estimated at each testing temperature using an optimization procedure based on the nonlinear least square method. Since it was found that the determined model parameters are temperature dependent, an approximation of this temperature dependence was given as

$$C = C_0 \exp\left(C_1\frac{T - T_0}{T_0}\right) \tag{8}$$

$$a = a_0\left(1 + a_1\frac{T - T_0}{T_0}\right) \tag{9}$$

$$b = b_0\left(1 + b_1\frac{T - T_0}{T_0}\right) \tag{10}$$

where, C_0, C_1, a_0, a_1, b_0 and b_1 = material constants; T = testing temperature; and T_0 = reference temperature which was taken as 273 K in the present paper.

The final values of six model constants for two filled silicone rubbers are listed as follows: For silicone rubber with hardness of 50, C_0 is 0.63 MPa, C_1 is −1.03, a_0 is −0.62 MPa, a_1 is −1.67, b_0 is 0.30 and b_1 is 3.99. For silicone rubber with hardness of 80, C_0 is 3.08 MPa, C_1 is −1.78, a_0 is −2.66 MPa, a_1 is −2.60, b_0 is 0.17 and b_1 is 1.23. Figure 5 shows the comparison of model correlations with the experimental data. Good agreement is obtained,

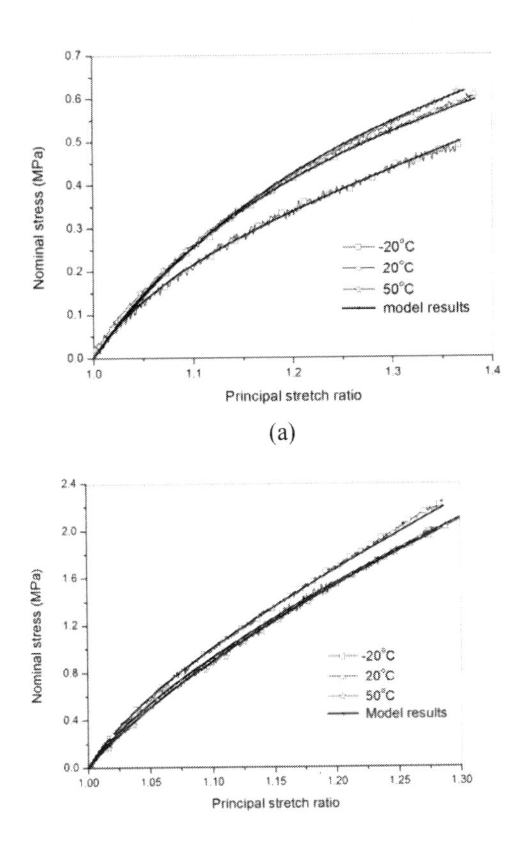

Figure 5. Correlation between the modified MR model and the experimental data for silicone rubbers with (a) hardness of 50 and (b) hardness of 80.

which indicates that such model is applicable to describe the temperature dependent deformation behavior of filled silicone rubbers subjected to uniaxial tension loading within the moderate strain range. For detailed investigation of the reliability and efficiency of the proposed model in complicated modes of deformation for engineering applications, it is necessary to perform more tests such as pure shear and biaxial tension tests at different temperatures in the future work.

4 CONCLUSIONS

The measurement and characterization of the stress-strain behavior for an engineering material is one of the most fundamental tasks for the analysis and de-sign of structures. In this paper, the effect of temperature on the tension stress-strain responses of two filled silicone rubbers with hardness (Shore A) of 50 and 80 was investigated within the temperature range of −20 to 50°C. The uniaxial tension experiments were performed under quasi-static loading conditions at the nominal strain rate of 0.001 s^{-1}. The automated grid method was used to accurately measure the full-field strain at moderate strains. The full-field strain contours indicate that the homogeneous deformation field can be obtained using the present specimen geometry. Furthermore, the Poisson effect was investigated and the incompressibility of the filled silicone rubbers was evaluated. The obtained nominal stress–strain results show that the tension behavior of filled silicone rubbers within the moderate deformation range exhibits obvious non-linear elastic characteristics and has the temperature sensitivity. The values of stiffness and nominal stress at a given elongation change with testing temperatures. According to the isotropic incompressibility hypothesis, a modified hyperelastic constitutive model was proposed to describe the temperature-dependent tension behavior of silicon rubbers. The model correlations agree well with the experimental data investigated within the present strain and temperature ranges.

ACKNOWLEDGEMENT

This research was supported through the NSAF foundation under Grant No. U1230103.

REFERENCES

Arruda, E.M. & Boyce, M.C. 1993. A three-dimensional constitutive model for the large stretch behavior of rubber elastic materials. *Journal of Mechanics and Physics of Solids,* 41: 389–412.

Bailly, L., Toungara, M., Orgéas, L., Bertrand, E., Deplano, V. & Geindreau, C. 2014. In-plane mechanics of soft architectured fibre-reinforced silicone rubber membranes. *Journal of the mechanical behavior of biomedical materials,* 40: 339–353.

Benevides, R.O. & Nunes, L.C.S. 2015. Mechanical behavior of the alumina-filled silicone rubber under pure shear at finite strain. *Mechanics of Materials,* 85: 57–65.

Boyce, M. & Arruda, E. 2000. Constitutive models of rubber elasticity: a review. *Rubber Chemistry and Technology,* 73: 505–523.

Buisson, G. & Ravi-Chandar, K. 1990. On the constitutive behaviour of polycarbonate under large deformation. *Polymer,* 31: 2071–2076.

Korochkina, T.V., Jewell, E.H., Claypole, T.C. & Gethin, D.T. 2008. Experimental and numerical investigation into nonlinear deformation of silicone rubber pads during ink transfer process. *Polymer Testing,* 27: 778–791.

Lion, A. 1997. A physically based method to represent the thermo-mechanical behaviour of elastomers. *Acta Mechanica,* 123: 1–25.

Machado, G., Chagnon, G. & Favier, D. 2012. Induced anisotropy by the Mullins effect in filled silicone rubber. *Mechanics of Materials,* 50: 70–80.
Meunier, L., Chagnon, G., Favier, D., Orgéas, L. & Vacher, P. 2008. Mechanical experimental characterization and numerical modelling of an unfilled silicone rubber. *Polymer Testing,* 27(6): 765–777.
Mooney, M. 1940. A theory of large elastic deformation. *Journal of Applied Physics,* 11(9): 582–592.
Nazarenko, S., Bensason, S., Hiltnert, A. & Baer, E. 1994. The effect of temperature and pressure on necking of polycarbonate. *Polymer,* 35: 3883–3891.
Nunes, L.C.S. 2010. Shear modulus estimation of the polymer polydimethylsiloxane (PDMS) using digital image correlation. *Materials and Design,* 31: 583–588.
Podnos, E., Becker, E., Klawitter, J. & Strzepa, P. 2006. FEA analysis of silicone MCP implant. *Journal of Biomech,* 39: 1217–1226.
Rastogi, P.K. 2000. Principles of holographic interferometry and speckle metrology. *Topics in Applied Physics,* 77: 103–150.
Rey, T., Chagnon, G., Le Cam J.B. & Favier, D. 2013. Influence of the temperature on the mechanical behaviour of filled and unfilled silicone rubbers. *Polymer Testing,* 32: 492–501.
Rivlin, R.S. 1948. Large elastic deformations of isotropic materials. IV. Further developments of the general theory. *Philosophical Transactions of the Royal Society A,* 241: 379–397.
Sirkis, J.S. & Lim, T.J. 1991. Displacement and strain measurement with automated grid methods. *Experimental Mechanics,* 31(4): 382–388.
Yeoh, O.H. 1990. Characterization of elastic properties of carbon-black-filled rubber vulcanizates. *Rubber Chemistry and Technology,* 63: 792–805.

Advanced Materials, Mechanical and Structural Engineering – Hong, Seo & Moon (Eds)
© 2016 Taylor & Francis Group, London, ISBN: 978-1-138-02908-8

A study on High Performance Fiber Reinforced Cementitious Composites containing polymers

G.J. Park, J.J. Park & S.W. Kim
Korea Institute of Civil Engineering and Building Technology, Goyang, Korea

ABSTRACT: This paper presents an experimental study on the improvement of the ductile behavior of High Performance Fiber Reinforced Cementitious Composites (HPFRCC) by admixing organic fiber together with polymers. The applicable types of polymer and minimum mix ratios are investigated by evaluating the physical properties like the compressive strength and tensile strength. In view of the influence of the admixing of polymer on HPFRCC reinforced by organic fiber, the best tensile strength was obtained when organic fiber and polymer are adopted concurrently than when reinforcement is provided by the organic fiber only. With regard to the type of polymer, it appeared that SBR and EVA were appropriate. The mix ratio of polymer was increased from 1% to 5% to find the minimum ratio and revealed that the compressive strength reduced gradually regardless of the type of polymer. Besides, the tensile and flexural strengths increased with larger mix ratio of polymer. Especially, the tensile and flexural strengths with polymer mix ratio of 5% were measured to be larger than when only organic fiber was used. Consequently, a minimum mix ratio of 5% is required for the polymer admixed with cement.

1 INTRODUCTION

The preference given to concrete as construction material in our modern society compared to other materials is due to its economy, its workability and the possibility to shape it easily as desired. Moreover, concrete is being adopted for the fabrication of numerous structures owing to its outstanding compressive strength and durability. However, its tensile strength reaching merely 10% of the compressive strength and its proneness to brittle failure even at small strain level remain still topics of material research.

Various approaches were adopted to improve the brittle nature and tensile strength of concrete. For example, High Performance Fiber Reinforced Cementitious Composites (HPFRCC) was reported to exhibit deformation hardening characteristics with the formation of numerous microcracks due to the tensile resistance of the fibers and to develop high tensile strength and ductile capacity compared to normal concrete (Kim et al. 2011). As another approach, there is also the method improving the ductility of the matrix itself. The combined use of cement mortar together with polymers that are high molecular organic compounds was reported to have improving effect on the tensile, flexural and bond strengths and on the ductility at low strength owing to the formation of a polymer film (Ohama 1995). However, the current lack of studies dedicated to HPFRCC using polymers is noteworthy and this topic needs to be addressed further.

Accordingly, this study intends to improve the tensile and flexural strengths of HPFRCC with compressive strength of 40 MPa by admixing polymers. To that goal, the physical properties of HPFRCC are evaluated in absence and presence of polymer. Based upon the evaluation, the physical performance of HPFRCC is examined according to the change in the type and mixing ratio of polymer so as to propose the minimum mixing ratio of polymer improving the tensile and flexural strengths.

2 TEST METHODS AND PURPOSES

2.1 *Materials and mix proportions*

This study adopts type 1 Portland cement and type 2 Fly Ash (FA), fine aggregates with average grain size of 0.7 to 1.2 mm, and polycarboxylate superplasticizer. PVA organic fiber is admixed with a volume fraction of 1% to improve the toughness of HPFRCC. In addition, latex-type polymer dispersions are used to increase the ductility of HPFRCC. Three types of polymers that are Styrene Butadiene Rubber (SBR), Polyacrylic Ester (PAE), and

Ethylene Vinyl Acetate (EVA) of which chemical properties are listed in Table 1 are adopted.

The physical performances are compared according to the individual or combined admixing of organic fiber and polymer. Table 2 arranges the compositions considered for the evaluation of the minimum mixing ratio of polymer with respect to the type of polymer and polymer-to-cement ratio (P/C).

2.2 *Test methods*

Tests were conducted for the compressive strength, flexural strength and tensile strength as shown in Figure 1 to evaluate the physical properties of HPFRCC according to the admixing of polymer. The flexural strength was measured through 3-point loading test on 40 × 40 × 160 mm cubes in compliance with ASTM C 348. Compressive strength test was performed by reusing both ends of the failed specimens with respect to ASTM C 349. However, due to the absence of regulations defined for the tensile strength, the tensile strength test was carried out on fabricated dog-bone specimens. Moreover, notches of 12.5 mm were prepared at both sides of the center of the specimens in order to induce tensile failure at identical locations, and the strain was measured on the specimens fixed by clip gages. All the specimens were subjected to wet curing at temperature of 20°C and humidity of 60%, and measurement was executed at 28 days. The tests were performed at speed of 1 mm/min.

Table 1. Chemical properties of polymers.

Composition	SBR	PAE	EVA
Solid (%)	48.5	49	55
Color	Milk	Milk	Milk
pH	9.7	8	7–8
MFT* (°C)	10	0	10
Tg** (°C)	0	5	0
Viscosity (mPa · s)	500–2000	<100	300–1500
Specific gravity (kg/l)	1.015	0.98	0.94

* MFT = Minimum film forming temperature.
** Tg = Glass transition temperature.

Figure 1. Test methods of HPFRCC containing polymer: (left) flexural strength; (center) compressive strength; (right) tensile strength.

3 TEST RESULTS AND DISCUSSION

3.1 *Effect of mixing method of organic fiber and polymer*

In order to verify the influence of the admixing of polymer on the physical properties of HPFRCC, compression test and bending test were conducted on specimens containing organic fiber (PVA) and polymer (SBR) individually and in combination.

Figure 2 compares the compressive strengths measured for the considered mixes. Considering that the compressive strength of Plain reaches 42 MPa, the admixing of organic fiber only is seen to increase the compressive strength by 5%. This increase is due to the partial constraint by the reinforcing fiber of the stress developed in the specimen. This observation is in agreement with the experimental of Song (2004). However, the compressive strength experiences decrease in all the mixes containing polymer. Especially, the combined admixing of polymer and fiber exhibits the largest decrease rate with a loss of about 7 MPa. This is in agreement with the experimental results of Ma (2013), which explained this loss by the delay in the generation of the cement hydrates by the polymer film.

Besides, Figure 3 reveals a pattern for the flexural strength different to that of the compressive strength. Considering that Plain develops flexural strength of approximately 6 MPa, it can be seen that the admixing of polymer only increases the flexural strength by 33% up to about 8 MPa. The adoption of fiber only results in a flexural strength of 9.6 MPa and the combined use of fiber and polymer produces the highest flexural strength with 10

Table 2. Matrix composition (in weight ratio).

Designation	Cement	FA	Sand	Water	Superplasticizer	P/C	PVA
Plain	1	0.2	0.8	0.4	0.05	0	0%
F1	1	0.2	0.8	0.4	0.05	0	1%
P5	1	0.2	0.8	0.4	0.05	0.05	0%
F1P1	1	0.2	0.8	0.4	0.05	0.01	1%
F1P3	1	0.2	0.8	0.4	0.05	0.03	1%
F1P5	1	0.2	0.8	0.4	0.05	0.05	1%

MPa. Accordingly, this verifies that the combined use of fiber and polymer increases further the tensile strength and ductility of HPFRCC.

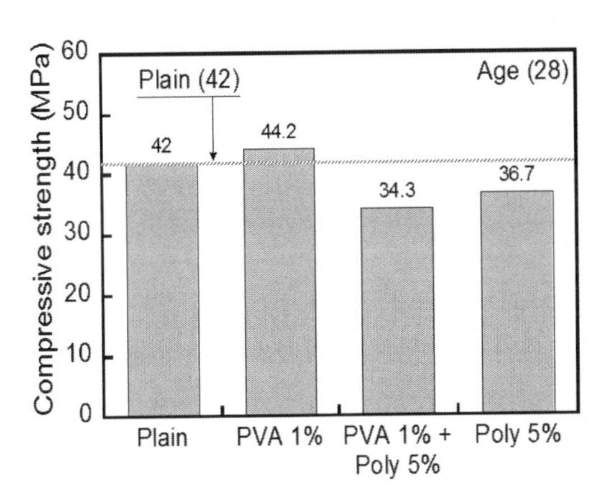

Figure 2. Compressive strength of HPFRCC containing polymer and PVA.

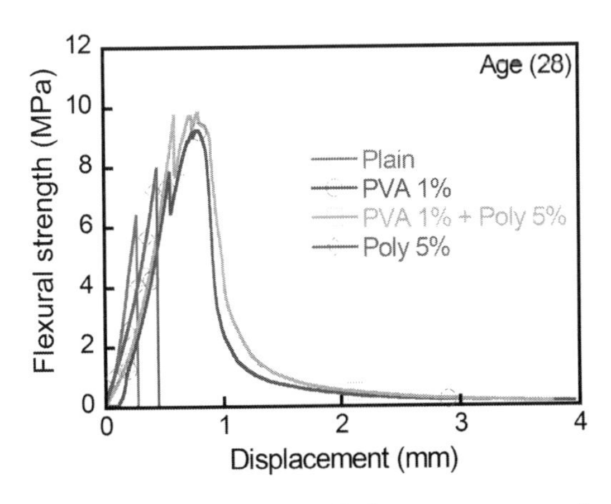

Figure 3. Flexural stress-displacement curves of HPFRCC containing polymer and PVA.

3.2 *Effect of mixing ratio and type of polymer*

Based upon the experimental results obtained in the previous section, the physical performance of HPFRCC reinforced by organic fiber is evaluated with respect to the type and mixing ratio of polymer. The corresponding compressive strengths are compared in Figure 4. It can be seen that the compressive strength tends to reduce when the polymer mixing ratio is increased from 1% to 5% and regardless of the type of polymer. On the contrary, the tensile strength in Figure 5 tends to increase by 10 to 28% with larger mixing ratio and regardless of the type of polymer. For SBR and EVA, the tensile

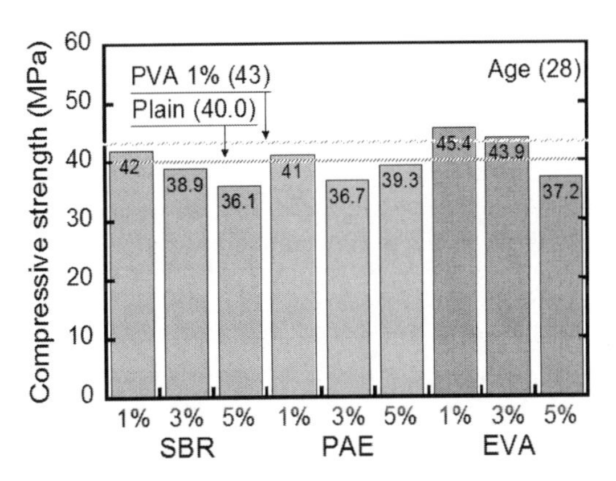

Figure 4. Compressive strength of HPFRCC according to type and mixing ratio of polymer.

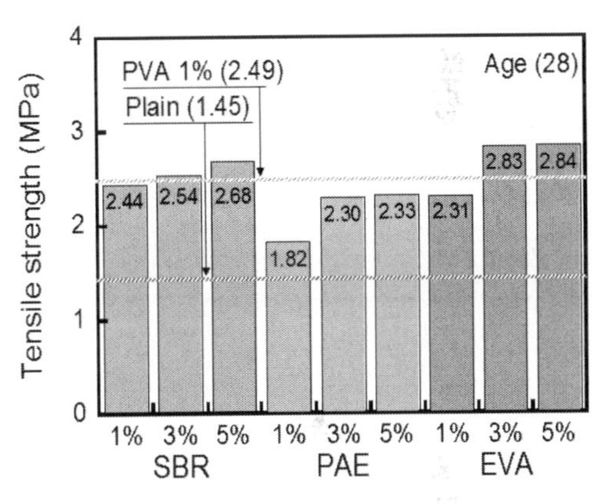

Figure 5. Tensile strength of HPFRCC according to type and mixing ratio of polymer.

strength increases at polymer mixing ratio larger than 3%. This can be explained by the improvement of the ductility of the matrix itself according to the admixing of polymer. However, for PAE, the tensile strength is seen to decrease by 6% compared to the mix using only organic fiber. Since clear pattern could not be identified in the tests involving specimens using PAE, need is for additional test and studies on the chemistry of PAE.

Figure 6 plots the flexural strength of HPFRCC with respect to the type and mixing ratio of polymer. For SBR and EVA, the flexural strength is measured to be higher by about 5% than in the mixes using only organic fiber when the polymer mixing ratio is larger than 5%. Ohama (1955) reported that, even if the compressive strength degrades due to the formation of the polymer film, the bond strength between the materials tends to increase. The results shown in Figure 6 confirm this tendency. However, similarly to the tensile strength, all the mixes using PAE are seen to develop lower

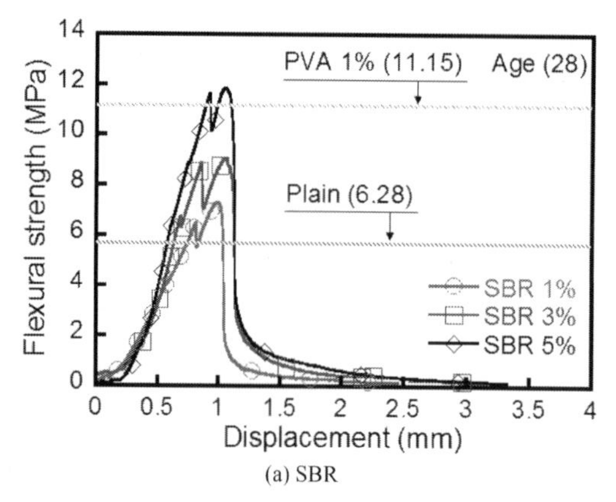

(a) SBR

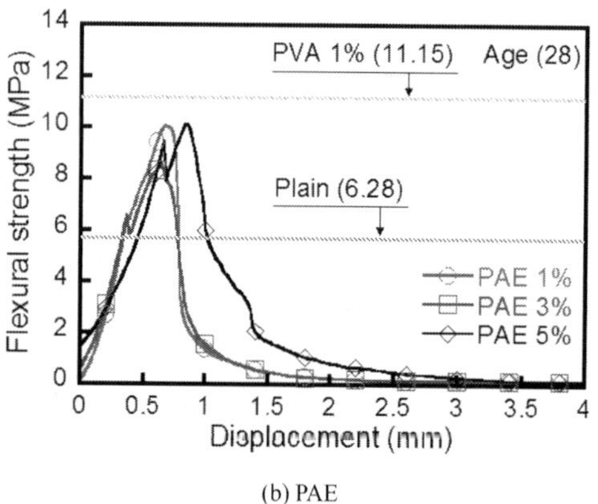

(b) PAE

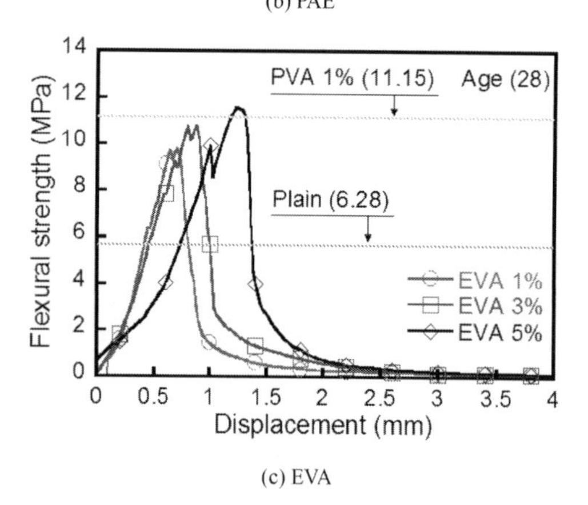

(c) EVA

Figure 6. Flexural strength of HPFRCC according to type and mixing ratio of polymer.

flexural strength than the mixes using only organic fiber. Accordingly, it appears that SBR and EVA are adequate as polymer applicable to the cementitious composite. Based on these results, the minimum mixing ratio of polymer shall be larger than 5% for the improvement of the flexural and tensile strength of HPFRCC reinforced by organic fiber.

4 CONCLUSIONS

This paper presented an experimental study on the admixing of polymer in a will to improve the tensile strength and ductility of HPFRCC. The following conclusions could be drawn from the results.

1. The combined use of polymer and organic fiber appeared to provide the best flexural and tensile strengths than the exclusive use of polymer or organic fiber.
2. The examination of the physical performance of HPFRCC according to the type of polymer revealed definite improvement of the properties according to the mixing ratio when using SBR and EVA.
3. Loss of the compressive strength up to a maximum of 18% occurred with the increase of the polymer mixing ratio from 1% to 5%. On the other hand, the flexural and tensile strengths appeared to increase with mixing ratio larger than 5%.
4. Consequently, a minimum mixing ratio of 5% can be recommended for the polymer to improve the flexural strength and ductility with regard to the mixing ratio of polymer.

ACKNOWLEDGEMENT

This research was supported by a Construction Technology Research Project 13SCIP502 (Development of impact/blast resistant HPFRCC and evaluation technique thereof) funded by the Ministry of Land, Infrastructure and Transport.

REFERENCES

ASTM C 348. 2014. Standard test method for flexural strength of hydraulic-cement mortars.

ASTM C 349. 2014. Standard test method for compressive strength of hydraulic-cement mortars (using portions of prisms broken in flexure).

Kim, D.J., Park, S.H., Ryu, G.S. & Koh, K.T. 2011. Comparative flexural behavior of hybrid ultra high performance fiber reinforced concrete with different macro fibers. *Construction and Building Materials,* 25(11): 4144–4155.

Ma, H & Li, Z. 2013. Microstructures and mechanical properties of polymer modified mortars under distinct mechanisms. *Construction and Building Materials,* 47: 579–587.

Ohame, Y. 1995. *Handbook of polymer-modified concrete and mortars properties and process technology.* Japan: Noyes Publications.

Ohama, Y. 1998. Polymer-based admixtures. *Cement and Concrete Composites,* 20(2): 189–212.

Song, P.S. & Hwang, S. 2004. Mechanical properties of high-strength steel fiber-reinforced concrete. *Construction and Building Materials,* 18(9): 669–673.

Advanced Materials, Mechanical and Structural Engineering – Hong, Seo & Moon (Eds)
© 2016 Taylor & Francis Group, London, ISBN: 978-1-138-02908-8

CVD growth of carbon nanofibers on copper particles

T.S. Koltsova, T.V. Larionova & O.V. Tolochko
Peter the Great St. Petersburg Polytechnic University, Polytechnicheskaya, St. Petersburg, Russia

ABSTRACT: Composite materials Cu-carbon nanofibers has a great potential as materials with high electrical conductivities, mechanical and tribological properties. Growth mechanism of Carbon Nanofibers (CNFs) by Chemical Vapor Deposition (CVD) on the surface of copper particles without using other catalysts is considered. The synthesized structures are analyzed by transmission electron microscopy. It is shown that the fiber structure is determined by the C: H ratio in the gas phase.

1 INTRODUCTION

Carbon nanostructures, such as Carbon Nanotubes (CNTs), Nanofibers (CNFs) and graphene, have extraordinary mechanical—stiffness, strength, and physical—thermal, electrical conductivities, properties. On this account, metal-based composites reinforced with CNTs or CNFs are being projected for use in structural applications for their high specific strength as well as functional materials for their exciting thermal and electrical characteristics (Bakshi et al. 2010). An effective method for synthesizing a composite material based on copper and carbon nanostructures was proposed in (Nasibulin et al. 2013). Carbon nanostructures—tubes, fibers, and multilayer graphene, were synthesized directly on copper particles without additional catalyst by deposition from the gas phase. This method allows to exclude such complex operations, as functionalization of carbon nanostructures and their introduction into the matrix. The Cu-CNF compact composite, prepared by the proposed method, have the wear reduced by more than twice compared to that of traditional Cu-graphite samples practically retaining the copper electrical properties (Larionova et al. 2014).

Synthesis of carbon nanostructures from the gas phase on metal powders of the iron subgroup was widely discussed in the literature (see, e.g. (Helveg et al. 2004, Hofmann et al. 2007)). At present, there exist several models of CNT and CNF growth on the surface of iron, cobalt, nickel, and other materials, but in the presence of a catalyst of the iron subgroup. It is believed that the growth of nanofibers is determined by the limited solubility of carbon and its diffusion in the metal phase and can occur with a carbide forming cycle as well as in its absence with the formation of a solid solution. However, these models are inapplicable for carbon nanostructures growth on copper particles in view of the zero solubility as well as diffusion of carbon in copper. In this study, we trace and analyze the regularities of formation of carbon nanostructures on copper particles.

2 EXPERIMENTAL

Carbon nanostructures were synthesized in accordance of the technique described earlier (Nasibulin et al. 2013) in acetylene–hydrogen media for the C:H ratio from 1:5 to 1:42. The total gas flow rate was about 400 cm^3/min. The temperature interval of synthesis was 600–950°C. After synthesis, cooling to room temperature was carried out in an inert atmosphere. For the copper matrix, we used copper powder of the dendrite form with a particle size of 5–10 μm and purity of 99.7%. Electron microscopic studies were fulfilled with JEOL2010 HRTEM and Philips CM200FEG microscopes. The Raman spectra were recorded at room temperature; for the source of monochromatic radiation, the YAG laser (with a wavelength of 532.25 nm and a power of 30 mW) was used.

3 RESULTS AND DISCUSSION

Figure 1 shows the dependences of the copper powder mass gain during the CVD on the synthesis temperature for various compositions of the gas phase. In view of active sintering of copper, the experiment was limited to a temperature of 940°C.

Typical temperature dependences are curves with maximums; the position of the peak is determined by the composition of the gas phase. In the temperature interval on the left of the peak, the amount of deposited carbon is determined by the thermodynamics of the catalytic decomposition of hydrocarbons; (the dependences

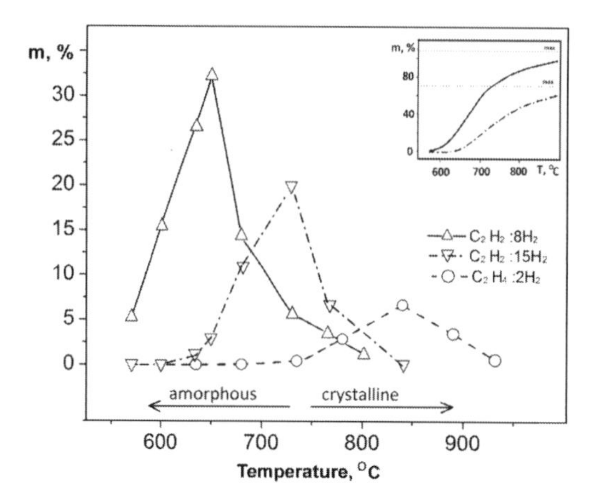

Figure 1. Experimental dependence of specimen weight gain on the temperature and gas content (synthesis duration 20 min). The inset shows the dependences calculated for C_2H_2:8H_2 and C_2H_2:15H_2 in accordance with thermodynamics.

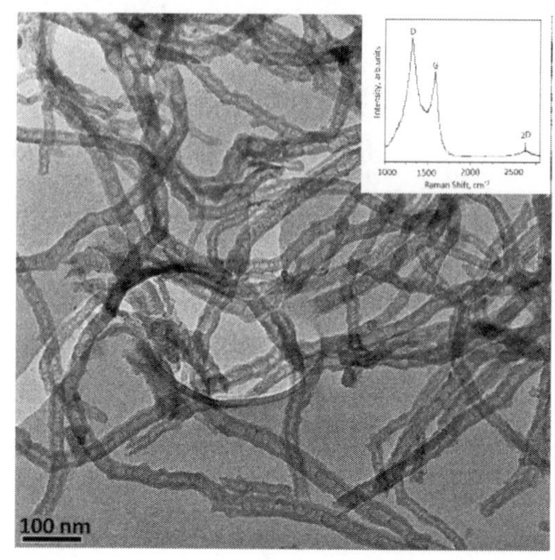

Figure 2. TEM image of CNFs synthesized at 765°C for 20 min in C_2H_2:15H_2. Inset: Raman spectrum.

calculated in accordance with thermodynamics are shown in the inset). Carbon deposited at these temperatures is in the amorphous state. A decrease in the mass increment of deposited carbon upon an increase of temperature (right of the peak) can be explained by a transition to the adsorption control regime of synthesis (Lebedeva et al. 2011). At these temperatures, carbon formed in the gas phase is predominantly carried out with the gas flow or deposited on the cold parts of the reactor. The carbon deposited on copper particles at these temperatures crystallizes forming carbon nanofibers or multilayer graphene containing from 5 to 8 graphene monolayers. The images of carbon deposits synthesized in this temperature region are shown in Figure 2a. The samples obtained in this way were tested by the Raman scattering method; typical spectrum is shown in inset of Figure 2.

Examples of CNF images obtained in a transmission electron microscope is shown in Figure 3. The CNF structure has two regions: the central region, which is formed by carbon layers located at an angle to the symmetry axis, and the periphery region in which carbon layers are parallel to the fiber axis. In addition, amorphous carbon is always present on the surface of the fiber.

According to the results of microscopy, an increase in the concentration of carbon in the gas phase leads to the formation of a large number of periphery layer and, as a consequence, to an increase in the outer diameter of the fiber and the total mass of the product. The diameter of the central region remains almost unchanged and is 45–50 nm.

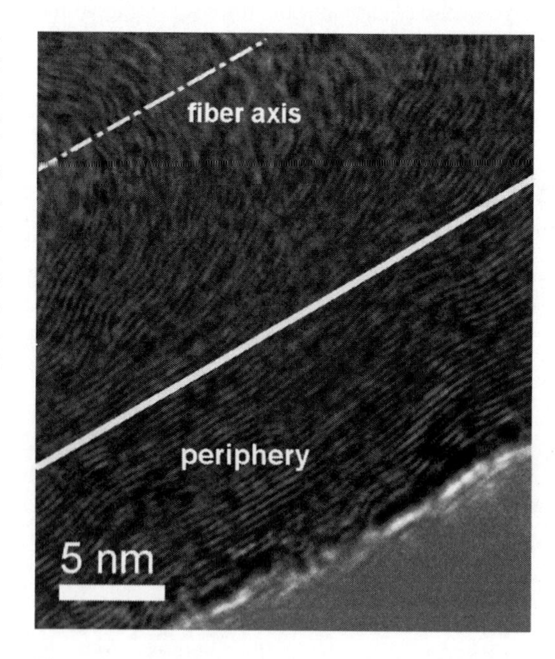

Figure 3. TEM image of carbon nanofiber, synthesized at 940°C for 20 min.

Apart from the number of periphery carbon layers, the carbon concentration in the gas phase determines the central structure of the CNFs. Figure 4 demonstrates a dependence of the outer diameter of the CNF and the angle between the graphene layers in the central part of the fiber on the composition of the gas phase. Upon a change in the concentration of carbon atoms in the gas

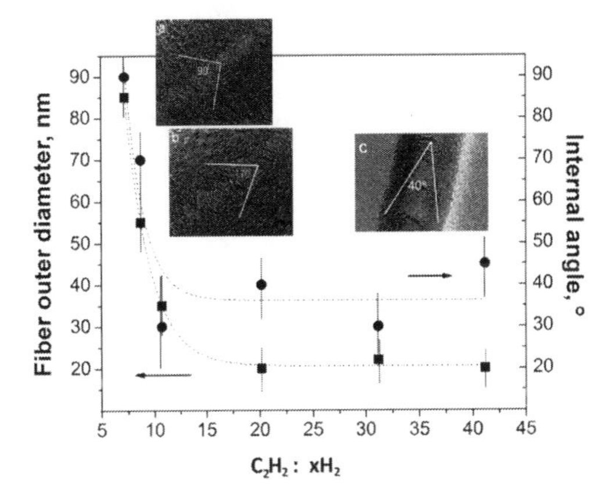

Figure 4. Dependences of the fibers internal angle and outer diameter on gas phase composition. The insets show TEM images of the fibers, synthesized at $C_2H_2:H_2$ = 1:7 (a), 1:8.5 (b) and 1:41 (c).

phase, the angle between carbon layers in the central part of the fiber changes from 90 to 40°. The insets to Figure 4 show the images of the fibers synthesized for different compositions of the gas phase: for C_2H_2: $7H_2$, the angle between carbon layers is 90°, and the structure of the fiber is something in between a "bamboo" fiber and a "stack of cups". Upon an increase in the hydrogen concentration, the fiber acquires a more clearly manifested "bamboo" structure, the angle between graphene layers decreases to 40°, and the total amount of the carbon product diminishes due to a decrease in the number of periphery carbon layers. A further decrease in the concentration of carbon in the gas mixture does not lead to visible changes in the structure of the carbon fiber, but decreases the amount of carbon deposit down to zero.

High concentration of carbon in gas phase results in more pronounced deposition and formation of less crystalline product.

Figure 5 shows the TEM image of a copper particle with carbon layers synthesized on it. The detachment of the carbon layer from the copper surface, which is observed in the upper part, apparently occurs during cooling and is due to the difference between the thermal expansion coefficients of copper and carbon (Li et al. 2009). The nucleation of the fiber is observed in the left conical part; there is no detachment of the carbon layer in this part in spite of the large thickness.

The interplanar distance measured from the TEM image for crystallographic planes of copper parallel to the surface in the conical part of the particle is 0.200 ± 0.01 nm, which closes to the Cu (111) planes (d = 0.208 nm). Arrangement of atoms in the Cu (111) plane is in conformity with the arrangement of atoms on the C (0002) plane.

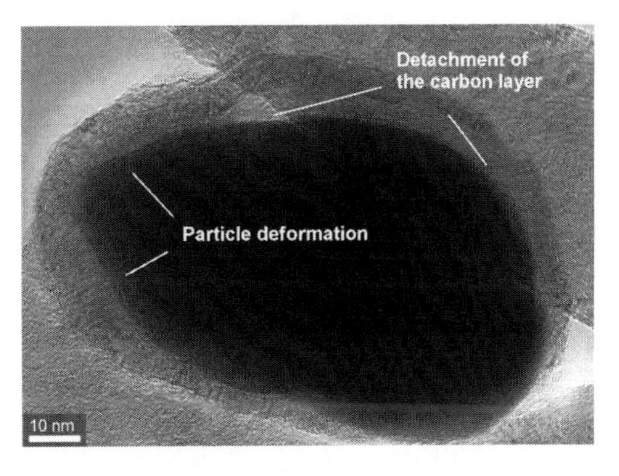

Figure 5. TEM images of the copper particles, carbon layers and fibers.

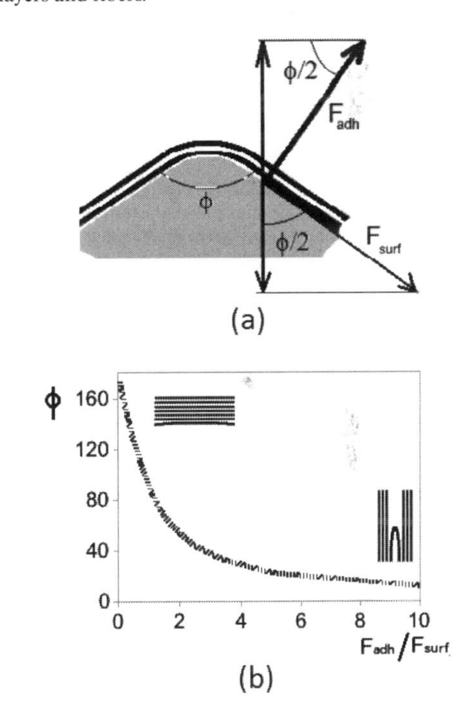

Figure 6. Scheme of the internal angle formation (a) and dependence of internal angle on surface tension and adhesion relation (b).

These planes may play the role of the substrate for coherent nucleation of graphene. The coherence of the interface is the reason for stronger adhesion of the carbon layer on the surface sites of Cu (111).

As can be seen from Figure 5, during the growth of the fiber, local deformation of the copper particle takes place in the region of its nucleation; the detachment of the nucleus from the particle is accompanied by the restoration of the particle shape and the formation of a new nucleus. Deformation is stronger, the smaller the surface tension of copper and the stronger the adhesion between the carbon layer and the copper surface. When the surface tension exceeds adhesion, the carbon layer is separated from the deformed surface, which causes the bam-boo structure of the fiber. The simplified scheme of this representation is shown in Figure 6a. Judging from the diagram in Figure 6a, the angle between graphene layers in the central part of the fiber is determined by the relation between adhesive forces between the graphene layer on the particle surface and surface tension of copper Figure 6b).

Apparently, high concentration of carbon in the gas phase results in more pronounced deposition and formation of more defective carbon layers with weak coherence and correspondingly low adhesion to Cu (111). Supposing the same Cu surface tension for all the particles as independent on the gas phase, the adhesion will determine the center structure of the fiber (Figure 6b).

4 CONCLUSION

We have investigated the synthesis of carbon nanostructures from the gas phase on the surface of copper particles. Typical temperature dependences are curves with maximums; the position of the peak is determined by the composition of the gas phase. In the temperature interval on the left of the peak, the amount of deposited carbon is determined by the thermodynamics of the catalytic decomposition of hydrocarbons. A decrease of deposited carbon upon an increase of temperature (right of the peak) can be explained by a transition to the adsorption control regime of synthesis.

The structure of the fiber is mostly determined by the C: H ratio in the gas phase: for higher concentrations of carbon, thicker fibers with a large internal angle (of the type of a stack of cups) are formed; upon an increase in the hydrogen concentration, the fiber acquires bamboo structure. We have proposed a model for the formation on the central angle in the fiber depending on the relation between the surface tension of copper and adhesion of the carbon layer.

REFERENCES

Bakshi, S.R. Lahiri, D. & Agarwal, A. 2010. Carbon nanotube reinforced metal matrix composites—a review. *International Materials Reviews,* 55(1): 41–64.

Helveg, S., LopezCartes, C., Sehested, J., Hansen, P.L., Clausen, B.S., RostrupNielsen, J.R., AbildPedersen, P. & Norskov, J.K. 2004. Atomic-scale imaging of carbon nanofibre growth. *Nature,* 427: 426–429.

Hofmann, S., Sharma, R., Du, G., Matteri, C., Cepek, C., Cantoro, M., Ducati, C., Pisana, S., Parvez, A., Ferrari, A.C., Dunin-Borkowski, R., Lizzit, S., Petaccia, L., Goldoni, A. & Robertson, J. 2007. In-situ observations of catalyst dynamics during surface-bound carbon nanotube nucleation. *Nano Letter,* 7: 602–608.

Larionova, T., Koltsova. T., Fadin, Yu. & Tolochko, O. 2014. Friction and wear of copper–carbon nanofibers compact composites prepared by chemical vapor deposition. *Wear,* 319: 118–122.

Lebedeva, I.V., Knizhnik, A.A., Gavrikov, A.V., Baranov, A.E., Potapkin, B.V., Aceto, S.J., Bui, P.-A., Eastman, C.M., Grossner, U., Smith, D.J. & Sommerer, T.J. 2011. First-principles based kinetic modeling of effect of hydrogen on growth of carbon nanotubes. *Carbon,* 49: 2508–2521.

Li, X., Cai, W., An, J., Kim, S., Nah, J., Yang, D., Piner, R., Velamakanni, A., Jung, I., Tutuc, E., Banerjee, S., Colombo, K. & Ruoff, L.R.S. 2009. Large-area synthesis of high-quality and uniform graphene films on copper foils. *Science,* 324: 1312–1314.

Nasibulin, A.G., Nasibulina, L.I., Anoshkin, I.V., Kauppinen, E.I., Koltsova, T., Semencha, A. & Tolochko, O.V. 2013. A novel approach to composite preparation by direct synthesis of carbon nanomaterial on matrix or filler particles. *Acta Materialia,* 61(6): 1862–1871.

Advanced Materials, Mechanical and Structural Engineering – Hong, Seo & Moon (Eds)
 ISBN: 978-1-138-02908-8

Evaluation of methacrylic acid based polymers as green inhibitors for Reverse Osmosis

A. Pervov, A. Andrianov, V. Chukhin & R. Efremov
Department of Water Supply, Moscow State University of Civil Engineering (National Research University), Moscow, Russia

ABSTRACT: The formation of calcium carbonate and calcium sulfate scales is a very important problem in reverse osmosis plants operation. To reduce scaling, inhibitor dosing is commonly used in RO desalination. In the present work several pilot samples of environmentally friendly low phosphorus inhibitors based on acrylic and methacrylic acids, maleic anhydride and sucrose allyl ether was prepared and tested in comparison with well-known RO chemicals. Scaling tests were conducted using commercial RO module in circulation mode with gradual concentrating of initial solution. The highest inhibitor efficiency was demonstrated by co-polymer of methacrylic acid and maleic anhydride at concentration values of 3 mg/L and higher than that. The rarely cross-linked co-polymer of methacrylic acid and sucrose allyl ether (marked as RPAC-4) and rarely cross-linked co-polymer of acrylic acid and sucrose allyl ether (CAAC) also showed good results that were similar with conventional inhibitors such as OEDP, NTP, and Aminat-K (based on phosphonic acids).

1 INTRODUCTION

Reverse Osmosis (RO) is an advanced membrane technique used as a unique tool to treat water from different sources as well as to recycle and reclaim wastewater; -due to its efficiency it gives way to reject various water contaminants. Meanwhile, membrane systems when applied into practice suffer certain operational problems (such as high chemical and power consumption, pretreatment costs and large effluent disposal) that could be attributed to membrane fouling and measures to prevent it. One of the major problems of RO is formation of calcium carbonate and calcium sulfate dihydrate scales during RO plant operation. To prevent scaling on membrane surface the inhibitor dosing is widely used in RO desalination. These chemicals are commonly based on phosphonic or phosphoric acid that remains in the rejected stream (the reverse osmosis concentrate) and are discharged into surface reservoirs, which creates a serious environmental problem associated with eutrophication (Lattemann & Höpner 2008; Feiner et al. 2015).

The trend of the last decade is synthesis of new generation of environmentally friendly inhibitors that do not contain phosphates and other nutrients, biodegradable in natural environment and efficient for scale prevention as well.

Nowadays a large number of antiscalants is produced all over the world. These chemicals can be classified into the following groups:

1. Polyphosphates;
2. Complexons including phosphonates;
3. Polymers (polyacrylic acid, polymethacrylic acid, polymaleic acid, polyaspartic acid, polymethyl methacrylate, polyethyl methacrylate etc.);
4. Co-polymers.

Polymers and co-polymers are used to synthesis a green scale inhibitors. Husson et al. (2011) states that the most promising green antiscalants are currently based on polyaspartic acid which does not contain nitrogen and phosphorus and is sufficiently biodegradable (Gao et al. 2009). Numerous studies that were carried out under various conditions also showed high efficiency of polyaspartic acid for calcium sulfate scale inhibition (Hasson et al. 2011; Ali et al. 2015; (Shemer & Hasson 2014; Quan et al. 2008; Chaussemier et al. 2015). To improve the performance of polyaspartic acid based on inhibitors are modified with the attachment of open-chain polysuccin-imide functional groups (Chen et al. 2015).

Another green scale inhibitor is a polyepoxysuccinic acid (Sun et al. 2009). Similarly to a polyaspartic acid it has the favorable characteristics: does not contain nitrogen and phosphorus and is readily biodegradable. This inhibitor can be used for various water compositions. In (Liu et al. 2012) it was indicated that polyepoxysuccinic acid gives the best effect with respect to $CaCO_3$ and $SrSO_4$ and polyaspartic acid gives the best effect with respect to $CaSO_4 \cdot 2H_2O$ and $BaSO_4$. Nevertheless, wide industrial tests of polyepoxysuccinic acid based inhibitors have not yet been conducted and the performance of these compounds requires further confirmation (Quan et al. 2008).

A third promising class of inhibitors is a biodegradable and non-toxic polysaccharide-based poly-carboxylates which are obtained from inulin by chemical synthesis (e.g. carboxymethylinulin) (Kir-boga & Öner 2012). These compounds have significant inhibition effect on calcium carbonate crystallisation due to the presence of carboxylic acid groups in its structure (Chaussemier et al. 2015).

A number of studies (Wang et al. 2014; Amjad & Koutsoukos 2014; Popuri et al. 2014) were devoted to maleic acid based scale inhibitors. Thus Amjad & Koutsoukos (2014) studied 14 inhibitors based on acrylic and maleic acid and obtained the highest inhibitory ability at low doses (about 2 mg/L) that showed a low molecular weight (~2000 MW) compounds—polyacrylic acid, polymaleic acid and other polymers with free carboxyl groups. More complex compounds created on the maleic acid base exhibit better scale control properties as well—for example hydrolysed polymaleic anhydride and its mixtures with 1-hydroxyethane-1, 1-diphosphonic acid and polyacrylic acid (Shen et al. 2012).

Other traditional scale inhibitors—polycarboxylates—also have high efficiency and environmental suitability. The most common are polyacrylates and methacrylates with a molecular weight ranging between 5000 and 6000 g/mol and polymers based on polyacrylates (Antonya et al. 2011). Polyacrylates are highly effective at the stage of nucleation and crystallization for scale-producing salts due to their adsorption on the growing seed crystals.

In addition to these simple compounds, a number of studies have been devoted to the development of complex mixed polymeric scale inhibitors (co-polymers). Using a set of polymers with different properties to synthesize inhibitors enhances its effectiveness and application range (Wang et al. 2014). Popuri et al. (2014) used synthesized green inhibitors—copolymers of maleic and citric acids to prevent precipitation of calcium phosphate and Ling et al. (2012) tested a new non-phosphorus inhibitor—Acrylic Acid (AA)-allyloxy poly(ethylene glycol) polyglycerol carboxylate copolymer—to prevent precipitation of calcium carbonate, sulfate, and phosphate in cooling systems.

Recent advances in the scale inhibitor development were in the field of synthesizing dendritic polymers with highly branched three-dimensional structure (rarely cross-linked polymers), some of which could be used as an environmentally friendly inhibitors (Demadis et al. 2005).

Based on state-of-the-art review, the availability of raw materials in Russia for antiscalant production and existing experience in polymer synthesizing, the selection has been made in favor of acrylic and methacrylic acid co-polymers with biodegradable fragments to be used as low phosphorus inhibitors.

2 EXPERIMENTAL

2.1 *Materials*

In order to obtain various spatial structure of polymers that meets the required properties of antiscalant, the polymerization reactions of acrylic and methacrylic acids used the following cross-linking agents: N, N'-methylene-bis-acrylamide; polyallyl ethers of pentaerythritol (the main component is a pentaerythritol triallyl ether) and polyallyl sucrose (the main component is hexa-polyallyl sucrose).

These cross-linking agents were chosen due to the peculiarities in formation of different spatial polymer structures. Since N, N'-methylene-bis-acrylamide has two reaction sites for the polymerization process and it is commonly used in the synthesis of polymeric hydrogels based on the acrylic acid. A higher branched structure (as for dendrimers synthesis) is achieved using polyallyl ethers of pentaerythritol, but the most extensive structure will be exhibited for polyallyl sucrose derivatives. The general advantage of the selected cross-linking agents is the presence of a central group which joints the polymer chains branches.

Six samples of new polymers were synthesized:

- RPAC-1 (rarely cross-linked polymer based on acrylic acid and N, N'-methylene-bis-acrylamide as a cross-linking agent);
- RPAC-2 (rarely cross-linked polymer based on methacrylic acid and N, N'-methylene-bis-acrylamide as a cross-linking agent);
- RPAC-3 (rarely cross-linked co-polymer based on methacrylic acid and allyl ether of pentaerythritol as a cross-linking agent);
- RPAC-4 (rarely cross-linked co-polymer of methacrylic acid and allyl ether of sucrose as a cross-linking agent);
- MAAC (co-polymer of maleic anhydride and methacrylic acid);
- CAAC (rarely cross-linked copolymer of acrylic acid and allyl ether of sucrose as a cross-linking agent).

The reference inhibitors were xyethylidenediphosphonic acid (OEDP), nitrilotrimethyl phosphon-ic acid (NTP) and Aminat-K (a mixture of sodium salts of nitrilotrimethyl-phosphonic and me-thyliminobis-methylenephosphonic acids).

All chemicals except Aminat-K were dry (powder or crystals) and were prepared for dosage as 10 or 20 mg/ml solutions using distilled water. Water solution of RPAC-4, MAAC, and CAAC were prepared by swelling in distilled water during these 24 hours.

All samples were synthesized and characterized (using the NMR and IR spectroscopy) by Fine Chemicals Research Center (Moscow, Russia). Scaling experiments was conducted in the Water treatment laboratory of Department of Water Supply, Moscow State University of Civil Engineering (National Research University).

2.2 *RO membrane scaling experiments*

Membrane scaling tests were carried out using the commercial 4040 spiral wound membrane module (model ERN-B-45-300, ZAO STC "Vladipor", Russia) assembled on laboratory membrane unit (Figure 1). The membrane element is rolled using reverse osmosis ESPA membranes with selectivity up to 98.5% (on 0.15% NaCl test solution).

The feed solution (tap water or model solution) is placed in the feed water tank (1) and delivered to membrane module via centrifugal multistage pump (2). The transmembrane pressure, cross-flow and recovery rate is adjusted by valves (10, 11, and 12) and controlled by pressure gauges (6) and flowmeters (7 and 9).

All scaling tests were conducted in circulation mode whereby reject flow (concentrate) was returned to the feed water tank (1) and permeate is collected in separate tank (4). The transmembrane pressure was maintained at 7.0 ± 0.2 bars. The

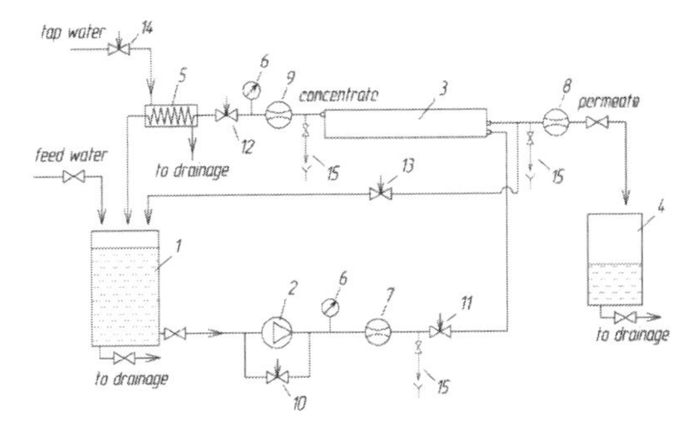

Figure 1. Schematic diagram of laboratory RO unit for membrane scaling tests: 1–feed water tank; 2–pump; 3–spiral wound membrane module; 4–permeate tank; 5–heat ex-changer; 6–pressure-gauge; 7–feed water rotameter; 8–permeate rotameter; 9–concentrate rotameter; 10–by-pass adjusting valve; 11–feed water adjusting valve; 12–concentrate adjusting valve; 13–cooling water adjusting valve; 14–sampler.

product flux, depending on tap water temperature, varied from 100 to 150 liters per hour. The virtual selectivity on the tap water was 97.5…98.0%. The volume of feed solution was 80 ± 2 liters. Concentrate flow was kept constant at 100 ± 10 l/h and the recovery rate was ranging from 50 to 60%. In order to expand scaling time and escape flux decrease caused by osmotic pressure raised the circulating solution as a part of product water was returned to feed water tank and the other part of the product flow was directed to the permeate tank constantly kept changing.

Experiments were carried out with Moscow tap water from April 2015 to May 2015. During this period the tap water had quite stable quality and its TDS was 246…266 ppm, the total hardness was 3.1…3.4 meq/L (155…170 ppm of $CaCO_3$), total alkalinity was 2.5…2.9 meq/L, calcium was 2.2…2.5 meq/L, pH was 7.75…8.2, sulphates were 10…13 mg/L, chlorides were 8…10 mg/L.

The samples are taken for initial feed solution from tank (1), for circulating solution—from tank (1) (for various concentration ratios) and for permeate—from tank (4) (one sample characterizing the averaged quality of product water). In all samples the temperature, TDS (conductivity), pH, the total hardness, the total alkalinity and calcium are determined.

To restore membrane element performance and to remove accumulated scales, every 10–15 tests chemical washing was conducted using citric acid or EDTA.

Scaling experiments were conducted for the series of new antiscalants and selected reference scale inhibitors. The dosages were: 0 mg/L (without antiscalant), 3 mg/L, 5 mg/L, and 10 mg/L (0; 3; 5 и 10 ml/L respectively for liquid Aminat-K). Most tests were repeated twice to improve the accuracy of the results.

The amount of scales of $CaCO_3$ and $CaSO_4$ (ex-pressed as Ca_2^+ in meq or mg) accumulated in membrane module were calculated as the difference between initial amount of calcium in feed solution and sum of amount of calcium in concentrate (circulating solution) and permeate (Pervov 1999):

$$M_{Ca} = V \cdot C_{Ca} - (V_c^t \cdot C_{cCa}^t + V_p^t \cdot C_{pCa}) \qquad (1)$$

where, M_{Ca} = amount of calcium accumulated in membrane module (meq); V = feed solution volume (l); C_{Ca} = concentration of calcium in feed solution (meq/L); V^t_c, V^t_p = volume of circulating solution and total permeate respectively for time *t* (l); C^t_c C_a, $C^t_{p\,Ca}$ = concentration of calcium in circulating solution and total permeate respectively for time *t* (meq/L).

This difference value was calculated for concentration ratios 2, 3, and 5. The total hardness and total alkalinity were determined to control the correctness of other parameters determination.

The amount of calcium can be converted to mass of calcium carbonate (mg):

$$M_{CaCO_3} = M_{Ca^{2+}} \cdot 50 \quad (2)$$

Antiscalant efficiency as a calcium carbonate inhibitor was calculated by using the following equation:

$$E(\%) = \frac{M_{CaCO_3}^{\text{blank}} - M_{CaCO_3}^{\text{antiscalant}}}{M_{CaCO_3}^{\text{blank}}} \cdot 100 \quad (3)$$

where, $M_{CaCO_3}^{\text{blank}}, M_{CaCO_3}^{\text{antiscalant}}$ = mass of calcium carbonate accumulated in membrane module in the absence of scale inhibitor and with inhibitor dosing respectively (meq or mg).

3 RESULTS AND DISCUSSION

The relationships between the accumulated scale amount, inhibitor concentration and recoveries (determined as concentration ratio) for selected antiscalants is shown in Figure 2. The summary results of inhibitors efficiency are listed in the Table 1.

The preliminary results showed that for a number of tested scale inhibitors (OEDP, NTP,

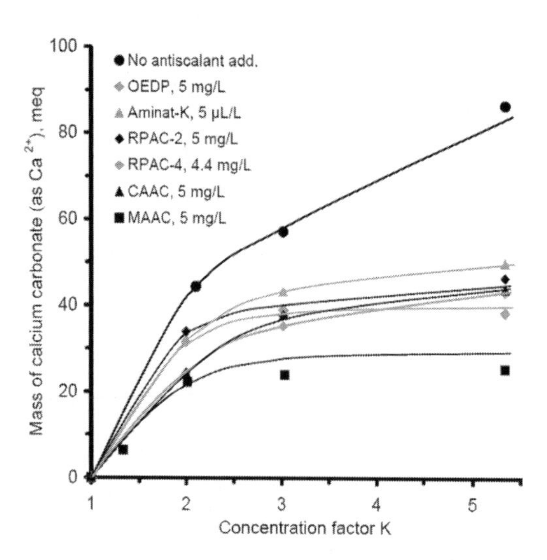

Figure 2. The inhibition efficiency of selected inhibitors for concentration 5 mg/L.

Table 1. The scale inhibitors efficiency in reverse osmosis tests (concentration factor is 5.3).

Inhibitor	Dose, mg/L (µL/L)	Inhibitor efficiency, %
Aminat-K	3	32 ± 2
	5	50 ± 3
	10	56 ± 3
OEDP	3	45 ± 3
	5	62 ± 4
	10	64 ± 4
NTP	3	46 ± 3
	5	62 ± 3
	10	49 ± 3
RPAC-1	5	47 ± 3
RPAC-2	5	50 ± 4
RPAC-3	5	40 ± 3
RPAC-4	4.4	57 ± 4
MAAC	3	62 ± 3
	5	75 ± 3
	10	69 ± 3
CAAC	5	56 ± 3

Aminat-K) the concentration of 3 mg/L is not sufficient to effectively inhibit calcium carbonate precipitation. The concentration of 10 mg/L has no significant impact on improving the inhibition efficiency compared with concentration of 5 mg/L, so most of the samples (CAAC and RPAC series) were tested at optimal concentration of 5 mg/L.

For dose of 5 mg/L all studied scale inhibitors can be arranged in the descending order of effectiveness: MAAC > OEDP > NTP ≈ RPAC-4 ≈ CAAC > Aminat-K ≈ RPAC-2 > RPAC-1 > RPAC-3. The highest efficiency was achieved for OEDP, RPAC-4, CAAC and MAAC. At the same time MAAC demonstrated the highest efficiency at the lowest dose of 3 mg/L (Figure 3a).

In the next step several tests were carried out for the best sample MAAC with lower concentrations –1, 2, 3, and 4 mg/L (Figure 3b). For concentration of 1 and 2 mg/L the MAAC efficiency was low, but comparable to that of other tested samples in range of 5...10 mg/L. Best results were obtained with doses of 3 mg/L or more.

The relationships between the mass of accumulated scale versus concentration ratio (Figure 3b) indicates that MAAC efficiencies were similar at relatively low feed water hardness (7...8 meq/L) and feed water concentration ratio up to 3, but with further increase in the concentration ratio and hardness values (up to 12...14 meq/L) the mass of accumulated scale becomes inversely proportional to the inhibitor concentration (Figure 3b).

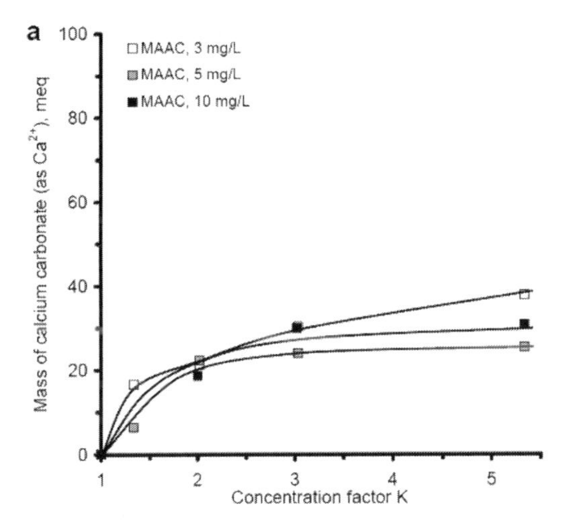

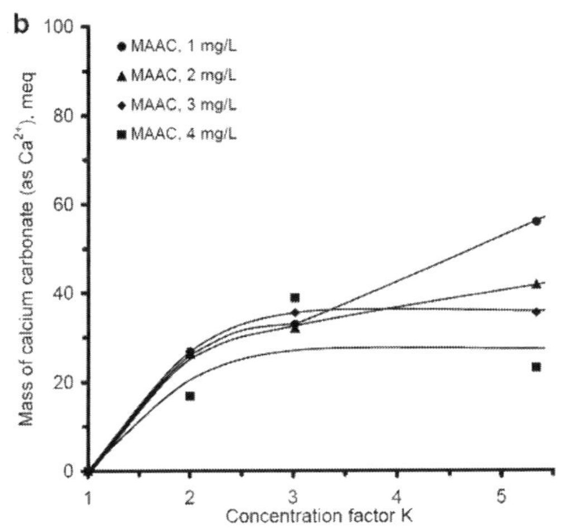

Figure 3. The inhibition efficiency of MAAC for high (a) and low (b) concentrations.

4 CONCLUSION

The trend of the last decade is development of environmentally friendly scale inhibitors: low phosphorus and biodegradable. The most promising compounds for synthesis of new inhibitors are polyaspatic acid, polyepoxysuccinic acid, polysaccharide-based polycarboxylates, polymaleic and maleic acid, methacrylic acid, etc.

Several samples of pilot low phosphorus inhibitors based on co-polymers of acrylic and methacrylic acid was prepared and tested to prevent calcium carbonate and calcium sulfate deposition in RO membrane module in comparison with three reference conventional inhibitors NTP, OEDP, and Aminat-K. The best performance was demonstrated by rarely cross-linked co-polymer of methacrylic acid and sucrose allyl ether (RPAC-4), rarely cross-linked co-polymer of acrylic acid and sucrose allyl ether (CAAC) and co-polymer of methacrylic acid and maleic anhydride (MAAC). The inhibition efficiencies of these polymers were equal or even higher than of conventional OEDP, NTP, and Aminat-K. At the lower doses (≤3 mg/L) MAAC showed the superior antiscaling efficiency.

Thus, several synthesized green inhibitors demonstrated quite good antiscaling efficiencies and could be considered very competitive with conventionally used antiscalants and in some cases (MAAK) even superior to them. Nevertheless, to create more efficient and economically attractive commercial scale inhibitors, further research studies are required.

ACKNOWLEDGMENT

This research was financed by Federal Targeted Program "Research and development in priority areas of advancement of Russian scientific and technological complex for 2014–2020" (Grant agreement No 14.582.21.0007).

REFERENCES

Ali, S.A., Kazi, I.W. & Rahman, F. 2015. Synthesis and evaluation of phosphate-free antiscalants to control $CaSO_4\,2H_2O$ scale formation in reverse osmosis desalination plants. *Desalination*, 357: 36–44.

Amjad, Z. & Koutsoukos, P.G. 2014. Evaluation of maleic acid based polymers as scale inhibitors and dispersants for industrial water applications. *Desalination*, 335: 55–63.

Antonya, A., Low, J.H., Gray, S., Childress, A.E., Le-Clech, P. & Leslie, G. 2011. Scale formation and control in high pressure membrane water treatment systems: A review. *Journal of Membrane Science*, 383: 1–16.

Chaussemier, M. et al. 2015. State of art of natural inhibitors of calcium carbonate scaling. A review article. *Desalination*, 356: 47–55.

Chen, J., Xu, L., Han, J., Su, M. & Wu, Q. 2015. Synthesis of modified polyaspartic acid and evaluation of its scale inhibition and dispersion capacity. *Desalination*, 358: 42–48.

Demadis, K.D., Neofotistou, E., Mavredaki, E., Tsiknakis, M., Sarigiannidou, E.-M. & Katarachia, S.D. 2005. Inorganic foulants in membrane systems: chemical control strategies and the contribution of green chemistry. *Desalination*, 179: 281–295.

Feiner, M., Beggel, S., Jaeger, N. & Geist J. 2015. Increased RO concentrate toxicity following application of antiscalants—acute toxicity tests with the amphipods *Gammarus pulex and Gammarus roeseli. Environ. Pollution*, 197: 309–312.

Gao, Y., Liu, Z., Zhang, L. & Wang Y. 2009. *Synthesis and performance research of biodegradable modified polyaspartic acid*. Presented on: 3rd International Conference on Bioinformatics and Biomedical Engineering, Beijing, China, Jun 11–12, 2009.

Hasson, D., Shemer, H. & Sher, A. 2011. State of the art of friendly green scale control inhibitors. *Industrial & Engineering Chemistry Research*, 50: 7601–7607.

Kirboga, S. & Öner, M. 2012. The inhibitory effects of carbox-ymethyl inulin on the seeded growth of calcium carbonate. *Colloids and Surfaces B: Biointerfaces*, 91: 18–25.

Lattemann, S. & Höpner, T. 2008. Environmental impact and impact assessment of seawater desalination. *Desalination*, 220: 1–15.

Ling, L. et al. 2012. Carboxylate-terminated double-hydrophilic block copolymer as an effective and environ-mental inhibitor in cooling water systems. *Desalination*, 304: 33–40.

Liu, D., Dong, W., Li, F., Hui, F. & Lédion, J. 2012. Comparative performance of polyepoxysuccinic acid and polyaspartic acid on scaling inhibition by static and rapid controlled precipitation methods. *Desalination*, 304: 1–10.

Pervov, A.G. 1999. A simplified RO process design based on understanding of fouling mechanisms. *Desalination*, 126: 227–247.

Popuri, S.R., Hall, C., Wang, C.C. & Chang, C.Y. 2014. Development of green/biodegradable polymers for water scaling applications. *International Biodeterioration & Biodegradation*, 95(Part A): 225–231.

Quan, Z., Chen, Y., Wang, X., Shi, C., Liu, Y. & Ma, C.F. 2008. Experimental study on scale inhibition performance of a green scale inhibitor polyaspartic acid. *Science in China Series B: Chemistry*, 51(7): 695–699.

Shemer, H. & Hasson, D. 2014. *Characterization of the inhibitory effectiveness of prominent environmentally friendly antiscalants*. Presented at the Conference on Desalination for the Environment: Clean Water and Energy, 11–15 May 2014, Limassol, Cyprus.

Shen, Z., Li, J., Xu, K., Ding, L. & Ren, H. 2012. The effect of synthesized hydrolyzed polymaleic anhydride (HPMA) on the crystal of calcium carbonate. *Desalination*, 284: 238–244.

Sun, Y.H., Xiang, W.H. & Wang, Y. 2009. Study on polyepoxysuccinic acid reverse osmosis scale inhibitor. *Journal of Environmental Sciences*, 21: 73–75.

Wang, H., Zhou, Y., Yao, Q., Ma, S., Wuc, W. & Sun, W. 2014. Synthesis of fluorescent-tagged scale inhibitor and evaluation of its calcium carbonate precipitation performance. *Desalination*, 340: 1–10.

Advanced Materials, Mechanical and Structural Engineering – Hong, Seo & Moon (Eds)
© 2016 Taylor & Francis Group, London, ISBN: 978-1-138-02908-8

A carbon coating cathode for high-performance Zinc-Bromine flow Battery

Y. Kim & J. Jeon
Department of Advanced Material and Energy, Dongguk-University, Seoul, Republic of Korea

ABSTRACT: Zinc-Bromine flow Batteries (ZBBs) require cathode materials having a high performance against bromine. This paper describes cathode electrode by a simple carbon coating method. This method is made of 3 steps. First step is making a carbon slurry (mixing 5 wt% of carbon powder, 4 wt% of Polyvinylidene Fluoride (PVDF) and solvent (NMP) for 8 hrs at 30°C). And the second step is to coat the surface of a graphite foil (cathode electrode) with the carbon slurry maintaining 100 μm of thickness. The final step is heating electrode covered with carbon slurry during 4 hrs at 120°C for decoupling solvent. The experimental results show that the proposed (carbon coating) electrode allows higher chemical stability and durability for long-time cell operations and also provides a high surface area for improving bromine-reduction reaction in a cathode half-cell. This work provides a new insight into the design of a high-performance cathode for cost effective and reliable ZBB.

1 INTRODUCTION

The Zinc-Bromine flow Battery (ZBB) is an attractive technology for large-scale energy storage due to its higher energy density and low cost (Kim & Jeon 2015). Electrolyte of ZBB is primarily composed of an aqueous zinc-bromide salt dissolved in distilled water. The electrolyte is stored externally in two tanks and is circulated through cell stacks constantly during the charge and discharge. During charge, the zinc-bromide salt is converted into zinc metal and elemental bromine. On discharge, the metallic zinc plated on the anode dissolves as zinc ion into the electrolyte and is available for platting again at the next charge cycle (Lai et al. 2013).

The reactive cell of a ZBB is divided into two half-cells by the separator (Cheng et al. 2014, Yang et al. 2015). In this case in the cathodic half-cell, there are various problems to be solved. During charge, bromine having low reaction rate evolves from two bromide ion at the cathode. Because of the low reaction rate of the bromine, some bromine molecular cannot convert bromide ion during the discharge. Therefore, it makes the coulombic efficiency and discharge capacity low. For overcoming this problem, the reaction surface of the cathode has to enlarge in order to enhance the reaction rate of the cation.

There is another problem. Toxic bromine (Xu & Zhao 2015, Cho et al. 2015, Coad & Lex 2012) make corrosion of the electrode. Thus, the cathode must have chemical resistance with the bromine to prevent corrosion.

In general, carbon coating on electrode materials is used to increase the electrode conductivity and surface area (Heon-Young Lee & Sung-Man Lee, et al. 2004, Min-Ho Seo & Mihee Park, et al. 2011). However, this attempt is limited to specific chemical environments in a bromine half-cell, requiring high stability and durability against bromine. Hence, choosing substrate and binder materials is very important for the design of a carbon coated electrode applicable to a cathode the ZBB. Also, the cost of the developed electrode has to be considered.

This paper describes a cost-effective and high performance cathode for ZBBs. The proposed cathode is coated with carbon powder on a graphite foil through the use of polyvinylidene fluoride [-(CF2-CH2)n-] as a binder. To demonstrate the effectiveness of the proposed electrode, Cyclic Voltammetry (CV) measurements and 6 cm^2 flow cell operations are carried out in 2.0 M $ZnBr_2$ electrolyte solutions. And electrode surface morphology, before and after cell operation (18 cycles), is also observed by Scanning Electron Microscope (SEM) for searching damaged electrode surface, respectively. Also, the experimental results demonstrate that the proposed electrode allows higher chemical resistance (Park et al. 2008, Park et al. 2015) against bromine for the durability improvement of ZBB cells than original cathode and provides a widely extended surface area to improve the reaction rate of bromine in the cathode half-cell.

2 EXPERIMENT

2.1 *CV measurements*

The CV is used to characterize electrochemical reactions of 2.0 M $ZnBr_2$ electrolyte according to using a cathode. And the CV measurements are performed for 10 cycles in a three-electrode cell (Wu et al. 2015). The measurements include test electrode (coated electrode), a platinum-wire electrode and a Saturated Calomel Electrode (SCE) are employed as a working electrode, a counter electrode and a reference electrode, respectively. The potential-scanning range is from −1.5 V to 1.5 V with the scan rate as 10 mVs^{-1} at 25°C.

2.2 *Proposed carbon coating technology*

Some studies related with coating technology are done to demonstrate utility of carbon coating for lithium ion batteries (Lee et al. 2004, Seo et al. 2011). But the proposed carbon coating method is the first technology for ZBB. This is prepared based on Carbon Powder (CP) Polyvinylidene Fluoride (PVDF) powder and N-Methyl-2-Pyrrolidone (NMP). Also, in this work, Graphite Foil (GF) is used as an electrode substrate for the carbon coating. The related mechanical data of these elements are indicated in Table 1.

The proposed carbon coating process for ZBB is according to the following three steps: (1) mixing of three-component (CP, PVDF, and NMP) for 8 hrs at 30°C to make a mixture (2) plating a mixture (maintain 100 μm of thickness) on the surface of GF (3) heating a mixture-paste GF during 4 hrs at 120°C for decoupling solvent.

The mixing ratio of the materials for carbon coating is determined by CV measurements to evaluate the bromine-reduction reaction of a carbon-coated electrode. So, four carbon-coated GFs are compared, which are coated with mixture slurries including 0.5 g, 1.0 g, 1.5 g, and 2.0 g of CP for a given 1 g PVDF and 30 mL NMP, respectively.

Table 1. Mechanical properties of coating materials.

Components/model	Properties
Carbon Powder/ ABG1005	LOI: 99.95%, Ash: 0.05%, moisture: <0.1%, true density: 2.15 g/cm, scoot density: 0.049 g/ccm, surface area BET: 0.5 sm/g, Oversize: <0.1%,
Polyvinylidene fluoride (PVDF)/Solef 6020	specific gravy: 1.75~1.8, heat of fusion: 57.0~66.0 J/g, crystallization heat: 47.0~52.0 J/g, melting temperature: 171~175°C
Graphite Foil/TF6	thickness: 0.6 mm, bulk density: 1.7 g/cm³, electrical resistivity: <10 Ω mm

Figure 1 shows CV curves when the four carbon-coated electrodes are used as an working electrode. It can be seen that 1.5 g carbon coating provides the highest cathode-reduction-current peak of about 45 mA at the potential voltage of 0.24 V, where, 1.5 g CP for a given 1 g PVDF and 30 mL NMP leads to a mixing ratio such that CP: PVDF: NMP = 5: 4: 91 wt%. Thus, it can be said that 5 wt% carbon coating has the most extended surface area than the others and can provide an improved reaction rate of bromine in the cathode half-cell. However, there still remains a durability problem (: chemical resistance against bromine) of the proposed electrode, so, charge and discharge test (operation test) and observing the use of SEM was occurred.

2.3 *Operation of flow cell*

To demonstrate the effectiveness of the proposed electrode, cell operation experiments are carried out on 2.0 M $ZnBr_2$ electrolyte with no additives. A flow cell (6 cm^2) employing a micro-porous-membrane (Asahi Kasei, SF600) (Wei et al. 2015, Busch & Schmitz 2010) is used as a separator. And the cyclic cell operation has been carried out for 18 cycles by using WBCS3000 (Won A tech Co., Korea). Also, the current density is kept constant at 20 mA/cm^2 during charge and discharge, and the charging time per cycle is 50 min.

2.4 *SEM observation*

The surface morphology of the cathode is observed by using JSM-7800F (JEOL Ltd., Japan) at an acceleration voltage of 10 KV for analysis damage of electrode against bromine. The samples are coated with platinum before the SEM analysis.

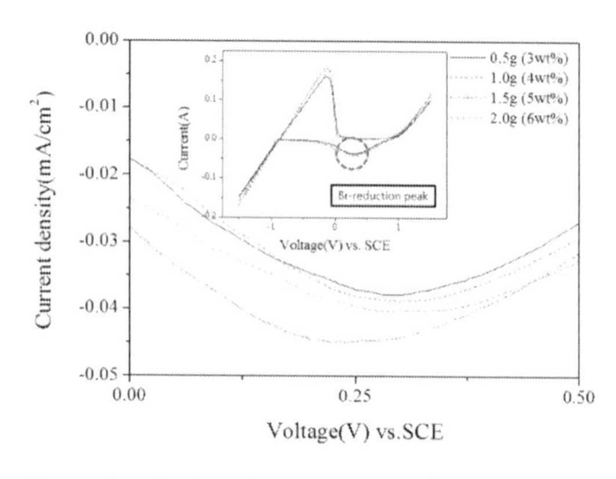

Figure 1. Cyclic voltagram curves for four carbon-coated electrodes to determine a mixture rate.

3 RESULTS AND DISCUSSION

Figure 2 shows CV-curve comparison of two electrodes: one is an original (uncoated) GF and the other is a proposed GF (5 wt% carbon coating). In the figure, I_{PA} (peak current of anode: zinc dissolution) and I_{PC} (peak current of cathode: bromine forming) is shown. I_{PA} has no visible change before and after coating, while I_{PC} is 50 mA/cm^2 higher after coating than before. This means that the carbon coating causes a high surface area that allows the improvement of bromine reduction reaction in the cathodic half-cell.

Figure 3 shows the operation curves of 18 cycles. In Figure 3 (a), the result of using original GFs as both anode and cathode of the ZBB cell is shown. Also, Figure 3 (b) shows the result of original GF and coated GF as the anode and cathode, respectively.

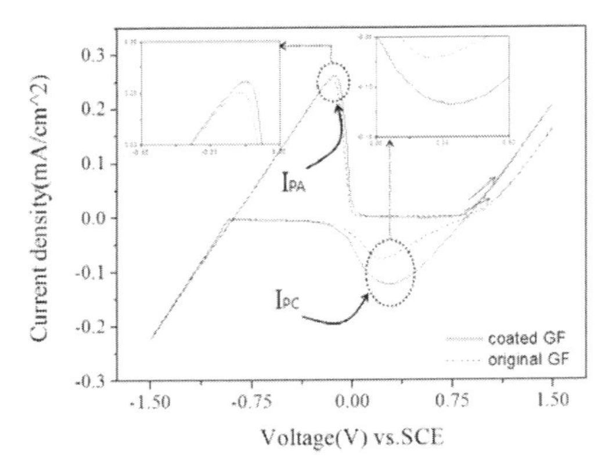

Figure 2. Cyclic voltagram curves for the comparison of uncoated and coated electrodes (based on GF).

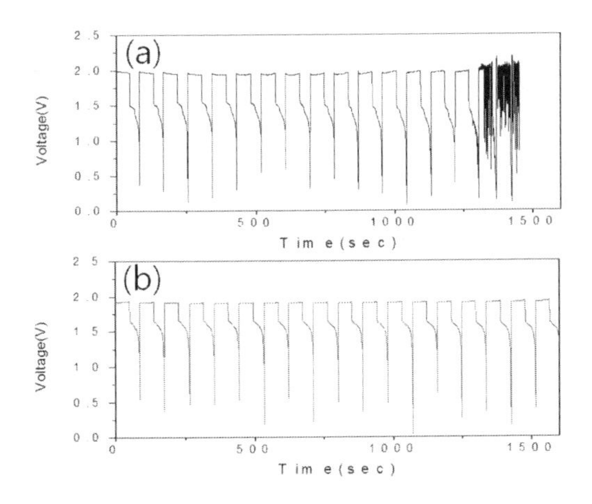

Figure 3. Charge and discharge curves of 18 cycles using (a) (uncoated) original GF as both anode and cathode (b) only coated GF as a cathode.

Figure 3 (a) that using the original GF as the cathode shows the unstable cell voltage. That is abnormally terminated at 16th cycle. Especially, the last 2 cycles before premature terminations are displayed, and the aberrant charging and discharging voltage curve have caused fatal electrode damage against bromine in the cathode half-cell. After operation test, this can be observed from the SEM image displayed in Figure 4. Figure 4 (a) and (b) are before and after the 18 cyclic operations, respectively. Inevitably, it can be found many tiny holes on the damaged surface of the cathode.

In contrast, as depicted in Figure 3 (b), the proposed cathode (5 wt% carbon, 4 wt% PVDF coating) leads to the improved durability more than about 11% for 18 cycles as compared with the original GF. Also, it appears that there is no damage on the surface of the proposed electrode when comparing the before and after, as displayed in Figure 5 (a) and (b), respectively. The reason is because PVDF used as a binder for carbon coating prevents the electrode surface damage against bromine in the cathode half-cell and consequently, provides high chemical stability for long-time cell operation.

Figure 6 shows the performance comparison of the two cells employing the original GF and the proposed GF as cathode electrode, respectively,

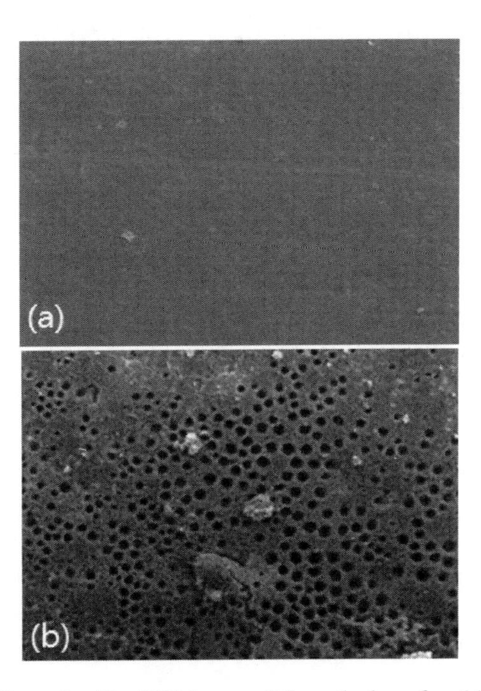

Figure 4. The SEM images of the cathode surface (a) before and (b) after the cyclic operation of 6 cm^2 cell employing an original GF as a cathode.

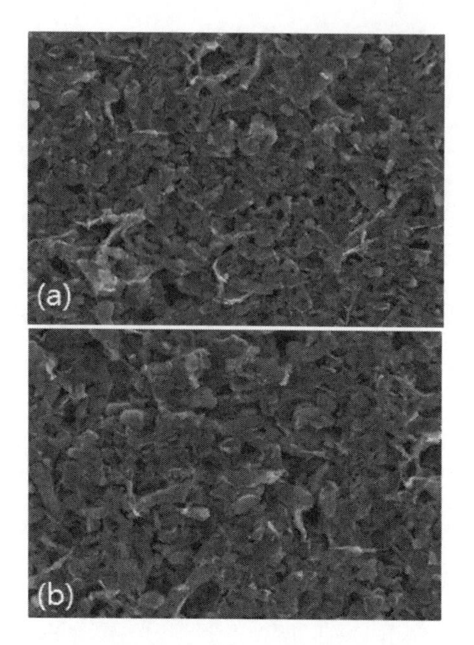

Figure 5. The SEM images of the cathode surface (a) before and (b) after the cyclic operation of 6 cm^2 miniature cell employing a coated GF (5 wt% carbon and 4 wt% PVDF) as a cathode.

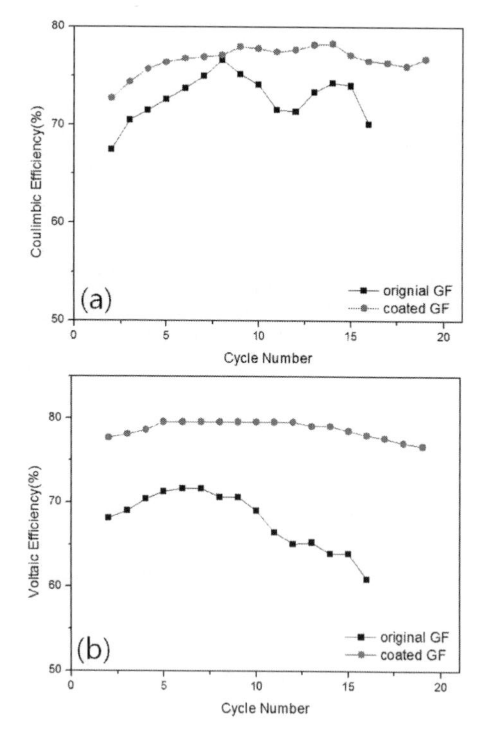

Figure 6. Performance comparison due to cell operation: (a) Coulombic Efficiency (CE), (b) Voltaic Efficiency (VE).

where CE and VE denote current efficiency and voltaic efficiency. It is shown that VE is averagely 10.5% higher after coating than before, and CE is averagely 4.0% higher. More remarkably, the VE is acceptably high by (77–80%) and is kept stable during the 18th cyclic operations. The main reason for these results is due to the fact that the carbon coating provides high conductivity as well as an extended surface area to improve bromine reaction in the cathode half-cell.

4 CONCLUSION

In this paper, a carbon coating technology has been proposed which allows a noble cathode electrode for high-performance ZBB. A mixing ratio for the carbon coating has also been derived such that CP: PVDF: NMP = 5: 4: 91 wt%. The experimental results have shown that the proposed carbon coating provides high stability and durability in a cathodic half-cell, and it improves the bromine reaction rate by increasing the reaction surface area of the cathode. It has been concluded that this advanced carbon coating arrangement of cathode electrodes gave rise to high conductivity as well as stable chemical reaction. Hence, a solution to the problem is found and encountered in the ZBBs.

ACKNOWLEDGMENT

This work has been supported by the Industrial Strategic Technology Development Program (10043787, Development of Ion Selective Membranes for Redox Flow Battery (RFB)) and funded by the Ministry of Trade, Industry & Energy (MI, Korea).

REFERENCES

Busch, D. & Schmitz, B. 2010. *Microporous foil for batteries having shutdown function*, US Patent NO. 8,927,135 B2.

Cheng, D., Xu, Q. & Yao, W. 2014. The Study of High Pressure Pulse Circuit in Ultrasonic Nondestructive Testing, *Advanced Materials Research,* 1070–1072: 449–455.

Cho, K.T., Tucker, M.C., Ding, M., Ridgway, P., Battaglia, V.S., Srinivasan, V. & Weber, A.Z. 2015. Cyclic Performance Analysis of Hydrogen/Bromine Flow Batteries for Grid-Scale Energy Storage. *ChemPlusChem,* 80: 402–411.

Coad, N. & Lex, P. 2012. *white paper, (ZBB Energy Corporation)*, 8–9.

Kim, D. & Jeon, J. 2015. Study on Durability and Stability of an Aqueous Electrolyte Solution for Zinc Bromide Hybrid Flow Batteries, *Journal of Physics: Conference Series*, 574: 012074.

Lai, Q., Zhang, H., Li, X., Zhang, L. & Cheng, Y. 2013. A novel single flow zinc–bromine battery with improved energy density. *Journal of Power Source*, 235: 1–4.

Lee, H.Y. & Lee, S.M. 2004. Carbon-coated nano-Si dispersed oxides/graphite composites as anode material for lithium ion batteries, *Electrochemistry Communications,* 6: 465–469.

Li, Y., Song, W., Fu, W., Tsang, D.C.W. & Yang, X. 2015. The roles of halides in the acetaminophen degradation by UV/H_2O_2 treatment: Kinetics, mechanisms, and products analysis. *Chemical Engineering Journal*, 271: 214–222.

Park, H.H., Deshwal, B.R., Kim, I.W. & Lee, H.K. 2008. Absorption of SO_2 from flue gas using PVDF hollow fiber membranes in a gas–liquid contactor. *Journal of Membrane Science*, 319: 29–37.

Park, J.W., Wycisk, R. & Pintauro, P.N. 2015. Nafion/PVDF Nanofiber Composite Membranes for Regenerative Hydrogen/Bromine Fuel Cells. *Journal of Membrane Science*, 490: 103–112.

Seo, M.H., Park, M.H., Lee, K.T., Kim, K.T., Kim, J.Y. & Cho, J.P. 2011. High performance Ge nanowire anode sheathed with carbon for lithium rechargeable batteries. *Energy Environ,* 4: 425–428.

Wei, X., Li, B. & Wang, W. 2015. Porous Polymeric Composite Separators for Redox Flow Batteries. *Journal of Macromolecular Science: Part C: Polymer Reviews*, 55(2): 247–272.

Wu, S., Zhao, Y., Li, D., Xia, Y. & Si, S. 2015. An asymmetric Zn//Ag doped polyaniline microparticle suspension flow battery with high discharge capacity. *Journal of Power Sources*, 275: 305–311.

Xu, Q. & Zhao, T.S. 2015. Fundamental models for flow batteries. *Progress in Energy and Combustion Science*, 49: 40–58.

Yang, J.H., Yang, H.S., Ra, H.W., Shim, J. & Jeon, J.D. 2015. Effect of a surface active agent on performance of zinc/bromine redox flow batteries: Improvement in current efficiency and system stability. *Journal of Power Sources*, 275: 294–297.

Advanced Materials, Mechanical and Structural Engineering – Hong, Seo & Moon (Eds)
 ISBN: 978-1-138-02908-8

Formation of the carbon microcoils dangled from the carbon nanofibers

G.H. Kang & S.H. Kim
Center for Green Fusion Technology and Department of Engineering in Energy and Applied Chemistry, Silla University, Busan, Republic of Korea

G.R. Kim
Department of Electronics Engineering, Silla University, Busan, Korea

ABSTRACT: Carbon nanofibers and carbon microcoils were synthesized using a thermal chemical vapor deposition system with C_2H_2 and H_2 as source gases and SF_6 as an additive gas. Formation of the carbon nanofibers and the double helix type carbon microcoils were investigated according to the injection stage and time of SF_6 flow during the reaction. The injection of SF_6 flow during the initial reaction stage gave rise to the continuous growth of the carbon microcoils. In this case, during the middle reaction stage and the injection of SF_6 flow, we could observe the formation of the carbon microcoils dangled from the carbon nanofibers. We suggest the growth mode of the dangled carbon microcoils from the carbon nanofibers. The density of the dangledcarbon microcoils seemed to be slightly high with the increasing injection time of SF_6 flow. The injection of SF_6 flow during the final reaction stage gave rise to the formation of the carbon nanofibers without the formation of the carbon microcoils.

1 INTRODUCTION

Because of the diversity in the morphology of carbon nanomaterials that can extend the practical application areas by using the morphology-correlated properties thus, many researchers have found that there are diverse unique forms of the carbon nanomaterials, such as carbon nanotubes, graphenes, carbon coils, and so on (Dresselhaus et al. 2001, Dupuis 2005, Shaikjee & Coville 2012). Therefore, the synthesis of the carbon nanomaterials having the peculiar morphologies is an initial primary step to develop the morphology-correlated materials properties.

Among the unique morphologies observed in the carbon nanomaterials, the carbon coils are particularly noticed because of their unique double helix type geometries. Carbon coils have been proposed as the suitable materials for field emitters, highly sensitive nano—and microsized detectors, reinforcing fillers for composites, and building blocks for the fabrication of nanodevices (Hokushin et al. 2007, Hemadi et al. 2001). Practically, carbon coils are noticed to be used as the electromagnetic absorbers (Coville 2012, Pan et al. 2001).

For the preparation method of the carbon coils, catalytic Chemical Vapor Deposition (CVD) technique using thermal CVD and the metal catalyst has been used because of its relative inexpensive and applicable feature. The iron family elements (Fe, Co, and Ni), especially Ni element, were known to be effective metal catalyst for the growth of the carbon coils on the substrates (Hou et al. 2003).

The addition of a small amount of the sulfur incorporated impurities during the reaction has also been noticed as one of the methods to enhance the carbon coils production yield and reproducibility (Motojima et al. 1990). Usually, sulfur was incorporated in the form of hydrogen sulfide (H_2S), carbon disulfide (CS_2), and thiophene (C_4H_4S) (Motojima et al. 1995, 1996, Chen et al. 1999). Previously, we introduced a trace of sulfur in the form of SF_6 (Kim et al. 2010) as an impurity to enhance the formation of the carbon coils because the fluorine incorporated species were known to enhance the nucleation sites for the carbon-related materials (Asmann et al. 1999, Wong et al. 1992, Corat et al. 1997).

Furthermore, we presented an in-situ cycling on/off modulation process of C_2H_2/H_2 flow to enhance the formation density of the Carbon Nanofilaments (CNFs) (Jeon et al. 2012). It can be simply achieved by turning a source gas flow on or off during the reaction stage. In general, the carbon coils have various geometries, such as double helix type microcoils, single helix type nanocoils, wavelike type nanocoils (Eum et al. 2012). By the in-situ cycling on/off modulation process employing SF_6 flow, we could obtain the formation of the geometrically controlled carbon microcoils (Eum et al. 2012, Jeon et al. 2012).

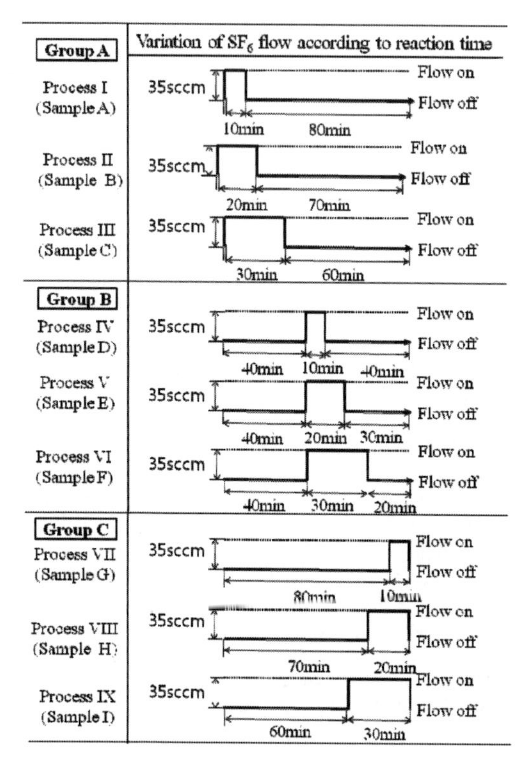

Figure 1. Different reaction processes with different samples.

In this work, we introduced a unique type of shape of the carbon nanomaterials, namely the carbon microcoils dangled from the carbon nanofibers. This morphology could be achieved by manipulating the injection stage of SF_6 flow during the reaction. We investigated the morphologies of the carbon nanomaterials according to the injection stage of SF_6 flow. In addition, we varied the injection time of SF_6 flow. Finally, we suggest the growth mode of the carbon microcoils dangled from the carbon nanofibers.

2 EXPERIMENTAL DETAILS

SiO_2-layered Si substrates were prepared by the thermal oxidation of 2.0 × 2.0 cm^2 p-type Si (100) substrates. The thickness of the SiO_2 layer on the Si substrate was estimated to be ca. 300 nm.

Ni powder (0.1 mg, ca. 99.7%) was evaporated for 1 min to form a Ni catalyst layer on the substrate using a thermal evaporator. The estimated thickness of the Ni catalyst layer was ca. 250 nm.

For the deposition of the carbon coils, a thermal CVD system was employed. C_2H_2 and H_2 as source gases and SF_6 as an additive gas were injected into the reactor. The flow rate of C_2H_2, H_2, and SF_6 were fixed at 250, 35, and 35 standard cm^3/min (sccm), respectively. Total reaction time (mainly by C_2H_2 + H_2 gases) was 90 minutes.

According to the different reaction processes, the injection of SF_6 flow was performed during the initial, the middle, and the final reaction stages of the process as our previous report (Park et al. 2013). As shown in Figure 1, samples A–C (group A), D–F (group B), and G–I (group C)

Table 1. Experimental conditions of the carbon coils deposition for the different samples.

Samples	Substrate Temp. (°C)	C_2H_2 flow rate (sccm)	H_2 flow rate (sccm)	SF_6 flow rate (sccm)	Total pressure (Torr)	Source gases flow time (min)			Injection stage of SF_6 flow
						C_2H_2	H_2	SF_6	
A	750	250	35	35	100	90	90	10	Initial
B	750	250	35	35	100	90	90	20	Initial
C	750	250	35	35	100	90	90	30	Initial
D	750	250	35	35	100	90	90	10	Middle
E	750	250	35	35	100	90	90	20	Middle
F	750	250	35	35	100	90	90	30	Middle
G	750	250	35	35	100	90	90	10	Final
H	750	250	35	35	100	90	90	20	Final
I	750	250	35	35	100	90	90	30	Final

were obtained after the different reaction processes (process I–IX). In these cases, C_2H_2 and H_2 flows were continuously injected during the reaction. The morphological details of the carbon coils deposited onto the different substrates were investigated using field emission scanning electron microscopy (FESEM, Hitach 4500). The detailed reaction conditions of the different processes are shown in Table 1.

3 RESULTS AND DISCUSSION

Figure 2 shows the representative sample in this work. We could observe the uniform growth of the carbon nanomaterials on the whole surface of the sample, indicating the growth uniformity of the carbon nanomaterials on the sample.

Figure 3 shows the magnified FESEM images for the samples of group A (samples A–C) carried out the processes I–III. Among the various types of the carbon coils-related geometries, the carbon microcoils were most frequently observed on these samples. In addition, a couple of the carbon nanofilaments and the wave-like type carbon nanocoils were occasionally observed as shown in Figure 3.

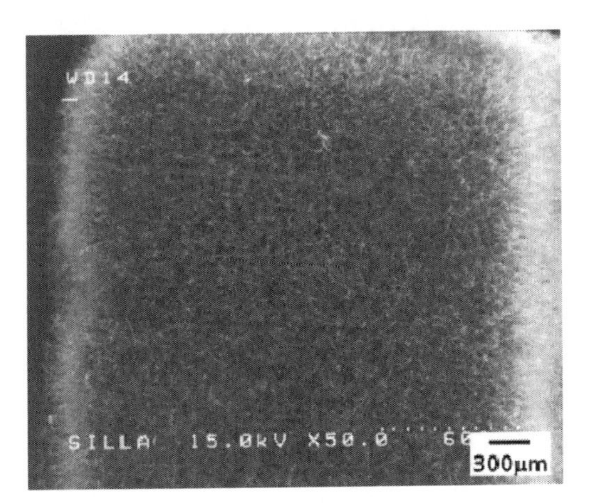

Figure 2. The representative FESEM image of group A.

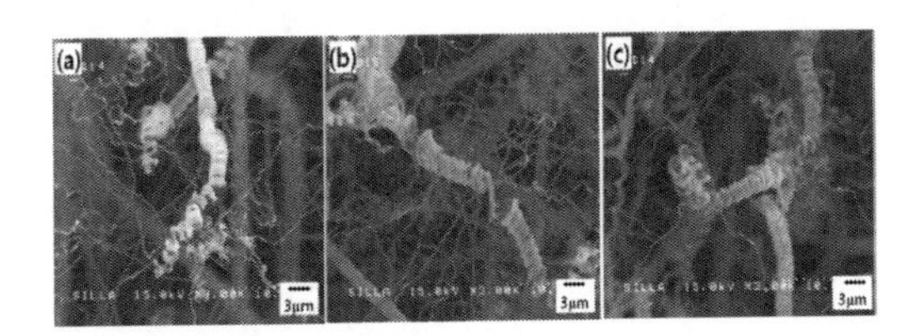

Figure 3. The FESEM images of samples A, B, and C.

Meanwhile, in the case of the samples of group B, we could clearly observe the formation of the carbon microcoils dangled from the carbon nanofibers as shown in Figure 4.

To identify the growth mode of the dangled carbon microcoils from the carbon nanofibers, we investigated the morphologies which focuses on the starting point of the carbon microcoils from the carbon nanofibers. Figure 5 shows FESEM images indicating the formation of the carbon microcoils from the carbon nanofibers for the samples D–F.

As shown in these Figures, we could observe the linear-type carbon nanofibers (see the inside of the dotted rectangles) and the curl-type carbon nanocoils (see the inside of the dotted circles) around the starting point of the carbon microcoils. The inset in Figure 5a clearly shows the initiation of the carbon microcoil from two curl-type carbon nanocoils. Therefore, we suggest the growth mode of the dangled carbon microcoils from the carbon nanofibers as follows.

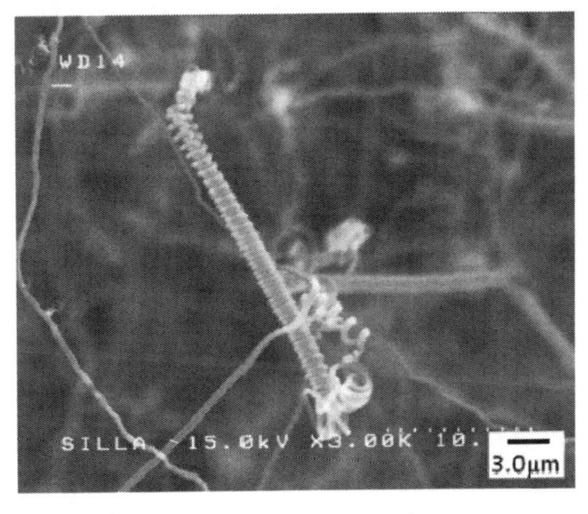

Figure 4. The representative FESEM image of the sample for group B showing the formation of the carbon microcoils dangled from the carbon nanofibers.

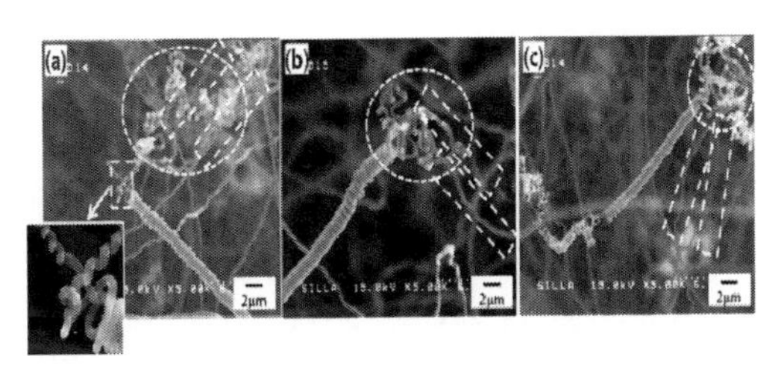

Figure 5. FESEM images of (a) sample D, (b) sample E, and (c) sample F with focusing on the transformation from the carbon nanofibers into the carbon microcoils.

1. Linear type carbon nanofibers were transformed into the curl-type carbon nanocoils by the injection of SF_6 flow.
2. The heads of two curl-type carbon nanocoils which is located at the neighboring sites met and formed the growth starting point of the carbon microcoils. In this case the Ni elements were located on the head of the curl-type carbon nanocoils. The conglomeration of the heads of two curl-type carbon nanocoils is considered as the main cause for the formation of the starting point, namely the head, of the carbon microcoils (Park et al. 2013).
3. Two attached carbon nanocoils would rotate in the same direction due to the collapsed balance of the head of the carbon microcoils as the previous report (Park et al. 2013). Finally, they can form the dangled carbon microcoils from the carbon nanofibers as shown in Figure 6.

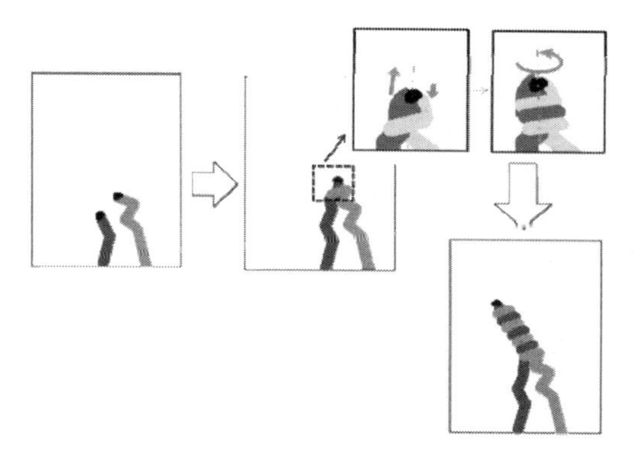

Figure 6. The schematic diagram showing the growth mode of the carbon microcoils from the linear-type carbon nanofibers.

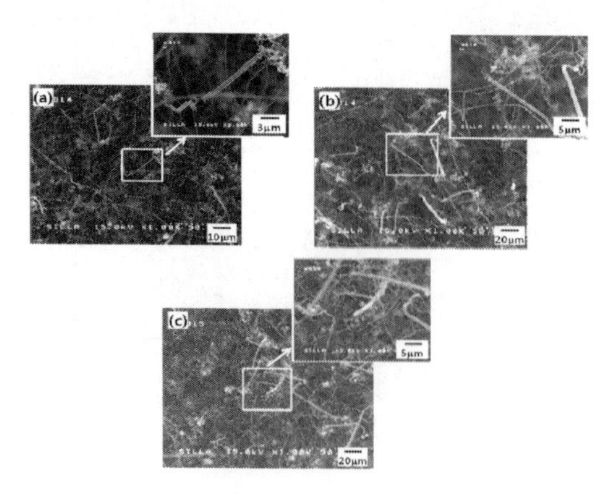

Figure 7. FESEM images of (a) sample D, (b) sample E, and (c) sample F. The insets show the magnified images of the dotted square in each sample.

As shown in Figure 7, the density of the dangled carbon microcoils seemed to be slightly high with the increasing injection time of SF_6 flow (compare the inset of Figure 7c with 7a and 7b). However the lengths of the dangledcarbon microcoils do not vary much according to the injection time of SF_6 flow. These results reveal that the geometries of the carbon microcoils would not be much affected by the injection time of SF_6 flow in the case of middle injection stage of SF_6 flow. The study of the cause for these results is underway.

Figure 8 shows the representative FESEM images for the samples of group C (samples G–I) carried out the processes VII–IX. In this case we could merely observe the linear-type carbon nanofibers without any kind of the carbon coils-related nanomaterials. However, at some position on the sample I, we could occasionally observe the under-developed carbon microcoils as shown in Figure 9.

Although, in this case, without the carbon microcoils we cannot fully explain the exact reason for the formation of the carbon nanofibers, but we

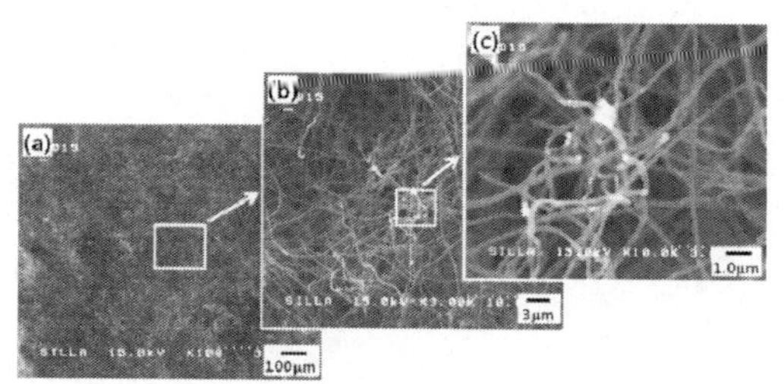

Figure 8. The representative FESEM images of group C under the magnification of (a) × 100, (b) × 3,000, and (c) × 10,000.

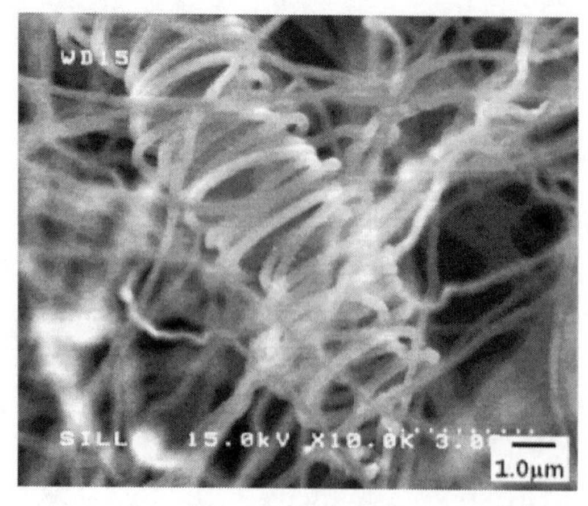

Figure 9. FESEM image of sample I showing the under-developed carbon microcoils.

could understand that the injection of SF_6 flow during the final stage of the reaction is not appropriate for the formation of the carbon microcoils on the sample.

4 CONCLUSION

By manipulating the injection stage of SF_6 flow during the reaction, we could obtain the dangled carbon microcoils from the carbon nanofibers. We suggest the growth mode of the dangled carbon microcoils. The injection of SF_6 flow during the final stage of the reaction is not appropriate for the formation of the well-developed carbon microcoils.

ACKNOWLEDGMENT

This work has been supported by the Human Resource Training Program for Regional Innovation and Creativity through the Ministry of Education and National Research Foundation of Korea (2014 H1C1 A1066859).

REFERENCES

Asmann, M., Heberlein, J. & Pfender, E. 1999. A review of diamond CVD utilizing halogenated precursors, *Diamond and Related Materials,* 8: 1–16.

Chen, X. & Motojima, S. 1999, Morphologies of carbon microcoils grown by chemical vapor deposition, *Journal of Materials Science*, 34: 5519–5524.

Corat, E.J., Trava-Airoldi, V.J., Leite, L.F., Nono, M.C.A. & Baranauskas, V. 1997, Diamond growth with CF_4 addition in hot-filament chemical vapor deposition, *Journal of Materials Science,* 32: 941–947.

Coville, N.J., Mhlanga, S.D., Nxumalo, E.D. & Shaikjee, A. 2011. A review of shaped carbon nanomaterials. *South African Journal of Science,* 2011; 107(3/4), Art. #418, 15 pages. DOI: 10.4102/ sajs.v107i3/4.418.

Dresselhaus, M.S., Dresselhaus, G. & Avouris, P. 2001. *Carbon nanotubes: synthesis, structure, properties, and applications,* Berlin: Springer-Verlag: 12–51.

Dupuis, A.C. 2005. The catalyst in the CCVD of carbon nanotubes-a review, *Progress in Materials Science,* 50: 929–961.

Eum, J.H., Kim, S.H., Yi, S.S. & Jang, K. 2012. Large-scale synthesis of the controlled-geometry carbon coils by the manipulation of the SF_6 gas flow injection time, *Journal of Nanoscience and Nanotechnology,* 12(5): 4397–4402.

Hernadi, K., Thien-Nga, L. & Forro, L. 2001. Growth and microstructure of catalytically produced coiled carbon nanotubes, *The Journal of Physical Chemistry B,* 105: 12464–12468.

Hokushin, S., Pan, L.J., Konishi, Y., Tanaka, H. & Nakayama, Y. 2007. Field emission properties and structural changes of a stand-alone carbon nanocoil, *Japanese Journal of Applied Physics*, 46: L565-L567.

Hou, H., Jun, Z., Weller, F. & Greiner, A. 2003. Large-scale synthesis and characterization of helically coiled carbon nanotubes by use of $Fe(CO)_5$ as floating catalyst precursor, *Chemistry of Materials,* 15: 3170–3175.

Jeon, Y.C., Eum, J.H., Kim, S.H., Park, J.C. & Ahn, S.I. 2012. Effect of the on/off cycling modulation time ratio of C_2H_2/SF_6 flows on the formation of geometrically controlled carbon coils, *Journal of Nanomaterials,* Article ID 908961.

Kim, K.D., Kim, S.H., Yi, S.S. & Jang, K. 2010. Effect of SF_6 incorporation in the cyclic process on the low temperature deposition of carbon nanofilaments, *Thin Solid Films,* 518: 6412–6416.

Motojima, S., Asakura, S., Kasemura, T., Takeuchi, S. &. Iwanaga, H. 1996, Catalytic effects of metal carbides, oxides and Ni single crystal on the vapor growth of micro-coiled carbon fibers, *Carbon*, 34: 289–296.

Motojima, S., Itoh, Y., Asakura, S. & Iwanaga, H. 1995, Preparation of micro-coiled carbon fibers by metal powder-activated pyrolysis of acetylene containing a small amount of sulfur compounds. *Journal of Materials Science*, 30: 5049–5055.

Motojima, S., Kawaguchi, M., Nozaki, K. & Iwanaga, H. 1990, Growth of regularly coiled carbon filaments by Ni catalyzed pyrolysis of acetylene, and their morphology and extension characteristics, *Applied Physics Letters*, 56: 321, 321–323.

Pan, L.J., Hayashida, T., Zhang, M. & Nakayama, Y. 2001. Field emission properties of carbon tubule nanocoils, *Japanese Journal of Applied Physics,* 40: L235–L237.

Park, S., Jeon, Y.C. & Kim, S.H. 2013. Effect of injection stage of SF_6 flow on carbon micro coils formation, *ECS Journal of Solid State Science and Technology*, 2 (11): M56-M59.

Shaikjee, A. & Coville, N.J. 2012. The synthesis, properties and uses of carbon materials with helical morphology, *Journal of Advanced Research*, 3(3): 195–223.

Wong, M.S. & Wu, C.H. 1992, Complications of halogen-assisted chemical vapor deposition of diamond, *Diamond and Related Materials,* 1: 369–372.

Advanced Materials, Mechanical and Structural Engineering – Hong, Seo & Moon (Eds)
© 2016 Taylor & Francis Group, London, ISBN: 978-1-138-02908-8

Study on the anodizing of AZ31 magnesium alloys in ethanol solution

S.A. Salman
EcoTopia Science Institute, Nagoya University, Furo–cho, Chikusa ku, Nagoya, Japan
Graduate School of Engineering, Al-Azhar University, Cairo, Egypt

ABSTRACT: Magnesium is the lightest structural metal materials in practical use, with a density equivalent to two-thirds, one-third and one quarter of Al, Zn and steel densities respectively. Magnesium alloys are therefore recognized as alternatives to Al alloys and steel in reducing the weight of structural materials. Furthermore, magnesium alloys have high strength-to-weight ratios, excellent castability, machinability, weldability, thermal stability and good damping capacity. Despite these good features, there are major limitations concerning magnesium alloys because of their extremely negative equilibrium potential and poor corrosion resistance. In order to improve the corrosion resistance, anodizing of AZ31 Mg alloy was carried out in ethanol solutions containing cerium nitrate ions. The corrosion resistance was examined using salt spray corrosion tests and polarization tests. The composition and structure were analyzed using SEM, Energy Dispersive X-ray Spectroscopy (EDS), and XRD. Anodic films were successfully formed on AZ31 Mg alloys in cerium-containing ethanol solutions at a constant potential of 10 V for 600 s. The produced anodic films had no cracks, and the corrosion resistance was improved.

1 INTRODUCTION

Magnesium is one of the most prevalent elements in the earth's crust. It is also present in large quantities in seawater. Magnesium has some promising properties, such as its combination of high strength and low density, and also its high dimensional stability, high thermal conductivity, good electromagnetic shielding characteristics, high damping, and good machinability (Salman & Okido 2013). These properties make it an attractive proposition and an excellent choice when weight reduction is needed. There are several applications of magnesium alloys, such as in vehicles, electronic parts, and aerospace equipment. Unfortunately, the high chemical and electrochemical activity of magnesium leads to a poor corrosion resistance and consequently prevents the widespread use of magnesium alloys in many applications. The internal galvanic corrosion from second phases or impurities and the weakness of the surface hydroxide film on magnesium are other reasons for the poor corrosion resistance of magnesium alloys. Furthermore, the oxide film produced is less stable than the anodic films formed on Al alloys and steel.

Various surface treatments have been carried out to achieve a good corrosion resistance, such as electrochemical plating, chemical conversion coating, anodizing, self-assembled monolayers and thermal spraying (Salman & Okido 2012, da Conceicao et al. 2010, Salman et al. 2007, Salman et al. 2008a,b, Salman et al. 2009). Of the previous processes, anodizing produces a thick, durable oxide film on Mg alloys, which improves the corrosion resistance. An organic coating pretreatment is often used to provide better adhesion between the anodic film and the substrate and to improve the corrosion resistance. A chromate bath is traditionally applied despite its being toxic to humans and difficult to recycle. In our previous research, we have investigated the anodizing of AZ31 magnesium alloy samples in alkaline solution at various applied potentials (Mizutani et al. 2003, Salman et al. 2010, Choi et al. 2013).

Virtanen et al. investigated the anodizing of pure magnesium in electrolytes composed of water-free ethanol and methanol containing ammonium nitrate ions. A black anodic layer with a flake-like nanoporous structure was produced (Brunner et al. 2009). Tannic acid-based conversion coatings have been formed on AZ91D magnesium alloy substrates, but irregular cracks formed on the surface (Hu et al. 2010). Organic coatings silane-based for AZ91D magnesium alloy have been synthesized as prospective surface treatments for AZ91D magnesium alloy by hydrolysis and condensation reaction of the different silanes in ethanol (Chen et al. 2008). Chemical conversion coatings have been obtained on magnesium alloys using cerium nitrate in aqueous (Salman & Okido 2012, Lin & Li 2006) and ethanol (Wang et al. 2009) solutions. In the present work, we have investigated the anodizing of magnesium alloys in ethanol solutions containing cerium nitrate ions.

The composition and structure of the anodic films were analyzed using SEM, EDS, and XRD. The corrosion resistance was examined using a salt spray corrosion test and a polarization test. The corrosion resistance of the resulting films was improved after the anodizing

2 EXPERIMENTAL

The chemical composition of the commercially available AZ31 Mg alloy substrate used is shown in Table 1. The Mg alloy samples were ground with emery paper up to 2000 grit, polished with 0.05 μm Al2O3 powder, washed with water, and dried in hot air. The working area of each sample (1 cm^2) was defined using hydrophobic adhesive masking tape.

The samples were then anodized in 200 ml of a 0.05 M $Ce(NO_3)_3$ ethanol solution at 298 K using a constant potential of 10 V for 600 s. After the samples were withdrawn from the anodizing solution, they were immediately rinsed with deionized water and dried in cool air using a handheld dryer. The surface and cross-sectional microstructures and corresponding chemical compositions of the samples were characterized using SEM, EDS, and XRD. The anticorrosion properties of the anodic coatings were examined using potentiodynamic polarization in a 0.1 M NaCl aqueous solution at 298 K with a typical three-electrode electrochemical cell: the Mg alloy samples served as the working electrode, a Pt coil as the counter electrode, and a saturated Ag/AgCl electrode (3.3 M KCl) as the reference electrode.

3 RESULTS

The current transients of the AZ31 magnesium alloy samples are shown in Figure 1.

During the initial stages of the anodizing process, the current density was high, but it decreased gradually with anodizing time after the initial five seconds. The higher current density observed at the start of anodizing occurred because the current could flow through the metallic Mg. The current density decreased after the formation of the anodic film, which exhibited a higher corrosion resistance than the bare metallic Mg. The current density then sharply increased up to a period of 30 s and subsequently decreased for a period of a few

Table 1. Chemical composition (mass%) of AZ31 Mg alloy.

Element	Mg	Al	Zn	Mn	Si	Cu	Ni	Fe
Conc.	Bal.	3.0	1.0	0.43	0.01	<0.01	<0.001	0.003

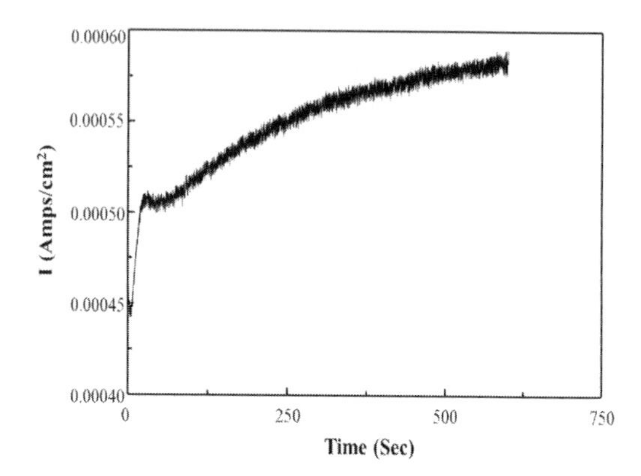

Figure 1. Current change during constant potential anodizing of AZ31 magnesium alloy in cerium contained ethanol solution at 298 K for 600 s.

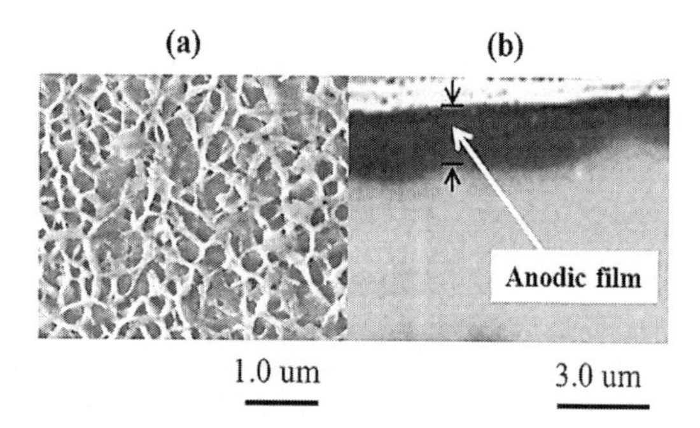

Figure 2. Surface morphologies and cross section of of the anodic films formed on AZ31.

seconds. These current contributions are probably related to a series of destruction, nucleation, and growth process of MgO and other oxides/hydroxides on the electrode. The current density gradually increased after a period of 50 s at an almost stable rate until the end of the anodizing process.

Figure 2 shows SEM images of the surface and cross-section of an anodic film.

The coating had a porous surface and showed a net-like structure, as shown in Figure 2(a); the formation of this porous structure is attributed to the Pilling–Bedworth ratio: the molar volume of MgO (Voxide) was 11.3 cm^3 mol^{-1}, whereas the molar volume of metallic Mg (Vmetal) was 14.0 cm^3 mol^{-1} (Choi et al. 2013, Blawert et al. 2006). From observation of the surface, we can see that this porous oxide layer did not extend deeply into the lower part of the anodic film. The thickness of the film was about 2 μm, with no cracks, as shown in Figure 2(b).

Figure 3 shows the XRD patterns of the anodized AZ31 magnesium alloy samples. The anodic films were mainly composed of magnesium and

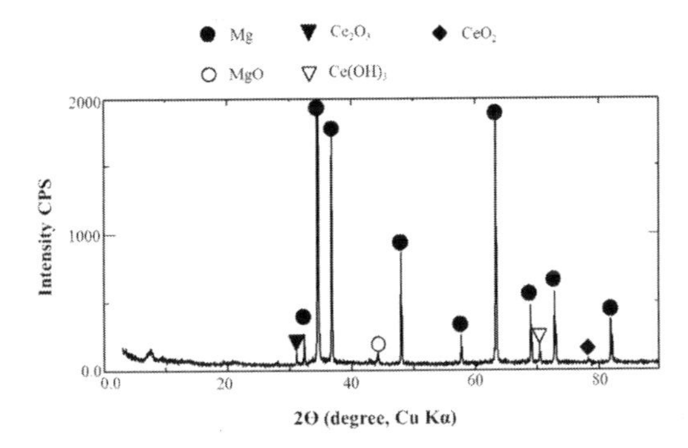

Figure 3. Typical X-ray diffraction pattern of the anodic films formed on AZ31 Mg alloy in cerium contained ethanol solution at 298 K for 600 s.

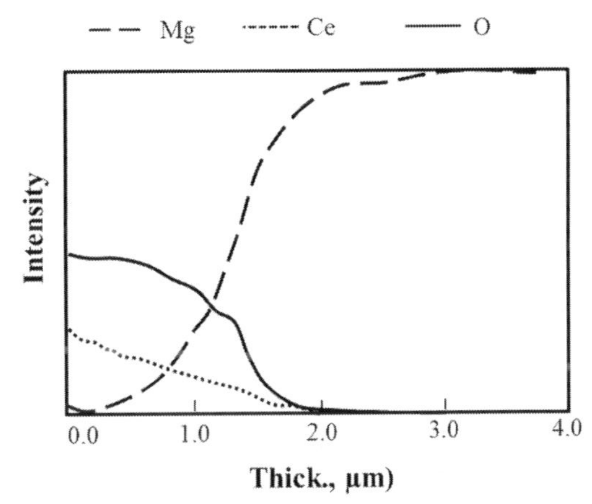

Figure 4. Schematic diagram of EDS line scanning throughout the thickness of the anodic film formed on AZ31 Mg alloy in cerium contained ethanol solution at 298 K for 600 s.

cerium oxides/hydroxides. It has been reported that MgO is only produced at high potential when anodizing in alkaline solutions.

However, based on our XRD results, MgO can form in the case of anodizing in an ethanol solution containing cerium ions. Magnesium oxide can form according to Reaction 1.

$$2\,Mg + O_2 \rightarrow 2\,MgO \tag{1}$$

It can be also form according to Reaction 2.

$$Mg(OH)_2 \rightarrow MgO + H_2O \tag{2}$$

CeO_2, Ce $(OH)_3$, and Ce_2O_3 were also detected in the films. We suggest that these compounds are formed according to following equations.

$$2H_2O + 2e^- \rightarrow 2OH^- + H_2 \tag{3}$$

$$Ce^3 + 3OH^- \rightarrow Ce(OH)_3 \tag{4}$$

$$2\,Ce(OH)_3 \rightarrow Ce_2O_3 + 3H_2O \tag{5}$$

$$Ce_2O_3 + O_2 \rightarrow 2CeO_2 \tag{6}$$

Figure 4 shows a schematic diagram of the EDS line scanning performed throughout the thickness of the anodic films. Oxygen and cerium show higher values on the top of the anodic films because of the formation of cerium and magnesium oxides. On traversing downward toward the substrate, cerium disappears and oxygen decreases notably.

On the other hand, magnesium exhibits a lower value on the top of the anodic films and increases on traversing in the lower substrate direction.

Figure 5 shows the polarization curves of AZ31 Mg alloy samples before and after the anodizing treatment. Tafel plots how that the non-anodized samples had a pitting potential (E_{pit}) of −1.39 ± 0.02 V Ag/AgCl and a corrosion current density (I_{corr}) of 4×10^{-4} Acm^2. On the other hand, the anodized samples showed a significant increase in their pitting potential of −0.9 V ± 0.01 V Ag/AgCl and a significant reduction in their corrosion current density of 9.9×10^{-7} Acm^2 compared with the non-anodized samples. These results show that the anodic coatings were much less affected by corrosive media because of the passivation action of the anodic films.

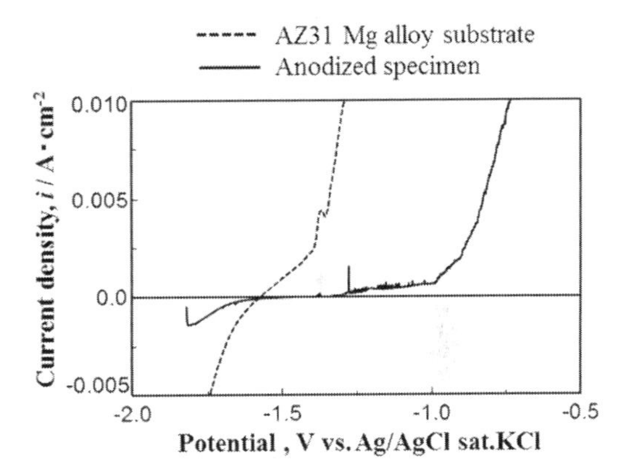

Figure 5. Potentiodynamic polarization curves of AZ31 Mg alloy before and after anodizing in cerium contained ethanol solution at 298 K for 600 s.

The surface appearance of the anodized and non-anodized AZ31 Mg alloy samples after exposure to a 5 wt% NaCl salt spray for 48 h is shown in Figure 6. The salt spray test results agree with the polarization results. The non-anodized samples exhibited many pits, and the surface was almost covered with the corrosion products after 48 h. On the other hand, there were fewer small pits observed on the surface of the anodized samples.

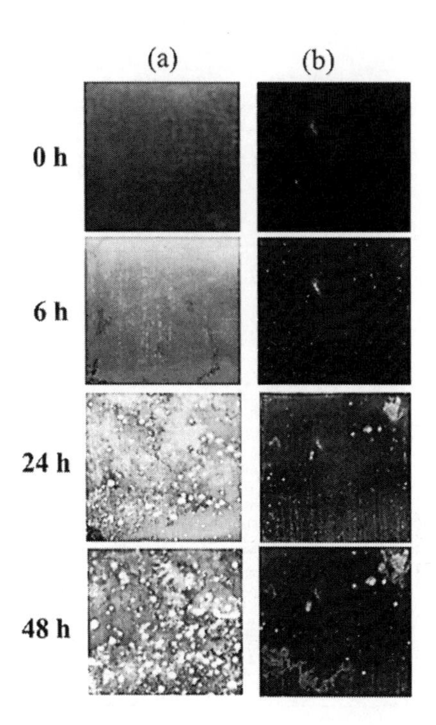

Figure 6. Surface morphologies of non-anodized AZ31 Mg alloy (a) and anodized specimen (b) and before and after salt spray test.

4 CONCLUSIONS

Anodic films were successfully formed on AZ31 Mg alloys in cerium-containing ethanol solutions at a constant potential of 10 V for 600 s. These films were composed mainly of Mg, Ce_2O_3 and $Ce(OH)_3$ cerium oxides. The corrosion resistance of AZ31 Mg alloy was improved with anodizing; A significant increase in the pitting potential and a decrease in the corrosion current density were obtained.

ACKNOWLEDGEMENT

The authors gratefully acknowledge the financial support of Aichi Center for Industry and Science Technology and the Ministry of the Education, Culture, Sports, Science and Technology, Japan. They wish also to acknowledge the Ministry of Higher Education and the Faculty of Engineering, Al-Azhar University for endless support.

REFERENCES

Blawert, C., Dietzel, W., Ghali, E. & Song, G. 2006. Anodizing Treatments for Magnesium Alloys and Their Effect on Corrosion Resistance in Various Environments, *Advanced Engineering Materials*, 8: 511–533.

Brunner, J.G., Hahn, R., Kunze, J. & Virtanen, S. 2009. Porosity Tailored Growth of Black Anodic Layers on Magnesium in an Organic Electrolyte, *Journal of the Electrochemical Society,* 156: 62–66.

Chen, X.M., Li, G.G., Lian, J.S. & Jiang, Q. 2008. An organic chromium-free conversion coating on AZ91D magnesium alloy, *Applied Surface Science*, 255: 2322–2328.

Choi, Y.I., Salman, S., Kuroda, K. & Okido, M. 2013. Enhanced Corrosion Resistance of AZ31 Magnesium Alloy by Pulse Anodization, *Journal of the Electrochemical Society,* 160: 364–368.

Choi, Y.I., Salman, S., Kuroda, K. & Okido, M. 2013. Synergistic corrosion protection for AZ31 Mg alloy by anodizing and stannate post-sealing treatments, *Electrochimica Acta,* 97: 313–319.

da Conceicao, T.F., Scharnagl, N., Blawert, C., Dietzel, W. & Kainer, K.U. 2010. Surface modification of magnesium alloy AZ31 by hydrofluoric acid treatment and its effect on the corrosion behaviour, *Thin Solid Films*, 518: 5209–5218.

Hu, J.Y., Li, Q., Zhong, X.K., Li, L.Q. & Zhang, L. 2010. Organic coatings silane-based for AZ91D magnesium alloy, *Thin Solid Films,* 519: 1361–1366.

Lin, C.S. & Li, W.J. 2006. Corrosion Resistance of Cerium-Conversion Coated AZ31 Magnesium Alloys in Cerium Nitrate Solutions, *Materials Transactions,* 47: 1020–1025.

Mizutani, Y., Kim, S.J., Ichino, R. & Okido, M. 2003. Anodizing of Mg alloys in alkaline solutions, *Surface and Coatings Technology*, 169: 143–146.

Salman, S.A. & Okido, M. 2012. Self-assembled monolayers formed on AZ31 Mg alloy, *Journal of Physics and Chemistry of Solids*, 73: 863–866.

Salman, S.A. & Okido, M. 2013. *Anodization of magnesium (Mg) alloys to improve corrosion resistance, in: G-L Song (Eds.), Corrosion prevention of magnesium alloys*, Woodhead Publishing, Cambridge, 197–231.

Salman, S.A., Ichino, R. & Okido, M. 2007. Development of Cerium-based Conversion Coating on ZA31 Magnesium Alloy, *Chemistry Letters*. 36: 1024–1025.

Salman, S.A., Ichino, R. & Okido, M. 2008a. Production of Alumina-Rich Surface Film on AZ31 Magnesium Alloy by Anodizing with Co Precipitation of Nano-Sized Alumina, *Materials Transactions,* 49: 1038–1041.

Salman, S.A., Ichino, R. & Okido, M. 2008b. Influence of calcium hydroxide and anodic solution temperature on corrosion property of anodising coatings formed on AZ31 Mg alloys, *Surface Engineering*, 24: 242–245.

Salman, S.A., Ichino, R. & Okido, M. 2009. Improvement of corrosion resistance of AZ31 Mg alloy by anodizing with co-precipitation of cerium oxide, *Transactions of Nonferrous Metals Society of China*, 19: 883–886.

Salman, S.A., Mori, R., Ichino, R. & Okido, M. 2010. Effect of Anodizing Potential on the Surface Morphology and Corrosion Property of AZ31 Magnesium Alloy, *Materials Transactions*, 51: 1109 - 1113.

Wang, C., Zhu, S.L., Jiang, F. & Wang, F.H. 2009. Cerium conversion coatings for AZ91D magnesium alloy in ethanol solution and its corrosion resistance, *Corrosion Science,* 51: 2916–2923.

Advanced Materials, Mechanical and Structural Engineering – Hong, Seo & Moon (Eds)
© 2016 Taylor & Francis Group, London, ISBN: 978-1-138-02908-8

CO_2 reduction from cement industry

D. Benghida
Department of Architecture, Dong-A University, South Korea

ABSTRACT: Arguably, concrete has been and will continue to be the most important substance that has endured for centuries. Approximately 32 billion tons of concrete are produced each year. Concrete has found its major use in dams, highways and many different kinds of building and constructions. In addition to longevity, the Roman recipe (used to construct Roman cities and landmarks) is reported to be a much greener material, requiring substantially less energy in the manufacturing process. However, many scholars argue that nowadays the making of concrete takes tons of energy; its production leads to a global carbon dioxide (CO_2) emission every year, which accounts to 13% of global carbon emissions in 2014 due to its huge demand, in quantity. This makes buildings partly responsible for energy use around the world, and this is the reason why the growing trend in the building industry today is towards eco-friendly designs. In the last decade, the concrete industries have spent significant resources to promote their material as the optimum solution for sustainable building design. Pressure to reduce CO_2 emissions continues to grow at the same time with levels of legislation and incentives to preserve our environment. The review in this study helps, first, in developing an insight about the role of buildings in global carbon emission with a focus on cement, and second, in understanding what major solutions are needed to reduce it with appropriate technologies and materials.

1 INTRODUCTION

Many parts of the world are going through an accelerated industrial expansion that has resulted in a rapid urban growth and a fast urbanization. This expansive growth demands more energy, more power, and more construction materials, hence it creates the need for raw material extraction, transportation and manufacturing, logistics, packaging, and waste management, among others aspects.

Figure 1, shows the worldwide cement production, by major producing countries; clearly China and India are the global top two cement producers respectively with 59.80% and 6.69% (U.S. Geological Survey, Mineral Commodity Summaries, 2015).

According to the U.S. Geological Survey, in a report published January 2015, the global cement output in 2014 is estimated at 4.18 billion tons, driven by growth in the top two producing countries. Therefore, the concrete industry is looking into various alternatives to reduce their environmental impact.

The building sector with its consumption and production patterns has gradually impacted on key environmental aspects such as energy, carbon, and waste.

In general terms, buildings contribute by approximately 30% to total global Greenhouse Gas (GHG) emissions (UNEP, 2009). Because energy consumption and greenhouse gas emissions from buildings have both become a known fact, a lot of efforts have been made to reduce the environmental impacts of buildings.

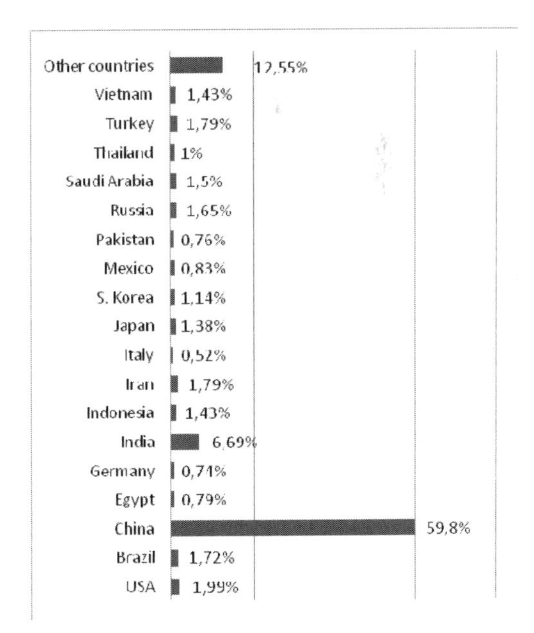

Figure 1. World cement production 2014 in %, by region and main countries, updated from USGS Mineral Program Cement Report (January 2015).

In this paper, I review the role of cement industry in global CO_2 emissions, describe the cement production process, and then compare cement to other construction materials. I will also provide a review of the opportunities for emission reduction by listing the current best approach to tackle concrete's CO_2 emissions in the building industry.

2 CONCRETE, A COMPLEX BALANCED MIX

The first concrete was used by Egyptians, then transmitted to Romans and then after some 2000 years, in the early nineteenth century, it was born again with the discovery of Portland cement; the key ingredient used in concretes today, i.e., a concrete made by burning grounded chalk and finely crushed clay in a limekiln till the carbon dioxide evaporates.

Historical evidence points that ancient Egyptians have used lime and gypsum cement for concrete. Their pyramids are built with Lime mortars and gypsums. Romans from their side are asked to use a primal mix consisting in Pozzolana, or small gravel and coarse sand mixed with hot lime and water, and sometimes even milk or animal blood.

One key element in the general properties of concrete nowadays is that it should possess certain physical and chemical properties, tensile strength, low-level of permeability to avoid moisture and retain chemical and volume stability. Essentially, it has a high level of compressional strength however because its tensile strength is weak (which may lead to cracking), it is further reinforced by steel bars or fiber and iron mesh.

It turns out that the Romans would have baked their ingredients at much lower temperatures, reducing the amount of fuel burned to make concrete. They have also included additives such as animal blood to improve concrete performance, and the Chinese added sticky rice to their mixtures when building the Great Wall during the Ming dynasty.

Scientists and engineers nowadays are actively looking into reducing its environmental impact. The one ingredient that is responsible for its high carbon emissions is cement. Indeed, nowadays there are several ways being researched to increase the workability of concrete and reduce its environmental impact. Workability of concrete means the ability of a concrete to fill the mould appropriately. This can be achieved with the water ratio, the level of hydration and shape and size of the aggregate. In order to produce strong varieties of concrete, additives called plasticizers are also used.

Ultra strong concrete = Ultra strong plasticizers + Maximum water reduction

3 WHAT IS THE DIFFERENCE BETWEEN CEMENT AND CONCRETE?

The terms "cement" and "concrete" are often used interchangeably which may lead to confusion because cement is actually an ingredient of concrete. Cement is the fine and gray powder that, when mixed with water, sand and gravel, forms the rock-like mass known as concrete.

Concrete or as named in Latin 'concretus', means compact and condensed. It is a mixture of concocted materials like sand, gravel, recycled concrete, or small rocks combined with any type of cement and water then allowed to dry and harden. In fact, water, sand, stone or gravel, and other ingredients make up about 90% of the concrete mixture by weight.

Figure 2, clearly shows that cement is just part of the concoction (11%). Actually, cement acts as a "binding agent" or "glue" that holds firmly all materials together. Varying the volume proportions of the concrete will deliver different types of concrete, hence different types of structures. (Goodfellow, 2011).

The product from the burning process during manufacture of cement Equation (1), called clinker, is then blended with other ingredients to produce the final cement product Equation (2).

The cement is manufactured following a whole process that goes from quarrying, crushing, or grinding the raw materials (often it is limestone [$CaCO_3$], chalk, and clay), which are then turned into either a dry powder or wet slurry in the kiln. The heat in the kiln (which can go up to 1450°C) will then fuse the raw materials into small pellets known as clinker. The latter is then left to cool down before it is turned into the fine powder known as Portland cement:

Figure 2. Typical volumetric proportions of concrete basic ingredients (Goodfellow, 2011).

$$CaCO_3 + heat(700-1000°C) \rightarrow CaO + CO_2 \uparrow \quad (1)$$

$$CaO + SiO_2 + Al_2O_3 + Fe_2O_3 + \text{other oxides} + heat(1350-1450°C) \rightarrow \text{Cement} \quad (2)$$

So during cement production, first, limestone ($CaCO_3$) is converted to lime (CaO) and second, CO_2 is emitted by fossil fuel combustion during clinker production. CO_2 is also emitted in a relatively small amount during the process of mining sand and gravel, crushing stone, combining the materials in a concrete plant and transporting concrete to the construction site. But it remains that CO_2 emissions from cement clinker production constitutes one of the largest source of CO_2 emissions in general. In 2013, the cement clinker CO_2 emission increased globally by 7.4% mainly due to a 9.3% increase in the production in China, which accounted for more than half of total global production (Olivier et al. 2014).

"The rule of thumb is that for every ton of cement you make, one ton of CO_2 is produced," says Marios Soutsos, who studies concrete at the University of Liverpool, UK (Crow 2008). Knowing that the global emissions of CO_2 stood at 32.30 billion tons in 2014 (International Energy Agency, 2015), and by using Soutsos rule, my calculations of the global CO_2 emissions from cement production during the year 2014 came to equal: 12.94%.

Marios Soutsos says that "Modern cement kilns are now more efficient, and produce about 800 kg of CO_2 per ton—but that is still a big emission." (Crow, 2008). Of those 800 kg of CO_2, around 530 kg is released by the limestone decomposition reaction itself.

So to resume, the production of cement releases greenhouse gas emissions at different steps while:

- Heating the limestone (chemical process called calcination). This process accounts for about 50% of all emissions from cement production.
- Burning fossil fuels to heat the kiln because coal, natural gas, or oil are used. This represents around 40% of CO_2 cement emissions.

Powering additional plant machinery through electricity and transporting cement accounts for 5 to 10% of the industry's emissions.

4 HOW DOES CEMENT COMPARE TO OTHER BUILDING MATERIALS?

A large amount of materials is used in buildings, which explains the high content of embodied energy in most buildings. Embodied energy refers to energy consumed by all of the processes associated with the production of building materials (mining, manufacturing of materials, etc.). A high level of embodied energy used implies a high level of pollution at the end of the production line, because as energy is consumed, CO_2 is emitted. The materials that have most likely the highest degree of CO_2 emission are aluminum and steel.

When comparing the energy of production for concrete and other building materials, a research study concluded that the energy required for producing one metric ton of reinforced concrete was 2.5 GJ/ton compared to 30 GJ/ton for steel and 2.0 GJ/ton for wood (Figure 3).

A similar and obvious difference was noted when the same study was compared with the CO_2 emissions of several different building materials per 1000 kg for residential construction (Table 1):

Concrete came to be accounted for 147 kg of CO_2, metals accounted for 3000 kg of CO_2, and wood accounted for 127 kg of CO_2 (Pentalla, 1997).

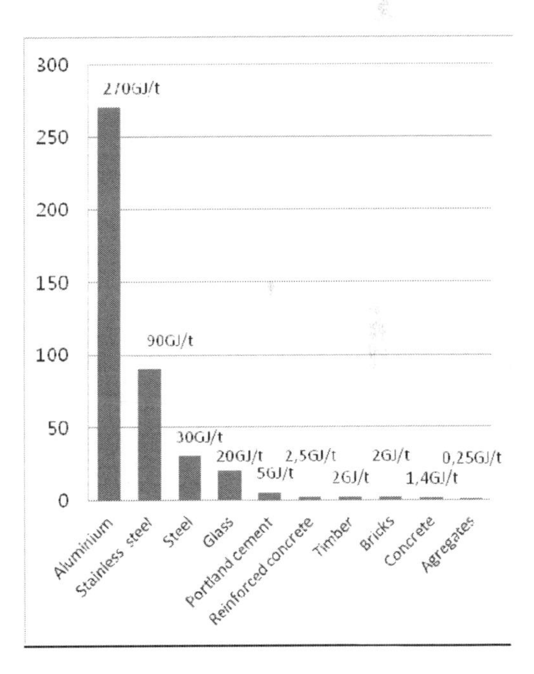

Figure 3. Production energy consumption of different building materials (Penttala, 1997).

Table 1. Production CO_2 emissions of building materials (Penttala, 1997).

Building material	CO_2 [Kg/t]
Wood	124
Concrete	147
Glass	2100
Oil & Plastic	6000
Metals	3000

5 STRATEGIES TO REDUCE CO_2 EMISSIONS FROM CEMENT INDUSTRY

One of the best approaches to tackle concrete's CO_2 emissions in the building industry is to focus on the Portland cement production process. Among the new technologies and production fuels requirement that are used to reduce the effects of Portland cement, there is a use of recycled scrap material to heat kiln like old tires for instance (Laboy-Nieves, 2014).

Replacing the Portland clinker, either partially or entirely, with alternative cements is also another approach. This can refer to any waste materials like slag (from blast furnaces) or fly ash (from coal-fired power stations). Nevertheless, clinker replacement remains a limited option when there is a high demand for cement. Replacing clinker with slag necessitates to mix alkali to activate the mixture, however, alkali-silica reaction is a problem in the future, because as time passes (which can mean over 60 years) more and more aggregates are reactive (Scrivener & Kirkpatrick, 2008).

Another approach to the carbon footprint problem is to reduce the amount of cement by taking out at least 10 per cent of the cement content while retaining durability. Ravindra Dhir, director of the concrete technology group at the University of Dundee in the United Kingdom, believes that if 20% of cement is replaced with a particular mixture, the durability of the final concrete will improve: "the less [cement] you use, the better the concrete should be" (Crow, 2008).

Nevertheless, even if there is a particular mixture of blended cements which has high ability to reduce the CO_2, the problem of uncertainty about their future performance and durability "in the future" remains. Similarly, even though the uptake of alternative cementitious materials has been positive, the demand for cement remains in force.

6 CONCLUSION

About 13% of the global CO_2 emissions is coming from cement production, mainly due to its production processes. Many scientific approaches and substantial strategies are made to achieve deep CO_2 emission reductions, to better the efficiency of cement materials' use and to create new cement materials that allow for higher energy efficiency during product use. For the time being, these efforts are not sufficient to meet the projected goal. The development of new cement materials, new cement substitutes or new processes are all limited either by the uncertainty of their future efficiency, by the capital stock turnover, norms and standards for alternative cements, or by saleable non-acceptance by the construction industry market.

From an energy use standpoint, concrete is preferable to aluminum or steel in the construction industry, and is generally more advantageous than glass, plastic, and metals with regard to generation of CO_2 and other greenhouse gases. So from an environmental impact point of view, considering that many approaches have been taken to reduce CO_2 emissions from cement industry, results consistently shows that concrete construction will keep being promoted as environmentally advantageous to alternative forms of construction. Therefore, it can be easily forecasted that cement use will keep increasing for decades to come.

ACKNOWLEDGMENT

This work was supported by the Dong-A University Research Fund.

REFERENCES

Crow, J.M. 2008. The Concrete Conundrum, *Chemistry World*, March, 62–66.

Goodfellow, R.J.F. 2011. *Concrete for Underground Structures Guidelines for Design and Construction.* Englewood, Colo: Society for Mining, Metallurgy, and Exploration.

International Energy Agency, 2015. Global energy-related emissions of carbon dioxide stalled in 2014, March. [online] Available at: http://www.iea.org/newsroomandevents/news/2015/march/global-energy-related-emissions-of-carbon-dioxide-stalled-in−2014.html [Accessed 7 Sep. 2015].

Laboy-Nieves, E.N. 2014. Energy Recovery from Scrap Tires: A Sustainable Option for Small Islands like Puerto Rico. Sustainability, 6: 3105–3121.

Olivier, J. 2014. *Trends in global CO2 emissions: 2014.* Report PBL Netherlands Environmental Assessment Agency The Hague, The Hague: PBL Netherlands Environmental Assessment Agency, Available at: http://jrc-2014-trends-in-global-co2-emissions-2014-report-93171.pdf [Accessed 7 Sep. 2015].

Pentalla, V. 1997. *Concrete and Sustainable Development, ACI Materials Journal*, September-October, American Concrete Institute, Farmington Hills, MI.

Scrivener, K.L. & Kirkpatrick, R.J. 2008. Innovation in use and research on cementitious material, *Cement and Concrete Research,* 38(2): 128–136.

U.S. Geological Survey, 2015. *Mineral Commodity Summaries: Cement publication news*, Virginia, available online at: http://minerals.usgs.gov/minerals/pubs/commodity/cement/mcs-2015-cemen.pdf [Accessed 7 Sep. 2015].

UNEP. 2009. *Buildings and Climate Change Summary for Decision Makers*. UNEP DTIE, Sustainable Consumption & Production Branch, Paris, France. Available at: http://www.unep.org/sbci/pdfs/SBCI-BCC-Summary.pdf [Accessed 7 Sep. 2015].

Advanced Materials, Mechanical and Structural Engineering – Hong, Seo & Moon (Eds)
© 2016 Taylor & Francis Group, London, ISBN: 978-1-138-02908-8

A study on the removal of arsenic from arsenopyrite in gold concentrate using thermal decomposition method

J.P. Wang
Department of Metallurgical Engineering, Pukyong National University, Busan, Korea

ABSTRACT: A study on the removal of arsenic from gold concentrate was conducted using thermal decomposition method under inert atmosphere (UHP Ar). Thermodynamically, the Arsenic (As) will be decomposed from arsenopyrite (FeAsS) at the temperature range of 600°C to 900°C. Thermogravimetry (TGA) was applied to measure weigh loss of the concentrate for decomposition of arsenic as a function of temperature and time. The content of arsenic (As) was successfully reduced to about 0.3 wt.% from 11.07 wt.% of the initial concentrate regardless of temperature and the arsenic sulfide (AsS) was also recovered as a form of powder with mean particle size of 5 μm.

1 INTRODUCTION

Gold (Au), a representative precious metal, is used across various industrial fields due to its excellent physicochemical properties. Particularly, gold is widely applied as a surface coating material and as a bonding wire material for electrical and electronic parts, and the industrial demand for gold is expected to increase even further in the future (Marsden & House, 2006, Adams 2005).

Various studies have been performed to extract gold from gold concentrates and tailings, while the environmental constraints have been encountered in the leaching process due to the use of cyanide and the presence of various impurities other than gold (Nam et al. 2008, Tongamp et al. 2010, Curreli et al. 2009, Yang et al. 2009). In particular, the gold concentrate contains impure elements such as iron (Fe), Copper (Cu), Zinc (Zn), lead (Pb), and Arsenic (As). In the case of high As content, the dissolution rate is low, which makes separation and recovery of gold in the refining process difficult, and even the high arsenic content causes a difficulty in arsenic treatment.

Kim et al. (2005) performed a study to remove arsenic and heavy metal from tailings by a soil washing process. The most effective detergents for arsenic removal were determined to be oxalic acid (72% removal efficiency) and phosphoric acid (65% removal efficiency), of which oxalic acid was proven to be very effective in removing heavy metals including Cu (Kim & Kim 2005). However, this process has limitations in commercialization as the removal rate of arsenic was still relatively low and the wastewater after the soil washing process would pose an environmental risk of secondary contamination.

A characteristic study by the electrokinetic technique was performed for removal of the arsenic from regosol in heavy metal mines, and the arsenic removal rate was measured using four types of electrolyte in this study. The results showed that the removal rate of arsenic was found to be 18.6% for citric acid + Sodium Dodecyl Sulfate (SDS), 8.1% for 0.1 NHNO, 7.4% for HAC, and 6.6% for distilled water. The highest removal rate was just under 20%, which demonstrated the technical potential of the process in question, but was too low to make the process viable for practical application (Shin & Yoon 2006).

Thus, this study was conducted with an aim to improve on removing arsenic from gold concentrate using an eco-friendly thermal-treatment method in contrast to the conventional processes of removing arsenic using wet methods and to maximize the recovery rate of gold in the refining process. The gold concentrate as a form of powder was placed inside an electric furnace with an inert atmosphere and the reaction temperature and time were varied to determine the conditions for efficient removal and collection of arsenic using a bag-filter.

2 MATERIALS

The material used in this study was the concentrate in a form of powder collected from a "Mine A" in Korea. The chemical composition of gold concentrate was analyzed using Inductively Coupled Plasma Optical Emission Spectrometry (ICP-OES) and its result is shown in Table 1. The concentrate was found to be a sulfide mineral which mainly contains Fe and Zn with S. The content of As of the total amount that should be removed was about 11% as well as minor amounts of Ca, Zn, Si, Mn and Al contained in the concentrate. In addition, the concentrate was also analyzed by the X-Ray Diffraction (XRD) shown in Figure 1. The patterns of peak illustrate that the gold concentrate is comprised of ZnS, FeS, and FeAsS and that As is observed to be only present in FeAsS (arsenopyrite). The FeAsS is known to be a representative ore containing As with an average As content of approximately 46% (Hurlburt & Klein, 1985).

Figure 2 is a schematic diagram of the experimental apparatus, which was constructed to allow the reactor to be inserted into the upper part of the box-type electric furnace. The reactor body and cover used in this experiment was made of the SUS310 material, which is durable and corrosion-resistant at high temperatures in order to minimize the potential thermal fatigue and failure defects that could take place during the experiment. To ensure perfect sealing, an O-ring was installed between the reactor body and cover to prevent the passage of internal and external gases. A cooling line was installed underneath the cover to ensure that the O-ring did not melt even at high temperatures. The sample powder was placed in an alumina boat and inserted into the center of the reactor to facilitate the release of gaseous arsenic through an off-gas line. The off-gas line was connected to an Erlenmeyer flask to collect gaseous arsenic, and a

Table 1. Chemical composition of gold concentrate analyzed by ICP.

	Chemical composition (wt.%)			
	Al	Si	S	Ca
Content	0.140	2.486	36.355	1.899
	Mn	Fe	Zn	As
Content	0.302	32.371	7.777	11.071

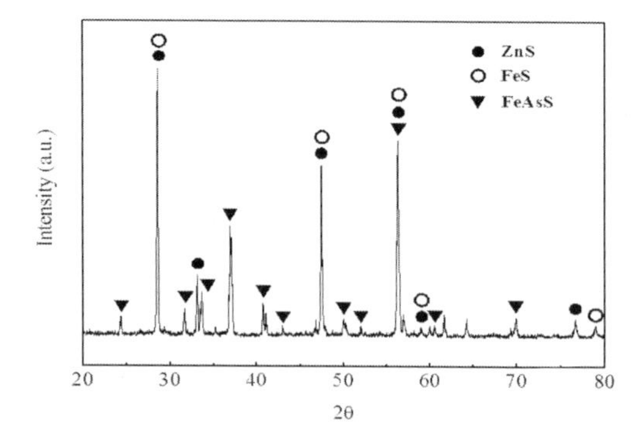

Figure 1. XRD patterns of gold concentrate.

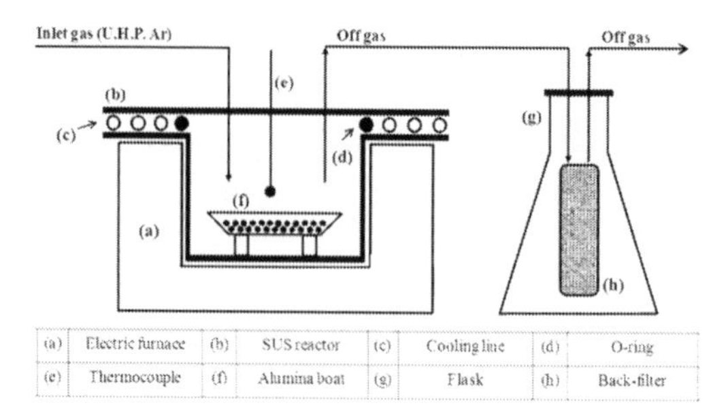

Figure 2. Experimental apparatus for removal of arsenic from gold concentrate.

bag-filter was placed at the end of the line to collect arsenic at room temperature.

As for the experimental method, the gold concentrate in a form of powder was spread out on an alumina boat in a 5 mm thick layer and the boat was put into the center of the reactor, after which the reactor cover was combined with the body. Then, the U.H.P argon gas was fed into the system at a rate of 0.2 liter/min and then temperatures were increased to 600°C, 700°C, 800°C, 900°C for 120 minutes. During the experiment off-gas generated was solidified to be powder form in the bag-filter at room temperature.

The chemical composition of the gold concentrate conducted at each temperature was analyzed by Inductively Coupled Plasma Optical Emission Spectrometry (ICP-OES) and the powder collected in the bag-filter was also examined through X-Ray Diffraction (XRD) and Energy Dispersive X-ray Spectrometer (EDX).

3 RESULTS AND DISCUSSION

This experiment was conducted with consecutive processes for the removal and recovery of arsenic from the gold concentrate, and the chemical reaction equations for the processes are shown in (1) and (2). The first-step (1) is a thermal decomposition reaction in which decomposition will take place spontaneously at high temperatures. As may be seen in Figure 1, As in the gold concentrate is present only in the FeAsS phase, which, thermodynamically speaking, may be thermally decomposed into FeS and As at the temperature range of 600°C to 900°C. Hence it may be seen that it is possible to remove arsenic as a way to cause a phase transition of arsenic from solid to gas by raising the temperatures under inert environment (UHP Ar). In particular, the boiling point of arsenic is lower than its melting point at standard atmospheric temperature, which means that solid arsenic will sublimate into the gas phase at about 614°C (887 K) and thus the temperature of 600°C for the thermal decomposition reaction is deemed sufficient for removal of arsenic from the concentrate. On the other hand, second-step (2) represents a phase change of arsenic from gas to solid at room temperature and the arsenic powder is collected using a bag-filter mounted at the end of the off-gas line in the long run.

$$FeAsS \rightarrow FeS + As(g)\ \Delta G_{600^\circ C} = -10.55\ kcal$$
$$FeAsS \rightarrow FeS + As(g)\ \Delta G_{700^\circ C} = -10.24\ kcal$$
$$FeAsS \rightarrow FeS + As(g)\ \Delta G_{800^\circ C} = -9.99\ kcal$$
$$FeAsS \rightarrow FeS + As(g)\ \Delta G_{600^\circ C} = -9.95\ kcal \quad (1)$$

$$As(g) \rightarrow As(s) \quad (2)$$

Based on thermodynamic considerations for the thermal decomposition method, Thermogravimetry (TGA) was used to examine weight loss of the gold concentrate at elevated temperatures and its result is shown in Figure 3. It can be seen that the pattern of weight loss was found to be almost same regardless of temperatures and weight loss of the sample was dramatically increased at about 90 minutes. It was found that the weight loss of (a) and (b) was relatively larger than (c) and (d), since Gibbs free energies at low temperatures of 600°C and 700°C have more negative values than 800°C and 900°C which might encourages thermal decomposition reaction to occur actively.

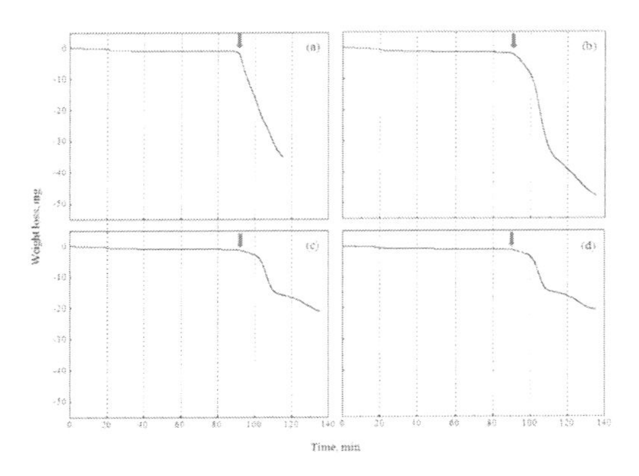

Figure 3. Weight loss with time at each temperature analyzed by TGA. (a) 600°C (b) 700°C (c) 800°C (d) 900°C.

The chemical composition of gold concentrate after thermal decomposition at each temperature was measured by Inductively Coupled Plasma Optical Emission Spectrometry (ICP-OES) and its results are summarized in Table 2. The As content of the initial sample are found to be about 11.071 wt.% and Fe and Zn are major composition of the concentrate. Once thermal treatment was finished at each temperature for 120 minutes, the content of As was dramatically decreased to 0.397 wt.% at 600°C, 0.298 wt.% at 700°C, 0.311 wt.% at 800°C, and 0.307 wt.% at 900°C, respectively. The arsenic in gold concentrate exists as a part of the complex compound (FeAsS), from which it decompose to be FeS and As to be removed in its gas form. Furthermore, the contents of other elements such as Al, Si, Ca, Mn, Zn remain almost unchanged.

The change of content of Fe, S, Zn, and As and the removal rate of As with temperatures are shown in Figure 4. It is noted that the content of S in concentrate was found to be decreased with similar trend of As due to its evaporation during thermal decomposition process, while the content of Fe was increased by the loss of As and S in FeAsS of the concentrate. In addition, the content of Zn was also increased by the loss of S in ZnS of the concentrate. The removal rates of As from the initial sample were calculated to 96.41% at 600°C, 97.31% at 700°C, 97.19% at 800°C, and 97.22% at 900°C, respectively. The minute amount of arsenic remaining inside the compound requires more reaction time than that in this experimental condition to be thermally decomposed and completely removed, which can be considered to be a limitation for removing arsenic by a dry method.

Figure 5 shows the results of XRD patterns of the arsenic powder collected at room temperature using a bag-filter mounted at the end of the off-gas line during thermal decomposition process. It can be found that only pure arsenic sulfide (AsS) was obtained with no detection of other impurity forms. The reason that the powder collected in the bag-filter was AsS than As in its single element form might be considered to be the fact that the sulfur contained in the sulfide ore was sublimated and

Table 2. Chemical composition of gold concentrate at each temperature.

	Chemical composition (wt.%)							
	Al	Si	S	Ca	Mn	Fe	Zn	As
Concentrate	0.140	2.486	36.355	1.899	0.302	32.371	7.777	11.071
600°C	0.182	2.054	23.901	1.743	0.402	43.548	9.524	0.397
700°C	0.166	1.817	26.918	1.685	0.323	44.039	8.398	0.298
800°C	0.183	1.755	21.227	1.604	0.386	47.296	9.112	0.311
900°C	0.192	1.974	20.952	1.888	0.382	44.586	10.200	0.307

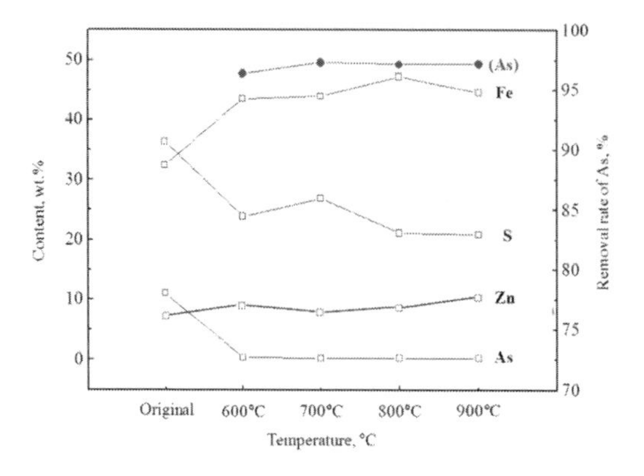

Figure 4. Change of contents of Fe, S, Zn, and As in gold concentrate and removal rate of As with temperatures.

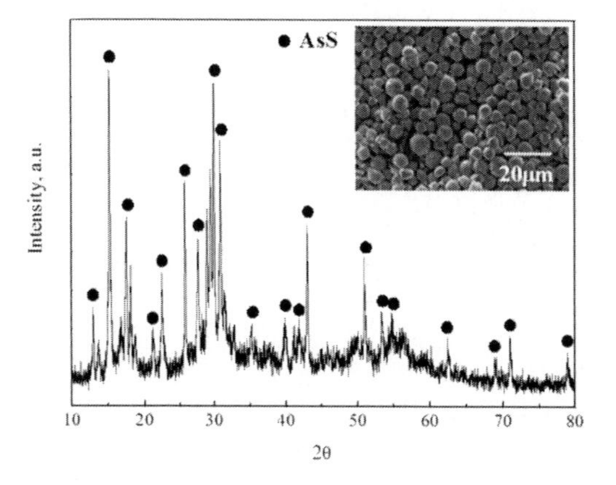

Figure 5. XRD patterns and shape of powder collected at 700°C for 120 minutes..

was released in the gas phase reacted with arsenic to form AsS before being collected in the bag-filter. The recovered arsenic sulfide (AsS) may be used as pigment or colorant and the arsenic may be used as an additive in the lead battery for automobiles and semiconductor materials (GaAs) as a means to minimize environmental problems caused by the arsenic gas emission and to enhance the added value of waste resources. As shown in the chemical reaction equation for As and S (3), the Gibbs free energy in this reaction is −43.83 kcal, which means that this reaction spontaneously occurs to produce AsS at 700°C.

$$As + S(g) = AsS\ \Delta G_{700°C} = -10.24\ kcal \quad (3)$$

Also, the recovered AsS was a form of powder in spherical shape with an average diameter of about 5 μm. The synthetic powder collected could be easily separated from the bag-filter with just a slight impact for easy recovery as a spherical powder shape. In summary, As from FeAsS was efficiently removed through thermal decomposition reaction and AsS in powder form could also be recovered using a bag-filter at the rear end of the off-gas line.

4 CONCLUSIONS

In this study, thermal decomposition method was carried out to remove arsenic (As) from arsenopyrite (FeAsS) in the gold concentrate. The content of arsenic in the initial sample was reduced to about 0.3 wt.% regardless of temperatures and the highest removal rage of arsenic was found to be about 97.22% at 900°C for 120 minutes. The arsenic released as the off-gas was effectively recovered as a form of arsenic sulfide (AsS) by a bag-filter.

ACKNOWLEDGEMENT

This work was supported by a Research Grant of Pukyong National University (2015 year).

REFERENCES

Adams, M.D. 2005. *Advances in Gold Ore Pressing*, Elsevier.

Curreli, L., Garbarino, C., Ghiani, M. & Orru, G. 2009. Arsenic leaching from a gold bearing enargite flotation concentrate, *Hydrometallurgy*, 96: 258–263.

Hurlburt, C.S. & Klein, C. 1985. *Manual of Mineralogy*, 20th edn. John Wiley & Sons, Inc.

Kim, T.S. & Kim, M.J. 2005. Remediation of Mine Tailings Contaminated with Arsenic and Heavy Metals: Removal of Arsenic by Soil Washing, *Journal of Korean Society of Environmental Engineers*, 30: 808–818.

Marsden, J.O. & House, C.I. 2006. *The Chemistry of Gold Extraction*, SME.

Nam, K.S., Jung, B.H., An, J.W., Ha, T.J., Tran, T. & Kim, M.J. 2008. Use of chloride-hypochlorite leachants to recover gold from tailing, *International Journal of Mineral Processing*, 86: 131–140.

Shin, H.M. & Yoon, S.S. 2006. Removal Characteristics of Arsenic from Abandoned Metal Mining Tailings by Electrokinetic Technique, *Journal of the Environmental Science*, 15: 279–286.

Tongamp, W., Takasaki, Y. & Shibayama, A. 2010. Selective Leaching of Arsenic from Enargite in NaHS-NaOH Media, *Hydrometallurgy*, 101: 64–68.

Yang, J.S., Lee, J.Y., Baek, K., Kwon, T.S. & Choi, J. 2009. Extraction behavior of As, Pb, and Zn from mine tailings with acid and base solutions, *Journal of Hazardous Materials*, 171: 443–451.

Advanced Materials, Mechanical and Structural Engineering – Hong, Seo & Moon (Eds)
© 2016 Taylor & Francis Group, London, ISBN: 978-1-138-02908-8

Process control system of Kazakhstan Tokamak for Material Testing

A.G. Korovikov & D.A. Olkhovik
National Nuclear Center, Kurchatov, Republic of Kazakhstan

ABSTRACT: In this experiment, the automation object of a control system that helps in the preparation of the vacuum chamber is being investigated; the configuration of the Process Control System (PCS) and software is being developed; and the equipment and the SCADA system are chosen. The tests of complex performance and the study of the PCS are specified on the real control object that is confirmed by compliance of technical characteristics along with its requirements.

1 INTRODUCTION

Currently, Kazakhstan Tokamak for Material Testing KTM is being constructed in Kurchatov. In tokamaks plasma parameters are defined not only by the quantity and volume of the composition of residual gas but also by the state of vacuum chamber surface which is a source of intaking various impurities such as vapors of water, oxygen, carbon, etc., into the plasma. The efficiency of investigations conducted at tokamaks is specified by the complications of the Automation System (AS) available and its capabilities. KTM AS is a high-productive system functionally consisting of Process Control System (PCS), Plasma Control System (PlCS), Digital Control Power Supply System (DCPSS), Emergency Shutdown & Synchronization System (ESSS) and Information & Measuring System (IMS). In this experiment, the given article describes about the PCS thus, ensuring for the preparation of the vacuum chamber.

2 AUTOMATION OBJECT

There are various process systems that are being used in order to decrease the impurities inside of KTM vacuum chamber. They are as follows (Azizov 2006):

- High-Vacuum Pumping-out System (HVPS)
- Water Cooling System (WCS), and
- Vacuum Chamber Heating System (VCHS).

To operate properly KTM needs synchronous and safe work from all in-built process systems. In such a way the process systems are required to be incorporated in a unique Program & Process Complex.

The KTM control system consists of three control subsystems.

The first VCHS is designed for heating and keeping the temperature properly inside the chamber and its branch pipes at a level of 200 ± 10ºC during the pumping-out and purifying.

Automation object includes:

- vacuum chamber divided into 6 heating regions
- ohmic heaters
- induction heater, and
- temperature sensors (E-type thermocouples).

The heaters installed on the vacuum chambers are divided into 5 groups which are connected to individual power supplies. There are 38 thermocouples that control the temperature on a surface of the chamber and heaters.

The internal surface of the vacuum chamber is an additional heating region which is heated to keep the temperature. The controlled power source ensures the induction supply by the current up to 250 A, voltage up to 700 V and frequency up to 50 Hz.

The second WCS subsystem is intended for heat removal from electromagnetic system, vacuum pumps and high-frequency plasma heating system.

In this subsystem, the key parameters for controlled and operated process are the temperature and coolant flow rate at the outlet from cooling the channels and its pressure in feeding the collectors, as well as the temperature in places of screw junctions of the coils, high-vacuum pumps and HF-generators.

The third HVPS subsystem has to pump out 5000 l vacuum chamber from atmospheric pressure and keep their background discharge to 10–5 pa while the heating of the chamber walls goes up to 200°C. The HVPS is operating accurately under established cyclogram.

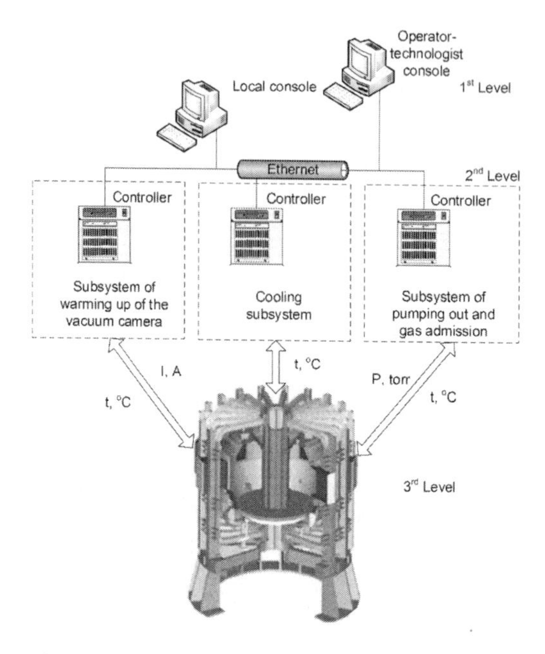

Figure 1. Structure of a system.

3 SYSTEM CONFIGURATION

Control system is a complex three-level process (Figure 1). The first level consists of the operator-technologist work station, the common-users' TV, and the functional keyboard which needs to input a command for process equipment control and give an indication of its state. Data processing and visualization is performed at this level.

The second level coordinates with running the units of the third level and is capable to directly control and monitor the separate parameters of the process system. Three microcontrollers and local console are also a part of the second level.

The third level controls separate blocks and units of the process system; it measures, collects, and processes the input data as well as provides interaction with controllers of the second level through the digital communication channels. This level includes ADAM-6000 series controllers, primary converters, pumps, shutters, etc.

4 SOFTWARE

The software includes SCADA-System (Supervisory Control and Data Acquisition) Trace Mode 6.04. Software operator's workstation (local console, operator-technologist workstation) is designed for Microsoft Windows platform using Trace Mode MRT-monitor real-time. The software of the controllers is designed for Microsoft DOS 6.22 platform (Figure 2).

In the course of each data collecting the cycle, the controller reads out the current values of pressure, temperature and vacuum that then makes preliminary data processing, control and logging in a RAM memory that separates each parameter according to Modbus TCP and ASCII protocol (Figure 3).

The synthesis of automatic control system resulted in obtaining PI-regulators' factors per each heating region in the vacuum chamber. For this the CHR approach was used which is based on the use of the maximum rise in speed in the absence of reregulation or in presence of not more than 20% reregulation foot (Chien 1972).

The program module of Warning & Emergency Protection System (WEPS) is developed. At the temperament of the module the algorithms of control process units are applied, which are involved in the process of high-vacuum pumping. The vacuum chamber and cooling systems are developed and investigated on using the mathematical apparatus—Petri's network (Cassandras 2008). The module allowed operating the long technological

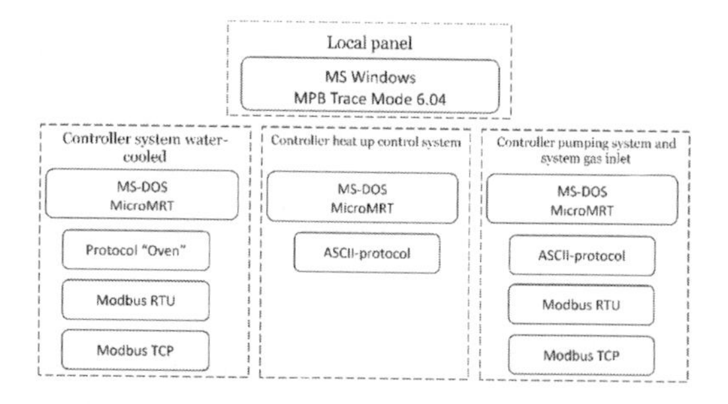

Figure 2. Structure of the software.

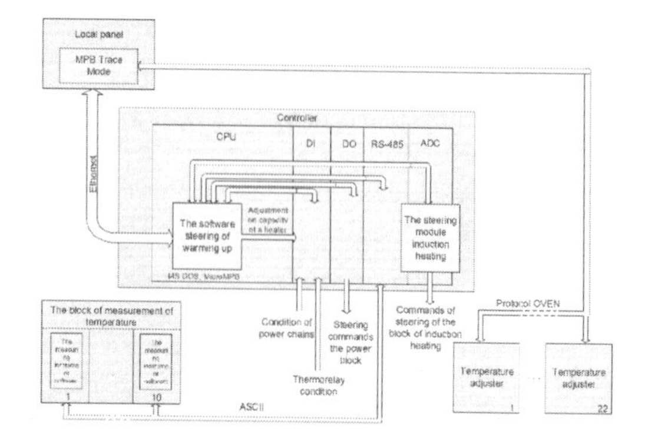

Figure 3. Data interchange in Vacuum chamber heat subsystem.

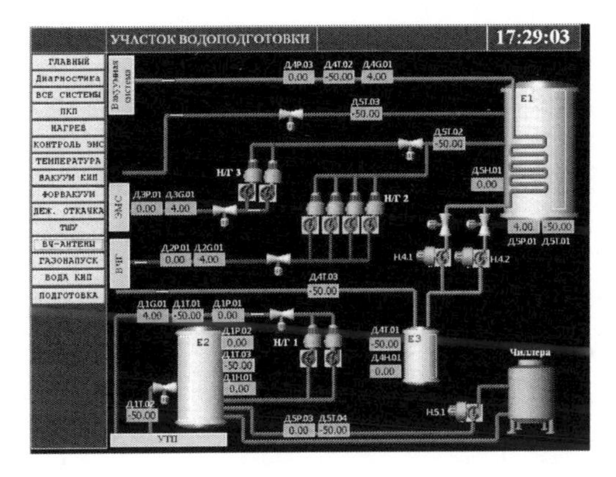

Figure 4. Screen of operator-technologist.

processes under the automatic mode and preventing the expensive equipment against failure in case of any emergency situations or human error.

In SCADA Trace Mode we've applied IEC standard 61131 programming languages like FBD (Function Block Diagram) and ST (Structured Text) (IEC61131-3) to program signal processing algorithms, control algorithms, and drivers of devices.

The technological data processing was performed by the means of human machine interface. Preventive and alarm signals (about an output of those parameters for admissible boundaries, run errors of commands) are expressed as sound, marked in color or displayed as text format (Figure 4).

5 CONCLUSIONS

1. The proposed architecture has made it possible to design territorially distributed system by enabling to the strengthening capacity and widening the functions. The chosen equipment bears high functional, dynamic, and precision specifications.
2. The conducted study of the system justifies the property of designed control algorithms and confirms that theses technical specifications comply with the requirements as follows:
    a. Experimentally received error estimates for measuring channels without primary converters, do not exceed calculated values of maximum permissible error
    b. Quality of heating algorithms completely meets the requirements, namely maximum control error does not exceed 10%, and not more than 2% under steady-state mode
    c. Operating the discrete-event systems correspond to working cyclogram of the high-vacuum pumping-out and cooling; control systems run under automatic mode
    d. No software failure or equipment fault was detected in the course of life tests of hardware-software complex that lasted for 72 hours
3. The work has been completed by introducing the system in KTM tokamak that is acknowledged by the Act of Introduction.

REFERENCES

Azizov, E.A. 2006. *Kazakhstan Materials Research tolamak KTM and challenges of thermonuclear fusion*, Almaty, 236.

Cassandras, C.G. 2008. *Introduction to discrete event*, Dordrecht: Kluwer Academic Publishers, 848.

Chien, K.L., Hrones, J.A. & Reswick, J.B. 1972. On automatic control of generalized passive systems, *Transactions of the ASME*. 74: 175–185.

IEC 61131-3 Ed.3 2012. *International standart. Programmable controllers*. International electrotechnical commission, 229.

Advanced Materials, Mechanical and Structural Engineering – Hong, Seo & Moon (Eds)
© 2016 Taylor & Francis Group, London, ISBN: 978-1-138-02908-8

Performance and reliability analysis of the robot centered FMS and its configuration optimization

D. Zhang & Y.J. Zhang
School of Mechanical Engineering, Xi'an Jiaotong University, Xi'an, P.R. China

ABSTRACT: A reliability analysis model for the robot centered Flexible Manufacturing System (FMS) is proposed by using PRISM, a probabilistic model checker. The test FMS system consists of four components, three machines and a robot. In this model, a novel reliability analysis method is proposed for the evaluation of machines or robots, where the actual operation time spent on equipment is considered to determine the generation of failures. Furthermore, the effect of system reliability on productivity is studied. It is concluded that the Mean Time Between Failures (MTBF) and Mean Time To Repair (MTTR) display a sensitive influence on both productivity and maintenance cost. Aiming at seeking an optimal configuration scheme by available machines and robots in inventory for a greatest benefit, an optimization configuration algorithm based on bottleneck shifting is proposed.

1 INTRODUCTION

The optimal design of FMS is still a critical issue and it is also a complex problem, the most common analytical method is based on mathematical programming (Kumar et al. 2006; Chan and Swarnkar 2006), but when reliability and safety evaluation are needed, analyses on the subsystems of FMS and their complicated inter-relationships turn to be necessary to accurately calculate the performance measures. Therefore, computer modeling and simulation are a more practical and accurate way to solve this problem.

Some vague understanding of reliability analysis for machining and manufacturing systems is clarified in this paper, which include the difference of reliability modelling for electronic devices and machining equipment, and state space representation for machines. Configuration optimization of a robot centered FMS is shown as a case study subsequently.

2 ANALYTIC MODELS

2.1 *Reliability consideration*

Reliability analysis has been done better on electronic equipment or system, in which reliability is defined as the ability of a system or component to perform its required functions under given conditions for a specified period of time, but the prerequisite "under a given condition" is very hard to identify in the manufacturing industry. If two machine tools in a system hold the same MTBF value, but different utilization rates, it is certain that a machine with higher utilization rate breaks down with a higher frequency. Besides, FMS makes it possible to automate production. The production capacity and utilization rate of machines in a FMS is much higher than the stand-alone machine because of the operator break time, material waiting time and tool management time decreases with great amount. Reliability computation in these two cases is different, and this issue rarely has analyzed in previous literatures. The given *conditions* change with the different elements, such as styles of parts, machining processes, load spectrum, utilization rates of machine tools' components and other processing conditions, and there may be a gap between the official MTBF values of machine tools and the actual one.

FMS can be modeled by using a Continuous-Time Markov Chain (CTMC), in which the machines and robots break down with their own failure rates (reciprocals of MTBFs), and come to use again with repair rates (reciprocals of MTTRs). Failure rates of the machines in a manufacturing system are generally considered as a fixed value in many literatures, and, the total time in the service period of the machines is summed up for the reliability calculation. Practically, different from electric equipment, jobs in manufacturing systems are discrete, and failure rates have a special and intimate relationship with the actual processing time of parts, the failure frequency increases with the processing time.

The manufacturing system in this scheme is repairable, machines fail with rate λ (the inverse of MTBF), and are repaired as new with rate μ(the inverse of MTTR). As shown in Figure 1(a), only

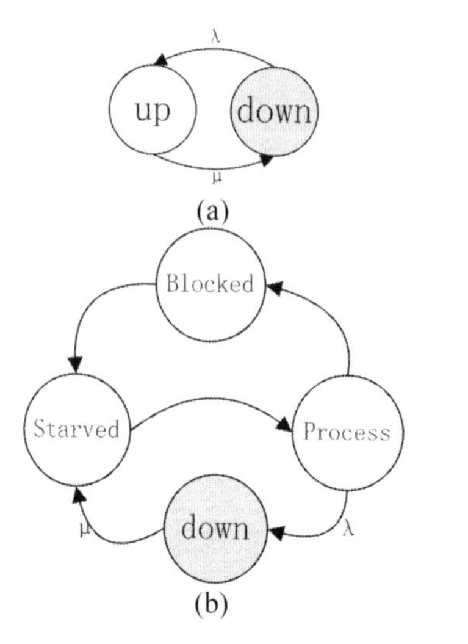

Figure 1. (a) Two states model in previous research. (b) The machine's working processes.

two states of a machine are considered according to reliability methods of Savsar and Aldaihani (M. Savsar 2000, Aldaihani & Savsar 2005, Mehmet Savsar & Aldaihani 2008), machines are either in failed state or in normal state. In Figure 1(b), the states of the machine should be divided into four parts, which are starved, processing, blocked, failed respectively. Only failures in machines' processing time are considered, and the idle failures in the starved and blocked time are neglected.

2.2 *PRISM modeling language*

PRISM is an open-source probabilistic model checker developed initially at the University of Birmingham (Heath et al. 2008), probabilistic models in PRISM are coded and analyzed by the PRISM language, and the model is described as m parallel modules, M = $\{M_1, \ldots, M_m\}$, each of which corresponded to a component of the system, such as a machine tool in FMS. Each *Mi* module is defined as the combination of (V_i, U_i), the set V_i defines the state space of the variables and *Ui* is a set of updates of V_i states.

The behavior of M_i can be described by the states of the V_i and U_i, in each module, the U_i is constructed as guarded commands. The *guard* is a pre-judging before an *update*, the *guard* is composed of several temporal logics joined with "|" or "&", each *update* occurs with the *rate* when the guard is *true* (i.e. the delay before this transition completed is sampled from a negative exponential distribution with parameter *rate*).
[*act*] (*guard*) -> (*rate*):(*update*).

3 BOTTLENECK DETECTION AND OPTIMIZED ALGORITHM

MTBF is the important assessment criteria of a device's reliability, processing time of a part varies little between different machines processing a single working procedure, and MTTR is mainly depend on the level of maintenance. The remaining parameter MTBF has a sensitive effect on the total benefit of the enterprise not only for the considerable extra production, but also the decrease of total maintenance cost. However, these benefits are at the expense of higher allocation cost. As a result, production managers tend to judge and weigh the benefits and costs to decide the appropriate system configuration, and the question to them is how to select an optimized combination of the robot and machines and achieve a highest economic efficiency.

Firstly, the damage of machines' failures should be evaluated, the states of a device can be divided into four parts, including starvation, processing, blockage and failed, the states. And their probabilities can be achieved by properties in PRISM as followings.

In Table 1, X can be replaced by M1, M2, M3 and R. The probabilities of M1, M2, M3 and the robot's failure being an obstacle to the total performance can be calculated respectively as follow.

SM1 = ?[(Raw_part = 1&R_idle = 1&M1_failed = m1)|(M1_failed = m1&M2_idle > 0&R_idle = 1)].----The raw parts conveyor and the robot are ready, or M2 is in starvation and the robot is ready, but M1 is failed.

SM2 = ?[(M1_finished > 0&R_idle = 1&M2_failed = m2)|(M2_failed = m2&R_idle = 1&M3_idle > 0)].----The M1 is blocked and the robot is ready, or M3 is in starvation and the robot is ready, but M2 is failed.

SM3 = ?[(M2_finished > 0&R_idle = 1&M3_failed = m3)|(M3_failed = m3&R_idle = 1&Finished_part = 0)].----The M2 is blocked and the robot is ready, or the finished parts conveyor is in starvation and the robot is ready, but M3 is failed.

SR = ?[(Raw_part = 1&M1_idle > 0&R_failed = 1)|(M1_fiished > 0&M2_idle > 0&R_failed = 1)| (M2_finished > 0&M3_idle > 0&R_failed = 1)| (M3_finished > 0&Finished_part = 0&R_failed = 1)]. ----The situations that the robot is failed to do its job.

Losses of devices' failures (*DX*) is computed as

$$D_X = S_X \times P_X, \quad X = M1, M2, M3, R. \tag{1}$$

where, *SX* is the probability of X being an obstacle, and *PX* is mean productivity loss under failures.

Table 1. Meanings of the properties.

Properties specification language	Meanings
S = ? [X_idle = 1]	The long-run probability of X's starvation.
S = ? [X_process = 1]	The long-run probability of X being processing a part.
S = ? [X_finished = 1]	The long-run probability of X's blockage.
S = ? [X_failed = 1]	The long-run probability of X being failed.

Device with the biggest *DX* brings the most damage, and it is considered as the bottleneck of the system, the next is the second bottleneck and so on.

An optimized algorithm for configuration is proposed based on bottleneck shifting. There are different definitions about a bottleneck; here we define it as the machine or robot's failure that brings the most damage to the total system, they can lead to the blockage of upstream machine or the starvation of the downstream machine. In Figure 2, the FMS configuration can be listed with the combination F(L, M, C, R), L for lathe code, M for Milling machine code, C for machining center code and R for robot code. The performance of F(L, M, C, R) is usually judged by indexes such as utilization rates of devices, availability and productivity of the system. But the most straightforward way to evaluate the manufacturing system is the benefit it brings, a Benefit-Cost model based on PRISM is proposed as follow.

$$TB = R_P - R_m - C_s.\ s.t. C_s \le C. \quad (2)$$

where, *TB* is the total benefit the system brings, *Rp* is the net income of finished parts, *Rm* is the maintenance cost, *Cs* is the initial setup cost of the configuration, and *C* is the planned maximal setup cost.

Based on the above points, an optimization algorithm can be described as followings:

1. Decide a relatively lower configuration.
2. Calculate *TB* and *DX* of the configuration combination and determine the bottleneck machine or robot based on the value *DX*.
3. Find the nearest machine or robot with a higher MTBF, and calculate the total benefit TB'.
4. Compare TB' with TB, if TB' is higher than TB and satisfying that Cs ≤ C, then choose the configuration combination with TB', and select it as the new configuration combination for iteration.

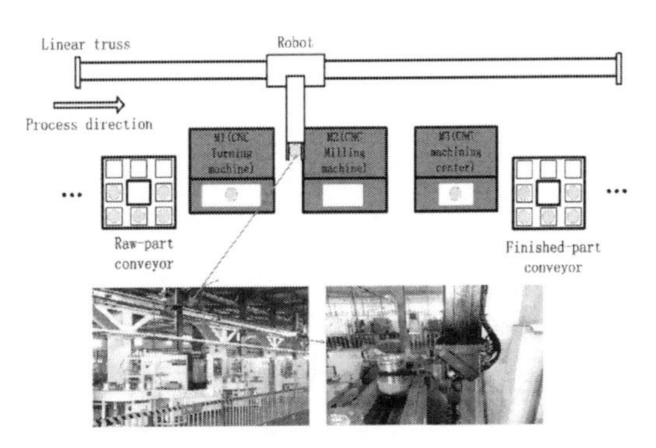

Figure 2. Piston flexible manufacturing line.

Table 2. Information of the machines and robot.

Device type	MTBF (min)	MTTR (min)	Processing time (min/part)	Sale price ($\times10^3$\$)
Robot	80	5	0.05	100
M1	100	10	0.5	150
M2	120	8	0.7	160
M3	150	12	0.55(Route1), 1.2(Route 2)	300

* The profit of each piston is 4 $.

5. If TB' is smaller than TB, then check that whether the bottleneck is the last one, If not, go to the next bottleneck and continue the iteration.
6. If the bottleneck is the last one, then an optimized configuration is achieved.

4 CASE STUDY

4.1 *Modeling for robot centered FMS*

Robot centered system is an important style of FMS, and it is applied widely in manufacturing industry, a robot centered system selected from the piston flexible manufacturing line (as shown in Figure 2) is modeled as a case study, three machines and one robot is contained, the parameters of the system are listed in Table 2. The behavior of FMS can be described as the movement of its working pieces, as in Figure 2, the robot picks up a part from the raw-part conveyor, and loads the M1 processing the first process of the part, then unloads M1 and goes to the M2, until the final process is finished by M3 and a finished part is collected by the finished-part conveyor, if M2 is failed, then the part is shifted to M3 directly in the second process route. It is obvious that the states of machines and the robot is very restricted, a machine

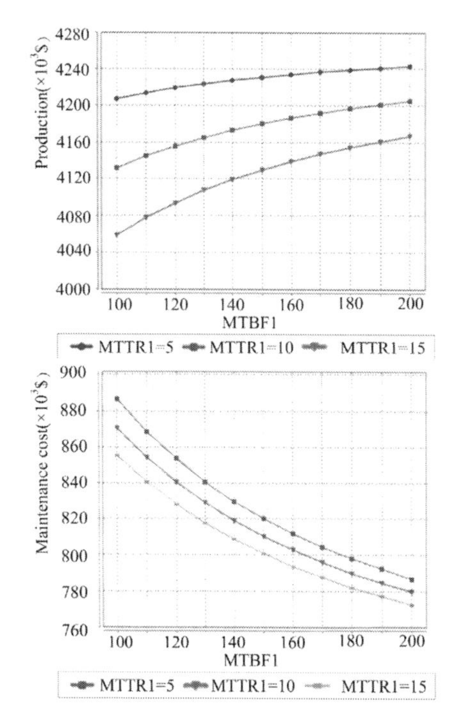

Figure 3. The influences of M1's MTBF and MTTR.

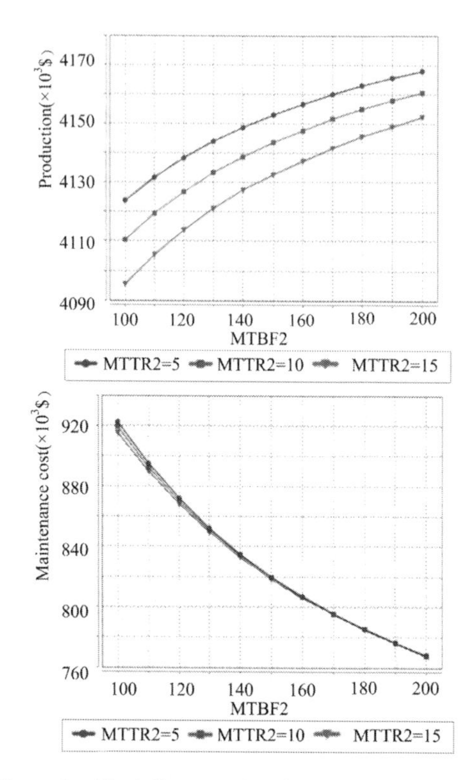

Figure 4. The influences of M2's MTBF and MTTR.

in the system can be divided into four states, including idle time, processing time, failed time, and waiting time, the system goes on working along with the machine and robot states change.

To further investigation of reliability influences, system throughputs under different MTBF, MTTR combinations are shown in Figure 3, Figure 4 and Figure 5, it is indicated that the reliability of the machine does have a great influence on the total productivity and maintenance cost.

The effects of MTBF values of M1 and M3 on the system performance are obvious by the above results, and slightly 10 min more of MTBF can improve a highest benefit of 80000 $, besides, the maintenance level of the team is very important to keep a high productivity.

Reliability analysis with the consideration of system productivity has been rarely studied in traditional configuration optimization methods. For this reason, the effects of reliability and maintainability have been shown with a real system. Besides, the proposed performance and reliability analysis method provides a basis for decision of maintenance resources allocation or reliability improving priorities. By contrast, maintenance resources like spare parts preparation or maintenance interval determination have been simply allocated according to equipment's MTBF and MTTR in former

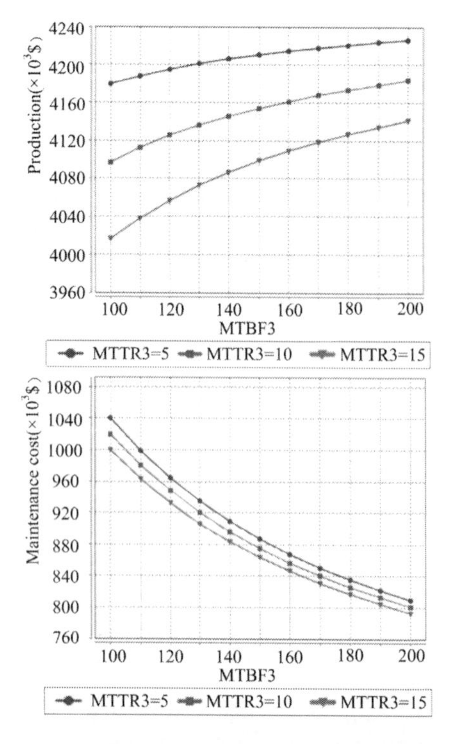

Figure 5. The influences of M3's MTBF and MTTR.

methods, which ignore the sensitivity analysis of these parameters.

5 CONCLUSION

In this paper, an analysis model to the robot centered FMS with three machines and a robot is proposed based on PRISM. A novel method processing reliability of equipment is introduced, in which the actual operation time of these devices is considered to estimate a failure. Effects of different reliability indexes on the system behavior are studied. It is figured that MTBF and MTTR display a sensitive influence on the system productivity and maintenance cost. Then a configuration algorithm based on bottleneck shifting is proposed to group the most suitable machines and robots available in inventory.

ACKNOWLEDGEMENT

This project is supported by National Natural Science Foundation of China (Grant No. 51375377).

REFERENCES

Aldaihani, M.M. & Savsar, M. 2005. A stochastic model for the analysis of a two-machine flexible manufacturing cell. *Computers & Industrial Engineering,* 49(4): 600–610.

Chan, F.T.S. & Swarnkar, R. 2006. Ant colony optimization approach to a fuzzy goal programming model for a machine tool selection and operation allocation problem in an FMS. *Robotics and Computer-Integrated Manufacturing,* 22(4): 353–362.

Heath, J., Kwiatkowska, M., Norman, G., Parker, D. & Tymchyshyn, O. 2008. Probabilistic model checking of complex biological pathways. *Theoretical Computer Science,* 391(3): 239–257.

Kumar, A., Prakash, Tiwari, M.K., Shankar, R. & Baveja, A. 2006. Solving machine-loading problem of a flexible manufacturing system with constraint-based genetic algorithm. *European Journal of Operational Research,* 175(2): 1043–1069.

Savsar, M. 2000. Reliability analysis of a flexible manufacturing cell. *Reliability Engineering & System Safety,* 67(2): 147–152.

Savsar, M. & Aldaihani, M. 2008. Modeling of machine failures in a flexible manufacturing cell with two machines served by a robot. *Reliability Engineering & System Safety,* 93(10): 1551–1562.

Advanced Materials, Mechanical and Structural Engineering – Hong, Seo & Moon (Eds)
© 2016 Taylor & Francis Group, London, ISBN: 978-1-138-02908-8

The device designed for clamped positioning error compensation based on passive compliance theory

H. Wang, L.D. Sun & X. Zhao
School of Mechanical Engineering, University of Jinan, Jinan, China

L. Tang
Miracle Automation Engineering Co. Ltd., Wuxi, China

ABSTRACT: In order to solve the problems of the centering difficulties for peg-in-hole assembly task and the stress injury of the key components, the clamping position error compensation device was designed. On the basis of passive compliance assembly theory, which used the flexible mechanisms consisted of the linear guide, V-block, compression spring, et al. Besides, the pressure spring being regarded as the key mechanical component, it was analyzed by numerical simulations. In addition to this, the device can realize the fine-tuning of the relative position in both, the X and Z direction and achieve the compensation for the position error during peg-in-hole assembly task. In general, it has a simple structure, small assembly force and it can improve the efficiency when the rope wheel was installed.

1 INTRODUCTION

The main function of the wire rope was checking of clean inspection maintenance system and maintaining the grid lines after tension stringing, in order to improve the service life of wire rope and ensure the safety of construction. The pay-off equipment was an important part of maintenance system, which can spread the wire rope by winding the wheel and ensuring inspection maintenance system that can work normally. When the pay-off equipment clamped the rope wheel, the main shaft needed to insert the center hole of the pay-off equipment. Especially the original clamped positioning device lacks flexible mechanism in the main shaft in the application process, which was difficult to make the rope wheel hole at the right position of the spindle, because of the deformation of stringing construction of rope wheel. At the same time, operation is time-consuming and high labor intensity for workers reduced the working efficiency of the whole maintenance system.

At present, in the study domain of shaft's hole assembly, we usually take a certain error compensation ability of flexible shaft hole assembly device to solve the problem of neutral, named compliance assembly technology (Xiao et al. 1994). In this paper, a clamping position error compensation device based on the passive compliance assembly theory was designed and studied to solve the problem of the centering difficulties for peg-in-hole assembly task of pay-off equipment.

2 COMPENSATION PRINCIPLE OF PASSIVE COMPLIANT ASSEMBLY ERROR

Compliant assembly can be divided into three types in generally:

1. Active compliant. Equipped with sensors to assembly equipment, and combined with the appropriate control method to realize assembly work.
2. Passive compliant. Assembly equipment added a few auxiliary compliant mechanisms, makes natural obedience to external forces in the assembly process, to compensate the alignment error, to complete the assembly.
3. Mixed compliant. The active compliant and passive compliant were combined and different compliant methods were used in different assembly stages (Wu et al. 2005).

This article mainly used the principle of passive compliant assembly error compensation. The passive compliant mechanism mainly used the elastic deformation of the flexible member to adapt the force and motion, to adjust the relative position of the shaft hole and complete the assembly. It had the advantage to reduce the assembly time and protect the assembly equipment effectively. The principle can be showed by the RCC (Remote the Compliance Center) (Dong et al. 2007). The structure diagram was shown in Figure 1.

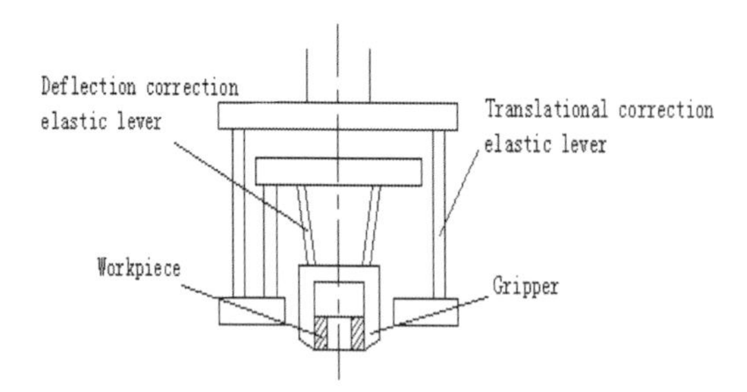

Figure 1. RCC structure diagram.

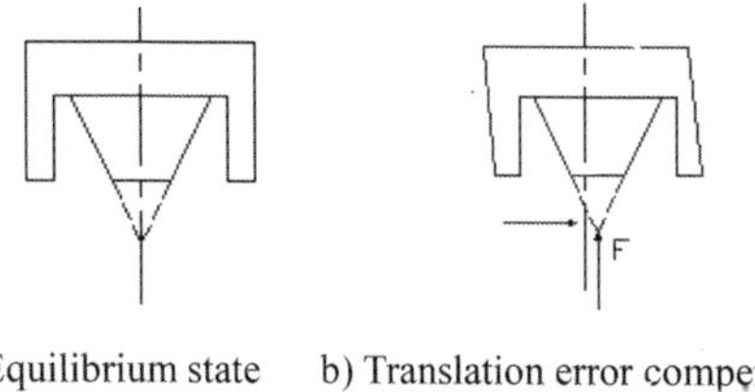

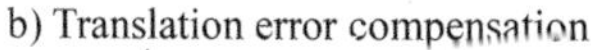

a) Equilibrium state b) Translation error compensation

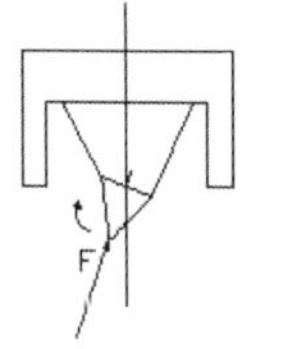

c) Angle error compensation

Figure 2. The neutral principle diagram of the error compensation.

The mechanism was characterized by using the RCC principle to determine a compliant center point (as shown in Figure 2a). When the hole in the center from the axis side force did not produce the angular displacement but lateral movement (as shown in Figure 2b), when the hole loaded from the shaft torque only produced angular displacement and did not produce the lateral movement (as shown in Figure 2c).

3 SCHEME DESIGN OF THE DEVICE FOR COMPENSATING FOR THE CLAMPING POSITION ERRORS

At present, there are some problems of the existing pay-off equipment which were shown in Figure 3 when it clamped the rope wheels. Because of the deformation of the rope wheels and the equipment lack of flexible regulation, it was difficult to insert the rope wheels by clamping spindle. Besides, the workers need to take a lot of effort to adjust the position of the rope wheel with the help of auxiliary device, which is inefficient. If the adjusted force to the pay-off equipment was too large, the main shaft, lifting device and other key parts would not be damaged easily, and ineffectively. This device for clamping position error compensation was presented based on the principle of the Passive compliant. In order to optimize and improve the design of existing equipment, the scheme design of the error compensation mechanism which was based on the passive compliant theory was proposed.

3.1 *Clamping function analysis of positioning error compensation device*

Based on the actual job requirements of the pay-off equipment, the device for compensating for the clamping position errors should have the following functions:

1. The function of transportation. The rope plate was transported from the shop doorway by artificial to pay-off equipment. It was an overloaded transport and had a fixed route.
2. Clamping and positioning function. Locating the position of the center of the rope to ensure that the central hole of the rope and the main shaft of the wire rope were not too large.

Figure 3. Rope plate of the clamping operation.

3. The error compensated function. The center position could be adjusted automatically through the flexible element when the rope wheel was clamped, and the coaxial degree error of the initial position of the shaft hole could be compensated. It needed to ensure the high efficiency and safety of the device for the rope clamping.

3.2 *Scheme design*

According to the functional requirements of the device for clamping and compensating the position errors, the function analysis method was used to carry out the system analysis of the device for clamping and compensating the position errors, as shown in Figure 4.

1. Realization of transport functions.

The rope wheel full of wire weighed about 1600 Kg, and the conveying path was fixed and determined by the selection of rail transportation (the small car) through analysis.

2. The realization of the positioning and the clamping and compensating position errors function.

The practical and economical characteristics of the device were considered based on the principle of the passive compliant assembly error compensation, selected the screw spring, the linear guide and the V-block to make up the compliant mechanism. The V-block had the center position function of the large diameter rope wheel, and the clamping capacity was also considered, with the pressure spring and guide rail achieving the error compensation function at the same time.

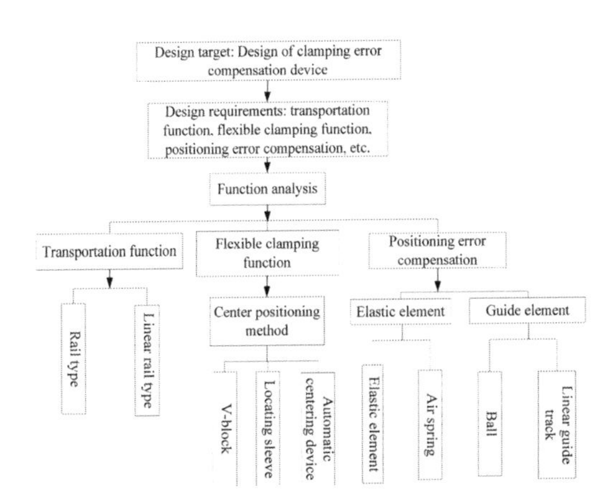

Figure 4. The function of the positioning error compensation device.

4 STRUCTURE DESIGN OF THE DEVICE FOR COMPENSATING FOR THE CLAMPING POSITION ERRORS

4.1 *The overall design of the device for compensating for the clamping position errors*

The overall structural design was discussed in detail, as shown in Figure 5. The device includes a conveying machinery, guiding machinery, and error compensation machinery.

4.2 *The key institutions design of the device for compensating for the clamping position errors*

4.2.1 *Design of guide mechanism*

During the rope wheel clamped, in order to realize the clamping operation of the Y-direction (along the wire disk spindle direction) and the adjustment of the error compensation along with the X-direction (along the track direction), it designed the guide mechanism as shown in Figure 6.

By using the Y-direction linear guide 1 which is connected to the frame and the tray, to implement the Y-direction to orient some features; to realize the error compensation function of

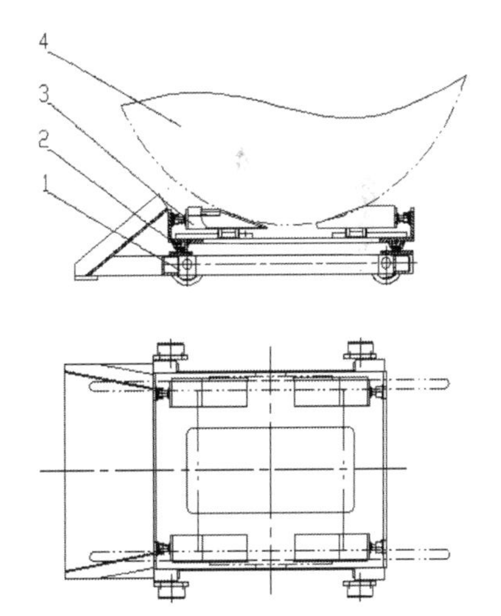

1-conveying mechanism; 2-guiding mechanism; 3-error compensation mechanism; 4-rope coil

Figure 5. The device for compensating for the clamping position errors.

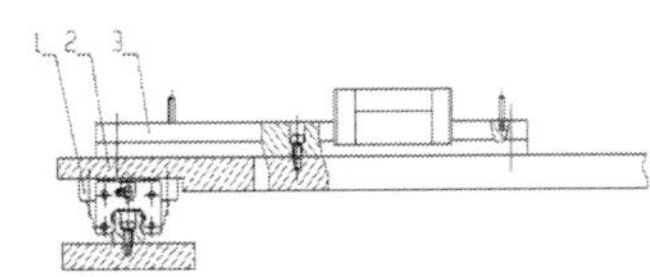

1-Y-direction linear guide; 2-pallet; 3-X-direction linear guide.

Figure 6. Guide mechanism.

the compensation movement in the X-direction, the tray was provided with a linear guide 3 on X-direction.

4.2.2 *Mechanism design for compensating for the clamping position errors*

The working part of the positioning error compensation mechanism (as shown in Figure 7) was mainly composed of a tray 2, a linear guide rail 3, a baffle 4, a press spring 5, and a V-block 6. The baffle plate and the linear guide rail were fixed on the pallet. The V-block was fixed on the slide block of the linear slide rail, and connected with the baffle by a compression spring.

The working principle: the wire disk was sandwiched in between four V-shaped blocks (shown as the Figure 5). At this moment, the cable disc was balanced by its own gravity and the pressure spring pressure. In other words, the position was the equilibrium point of the passive error compensation principle. Adjusting the position of the spindle of the wire equipment roughly and aligning it with the center bore of the rope wheel. Because of various deformations and operating error of the rope wheel, the central hole of the spindle was difficult to achieve the rope wheel. The balance would be broken up, when the spindle was put into the center hole of the rope wheel. According to the principle of passive compliant error compensation, when the hole in the center from the axis side force did not produce the angular displacement

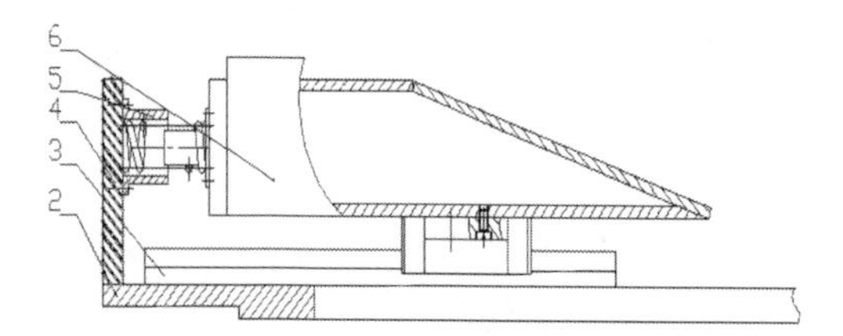

2-tray; 3-linear guide; 4-baffle; 5-the spring; 6-V-block ;

Figure 7. Structure view (partial) for error compensation device.

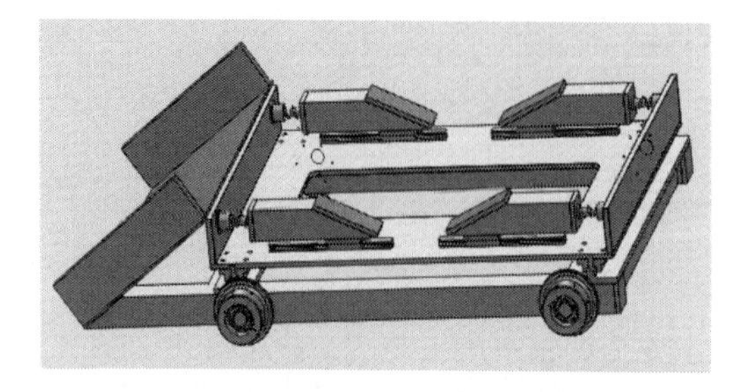

Figure 8. The 3D model of the error compensation device for clamping position.

but the direction of lateral, when the hole from the shaft torque only produced angular displacement and did not produce with the lateral. The orientation error of the spindle X-direction would generate the pressure in the X-direction. This pressure would affect the rope wheel hole, through the role of rope wheel to a V-shaped block, which made the V-shaped block on the X-direction of the linear guide. Finally, the pressure was balanced between the two ends of the V-block, and the positioning error of the X direction was compensated.

The error compensation principle of the Z-direction (vertical direction) is the same as above. The error compensation of the Z-direction is realized mainly through the position adjustment of the two V block in the X-direction. Spindle and rope tray hole deviation angle, will produce a torque on the wire disk and eventually installing clip positioning error compensation device of V-type block, because four V-shaped block independent of each other, resulting in the result is that four V-shaped blocks are respectively along the X-direction relative position adjustment, so as to realize the angle error compensation.

The error compensation principle of Z-direction (vertical direction) was the same as above. The error compensation of the Z-direction was realized mainly through the adjustment of the two V-blocks in the X-direction indirectly. The deviation angle of spindle and rope tray hole would produce a torque on the wire disk and eventually to the V-block of the device for compensating for the clamping position errors. Since the four V-shaped blocks were independent of each other, the result was that the relative position of the 4 V-shaped blocks were adjusted along the X-direction respectively, so as to compensate the angle error.

5 FEA OF THE KEY PARTS OF ERROR COMPENSATION DEVICE

By the analysis of the principle of the error compensation for clamping and structure designed, four pressure springs which were connected with the V-block was important for the error compensation of three direction and angle positioning. Therefore, it was necessary to analyze the finite element of the spring.

The error compensation device for clamping the clip was shown in Figure 5.

According to the design requirements of the error compensation device for clamping position, choosing the θ angle of the two wedge block making up the V-block, $Tan\ \theta = 1/3$, the plateful wire rope $G = 16000\ N$. The external force generated in the misalignment can be ignored, because it was smaller compared to the misalignment of external force. The stress analysis of a single V-block was shown in Figure 9.

$Fz = G/4 = 4000$ N, simplified to the following analysis chart:

According to the stress of the spring, the specifications of the spring were YA6 × 32 × 80 GB/T2089. To judge whether the pressure spring satisfies the use requirements, the stress analysis of

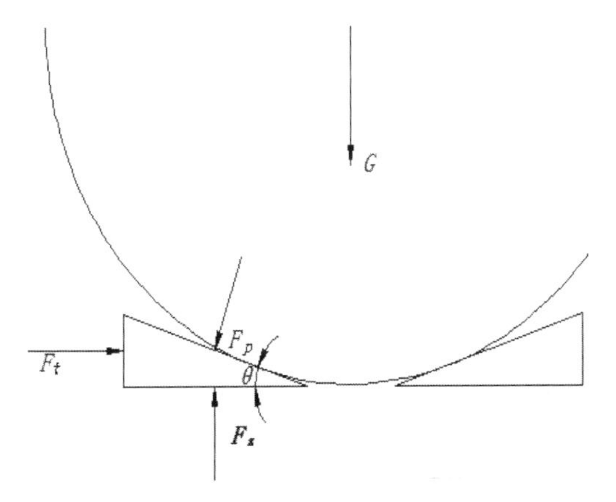

Figure 9. Stress model chart.

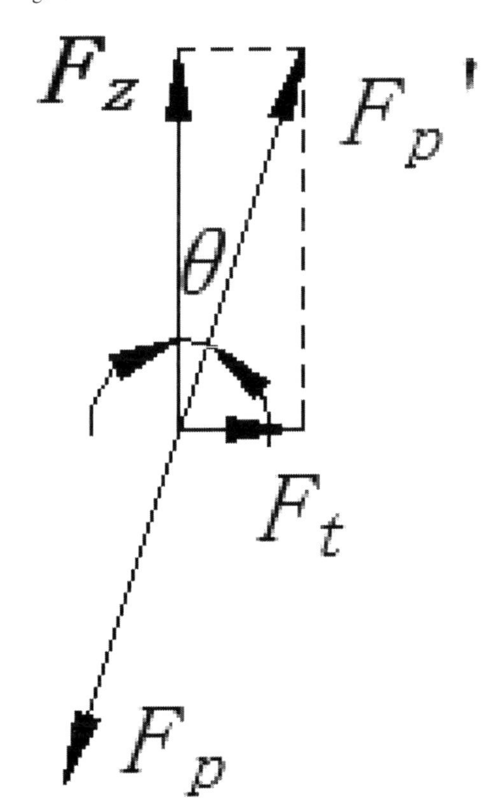

Figure 10. Stress diagram.

simulation module was carried out by solid works, exerted the $F_{t\,(rated)} = \mu\ F_t = 1466.6$ N ($\mu = 1.1$, is the Safety factor). Results were shown in Figure 10.

From the deformation distribution, the maximum displacement could be seen was about 23 mm, which was less than the maximum working deformation amount used under the request of 25 mm, the effect to the maximum stress point could be seen in the inner pressure spring produce about 1.32×10^9 N/m^2, lower than the 60Si2MnA's Yield strength 1.37×10^9 N/m^2. From this it could be seen that the pressure spring meets the use of the requirements.

Through above of the design and the theoretical analysis, we designed the rope wheel clamped positioning error compensation device based on the

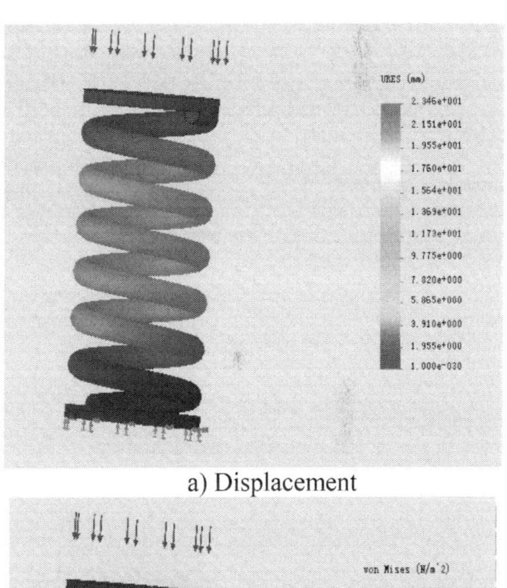

a) Displacement

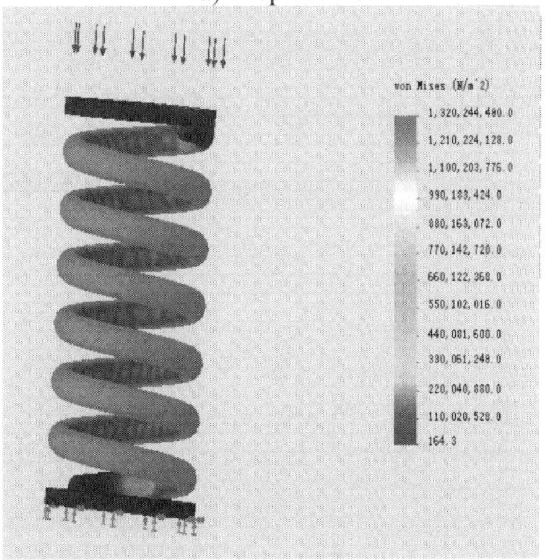

b) Stress

Figure 11. The contours of stress and deformation distribution of pressure spring.

passive compliance theory and the spring flexible device, which could satisfy the need of the displacement about the problem that the centers of the rope wheel and the clamping device were not right when it stood the rope wheel's weight. So the new device can achieve the positioning error compensation.

6 CONCLUSION

Aimed at the problem of the shaft hard coincidence to the hole during the installation of wire rope wheel, a compensation device for the clamped positioning error of pay-off equipment for the wire rope mounting was designed, which was based on the passive compliant assembly theory. The device was composed of V-block, screw spring and linear guide, the displacement deviation and the angular deviation of the direction of X and Z were converted to the displacement deviation of X-direction, then realized the function of the error compensation. The transport function of the rope wheel was realized by the earth rail transport mechanism, the guide function of the rope wheel movement in the process of the rope wheel loading was realized by the linear guide mechanism. Based on this, a numerical simulation analysis of the key pressure spring was carried out to ensure the reliability of the design.

With the device's application, it could reduce the difficulty of rope disk clamping greatly, and reduce the labor intensity of the workers, also it could eliminate the damage of the rope wheel installed on line equipment. So as to improve the efficiency of the wire rope inspection and maintenance line, and improve the economic benefit of the enterprise.

REFERENCES

Deng, X.X., Li, B.Q. & Chen, Z.Q. 1994. A flat compliant gripper. *Chinese Mechanical Engineering Society*, 6: 48–49.

Wu, G.S., Zhang, S.J., Zheng, W. & Zheng, Y.G. 2005. CMAM system of passive composite compliant for robot assembly operation. *Combined Machine Tool and Automatic Machining Technology*, 6: 25–26.

Xin, P., Cheng, D., Quan, Li. & Dai, S.X. 2007. ANSYS test method for compliant wrist compliance center of RCC. *Ship Science and Technology*, 29: 97–101.

Advanced Materials, Mechanical and Structural Engineering – Hong, Seo & Moon (Eds)
© 2016 Taylor & Francis Group, London, ISBN: 978-1-138-02908-8

The susceptibility of forming efflorescence on concrete depending on the mould-releasing agent

R. Pernicová
Klokner Institute, Czech Technical University, Prague, Czech Republic

ABSTRACT: The susceptibility of forming efflorescence on concrete with different types of cement depending on the used mould-releasing agent is presented in this paper. This article is focused on primary efflorescence, specifically lime-based, which cannot be completely avoided. The measurement involves qualitative and quantitative efflorescence tests and also includes the determination of chemical composition of leaching salt crystals. These properties were tested depending on the selected mould-releasing agent and used cement. On the basis of the performed experiments, it can be concluded that the different kind of separators as well as type of cements impact the efflorescence of concrete. Data also shows that the amount of salt crystal in the mixture with the mineral oil is 20% lower than other mixtures.

1 INTRODUCTION

Efflorescence on concrete elements is based on the principle of delivery of efflorescence-forming substances to the surface. The mechanism by which salt gets transported to the surface of structures by moisture through porous materials is well known (Pel 2003). Efflorescence is natural process that appears mostly at the time of forming samples and always with high humidity (Charola 1979). This leads to the requirement to avoid climatic influences, especially water (such as rain, moisture, and others). This efflorescence cannot be completely avoided.

Chemical-mineralogically, efflorescence frequently formed hardly soluble calcium carbonate on concrete surfaces (Pernicova 2014). This is only a temporary aesthetic defect that does not affect the usable properties of the product and mostly disappears within three years via the surrounding environment. Despite this, the current trend in civil engineering is renovating such construction so that efflorescence is not visible.

Primary efflorescence forms for only a limited time after the production of concrete elements. Therefore, it is important to examine the processes occurring in early stages when the structure of the material is formed (Ďemo 2012). These measurements can point to the effect of chemical composition on the concrete element during the formation of efflorescence. Selected type of separator, additives, or admixtures can significantly influence efflorescence (Pernicova 2014). Using appropriate means cannot avoid efflorescence entirely; however, it can help reduce the risk of their occurrence (Higgins 1982).

2 PREPARATION

2.1 *Materials*

Two port-land cements from different sides of the origin were used for preparation of concrete. The first one is from cement plant "Schwenk" (denoted as "S"), the second one is from "Holcim" (denoted as "H").

Two different kinds of material were used as mould-releasing agents. The base of the first separator is vegetable and mineral oil soluble in water ("MO"). This emulsion is called ecological for its other properties. Second is wax dissolved in petrol ("WP").

2.2 *Samples*

Concrete was prepared in mass of ingredients: a binder (cement, micro silica, including additives) and standardized sand in a ratio of 1:4. Portland cement was used for this mixture. Samples were mixed in a mixer according to ČSN EN196-1. Specimens (150 × 150 × 150 mm in dimension) were cut from bigger concrete elements, which was removed from the mould after 24 hours and then was cured 7 days at temperature 20 ± 1°C. Samples from the front side have a specially-treated surface that will not be further modified.

Table 1. Composition of samples.

No. sample	Cement	Mould-releasing agent
1	S	MO
2	S	WP
3	H	MO
4	H	WP

3 EXPERIMENTAL METHOD AND RESULTS

3.1 *Qualitative test of efflorescence*

A qualitative test involves applying a puddle of dis-tilled water on the surface of the concrete. Concrete has been visually assessed for formation of efflorescence right after water has evaporated from the surface. A Standard Test is in an American norm ASTM C67. The test for efflorescence involves partial immersion of the samples in distilled water in a climatic chamber. The dried samples were placed in a chamber at a temperature of 24 ± 1°C and alternating humidity. Humidity has been tested strictly at 30% and 70% alternating about 4 hours during the 7-day period. Each specimen has been placed in a shallow pan containing enough water (water level was 5 ± 1 mm on each side). Then the elements were drying in the oven at 110 ± 5°C for additional 24 hours. Finally, formed efflorescence was compared to control samples that were in the same drying room except there was no contact with water.

Due to contrary conditions of the storage, crystallization began even before the starting measurement. Concrete samples with cement H were more inclined to efflorescence (Figure 1). White efflorescence appeared on all samples after the test no matter which mould-releasing agent or cement was used.

3.2 *Quantitative test of efflorescence*

A quantitative test focuses on the extraction of lime by distilled water. A Swedish company developed this test back in 1978 and named it after the founder Mr Marcus Samuelsson. The Samuelsson test involves controlled amount of distilled water,

Figure 1. Efflorescence on concrete made using Holcim cement.

especially one liter, which acts for 16 hours on the surface of the specimen. After this time, the water was placed in the climate chamber until the specimen dried, and the formation of salt crystals was observed. The minimum quantity of lime necessary to cause efflorescence (visible discoloration) was of the order of 1 g/cm^2.

Figures 2, 3, and 4 show crystals from the distilled water after quantitative test. These two samples had the same composition, were prepared in the same way, and stored in the same external conditions but with different kinds of cement.

Table 2 shows the amounts of salt crystals after the quantitative test. It can be observed that the mould-releasing agent MO caused less salt leaching from concrete structure by approximately 20%. It is likely that crystals started to form even before the measurement began. Probably, this is the reason why lime efflorescence was in lesser amounts

Figure 2. Detail of salt crystals from concrete made using Holcim cement.

Figure 3. Detail of salt crystals from concrete made using Schwenk cement.

by the quantitative test. Therefore, it can be concluded that the mixture with cement Schwenk forms more crystals but is more resistant to the initial occurence of efflorescence.

3.3 *Chemical composition*

The chemical nature of the crystal and propositions for their creation from the stored materials in the structure were very important in addition to the abovementioned two properties. To determine the chemical composition we used an X-ray fluorescent spectrometer. Amount of salt ions was determined from the solution, which was obtained by leaching samples in distilled water. A strip has been cut from the sample (about 2 cm thick). All of its sides, except tested surface, are coated with a resin to prevent ingress of fluid into sample during the test and subsequently dried at a temperature of 40 ± 5°C to obtain a constant weight. The analytical sample left leached at ratio of 1:2.5 (surveyed surface: distilled water) for 7 days at a temperature of 20 ± 5°C.

Subsequently, it was filtered in smaller samples for chemical analysis. The result of this measurement was an amount of salt ions contained in the extract, which indicates the nature and composition of the efflorescence. Crystals can be based on sulfates, chlorides, nitrates, carbonates, etc. (Brocken 2004). Table 3 lists the measured values of water-soluble salts.

Figure 4. Amount of salt crystals from concrete made using different cement: Holcim (left), Schwenk (right).

Table 2. Amount of salt crystals after test.

	1	2	3	4
[g/m^2]	48.24	59.33	11.33	14.07

Table 3. Chemical composition of leaches.

Water-soluble salts	1	2	3	4
Ca [g/m^2]	6.00	6.625	0.923	1.925
Na [g/m^2]	0.948	0.615	0.840	0.933
K [g/m^2]	5.93	4.85	4.10	3.90
ΣCa+Mg [mmol/m^2]	0.15	0.165	0.023	0.048
$KNK_{4.5}$ [mmol/m^2]	0.425	0.425	0.168	0.198

4 CONCLUSION

The possible potential of lime efflorescence can be caused by many factors such as: extremely unsuitable climatic conditions, packaging product damage, long improper storage, incorrectly carried out construction of the building, and many more. An appropriate concept of concrete mix composition helps to reduce the risk. The most dangerous one is moisture.

Holcim cement samples were more sensitive on the presence of humidity and efflorescence process started only few days after the preparation of the sample. Very likely, crystals started to form even before the measurement began. Therefore, it can be concluded that the mixture with cement Schwenk forms more crystals while it was more resistant to the initial formation of efflorescence.

Effect of separator on efflorescence is noticeable. Measured data on mixture with the product MO shows 20% lower amount of salts crystal obtained by quantitative test than that from mixture with WP. A similar effect can be observed on the values of leaching salt ions, except leaching calcium. The surface of sample treated by the separator MO shows a higher value of soluble calcium salts.

A combination of suitable materials can achieve a positive influence on the efflorescence properties of concrete. Further studies would have to be done to verify the possibility of applications of hydrophobic impregnation to improve the situation and limitations of efflorescence. Behavior of hydrophobic samples during the time must be studied as well to find out exactly the best time for application or effective service life.

ACKNOWLEDGMENT

The research was supported by grant project No. P105/12/G059.

REFERENCES

Brocken, H. & Nijland, T.D. 2004. White efflorescence on brick mansory and concrete mansory blocks, with special emphasis on sulfate efflorescence on concrete blocks. *Construction and Bulding Materials*, 18(5): 315–323.

Charola, A.E. & Lewin S.Z. 1979. Efflorescences on building stones—SEM in the characterization and elucidation of the mechanisms of formation. *Scanning Electron Microscopy,* (1): 379–386.
Demo, P. et al. 2012. Physical and chemical aspects of the nucleation of cement-based materials. *Acta Polytechnica,* 52(6): 15–21.
Higgins, D.D. 1982. Efflorescence on concrete. *Appearance Matters*, 4: 1–8.
Pel, L., Huinink, H. & Kopinga, K. 2003. Salt transport and crystallization in porous building materials. *Magnetic Resonance Imaging,* 21(3–4): 317–320.
Pernicova, R. 2014. Analysis of formation and testing of efflorescence on concrete elements. *Advanced Materials Research,* 1025–1026: 641–644.
Pernicova, R. & Kostelecka M. 2014. Influence of Mould Release Agent on Surface Properties of Concrete with Special Matrix. *Advanced Materials Research*, 1025–1026: 637–640.

Advanced Materials, Mechanical and Structural Engineering – Hong, Seo & Moon (Eds)
© 2016 Taylor & Francis Group, London, ISBN: 978-1-138-02908-8

Evaluation of the durability performance of the HPC on its strength level to apply it to the containment buildings in nuclear power plants

D.G. Kim, H.J. Lee, N.W. Yang & J.H. Lee
Korea Institute of Civil Engineering and Building Technology, Goyang, South Korea

M.S. Cho
Korea Electric Power Research Institute, Daejeon, South Korea

ABSTRACT: Most nuclear power plant structures which require safety measures are made of concrete which is a mixture of various materials such as cements, aggregates, and admixtures. Concrete has a material characteristic whose applicability should be verified according to the required features and use the environment of the structures. Moreover, considering that construction duration of structures during nuclear power plant construction is proportionally longer, enormous material amounts are required in the structure construction, it is necessary to develop high performance technology of concretes as major materials of nuclear power plants to improve economic feasibility and shorten the construction period. Thus, this study evaluated construction properties of HPC mixes of 6,000 psi, 8,000 psi, and 10,000 psi as well as salt damage resistance performance, sulfate attack resistance performance, freeze–thaw attack resistance performance, and carbonation resistance performance thereby evaluating a service life of concrete.

1 INTRODUCTION

Many nations including South Korea have been interested in the generation of nuclear power to produce and utilize energy efficiently and much social attention has been paid to safety of nuclear power plant structures after Fukushima nuclear accident. Therefore, it is time to verify the structural materials to ensure the safety of nuclear power plant structures primarily along with efficient designs.

In particular, most nuclear power plant structures which require safety measures are made of concrete which is a mixture of various materials such as cements, aggregates, and admixtures. Concrete has a material characteristic whose applicability should be verified according to the required features and use environments of structures.

Moreover, considering that the construction duration of the structures during nuclear power plant construction is proportionally long and enormous materials are required in structure construction, it is necessary to develop high performance technology of concretes as major materials of nuclear power plants to improve economic feasibility and shorten a construction period.

The ACI 349 Design Standard Committee has completed internal review on ACI 349-12 design standards, which is a revised edition in 2012, and now in the process of approval process to publish the standards. This standard is based on ACI 318-08 and is expected to have a significant change such as the use of high strength reinforced bars compared to ACI 349-06. In ASME Section III, Division 2, which is based on the Allowable Stress Design (ASD) Act, no modifications in relation to high performance reinforced concretes were found.

In ASME Section III, Division 2, no particular restriction on concrete strength was also found. The corresponding design of the standard section was of Subsection CC-3421, which also has no additional restriction on the concrete strength. In the shear strength section, an allowable value of concrete tensile strength was only restricted to 3.5 (fc′)1/2. In summary, it is difficult to say that concrete strength has been restricted based on only this section because ACI 349 did not provide any clear numerical figures on concrete strength.

According to the Standard Review Plan (SRP) 3.8.1 (Rev. 2, 2007), allowable tensile strength of concretes in relation to shear strength was restricted to a range of 40~60 psi. This is a basic position where NRC stands in relation to authorization and approval. According to the SRP, concrete compressive strength is restricted to 6,000 psi. However, in the recently published SRP 3.8.1

(Rev. 3), review items of Subsection CC-3421.5 have been revised significantly in which restriction of allowable tensile strength range of concrete has been removed.

Thus, based on the above review results, there was no restriction on concrete compressive strength in the ASME standards. However, the standpoint of the NRC on safety related authorization and approval is still based on 6,000 psi compressive strength. This standpoint was firm until the SRP was published in 2007 at least. Although, recent standards have removed the related items, it is still not clear that NRC has changed their standpoint. Hence, until the standpoint of the NRC on the authorization and approval is clearly represented, it is quite reasonable to accept the restriction of compressive strength as 6,000 psi.

Thus, this study reviewed compressive strength which can represent concrete performance among mechanical properties of concretes to validate the secured safety of concretes and cement setting test that was conducted to understand the basic characteristics of fresh concretes and hardening characteristics by mix were reviewed by comparing initial and final set times according to mixes. Furthermore, this study evaluated construction properties of HPC mixes of 6,000 psi, 8,000 psi, and 10,000 psi as well as salt damage resistance performance, sulfate attack resistance performance, freeze–thaw attack resistance performance and carbonation resistance performance, thereby, evaluating a service life of concrete.

2 EXPERIMENTAL DESIGN TO EVALUATE DURABILITY PER STRENGTH LEVEL OF HPC

As a binder to manufacture concretes, Ordinary Portland Cement (OPC), Fly Ash (FA), Ground Granulated Blast furnace Slag (GGBS), Silica Fume (SF) were used.

In this study, preparations for density and components of each material were carried out prior to mixes to estimate qualities and basic physical properties of manufactured concretes.

The physical properties of materials that were used in this study have successfully satisfied the local and international standards including the KCI as shown in Tables 1, 2.

Table 3 shows the mix proportion of 6,000 psi The HPC used in the experiment and Tables 4 and 5 shows the mix proportions of 8,000 psi and 10,000 psi (Table 5).

Table 1. Origin of binder and density measurement.

Item	Origin and manufacturer	Density (g/cm^3)
NQA©	Shin Uljin	3.15
Slag (SP)	Youngjin Global	2.94
Fly ash (FA)	Shin Uljin	2.35
Silica fume (SF)	Elkem	2.32

Table 2. Origin of aggregates and physical characteristic measurements.

	Chemical characteristic								
Item	SiO_2	Al_2O_3	Fe_2O_3	CaO	MgO	SO_3	K_2O	Na_2O	LOI
OPC	21.55	5.31	3.56	61.23	3.74	1.95	1.08	0.13	1.24
FA	64.02	19.89	4.45	3.82	1.09	0.00	1.13	1.04	6.76
SP	36.04	15.79	0.45	42.16	3.94	1.95	0.50	0.22	–0.70

Table 3. Determination of mix design of 6000 psi-grade HPC.

6000 psi	W/B (%)	Water	Cement	FA	BS	SF	Coarse Agg.	Fine Agg.	Use environment
FA20	40	162	324	81	0	0	1027	750	Current(APR1400)
SF5	40	155	368	0	0	19	1057	772	Severe cold
FA25	40	155	291	97	0	0	1044	762	General
BS25FA25	40	155	194	97	97	0	1041	760	
BS30FA30	40	155	155	117	116	0	1037	757	
BS50	40	155	194	0	194	0	1054	769	
BS65SF5	40	155	116	0	252	20	1048	765	Scorching heat

※ FA: Fly Ash, BS: Blast furnace slag, SF: Silica Fume Notation example: BS30FA25SF5 (blast-furnace slag 30%, Fly Ash 25%, and Silica Fume 5% in a binder).

Table 4. Determination of mix design of 8000 psi-grade HPC.

8000 psi	W/B(%)	Water	Cement	FA	BS	SF	Coarse Agg.	Fine Agg.	Use environment
SF5	34	155	433	0	0	23	1094	677	Severe cold
FA25	34	155	342	114	0	0	1078	667	
BS25FA25	34	155	228	114	114	0	1073	664	General
BS30FA30	34	155	182	137	137	0	1069	661	
BS50	34	155	228	0	228	0	1090	674	
BS65SF5	34	155	137	0	296	23	1083	670	Scorching heat

Table 5. Determination of mix design of 10000 psi-grade HPC.

10000 psi	W/B (%)	Water	Cement	FA	GS	SF	Coarse Agg.	Fine Agg.	Use environment
SF5	28	155	526	0	0	28	1041	644	Severe cold
FA25SF5	28	155	388	138	0	28	1016	628	
BS25FA20SF5	28	155	277	111	138	28	1016	628	General
BS30FA25SF5	28	155	222	138	166	28	1009	624	
BS45 SF5	28	155	277	0	249	28	1032	638	
BS65 SF5	28	155	166	0	360	28	1027	635	Scorching heat

3 ANALYSIS ON PHYSICAL AND CHEMICAL CHARACTERISTICS OD USED MATERIALS

3.1 *Evaluation on the HPC construction property*

For slump experiments, KS F 2594 and ASTM C 143 measurement methods were followed. The evaluation result showed that all mixes of 6,000 psi and 8,000 psi concretes satisfied a target value 7.0 ± 1.5 inch and 10,000 psi slump flow was 7.0 ± 1.5 inch which also satisfied a target value 500 ± 100 mm.

For air content experiments, KS F 2421 and ASTM C173 measurement methods were employed and air content of concretes was calculated according to the following equation.

$$A = A1 - G \tag{1}$$

where A is air content of concrete (volume percentage of air contents in concrete), A1 is apparent air content (volume percentage of air contents in concrete), and G is an aggregate correction factor (volume percentage of air contents in concrete).

The evaluation result showed that a trend of reduction in air contents was revealed as a replacement ratio was increased overall but air contents in all mixes satisfied a target air content 4.5 ± 1.5% within a range of 3.3~4.5%.

An admixture usage amount had about two times difference between mixes using only FA and only slag despite of the same admixture that was used. The required flowability was achieved with a smaller amount of admixture material than other admixture materials due to the improved flow of the dispersed slag.

For the setting properties, experiments were conducted in accordance with the ASTM C191 evaluation method and a setting time was 8 hours and 45 minutes to 16 hours and 30 minutes was set.

A final setting time showed FA25SF5 (10 ksi) > FA25 (6 ksi) > SF5 (10 ksi) > FA20 (6 ksi) > SF5 (6 ksi) > BS65SF5 (6 ksi) > BS65SF5 (10 ksi) and two hour earlier setting property than the current local mix (FA20) was developed in case of BS65SF5 (scorching heat).

3.2 *Evaluation on salt damage resistance performance*

In this evaluation, analysis on changes in salt diffusion coefficient and estimation on service life by age (28, 91, and 180 days) were conducted through the accelerated test method (NT Build 492 method). After a penetration test, specimens were split and AgNO3 solution was sprayed to determine a salt penetration depth. The evaluation result showed that the diffusion coefficient was decreased compared to the existing ones in all mixes except for FA25 mix for 6,000 psi while diffusion coefficient of 8,000 and 10,000 psi was significantly decreased compared to that of 6,000 psi. This reduction in diffusion coefficient contributes to the prevention of the internal movements of salt damage and other deterioration factors as major life degradation factors against nuclear power structures into concretes, which indicates proof of increases in service life of concretes.

3.3 *Evaluation on freezing and thawing resistance performance*

For this evaluation, specimens were manufactured according to ASTM C666 and 150 and 300 cycles of freezing and thawing that were applied. A mass

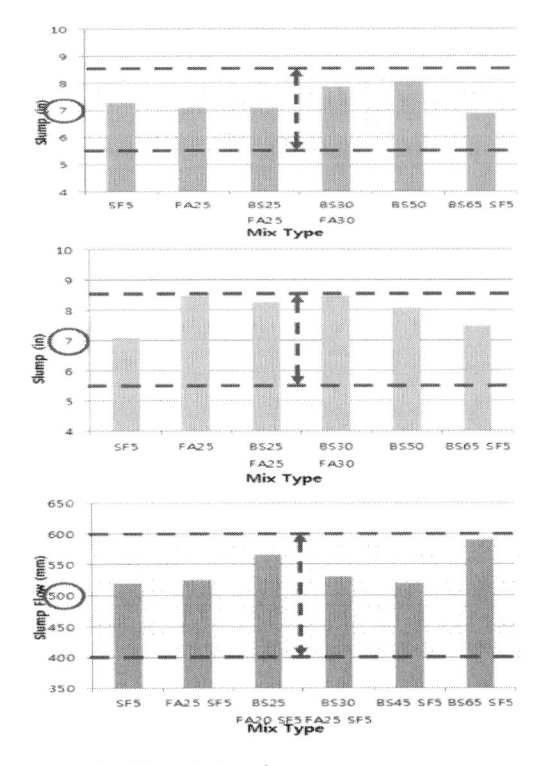

Figure 1. Slump test results.

loss rate means mass reduction due to spalling in the concrete surface such as scaling caused by freezing and thawing. The lower the mass loss rate, the better is the resistance against freezing and thawing. Most specimens of standard mix by strength level showed a lower mass loss rate than that of the existing nuclear power plant mix and ternary mixes such BS25 and FA25 which had an increase in mass loss rate.

Durability factor is the result that is calculated through the relative dynamic modulus of elasticity measurement. The higher the durability factor, the better is the resistance against freezing and thawing. In general, freezing and thawing resistance was evaluated as good if the relative dynamic modulus of the elasticity is over 60 after 300 cycles of freezing and thawing. Except for ternary mixes, durability factor was good compared to that of existing nuclear power plant mixes.

3.4 *Evaluation on carbonation resistance performance*

For this evaluation, carbonation acceleration test was conducted in accordance with KS F2584 and carbonation depth was measured using 1% phenolphthalein solution. The evaluation result showed that carbonation did not occur except for four

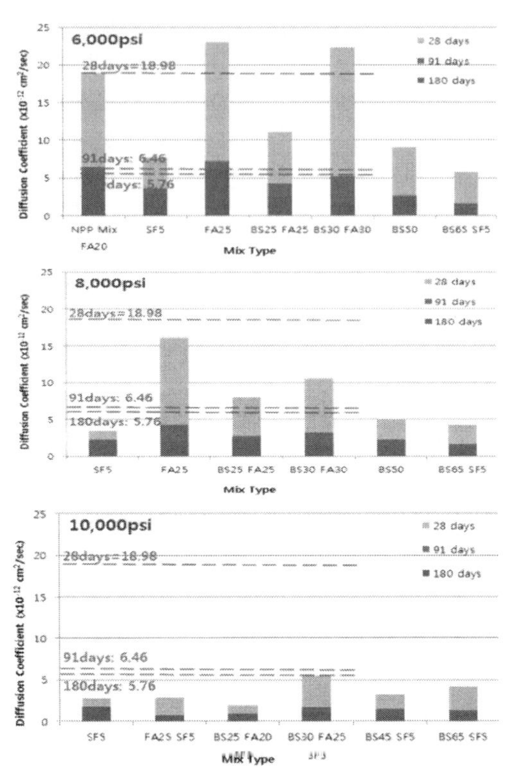

Figure 2. Evaluation on salt damage resistance performance.

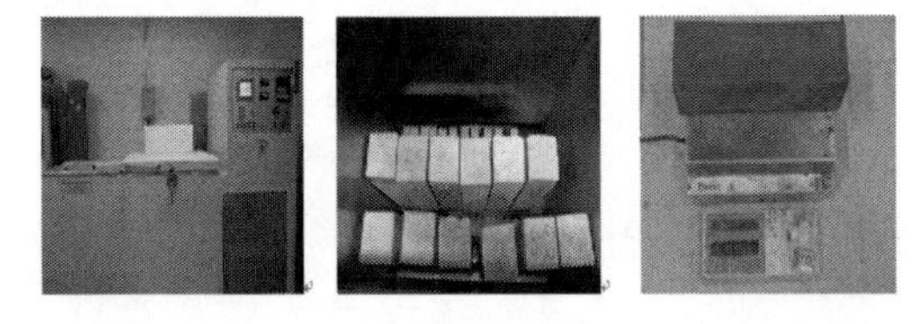

Figure 3. Freezing and thawing tester.

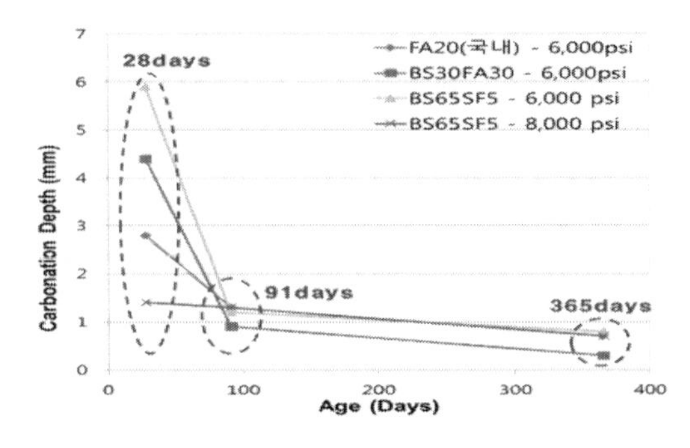

Figure 4. Evaluation on carbonation resistance performance.

mixes (three 6,000 psi strength) and carbonation resistance performance was improved as strength increased. Thus, high strength over 8,000 psi needs not to concern about damages due to carbonation.

4 CONCLUSION

This study evaluated the construction properties of HPC mixes of 6,000 psi, 8,000 psi, and 10,000 psi as well as salt damage resistance performance, sulfate attack resistance performance, freeze–thaw attack resistance performance and carbonation resistance performance, thereby, evaluating a service life of concrete.

An evaluation on service life due to salt damage assessed a staring time of reinforcement corrosion due to chloride and considered concrete cover thickness (1.5 and 2.0 inch). A salt damage service life evaluation was conducted through determination of reach time of critical chloride contents on reinforcement surface using the ACI Life 365 program. Mixes over 8,000 psi strength satisfied service life 100 years in all cover thickness conditions while durability performance of BS50 had an excellent durability performance compared to that of current FA20 mixes in the case 6,000 psi.

An evaluation on service life due to freezing and thawing was conducted to calculate service life according to Finland Code (BY50) and a service life due to freezing and thawing was determined as 260 years approximately.

From the above experimental results, most mixes except for some ternary mixes had improved durability performance according to increase in strength against major life degradation factors of nuclear power plant structures such as salt damage, sulfate attack, freezing and thawing, and carbonation.

With 6,000 psi strength, BS50 had better durability performance than that of current FA20 mix, which indicates utilization of blast furnace slag that shall be considered to improve the safety of nuclear power plants in Korea.

ACKNOWLEDGMENT

This work was supported by the Nuclear Power R&D Program of the Korea Institute of Energy Technology Evaluation and Planning (KETEP) grant funded by the Korea government Ministry of Knowledge Economy (No. 2014151010169 A).

REFERENCES

ACI Committee 318, 2005. *Building Code Requirements for Reinforced Concrete*, Farmington Hills, MI: American Concrete Institute.

ACI Committee 318, 2008. Building Code Requirements for Structural Concrete and Commentary, Farmington Hills, MI: American Concrete Institute.

Hassoun, M.N. & Al-Manaseer, A. 2005. *Structural Concrete: Theory and Design*, John Wiley & Sons, 15–22.

Patrick, P. & Denis, M. 2003. Code Provisions for High-Strength Concrete-an International Perspective. *Concrete International*, 5: 76–90.

Uchiyama, T., Ishimura, K., Takahashi, T. & Hirade, T. 1991. *Study on Reactor Building Structure Using Ultrahigh Strength Materials Part 4 Bending Shear Tests of RC Shear Walls,* Trans. of the Internal Assoc. for Struct. Mech. in Reactor Tech. Conf., SMiRT-11, Paper No. H13/4.Tokyo, Japan, Aug. 1991.

Advanced Materials, Mechanical and Structural Engineering – Hong, Seo & Moon (Eds)
 ISBN: 978-1-138-02908-8

A finite element analysis on roll forming and stamping with convex sheet

Y. Zhang, D.H. Yoon, D.H. Kim & D.W. Jung
Department of Mechanical Engineering, Jeju National University, Jeju-do, Republic of Korea

ABSTRACT: The roll-forming process of the metal sections is a significant field in the advanced formation of strip metal, and it is influenced by many factors. Before we used the stamping method to do the blank's products of SGARC440 alloy, and now we are using the roll forming process. Compared with the stamping process, the roll forming process has some advantages and differences in some areas such as the internal energy and the kinetic energy. In this paper we use the finite element method to do the comparison between roll forming process and stamping process with the ABAQUS software. Though the results in the two processes have various differences, we can analyze which process is better.

1 INTRODUCTION

Roll forming process has been used to describe a large class of continuous manufacturing process, where a long strip of sheet-metal is deformed into products of desired geometry by passing through a series of rotating mills arranged in tandem (Neugebauer 2007). Since the deformation process is complex, usually cold roll forming has to be developed on an empirical basis from an experimental knowledge (Palumbo 2008). The increased use of advanced high strength steels in the automotive industry has resulted in an increase in formability and wear issues during stamping production. For the successfully stamping experience, the stamping press shop shows that press speeds and production rates often need to be reduced to a minimum. Nowadays, the Finite Element Method (FEM), using ABAQUS is considered to be an effective tool to simulate the roll forming process and stamping process. SGARC 440 has been widely used in many areas (Nefussi 1993). Many researchers have studied the problems that result in forming defects during the manufacturing process (sheet bending). However, which method is better in this kind of sheet bending is still a problem for manufacturing. In order to obtain a deeper understanding of the sheet bending process, many experiments and Finite Element Method (FEM) analyses have been carried out. Based on these outcomes, we analyzed the differences between roll forming and stamping, and its relation to the mechanical properties of the material (McClure 1995).

2 EXPERIMENT DETAILS

2.1 *Material properties+*

The sheet used in the roll forming experiments was SGARC 440. Table 1 shows the parameters of this material which is got from the material's experiments. Figure 1 show the plastic strain-Yield strain of the sheet, whose thickness was 1 mm.

2.2 *Forming conditions*

In both of these simulations we used the ABAQUS/Explicit. ABAQUS/Explicit is capable of handling difficult contact analyses more readily and with fewer manipulations of steps and boundary conditions (Kleiner 2003). The material formed is a concave sheet and the dimensions are shown in Figure 2. The sheet has a length of 900 mm and the convex is a part of circle whose radius is 3000 mm. The width is 215 mm.

In these roll forming processes, the deformed sheet progressively passes through a series of rolls. The angles of the rolls from left to right are 15°, 30°, and 45°. Figure 3 shows the continual roll

Table 1. Mechanical properties of the SGARC 440 sheet.

	Young's modulus	Tensile strength	Yield strength	Poisson's ratio
SGARC 440	207000 [MPa]	440 [MPa]	300 [MPa]	0.28

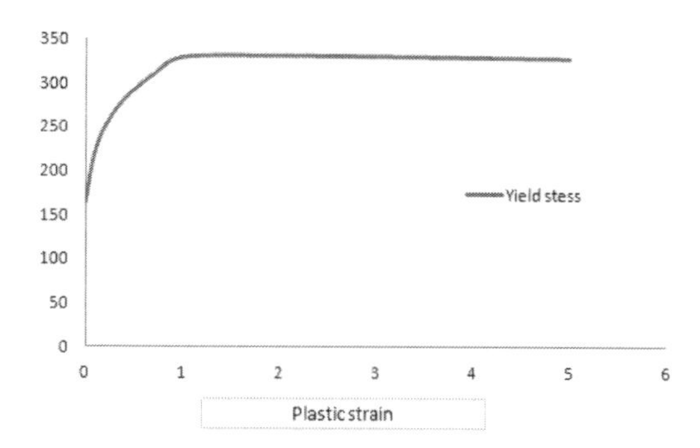

Figure 1. The plastic strain-yield stress of SGARC440.

Figure 2. The sheet of SGARC440.

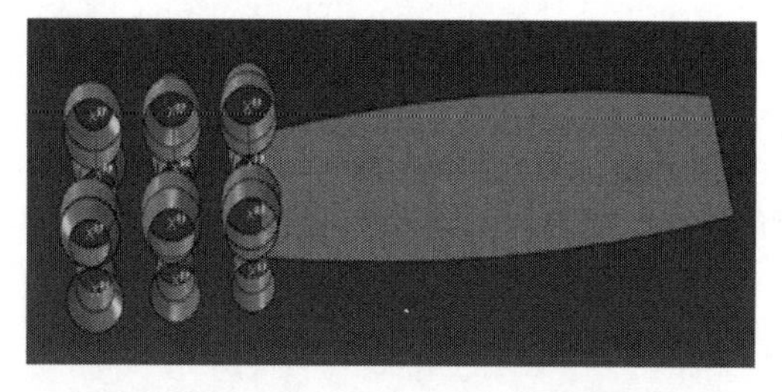

Figure 3. The roll forming process in ABAQUS.

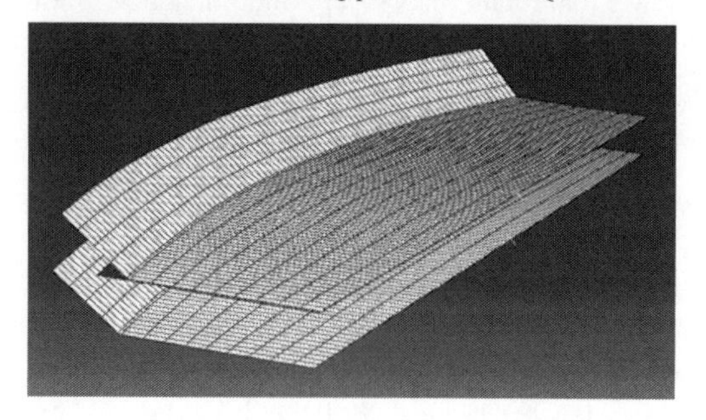

Figure 4. Stamping process of SGARC440 in ABAQUS.

forming process. The sheet moves from the left side to the right side, through the rollers in turns. The roll forming direction is shown in Figure 3.

The sheet used in stamping is the same with the one in the roll forming process. Figure 4 shows the stamping process. The distance between the upper die and sheet is 0.5 mm and the distance between sheet and lower die is 25.5 mm, while the sheet here uses shell element. The bending length is 15 mm and bending angle is 45°.

3 FINITE ELEMENT MODELING

We use ABAQUS as the simulation tool to analyze the roll forming and stamping process. The simulation conditions were chosen with the objective of reducing computer time where possible, while retaining the important features of the experiment (Mori 2007). It was assumed that the rolls and dies were completely rigid bodies, and the sheets were plastic bodies. All modeling of this process was undertaken using the ABAQUS finite-element package. As the channel section is longitudinally symmetric, only half of the strip width is included in the model. There are no complicated numerical procedures required in this model.

3.1 *Geometry and mesh*

All the modeling of this process was undertaken using the ABAQUS finite-element package (Version 6.13). Based on this consideration, the finite-element model was set up to include important features of the experiment, while reducing computation time where possible. This finite model allows describing the deformation and stress-strain within the simulation. To perform a better simulation, the three roll stations were modeled using the "discrete rigid surfaces" option of ABAQUS. Also in stamping, the die and punch were modeled using the four node shell element "S4R."

3.2 *Boundary conditions*

For the roll forming strip model, the symmetry boundary condition was applied to the strip model.7 U1 = UR2 = UR3 = 0. The boundary conditions were applied to the plane to prevent any translation and rotation in any direction without the bending direction. The interaction is set to be a general interaction (Explicit). In view of the existing works, we realize a theoretical problem is that the friction will have influence on the simulation. Taking this into consideration, we assumed that the friction between the rolls and sheet is small enough to omit from the simulation. So the sheet is fixed and only the roll is moving along with the edge direction. The steps in roll forming simulation are set to be four steps. First step is Roll-1 approach to the sheet. Second is Roll-1 moving on the sheet and Roll-2 approaching the sheet. Third is Roll-2 moving on the sheet and Roll-4 approaching the sheet, and the last one Roll-4 moving with the sheet (Wen 1994).

Figure 5. The mesh of roll forming in ABAQUS.

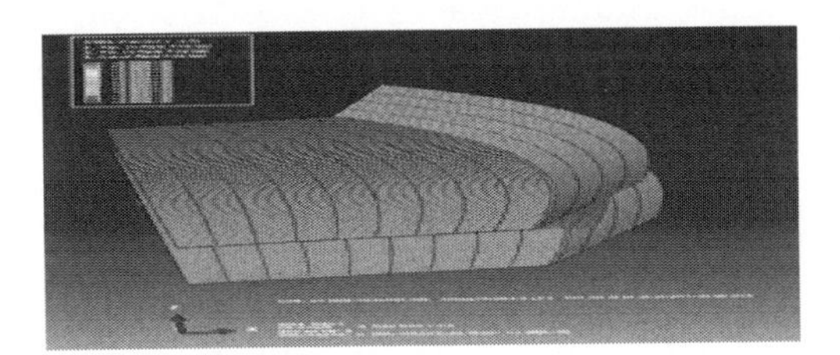

Figure 6. The mesh of stamping in ABAQUS.

For the stamping process, the symmetry condition is same to the roll forming. For the punch, the boundary condition was 25 mm displacement in the vertical direction. A friction coefficient of 0.1 was assumed. The interaction is set to be surface to surface (Explicit) and the rigid parts are master surfaces. The master surfaces are rigid bodies. The simulation time for the stamping is 1 s, which is an important factor commonly used in the stamping process design. Using a proper simulation time will lead to a better result (Han 2007). The boundary condition for the bunch is displacement and rotation and U2 = 25. Other directions are fixed to prevent any translation and displacement (Sieger 2003).

4 ANALYSIS SEQUENCE

The objective of this study is to investigate the differences between the roll forming process and stamping process. The geometries of the strip after stamping deformation and roll forming deformation are shown in Figure 7 and Figure 8. The geometries of the sheet after roll forming and stamping have some differences. Considering the details of the roll forming and stamping simulations, it is obvious that there are differences between the two geometries. Figures 9 and 10 show the productions after the experiment.

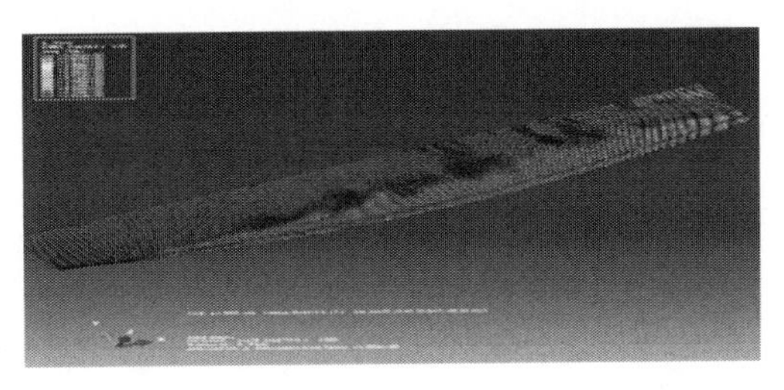

Figure 7. The sheet after roll forming in ABAQUS.

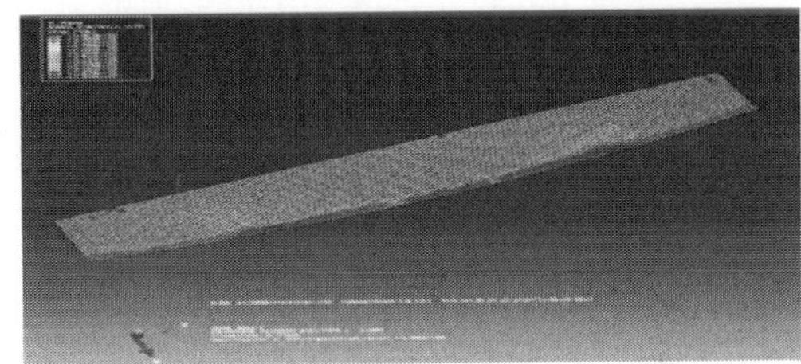

Figure 8. The sheet after stamping in ABAQUS.

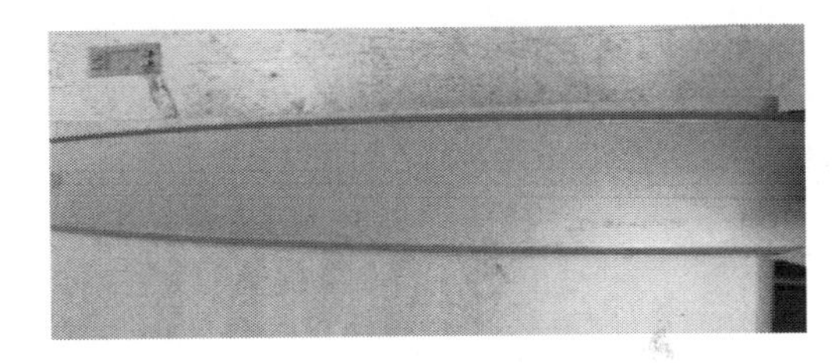

Figure 9. The sheet after roll forming in experiment.

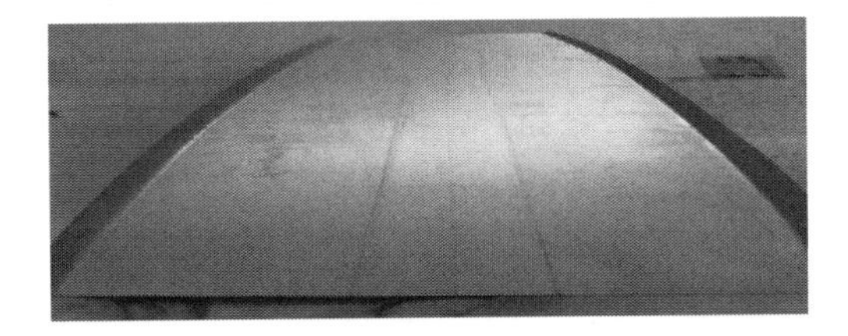

Figure 10. The sheet after stamping in the experiment.

4.1 *History of internal energies*

Representative simulation energies are shown in Figure 11. Comparing the two pictures in Figure 11 shows that the internal energies in stamping and roll forming are a small fraction (less than 1%) of the internal energy through all.

Figures 13 and 14 are a comparison of the kinetic energies in roll forming and stamping from the simulation. The kinetic energy history of roll forming shown in Figure 12 oscillates significantly.

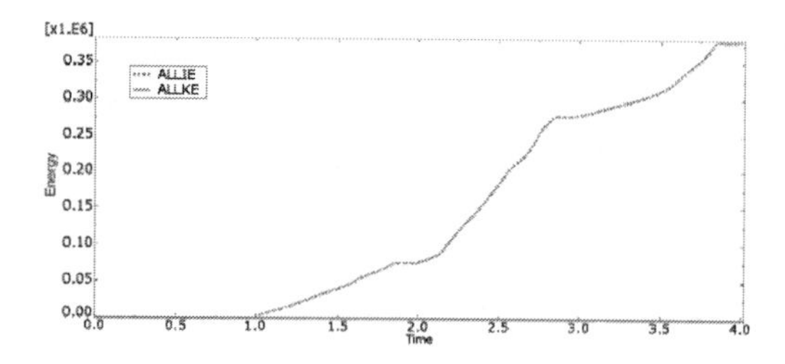

Figure 11. The history of internal energies in roll forming.

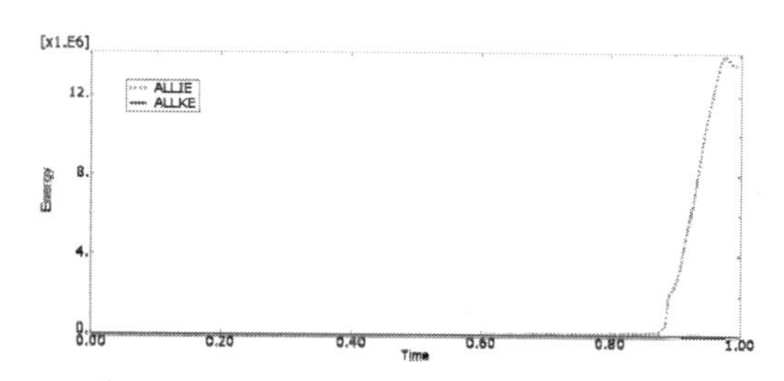

Figure 12. The history of internal energies in stamping.

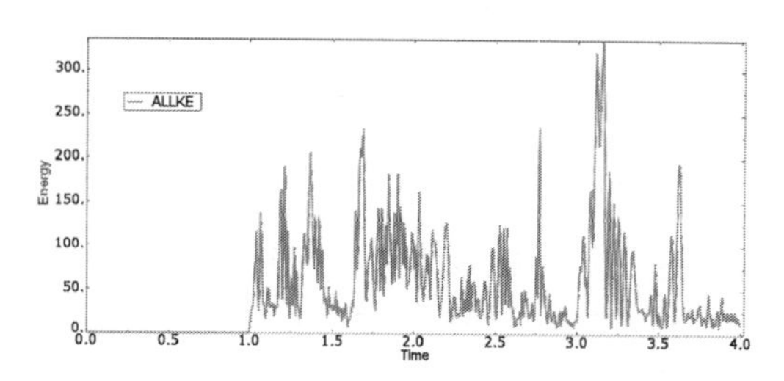

Figure 13. The history of kinetic energies in roll forming.

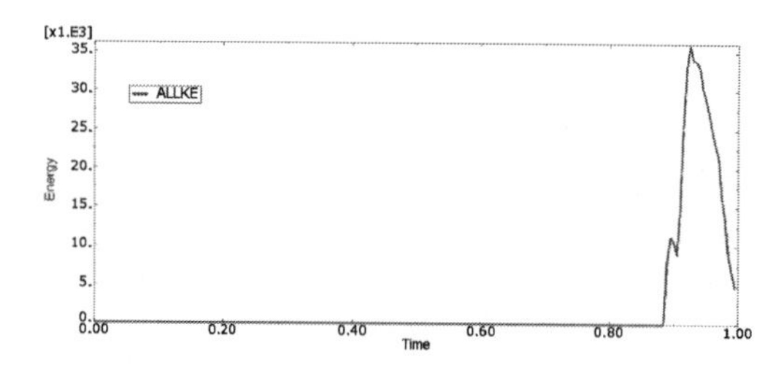

Figure 14. The history of kinetic energies in stamping process.

5 CONCLUSION

This paper does the research on the roll forming and stamping, and compares the two processes to find which method is better. With the finite element analysis, we get the results of the roll forming and stamping. The cures of the internal energy and kinetic energy of the roll forming and stamping process have been figured out. Comparing the results we get some conclusions as follow.

1. From comparing the results of internal energies between roll forming and stamping in ABAQUS we find there aren't so many differences in roll forming and stamping.
2. From comparing the results of kinetic energies between roll forming and stamping in ABAQUS we also find there aren't so much differences in roll forming and stamping.
3. We get the results about the process of roll forming and stamping, the quality of the roll forming and stamping is almost the same. But the roll forming process can improve the speed of the efficiency and utilization, so the roll forming process is a better method compared to stamping.

ACKNOWLEDGMENT

This research was supported by the Basic Science Research Program through the National Research Foundation of Korea (NRF) funded by the Ministry of Education (NRF-2014R1A1A4A01009199).

REFERENCES

Han, Z.W., Liu, C., Lu, W.P. & Ren, L.Q. 2002. Simulation of a multi-stand roll forming process for thick channel section. *Journal of Materials Processing Technology,* 127: 382–387.

Kleiner, M., Geiger, M. & Klaus, A. 2003. Manufacturing of lightweight components by metal forming, *CIRP Annals,* 52(2): 521–542.

McClure, C.K. & Li, H. 1995. Roll forming simulation using finite element analysis. *Manufacturing Review,* 8(2): 114–122.

Mori, K. & Tsuji, H. 2007. Cold deep drawing of commercial magnesium alloy sheets, *CIRP Annals,* 56(1): 285–288.

Nefussi, G. & Gilormini, P. 1993. A simplified method for the simulation of cold rolls forming. *International Journal of Mechanical Sciences*, 35(10): 867–878.

Neugebauer, R., Altan, T., Geiger, M., Kleiner, M. & Sterzing, A. 2006. Sheet metal forming at elevated temperatures, *CIRP Annals,* 55(2): 793–816.

Palumbo, G., Sorgente, D. & Tricarico, L. 2008. The design of a formability test in warm conditions for an AZ31 magnesium alloy avoiding friction and strain rate effects, *International Journal of Machine Tools and Manufacture,* 48(14): 1535–1545.

Siegert, K., Jager, S. & Vulcan, M. 2003. Pneumatic bulging of magnesium AZ31 sheet metals at elevated temperatures, *CIRP Annals,* 52(1):241–244.

Wen, B. & Pick, R.J. 1994. Modelling of skelp edge instabilities in the roll forming of ERW pipe. *Journal of Materials Processing Technology,* 41: 425–446.

Advanced Materials, Mechanical and Structural Engineering – Hong, Seo & Moon (Eds)
© 2016 Taylor & Francis Group, London, ISBN: 978-1-138-02908-8

Effect of free air and underground blast on the concrete structure

Y.H. Yoo
Agency for Defense Development and University of Science and Technology, Daejeon, Republic of Korea

Y.S. Choi & J.W. Lee
School of Mechanical Engineering, Chung-Ang University, Seoul, Republic of Korea

ABSTRACT: This study investigates the effect of free air and underground blast on the concrete structure. The propagation of the blast wave and its effect on the concrete structure was simulated. Due to the difference of the medium, the speed of the pressure propagation from free air blast was faster than the underground blast. The peak pressure in free air blast was three times higher than that in underground blast. Also, arrival time of the peak pressure was 26 times faster in the free air blast. Pressure rise time, positive duration time, and negative duration time in the underground blast were longer than those in the free air blast. The behavior of the concrete structure was also predicted. In the case of free air blast, the top and side surfaces of the concrete structure failed. However, in the case of underground blast, only the top surface of the concrete structure failed.

1 INTRODUCTION

The protection of various structures has been extensively studied in recent years due to the terrorism. The terrors by conventional weapons are more destructive toward structures in direct hit or near field blast. In case of the near field blast, the blast pressure passing through the medium such as air and soil is the main issue which influences the structures.

Free air blast phenomenon is well known due to the massive laboratory and full scale experiments (Goodman 1960). The focus on these massive experiments is to predict the pressure distribution and the blast wave due to the explosion. The blast wave generates overpressure which is critical to the structures.

The influence of the blast wave on the structure has been investigated by many researchers. These studies have focused on free air blast with laboratory experiments. Steel plate's exposure because of the blast load was experimented to predict the behavior of steel plates (Jacob et al. 2007).

Most blast proof structures are made of concrete or reinforced concrete material. Similar to steel plate experiments, these materials have been investigated on their behavior due to the blast pressure by several researchers (Tabatabaei et al. 2013).

Soil is another important medium for the pressure or shock wave. Most important infrastructures such as the bunker are buried in soil. Small scale experiments were widely performed in case of the mine blast which was buried shallow (Anderson et al. 2011). However, most real bunkers are buried deep and it is very difficult to carry out full scale experiments on this case.

Development of the numerical or finite element analysis has become a major solution to replace these full scale experiments. Due to the limitation and risk, finite element analysis has become one of the most popular methods in blast simulations. Hence, manifold blast simulations using finite element analysis have been performed.

There are several methods used in finite element analysis: Lagrangian, Eulerian, and arbitrary Lagrangian Eulerian method. These methods are based on nodes and meshes, which are created in the model. Lagrangian method is well fit in predicting the behavior of the soil structures, whereas Eulerian method is good in predicting the fluid behavior. The arbitrary Lagrangian Eulerian method is the combination of the advantages of the Lagrangian and Eulerian methods.

In this study, the pressure wave in the free air and underground blast are simulated using finite element analysis. The influence of medium on the pressure propagation is investigated. Subsequently, the effect of the pressure on the concrete structure is investigated to compare the difference between the free air and underground blast.

2 SIMULATION SETUP

Figure 1 shows the schematic view and dimensions of underground blast model. The model which contains the soil has 10 cm thickness concrete structure inside. A yz-plane symmetric condition was applied to reduce simulation time. The mass of the TNT is 26 kg and buried under 1 m from the surface of the soil. The pressure sensor is placed above the concrete structure to calculate the pressure distribution. The element size is 50 mm and the total number of elements is 327376.

To compare free air and underground blast conditions, another model was created. Similar to the underground blast model, all dimensions are same but free air blast model does not have soil inside. Figure 2 shows the two models.

The soil model used the deviatoric perfectly plastic yield function ϕ is shown in Equation 1.

$$\phi = J_2 - \left[a_0 + a_1 p + a_2 p^2\right] \tag{1}$$

where, J_2 = second invariant; a_0, a_1, and a_2 = constants; p = pressure. This model is the simplest function to simulate in LS-DYNA (Hallquist 2007). Input parameters for the soil model are shown in Table 1 (Thomas et al. 2008).

The soil model requires a pressure versus volumetric strain curve. For the soil model, ten data points were selected. The selected soil data were from triaxial compression, hydrostatic compression, and uniaxial strain tests. The data points for soil model are shown in Table 2 (Thomas et al. 2008).

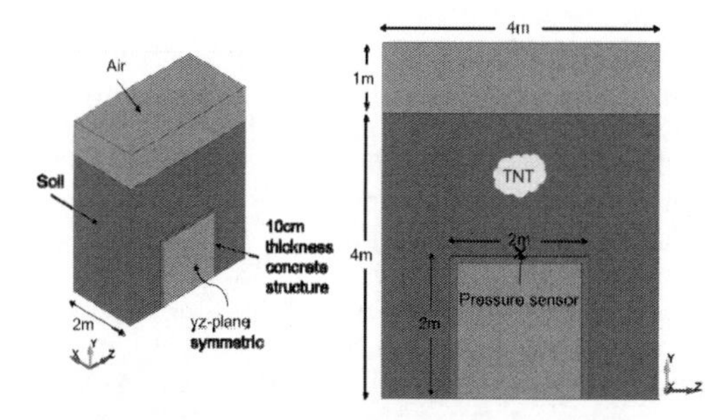

Figure 1. Underground blast model.

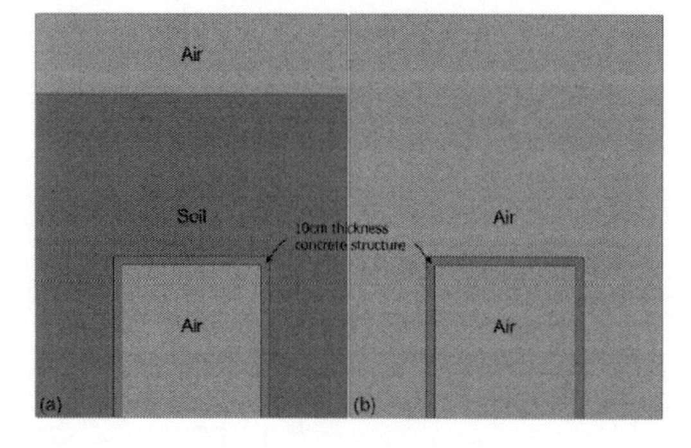

Figure 2. Models (a) underground blast (b) free air blast.

TNT used Jones–Wilkins–Lee (JWL) equation of state where pressure, p, is defined as shown in Equation 2 (Hallquist 2007).

$$p = A\left(1 - \frac{\omega}{Q_1}\right)e^{-Q_1} + B\left(1 - \frac{\omega}{Q_2}\right)e^{-Q_2} + \frac{\omega E}{V} \tag{2}$$

$$Q_1 = R_1 V$$
$$Q_2 = R_2 V$$
$$V = \rho_0/\rho$$

where, A, B, R_1, R_2 and ω = function constants; E = internal energy; ρ = density. Input parameters for the TNT model are shown in Table 3 (Adamik et al. 2005).

Linear polynomial equation of state is used for the air model. The pressure, p, is shown in Equation 3 (Hallquist 2007).

$$p = C_0 + C_1\mu + C_2\mu^2 + C_3\mu^3 + \left(C_4 + C_5\mu + C_6\mu^6\right)E \tag{3}$$

$$\mu = \frac{\rho}{\rho_0} - 1$$

Table 1. Input parameters for soil model.

	Symbol	Value
Density (kg/mm³)	RO	14.534×10^{-6}
Shear modulus (GPa)	G	0.017237
Bulk unloading modulus (GPa)	K	0.119969
Function constant (GPa²)	A0	285.67
Function constant (GPa)	A1	1.052×10^{-4}
Function constant (–)	A2	0.75
Pressure cutoff (GPa)	PC	-1.379×10^{-5}

Table 2. Data points for soil model.

	Volumetric strain		Pressure	
	Symbol	Value (–)	Symbol	Value (MPa)
Point 1	EPS1	0.0	P1	0
Point 2	EPS2	−0.0089	P2	0.2068
Point 3	EPS3	−0.0104	P3	0.2413
Point 4	EPS4	−0.0120	P4	0.2758
Point 5	EPS5	−0.0138	P5	0.3103
Point 6	EPS6	−0.0155	P6	0.3447
Point 7	EPS7	−0.0190	P7	0.4137
Point 8	EPS8	−0.0226	P8	0.4826
Point 9	EPS9	−0.0263	P9	0.5516
Point 10	EPS10	−0.0291	P10	0.6054

Table 3. Input parameters for TNT model.

	Symbol	Value
Density (kg/mm^3)	RO	1.590×10^{-6}
Detonation velocity (m/s)	D	6930
Chapman–Jouget pressure (GPa)	PCJ	21
Function constant (GPa)	A	371.2
Function constant (GPa)	B	3.23
Function constant (–)	R1	4.15
Function constant (–)	R2	0.95
Function constant (–)	OMEG	0.30
Function constant (–)	E0	7.0

Table 4. Input parameters for air model.

	Symbol	Value
Density (kg/mm^3)	RO	1.29×10^{-9}
Equation coefficient (–)	C0	0
Equation coefficient (–)	C1	0
Equation coefficient (–)	C2	0
Equation coefficient (–)	C3	0
Equation coefficient (–)	C4	0.4
Equation coefficient (–)	C5	0.4
Equation coefficient (–)	C6	0
Function constant (–)	E0	2.5×10^{-4}

where, C_0 to C_6 = equation coefficients; E = internal energy; ρ = density. Input parameters for the air model are shown in Table 4 (Schwer 2010).

In the case of concrete, continuous surface cap model was selected with density of 2.34×10^{-6} kg/mm^3.

3 RESULT

Figure 3 shows the incident pressure propagation of the free air and underground blast. The incident pressure propagation due to underground blast is denser than that due to free air blast. In Figure 3, the distance where pressure travels through each medium is equal. However, the blast wave due to underground blast took 7.63 ms to reach the concrete structure and 0.25 ms in the case of free air blast: the blast wave due to free air blast is 13 times faster than that due to underground blast.

Figure 4 shows the pressure distribution at the sensor. The pressure sensor was placed above the concrete structure. In case of free air blast, the pressure rise time which is the time for reaching the peak pressure was very small (less than 0.1 ms). The positive duration time which is the time when the peak pressure starts to rise to its peak then decreases to reach the initial pressure (i.e. reference pressure), which is 2.25 ms. After the pressure reaches the reference pressure, it continuously decreases then returns to the reference pressure. This time frame is called the negative duration time, which is 0.76 ms.

In the case of underground blast, it took more time to reach the concrete structure. The pressure rise time is 2.38 ms, which is longer than free air blast. The positive duration time is 3.59 ms, while the negative duration time is 1.11 ms. The results are listed in Table 5.

Figure 5 shows the behavior of concrete structure due to underground blast pressure. As the pressure travels slower than free air blast, the concrete structure starts to crack after 9.39 ms. At 9.74 ms, the shape of the cracks form a circular hole: due to the symmetric condition, a semicircle crack

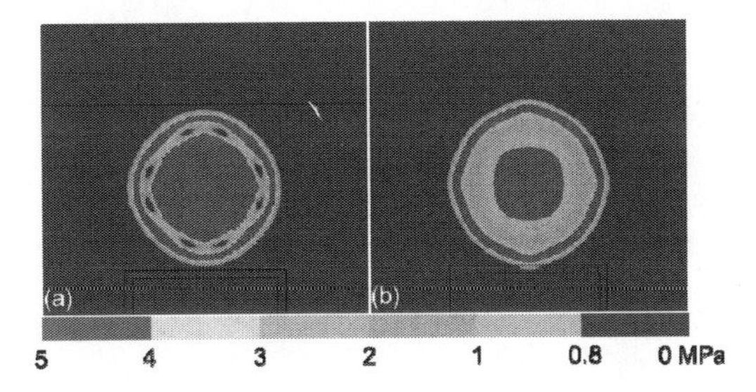

Figure 3. Pressure propagation (a) underground blast at t = 7.63 ms and (b) free air blast at t = 0.25 ms.

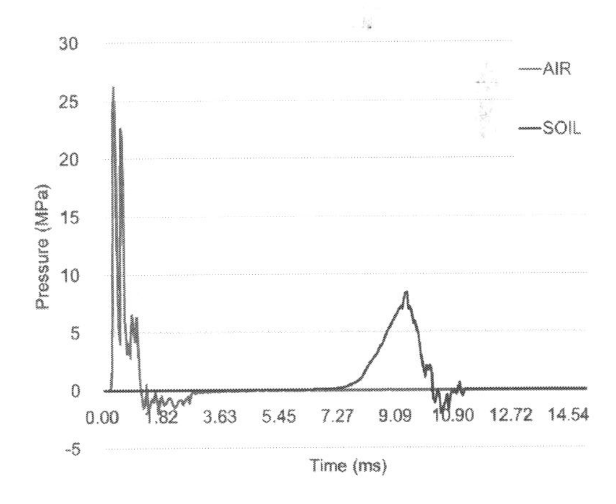

Figure 4. Pressure distribution at sensor.

Table 5. Sensor results.

	Air	Soil
Peak pressure (MPa)	26.24	8.4
Time at peak (ms)	0.35	9.39
Pressure rise time (ms)	<0.1	2.38
Positive duration time (ms)	2.25	3.59
Negative duration time (ms)	0.76	1.11

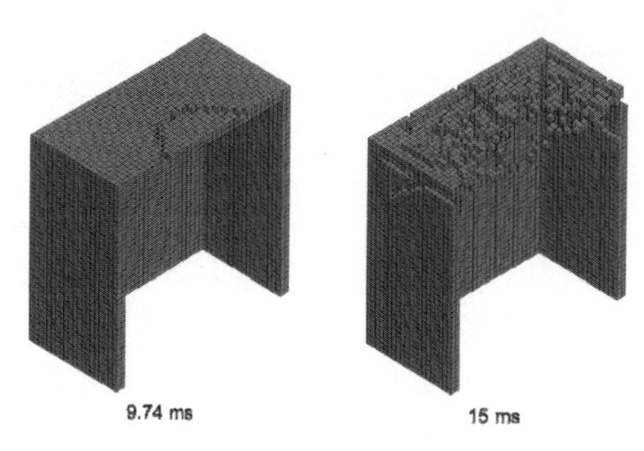

Figure 5. Structural behavior due to underground blast.

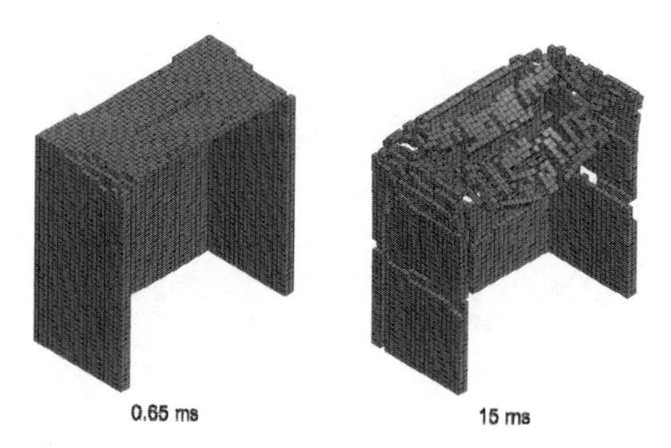

Figure 6. Structural behavior due to free air blast.

is shown at 9.74 ms. After the peak pressure passes over the top surface of the concrete, the structure fails in the wide range.

Figure 6 shows the behavior of concrete structure due to free air blast pressure. When compared with underground blast effect, the concrete structure starts to crack faster (0.65 ms). The shape of the cracks forms a straight line. After the pressure passes through, the structure starts to crack and fail at the top surface as well as the side surface. At 15 ms, the top surface totally collapsed and the side surface also failed with a horizontal crack line.

4 CONCLUSION

Two different media, air and soil, were simulated to compare the effect of the free air and underground blast on the concrete structure. Due to the difference of the media, the characteristics of the pressure propagation were different significantly. Free air blast propagated faster and resulted in a higher peak pressure at the same sensing point when compared with underground blast.

The peak pressure was 26.24 MPa for free air blast and 8.4 MPa for underground blast. The time for the peak pressure arrival in case of the free air and underground blast was 0.35 and 9.39 ms, respectively.

The pressure rise time for underground blast was 2.38 ms, while that for free air blast was very small (less than 0.1 ms). The positive duration time for free air blast and underground blast was 2.25 and 3.59 ms, respectively. The negative duration time for free air blast was 0.76 ms, whereas that for underground blast was 1.11 ms.

In case of the free air blast, the concrete structure started to crack with a straight line at the top surface. Subsequently, the top surface failed and the side surface failed as well. In case of the underground blast, the concrete structure started to crack in a circular form. Consequently, the top surface failed.

ACKNOWLEDGMENT

This research was supported by the Agency for Defense Development (Grant Nos. ADD-13-70-04-05 and ADD-UE145116GD).

REFERENCES

Adamik, V., Trzcinski, W., Trebinski, R. & Cudzilo, S. 2005. *Investigation of the behaviour of steel and laminated fabric plates under blast wave load.* V International Armament Conference, Waplewo.

Anderson, C.E., Behner, T. & Weiss, C.E. 2011. Mine blast loading experiments. *International Journal of Impact Engineering,* 38(8): 697–706.

Goodman, H.J. 1960. *Compiled free-air blast data on bare spherical pentolite (No. BRL-1092).* Army Ballistic Research Lab Aberdeen Proving Ground Md.

Hallquist, J.O. 2007. *LS-DYNA keyword user's manual.* Livermore Software Technology Corporation 970.

Jacob, N., Nurick, G.N. & Langdon, G.S. 2007. The effect of stand-off distance on the failure of fully clamped circular mild steel plates subjected to blast loads. *Engineering Structures,* 29(10): 2723–2736.

Schwer, L. 2010. *A brief introduction to coupling load blast enhanced with Multi-Material ALE: the best of both worlds for air blast simulation.* LS-DYNA Forum, Bamberg.

Tabatabaei, Z.S., Volz, J.S., Baird, J., Gliha, B.P. & Keener, D.I. 2013. Experimental and numerical analyses of long carbon fiber reinforced concrete panels exposed to blast loading. *International Journal of Impact Engineering,* 57: 70–80.

Thomas, M.A., Chitty, D.E., Gildea, M.L. & T'Kindt, C.M. 2008. *Constitutive Soil Properties for Cuddeback Lake,* California and Carson Sink, Nevada. NASA CR 215345.

Advanced Materials, Mechanical and Structural Engineering – Hong, Seo & Moon (Eds)
 ISBN: 978-1-138-02908-8

Tests on moist-cured lightweight foamed concrete with slag as a supplementary cementitious material

K.H. Lee
Department of Architectural Engineering, Kyonggi University Graduate School, Seoul, Korea

K.H. Yang
Department of Plant and Architectural Engineering, Kyonggi University, Suwon, Korea

ABSTRACT: The present study tested the fundamental properties of foamed concrete with the addition of Ground Granulated Blast-furnace Slag (GGBS) as a partial replacement of ordinary portland cement. Five foamed concrete mixes were prepared at the different replacement levels (R_G) of GGBS, which varied from 0% to 90%. Test results showed that the flows of the foamed concrete increased as R_G increases up to 30%, beyond which the flow remained constant. With the increase in R_G, the depth of defoaming tended to increase, whereas the oven-dried density and porosity of hardened concrete decreased slightly. Considering the segregation of cementitious materials and strength development corresponding to the requirements of AAC-2 class specified in ASTM provision, the optimum R_G for foamed concrete could be recommended to be 50%.

1 INTRODUCTION

Foamed concrete (LFC) is characterized by a high lightweight potential because of its porous structure without aggregates. It is primarily used for thermal and acoustic insulation purposes in buildings (Pan et al. 2014, Yang et al. 2014). According to the manufacture process, the foamed concrete is classified into two groups: high-pressure steam-cured Autoclaved Aerated Concrete (AAC) and moist-cured in-situ concrete. It is commonly recognized that the strength development of AAC is higher than that of moist-cured foamed concrete with a same density because the precast AAC is a special curing process under high temperature and high pressure. However, the AAC production is accompanied by a high-energy consumption process (Yang & Lee 2015).

A large proportion of the foamed concrete industry is very interested in alternating the AAC block. As a result, deliberate attempts to develop the sustainable foamed with low CO_2 emission and low energy consumption have been increasingly made. In particular, considerable efforts for safe and sound recycling of by-products including Ground Granulated Blast-furnace Slag (GGBS) have been enacted in the foamed concrete industry because the CO_2-emission coefficient of Ordinary Portland Cement (OPC) from a Life Cycle Assessment (LCA) is commonly evaluated to be 0.8 to 1.0 ton-CO_2/ton. To address these issues, the present study examined the fundamental properties of foamed concrete with the addition of GGBS as a partial replacement of OPC. The measured properties of the prepared foamed concrete were compared with the ASTM strength class (ASTM 2012).

2 EXPERIMENATAL DETAILS

2.1 *Materials*

Table 1 presents the chemical compositions of the cementitious materials used. The GGBS had a high CaO content of 42.2% and SiO_2-to-Al_2O_3 ratio by mass of 2.3, which conforms to ASTM C989. The OPC (ASTM Type I) was mainly composed of CaO of 60.57% and SiO_2 of 22.13%. The dry density and specific surface area of the cementitious materials were also measured as explained in the follows: 2.91 g/cm^3 and 4400 cm^2/g, respectively, for GGBS, and 3.15 g/cm^3 and 3000 cm^2/g for OPC. The foaming agent used to produce the pre-formed foam was based on protein with enzymatic active components, which does not induce a chemical reaction. Up to 2000 L of foam can be produced from 1 kg of agent. To achieve the designed initial flow, a specially modified polycarboxylate-based high-range water-reducing agent was added by 0.5% of the binder weight.

2.2 *Mixture proportions*

The main parameter investigated to produce a sustainable foamed concrete was the replacement level (R_G) of GGBS, which varied from 0%

Table 1. Chemical composition of cementitious materials (% by mass).

Material	SiO_2	Al_2O_3	Fe_2O_3	CaO	MgO	K_2O	Na_2O	TiO_3	SO_3
OPC	22.13	3.68	5.21	60.57	3.02	0.78	0.13	–	2.08
GGBS	32.63	14.18	0.43	42.17	4.87	–	0.23	–	4.04

to 90% (see Table 2). The foam volume required depended on the binder content and water-to-binder (*W*/*B*) ratio. The unit binder content and *W*/*B* ratio were fixed to be 500 kg/m³ and 30% for all mixes. To achieve the self-compactability requirement of the foamed concrete, targeted flow was selected to be above 200 mm. The compressive strength and dry density of concrete specimens were designed to meet the requirement of class AAC-2 specified in ASTM C1693. The mixture proportioning of the foamed concrete for the targeted requirements was determined in accordance with the procedure proposed by Lee (2014). All concrete mixes did not include aggregates or fillers.

2.3 *Casting and testing*

In accordance with the mixing procedure recommended by Lee (2014), pre-formed foam was added to the cementitious slurry produced by mixing the pre-blended binder and water in a mixer pan of 0.12 m³ capacity, equipped with rubber wiper blades. The foaming agent was diluted with water in 1:32 ratio by volume and thereafter aerated to 40 kg/m³ density using a foam generator connected with a compressed air source, to produce foam. The water content was considered to be a part of the water in the overall mixture. Fresh concrete was cast in various steel molds lined with stiff vinyl to prevent their interaction with the mould release oil. Most specimens were then sealed using a plastic bag to prevent evaporation and cured at room temperature until testing. The specimens intended for measuring dry density were cured under wet condition after one day.

The initial flow and depth of defoaming in fresh concrete were measured. The flow was tested without raising and dropping the flow table to ascertain self-compactability. The actual foam volume in the fresh concrete was also recorded using a mess cylinder and methyl alcohol. The compressive strength of the concrete was measured using cylindrical specimens of 100 mm diameter and 200 mm height at ages of 7 d and 28 d. The dry density of hardened concrete were measured in accordance with the KS provision. For the oven-dried concrete, the pore size distribution and pore morphology of the specimens were recorded by image analysis.

Table 2. Details of mixture proportions.

Specimens	Composition of binder (%) OPC	GGBS	Unit binder content (kg/m³)	*W*/*B* ratio (%)	SP (%)	Air content by volume (%)
O10	100	0				63.6
O7G3	70	30				63.2
O5G5	50	50	500	30	0.5	62.9
O3G7	30	70				62.7
O1G9	90	10				62.4

(Specimen notations were identified using the type and composition of cementitious materials to produce binder. For example, O7G3 indicates a foamed concrete with 70% OPC and 30% GGBS.).

3 EXPERIMENTAL RESULTS AND ANALYSIS

3.1 *Flow and depth of defoaming*

Figures 1 and 2 represent the initial flow and depth of defoaming of fresh concrete, respectively. Test results are also summarized in Table 3. All concrete specimens achieved the targeted initial flow. The flows of the foamed concrete increased with the increase in R_G up to 30%, beyond which the flow remained constant. The initial flow increased from 200 mm to 270 mm when R_G varied from 0 to 30%, whereas the flow was measured to be 280 mm for the foamed concrete with R_G of 70%. The depths of the defoaming tended to increase with the increase in R_G. No defoaming was observed for foamed concrete with R_G of 30%, whereas the specimen with R_G of 90% was defoamed by 6 mm. For foamed concrete specimens with R_G above 70%, the segregation of cementitious materials was observed, as shown in Figure 3. The segregation accelerated the defoaming.

3.2 *Oven-dried density and compressive strength*

The oven-dried density of hardened foamed concrete tended to decrease slightly as R_G increases, as shown in Figure 4. The oven-dried density of specimens decreased from 570 kg/m³ to 480 kg/m³ when R_G increased from 0% to 90%. The density

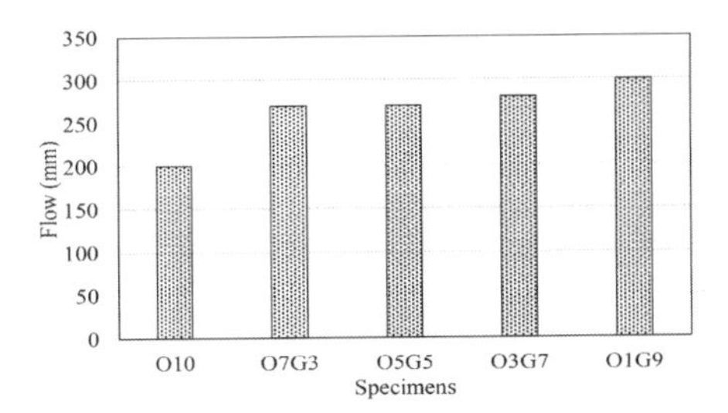

Figure 1. Initial flow of foamed concrete tested.

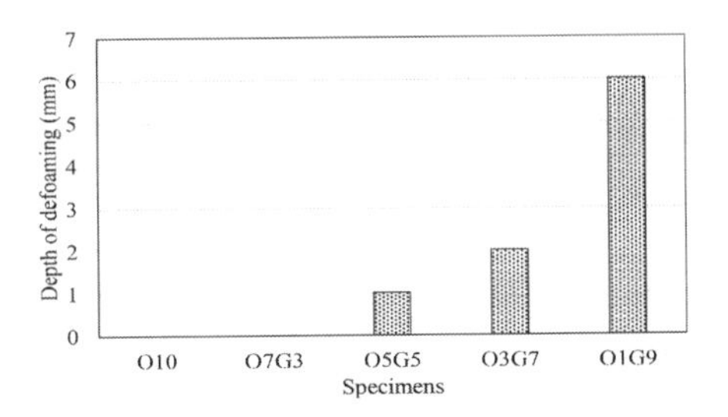

Figure 2. Depth of defoaming of concrete specimens.

Table 3. Summary of test results.

Speci-mens	Flow (mm)	Defoamed depth (mm)	Compressive strength (MPa)		Oven-dried density (kg/m^3)	Air void (%)
			7 day	8 day		
O10	200	0	2.8	3.6	550	54.8
O7G3	270	0	1.6	2.5	500	53.6
O5G5	270	1	1.6	2.3	520	41.4
O3G7	280	2	1.0	2.1	510	27.2
O1G9	300	6	0.6	1.5	480	41.0

of GGBS is lower by approximately 8% than that of OPC. For foamed concrete not including aggregates, the density of cementitious materials affects the oven-dried density of hardened concrete.

Figure 5 shows the compressive strength of the specimens at an age of 7 d and 28 d. The 7-day compressive strength ranged between 77% and 40% of the 28-day strength. This ratio decreased as R_G increased. This may be attributed to the pozzolanic reaction of GGBS (Shi et al 2006). The compressive strength of foamed concrete decreased with the increase in R_G.

Compared with the compressive strength of foamed concrete (specimen O10) without GGBS, O7G3 mix had lower strength by 56% and 67% at an age of 7 d and 28 d, respectively. The 28-day strength of O5G5 was lower by 92% than that of O7G3 mix.

Comparing the 28-day oven-dried density and compressive strength with those requirements

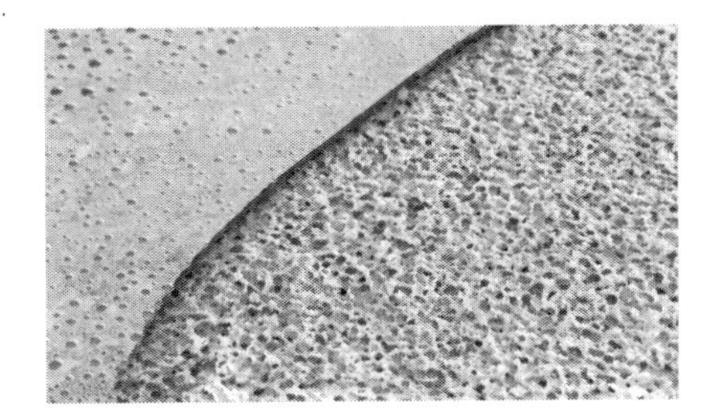

Figure 3. Segregation of cementitious materials observed for specimens with R_G above 70%.

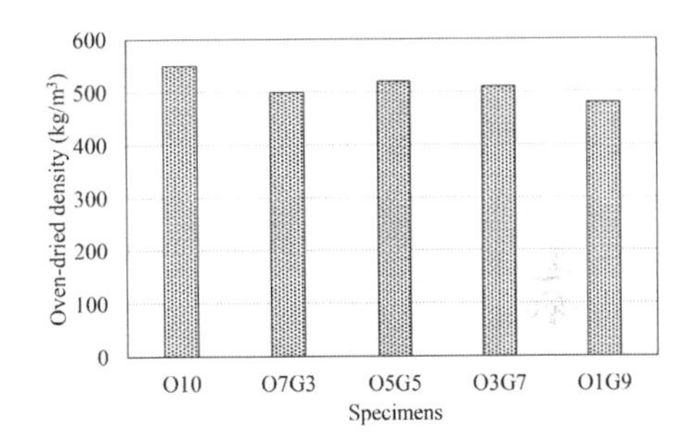

Figure 4. Oven-dry density of specimens.

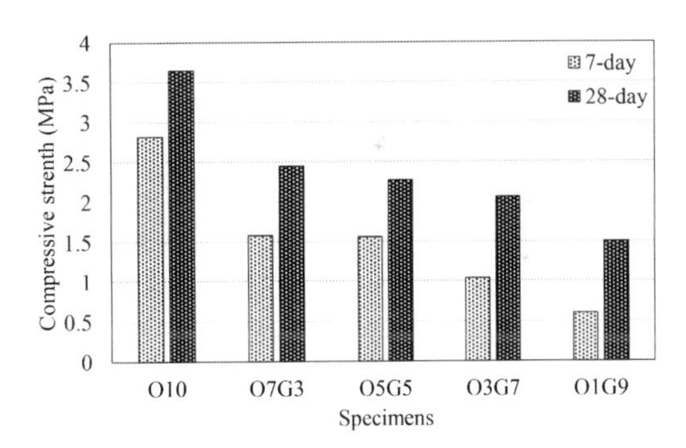

Figure 5. Compressive strength of specimens.

specified in ASTM, it was found that the foamed concrete mixes with R_G below 70% belong to the AAC-2 class (density limits: 350–550 kg/m^3; minimum compressive strength: 2 MPa). The specimen O1G9 had lower compressive strength than the minimum requirement because of the segregation of cementitious materials.

3.3 *Relationship of R_G and compressive strength*

For the further development of foamed concrete with high-volume GGBS, simple equations to predict 28-day compressive strength ($f'c$) was proposed based on a multiple regression analysis using the test data. The basic models were adjusted based on trial-and-error approach until a relatively high value of a coefficient of correlation was obtained. Overall, the following equation could be determined for foamed concrete with GGBS (see Figure 6):

$$f_c' = 3.52e^{-0.009R_G} \quad (1)$$

Noted that the above equation needs to be examined for foamed concrete with different densities. According to Equation (1), the maximum R_G can be recommended to be 62% to satisfy the limitation for the AAC-2 class of ASTM C1693. However, considering the segregation of cementitious materials, the optimum R_G could be recommended to be 50%.

3.4 *Porosity and pore-size distribution*

The pore size of the foamed concrete was measured to be between 0.03 mm and 6 mm (Figure 7). From the pore distribution, porosity of each specimen was calculated to be 54.8, 53.6, 41.4, and 27.2 for foamed concrete mixes with R_G of 0, 30, 50, and 70, respectively, indicating that the porosities decreased with the increase in GGBS content. Specimen O1G9 revealed high void share at the pore size of 1.0–2.5 mm. This may be attributed to the severe segregation of cementitiouls materials. Figure 7 also showed a uniform pore distribution in all mixes when the maximum pore size was less than 1 mm, beyond which the pore distribution became slightly irregular because of the artificial introduction of external air.

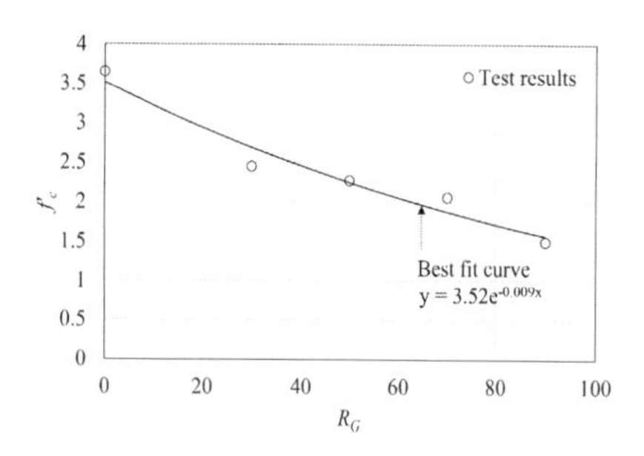

Figure 6. Effect of R_G on 28-day compressive strength of foamed concrete.

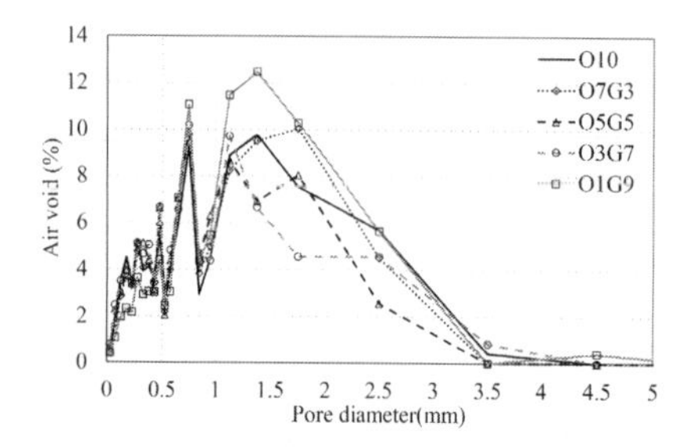

Figure 7. Pore distribution of foamed concrete.

4 CONCLUSION

From the current tests using five foamed concrete mixes with different GGBS contents (R_G) as a partial replacement of OPC, the following conclusions may be drawn:

1. The flows of the foamed concrete increased with the increase in R_G up to 30%, beyond which the flow remained constant.
2. No defoaming was observed for foamed concrete with R_G of 30%, whereas the specimen with R_G of 90% was defoamed by 6 mm because of the segregation of cementitious materials.
3. The oven-dried density and porosity of hardened foamed concrete tended to decrease slightly as R_G increases.
4. Considering the segregation of cementitious materials and compressive strength development corresponding to the requirements of AAC-2 class specified in ASTM provision, the optimum R_G for foamed concrete could be recommended to be 50%.
5. A uniform pore distribution was observed when the maximum pore size was less than 1 mm, beyond which the pore distribution became slightly irregular.

ACKNOWLEDGEMENT

This research was supported by a grant (14CTAP-C078666-01) from the Infrastructure and Transportation Technology Promotion Research Program funded by the Ministry of Land, Infrastructure and Transport of the Korean government.

REFERENCES

ASTM C 1693. 2012. *Standard Specification for Precast Auto-Claved Aerated Concrete (AAC) Wall Construction Units.* Annual Book of ASTM standards: V. 4.02. ASTM International.

Lee, K.H. 2014. *Development of Mixture Proportioning Model for Low-Density High-Strength Foamed Concrete.* Master's Dissertation−Kyonggi University.

Neville, A.M. 1995. *Properties of concrete*. Addison Wesley Longman; Edinburgh Gate England.

Pan, Z., Li, H. & Liu, W. 2014. Preparation and Characterization of Super Low Density Foamed Concrete from Portland Cement and Admixtures. *Construction and Building Materials,* 72(15): 256–261.

Shi, C., Krivenko, P.V. & Roy, D. 2006. *Alkali-activated Cements and Concretes*. New York: Taylor and Francis.

Yang, K.H., Lee, K.H., Song, J.K. & Gong, M.H. 2014. Properties and Sustainability of Alkali-Activated Slag Foamed Concrete. *Journal of Cleaner Production,* 68(1): 226–233.

Yang, K.H. & Lee, K.H. 2015. Tests on High-Performance Aerated Concrete with a Lower Density. *Construction and Building Materials,* 74(15): 109–117.

Advanced Materials, Mechanical and Structural Engineering – Hong, Seo & Moon (Eds)
© 2016 Taylor & Francis Group, London, ISBN: 978-1-138-02908-8

Flexural tests on Reinforced Concrete rectangular columns with supplementary V-ties

M.K. Kwak
Department of Architectural Engineering, Graduate School, Kyonggi University, Gyeonggi-Do, South Korea

K.H. Yang
Department of Plant and Architectural Engineering, Kyonggi University, Gyonggi-Do, South Korea

J.I. Sim
R&D Team, GL Construction Co. Ltd., Gwangju, South Korea

ABSTRACT: The objective of this study is to examine the practical potentials of the simplified V-tie arrangement method in reinforced concrete columns as an alternative to the conventional crossties specified in ACI 318-11. Six columns were tested to failure under constant axial load and reversal lateral loads. The axial load level varied to be 0.25, 0.4, and 0.55 for columns with V-ties or crossties. Test results ascertained that 90-degree hooks of crossties were gradually opened after the peak lateral load, resulting in the premature buckling of longitudinal bars. Meanwhile, no V-ties were extracted from concrete core until the columns failed. As compared with the work damage indicator of the companion crosstie columns, the V-tie columns had higher value as much as 249%, 249%, and 548% for the axial load level of 0.25, 0.4, and 0.55, respectively.

1 INTRODUCTION

Reinforced Concrete (RC) columns carrying axial compressive loads with or without moment require enough ductile response to withstand an unexpected excessive-lateral deformation. The ductility of RC columns significantly depends on the lateral confining pressure provided by transverse reinforcement (ACI-ASCE, 1997). In particular, the supplementary ties are effective in preventing the premature buckling of unsupported longitudinal middle bars and widening the effectively confined area of column core (Sheikh & Khoury 1993). It is also commonly known (Ozcebe & Saatcioglu 1987; Sakai and Sheikh, 1989) that the selection of a proper configuration of supplementary ties is a more feasible approach than a reduction of spacing of peripheral ties to achieve the same level of ductility of columns. Based on extensive test result, ACI 318-11 provision (2011) specifies the consecutive crossties engaging the same longitudinal middle bars as a supplementary ties. Considering practical installation of crossties onto longitudinal bars, ACI 318-11 provision permits the hook angle of the crossties to be 90-degree at one end and 135-degree at the other end, although most experimental studies (Mander et al. 1988; Saatcioglu & Razvi, 1992) used 135-degree hooks or 180-degree hooks for extension of the crossties into the core concrete.

Yang & Kim (2015) pointed out that the 90-degree hook of the crossties are frequently opened with the loss of concrete cover. Sheikh & Khoury (1993) also mentioned that the conventional crossties with a 90-degree hook would be insufficient to achieve the ductility of columns under high axial load level because the opening of the 90-degree hook accelerates the premature buckling of longitudinal bars and thereby load transfer capacity of column is suddenly dropped.

The present study proposed a simplified V-shaped tie arrangement method as an alternative for satisfying the ACI 318-11 provision specified for crossties of RC columns. The mechanical contribution of V-ties to the concrete confinement and prevention of premature buckling of longitudinal bars were established using a equivalent lateral force concept derived from the lateral displacement of longitudinal bars owing to the buckling. Six columns were tested under constant axial load and reversal lateral loads to examine the effectiveness of the proposed V-tie arrangement approach on the flexural ductility of columns.

2 MECHANICAL CONTRIBUTION OF V-TIES

Yang & Kim (2015) established the mechanical role of V-shaped bars as the supplementary ties and verified through the extensive axial tests of columns. The mechanical contribution of V-ties to the ductility of columns can be summarized as follows. The supplementary ties engaging the unsupported longitudinal middle bars in columns minimizes the area of the unconfined concrete through the quasi-uniform distribution of confining pressure and prevents the premature buckling of longitudinal bars at a section midway. The opening of 90-degree hooks of the conventional crossties gradually reduces the confining pressure, whereas V-ties sufficiently anchored in the concrete core maintains the confinement until the failure of columns, as shown in Figure 1. Based on Bae (2005) approach, the relationship between axial stress (f_s) and transverse displacement (Δ_{tra}) of longitudinal reinforcing bars due to buckling under axial concentric loads can be obtained. From the load–displacement relationship of a simply supported beam, the transverse displacement due to buckling can be converted into an equivalent concentric one-point load (N_{ea}) applied at the mid-span of longitudinal bars laterally supported by upper and lower transverse reinforcements, as shown in Figure 2. The equivalent force can be expressed as follows:

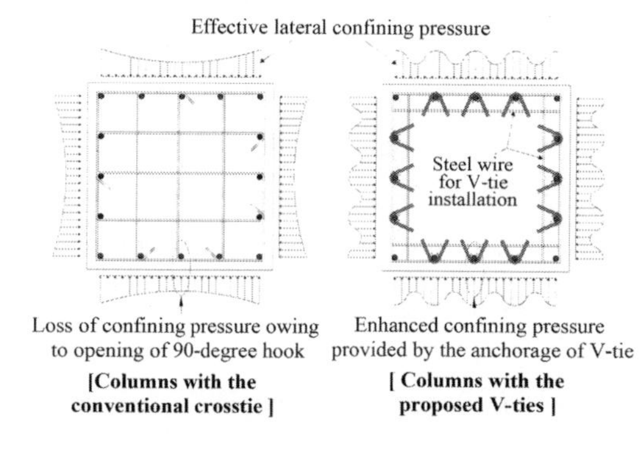

Figure 1. Effective confining pressure by transverse reinforcement.

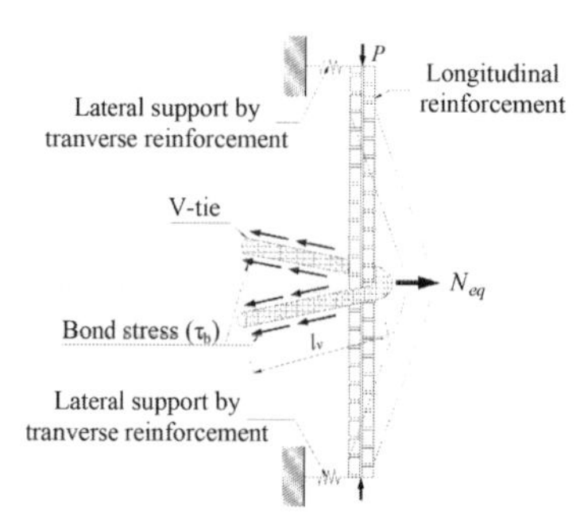

Figure 2. Bond resistance of V-tie against the equivalent force due to buckling of a longitudinal bar.

$$N_{eq} = \frac{48 E_{it} I_s \Delta_{tra}}{s_v^3} \tag{1}$$

where s_v is the spacing of the transverse reinforcement, I_s and E_{it} are the moment of inertia and tangential modulus of the longitudinal bar, respectively. The equivalent force should be resisted through the bond stresses along the V-tie legs extended into core concrete (Figure 2). Therefore, to prevent the extraction of a V-tie from the buckling of a longitudinal bar, the V-tie leg should be extended into concrete core to be more than the followings:

$$l_v \geq \frac{N_{eq}}{2\tau_b \sum_0 \cos(\theta/2)} \tag{2}$$

where τ_b is the bond strength of concrete, Σ_0 is the perimeter of V-tie bar, and θ is the angle between the two legs of the V-tie bar. The angle between two legs of V-tie bars is also recommended to be 45-degrees in order to enhance the bond resistance of the V-tie and to avoid their mutual interference during installation.

3 EXPERIMENTAL DETAILS

3.1 *Specimen details and materials*

Three columns with the conventional crossties specified in ACI 318-11 and the companion three columns with the proposed V-ties were prepared.

Table 1. Detail of specimens.

	Details of transverse reinforcement			Axial load	
Specimen[1]	s_v	ρ_v	Supplementary ties	N_n	$\frac{N_n}{f'_c A_g}$
C-0.25	130	1.03	Crossties	735	0.25
C-0.4				1176	0.4
C-0.55				1617	0.55
V-0.25		0.89	V-ties	735	0.25
V-0.4				1176	0.4
V-0.55				1617	0.55

[1]The specimen notation includes two parts. The first letter refers to the type of supplementary ties: C for the conventional crossties and V for the proposed V-shaped ties. The second part indicates the applied axial load level. For example, specimen C-0.25 indicates a crosstie column under the constant axial load of 0.25. $f'_c A_g$ Note: S_v, and ρ_v = spacing, and volumetric ratio of transverse re-inforcement, respectively, N_n = applied axial load, and A_g = area of column section.

The main parameter selected for each column group was the axial load level (f_c'), as given in Table 1, where N_n is the applied axial load, f_c' is the compressive strength of concrete, and A_g is the gross area of column section. The axial load level was 0.25, 0.4, and 0.55. Figure 3 shows the geometrical dimensions of the column sections and the arrangement details of longitudinal and transverse reinforcing bars. The column section size was 350 × 350 mm. All columns were cast integrally with a 550 × 1050 × 600 mm top stub and a 800 × 1350 × 600 mm bottom stub representing the column base. For longitudinal reinforcement, symmetrical arrangement was provided using eight deformed bars with a diameter of 19 mm, producing a longitudinal reinforcement ratio of 1.8%. Deformed bars of 10 mm diameter were used as transverse reinforcement at spacing (s_v) of 130 mm. The conventional crossties had a 90-degree hook at the one end and a 135-degree hook at the other end. The hooks were extended by 75 mm into core concrete. The 90-degree hook end of the crossties was alternated along the length column. The angle and extension length of V-tie legs were 45-degree and 75 mm, respectively. The V-tie was individually fixed onto a longitudinal bar and placed on the steel wire joining the peripheral ties, as shown in Figure 3. The average compressive strength of concrete measured from testing three control cylinders was 24 MPa. The yield strengths of mild steel bars were 516.3 MPa and 473.3 MPa for 19 mm and 10 mm diameter bars, respectively.

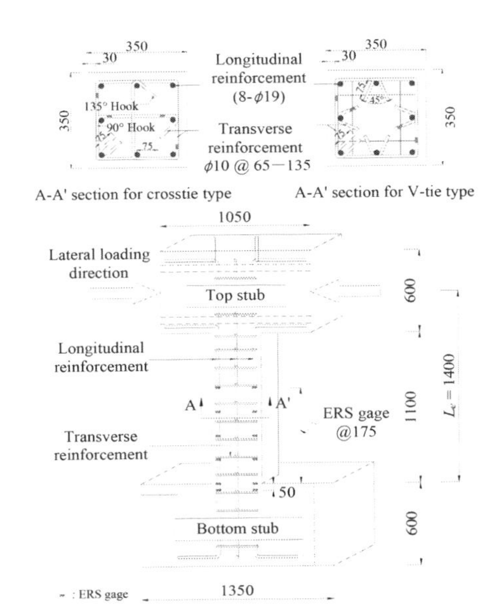

Figure 3. Specimen details and arrangement of reinforcement (all dimensions are in mm).

Figure 4. Test set-up.

3.2 *Test set-up*

All the specimens were tested under constant axial load and reversed cyclic lateral loads, as shown in Figure 4. The test strong frame was designed with two assemblies for axial loads and an assembly for lateral loads. The axis of rotation of the hinge for the axial load assembly coincided with the central axis of the column specimens. Axial compressive force was applied by pulling the load transfer assembly down using two 2000 kN capacity hydraulic jacks. After applying the full axial load, lateral load reversals were applied at the center of the top stub using a 2000 kN capacity hydraulic jack with a lateral displacement rate of 2 mm/min. The applied lateral load was controlled by the predetermined displacement history calculated as the function of the yield displacement (Δ_y) of each column.

4 TEST RESULTS AND DISCUSSIONS

4.1 *Lacuteral load-displacement relationship*

Figure 5 shows the lateral load–lateral displacement relationship of the column specimens. Table 2 also summarizes test results. The initial stiffness of columns increased with increasing λ_n. The pre-peak behavior and ultimate strength of columns were independent of the type of supplementary ties. The flexural capacity of V-tie columns was comparable to that of the companion crosstie columns. However, the slope at the descending branch of the lateral load-displacement curve was dropped rapider for crosstie columns than for V-tie columns. Both the stiffness degradation and strength reduction rate with every loading cycle were also much lower for the V-tie columns than for the crosstie columns. This observation was more pronounced for columns under a higher axial load level. After the columns reached the peak lateral load capacity, 90-degree hooks of

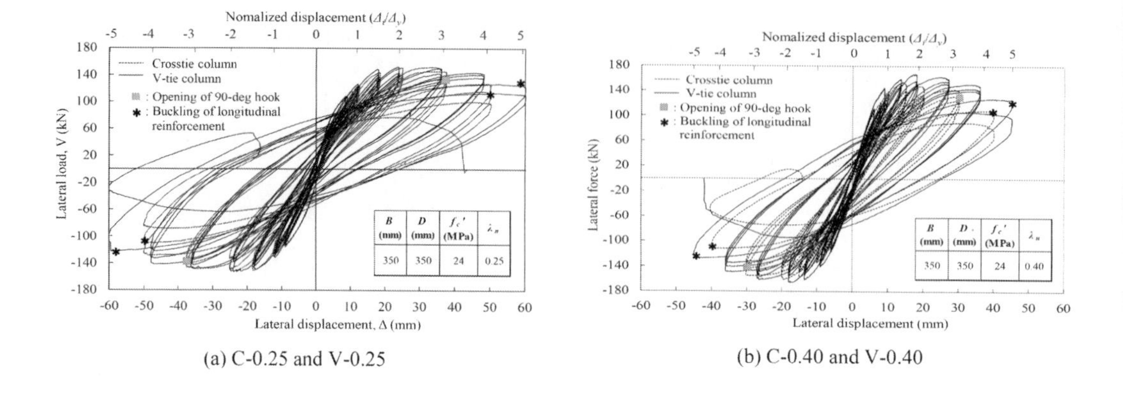

(a) C-0.25 and V-0.25

(b) C-0.40 and V-0.40

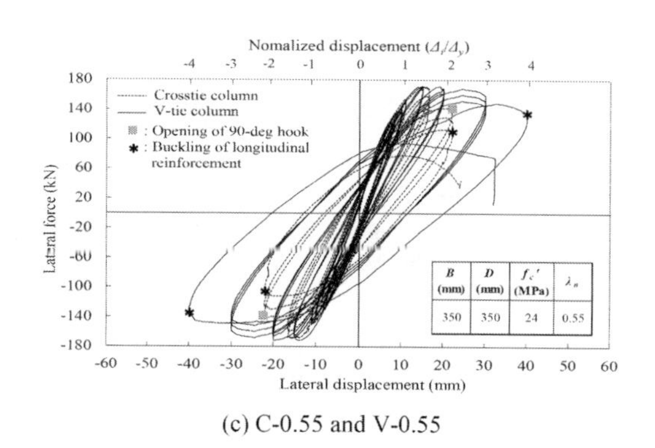

(c) C-0.55 and V-0.55

Figure 5. Lateral load-displacement relationship of column specimens.

Table 2. Summary of test results.

Specimen	Experimental results							Predicted V_n (kN)	$(V_n)_{Exp.}/(V_n)_{Pre.}$
	Peak lateral load (kN)								
	V_n^+	V_n^-	Ave	Δ_y (mm)	Δ_{80} (mm)	μ_Δ	W_{80}	ACI 318-11	
C-0.25	157.5	157.9	157.7	12.49	50.11	4.01	48.01	142.07	1.11
C-0.4	166.6	167.2	166.9	10.03	29.39	2.93	33.17	151.93	1.10
C-0.55	171.6	168.8	170.2	11.00	22.14	2.01	10.12	150.54	1.13
V-0.25	156.8	152.6	152.5	12.05	59.82	4.96	119.79	142.07	1.07
V-0.4	167.6	165.7	166.7	9.06	36.05	3.98	82.73	151.93	1.10
V-0.55	171.6	171.4	171.5	10.05	30.33	3.02	55.42	150.54	1.14

Note: Δ_y = yield displacement of column as an average of both loading direction, Δ_{80} = lateral displacement of column at 0.8 V_n on the descending branch of lateral load-displacement (V–Δ) curve, as an average of both loading direction, μ_Δ = displacement ductility ratio, W_{80} = work damage indicator until 0.8 V_n.
– Superscripts + and—mean that the positive and the negative loading directions, respectively.
– Parentheses indicate the V–Δ loop of incremental yield displacement at which specified features given in the table occurred.

crossties were gradually opened, which eventually caused a premature buckling of longitudinal reinforcement and a severe crushing of core concrete. The ductility deterioration due to the opening of 90-degree was more severe as λ_n increased.

4.2 *Flexural strength*

All columns had higher flexural capacity than the nominal strength predicted using the stress block specified in ACI 318-11. The ratio between measured strengths and predictions was more than 1.1, as given in Table 2. The flexural capacity of columns was independent of the type of the supplementary ties.

4.3 *Flexural ductility*

As compared with the value of work damage indicator (W_{80}) of the companion crosstie columns, the V-tie columns had higher value as much as 249.5%, 249.4%, and 547.6% for axial load level of 0.25, 0.4, and 0.55, respectively, as given in Table 2. The effect of the axial load level on μ_Δ was similar to the trend observed in the work damage indicator. The superiority of V-ties to the conventional crossties in enhancing the flexural ductility of columns is more pronounced as λ_n increase.

5 CONCLUSIONS

From the experimental program to examine the effectiveness of the proposed V-tie arrangement method on the flexural ductility of RC columns, the following conclusions may be drawn:

1. Both the stiffness degradation and strength reduction rate with every loading cycle were also much lower for the V-tie columns than for the crosstie columns. This observation was more pronounced as axial load level increased.
2. After the columns reached the peak lateral load capacity, 90-degree hooks of crossties were gradually opened, which eventually caused a premature buckling of longitudinal reinforcement and a severe crushing of core concrete.
3. All columns had higher flexural capacity than the nominal strength predicted using the stress block specified in ACI 318-11.

As compared with the value of work damage indicator (W_{80}) of the companion crosstie columns, the V-tie columns had higher value as much as 249.5%, 249.4%, and 547.6% for axial load level of 0.25, 0.4, and 0.55, respectively.

ACKNOWLEDGEMENTS

This work was supported by the National Research Foundation of Korea (NRF) grant funded by the Korean Government (MSIP) and the Small and Medium Business Administration (SMBA) grant funded by the Ministry of Trade, Industry & Energy, Republic of Korea (No. NRF-2014R1 A2 A2 A09054557 and No. S2187208).

REFERENCES

ACI-ASCE (ACI-ASCE Committee 441) (1997). High-Strength Concrete Columns: *State of the Art, ACI Structural Journal*, 94(5): 323–335.

ACI 318-11, *Building Code Requirements for Structural Concrete (ACI 318-11) and Commentary (ACI 318R-11)*. 2011. American Concrete Institute, Farmington Hills, MI, USA, 503.

Bae, S. 2005. *Seismic Performance of Full-Scale Reinforced Concrete Columns*, Ph.D. Dissertation, The University of Texas at Austin, USA, 312.

Mander, J.B., Priestley, M.J.N. & Park, R. 1988. Observed Stress–Strain Behavior of Confined Concrete, *Journal of Structural Engineering ASCE*, 114(8): 1827–1849.

Ozcebe, G. & Saatcioglu, M. 1987. Confinement of Concrete Columns for Seismic Loading, *ACI Structural Journal*, 84(4): 308–315.

Saatcioglu, M. & Razvi, S.R. 1992. Strength and Ductility of Confined Concrete, *Journal of Structural Engineering ASCE*, 118(6): 1590–1607.

Sakai, K. & Sheikh, S.A. 1989. What Do We Know about Confinement Reinforced Concrete Columns? A Critical Review of Previous Work and Code Provisions, *ACI Structural Journal*, 86(2): 192–207.

Sheikh, S.A. & Khoury, S.S. 1993. Confined Columns with Stubs, *ACI Structural Journal*, 90(4): 414–431.

Yang, K. H. & Kim, W.W. 2015. Axial Compression Performance of RC Short Columns with Supplementary V-Shaped Ties, *ACI Structural Journal*, Under Review.

Advanced Materials, Mechanical and Structural Engineering – Hong, Seo & Moon (Eds)
 ISBN: 978-1-138-02908-8

Shear creep of epoxy adhesive joints: Experiment and fractional derivatives of rheological modeling

X.M. Wang & C.W. Zhou
Nanjing University of Aeronautics and Astronautics, Nanjing, China

ABSTRACT: Epoxy adhesive joints have been widely used in the field of aeronautics, civil engineering, etc. Long term performance of epoxy adhesive joints is one of the key reasons for its successful engineering application. However, it has not been well studied. In this paper, experiments were carried out to investigate the shear creep behavior of epoxy adhesive joints under different stress levels. A fractional derivative rheological model is proposed to characterize the complicated creep deformation of epoxy adhesive joints with a simple expression. The results revealed that this model is capable of predicting the creep behavior of epoxy adhesive joints up to a sustained stress level of 30–70% of shear strength with a good accuracy.

1 INTRODUCTION

Epoxy adhesive joints have been extensively used in many fields such as aeronautics, astronautics, automotive, and civil engineering because of high strength-to-weight ratio, fewer sources of stress concentrations and better fatigue properties (Magalhaes & Moura 2005).

A number of researches (Mendoza–Navarro et al. 2013, Balzani et al. 2012, Rapp 2015) have been carried out on the short-term performance of epoxy adhesive joints such as the strength and failure mechanism. As for the long-term performance, studies (Zehsaz et al. 2014, Khalili et al. 2009, Majda & Skrodzewicz 2009) have shown that epoxy adhesive joints exhibits obvious viscoelasticity, especially in high-stress levels. The experiments carried out by Zehsaz et al. (2014) reveal that creep deformation increases with adhesive thickness. Khalili et al. (2009) investigated the effect of fiber reinforcing in epoxy adhesive joints, and the results show that the creep deformation of the specimens decrease significantly with fiber reinforcing. Experiments have been taken out by Feng et al. (2005) to determine the long-term time-dependent performance of polymers under different environmental conditions. Zehsaz and Majda, etc. (Zehsaz et al. 2014, Majda & Skrodzewicz 2009) propose different kinds of rheological models to simulate creep behavior of adhesive joint, but these models usually need a lot of parameters to get a more accurate result. Hence, an effective model is needed to describe the creep behavior of epoxy adhesive joints.

The fractional derivative of the rheological model is proposed in this paper to simulate the shear creep behavior of the epoxy adhesive. Its parameters are determined by shear creep tests of double-lap joint specimens.

2 EXPERIMENTAL METHODS

2.1 *Materials and specimen*

In recent years, Hybrid Fiber-Reinforced Polymer (HFRP) has been widely utilized in strengthening structures such as beams and columns for its mechanical and economic benefits. In HFRP two or more reinforcing fibers are usually employed in one FRP sheet, and the advantages of different fibers could be fully used. However, in some circumstances, HFRP needs to be overlapped. Therefore, HFRP is used to make double-lap joints in this paper.

The epoxy used in this test is produced by Nantong Xingchen Synthetic Material Co., LTD. in China. The epoxy is made by mixing the resin and hardener with a ratio of 1:1. The fiber-reinforced polymer is C/GFRP (Carbon and Glass Fiber Reinforced Polymer) sheet which is produced by China Nanjing Haituo Composite Co., LTD.

Geometry of double-lap joint specimen is shown in Figure 1. The whole length of the specimen is 250 mm, is as recommended by the national standards of P.R.C. (2008). The thickness and width of C/GFRP sheets on both sides are 0.65 mm and 25 mm, respectively. The length of cover lap is 41 mm, with a thickness of 0.65 mm. Both ends of the specimen are reinforced with aluminum sheets. The length and thickness of aluminum sheets are 50 mm and 2 mm, respectively, as shown in Figure 2(a). The tensile creep behavior of C/GFRP sheet

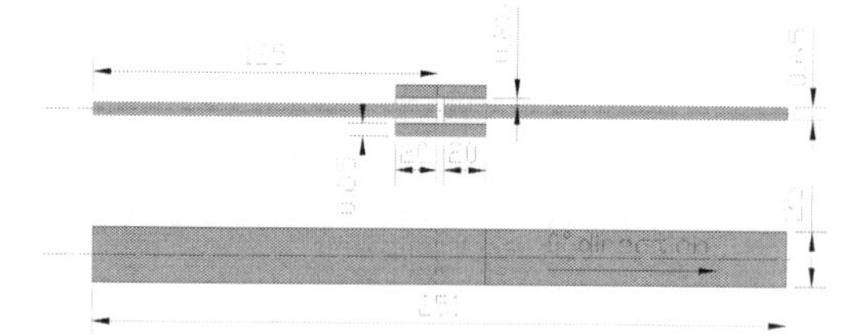

Figure 1. Geometry of double-lap joint specimen (unit: mm).

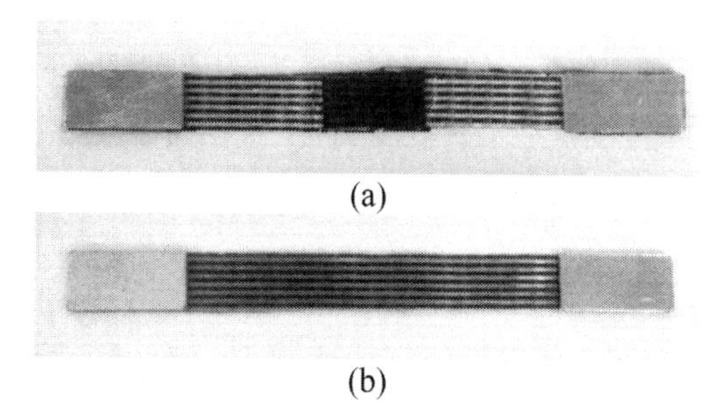

Figure 2. Specimens.

in 0° direction was also tested and the specimen is shown in Figure 2(b).

2.2 *Experiment method*

Tensile strength tests were carried out first and the average ultimate shear strength of epoxy adhesive is 12.1 MPa. The shear failure mode is shown in Figure 3. The sustained stress in creep tests is 4 MPa, 6 MPa, and 8 MPa, which corresponds to 30%–70% of shear strength.

The shear creep tests have been carried out using CSS-3910 electronic creep testing machine (Figure 4). CSS-3910 consists of EDC-60R (electronic digital controller) produced by Germany DOLL company, chucks, deformation measurement extensometer with the accuracy of 1 μm, computer systems and other accessories. The shear creep deformation is evaluated by measuring the distance between fixed point A/A' and B/B' which is as shown in Figure 4. The specimens were tested at 20 ± 5°C under dry condition (RH = 30%).

The creep test of C/GFRP sheet was carried out and no significant creep strain was observed. Therefore, the creep of C/GFRP sheet is neglected.

The total shear deformation of specimen consists of instantaneous elastic shear deformation and creep deformation. Instantaneous elastic shear deformation happens shortly after the sustained stress σ_0 is applied at time t_0. As shown in Figure 5, the instantaneous elastic shear strain is:

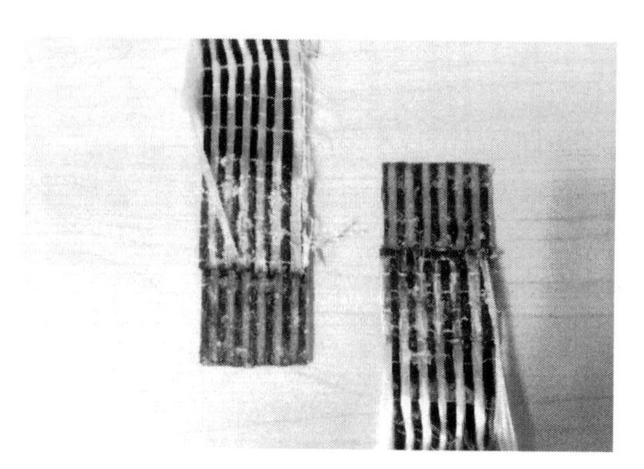

Figure 3. Failure mode of double-lap joint.

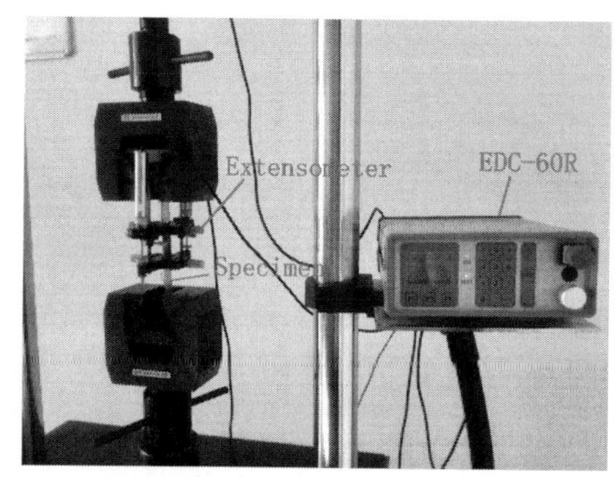

Figure 4. Specimen on jig of CSS-3910.

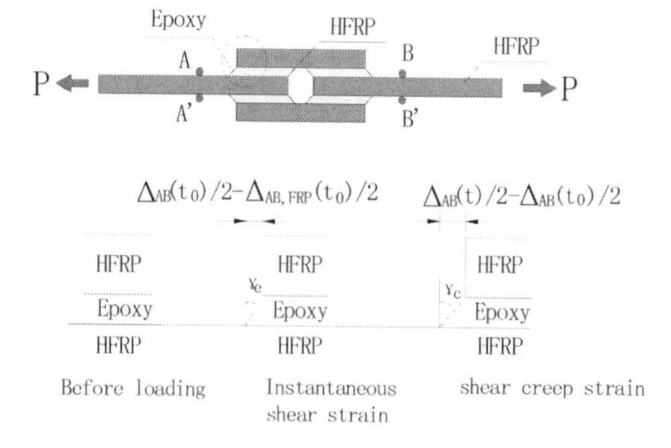

Figure 5. Calculation schematic of shear creep strain.

$$\gamma_e = \frac{\Delta_{AB}(t_0) - \Delta_{AB,FRP}(t_0)}{2h} \tag{1}$$

where, h is the thickness of epoxy adhesive.

The shear creep strain at the time t is

$$\gamma_c(t)=\frac{\Delta_{AB}(t)-\Delta_{AB}(t_0)}{2h} \tag{2}$$

The creep compliance $J(t)$ is defined as:

$$J(t)=\frac{\gamma_c(t)}{\sigma_0} \tag{3}$$

3 FRACTIONAL DERIVATIVE RHEOLOGICAL MODELING

Elastic spring and linear dashpot are the most common components in the rheological models. Fractional derivative dashpot is another kind of non-integer order approach for modeling creep behavior. The definition for Riemann-Liouville fractional calculus (Barpi & Valente 2004) is:

$$D^{-\alpha}f(t)=\frac{d^{-\alpha}f(t)}{dt^{-\alpha}}=\int_0^t\Phi_{1-\alpha}(t-\tau)f(\tau)d\tau \tag{4}$$

where $\Phi_\alpha(t)=\begin{Bmatrix}1/\left[t^{-\alpha}\Gamma(1-\alpha)\right], & t>0\\ 0, & t<0\end{Bmatrix}$, and Γ represents Gamma function: $\Gamma(z)=\int_0^\infty t^{z-1}e^{-t}dt$. Rearrange Equation (4), and the expression for the α-order fractional derivative of function f(t) turns into:

$$\begin{aligned}D^\alpha f(t)&=\frac{d\left[D^{-(1-\alpha)}f(t)\right]}{dt^\alpha}\\&=\frac{1}{\Gamma(1-\alpha)}\frac{d}{dt}\int_0^t\frac{f(\tau)}{(t-\tau)^\alpha}d\tau\end{aligned} \tag{5}$$

The constitutive relation defining the fractional derivative dashpot is given as:

$$\sigma(t)=E\tau^\alpha D^\alpha\varepsilon(t) \tag{6}$$

where E, τ, and, α are the parameters, and $0<\alpha<1$. This indicates fractional derivative dashpot representing the character between elastic spring ($\alpha=0$) and linear dashpot ($\alpha=1$).

Figure 6 shows some basic combinations of rheological models, which consist of springs and fractional derivative dashpots. Figure 6(a) shows fractional derivative Maxwell model and Figure 6(b) shows fractional derivative Kelvin model which are four-parameter models and Figure 6(c) shows

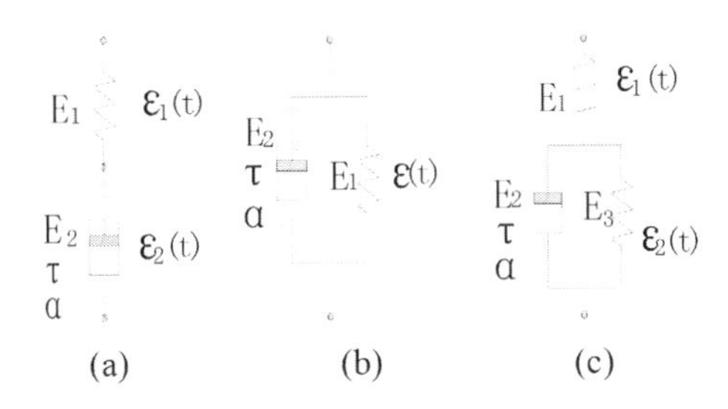

Figure 6. Various combinations of rheological models.

fractional derivative Kelvin–Voigt model which is the combination of Figure 6(a) and Figure 6(b).

The creep strain $\varepsilon(t)$ under sustained stress σ_0 for fractional derivative Maxwell model can be described as follows:

$$\varepsilon(t)=\varepsilon_1(t)+\varepsilon_2(t)=\frac{\sigma_0}{E_1}+\frac{\sigma_0}{E_1\tau^\alpha D^\alpha} \tag{7}$$

Rearranging Equation (7) we get the expression for σ_0 which is:

$$\sigma_0=\frac{E_1E_2\tau^\alpha D^\alpha}{E_1+E_2\tau^\alpha D^\alpha}\varepsilon(t) \tag{8}$$

In order to simplify the calculation of creep compliance $J(t)$, Laplace transfer technique is used. By using Laplace and inverse Laplace method, the creep compliance $J(t)$ of fractional derivative Maxwell model can be obtained:

$$\begin{aligned}J(t)&=L^{-1}\left(\frac{E_1+E_2\tau^\alpha s^\alpha}{E_1E_2\tau^\alpha s^\alpha}\frac{1}{s}\right)\\&=\frac{1}{E_1}\left[1+\frac{E_1t^\alpha}{E_2\tau^\alpha\Gamma(\alpha+1)}\right]\end{aligned} \tag{9}$$

In the same way we can get creep compliance of fractional derivative Kelvin model and Kelvin-Voigt model, respectively. However, the Kelvin Model fails in simulating the instantaneous deformation, while Kelvin-Voigt model contains infinite summation term. Therefore, fractional derivative Maxwell model is considered in this study.

4 RESULTS AND DISCUSSION

The shear creep test of epoxy adhesive joints under different sustained stress was shown in Figure 7. Obvious shear creep strain is observed in the test, and the shear creep strain increases with time and sustained stress.

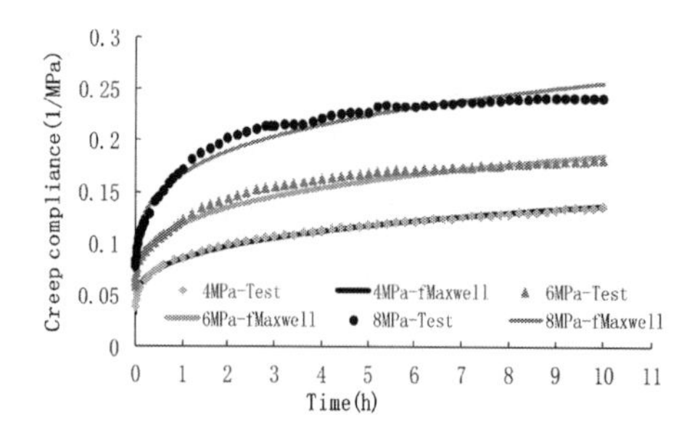

Figure 7. Creep curves of epoxy from the experiment and fractional Maxwell model under various sustained stresses.

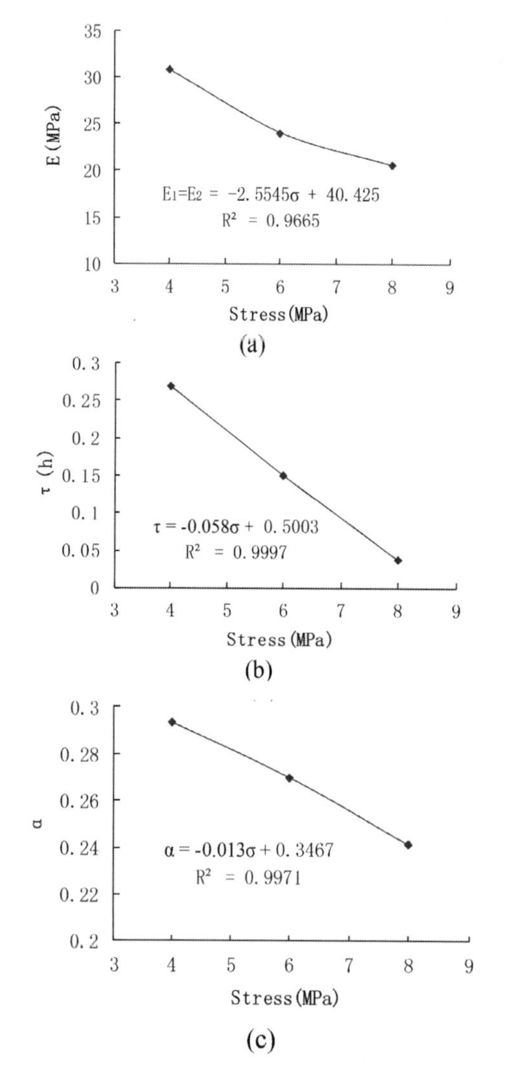

Figure 8. Variations of parameters in fractional derivative Maxwell model with sustained stress.

It can be seen from Figure 7 that creep curves provided by fractional derivative Maxwell model agrees well with the test results. For simplification moduli E_1 and E_2 are postulated as the same: $E_1 = E_2 = E$. The three independent parameters in Equation (9) are: E, α, and τ, which are formulated by fitting the experimental curves as shown as in Figure 8.

As demonstrated in Figure 8, the equations of parameters in fractional derivative Maxwell model can be obtained as a function of sustained stress σ_0 (in MPa):

$$E_1(\sigma) = E_2(\sigma) = -2.5545\sigma + 40.425 \tag{10}$$

$$\tau(\sigma) = -0.058\sigma + 0.5003 \tag{11}$$

$$\alpha(\sigma) = -0.013\sigma + 0.3467 \tag{12}$$

Finally, a nonlinear creep model defined as Equation (13) can be got by replacing the constant parameters in fractional derivative Maxwell model in Equation (9) with the stress-dependent functions in Equation (10–12).

$$J(t) = \frac{1}{E(\sigma)}\left[1 + \frac{t^{\alpha(\sigma)}}{\tau(\sigma)^{\alpha(\sigma)}\Gamma(\alpha(\sigma)+1)}\right] \tag{13}$$

5 CONCLUSION

Shear creep experiments of epoxy adhesive joints are performed on double lap shear test specimens under different sustained stresses. Obvious shear creep strain is observed, and the creep strain increases significantly with sustainable stress and time. A Maxwell type fractional derivative rheological model is proposed to predict the nonlinear shear creep behavior of epoxy adhesive joints. Compared with traditional rheological models, it can fit complicated experimental curves well with fewer parameters.

ACKNOWLEDGMENT

This work is supported by the Fund of Jiangsu Innovation Program for Graduate Education (KYLX_0220) and the Fundamental Research Funds for the Central Universities, National Natural Science Foundation of China (11272147, 10772078), Aviation Science Foundation (2013ZF52074), Fund of State Key Laboratory of Mechanical Structural Mechanics and Control

(0214G02), and the Project Funded by the Priority Academic Program Development of Jiangsu Higher Education Institutions (PAPD).

REFERENCES

Balzani, C., Wagner, W., Wilckens, D., Degenhardt, R., Büsing, S. & Reimerdes, H.G. 2012. Adhesive joints in composite laminates-A combined numerical/experimental estimate of critical energy release rates. *International Journal of Adhesion and Adhesives*, 32: 23–38.

Barpi, F. & Valente, S. 2004. A fractional order rate approach for modeling concrete structures subjected to creep and fracture. *International Journal of Solids and Structures,* 41: 2607–2621.

Feng, C.W., Keong, C.W., Hsueh, Y.P., Wang, Y.Y. & Sue, H.J. 2005. Modeling of long-term creep behavior of structural epoxy adhesives. *International Journal of Adhesion & Adhesives*, 25: 427–436.

Khalili, S.M.R., Jafarkarimi, M.H. & Abdollahi, M.A. 2009. Creep analysis of fiber reinforced adhesives in single lap joints-Experimental study. *International Journal of Adhesion & Adhesives*, 29: 656–661.

Magalhaes, A.G. & de Moura, M.F.S.F. 2005. Application of acoustic emission to study creep behavior of composite bonded lap shear joints. *NDT & E International,* 38: 45–52.

Majda, P. & Skrodzewicz, J. 2009. A modified creep model of epoxy adhesive at ambient temperature. *International Journal of Adhesion & Adhesives*, 29: 396–404.

Mendoza-Navarro, L.E., Diaz-Diaz, A., Castañeda-Balderas, R., Hunkeler, S. & Noret, R. 2013. Interfacial failure in adhesive joints: Experiments and predictions. *International Journal of Adhesion and Adhesives*, 44: 36–47.

National standards of P.R.C. *Plastics-Determination of tensile properties-Part 5: Test conditions for unidirectional fiber-reinforced plastic composites.* (GB/T 1040.5-2008/ISO 527-5:1997), 2008.

Rapp, P. 2015. Mechanics of adhesive joints as a plane problem of the theory of elasticity. Part II: Displacement formulation for orthotropic adherents. *Archives of Civil and Mechanical Engineering*, 15(2): 603–619.

Zehsaz, M., Vakili-Tahami, F. & Saeimi-Sadigh, M.A. 2014. Parametric study of the creep failure of double lap adhesively bonded joints. *Materials and Design*, 64: 520–526.

Advanced Materials, Mechanical and Structural Engineering – Hong, Seo & Moon (Eds)
© 2016 Taylor & Francis Group, London, ISBN: 978-1-138-02908-8

Experimental study on shaking table tests of Dougong model of Tianwang hall, Luzhi

Z.R. Li, T.Y. Hou, X.X. Zhao, X.L. Zhang, Q.Y. Chen, Z.L. Que & F. Wang
College of Materials Science and Engineering, Nanjing Forestry University, Nanjing, Jiangsu, China

ABSTRACT: Taking a "Dougong," the corbel bracket of the Tianwang hall in Baosheng Temple – Luzhi, as the research object, an experimental study is carried out on shaking table tests of the full-scale pine corbel bracket model. Through the analysis of acceleration and dynamic magnification coefficient and the process of displacement characteristics of corbel bracket in response to the changes of vibration, major conclusions are as follows. Seismic acceleration shows the response of vibrational energy instead of the maximum deformation of the bracket specimens. The vibration frequency plays an important role in the degree of rotation deformation, while the amplitude plays a decisive role in the sliding displacement of each component. The rotary displacement of cap block and flower arm occupies a dominant position among all the components. Flower arm with Xia-ang with one special part on the bracket, which mainly plays the role of decoration, is weak on the connection node position which needs more attention and relevant reinforcement measures.

1 INTRODUCTION

The evolution of corbel bracket is the significant symbol of Chinese traditional timber frame architectural evolution. As there is a transition between the roof structure and uprights, corbel bracket plays the dual function of force and decoration, which is more prominent as a shock absorber.

As a whole, corbel brackets can be regarded as an inverted elasticity fixed hinge support. Overhanging beam formed by overlapped crossbars makes it a prominent shock absorber as well as an elegant decoration. Corbel bracket overhangs in two directions, which reduces the span of upper structure, and enhances the carrying capacity of girder as well as square-column. Constructional elements of corbel bracket are formed by the mortise and tenon connection of timber components which brings the whole system a good ductility, and increases the ability of energy dissipation through friction between the components.

The study of historic Chinese building has transformed from the era only depended on minority pioneers' hard work into a number of fierce ideological collisions everywhere. Studies mainly focus on the outstanding anti-seismic ancient timber frame architecture, not a few of them did some tests on corbel bracket unit, including vertical monotonic loading, horizontal low cyclic loading, dynamic characteristics, etc. Mechanics models were built as well. Whereas, domestic scholars focus more on Song type and Qing type and use simplified symmetrical corbel bracket models, which can't fully reflect on the actual force condition; a few of the tests used full scaled models nor tested the corbel bracket as an individual unit.

2 MATERIAL AND METHODS

The original material of corbel brackets in the Tianwang hall was Chinese-fir. Considering the difference between ancient Chinese fir and modern local fast-growing Chinese Fir, we choose hard pine with stronger intensity over modern Chinese fir to simulate as far as possible. The result of essential mechanical properties determination of model test specimen is shown Table 1.

The model constructional element is constituted by corbel brackets that cover the bottom of bucket to its girder; each structure of the sheaf adopts

Table 1. Mechanical properties of materials.

Trees	Moisture content (%)	Compression strength parallel to grain (MPa)	Flexural strength (MPa)	Modulus of elasticity (MPa)
China Fir	11.0	44.9±1.4*	55.6±3.5	5601.7±52.1
Hard Pine	13.4	47.1±3.6	75.9±4.6	7153.1±75.2

* is standard deviation, similarly here in after.

straight tenons for their connection according to their actual service requirements; small blocks in the upper part were installed with beams and loading plates to find a balance individually. Due to the limited load capacity of the shaking table, our test is going to be carried out in an approximately no-load state, with their counterweight being 25 kilograms for seeking a balance during the vibrating process.

After packaging the test specimen, the cap block was fixed on the table-board. Lateral acceleration sensors are installed on the table-board, the bottom of the cap block and the top of corbel bracket are connecting them to the seismic shear shaking table test analysis system (as it illustrated in Figure 1: A,B,C). Gauging points for rotary displacement and sliding displacement are as shown in Figure 1.

Fifteen groups of tests are conducted in the overhanging direction and they lasted for 30 s each time. The equipment can output Sinusoidal waveform only, so we determined the frequency and amplitude according to El-centro. Lateral load shock was loaded in the vibrational frequency (frequency of electromotor) of 10 Hz, 15 Hz, 20 Hz, 25 Hz, and 30 Hz (the corresponding frequency of the shaking table is 1.05 Hz, 1.59 Hz, 2.10 Hz, 2.65 Hz, and 3.12 Hz) and in amplitude of 10 mm, 20 mm, and 30 mm. P1520 stands for the frequency of 15 Hz and amplitude 20 mm.

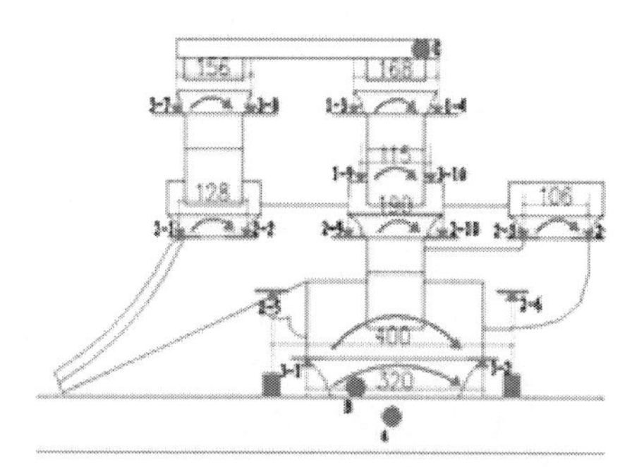

(a) Gauging point for rotary displacement

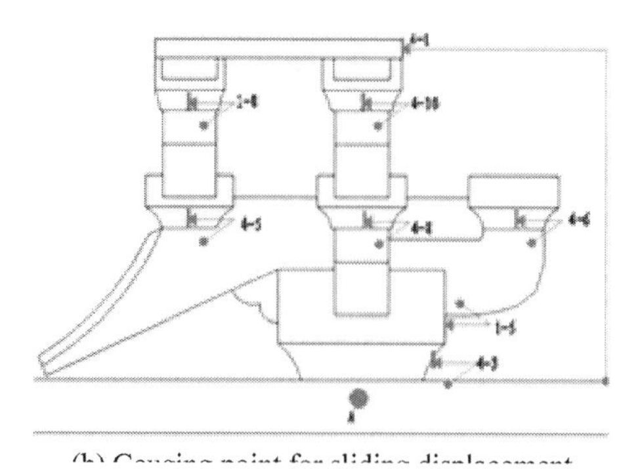

(b) Gauging point for sliding displacement

Figure 1. Arrangement of rotary and slip deformation measuring.

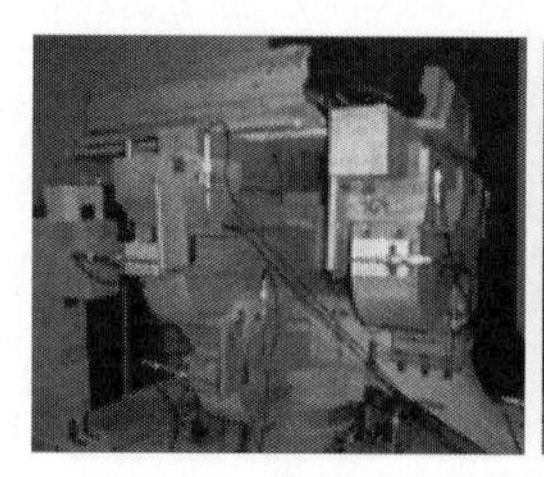

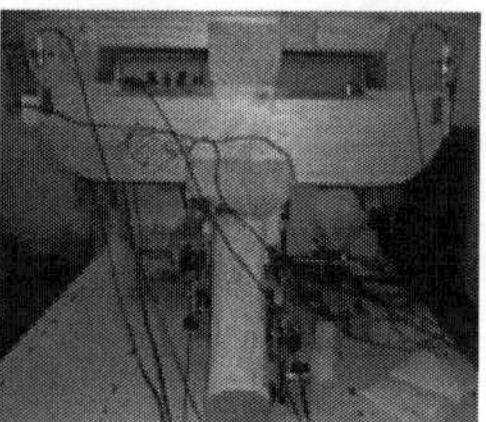

Figure 2. The model and shaking table.

The shaking table equipment and the dougong sample are as shown in Figure 2. The equipments: four-free-angles simulation seismic experiment platform, HPDZ-1 and acquisition, analysis system for vibration and dynamic signal, and HPU100-F are both from Nanjing Ho & P Technology Co., Ltd.

3 TEST RESULTS

The sample vibrates slightly in low frequency and amplitude. When amplitude increases to 20 millimeters, it will shake visibly. Moreover, dip angles and slip can be seen in connection to the joints between flower arm and Xia-ang especially in P1350 with squeak because of extrusion between ear of cap block and the joint of flower arm and Xia-ang. Along with the increasing frequency, the amplitude of the massive structure decreases, and its vibration is more disciplinary. On the macroscopic perspective, the model did not slip, and its chief transformation is based on the component node corner. Component part failure did not happen during the whole process. Whereas, after angle displacement and horizontal sliding, the level of absolute displacement in the long arm, underneath beam increased significantly, and the maximum slip quantity reached at 12 mm.

4 EXPERIMENTAL ANALYSES

4.1 *Structural dynamic characteristic*

With the measured acceleration values on the bottom and top of the bracket, calculate them and get the dynamic amplification factor of test specimen in different operating modes-β, $\beta = a_1/a_0$, they can reflect their vibration isolation effect (vide Figure 2).

Simulating seismic intensity with the vibrational energy corresponded to them by their accelerational response. Nonetheless, at the same prompting acceleration operating mode, the maxi-

Table 2. The acceleration and dynamic magnification coefficient.

Condition number	Vibration excitation (g)	Intensity of corresponding earthquake	Acceleration of the bottom a_0 (g)	Acceleration of the top a_1 (g)	Dynamic magnification factor β
P1010	0.048	IV	0.052	0.052	1.0
P1020	0.07	IV	0.12	0.145	1.208
P1030	0.12	VII	0.15	0.255	1.7
P1510	0.034	V	0.044	0.045	1.022
P1520	0.1	VII	0.16	0.17	1.063
P1530	0.13	VII	0.16	0.2	1.25
P2010	0.05	VI	0.06	0.045	0.75
P2020	0.18	VIII	0.22	0.19	0.863
P2030	0.23	VIII	0.27	0.242	0.896
P2510	0.07	VI	0.09	0.032	0.355
P2520	0.32	VIII	0.41	0.22	0.536
P2530	0.45	IX	0.50	0.24	0.48
P3010	0.11	VII	0.14	0.034	0.243
P3020	0.55	IX	0.56	0.17	0.303
P3030	0.75	X	0.77	0.27	0.351

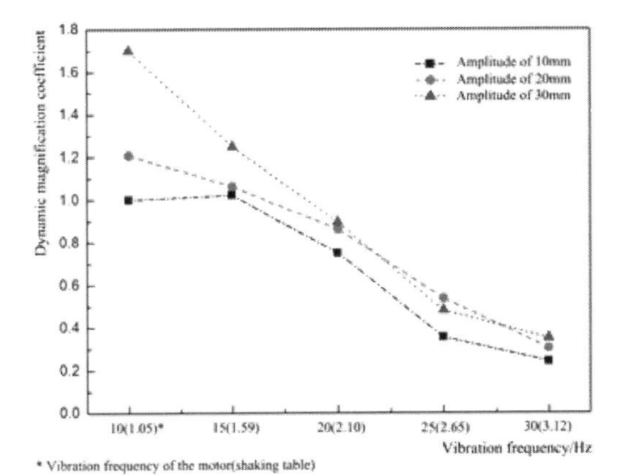

Figure 3. The change of dynamic magnification coefficient under same amplitude.

mum deformation of the test specimen will be different if the frequency and amplitudes differ. So the acceleration can be reflected mostly by the energy of vibration, not deformation maximum of test specimen directly. Then dispose the variation tendency of model dynamic amplification factor with the same amplitude but different frequencies (Figure 3).

1. Same inputted amplitude, model dynamic amplification factor will decrease with the increase of frequency. Static friction force, slip energy consumption, compressional deformation of occlusions in occlusions of arch and bucket, mortise and tenon joint between constructional elements will help shock absorption if the frequency decreases.
2. Same inputted frequency, model dynamic amplification factor amplifies with amplitude increase. When vibration frequency of tableboard is greater than 1.59 Hz (15 Hz for electromotor), the amplification of increasing dynamic amplification factor diminishes distinctly. It will almost remain at the frequency of 2.65 Hz and 3.12 Hz (25 Hz and 30 Hz for electromotor).
3. Below 1.59 Hz, dynamic amplification factor enlarges with the increase of amplitude, and no less than 1. Main reason is that the vertical displacement limits can be hardly found in a no-load condition, with limited friction between adjacent components. The bracket is easily going to swing widely; also at that time, frequency is not high and swing is sufficient. These will not happen in reality.

4.2 *Change features of displacement response*

Dispose measured values of each layer displacement in every mode, and draw its corresponding curve (take the operating mode 2.10 Hz–2.65 Hz as an example, vide Figure 4), choosing cap block, small arm, flower arm and long arm for instance. Then analyze the change law of bottom, central part and top of corbel bracket.

As shown in this figure:

1. When repetitive loading, vibration displacement curves of bottom, central part and top are in synch, it indicates that the constructional elements have good rigidity and keep tight well.
2. Displacement amplitude of fluctuation of the bottom is small, and it enlarges with its increase in height, until it reaches the maximum on

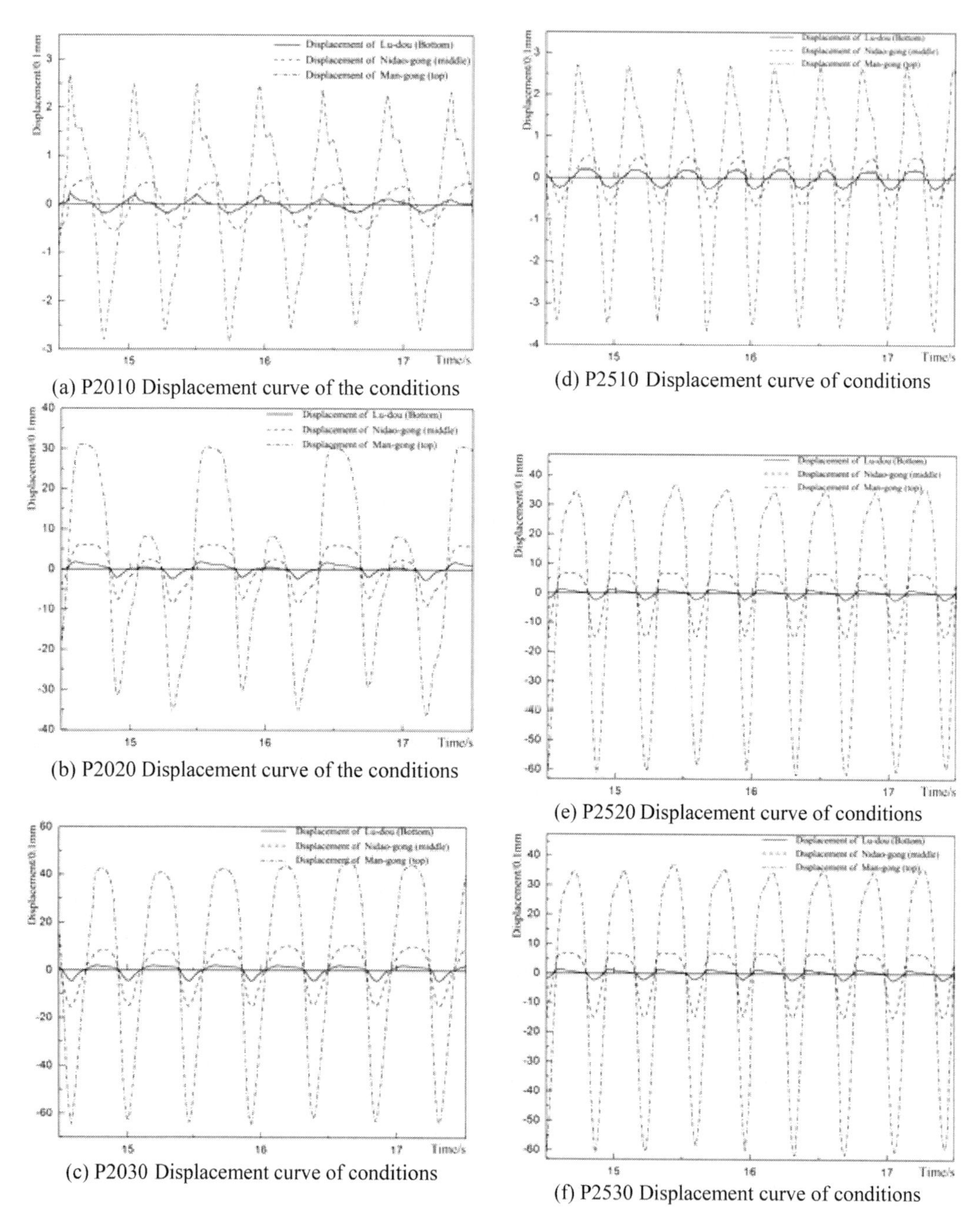

(a) P2010 Displacement curve of the conditions

(b) P2020 Displacement curve of the conditions

(c) P2030 Displacement curve of conditions

(d) P2510 Displacement curve of conditions

(e) P2520 Displacement curve of conditions

(f) P2530 Displacement curve of conditions

* Displacement in the graph all refers to the absolute horizontal displacement.

Figure 4. Changing curve of displacement under 20 Hz and 25 Hz.

the top. It is because the corners will occur in some constructional element in the central part, like flower arm and small arm, and they have distinct superposition to ensemble upward displacement. Besides, owing to the light top load, the center-of-gravity position shifts with the swing of corbel bracket while in vibration. To be exact, it is comparatively slow, and it provides an opposite acting force for horizontal displacement of top.

3. Same amplitudes, the large frequency input, the more obvious sine wave characteristics reflected by horizontal absolute displacement waveform are; horizontal displacement of each layer reach at maximum of P1530, whereas, it will decrease with increase of frequency then. It shows that if vibration frequency increases, force of friction between layers, compressional deformation energy consumption will help shock absorption and restrict the swing of bracket effectively.
4. Due to the same frequency, horizontal absolute displacement of small arm and flower arm will enlarge along with the increase of inputted amplitude. Almost unchanged displacement with same amplitude and different frequency can prove that amplitude is a vital factor to horizontal displacement peak; because cap block is fixed on the table-board, its horizontal displacement is restricted, and as there is a certain distance between the table-board and a horizontal displacement measuring to a point on the cap block, all these gather together, so as to create a displacement that is mostly caused by its corner displacement.
5. The connections of flower arm and Xia-ang are formed with joints by mortise and tenon, not just by an element. This combination does not contribute to any energy dissipation and could increase the risk of its lateral instability. Whereas, so many open joints bring extra burden to the whole bracket. We can see the shape and structure of the bracket in the late Ming Dynasty were transforming from structural element to architectural element, and its energy dissipation gets weakened.

Ultimately, it is here that we need to explain that our no-load vibration test leads to limits of friction and elastoplasticity, as well as a large swing. Therefore, the increasing displacement of the top of the bracket in the figures can't literally stand for the amplification of seismic dynamic in reality.

5 CONCLUSION

1. During vibration, static friction force, slip energy consumption, compressional deformation of occlusions in occlusions of arch and bucket, and mortise and tenon joints between constructional elements contribute to seismic energy absorption, stronger seismic shock, and more energy absorbed.
2. Maximum deformation of each constructional element model is related closely to the corbel bracket. Rotary deformation of cap block and flower arm govern an ensemble deformation.
3. For the horizontal sliding peak, amplitude is the determining factor; while vibrating, the changes of the frequency are important to the changes of the rotary deformation, but related little to the sliding displacement.
4. The component combined with flower arm and Xia-ang works almost as a decoration, both mortise and tenon joints in the front and back of it can't weight too much, which needs attention while its maintenance.
5. The key points of formation characteristics and aseismatic mechanism for the ensemble bracket experiencing geological process may be 1.59 Hz and 2.65 Hz, also for the rotary and sliding deformation of the cap block and flower arm, they are to be verified in subsequent tests and researches.

There being shortages, for example, using single unit bracket will cause structural imbalance easily. Besides, under no-load condition, rotational restraints to brackets are small, which makes seismic energy consumption not that into its full play. So it is accessible to some errors. In subsequent tests, we should face the reality to the best of our abilities, and restrain the rotation of the bracket appropriately; to assay the natural vibrational frequency of the test specimen, and use contrastive analysis to forecast the effect that the resonance may bring.

ACKNOWLEDGMENT

This research was financially supported by the National Undergraduate Training Programs for Innovation and Entrepreneurship Foundation (201410298016Z).

REFERENCES

Chen, C.Z. 1955. The Tianwang palace of Baosheng Temple in Luzhi. *Cultural Relics*, 8: 103–110.

Chen, Z.Y., Zhu, E.C. & Pan, J.L. 2012. Mechanics researching advance of Chinese ancient timber structure buildings. *Advance in Mechanics*, 42(5): 645–653.

Fan, J., Lv, C. & Zhang, H. 2008. *Study on the time—frequency characteristic of ground motions and its effect on structural earthquake response*. Proceedings(III) of the 17th National Conference on Structural Engineering, 014–018.

Fang, D.P., Yu, M.H., Miyamoto, Y., Iwasaki, S. & Hikosaka, H. 2001. Experimental studies on structural characteristics of ancient timber architectures. *Engineering Mechanics*, 18(1): 137–144.

Feng, J.L., Zhang, H.Y., Wang, H. & Zhou, H.D. 2009. Vibration analysis of Dougong layer in ancient timber architectures. *Sichuan Architecture*, 29(4): 132–133.

Gao, D.F., Zhao, H.T., Xue, J.Y. & Zhang, P.C. 2003. Experimental study on structural behavior of Dougong under the vertical action in Chinese ancient timber structure. *World Earthquake Engineering*, 19(3): 56–61.

Kyuke, H., Kusunoki, T., Yamaoto, M. et al. 2008. *Shaking table tests of 'MASUGUMI' used in traditional wooden architectures*. In: 10th World Conference on Timber Engineering, Miyazaki.

Pan, J.Z. 2008. *Study on the anti-seismic mechanism of Chinese ancient Timber Structure Buildings*. Shanghai: Tongji University.

Shao, Y., Qiu, H.X., Yue, Z. & Chun, Q. 2014. Experimental study of low-cycle loading test on Song-style and Qing-style dougong. *Building Structure*, 44(9): 79–82.

Sui, Y., Zhao, H.T. & Xue, J.Y. et al. 2010. Experimental study on lateral stiffness of Dougong layer in Chinese historic buildings. *Engineering Mechanics*, 27(3): 74–78.

Yuan, J.L., Chen, W., Wang, J. & Shi, Y. 2011. Experimental research on bracket set models of Yingxian Timber Pagoda. *Journal of Building Structures*, 23(7): 66–72.

Zhao, J.H., Yu, M.H., Yang, S.Y. et al. 1999. Dynamic experimental study on Dougong of ancient timber structures. *Journal of Experimental Mechanics*, 14(1): 106–112.

Advanced Materials, Mechanical and Structural Engineering – Hong, Seo & Moon (Eds)
 ISBN: 978-1-138-02908-8

Tensioned Fabric Structures with surface in the form of Monkey Saddle surface

H.M. Yee, M.N.A. Hadi & K.A. Ghani
Faculty of Civil Engineering, Universiti Teknologi MARA, Pulau Pinang, Malaysia

N.H.A. Hamid
Faculty of Civil Engineering, Universiti Teknologi MARA, Selangor, Malaysia

ABSTRACT: Form-finding has to be carried out for Tensioned Fabric Structure (TFS) in order to determine the initial equilibrium shape under prescribed support condition and pre-stress pattern. Tensioned fabric structures are normally designed to be in the form of equal tensioned surface. TFS is highly suited to be used for realizing surfaces of complex or new forms. However, research study on a new form as a TFS has not attracted much attention. Another source of inspiration minimal surface which could be adopted as form for TFS is very crucial. The aim of this study is to propose initial equilibrium shape of tensioned fabric structures in the form of Monkey Saddle. Computational form-finding is frequently used to determine the possible form of uniformly stressed surfaces. A TFS must curve equally in opposite directions to give the resulting surface a three dimensional stability. In an anticlastic doubly curved surface, the sum of all positive and all negative curvatures is zero. This study provides an alternative choice for structural designer to consider the Monkey Saddle applied in tensioned fabric structures. Cable pretension for cable reinforced Monkey Saddle TFS model $u = v = 0.6$ and $u = v = 1.2$, the four and two cables has been studied. The results on factors affecting initial equilibrium shape can serve as a reference for proper selection of surface parameter for achieving a structurally viable surface. Such in-sight will lead to improvement of rural basic infrastructure, economic gains, sustainability of built environment and green technology initiative.

1 INTRODUCTION

Tensioned Fabric Structure (TFS) is a suitable structure to be applied in large space area. Actually, TFS has been used over the past 50 years ago. Tensioned fabric structures are structures that are composed of tensioned fabric as structural members. Fabric patterns are joined together at seams and are tensioned through mechanical means or cables to rigid supporting system to typically provide a roofing structure as mentioned by Yee (2011).

The tensioned fabric structures are normally designed to be in the form of equal tensioned surface Yee (2011). Tensioned fabric structures employed membranes that have support geometry resulting in anticlastic membrane curvature, wherein outward loads are resisted by an increase in stress in the hogging warps about one axis of the membrane and inward forces are resisted by an increase in stress in the sagging warps about the other axis. Tensioned fabric structures are considered as form-resistant structures with doubly curved surfaces which must be pretension in such a way in order to resist applied environmental loading such as wind and snow. Form-finding using suitably formulated computational method has to be carried out for TFS in order to determine the initial equilibrium shape under prescribed support condition and pre-stress pattern. The principle of nonlinear analysis method is based on the large displacement finite element formulation used for analysis of structural behaviour under external loads. It can be used with suitable strategy proposed by Yee (2011) for form-finding purpose.

From Meadows (2015), in 1829, Anatomy Joseph Plateau investigated the physical and geometrical properties of soap film surfaces, which are elastic in the sense that they have smallest possible area. In the 1930s, Jesse Douglas and TiborRadó finally showed mathematically that no matter what shape the curve in, there is always a least-area spanning the curve. Further study, including work in the field of Geometric Measure Theory, showed that there is always a least area surface spanning the curve taken from Meadows (2015).

Most tension structures are designed to have a uniform prestress in their fabrics. In this condition, there is no shear stress in the fabric. This is also the condition which minimizes the fabric surface area for a given set of initial conditions. Pauletti & Pimenta (2008) presented an extension of the force density method called the natural force density method for the initial shape finding of cable and membrane structures, which led to the solution of a system of linear equations. With reference to a Helicoid soap film surface, the minimal surface associated with the prescribed boundary is obtained. Bletzinger et al. (2005) presented numerical methods to simulate soap film experiments as well as how they could be integrated among themselves and with structural optimization. Lewis & Gosling (1993) presented a study of minimum energy forms of prestressed cable nets and membranes in numerical form-finding and soap film models. Tensioned fabric structures are normally designed to be in the form of equal tensioned surface. Minimal surface such as classical minimal surfaces, Costa and Möbius strip or their variation have been studied as possible choice of surface form for TFS by Yee (2011), Yee & Samsudin (2014), Yee et al. (2011) and Yee & Choong (2013).

TFS is highly suited to be used for realizing surfaces of complex or new forms. However, none of the new examples mentioned present any results on the Monkey Saddle as load carrying members. Understanding of the possible Monkey Saddle initial equilibrium shapes to be obtained will provide alternative shapes for designers to consider. Form-finding using nonlinear analysis method can be used for both the initial equilibrium problem and load analysis for a new form of Monkey Saddle.

2 GENERATION OF MONKEY SADDLE IN TENSIONED FABRIC STRUCTURE

For this paper, the software ADINA (2003) has been used for the purpose of model generation. Aspect of modelling of surface of Monkey Saddle and form as well pre-stress pattern of the resulting TFS through form-finding using nonlinear analysis method is studied. Monkey Saddle as shown in Figure 1 can be represented parametrically by the following set of equations Weisstein:

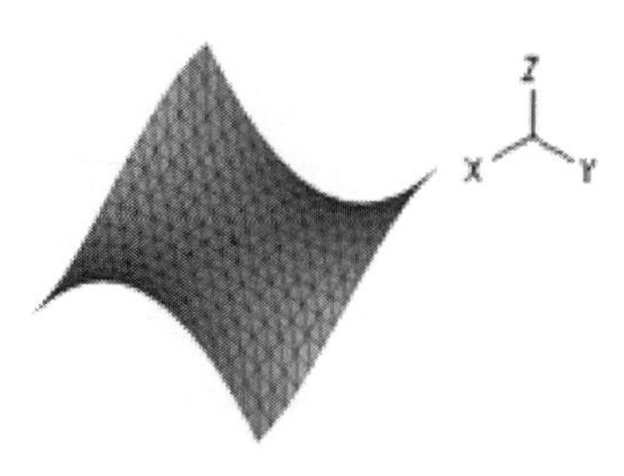

Figure 1. Monkey Saddle model (3-D view).

$$X = u, \quad Y = v, \quad Z = u^3 - 3uv^2 \tag{1}$$

The convergence criteria of form-finding adopted is that Least Square Error (LSE) of total warp and fill stress deviation should be < 0.01 based from Yee (2011). Figure 1 shows the Monkey Saddle model, this model is minimal surface.

3 COMPUTATIONAL ANALYSIS OF MONKEY SADDLE IN TENSIONED FABRIC STRUCTURE

The principle of proposed nonlinear analysis method by Yee (2011) is based on the large displacement finite element formulation used for analysis of structural behaviour under external loads is used in this study. Since the method can be used for both the initial equilibrium problem and load analysis, the approach using nonlinear analysis is quite common. As a first shape for the start of form-finding analysis procedure adopted in this study, initial guess shape is needed. The software ADINA (2003) has been used for the purpose of model generation of Monkey Saddle with variable $u = v = 0.6$ and $u = v = 1.2$. For the generation of such initial guess shape, anticlastic feature is incorporated into the model in order to produce a better initial guess shape. Such anticlastic feature has been incorporated by means of specification of selected sag, Δ, relative to two suitably chosen points on the fixed boundary with span, L. Using this Δ/L sag ratio, geometry for approximates the surface of TFS is then generated. Such geometry is then meshed to produce initial guess shape with anticlastic feature.

The proposed computational strategy involves two stages of analysis in one cycle. The first stage (denoted as SF1) is an analysis which starts with an initial assumed shape in order to obtain an updated shape for initial equilibrium surface. This is then followed by the second stage of analysis (denoted as SS1) aimed at checking the convergence of updated shape obtained at the end of stage SF1. During stage SF1, elastic modulus E with very small values, are used. Warp and fill stresses, σ_W and σ_F are kept constant. In the second stage SS1, the actual tensioned fabric properties values are used. Resulting warp and fill stresses are checked at the end of the analysis against prescribed stresses. Iterative calculation has to be carried out in order to achieve convergence. The resultant shape at the end of iterative step n (SSn) is considered to be in the state of initial equilibrium under the prescribed warp and fill stresses and boundary condition if difference between the obtained and the prescribed tensioned fabric stresses relative to the prescribed stress satisfied the specified criteria. Such checking of difference in the obtained and prescribed stresses has been presented in the form of total stress deviation in warp and fill direction versus stress analysis stage.

Monkey Saddle model with $u = v = 0.6$ and $u = v = 1.2$ have been chosen for discussion. The surface of a Monkey Saddle generated by Equation (1) for Monkey Saddle TFS models. The member pretension in warp and fill direction, denoted as σ_W and σ_F respectively, is 2000 N/m. The shear stress is zero. In this study, form-finding using nonlinear analysis method proposed by Yee (2011) has been carried out.

3.1 *Monkey Saddle model, u = v = 0.6*

Form-finding Monkey Saddle for $u = v = 0.6$ has been carried out. Figure 2 shows initial equilibrium shape of Monkey Saddle model of $u = v = 0.6$.

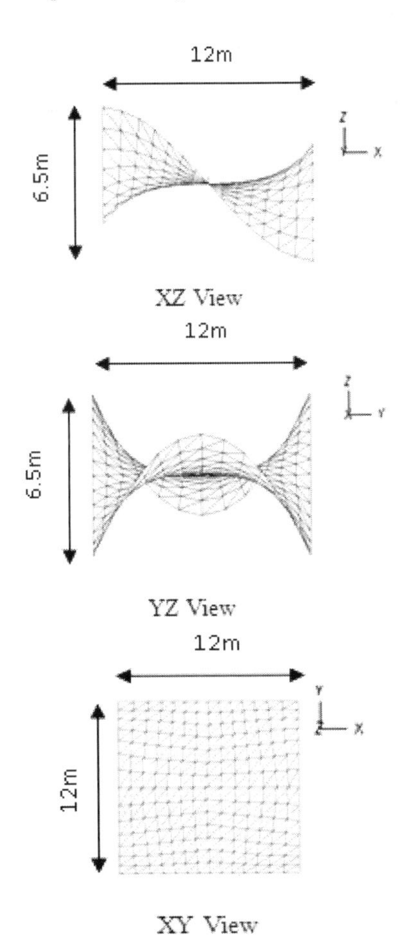

Figure 2. Different views for Monkey Saddle model, $u = v = 0.6$ after form-finding.

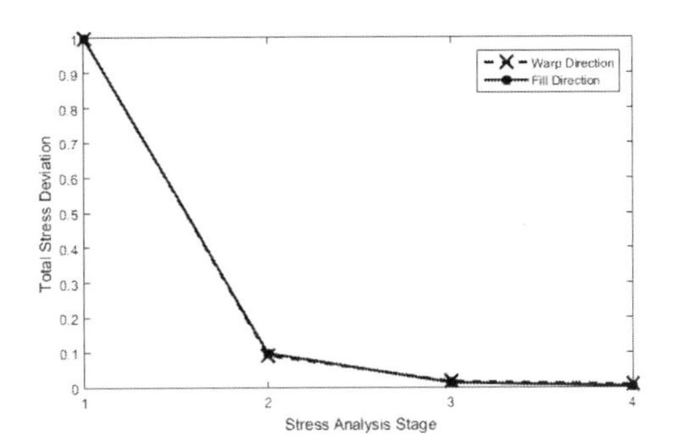

Figure 3. Variation of total stress deviation in warp and fill direction versus stress analysis stage for the Monkey Saddle model with $u = v = 0.6$.

Total number of nodes and elements of Monkey Saddle for $u = v = 0.6$ is 225 and 392, respectively. The warp and fill direction for Monkey Saddle model of $u = v = 0.6$ is 0.006884 and 0.002457, respectively. Figure 3 shows convergent curve of total warp and fill stress deviation for Monkey Saddle of $u = v = 0.6 < 0.01$.

3.2 *Monkey Saddle model, u = v = 1.2*

Form-finding Monkey Saddle model for $u = v = 1.2$ has been carried out. Total number of nodes and elements of Monkey Saddle for $u = v = 1.2$ is 225 and 392, respectively. The warp and fill direction for Monkey Saddle model of $u = v = 1.2$ is 0.009935 and 0.003556, respectively. The total warp and fill stress deviation for Monkey Saddle model with $u = v = 1.2 < 0.01$.

4 CABLE REINFORCED MONKEY SADDLE TFS MODEL

Cable reinforced of Monkey Saddle TFS model has been carried out. Area of cable and pretension of cable used for Monkey Saddle TFS model are 0.005 m^2 and 150000 N/m, warp and fill directions for fabric prestress is 2000 N/m. In this study, two variable has been used in cable reinforced of Monkey Saddle are $u = v = 0.6$ and $u = v = 1.2$. Figure 4 shows the location of reinforcing cables for Monkey Saddle $u = v = 0.6$ and $u = v = 1.2$. Cable reinforced for Monkey Saddle TFS models $u = v = 0.6$ with four cable and Monkey Saddle TFS models $u = v = 0.6$ with two cable.

4.1 *Cable reinforced Monkey Saddle TFS model, u = v = 0.6*

Form-finding cable reinforced of Monkey Saddle for $u = v = 0.6$ has been carried out. In this study,

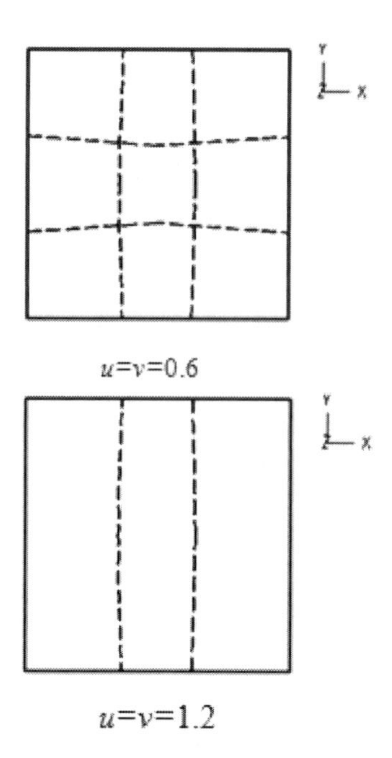

Figure 4. Location of reinforcing cables for Monkey Saddle $u = v = 0.6$ and $u = v = 1.2$.

two cable has been used for cable reinforced of Monkey Saddle for $u = v = 1.2$. The total number of nodes and elements cable reinforced of Monkey Saddle for $u = v = 0.6$ is 225 and 392, respectively. Dimension of cable reinforced for Monkey Saddle TFS models $u = v = 0.6$ is $12 \times 12 \times 6.5$ m. Figure 5 shows cable reinforced Monkey Saddle model and different views for cable reinforced Monkey Saddle TFS model for $u = v = 0.6$ after form-finding. In this result, cable pretension of Monkey Saddle TFS model for $u = v = 0.6$ is 14487.93 N. The warp to fill stress ratio of cable pretension for cable reinforced Monkey Saddle TFS model for $u = v = 0.6$ is 0.03418.

4.2 *Cable reinforced Monkey Saddle TFS model, $u = v = 1.2$*

Cable reinforced of Monkey Saddle for $u = v = 1.2$ has been carried out. In this study, two cable has been used for cable reinforced of Monkey Saddle for $u = v = 1.2$. The total number of nodes and elements cable reinforced of Monkey Saddle for $u = v = 1.2$ is 225 and 392, respectively. Dimension of cable reinforced for Monkey Saddle TFS models $u = v = 1.2$ is $24 \times 24 \times 51.8$ m. In this result, cable pretension for Monkey Saddle TFS model for $u = v = 1.2$ is 11393.32. The warp to fill stress ratio of cable pretension for cable reinforced Monkey Saddle TFS model for $u = v = 1.2$ is 0.24045.

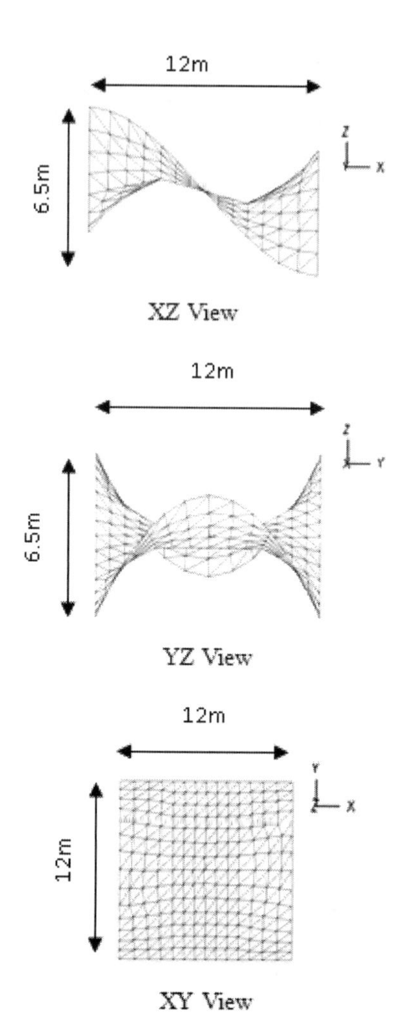

Figure 5. Different views for cable reinforced Monkey Saddle TFS model, $u = v = 0.6$ after form-finding.

5 CONCLUSIONS

Form-finding with surface in the form of Monkey Saddle with variables $u = v = 0.6$ and $u = v = 1.2$ for initial equilibrium shape and cable reinforced Monkey Saddle TFS model have been carried out successfully using the procedure adopted which is based on nonlinear analysis method. The Monkey Saddle to be obtained in this study will provide an alternative shapes for designers to be considered for adoption in tensioned fabric structures.

ACKNOWLEDGEMENT

The researchers wish to thank to Ministry of Education, Malaysia for funding the research project through the fund [Ref. No RAGS/1/2014/ TK02/

UITM//2] and the Research Management Institute (RMI), Universiti Teknologi MARA (UiTM) for the administrative support.

REFERENCES

ADINA. 2003. System 8.1. R&D Inc.

Bletzinger, K.U., Wüchner, R., Daoud, R. & Camprubi, N. 2005. Computational methods for form-finding and optimization of shells and membranes, *Computer Methods in Applied Mechanics and Engineering*, 194(30–33): 3438–3452.

Lewis, W.J. & Gosling, P.D. 1993. Stable minimal surfaces in form-finding of lightweight tension structures, *International Journal of Space Structures*, 8: 149–166.

Meadows, A. 2015. *Soap film geometry-minimal surfaces and isoperimetric problems in Mathematics*. Retrieved March 18, 2015. from http://faculty.smcm.edu/ammeadows/create/.

Pauletti, M.O. & Pimenta, P.M. 2008. *Shape finding of membrane structures by the natural force density method*, Proceedings of the Sixth International Conference on Computation of Shell and Spatial Structures IASS-IACM: 31–34.

Weisstein, EW. "Monkey Saddle." From Math-World—A Wolfram Web Resource. http://mathworld.wolfram.com/MonkeySaddle.html.

Yee, H.M. 2011. *A computational strategy for form-finding of tensioned fabric structure using nonlinear analysis method*, Ph.D. dissertation, School of Civil Engineering, Universiti Sains Malaysia, Pulau Pinang, Malaysia.

Yee, H.M & Choong, K.K. 2013. Form-Finding of Tensioned Fabric Structure in the Shape of Möbius Strip. *Iranica Journal of Energy & Environment. Geo-hazards and Civil Engineering*, 4(3): 247–253.

Yee, H.M., Choong, K.K. & Kim, J.Y. 2011. Form-Finding Analysis of Tensioned Fabric Structures using Nonlinear Analysis Method, *Advanced Materials Research*, 243–249: 1429–1434.

Yee, H.M. & Samsudin, M.N. 2014. Development and Investigation of the Moebius Strip in Tensioned Membrane Structures, *WSEAS Transactions on Environment and Development*, 10: 145–149.

Advanced Materials, Mechanical and Structural Engineering – Hong, Seo & Moon (Eds)
 ISBN: 978-1-138-02908-8

Analytical model for 3D space frame thin-walled structural joint

Y. Mohd Shukri
Faculty of Engineering, Universiti Selangor (UNISEL), Malaysia

M. Shuhaimi & S. Razali
Faculty of Mechanical Engineering, Universiti Teknologi Malaysia (UTM), Malaysia

ABSTRACT: In the automotive industry, thin-walled beam and structure are widely used to build light weight vehicles. The application of the thin-walled structure in the vehicular design manages to produce a high stiffness-to-weight ratio. However, the application of the thin-walled structure will expose the structure to the buckling and joint flexibility effects. Once the structure experiences buckling and flexibility effects, its stiffness is difficult to predict. In this study, a 3D space frame located at the frontal bulkhead will be used as a case study. 3D space frame is widely used by joining thin walled beams to other beams using a welding technique. In previous works, validated finite element model for 3D space frame thin walled structural joint was produced. Meanwhile, the intention of this research work is to develop an analytical model to predict the stiffness of the 3D space frame thin-walled structural joint. For automotive application, the thickness of the thin-walled structure will be in the range between 0.9 mm and 1.6 mm. The analytical model produced will be able to predict the equivalent stiffness of the 3D space thin-walled structural joint for that particular range of thickness. By producing an analytical model for 3D space frame thin-walled structural joint, design engineers can decide the size of all members to be more effective and efficient without depending on the experimental work. This can help the structural engineering designers to accelerate the process of designing vehicular structure.

1 INTRODUCTION

According to Megson (1974), there are no clear defined lines between sections which may be regarded as thin and those which may be considered as thick. However, it is suggested in the thin-walled theory, it may be applied with a reasonable accuracy if the sections' thickness is at least ten times thinner than the cross-section dimensions. Meanwhile, according to Kuang (1984), when the cross-section-to-thickness ratio exceeds 50, this beam is categorized as a thin-walled beam or structure. Another definition of the thin walled beam or structure is based on slenderness ratio (American Iron and Steel Institute, 2002).

In an automotive design, the weight of a vehicle will play an important role to improve the overall performance of a vehicle. Application of the thin-walled beam in vehicular technology offers a high strength-to-weight ratio (Domagoj et al. 2015, Claudio & Marco 2015).

However, the application of the thin-walled structure will cause a few drawbacks in designing the vehicles' structure. The main problem caused by the thin-walled structure is the buckling effect. Buckling is also known as a structural instability. Once this happens, the structures will experience a large deformation and it may lose its ability to carry the load (Wang et al. 2005). If this axial stiffness is estimated without considering the influence of the local buckling, the weaker structure is designed (Yasuaki et al. 2004). Another problem is the joint flexibility effect. Most body connections of necessity are not rigid, and their flexible characteristics contribute to one of the key factors dominating the response of the body structures (David 1974). Global torsion stiffness of a body-in-white greatly depends on joint flexibilities (Yasuaki et al. 2004).

In predicting the behavior of the thin-walled structure accurately, a finite element model which takes into consideration the buckling and joint flexibility effect must be developed. Modeling techniques take the additional flexibility into account using artificial stiffness at the joint area whereas the adjacent structures are modeled by beam elements (Jin et al. 2002). In the previous works, a validated finite element model to predict the equivalent stiffness of 3D space frame thin-walled structural joint has managed to produce satisfactory results (Shukri et al. 2013a, Shukri et al. 2013b, Shukri et al. 2014a, Shukri et al. 2014b).

However, the complexity of the finite element model will restrict the structural engineer designers to absorb this method in helping their work. As structural engineer designers would prefer to use engineering tools that are simple and

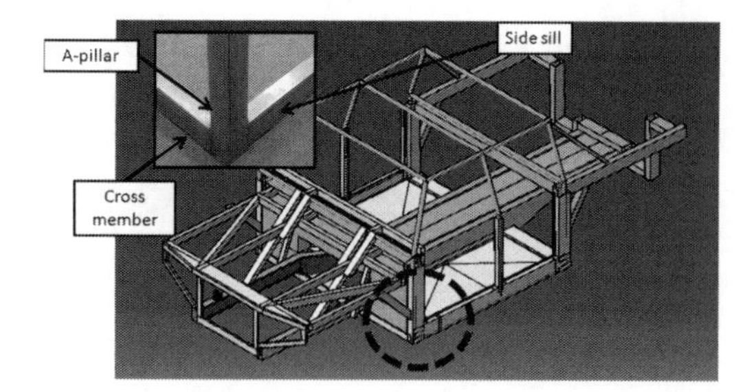

Figure 1. Application of 3D space frame structure joint in vehicle structure.

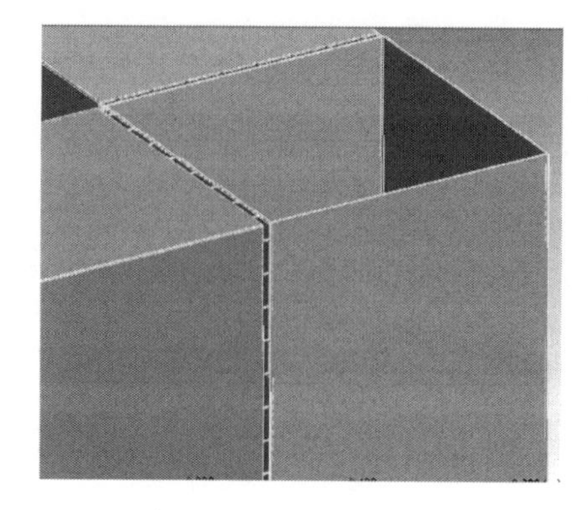

Figure 2. Application of circular beam element at joint area for 3D space frame thin walled structural joint.

reliable and can assist them to make important decisions at the onset of the design stage. Applications of this tool can help to produce a better design while saving time in the designing process (Hong et al. 2002).

In this research work, an analytical model that can predict the equivalent stiffness of 3D space frame will be developed. This will help the structural engineer designers to accelerate the designing process without carrying out any experimental testing or finite element model which is costly, time consuming, and requires special skills to be conducted. The application of 3D space frame thin-walled structural joint in the vehicular structure is shown in Figure 1.

2 METHODOLOGY

In previous works, a validated finite element model to predict equivalent stiffness of 3D space frame thin-walled structural joint was developed. Using this model, circular beam as shown in Figure 2 is introduced at the joint area to take care of buckling and joint flexibility effects. Using this model, accuracy of finite element model produces 97% accuracy (Shukri et al. 2014b).

In developing an analytical model for 3D space frame thin-walled structural joint, reduction member method is applied to determine the relation between all members of the 3D space frame thin-walled structural joint. Figure 3 shows the schematic diagram of this method.

When the joint effect is taken into consideration, the deformation of this structure is given by Equation 1.

$$\theta_{3D} = \theta_{cm} + \theta_{ap\&ss} + \theta_{joint} \tag{1}$$

From this relationship, the stiffness of this structure is given in Equation 2.

$$\frac{1}{K_{3d}} = \frac{1}{K_{cm}} + \frac{1}{K_{ap} + K_{ss}} + \frac{1}{K_{joint}} \tag{2}$$

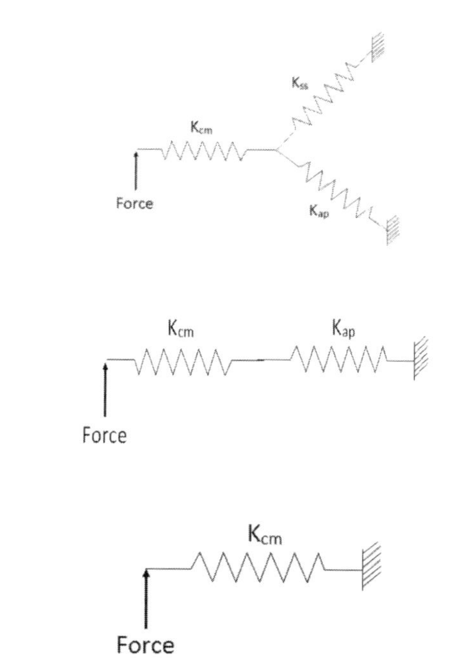

Figure 3. Schematic diagram of the structure.

K_{3D} = Stiffness of 3D space frame thin-walled structural joint (Nm/rad)
K_{cm} = Stiffness of cross member (Nm/rad)
K_{ap} = Stiffness of A pillar (Nm/rad)
K_{ss} = Stiffness of sidesill (Nm/rad)

While developing an analytical model, the joint stiffness of the 3D space frame thin-walled structural joint needs to be extracted from this model. By doing this, the relationship between joint stiffness and size of the thin-walled beam can be obtained. Using finite element model, the joint area is replaced with a solid and rigid element as shown in Figure 4.

By doing this, the stiffness of the joint will approach infinity.

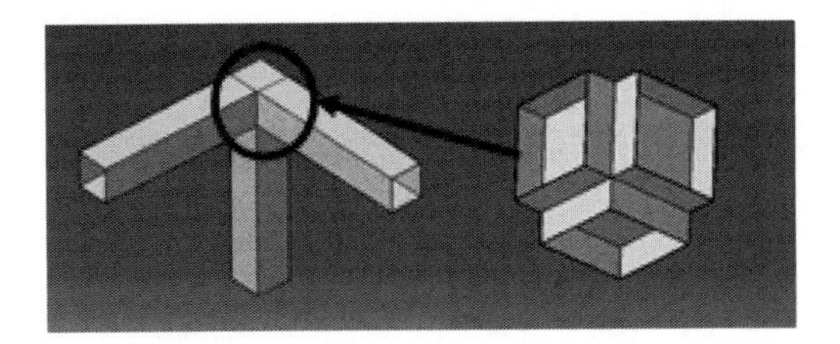

Figure 4. Solid and rigid elements at joint area.

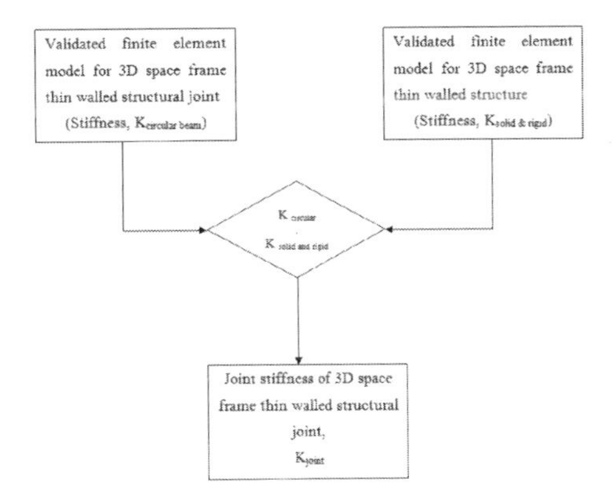

Figure 5. Flow chart to find joint stiffness of 3D space frame thinwalled structural joint.

$$K_{joint} = \infty$$
$$\frac{1}{K_{joint}} >>> 0$$

So, the analytical model for 3D the space frame thin-walled structure (without joint effect) is given in Equation 3.

$$\frac{1}{K_{3d}} = \frac{1}{K_{cm}} + \frac{1}{K_{ap} + K_{ss}} \tag{3}$$

By comparing the results using a circular beam element and a solid and rigid element, joint stiffness of 3D space frame thin-walled structural joints can be obtained as shown in Figure 5.

3 RESULTS AND DISCUSSIONS

By using finite element model with solid and rigid element at the joint area, the accuracy of the analytical model is tested. The accuracy of the analytical model for the size of 50 mm, 100 mm, 150 mm, and 200 mm is given in Figure 6.

From this result, the accuracy of analytical model will produce 99% accuracy if the length-to-base of the thin-walled beam is around 5 to 6 as shown as in Equation 4.

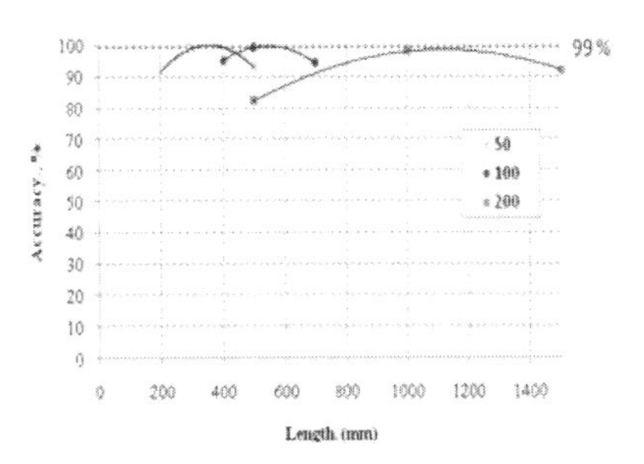

Figure 6. Accuracy of the analytical model for 3D space frame thinwalled structure.

Table 1. Ratio of physical properties for thin-walled beam.

Thickness, t (mm)	Base, B (mm)	Second moment of area, I (m^4)	Torsional constant, J (m^4)	$I_{t=x}/I_{t=0.9}$	$J_{t=x}/J_{t=0.9}$	$t_x/t_{0.9}$
1.2	50	9.3E-08	1.5E-07	1.31	1.33	1.33
	100	7.72E-07	1.2E-06	1.32	1.33	1.33
	150	2.64E-06	4.05E-06	1.33	1.33	1.33
	200	6.29E-06	9.6E-06	1.33	1.33	1.33
1.6	50	1.21E-07	2E-07	1.70	1.78	1.78
	100	1.02E-06	1.6E-06	1.74	1.78	1.78
	150	3.49E-06	5.4E-06	1.75	1.78	1.78
	200	8.33E-06	1.28E-05	1.76	1.78	1.78

$$L/B = 5 \text{ to } 6 \tag{4}$$

L = Length (m)
B = Base (m)

In the previous work, validated finite element model for 3D space frame thin-walled structural joint is only developed for the thickness of 1.2 mm only. However, in automotive industry, the thickness of thin walled beam that will be used is in the range of 0.9 mm to 1.6 mm.

In order to determine the factor that will contribute to the stiffness of 3D space frame thin-walled structure, the physical properties of the thin-walled beams are studied and compared. From this analysis, it is found that the ratio between the second moment of the area, torsional constant, and thickness are the same. These ratios are given in Table 1.

By using the relation between stiffness and second moment of area/torsional constant/thickness of thin-walled beam, the stiffness of 3D space frame thin-walled structural joint for the thickness of 0.9 mm and 1.6 mm can be generated using Equation 5.

$$\frac{K_{t=1.2}}{t} = \frac{K_{t=x}}{t_x} \tag{5}$$

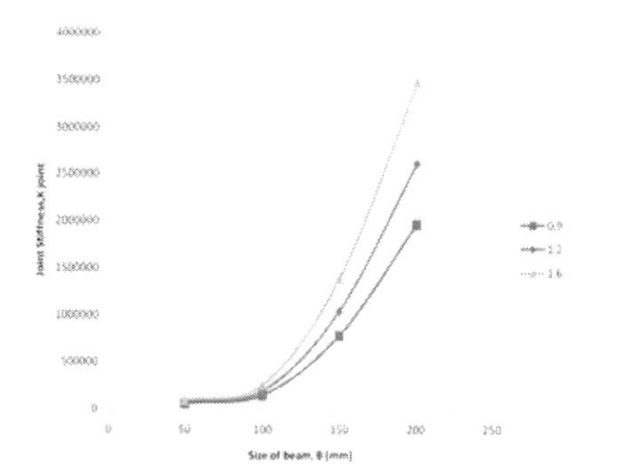

Figure 7. Joint stiffness of 3D space frame thin-walled structural joint for the size of 50 to 200 mm and thickness of 0.9 mm, 1.2 mm, and 1.6 mm.

K_t = joint stiffness (Nm/rad)
t = thickness (mm)

From this equation, the stiffness of 3D space frame thin-walled structural joint for the size of thin-walled beam from 50 to 200 mm and thickness from 0.9 to 1.6 mm manages to get generated as shown in Figure 7.

Finally, by combining the tables in Figure 7 and Equation 6, joint stiffness of the 3D space frame thin-walled structural joint can be predicted analytically.

$$\frac{1}{K_{3d}} = \frac{1}{K_{cm}} + \frac{1}{K_{ap} + K_{ss}} + \frac{1}{K_{joint}} \tag{6}$$

$$K_{cm} = \frac{3EI}{L}$$

$$K_{ap} = \frac{EI}{L}$$

$$K_{ss} = \frac{GJ}{L}$$

E = Young's Modulus (Pa)
I = Second moment of area (m^4)
L = Length of members (m)
J = Torsional constant (m^4)
G = Shear modulus of elasticity (Pa)

4 CONCLUSION

At the early of stage design, design engineers need to decide the preliminary sizes of all members without conducting an experimental work or running a finite element. Most of the time, design engineers will use their experience and also trial and error approach to decide the size of all members. By developing this model, it will be used as a guideline for design engineers to decide the size of all members in designing a 3D space frame thin-walled structural joint members for vehicular structural application.

ACKNOWLEDGMENT

I would like to thank Universiti Selangor (UNISEL) and Universiti Teknologi Malaysia (UTM) for their encouragement and support in this research work.

REFERENCES

American Iron and Steel Institute. 2002. *Automotive Steel De-sign Manual.* (6th ed.) (2002). Michigan: American Iron and Steel Institute.

Claudio, B. & Marco, S. 2015. European De-sign Approaches For Isolated Cold-Formed Thin-Walled Beam–Columns With Mono Symmetric Cross-Section. *Engineering Structures*, 86: 225–241.

David, C.C. 1974. Effect of Flexibility Connections on Body Structural Reponse. *SAE Technical Paper Series*, 740041.

Domagoj, L., Thuc, P.V., Goran, T. & Lee, J.H. 2015. Buckling Analysis of Thin-Walled Functionally Graded Sandwichbox Beams. *Elsevier*, 86: 148–156.

Kim, J.H., Kim, H.S., Kim, D.W. & Kim, Y.Y. 2002. New Accurate Efficient Modeling Techniques for the Vibration Analysis of T-Joint Thin-Walled Box Structures. *International Journal of Solids and Structures*, 39: 2893–2909.

Lin, K.H. 1984. Stiffness and Strength of Square Thin Walled Beam. *SAE Technical Paper Series,* 840734.

Megson, T.H.G. 1974. *Linear Analysis of Thin Walled,* Elastic Structures, Surrey University.

Mohd Shukri, Y., Shuhaimi, M. & Razali, S. 2013a. *Modelling of Equivalent Stiffness of Thin walled structural Joint.* International Conference on Mechanical, Automotive and Aerospace Engineering (ICMAAE' 13).

Mohd Shukri, Y., Shuhaimi, M. & Razali, S. 2013b. Finite Element Modelling to Predict Equivalent Stiffness of 3D Space Frame Structural Joint Using Circular Beam Element. *Applied Mechanics and Materials*, 431: 104–109.

Mohd Shukri, Y., Shuhaimi, M. & Razali, S. 2014a. Individual Stiffness of 3D Space Frame Thin Walled Structural Joint Considering Local Buckling Effect. *Applied Mechanics and Materials*, 554: 411–415.

Mohd Shukri, Y., Shuhaimi, M. & Razali, S. 2014b. Joint Stiffness of 3D Space Frame Thin Walled Structural Joint Considering Local Buckling Effect. *Applied Mechanics and Materials,* 660: 773–777.

Wang, C.M., Wang, C.Y. & Reddy, J.N. 2005. *Exact Solutions for Buckling of Structural Members CRC Press,* Boca Raton, Florida.

Yasuaki, T., Hidekazu, N., Toshiaki, N., Tatsuyuki, A. & Katsuya, F. 2004. First Order Analysis for Automotive Body Structure Design, Part 2: Joint Analysis Considering Nonlinear Behavior. *SAE Technical Paper Series*, 2004-01-1659.

Yim, H.J., Lee, S.B. & Pyun, S.D. 2002. A Study on Optimum Design for Thin Walled Beam Structures of Vehicles. *SAE Technical Paper Series,* 2002-01-1987.

Advanced Materials, Mechanical and Structural Engineering – Hong, Seo & Moon (Eds)
© 2016 Taylor & Francis Group, London, ISBN: 978-1-138-02908-8

Mathematical modelling of glass-metal composite cylindrical shell forming

A.A. Bocharova & A.A. Ratnikov
Far-Eastern Federal University, Vladivostok, Russia

ABSTRACT: The process of glass-metal composite manufacturing, providing the durable connection of glass and metal layers without surface micro cracks, is the cooling down of glass melt filling up the space between metal shells. Mathematical model of glass-metal composite shell deformation has been specified in the present article and a new approach to numerical solution of this problem has been proposed. The analytical method of residual strains determination on joints of layers of a glass-metal cylindrical shell arising in the course of its cooling description is provided in article. It was determined that maximal shear and detach-able strains on the layers conjunctions are dramatically lower than a breaking one.

1 INTRODUCTION

Glass-metal composite is a composition material consisting of a glass layer compressed between met-al shells (Pikul 2008). In the process of its manufacturing a durable connection of glass layer and shells is provided and necessary conditions for forming glass layer without surface micro cracks are created. As a result, glass layer and glass-metal composite obtain in general an exceptionally high shock resistance and durability having a relatively low weight. What is very important there are no real obstacles for manufacturing large ME equipment from glass-metal composite. Studies of glass-metal composite effectiveness in deepwater equipment have shown that durable hulls can be made of it and they are able to provide a necessary floatability of the deepwater equipment even at the final depth of the World ocean (Pikul & Goncharuk 2009).

There are five Russian Federation patents for glass-metal composite products manufacturing methods (Pikul 2010a,b, 2012a,b,c). The main idea of the manufacturing method is to fill the space between the metal shells with glass melts and to create the conditions for a durable connection of the forming glass layer with shells and to exclude the formation of surface microcracks on glass.

2 THE CYLINDRICAL SHELL FORMING

In the process of forming a cylindrical shell from glass-metal composite the glass melts having the temperature of more than 1000°C cools down to the normal temperature of 20°C. During the temperature decreasing the glass melt transfers from flowable state into viscous plastic, then into viscous elastic and finally into firm state. Forming of cylindrical composite shell happens under the influence of axially symmetric field of temperature change in time. As a consequence, the cylindrical shell will undergo axially symmetric deformation. The process of the shell formation goes on so slowly that inertia strength can be ignored. The temperature changes caused by glass-metal composite deformation in the process of cooling are small to negligible in comparison to the temperature of glass melt. In the consequence of it, the general task of the cylindrical shell formation falls into two parts: a temperature one and a deformation one. The temperature part of the task is solved independently from the deformation one.

For the numerical implementation of the proposed model modern numerical and digital-analytical methods were used. In connection with the exponential change of the required functions close to butt ends of the shell numerical methods lead to instability of calculation. In order to overcome this obstacle, Godunov's method of discrete orthogonalization was applied. It has significantly leveled disturbance peaks of strains on the surfaces conjunction of the composite shell layers.

Mathematical model of glass-metal composite shell deformation has been specified in the present article and a new approach to numerical solution of the glass-metal composite shell deformation problem has been proposed, in the basis of which an analytical solution has been taken and two deformation problems solutions have been present.

The deformation part of the problem of the glass-metal composite cylindrical shell formation is solved beginning from the temperature of glass transition to the temperature of complete cooling (20°C). In this case the results of the temperature part of the problem are used. The processing method of glass-metal composite formation includes the time of full relaxation of disturbances during the temperature of glass transition. As a consequence, the initial conditions of the deformation part of the problem solution are the conditions of the total absence of disturbances in the layers of glass-metal composite shell. Counting of the displacements and deformations also starts from the time of the absolute relaxation of the disturbances. That is why the initial conditions of the deformation part of the shell formation include the conditions of the total absence of displacements, deformations and the speeds of their change.

3 MATHEMATICAL MODEL

Mathematical model of glass-metal composite cylindrical shell formation is set up with the help of the deformed solid body three-dimensional equation physical lumping method (Pikul 2005). The essence of this method is that every layer is considered as an independent shell of constant depth, which is influenced by adjoining layers or environment. The obtained system of ordinary differential equations is in full accordance with continuum mechanics fundamental equations, which clearly reflects physical patterns of the shell material deformation. At the same time the accuracy of assignment of boundary conditions is increasing. In the first deformation problem all the materials parameters are taken as integrally-averaged throughout the whole temperature scope, in the second problem the temperature interval is subdivided into 10 equal segments. For this purpose the materials parameters are taken as integrally-averaged for a definite temperature segment. The final solution is found by summing up all the solutions found in each interval.

The deformation of each shell layer is considered in the local coordinate system under the alignment of the basic coordinate surfaces with the middle surfaces of the layers. The basic coordinate surfaces are circular cylindrical. Distance s along generatix and angle φ—in circumferential direction are taken as their coordinates. The third coordinates z are measured on normal to basic coordinate surfaces in the direction of outer normal. Such coordinates are accepted to use in the theory of shells. They enable to trace boundary value problem of shell mechanics to the consideration of the middle surface deformation and, in doing so, to decrease the complexity of solution no less than three orders less.

Figure 1 shows in its local coordinates systems a tentative scheme of layers force interaction of a three-layer glass-metal composite shell between each other and the environment.

Mathematical model of the cylindrical shell deformation includes fundamental equations suitable for every shell layer, equations of condition characterizing individual mechanical and thermophysical properties of layers materials, conditions of layers conjunction and boundary conditions:
– of balance

$$\frac{dN_{11}}{ds} = -\left(\sigma_{31}^{+} - \sigma_{31}^{-}\right);$$

$$\frac{d^2M_{11}}{ds^2} - \frac{N_{22}}{R} = -\frac{h}{2}\frac{d}{ds}\left(\sigma_{31}^{+} + \sigma_{31}^{-}\right) - \left(\sigma_{33}^{+} - \sigma_{33}^{-}\right);$$

$$Q_1 = \frac{dM_{11}}{ds} + \frac{h}{2}\left(\sigma_{31}^{+} + \sigma_{31}^{-}\right), \tag{1}$$

– of geometry

$$\varepsilon_{11} = \frac{du}{ds} + z\frac{d\psi_1}{ds}; \quad \varepsilon_{22} = \frac{w}{R}, \tag{2}$$

– of layers conjunction conditions

$$u^{(1)} + \frac{h_1}{2}\psi_1^{(1)} = u^{(2)} - \frac{h_2}{2}\psi_1^{(2)};$$

$$u^{(2)} + \frac{h_2}{2}\psi_1^{(2)} = u^{(3)} - \frac{h_3}{2}\psi_1^{(3)}, \tag{3}$$

$$w^{(1)} + \Delta w_1^{+} = w^{(2)} + \Delta w_2^{-};$$

$$w^{(2)} + \Delta w_2^{+} = w^{(3)} + \Delta w_3^{-}, \tag{4}$$

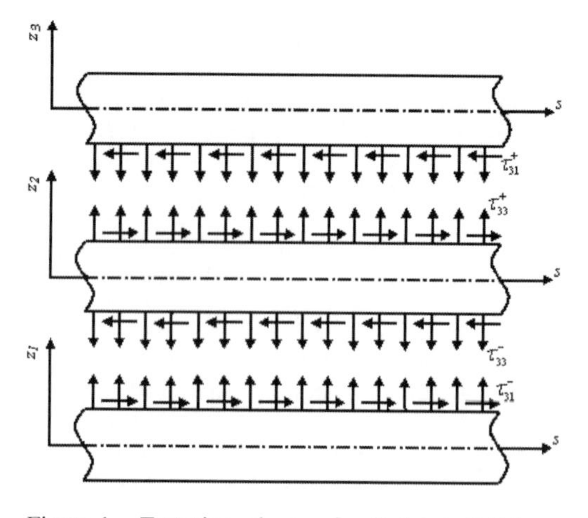

Figure 1. Tentative scheme of cylindrical shell layers force interaction between each other and environment.

– of boundary conditions

$$s=0, L: Q_1^{(i)}=0; \quad N_{11}^{(i)}=0; \quad M_{11}^{(i)}=0, \qquad (5)$$

where, N_{11} and N_{22} – specific tangential forces relegated to the unit of distance; M_{11} – specific moment of flection relegated to the unit of distance; σ_{33}^- and σ_{33}^- – tangential and normal strains respectively acting on the lower surface of the calculated layer; σ_{31}^+ and σ_{33}^+ – tangential and normal strains respectively acting on the upper surface of the calculated layer; Q_1 – specific lateral force relegated to the unit of distance; u and w – displacements of middle surface along the axes s and z correspondingly, depending only on s; ψ_1 – turning angle of normal to middle surface; ε_{11} – normal deformations in the direction of the cylindrical layer generatix, ε_{22} – normal deformations in the circumferential direction, R – radius of middle surface, h – layer depth, L – length of half-shell, Δw – additional deflections of the layers conjunction surfaces caused by a cramped part of lateral deformations:

$$\Delta w_i^{\pm} = \frac{1}{E_i} \int_0^{\pm h_i/2} \sigma_{33}^{(i)} dz.$$

Fundamental equations are supplied with the material condition equations:

– for external shells

$$\sigma_{11} = \frac{E}{1-\nu^2}\left[(\varepsilon_{11}+\nu\varepsilon_{22})-(1+\nu)\alpha^+\Delta T\right];$$

$$\sigma_{22} = \frac{E}{1-\nu^2}\left[(\varepsilon_{22}+\nu\varepsilon_{11})-(1+\nu)\alpha^+\Delta T\right],$$

– for glass layer

$$\sigma_{11} = \frac{E}{1-\nu^2}e^{-\mu t}$$

$$\times\int_{t_0}^{t}\left[(\dot{\varepsilon}_{11}+\nu\dot{\varepsilon}_{22})+(1+\nu)\dot{\alpha}^+\Delta T\right]e^{\mu t}dt;$$

$$\sigma_{22} = \frac{E}{1-\nu^2}e^{-\mu t}$$

$$\times\int_{t_0}^{t}\left[(\dot{\varepsilon}_{22}+\nu\dot{\varepsilon}_{11})+(1+\nu)\dot{\alpha}^+\Delta T\right]e^{\mu t}dt,$$

where, σ_{11} – normal strains in the direction of the cylindrical layer generatrix, σ_{22} – normal strains in the circumferential direction, E – temperature-averaged elasticity module, ν – the Poisson ratio, α^+ – temperature-averaged thermal-expansion coefficient, ΔT – temperature difference; T_0, T – initial and final temperatures of the layers materials; a dot above signifies time derivatives $t \in [t_0, t]$; μ – coefficient taking into account visco-elastic properties of the glass layer,

$$\mu = \frac{E}{2(1+\nu)\eta},$$

where, η – coefficient of glass dynamic viscosity.

The deformation process is symmetric in relation to the middle lateral section of the shell and axisymmetric in relation to symmetry axis. Due to this fact only half of the shell length is considered. While doing it, displacements, deformations and strains have proven to be independent from coordinate φ.

Specific forces and deflection moments can be defined through displacements:

$$Q_1 = \int_h \left(1+\frac{z}{R}\right)\sigma_{13}dz = \frac{Eh}{2(1+\nu)}\gamma_1;$$

$$N_{11} = \int_h \left(1+\frac{z}{R}\right)\sigma_{11}dz$$

$$= \frac{Eh}{1-\nu^2}\left[\frac{du}{ds}+\nu\frac{w}{R}+\frac{h^2}{12R}\frac{d\psi_1}{ds}\right]-\sigma_t h;$$

$$N_{22} = \int_h \sigma_{22}dz = \frac{Eh}{1-\nu^2}\left[\frac{w}{R}+\nu\frac{du}{ds}\right]-\sigma_t h;$$

$$M_{11} = \int_h \left(1+\frac{z}{R}\right)z\,\sigma_{11}dz$$

$$= \frac{Eh^3}{12(1-\nu^2)}\left[\frac{d\psi_1}{ds}+\frac{1}{R}\frac{du}{ds}\right]-\sigma_t\frac{h^3}{12R};$$

$$M_{22} = \int_h z\,\sigma_{22}dz = \frac{E\nu h^3}{12(1-\nu^2)}\frac{d\psi_1}{ds},$$

where, σ_t – strains temperature constituent.

$$\sigma_t = \frac{E}{1-\nu}\alpha^+\Delta T.$$

In order to get an analytical solution of the problem under consideration it is necessary to exclude specific forces N_{22} from the equilibrium Equations (1). In this regard let us refer to the conditions of internal forces self-balancing in the shell lateral sections:

$$\sum_{i=1}^{3} R_i N_{11}^{(i)} = 0; \quad \sum_{i=1}^{3} R_i Q_1^{(i)} = 0; \quad \sum_{i=1}^{3} N_{22}^{(i)} = 0 \tag{6}$$

From where it can be stated:

$$\tau_{31}^{-} = R_\tau \tau_{31}^{+}; \quad \tau_{33}^{-} = R_\tau \tau_{33}^{+}, \tag{7}$$

where, $R_\tau = \frac{R_3 - R_2}{R_1 - R_2}$.

Specific force N_{11} and moment M_{11} for every layer are stated from integrated equilibrium equations through layers junctions strains. In order to state specific force N_{22} through layers junctions strains and exclude it from consideration it is necessary to solve a system of equations algebraically. This system includes differentiated conditions of layers conjunction (3), written in deformations ε_{11}, and the last cylindrical shell self-balancing Equation (6). In general terms the solution is set down the following way:

$$N_{22}^{(i)} = a_i + b_i \int \tau_{31}^{+} ds \tag{8}$$

where, a_i И b_i – constants stated through the shell measures, materials parameters and constants of equilibrium equations integration $c_{(k)}^{(i)}$:

$$N_{11}^{(i)} = \int \left(\sigma_{31}^{-(i)} - \sigma_{31}^{+(i)} \right) ds + c_1^{(i)};$$

$$Q_1^{(i)} = \int \left(\frac{N_{22}^{(i)}}{R_i} + \sigma_{33}^{-(i)} - \sigma_{33}^{+(i)} \right) ds + c_2^{(i)};$$

$$M_{11}^{(i)} = \int \left(\int \left(\frac{N_{22}^{(i)}}{R_i} + \sigma_{33}^{-(i)} - \sigma_{33}^{+(i)} \right) ds \right) ds$$

$$- \frac{h_i}{2} \int \left(\sigma_{13}^{-(i)} + \sigma_{13}^{+(i)} \right) ds + c_2^{(i)} s + c_3^{(i)}, \tag{9}$$

where, $N_{22}^{(i)}$ is defined by the Formula (8).

4 THE EXAMPLE OF CALCULATIONS

The geometrical measures of the example shell are: glass layer middle surface radius $R_2 = 1$ m, its depth $h_2 = 0.1$ m; internal metal layer middle surface radius $R_1 = R_2 - (h_1 + h_2) / 2 = 0.9495$ m, its depth $h_1 = 0.001$ m; external metal layer middle surface radius $R_3 = R_2 + (h_3 + h_2) / 2 = 1.055$ m, its depth $h_3 = 0.01$ m; half-shell length $L = 4$ m.

We can manage to trace mathematical model of cylindrical shell deformation to the solution of one sextic differential equation with constant coefficients. The coefficients of this equation are determined on the basis of mechanical, thermophysical and geometrical parameters, which enables to find its solution. Boundary conditions are used to discover integration constants (5).

Graphs of tangential and detachable strains distribution acting on the border of metal layer and glass conjunction are present in Figure 2 and 3. The strains acting on the border of internal metal layer and glass conjunction are determined with the help of the Formulas (6).

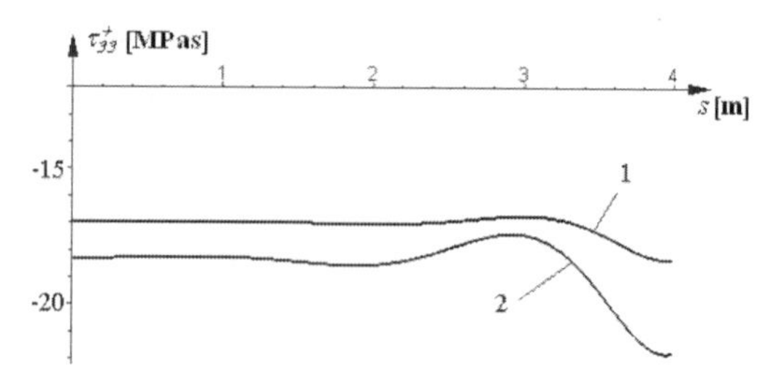

Figure 2. Distribution of normal strains along the length of the shell: 1 – solution under integral averaging of the whole temperature range; 2 – solution under recursive integral averaging.

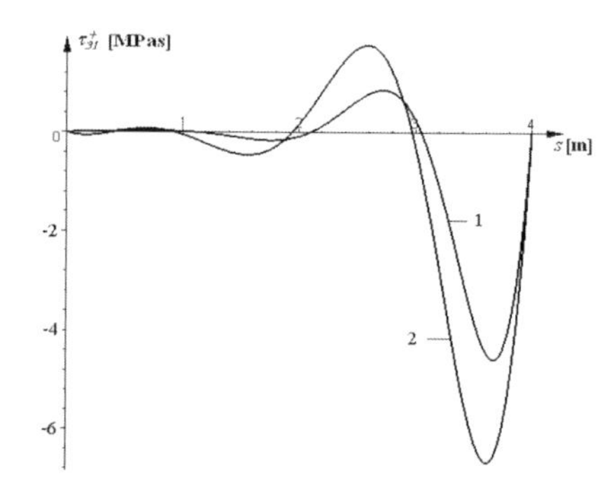

Figure 3. Distribution of shearing strains through the shell length: 1 – solution under integral averaging of the whole temperature range; 2 – solution under recursive integral averaging.

5 THE MAIN RESULTS AND CONCLUSIONS

1. On the basis of physically relevant theory of shells a mathematical model of glass-metal composite cylindrical shell deformation in the process of its manufacturing was developed.
2. The test has shown that the solution found fully satisfies all the fundamental equations of the mathematical model.
3. It was determined that maximal shear and detachable strains on the layers conjunctions are dramatically lower than a breaking one. This fact justifies the possibility of creating this kind of constructions.

4. While having a stepwise refinement of the solution there was marked a growth of shearing strain peak values by 46% and detachable strain by 18% in absolute terms.

ACKNOWLEDGEMENT

This work was supported by the Ministry of Education and Sciences of the Russian Federation. Applied scientific researches (project) unique identifier RFMEFI57814X0024.

REFERENCES

Pikul, V.V. & Goncharuk, V.K. 2009. The Composite Nano-material on the basis of glass—glass-metal composite, All materials. *Encyclopedic Reference Book*, 2009: 5–9.

Pikul, V.V. 2005. *Modern scientific problems in the field of applied mechanics: textbook in two parts*, Part 2, Mechanics of shells, Vladivostok, 524.

Pikul, V.V. 2008. About The Creation of Composite Nano-material on the basis of glass, *Non-Conventional Materials*, 2008: 78–83.

Pikul, V.V. 2010(a). *Waterproof durable hull of underwater device made of glass-metal composite*, R.F. Patent 2, 425, 776.

Pikul, V.V. 2010(b). *Method of manufacturing of glass-metal composite tube,* R.F. Patent 2, 433, 969.

Pikul, V.V. 2012(a). *Method of manufacturing of glass-metal composite cylindrical shell for a durable hull of underwater device,* R.F. Patent 2, 491, 202.

Pikul, V.V. 2012(b). *Method of making cylindrical shell of submarine strong hull of glass-metal composite,* R.F. Patent 2, 497, 709.

Pikul, V.V. 2012(c). *Method of manufacturing of sheet glass-metal composite,* R.F. Patent 2, 505, 495.

Advanced Materials, Mechanical and Structural Engineering – Hong, Seo & Moon (Eds)
© 2016 Taylor & Francis Group, London, ISBN: 978-1-138-02908-8

Nidltrusion as the method for composite reinforcement production. Part I: Mathematical model of heat transfer and cure

S.N. Grigoriev
Moscow State Technological University "STANKIN", Moscow, Russia

I.A. Kazakov & A.N. Krasnovskii
Department of Composite Materials, Moscow State Technological University "STANKIN", Moscow, Russia

ABSTRACT: Nidltrusion is a continuous manufacturing process used to produce high strength composite reinforcement rebars for construction industry. During this process, the solidification of composite material occurs in the heating chamber without touching its walls. The present work is focused on the determination of the heat transfer and curing of 8-mm diameter reinforcement using a two-dimensional implicit finite-difference analysis model. The infrared radiant heating theory is used for the analysis and the heat transfer between the composite rod and the surrounding air is taken into account. Final results allow concluding that the numerical results are stable and model can be further applied in other kinds of composite materials.

1 INTRODUCTION

The Fiber Reinforced composite rods (FRP rods), presenting a high strength to weight ratio, resistant to repeated loading and not subjected to corrosion. Thanks to these features they are widely used in the construction industry as reinforcement for concrete structures. The composite reinforcement is usually made of glass or basalt fibers and has special outer relief providing a good adhesion to concrete. The carbon fiber has higher characteristics but rarely used because of its high cost.

Among the several available techniques to produce composite reinforcement, nidltrusion process is becoming the most widespread and cost-effective continuous processing technique in Russia. Nidltrusion means molding on the needle and, unlike the braiding technology differing from, the fillers in composite rebar remain their unidirectional orientation (Figure 1).

A schematic overview of nidltrusion process and the equipment employed is shown in Figure 2.

In this process, dry reinforcement fibers in the form of continuous strands (roving) are pulled through a resin bath for impregnation. After the wetting process, the reinforcement pack is collimated in the forming taper into a performed shape before entering the heating chamber where it cured. The forming taper is typically characterized by a conical convergent shape, in order to promote compaction of the material, the removal of the air and excess resin. The circle shape of the rod cross section is fixed by spiral protrusions formed by a helical winding device. The forming taper and helical winding devices are practically always combined in one.

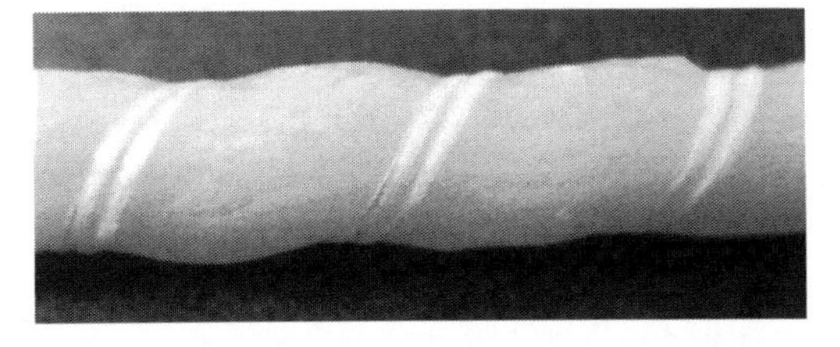

Figure 1. The part of 8-mm fiberglass reinforcement consisting of unidirectional and spiral protrusions made by nidltrusion.

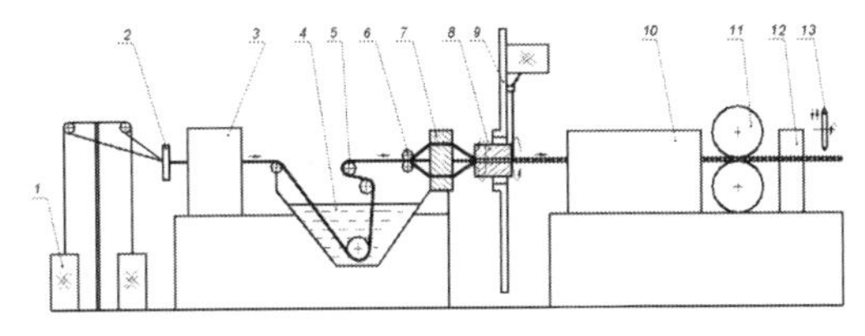

Figure 2. Schematic diagram of nidltrusion process: 1–roving, 2,7–fiber guiders, 3–drying chamber, 4–impregnation bath, 5–tension rollers, 6–resin streak device, 8–forming taper, 9–helical winding device, 10–heating chamber, 11–pulling mechanism, 12–footage counter, 13–moving cut-off saw.

After the winding the material pass through the heating chamber, where the material changes its status from liquid to solid. The heating chamber may be equipped with electrical tubular or Infrared (IR) heaters, but electrical tubular heaters are the most common ones. The composite reinforcement is cured in the heating chamber without touching its walls in contrast to conventional pultrusion process. Finally, outside the chamber, the composite reinforcement already polymerized and is pulled by a pulling sys-tem. Then a cut-off saw cuts the part into a desired length.

The heat provided by means of electrical tubular or IR heaters activates the exothermic cure reaction of the thermoset resin. In the first case the air as intervening medium is warmed in the oven which serves to heat the product through convection and radiation. In the second case the composite takes heat from the IR heaters directly and loses part of heat through convection and radiation to the surrounding medium. The composite material is cured in the heating oven without touching its walls in contrast to traditional Pultrusion. As the fiber-resin system enters and proceeds along the length of the heated chamber, the exothermic reaction of polymerization starts, thereby releasing heat and raising the temperature of the composite. Therefore, the temperature and degree of cure profiles in the composite material inside the heating chamber is a balance of heat transfer, generated by heating system and exothermic reaction of resin.

There have been several models (Joshi 2001, Krasnovskii & Kazakov 2012–2014, Carlone 2013, etc.) presented in the literature concerning the temperature and degree of cure relationships of pultrusion process. However, these models are based on the assumption that the outer surface of the composite rod is in direct contact with the walls of a heated die. Therefore, in relation to the nidltrusion process such models are not appropriate.

In the present work, the special mathematical model based on the mutual interactions between heat transfer, cure reaction and variation in the material properties was developed in order to predict the temperature and degree of cure of composite reinforcement for the nidltrusion process.

2 STATEMENT OF THE PROBLEM

Figure 3 shows a longitudinal section of the heating chamber with a composite reinforcement located inside. The cured part of the rod marked by gray color. The heating chamber is divided into several sections. Each section is set to strictly defined power output and left in this state for the duration of the process. The temperature for each section should be chosen so as to ensure the quality products production.

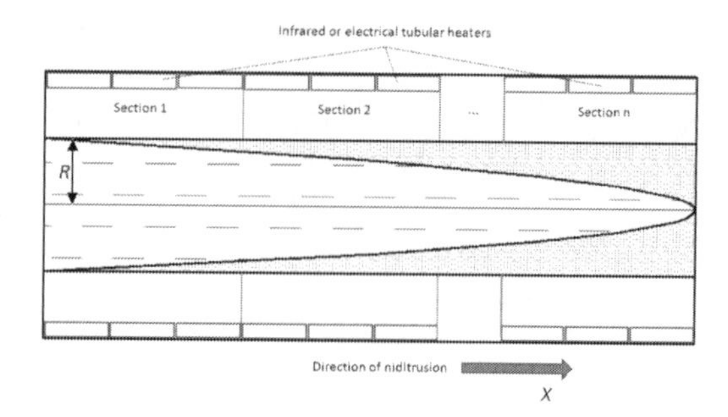

Figure 3. Longitudinal section of the heating chamber with a composite rod inside.

The problem is axisymmetric because the composite rod has a cylindrical shape and temperature of the heating chamber has an axisymmetric distribution.

In general, heat transfer equation for the cylindrical coordinate system is given by

$$\rho c_v \frac{dT}{dt} = \rho_r H \frac{d\alpha}{dt} + \frac{1}{r}\frac{\partial}{\partial r}\left(r\lambda_1 \frac{\partial T}{\partial r}\right) + \frac{\partial}{\partial x}\left(\lambda_2 \frac{\partial T}{\partial x}\right) + 2\mu\left(\left(\frac{\partial v_r}{\partial r}\right)^2 + \left(\frac{\partial v_x}{\partial x}\right)^2 + \frac{1}{2}\left(\frac{\partial v_r}{\partial x} + \frac{\partial v_x}{\partial r}\right)^2 + \left(\frac{v_r}{r}\right)^2\right), \tag{1}$$

where, ρ–density of composite, ρ_r–density of resin, μ–viscosity, α–degree of cure, H–the total heat generated by the reaction per unit mass of resin, λ_1, λ_2–heat-transfer coefficients, c_v–specific heat, x–axial direction, r–radial direction, v_r–radial component of the velocity, v_x–axial component of the velocity.

Here and below, nidltrusion process is considered as steady-state. In this case

$$\frac{dT}{dt} = \frac{\partial x}{\partial t}\frac{\partial T}{\partial x} + \frac{\partial r}{\partial t}\frac{\partial T}{\partial r} = U\frac{\partial T}{\partial x}, \tag{2}$$

where, U is the pull speed.

Assuming the thermal conductivity along the radius is a constant, and is absent along the die due to the dimensions of composite rod in longitudinal direction is much greater than the dimensions in the transverse direction. Then, Equation (1) can be written as

$$\rho c_v U \frac{\partial T}{\partial x} = \rho_r H \frac{d\alpha}{dt} + \frac{\lambda_1}{r}\frac{\partial}{\partial r}\left(r\frac{\partial T}{\partial r}\right). \tag{3}$$

Since the composite material is considered as a fiber/resin system, the total heat generated during the exothermic reaction per unit weight can be calculated as

$$H=\left(1-v_f\right)\cdot H_{tot}, \tag{4}$$

where, H_{tot}–the total heat generated by the exothermic reaction of resin, v_f–fiber volume fraction.

The density of composite, ρ and resin density, ρr can be expressed as follows:

$$\rho=\left(1-v_f\right)\cdot\rho_r+v_f\cdot\rho_f, \tag{5}$$

$$\rho_r=\alpha\cdot\rho_r^c+\left(1-\alpha\right)\cdot\rho_r^u, \tag{6}$$

where, ρf is the fiber density, ρ_r^u, ρ_r^C are uncured and cured resin densities correspondingly.

The specific heat capacities of composite material and resin are calculated in a similar manner:

$$c=\left(1-v_f\right)\cdot c_r+v_f\cdot c_f, \tag{7}$$

$$c_r=\alpha\cdot c_r^c+\left(1-\alpha\right)\cdot c_r^u, \tag{8}$$

where, cf is the specific heat of the fiber, C_r^u, C_r^C are uncured and cured resin specific heat correspondingly.

In the present investigation the well-established Kamal model has been adopted to describe the evolution of the cure reaction (Kamal 1974):

$$\frac{d\alpha}{dt}=\left(k_1+k_2\cdot\alpha^m\right)\cdot\left(1-\alpha\right)^n, \tag{9}$$

$$k_i=A_i\cdot\exp\left(-\frac{E_{ai}}{R\cdot T}\right),\quad (i=1,2), \tag{10}$$

where, k_i is the reaction rate, A_i is the pre-exponential factor, E_{ai} is the activation energy, R is the gas constant, n, m are the equation superscripts.

If the heating chamber is equipped with electrical tubular heaters the boundary conditions for the governing equation are:

$$T\,|_{x=0}=T_0,\ -\lambda_1\frac{\partial T}{\partial r}\,|_{r=R}=\gamma,\ \frac{\partial T}{\partial r}\,|_{r=0}=0, \tag{11}$$

where, T_0 is initial composite temperature at the computational domain inlet.

The net heat transfer through radiation is calculated using the Stefan-Boltzmann law (Bai, Vallee & Keller 2007):

$$\gamma=\varepsilon_r\cdot C_0\cdot\left(T^4\,|_{r=R}-T_{out}^4\right), \tag{12}$$

where, εr is the emissivity of the surface, C_0 is the Stefan-Boltzmann constant (5.67×10^{-8} W/(m$^2\cdot$K^4)), T_{out}–the steady temperature of the air in heating chamber sections.

In the case of heat transferred through radiation from IR heaters to outer surface of the composite rod, the boundary conditions for the governing equation are given by:

$$T\,|_{x=0}=T_0,\ -\lambda_1\frac{\partial T}{\partial r}\,|_{r=R}=\gamma_1+\gamma_2,\ \frac{\partial T}{\partial r}\,|_{r=0}=0, \tag{13}$$

where, the net heat transfer through radiation is calculated as

$$\gamma=\varepsilon_r\cdot C_0\cdot\left(T^4\,|_{r=R}-T_H^4\right), \tag{14}$$

where, T_H–the IR heater temperature.

The implicit finite difference method was used to solve the system of governing Equations (9–10). This method is based on discretizing the space and time domain through transformation into a finite difference form and solving the subsequent system of algebraic equations for the temperature and cure fields. The algorithm for solving the problem is as follows. For each time step, starting from the second, the unknown temperature is determined by using Equation (9) taking into account the boundary conditions, temperature and the degree of cure at the previous time step. Then, the degree of cure is determined from the Equation (10) for the current time step. A detailed description of the method has been shown in earlier studies (Grigoriev, Krasnovskii & Kazakov, 2012–2014) and will not be shown here.

A computer program is written using VBA to conduct the temperature and cure calculation.

3 APPLICATION AND DISCUSSION

The curing of fiberglass/epoxy reinforcement rod during nidltrusion is simulated to compare the experimental results which will be presented in the next part of this study. The numerical example for the processing rod with radius of 8 mm is given below. The heating chamber has 6 m length and consists of 8 sections. The temperature setting (by sections) is: [310; 310; 190; 130; 110; 120; 150; 150]°C. The inlet temperature is assumed to be equal to the resin bath temperature (24°C), while the epoxy matrix material is assumed to be totally uncured at the same cross section. The speed of nidltrusion has

Table 1. Values of physical quantities.

Property	Value
Resin Type: CYD-128/Vestamin ID	(388/100)
Heat of reaction, H_{tot}, J/g	456.44
Activation energy, E_{a1}, J/mol	23820
Pre-exponential factor, A_1, sec^{-1}	18
Activation energy, E_{a2}, J/mol	38400
Pre-exponential factor, A_2, sec^{-1}	2850
Equation superscript, n	1.9
Equation superscript, m	0.85
Fiber density, g/cm^3	2.56
Uncured resin density, g/cm^3	1.06
Completely cured resin density, g/cm^3	1.27
Heat-transfer coefficient, λl, W/(m · K)	0.25
Specific heat capacity of the fibers, J/(g · K)	0.84
Specific heat capacity of uncured resin, J/(g · K)	1.69
Specific heat capacity of cured resin, J/(g · K)	1.36
The emissivity of the surface, ε_r	0.7
The IR heater temperature, TH, °C	300

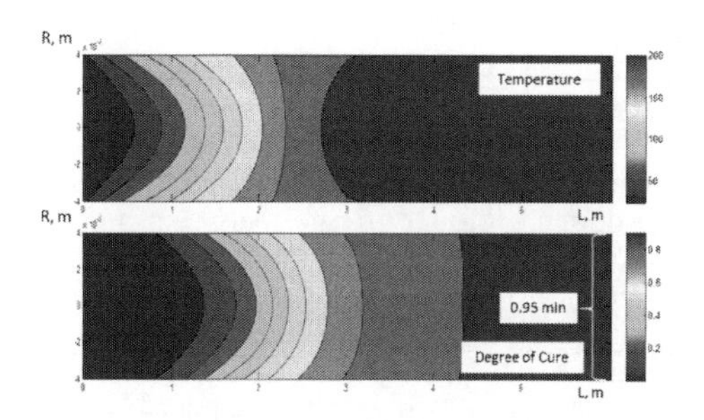

Figure 4. Temperature and degree of cure distributions within the rod (nidltrusion speed is 2.5 m/min).

been defined as 2.5 m/min. The fiber volume fraction is $v_f = 0.6$. The properties of components and the resin kinetic parameters are listed in Table 1.

The computed temperature field and the degree of cure distributions for the longitudinal section of the reinforcement are shown on Figure 4. It can be seen that the maximum temperature of the rod is approximately 200°C due to the high speed of the nidltrusion process although the first two sections of the heating chamber have temperatures exceeding of 300°C. The composite material near the centerline reaches the higher temperature faster than the outer surface due to the internal heat generation of the epoxy resin. The product is completely solidified at a distance of 6 m from the heating chamber inlet. It is assumed for the finished product $\alpha \geq 0.95$.

4 CONCLUSIONS

The theoretical work was presented regarding the mathematical modeling based on the mutual interactions between heat transfer, cure reaction and variation in the material properties for the nidltrusion process. The following conclusions were drawn:

1. Two-dimensional heat transfer and curing model is used to predict the temperature and degree of cure of composite reinforcement.
2. Complex boundary conditions are considered, including prescribed temperature and heat radiation. The model uses the IR radiant heating theory and takes into account the heat transfer between the composite rod and the surrounding air.
3. The numerical results are stable due to an implicit finite difference method was used.
4. The model can be further applied in other kinds of composite materials, if the necessary material parameters are determined.

ACKNOWLEDGEMENT

This work is a part of the State project in the field of scientific activity which has been granted by the Education and Science Ministry of the Russian Federation.

REFERENCES

Bai, Y. et al. 2007. Modeling of thermal responses for FRP composites under elevated and high temperatures. 2008. *Composites Science and Technology*, 68: 47–56.

Carlone, P. et al. 2013. Computational Approaches for Modeling the Multiphysics in Pultrusion Process. 2013. *Advances in Mechanical Engineering*, 2013: 1–14.

Grigoriev, S.N. et al. 2013. An analytic definition of the border polymerization line for axisymmetric composite rods. 2013. *Applied Composite Materials*, 20(6): 1055–1064.

Grigoriev, S.N. et al. 2014. The friction force determination of large-sized composite rods in pultrusion. 2014. *Applied Composite Materials*, 21(4): 651–659.

Grigoriev, S.N. et al. 2015. The Impact of Pre-heating on Pressure Behavior in Tapered Cylindrical Die in Pultrusion of Large-sized Composite Rods. 2015. *Advanced Materials Research*, 1064: 120–127.

Joshi, S.C. & Lam, Y.C. 2001. Three-dimensional finite-element/nodal-control-volume simulation of the pultrusion process with temperature-dependent material properties including resin shrinkage. 2001. *Composites Science and Technology*, 61(11): 1539–1547.

Kamal, M.R. 1974. Thermoset Characterization for Moldability Analysis. 1974. *Polymer Engineering and Science, March*, 14: 231–239.

Krasnovskii, A.N. & Kazakov, I.A. 2012. Determination of the optimal speed of pultrusion for large-sized composite rods. 2012. *Journal of Encapsulation and Adsorption Sciences*, 2(3): 21–26.

Kutin, A.A. et al. 2014. The Fiber Orientation Angle Determination for a Composite Anisotropic Solid Rod in Pultrusion. *Advanced Materials Research*, 941–944: 2273–2278.

Advanced Materials, Mechanical and Structural Engineering – Hong, Seo & Moon (Eds)
 ISBN: 978-1-138-02908-8

Experimental study on differential pressure control for a central heating system with a variable speed pump

S.W. Song
Korea Institute of Civil Engineering and Building Technology, Gyeonggi-do, South Korea

ABSTRACT: Variable Frequency Drives (VFDs) for pumps have been used to save pumping energy by modulating water flow based on a differential Pressure (dP) setpoint for a secondary water circulation loop system. However, the dP setpoint is normally set high and continues to run at the design conditions. This study analyzes the operational characteristics of a central heating system with a variable speed pump by changing the differential pressure setpoint from the design value (i.e., 0.3 bar) to the minimum value (i.e., 0.2 bar) under partial load conditions at the end of secondary water circulation loop of a multi-family residential building in Korea. As a result, pumping energy was estimated to be significantly reduced by 62.86% as the invert frequency was reduced by 23.40%, but the HW supply and return temperature difference increased by 3.29°C from 15.24°C to 18.53°C as the Hot Water (HW) flow rate decreased by 23.2% during selected experimental periods. Consequently, differential pressure setpoint could be reset lower than design value to minimize pumping energy for central heating (or cooling) systems within an acceptable range of water flow rate.

1 INTRODUCTION

Nearly zero or net Zero Energy Building (ZEB) is becoming the most important issue with respect to the global climate change. Several pilot projects of the ZEB have been implemented with various Energy Conservation Measures (ECMs) and renewable energy systems for residential and commercial buildings in Korea. However, it has often been limited to deliver the ZEB as expected without commissioning. Ongoing (or continuous) commissioning is one of the most promising technologies to achieve such ZEB by optimizing building and system performance with a minimum operational cost.

High-rise multi-family residential buildings with the radiant floor heating system served by district heating or central heating plant have widely been constructed and also designed for passive or zero energy buildings in Korea. Such high-rise multi-family residential buildings are often partially unoccupied during day time. However, most of the hot water circulation pumps run at a constant speed along with the by-pass control valve even during off-peak time. Variable Frequency Drive (VFD) for such central water circulation pumps can be used to save pumping used by modulating water flow rate based on a differential pressure setpoint.

When VFDs are applied for water circulation pumps (Tillack & Rishel 1998), a differential pressure sensor is generally installed at the end of the water circulation loop system but sometimes placed at the wrong spot near the pumps even though it could not ensure potential energy savings. In addition, differential pressure setpoint is normally set high and continues to run at design conditions even if it could be reset under partial load conditions to minimize the pumping energy (Lu & Cai 2002).

This study analyzes the operational characteristics of a central heating system with a variable speed pump, in terms of invert frequency, water flow rate, and supply and return water temperature difference, and then estimates energy savings based on adjusted baseline (EVO 2012); when the differential pressure at the end of the secondary water circulation loop is set to the minimum value (i.e., 0.2 bar) from the design value (i.e., 0.3 bar) for a case-study building in Korea.

2 CASE STUDY BUILDING

2.1 *Building description*

The case-study building is an eight-story (1,806 m^2), multi-family residential building including 15 residential units served by a central heating system. The building was constructed in December, 2012 as a pilot project to be a nearly zero energy building with several Energy Conservation Measures (ECMs), including: exterior super insulation, high-performance windows, movable exterior shadings

Table 1. Description of the case-study building.

Items	Description
Building Location	South Korea
Building Type	Multi-family Building
Construction Year	2012
Construction Type	Reinforced Concrete
Number of Floors	8 Floors
Total Floor Area	1,806 m^2
Roof U-value	0.11 W/m^2 K
Exterior Wall U-value	0.15 W/m^2 K
Floor U-value	0.10 W/m^2 K
Window U-value	1.0 m^2 K (SHGC 0.37)
System & Plant for Heating	Radiant Floor Heating Pellet Boiler (50 kW)
Others	Photovoltaic (PV) Panels

Figure 1. South façade of the case-study building.

(blinds), two pellet boilers for heating and Domestic Hot Water (DHW), and Photovoltaic (PV) panels, as shown in Table 1. Figure 1 shows the south façade of the case-study building.

2.2 *Central heating system and differential pressure control*

The case-study building includes 15 residential units served by a central heating plant (i.e., pellet boiler) providing hot water with a radiant floor heating system in each residential unit, which was controlled by the individual room thermostat to maintain 20°C of the heating temperature setpoint during selected experimental periods in 2014. A whole building energy monitoring and control system for the case-study building was installed to implement the on-going commissioning after construction. Table 2 shows the design conditions of the central heating system, and Figure 2 shows a schematic diagram of the energy monitoring and control system with measurement points for the central heating system of the case-study building.

For this study, a differential Pressure (dP) sensor was installed at the end of the Hot Water (HW) circulation loop system in order to control the circulation pump (P1) speed by a Variable Frequency Drive (VFD) based on a differential pressure setpoint. Two Resistance Temperature Detectors (RTDs) were installed to measure the HW supply (Ts) and return (Tr) temperature for the secondary HW circulation loop of the case-study building. An RTD sensor was also installed to detect the buffer tank's water temperature for the primary pumps (i.e., P2 and P3) and the pellet boiler operation (e.g., on/off) using the energy monitoring and control system installed for this study. In addition, a differential

Table 2. Design conditions of the central heating system.

Items		Design conditions
Heating Plant	Type Efficiency	Pellet Boiler (50 kW) 87%
Heat Exchanger	Type	Plat
HW Buffer Tank	Capacity	1 Ton
HW Circulation loop	Control	Differential Pressure (dP) Control
HW Circulation Pump (P1)	Type Control Capacity Head Flow Efficiency	Variable Speed Invert (VFD) 0.75 kW (1750 RPM) 12 m 71.6 l pm (4.3 CMH) 41.6%
Circulation Pump (P2)	Type Capacity	Single Speed (On/off) 0.55 kW
Boiler Pump (P3)	Type Capacity	Single Speed (On/off) 0.35 kW

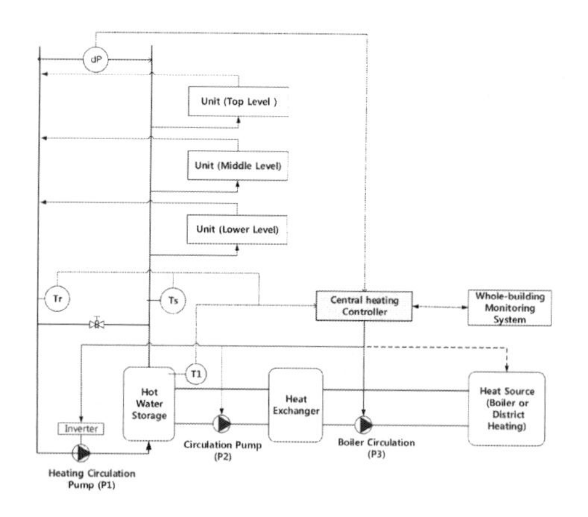

Figure 2. Schematic diagram of the energy monitoring and control system with measurement points.

pressure control valve was installed at the bottom of the secondary water circulation loop just in case of applying for constant speed pump control, which was initially setup at the design stage but replaced with variable speed pump control for this study.

3 RESULTS AND DISCUSSION

3.1 *Operational characteristic*

This study analyzes the operational characteristics of the secondary water circulation loop system with a variable speed pump when differential pressure setpoint was set to the minimum value (i.e., 0.2 bar) from the design value (i.e., 0.3 bar), in terms of differential pressure, invert frequency, hot water flow rate, and supply and return water temperature difference. All the data used in this study were measured every 15 minutes.

Figure 3 represents measured invert frequency of HW circulation pump against measured differential pressure. The invert frequency of HW pump was within the range of 25 Hz and 30 Hz when the dP was set to 0.3 bar, while it ranged from 20 Hz to 25 Hz when the dP setpoint was set to 0.2 bar, which was determined for the pump safety. As a result, it is identified that dP setpoint has a significant impact on the minimum operational range of variable frequency drive, which modulates the pump speed that is theoretically proportional to the water flow rate referred to as affinity laws of pump.

Figure 4 shows the relationship of measured invert frequency against measured HW flow rate. The invert frequency with 0.2 bar of dP setpoint began with 20 Hz and was almost constant until a certain range of HW flow rate (i.e., 0.2 m^3/15 min) and then increased in proportionality to the water flow rate, while the invert frequency with 0.3 bar of dP setpoint began with 25 Hz and then increased almost in proportionality to HW flow rate.

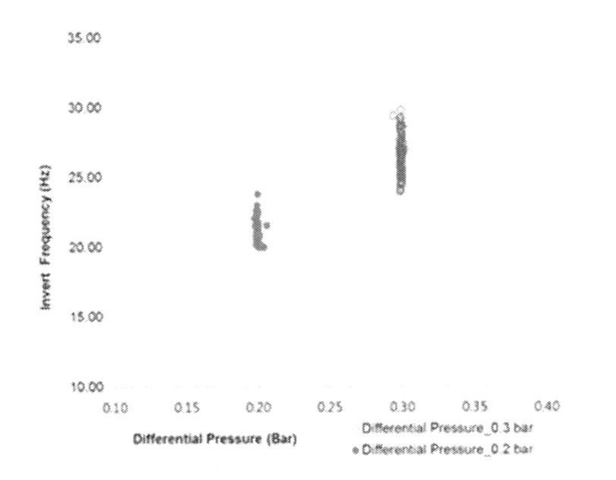

Figure 3. Invert frequency against HW temperature difference (supply—return).

Figure 5 shows the relationship of measured supply and return water temperature difference against the measured HW flow rate. The supply and return water temperature difference had a tendency to decrease within a certain range of temperature difference of about 15°C when HW flow rate increased. Overall, the supply and return water temperature difference with 0.2 bar of dP setpoint was relatively higher than that of 0.3 bar of dP setpoint as expected.

3.2 *Energy savings*

This study estimated the average pumping energy and savings based on the adjusted energy baselines (EVO 2012) when the differential Pressure (dP) setpoint after change) was changed from 0.3 bar to 0.2 bar, during the selected experimental periods. The adjusted energy baseline models are developed based on the linear regression models using 0.3 bar of dP setpoint (i.e., before change) but have the same outdoor temperature as that of the reporting period (i.e., after change). Table 3 shows daily average pumping energy and savings (%) based on the adjusted energy baselines during the experimental periods, including: HW flow rate, invert frequency, and pump power.

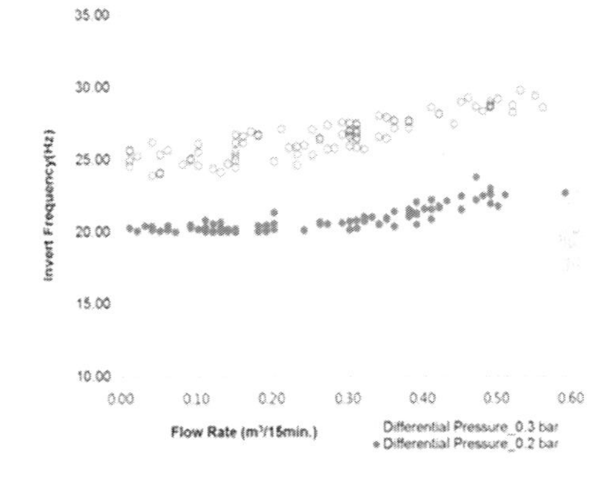

Figure 4. Invert frequency against HW flow rate.

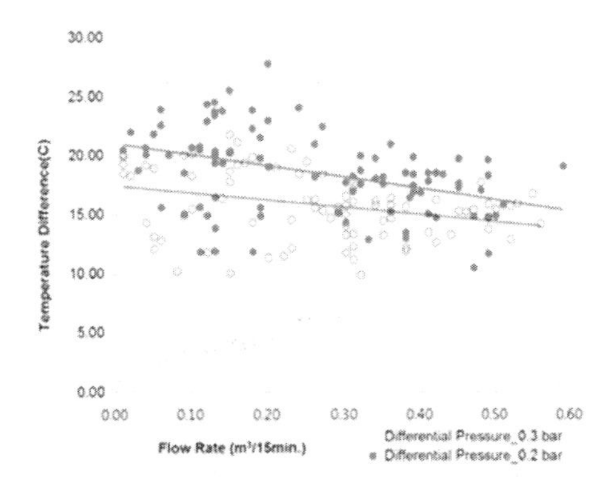

Figure 5. Temperature difference (supply—return) against HW flow rate.

Table 3. Daily average energy savings between 0.3 bar and 0.2 bar of dP setpoint.

	Before change	Adjusted baseline	After change	Energy savings	
Items	(dP 0.3 Bar)	(dP 0.3 Bar)	(dP 0.2 Bar)	Difference	Savings (%)
Measurement Period (Date)	12/12	–	12/22	–	–
Outdoor Temp. (°C)	−3.66	−4.95	−4.95	–	–
HW Flow Rate (m^3/15 min)	0.27	0.34	0.26	0.08	23.2
Invert Frequency (Hz)	26.60	27.18	20.83	6.35	23.4
Temp. Difference (°C)	15.81	15.24	18.53	3.29	
Pump Power (kW/15 min.)	0.26	0.35	0.13	0.22	62.86

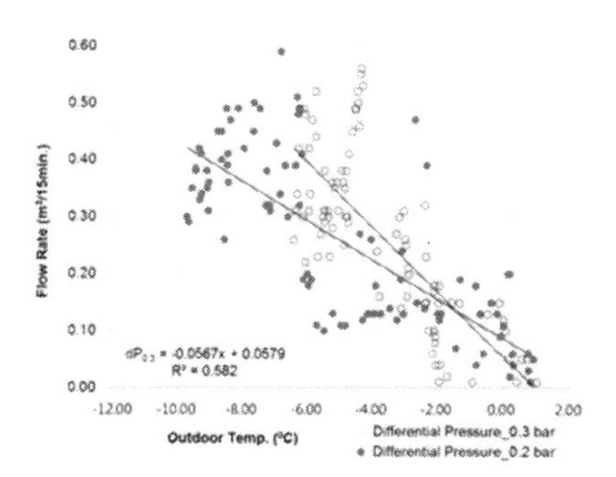

Figure 6. HW flow rate against outdoor temperature.

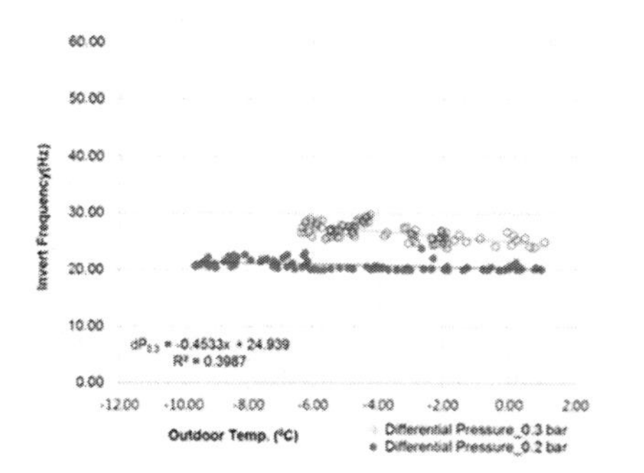

Figure 7. Invert frequency against outdoor temperature.

Figure 6 and Figure 7 shows measured HW flow rate and measured invert frequency against outdoor temperature, respectively. As a result, pump electricity use was estimated to be significantly reduced by 62.86% as the invert frequency was reduced by 23.40%, but HW supply and return temperature difference increased from 15.24°C to 18.53°C as the Hot Water (HW) flow rate decreased by 23.2% when the differential pressure at the end of water circulation loop was set to 0.2 bar from 0.3 bar during the selected experimental periods.

4 CONCLUSION

This study analyzes the operational characteristics of a central heating system with a variable speed pump, in terms of differential pressure, invert frequency, HW flow rate, and HW supply and return temperature difference, and then estimates the pumping energy savings when the differential pressure at the end of the secondary water circulation loop system was set to minimum value (i.e., 0.2 bar) from design value (i.e., 0.3 bar) for a multi-family residential building in Korea. As a result, pump electricity use was estimated to be significantly reduced by 62.86% as the invert frequency reduced by 23.40%, but HW supply and return temperature difference increased by 3.29°C from 15.24°C to 18.53°C as the Hot Water (HW) flow rate decreased by 23.2% during the selected experimental periods.

Consequently, differential pressure setpoint could be reset lower than the design value to minimize the pumping energy for central heating (or cooling) systems within an acceptable range of water flow.

ACKNOWLEDGMENT

This study was supported by the key R&D Project (No. 2015-0181) of Korea Institute of Civil Engineering and Building Technology.

REFERENCES

EVO, 2012. *International Performance Measurement and Verification Protocol: Concepts and Options for Determining Energy and Water savings*, Efficiency Valuation Organization (EVO), Washington, USA. 1.

Lu, L. & Cai, W. 2002. *A New Approach to Set Point Control in Chilled Water Loops*. International Refrigeration and Air Conditioning Conference, Purdue University, USA.

Tillack, L. & Rishel, J.B. 1998. Proper Control of HVAC Variable Speed Pumps, *ASHRAE Journal*, 41–47.

Advanced Materials, Mechanical and Structural Engineering – Hong, Seo & Moon (Eds)
 ISBN: 978-1-138-02908-8

An enhanced differential pressure reset control for a central heating system with a variable speed pump

S.W. Song
Korea Institute of Civil Engineering and Building Technology, Gyeonggi-do, South Korea

ABSTRACT: This study proposes an enhanced differential Pressure (dP) reset control, which can automatically reset a dP setpoint based on the measured supply and return water temperature difference within a certain range of differential pressure under partial load conditions, and demonstrates its use for the central heating system with a variable speed pump of a nearly zero energy, for a multi-family residential building in Korea. As a result, pumping energy was estimated to be reduced by 31.27% as the Hot Water (HW) flow rate decreased by 12.60% when the differential pressure at the end of a secondary water circulation loop system was automatically reset between the minimum value (0.2 bar) and the design value (0.3 bar). In addition, the average HW supply and return temperature difference also increased by only 1.19°C from 15.17°C to 16.36°C during selected measurement periods. Consequently, the differential pressure reset control proposed in this study can be used as an effective way to optimize the overall system's performance and save pumping energy under partial load conditions within an acceptable range of the water flow rate and supply and return temperature difference.

1 INTRODUCTION

Nearly Zero Energy Building (ZEB) is becoming the most important issues with respect to global climate change. Several pilot projects of the ZEB have been implemented with various Energy Conservation Measures (ECMs) and renewable energy systems for residential and commercial buildings in Korea. However, it has often been limited to deliver the ZEB as expected due to improper system operation without commissioning. Ongoing (or continuous) commissioning is one of the promising technologies to achieve such ZEB by optimizing building and system performance with a minimum operational cost.

High-rise multi-family residential buildings with the radiant floor heating system served by the district heating or central heating plant have widely been constructed and recently tried to be designed for passive or zero energy levels in Korea. Such high-rise multi-family residential buildings are often partially unoccupied during daytime. However, most of the hot water circulation pumps run at a constant speed along with the bypass control valve even during off-peak hours. Variable-Frequency Drives (VFDs) for water circulation pumps can be used to save pumping use by modulating water flow based on a differential pressure setpoint.

When VFDs are applied for water circulation pumps (Tillack & Rishel 1998), a differential pressure sensor for variable speed pumps is generally installed at the end of the water circulation loop system, and a differential pressure setpoint is normally set at a maximum value and continues to run at design conditions even if it could be reset under partial load conditions to minimize the pumping energy (Lu & Cai 2002). However, it is difficulty in determining a proper dP setpoint during partial load conditions.

This study proposes an enhanced differential Pressure (dP) reset control, which can automatically reset a dP setpoint based on the measured supply and return water temperature difference within a certain range of dP setpoint from the minimum dP value to the maximum dP value. It also analyzes the operational characteristics of a central heating system with a variable speed pump, in terms of invert frequency, water flow rate, and supply and return water temperature difference, when the differential pressure at the end of the water circulation loop system is automatically reset between the minimum value (0.2 bar) and the design value (0.3 bar) for a case-study building in Korea.

2 CASE-STUDY BUILDING

2.1 *Building description*

The case-study building is an eight-story (1,806 m^2), multi-family residential building including 15 residential units served by a central heating system. The building was constructed in December, 2012 as a pilot project to be a nearly zero energy building with several Energy Conservation Measures (ECMs), including: exterior super insulation, high-performance windows, movable exterior shadings (blinds), two pellet boilers for heating

Table 1. Description of the case-study building.

Items	Description
Building location	South Korea
Building type	Multi-family building
Construction year	2012
Construction type	Reinforced concrete
Number of floors	8 Floors
Total floor area	1,806 m^2
Roof U-value	0.11 $W/m^2 K$
Exterior wall U-value	0.15 $W/m^2 K$
Floor U-value	0.10 $W/m^2 K$
Window U-value	1.0 $m^2 K$ (SHGC 0.37)
System & Plant	Radiant floor heating
For heating	Pellet boiler (50 kW)
Others	Photovoltaic (PV) panels

and Domestic Hot Water (DHW), and Photovoltaic (PV) panels, as presented in Table 1.

2.2 *Central heating system and differential pressure reset control*

The case-study building includes 15 residential units served by a central heating plant (i.e., pellet boiler) providing hot water with a radiant floor heating system in each residential unit, which is controlled by individual room thermostat to maintain 20°C of heating setpoint during the experimental periods in 2014. A whole-building energy monitoring and control system for the case-study building was installed to implement ongoing commissioning after construction. Table 2 presents the design conditions of the central heating system, and Figure 1 shows a schematic diagram of the energy monitoring and control system with measurement points for the central heating system of the case-study building.

For this study, a differential Pressure (dP) sensor was installed at the end of the Hot Water (HW) circulation loop system in order to control the circulation pump (P1) speed by a VFD based on a differential pressure setpoint. Two Resistance Temperature Detector (RTD) sensors were installed to measure the HW supply (Ts) and return (Tr) temperature for the secondary HW circulation loop of the case-study building. An RTD sensor was also installed to detect the buffer tank's water temperature for the primary pumps (i.e., P2 and P3) and the pellet boiler operation (e.g., on/off) using the energy monitoring and control system installed for this study. In addition, a differential pressure control valve was installed at the bottom of the secondary water circulation loop just in case of applying for constant speed pump control, which was initially set up at the design stage but replaced with variable speed pump control for this study.

Table 2. Design conditions of the central heating system.

Items		Design condition
Heating plant	Type	Pellet boiler (50 kW)
	Efficiency	87%
Heat exchanger	Type	Plat
HW buffer tank	Capacity	1 Ton
Circulation loop	Control	Differential Pressure (dP) Reset control
HW Circulation pump (P1)	Type	Variable speed
	Control	Inverter (VFD)
	Capacity	0.75 kW (1750 RPM)
	Head	12 m
	Flow	71.6l pm (4.3 CMH)
	Efficiency	41.6%
Circulation pump (P2)	Type	Single speed (On/off)
	Capacity	0.55 kW
Boiler pump (P3)	Type	Single speed (On/off)
	Capacity	0.35 kW

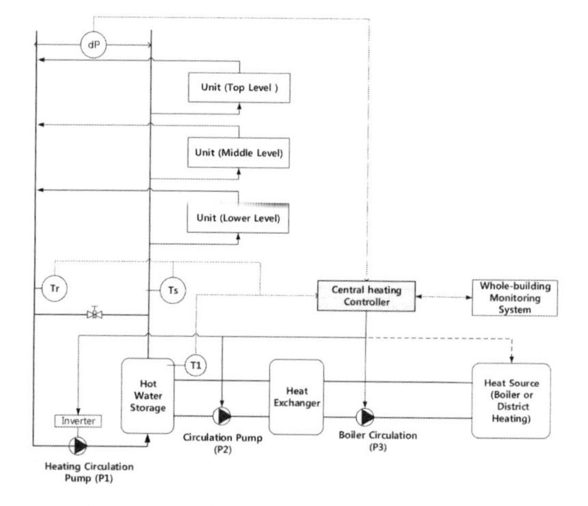

Figure 1. Schematic diagram of the energy monitoring and control system with measurement points.

3 RESULTS AND DISCUSSION

3.1 *Operational characteristic*

The case-study building was mostly unoccupied except for a few residential units, so that the hot water circulation pump (P1) was operating under part load conditions as well. This study analyzes the operational characteristics of a secondary water circulation loop system with a variable speed pump when the differential pressure setpoint is automatically reset based on the measured HW supply and return temperature difference within a certain range of differential pressure between the minimum dP setpoint (0.2 bar) to maximum dP setpoint (0.3 bar), as shown in Figure 2, in terms of differential pressure, invert frequency, hot water flow rate, and supply and return temperature difference. All the data used in this study were measured every 15 minutes.

Figure 2 shows the relationship of the measured differential pressure against the measured HW supply and return temperature difference. The differential pressure with 0.3 bar of dP setpoint was constant regardless of the temperature difference, while the differential pressure with dP reset control was proportional to the HW flow rate within a certain range of dP setpoint from the minimum dP setpoint (0.2 bar) to the maximum dP setpoint (0.3 bar) when the temperature difference ranged from the minimum value of 10°C to the maximum value of 20°C.

Figure 3 shows the measured invert frequency of the HW circulation pump against the measured HW flow rate. The invert frequency of HW pump ranged between 25 Hz and 30 Hz when the dP setpoint was set to 0.3 bar, while it ranged from 20 Hz to 30 Hz when the dP setpoint was reset. As a result, it was found that the dP setpoint had a significant impact on the minimum operational range of a VFD, which modulated the pump speed that

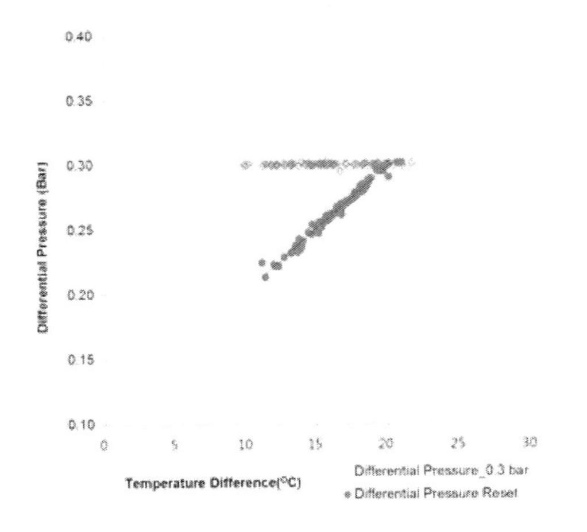

Figure 2. Invert frequency against HW temperature difference (supply—return).

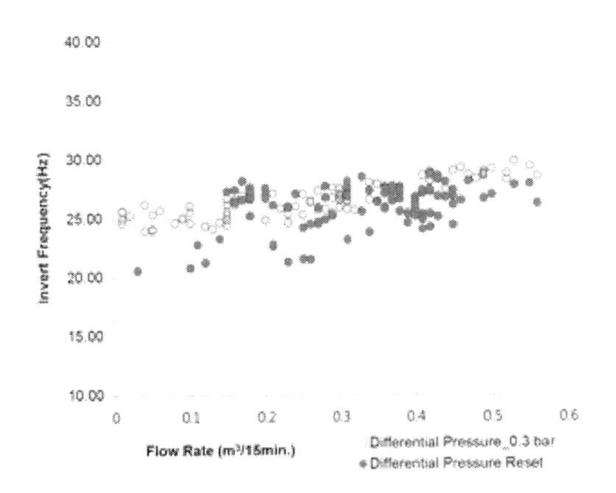

Figure 3. Invert frequency against HW flow rate.

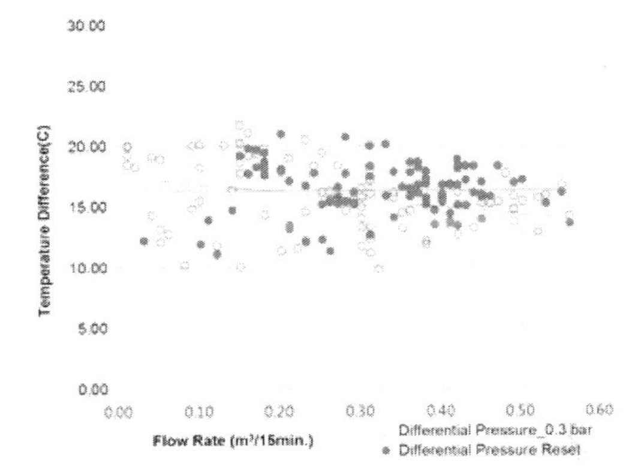

Figure 4. Temperature difference (supply—return) against HW flow rate.

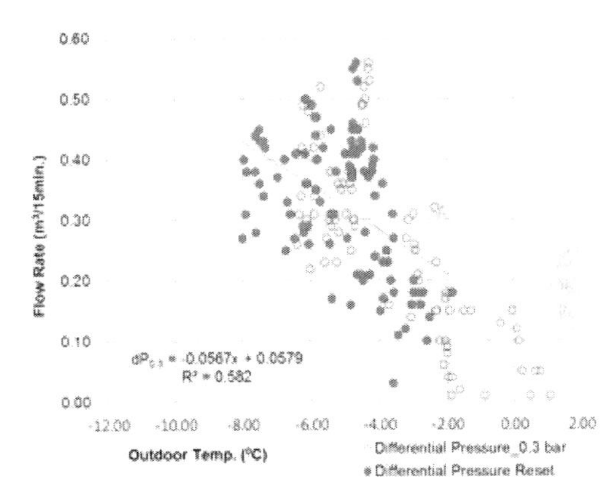

Figure 5. HW flow rate against outdoor temperature.

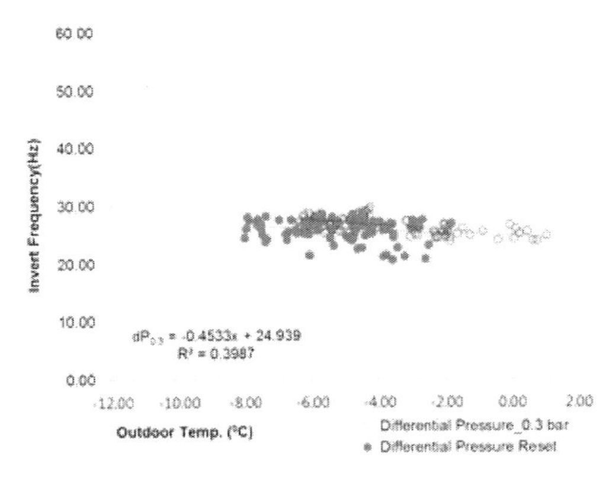

Figure 6. Invert frequency against outdoor temperature.

Table 3. Average energy savings between design values (0.3 bar) and reset control of dP setpoint.

	Before Change	Adjusted Baseline	After Change	Energy Savings	
Items	(dP 0.3 Bar)	(dP 0.3 Bar)	(dP Reset)	Difference	Savings (%)
Date	12/12	–	12/21	–	–
Outdoor Temp. (°C)	–3.66	–5.12	–5.12	–	–
Flow Rate (m^3/15min)	0.27	0.38	0.33	0.05	12.60
Invert Frequency (Hz)	26.60	27.26	26.01	1.25	4.59
Temp. Difference (°C)	15.81	15.17	16.36	1.19	
Pump Power (kW/15min.)	0.26	0.36	0.25	0.11	31.27

was theoretically proportional to the water flow rate referred to as affinity laws of pump.

Figure 4 shows the measured supply and return water temperature difference against the measured HW flow rate. As expected, the HW water temperature difference with 0.3 bar of dP setpoint slightly decreased when the HW flow rate increased. In addition, the temperature difference with 0.3 bar of dP setpoint was more scattered than that with dP reset control within a certain range of temperature difference of about 10°C.

3.2 *Energy savings*

This study estimated the average pumping energy and savings based on adjusted energy baselines (EVO 2012) when the differential Pressure (dP) setpoint was automatically reset between the minimum dP setpoint (0.2 bar) to the maximum dP setpoint (0.3 bar) under partial load conditions. The adjusted energy baselines were developed based on the linear regression models with 0.3 bar of dP setpoint (i.e., before change) but had the same outdoor temperature as the reporting period (i.e., after change). Figure 5 and Figure 6 show the HW flow rate and the invert frequency against outdoor temperature, respectively. Table 3 presents the daily average pumping energy and savings (%) based on the adjusted baselines, including: HW flow rate, invert frequency, and pump power. As a result, the average pump power was estimated to be reduced by 31.27% as the Hot Water (HW) flow rate reduced by 12.60%. In addition, the daily average HW supply and return temperature difference increased by only 1.19°C within an acceptable range of temperature difference from 15.17°C to 16.36°C.

4 CONCLUSION

This study proposes an enhanced differential Pressure (dP) reset control, which can automatically reset a dP setpoint based on the measured supply and return temperature difference within a certain range of differential pressure between the minimum dP setpoint (0.2 bar) to the maximum dP setpoint (0.3 bar), and demonstrates its use for the secondary water circulation loop system with a variable speed pump of a nearly zero energy, for a multi-family residential building in Korea.

As a result, the average pump power was estimated to be reduced by 31.27% as the Hot Water (HW) flow rate decreased by 12.60% when the differential pressure at the end of the secondary water circulation loop system was automatically reset between the minimum value (0.2 bar) and the design value (0.3 bar). In addition, the daily average HW supply and return temperature difference increased by only 1.19°C from 15.17°C to 16.36°C during the selected measurement periods.

Consequently, the differential pressure reset control proposed in this study can be used as an effective way to optimize the overall system's performance and save the pump's electricity use under partial load conditions within an acceptable water flow rate and supply and return temperature difference.

ACKNOWLEDGMENT

This study was supported by the key R&D Project (No. 2015-0181) of Korea Institute of Civil Engineering and Building Technology.

REFERENCES

EVO, 2012. *International Performance Measurement and Verification Protocol: Concepts and Options for Determining Energy and Water savings*, Efficiency Valuation Organization (EVO), Washington, USA. 1.

Lu, L. & Cai, W. 2002. *A New Approach to Set Point Control in Chilled Water Loops*. International Refrigeration and Air Conditioning Conference, Purdue University, USA.

Tillack, L. & Rishel, J.B. 1998. Proper Control of HVAC Variable Speed Pumps, *ASHRAE Journal*, 41–47.

Advanced Materials, Mechanical and Structural Engineering – Hong, Seo & Moon (Eds)
© 2016 Taylor & Francis Group, London, ISBN: 978-1-138-02908-8

Microstructure and fracture behavior of Ti-aluminide-reinforced Ti matrix laminates

W.Q. Zhou
Department of Mechanical Engineering, Tsinghua University, Beijing, China

K. Zhu & T. Jing
School of Materials Science and Engineering, Tsinghua University, Beijing, China

H.L. Hou & Y.J. Xu
Metal Forming Technology Department, Beijing Aeronautical Manufacturing Technology Research Institute, China

ABSTRACT: Ti-aluminide-reinforced Ti matrix laminated composites are produced from alternating titanium and aluminum foils by vacuum hot pressing. Microstructure characterization by Scanning Electron Microscopy (SEM), Energy Dispersive Spectroscopy (EDS), and X-Ray Diffraction (XRD) showed that the laminated composites were comprised of Ti, $TiAl_3$, and TiAl. Three-point bending tests with the load direction perpendicular and parallel to the laminates showed that the composites were anisotropic, and that the composites displayed better flexural strength when the load was perpendicular to the laminates. The three-point bending tests with edge-notched laminated composites were conducted to study the crack propagation. The unreacted Ti layers arrested the crack propagation and enhanced the toughness.

1 INTRODUCTION

Titanium aluminides have been promising structural materials in aviation and automobile industries because of their low density, high specific strength and mechanical behavior under high temperatures. Among various titanium aluminides, such as Ti_3Al, TiAl, $TiAl_2$, and $TiAl_3$, the formation of $TiAl_3$ is thermodynamically and kinetically favored over other aluminides (Soboyejo et al. 1996). Moreover, with the same percentage of the ductile reinforcing phase, the laminated ductile phase has the maximum toughness efficiency followed by fiber and particulate morphology (Soboyejo et al. 1996).

The fabrication methods of titanium aluminides laminated composites are available in the literature, including hot pressing (Harach & Vecchio 2001, Price et al. 2011, Patselov et al. 2012), vacuum hot pressing (Peng et al. 2005, Xu et al. 2006, Jiangwei et al. 2002, Wei et al. 2008), rolling and annealing (Sun et al. 2011, Cui et al. 2012, Kevorkijan & Škapin 2009), microwave activated combustion (Naplocha & Granat 2009), explosive welding and annealing (Bataev et al. 2012), pulsed current hot pressing (Mizuuchi et al. 2004), and electron beam physical vapor deposition (EB-PVD) (Ma et al. 2008). According to the study of Harach & Vecchio (2001), when Ti-$TiAl_3$ laminated composites were fabricated through reactive foil sintering in air, some oxides would exist in the composites (2001). While through vacuum hot pressing, Ti-$TiAl_3$ laminated composites can be manufactured without oxides between layers (Peng et al. 2005).

In the process of vacuum hot pressing, solid Ti and liquid Al would react directly; hence, $TiAl_3$ is obtained first, followed by TiAl (Sun et al. 2011). In addition, because Al is in the liquid state and the diffusion rate and the solid solubility of Al in Ti are larger than those of Ti in Al, Al is the main diffusion element in the reaction.

In this study, Ti-aluminide-reinforced Ti matrix laminated composites were synthesized by vacuum hot pressing. The flexural strength and fracture behavior were studied through a three-point bending test.

2 EXPERIMENTAL PROCEDURES

2.1 *Composite processing*

The commercial purity aluminum and titanium foils were used in this study. The dimensions of Ti foils were 35 mm * 35 mm * 0.1 mm and those of Al foils were 35 mm * 35 mm * 0.04 mm. First, the oxide films on the surface of Ti foils and Al foils were removed by 2%HF and 4%NaOH solutions, respectively. Then, 36 layers of Ti foils and 35 layers of Al foils were piled up with staggered stacking. Finally, the foil stacks were placed into a graphite mold in the vacuum furnace for vacuum hot pressing. It is worth noting that the vacuum degree reached to 0.01 MPa before heating.

The processing parameters are shown in Figure 1. First, the temperature was raised to 525°C in 50 minutes and then stayed at 525°C for 1 h with the pressure of 30 MPa. This period was for the layer diffusion. Then, the temperature was raised to 645°C for 3 h with the pressure reduced to 8 MPa to prevent the expulsion of liquid Al. This period was for the reaction between layers. Finally, the temperature was raised to 1050°C with the pressure of 40 MPa to reduce the pores in the composites.

2.2 *Microstructure, mechanical behavior, and crack propagation*

Scanning Electron Microscopy (SEM) was used to obtain the microstructure and the thickness of each layer of the sample. Energy Dispersive Spectroscopy (EDS) was used to quantify the contents of each element. X-Ray Diffraction (XRD) was used to analyze the phase composition.

The flexural strength was measured by a three-point bending test with two loading directions: one was perpendicular and the other was parallel to the laminates, as shown in Figure 2. As to the sample

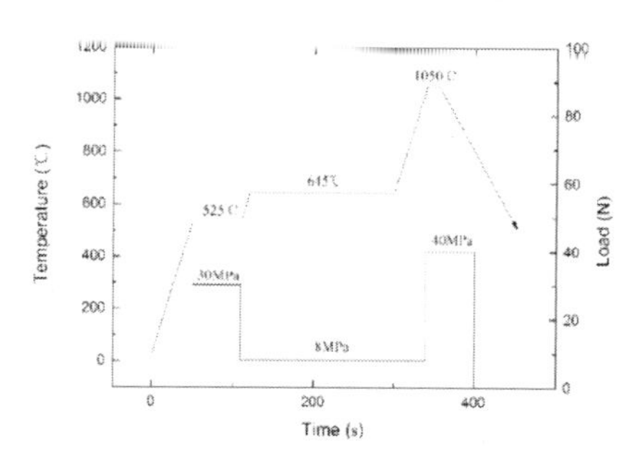

Figure 1. Processing parameters for Ti-aluminide-reinforced Ti matrix composites.

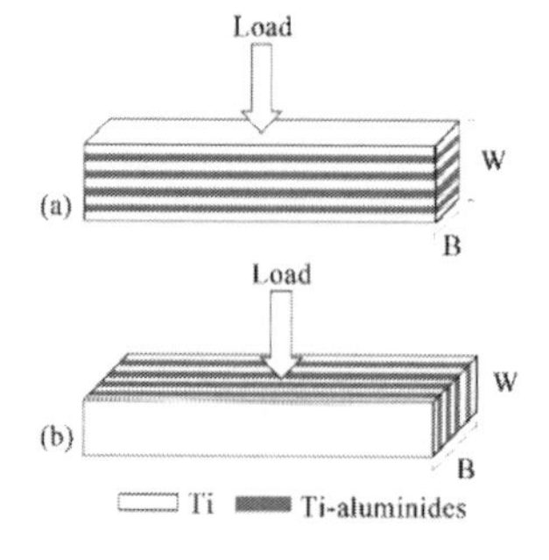

Figure 2. Samples for the three-point bending test. (a) Load direction perpendicular to the laminates, (b) load direction parallel to the laminates.

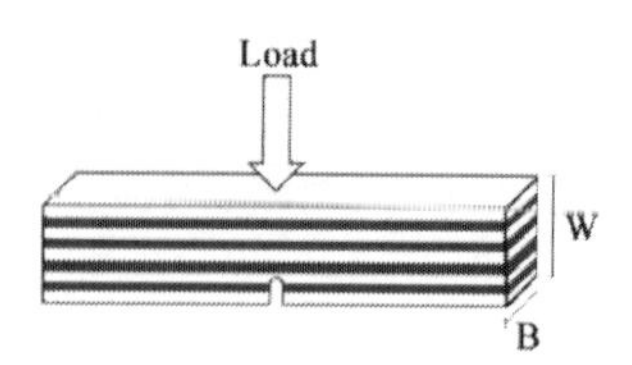

Figure 3. Sample for studying crack propagation.

with perpendicular load, the width and the thickness were 2.5 mm and 2.8 mm, respectively. As to the sample with parallel load, the width and the thickness were 2.8 mm and 2.5 mm, respectively.

The sample used for the crack propagation tests had a notch of 460 μm length and 300 μm width, as shown in Figure 3. The loading rate was 0.5 mm/min.

3 RESULTS AND DISCUSSION

3.1 *Microstructure and reaction mechanism*

The laminated composites consisted of unreacted Ti and Ti-aluminides, as shown in Figure 4. The thicknesses of Ti and Ti-aluminides were 51.52 μm and 25.14 μm, respectively. The analysis of XRD illustrated that these laminated composites consisted of Ti, $TiAl_3$, and TiAl, as shown in Figure 5.

According to the analysis of EDS, the content of Ti decreased gradually from the Ti layer to the Ti-aluminide layer, as shown in Figure 6(b). Table 1 presents the contents of Ti and Al from point 1 to point 12. Point 1 located at the original central line of the Ti layer contained 93.62% Ti, indicating that a little amount of Al diffused there. The atom ratio of Ti and Al in point 7 located at the original central line of the Al layer was about 1:3, indicating that the Ti-aluminide layer in this area was $TiAl_3$. In addition, the atom ratio of Ti and Al in point 4 was between 1:1 and 1:3, indicating that the Ti-aluminides in this area were TiAl and $TiAl_3$.

Thus, it could be concluded that during the reaction between Ti foils and Al foils, the Al in original Al layers mainly turned into $TiAl_3$. From the thermodynamic and kinetic viewpoints, the $TiAl_3$ was proved to be easier to form over the formation of other Ti-aluminides (Harach & Vecchio 2001). In addition, there were TiAl intermetallic compounds between the Ti layer and the $TiAl_3$ layer, indicating that some amount of TiAl may be formed after $TiAl_3$.

3.2 *Mechanical behavior*

Figure 7 shows the load–displacement curves with the load direction perpendicular and parallel to the laminates. The flexural strength, σ_f, can be calculated according to the following equation:

Figure 4. SEM micrographs of Ti and Ti-aluminide laminated composites.

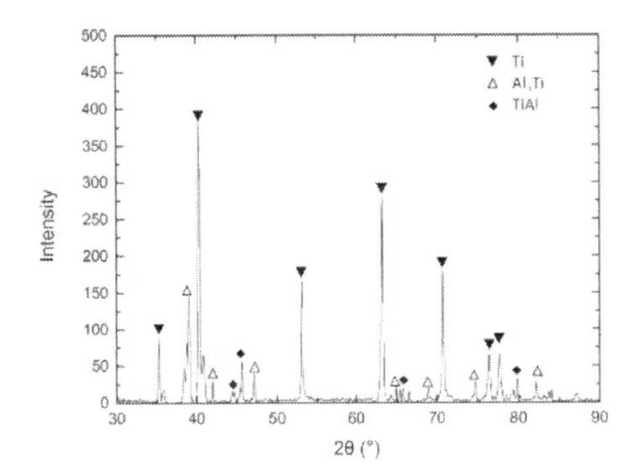

Figure 5. XRD patterns of Ti and Ti-aluminide laminated composites.

Table 1. Atomic ratio of Ti and Al.

Position	1	2	3	4	5	6
Ti (at.%)	93.6	92.5	87.6	59.3	34.9	26.4
Al (at.%)	6.4	7.5	12.4	40.7	65.1	73.6
Position	7	8	9	10	11	12
Ti (at.%)	26.2	26.1	33.7	64.6	89.8	91.5
Al (at.%)	73.8	73.9	66.3	35.4	10.2	8.5

$$\sigma_f = \frac{3P_{\max L}}{2BW^2} \tag{1}$$

where P_{max} is the maximum load; L is the supporting span; B is the specimen width; and W is the specimen thickness. The flexural strength of the samples with perpendicular and parallel loads was 156.10 MPa and 125.32 MPa, respectively.

These laminated composites showed good toughness, which was because the Ti layer can improve the toughness. These composites also displayed superior flexural strength when the load direction was perpendicular to the laminates.

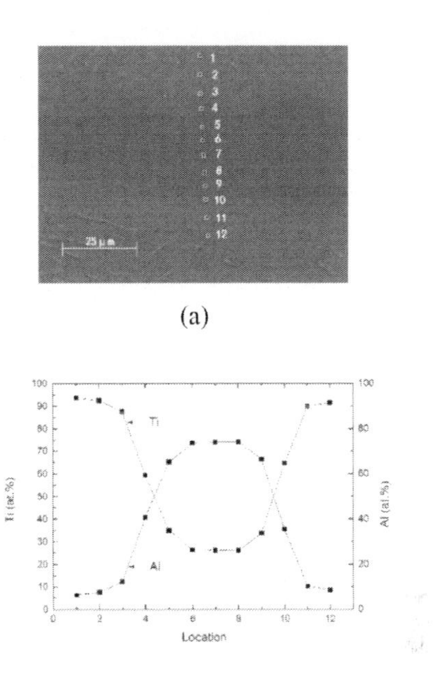

Figure 6. Distribution of Ti and Al: (a) position, (b) distribution curve.

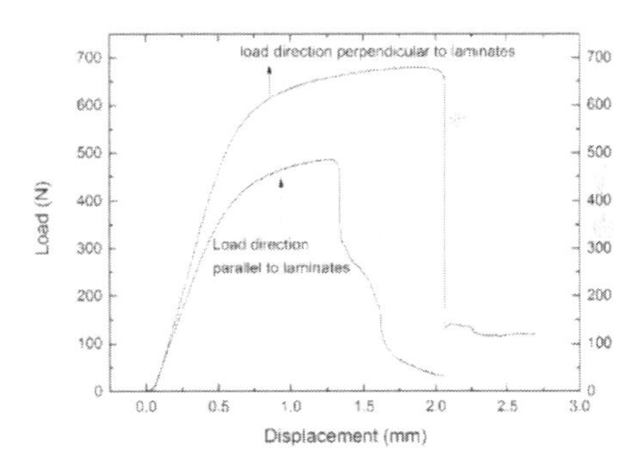

Figure 7. Load–displacement curve of Ti-aluminide-reinforced Ti matrix composites.

3.3 *Crack propagation*

The three-point bending test of the notched sample was used to study the crack propagation in the Ti-aluminide-reinforced Ti matrix composites. The load direction was perpendicular to the laminates. The load–displacement curve of the whole process is shown in Figure 8.

The whole crack propagation process is shown in Figure 9. Figure 9(a) shows the micrograph of the notched sample before loading. Figure 9(b) shows the crack after the first loading, and the crack first formed in the Ti-aluminide layer. The crack extended horizontally and then vertically.

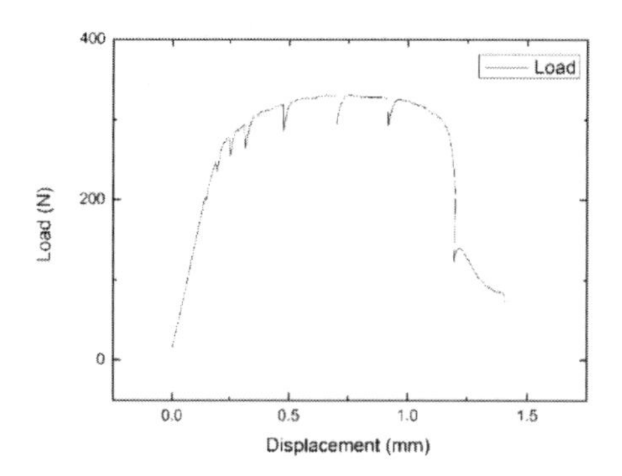

Figure 8. Load–displacement curves of notched Ti and Ti-aluminides laminated composites.

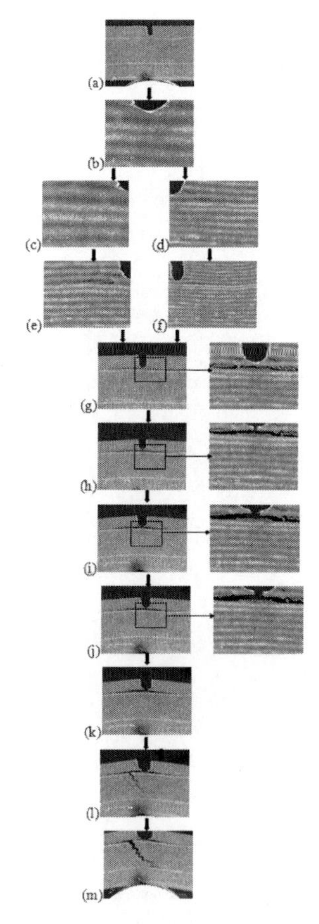

Figure 9. Micrographs of crack propagation. (a) Initial, (b) crack after the 1st loading, (c) left crack after the 2nd loading, (d) right crack after the 2nd loading, (e) left crack after the 3rd loading, (f) right crack after the 3rd loading, (g) crack after the 4th loading, (h) crack after the 5th loading, (i) crack after the 6th loading, (j) crack after the 7th loading, (k) crack after the 8th loading, (l) crack after the 9th loading, (m) crack after the 10th loading.

Figure 9(c), (d) shows the left crack and the right crack after the second loading, respectively. Crack nucleated and grew in the next Ti-aluminide layer. Figure 9(e), (f) shows the left crack and right crack after the third loading, respectively. Cracks grew along the Ti-aluminide layer and the width of the cracks became smaller. Figure 9(g), (h) further shows the cracks after the fourth and the fifth loading, respectively. Crack grew forward continually and some tiny vertical crack formed in the middle area of the sample. Figure 9(i), (j), (k) shows the cracks after the sixth, seventh, and eighth loading, respectively. Cracks in the middle area of the sample became more and more distributed in a triangle zone. Figure 9(l) shows the crack after the ninth loading, where the cracks in the middle area linked together and formed a main step-like crack. Finally, Figure 9(m) shows the crack after the tenth loading, where the vertical main crack grew bigger and the composites were damaged. Also, the Ti layer showed great plastic deformation in this process.

4 CONCLUSION

In this study, Ti-aluminide-reinforced Ti matrix laminated composites were produced by vacuum hot pressing. Then, SEM, XRD, and DES were used to study the microstructure, phase, and element distribution in the laminated composites. Moreover, three-point bending tests were conducted to study the flexural strength in different directions. In addition, crack propagation in the laminated composites was studied by the three-point bending test with notched samples. The conclusion drawn is as follows:

1. The composites consisted of the Ti layer and the Ti-aluminide layer. The original central line of the Ti layer is now the Ti layer with a little amount of Al that diffused into the Ti layer in the reaction. The original central line of the Al layer translated into $TiAl_3$ after the reaction. Between the Ti layer and Ti-aluminide layer are TiAl and $TiAl_3$, because TiAl formed after $TiAl_3$.
2. These laminated composites exhibited anisotropic features. The flexural strength with the load direction perpendicular to the laminates was superior to those with the parallel load direction. These composites also displayed some toughness because of the Ti layer.
3. When the load was put on these laminated composites, cracks were first formed in the Ti-aluminide layer and then grew along this layer. Cracks in the Ti-aluminide layer were bridged by the Ti layer and the cracks nucleated in the next Ti-aluminide layer. With increasing load put on the laminates, vertical cracks were

formed in Ti-aluminide layers. When the cracks reached to a certain amount, the Ti layers were damaged and the laminates failed. The Ti layer had plastic deformation before the damage.

REFERENCES

Bataev, I.A., Bataev, A.A. & Mali, V.I. et al. 2012. Structural and mechanical properties of metallic-intermetallic laminate composites produced by explosive welding and annealing. *Materials & Design*, 35: 225–234.

Cui, X., Fan, G. & Geng, L. et al. 2012. Fabrication of fully dense TiAl-based composite sheets with a novel microlaminated microstructure. *Scripta Materialia*, 66(5): 276–279.

Harach, D.J. & Vecchio, K.S. 2001. Microstructure evolution in metal-intermetallic laminate (MIL) composites synthesized by reactive foil sintering in air. *Metallurgical and Materials Transactions A*, 32(6): 1493–1505.

Jiangwei, R., Yajiang, L. & Tao, F. 2002. Microstructure characteristics in the interface zone of Ti/Al diffusion bonding. *Materials Letters*, 56(5): 647–652.

Kevorkijan, V. & Škapin, S.D. 2009. Fabrication and characterization of TiAl/Ti3 Al-based intermetallic composites (IMCs) reinforced with ceramic particles. *Archives of Materials Science and Engineering*, 40(2): 75–83.

Ma, L., Sun, Y. & He, X. 2008. Preparation and Performance of Large-Sized Ti/Ti-Al Microlaminated Composite. *Rare Metal Materials and Engineering*, 37(2): 325–329.

Mizuuchi, K., Inoue, K. & Sugioka, M. et al. 2004 Microstructure and mechanical properties of Ti-aluminides reinforced Ti matrix composites synthesized by pulsed current hot pressing. *Materials Science and Engineering: A*, 368(1): 260–268.

Naplocha, K. & Granat, K. 2009. Microwave activated combustion synthesis of porous Al-Ti structures for composite reinforcing. *Journal of Alloys and Compounds*, 486(1): 178–184.

Patselov, A., Greenberg, B. & Gladkovskii, S. et al. 2012. Layered metal-intermetallic composites in Ti-Al system: strength under static and dynamic load. *AASRI Procedia*, 3: 107–112.

Peng, L.M., Wang, J.H. & Li, H, et al. 2005. Synthesis and microstructural characterization of Ti-Al_3Ti metal–intermetallic laminate (MIL) composites. *Scripta materialia*, 52(3): 243–248.

Price, R.D., Jiang, F. & Kulin, R.M. et al. 2011. Effects of ductile phase volume fraction on the mechanical properties of Ti-Al_3Ti metal-intermetallic laminate (MIL) composites. *Materials Science and Engineering: A*, 528(7): 3134–3146.

Soboyejo, W.O., Ye, F., Chen, L.C. & Bahtishi, N. et al. 1996. Effects of reinforcement morphology on the fatigue and fracture behavior of MoSi2/Nb composites. *Acta Materialia*, 44:2027–41.

Sun, Y.B., Zhao, Y.Q. & Zhang, D. et al. 2011. Multi-layered Ti-Al intermetallic sheets fabricated by cold rolling and annealing of titanium and aluminum foils. *Transactions of Nonferrous Metals Society of China*, 21(8): 1722–1727.

Wei, Y., Aiping, W. & Guisheng, Z. et al. 2008. Formation process of the bonding joint in Ti/Al diffusion bonding. *Materials Science and Engineering: A*, 480(1): 456–463.

Xu, L., Cui, Y.Y. & Hao, Y.L. et al. 2006. Growth of intermetallic layer in multi-laminated Ti/Al diffusion couples. *Materials Science and Engineering: A*, 435: 638–647.

Advanced Materials, Mechanical and Structural Engineering – Hong, Seo & Moon (Eds)
© 2016 Taylor & Francis Group, London, ISBN: 978-1-138-02908-8

Stress distribution on endodontic post, gutta-percha, and dentin with different post materials

W.M.K. Helal
College of Mechanical and Electrical Engineering, Harbin Engineering University, Harbin, P.R. China
Department of Mechanical Engineering, Faculty of Engineering, Kafrelsheikh University, Kafrelsheikh, Egypt

D.Y. Shi
College of Mechanical and Electrical Engineering, Harbin Engineering University, Harbin, P.R. China

ABSTRACT: Choosing a suitable post material is a very important factor to obtain a satisfactory degree in the treatment process of the patient's tooth. The present study aims to analyze the stress distribution on the Endodontic Prefabricated Post (EPP), gutta-percha, and dentin with different post materials. Six EPP materials were considered in the present investigation in order to analyze the effect of the post material on the stress distribution through the Finite Element Analysis (FEA). The results obtained from the present analysis found that when comparing the maximum von Mises stress occurring on the dentin structure with six different post materials, the variation between the least and highest values was about 14%.

1 INTRODUCTION

For thousands of years, the replacement of the missing tooth structure using a post has been practiced (Trabert & Cooney 1984). In clinical dentistry, tooth loss is often due to caries or trauma. Nowadays, Endodontic Treatment (ET) is one of the most widely used techniques in order to save teeth.

According to post design, its designs can be divided into two types (Musikant & Deutsch 1984): 1- parallel and 2- tapered.

The ability of a post to resist vertical dislodging forces refers to post retention (Richard et al. 2004). The dentist must weigh retention against stress distribution while considering any post (Fernandes et al. 2001). A number of investigators have reported that a parallel post is more retentive than a tapered post (Standlee et al. 1978, Felton et al. 1991, Johnson & Sakamura 1978, Qualtrough et al. 2003). Thus, in the present investigation, a parallel post will be considered.

Nowadays, when analyzing structures, Finite Element Analysis (FEA), which has been proved to be a useful tool, is adopted (Asmussen et al. 2005, Chen & Xu 1994). A number of investigators have concluded that the E, and Poisson's ratio (υ) play vitally important roles at stress distribution. The present study aims to investigate and analyze the stress distribution on the EPP, gutta-percha, and dentin with different post materials. Six post materials were considered: (1) Carbon Fiber Post (CFP), (2) Gold (G), (3) Gold Alloy (GA), (4) Titanium (T), (5) Titanium Alloy (TA), and (6) Stainless Steel (SS).

2 MATERIALS AND METHODS

In order to investigate and analyze the stress distribution on the EPP, gutta-percha, and dentin, the following six different EPP materials were considered in the present investigation: (1) CFP, (2) G, (3) GA, (4) T, (5) TA, and (6) SS. Mechanical properties, E (GPa) and υ (non-dimensional) of the materials used were taken from the literature and are summarized in Table 1.

Nowadays, FEA is widely used in order to solve problems in biological structures. For the present investigation, the ANSYS package was used to perform the FEA model. The model was built in ANSYS Rel.12.1 by means of the Parametric Design Language. The structure of a simplified dentin FE model includes: dentin post, dentin, and gutta-percha. It was taken from the literature, as shown in Figure 1. It should be noted that 2/3 of

Table 1. Mechanical properties of the materials used in the present analysis.

Material	E (GPa)	υ	References
CFP	125	0.25	[12]
G	77	0.33	[10]
GA	93	0.33	[13]
T	112	0.33	[14]
TA	54	0.35	[15]
SS	210	0.3	[16]
Gutta-Percha	0.96E-3	0.3	[17]
Dentin	18.6	0.4	[17]

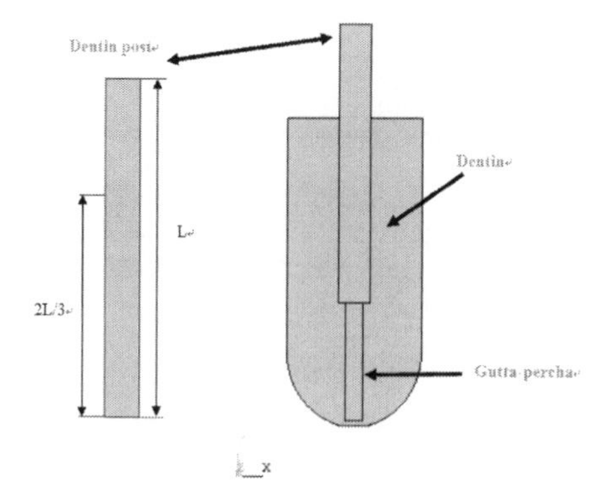

Figure 1. Structure of a simplified dentin FE model.

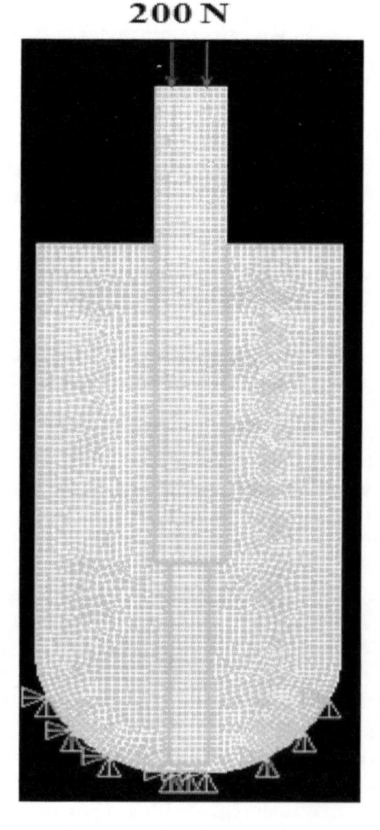

Figure 2. Boundary condition and external loads.

the EPP length was impeded inside the dentin cavity, as shown in Figure 1.

In the present work, in order to generate the model in the x-y plane, plane elements (PLANE 82) were used. The element (PLANE 82) was used as a plane stress. The element was defined as eight nodes having two degrees of freedom at each node.

It should be noted that all components of the FE model were considered as complete bonding. All exterior nodes on the curved bottom surfaces of the dentin structure were considered as the fixed boundary condition. In order to simulate the bruxism case, a compressive pressure of a magnitude of 200 N was applied on top of the EPP surface, as shown in Figure 2. It should be noted that, in the present investigation, all materials used were considered to be homogeneous, linear elastic, and isotropic.

3 RESULTS AND DISCUSSION

The main purpose of the present study was to determine the stress distribution surrounding the EPP, gutta-percha, and dentin with the six different post materials.

The surface of each component of the FE model was divided into a number of key points, numbered from the top right corner (clock wise), in order to present the stress distribution on the EPP, gutta-percha, and dentin.

3.1 *The stress distribution on the EPP*

Figure 3 shows the stress distribution on the EPP. The results found that at key points 2 and 4, higher stresses were observed. Due to key point 2, which is the last point that connects the EPP with the inner surface of the dentin, the stress concentration was observed at this key point. Also, due to the connecting between the EPP edge with the inner surface of the dentin, and due to the variations between the

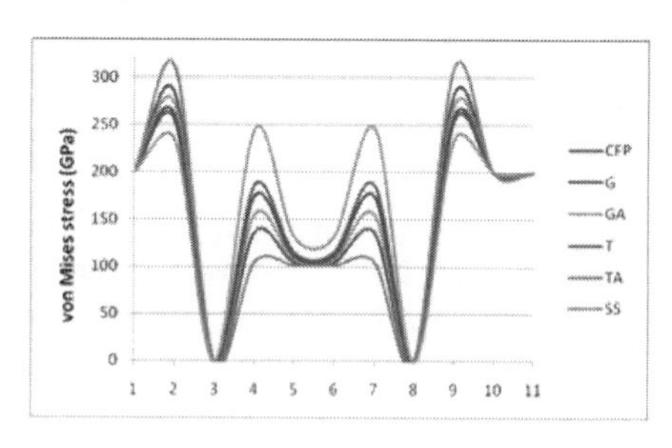

Figure 3. The stress distribution on the EPP.

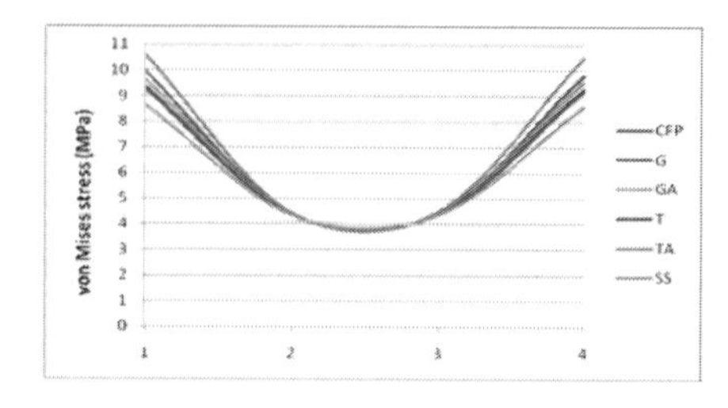

Figure 4. The stress distribution on the gutta-percha.

values of E of EPP and gutta-percha, the stress concentration was observed at key point 4.

When comparing the maximum von Mises stress occurring on the EPP surface with six various post materials, the variation between the least and highest values was about 24%, as shown in Figure 3.

3.2 *The stress distribution on the gutta-percha*

Figure 4 shows the stress distribution on the gutta-percha. The results found that at key point 1, higher stress was observed. Due to key point 1,

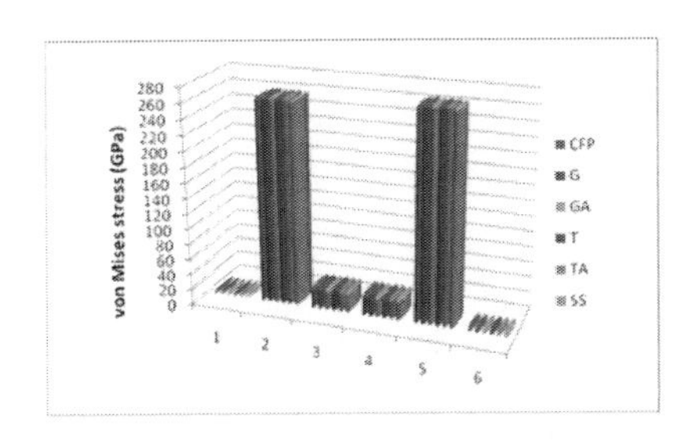

Figure 5. The stress distribution on the dentin.

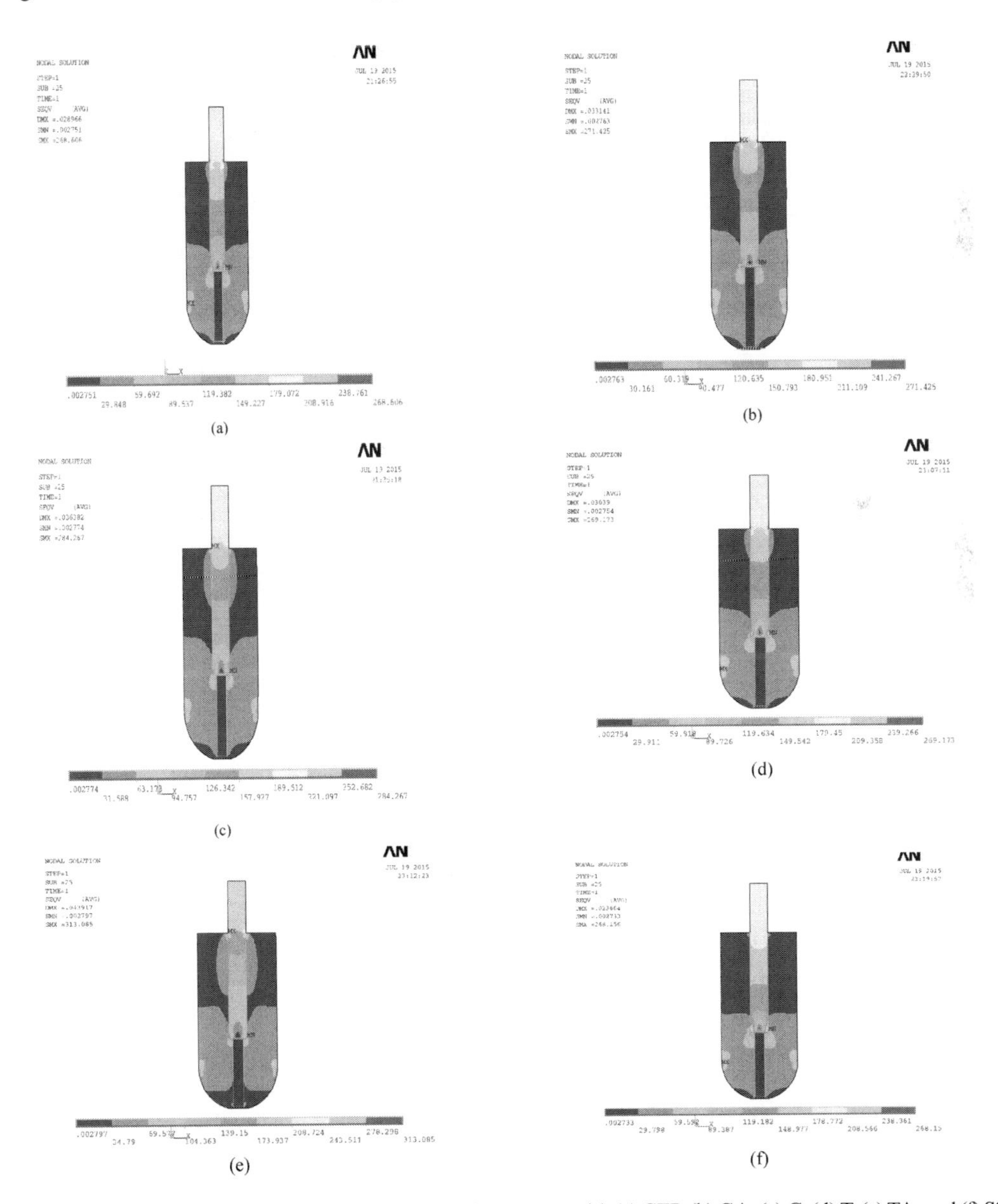

Figure 6. Simulation cases, von Mises stress, of the dentin structure with (a) CFP, (b) GA, (c) G, (d) T, (e) TA, and (f) SS.

which is the point that connects the EPP surface with the gutta-percha edge, the stress concentration was observed. The value of the stress at key point 2 is lower than that at key point 1 because the variation of the E in the post-gutta-percha connecting case is higher than that in the gutta-percha's inner surface of the dentin case.

We compared the maximum von Mises stress occurring on the gutta-percha surface with six different post materials. It was observed from the results that the variation between the least and highest values was about 17%.

3.3 *The stress distribution on the outer surface of the dentin*

Figure 5 shows the stress distribution on the outer surface of the dentin. Due to the top border curve of the boundary constraint, as shown in Figure 2, the maximum value of von Mises stress was observed at key point 5.

When comparing the maximum von Mises stress occurring on the outer surface of the dentin with six different post materials, it was observed that the variation between the least and highest values did not exceed 0.6%.

When comparing the maximum von Mises stress occurring on the dentin structure with six different post materials, it was observed that the highest value of von Mises stress was found at the TA followed by G, GA, T, and CFP, and the least value was found at the SS. The variation between the least and highest values was about 14%, as shown in Figure 6.

4 CONCLUSION

From the outcomes of the present investigation, it was observed that the highest value of von Mises stress was found at the TA followed by G, GA, T, and CFP, and the least value was found at the SS. The variation between the least and highest values was about 14%.

ACKNOWLEDGMENT

The paper was funded by the International Exchange Program of Harbin Engineering University for Innovation-oriented Talents Cultivation. The authors would like to thank Kafr El-Sheikh University, Egypt, for providing the facilities.

REFERENCES

Asmussen, E., Peutzfeldt, A. & Sahafi 2005. Finite element analysis of stresses in endodontically treated, dowel-restored teeth, *Journal of Prosthetic Dentistry*, 94(4): 321–329.

Chen, J. & Xu, L.J. 1994. A Finite Element Analysis of the Human Temporomandibular Joint, *Journal of Biomechanical Engineering*, 116(4): 401–407.

Felton, D.A., Webb, E.L., Kanoy, B.E. & Dugoni, J. 1991. Threaded endodontic dowels: Effect of post design on incidence of root fracture, *Journal of Prosthetic Dentistry*, 65(6): 179–187.

Fernandes, A., Rodrigues, S., SarDessai, G. & Mehta, A. 2001. Endodontology, *Endodontology*, 13: 11–18.

He, G. & Hagiwara. M. 2005. Bimodal structured Ti-base alloy with large elasticity and low Young's modulus, *Materials Science and Engineering C*, 25(3): 290 −295.

Johnson, J.K. & Sakamura, J.S. 1978. Dowel form and tensile force, *Journal of Prosthetic Dentistry*, 40(6): 645–649.

Joshi, S., Mukherjee, A., Kheur, M. & Mehta, A. 2001, Mechanical performance of endodontically treated teeth, *Finite Elements in Analysis and Design, 37*(8): 587–601.

Lanza, A., Aversa, R., Rengo, S., Apicella, D. & Apicella, A. 2005. 3D FEA of cemented steel, glass and carbon posts in a maxillary incisor, *Dental Materials*, 21(8): 709–715.

Maceri, F., Martignoni, M. & Vairo, G. 2007. Mechanical behaviour of endodontic restorations with multiple prefabricated posts: A finite-element approach, *Journal of Biomechanics*, 40: 2386–2398.

Muslkant, B.L. & Deutsch, A.S. 1984. A new prefabricated post and core system, *Journal of Prosthetic Dentistry*, 52(5): 631–634.

Pegoretti, A., Fambri, L., Zappini, G. & Bianchetti. M. 2002. Finite element analysis of a glass fibre reinforced composite endodontic post, *Biomaterials*, 23(13): 2667–2682.

Qualtrough, A.J., Chandler, N.P. & Purton, D.G. 2003. A comparison of the retention of tooth-colored posts, *Quintessence International*, 34(3): 199–201.

Robert G., Craig, Marcus L. 1997. *Restorative dental materials,* 10th Ed. Mosby, St. Louis, Missouri, USA, 1997.

Schwartz, R.S. & Robbins, J.W. 2004. Post Placement and Restoration of Endodontically Treated Teeth: A Literature Review, *Journal of Endodontics*, 30(5): 289–301.

Standlee, J.P., Caputo, A.A. & Hanson, E.C. 1978. Retention of endodontic dowels: Effects of cement, dowel length, diameter, and design, *Journal of Prosthetic Dentistry*, 39: 401–405.

Trabert, K.C. & Cooney, J.P. 1984. The endodontically treated tooth: restorative concepts and techniques, *Dental Clinics of North America*, 28(4): 923–951.

Advanced Materials, Mechanical and Structural Engineering – Hong, Seo & Moon (Eds)
© 2016 Taylor & Francis Group, London, ISBN: 978-1-138-02908-8

Development of organic/inorganic hybrid coating solution of steel sheet for automobiles

S.Y. Lee
Materials Science and Engineering of Graduate School, Pukyong National University, Busan, Korea

H.R. Jeong
Prepoll Co. Ltd., Gangseo-gu, Busan, Korea

K.W. Nam
Materials Science and Engineering, Pukyong National University, Busan, Korea

ABSTRACT: This study developed seven types of organic/inorganic hybrid coating solutions containing Si, and their various characteristics were evaluated. The coating solution with SiO_2 polysilicate 7 wt.% and melamine 3 wt.% shows superior corrosion resistance at 3 minutes of hardening time. In the salt spray test conducted for 7 hours, the corrosion area rate in SPCC and SPFC590 were 7.7% and 1.5%, respectively. The seven types of solutions were also excellent without cracking and peeling after cross-cutting, and showed excellent boiling water, rubbing, and bending resistance. It is judged that the melamine curing agent plays a cross-linking role in the synthesis of SiO_2 polysilicate and urethane resin. US_7M_3 is the optimal composition ratio in the range of solutions in this study.

1 INTRODUCTION

For recent automobiles, higher performance under high-speed impact is demanded for the enhancement of passengers' safety. Especially, in order to ensure the safety of the passengers under the lightweight and side collision, high strength material is applied to the upper part of the center pillar, and an even higher impact rigidity should be maintained than at the lower part. However, while car safety is extremely important, the appearance of the car is also important. Because of the cold-rolled steel sheet process applied in car manufacturing, the center pillar is susceptible to corrosion, and needs to be painted for corrosion resistance in a variety of ways.

Today, the use of Cr^{+6} contained in the chromate-treatment solution for corrosion resistance has been prohibited because of human risk. Recently, in response to this regulation, the development of an environmentally-friendly coating solution is being actively carried out. Studies on an eco-friendly anti-corrosive technology, such as an inorganic or organic coating treatment, are being actively investigated. (Zheludkevich et al. 2005) In this paper, the effects of coating zinc-coated steel sheet and cold-rolled steel sheet with an organic/inorganic hybrid solution were studied by analyzing the characteristics such as corrosion resistance. (Seo et al. 2010a, Seo et al. 2010b, Seo et al. 2010c).

Seven types of organic/inorganic hybrid coating solutions were developed in this study, and were coated onto two types of cold rolled steel sheet used for automobiles, which we named SPCC and SPFC590. The characteristics of the coating solutions were evaluated for their corrosion resistance, adhesion, boiling water resistance, and rubbing and bending resistance.

2 MATERIALS AND EXPERIMENTAL METHOD

An organic/inorganic hybrid coating solution was prepared using distilled water, ethanol, a urethane resin, SiO_2 polysilicate, and melamine (curing agent). The compositions of the coating solution are shown in Table 1. This composition was obtained through analysis and testing by authors. (Nam, unpubl.).

The organic/inorganic hybrid solution can create a polysilicate solution by hydrolysis reaction as shown in Figure 1. If a metal alkoxide represents to Me(OH)n, Me(OH)n produced by hydrolysis as the reaction of Formula (1) proceeds the reaction as Formula (2) in the solution.

$$Me(OR)n + nH_2O \rightarrow Me(OH)n + nROH \quad (1)$$

Here, Me refers to the metal group of Si, Ti, Al, Ba etc., R refers to the alkyl group of CH_3, C_2H_5,

Table 1. Compositions of coating solution (wt.%).

	Distilled water	EtOH	SiO_2 polysilicate	Urethane resin	Melamine
U	53	28.0	–	21	–
US_7	49.5	22.5	7	21	–
US_3M_3	50	23.0	3	21	3
US_7M_3	48	21.0	7	21	3
$US_{11}M_3$	46	19.0	11	21	3
US_7M_1	49	22.0	7	21	1
US_7M_5	47	20.0	7	21	5

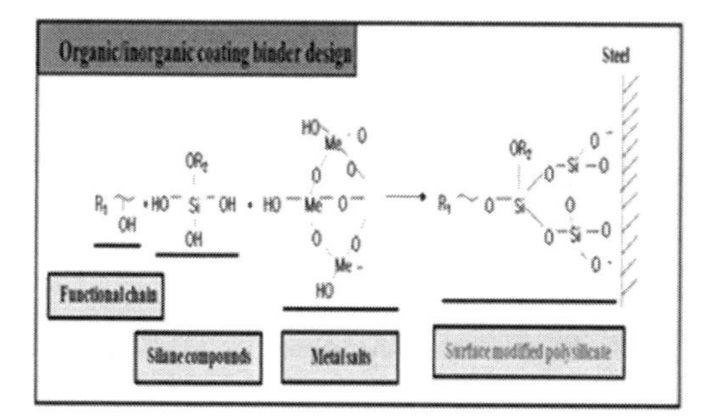

Figure 1. Organic/inorganic coating binder design.

C_3H_7 etc., and n represents the oxidation number of the metal.

$$Me(OH)n \rightarrow MeOn/2 + n/2H_2O \quad (2)$$

Polycondensation occurs with Me(OH)n from Formula (2). It forms oxide fine particles of -Me-O-Me-O- in the solution, and becomes a dense film after drying.

The specimens were made with cold-rolled steel sheet typically used for automobiles, named SPCC and SPFC590, produced by POSCO. The thickness of the material is 1 mm. Tables 2 and 3 show the mechanical properties and chemical compositions of the material used.

Salt spray specimens were made according to KS D9502, and were treated with taping to prevent corrosion on the edges. The coating was carried out using the bar-coater No. 3 (wet film thickness; 6.86 mm). The hardening treatment was performed for 1, 2, and 3 minutes at 463 K. (Seo et al. 2010a).

The specimens were tilted at about 45° in the salt water spay tester, and neutral salt water spray tests were conducted at 35 ± 2°C for 7 hours. The specimen was observed at 1 hour intervals. The evaluation of corrosion resistance of the test specimens was assessed according to the time at which the initial white rust and the area of white rust appeared, as captured by photography. Figure 2 shows a flowchart of the salt water spray test.

Table 2. Mechanical properties.

	σ_y (MPa)	σ_U (MPa)	ε (%)
SPCC	167.7	306.0	47
SPFC590	512.9	640.4	25

Table 3. Chemical compositions (%).

	C	Si	Mn	P	S
SPCC	0.0023	0.003	0.1	0.015	0.005
SPFC590	0.0836	0.185	1.863	0.02	0.005

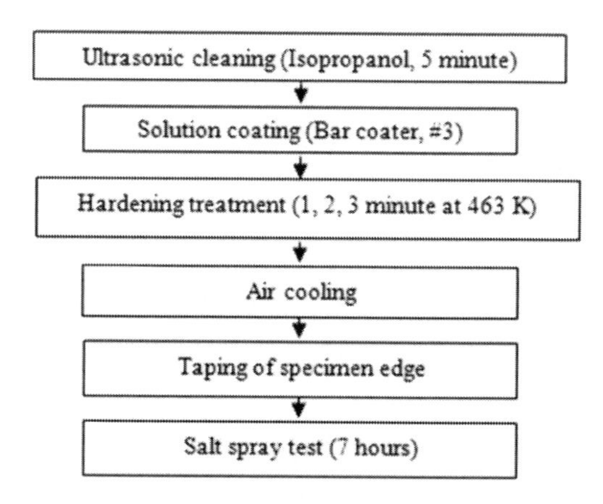

Figure 2. Flow chart of salt spray test.

The coating adhesion was evaluated by placing 3M tape on the cross-cuts at 1 mm intervals on an area of 10 × 10 mm. The coating adhesion was carried out according to the ASTM D3359-09 method. After immersion for 1 hour in boiling water, the boiling water resistance was observed for the surface. With rubbing, changes in the color of the surface were observed when rubbing 30 times using a cotton stick under wet conditions. Cracking and peeling of the coating layer were observed on the surface after 180° bending. These properties were evaluated with a hardening time of 3 minutes after coating.

3 RESULTS AND DISCUSSION

The appearances of the SPCC and SPFC590 specimens after a salt spray test of 7 hours are shown in Figure 3. Figure 3(a) shows the U solution,

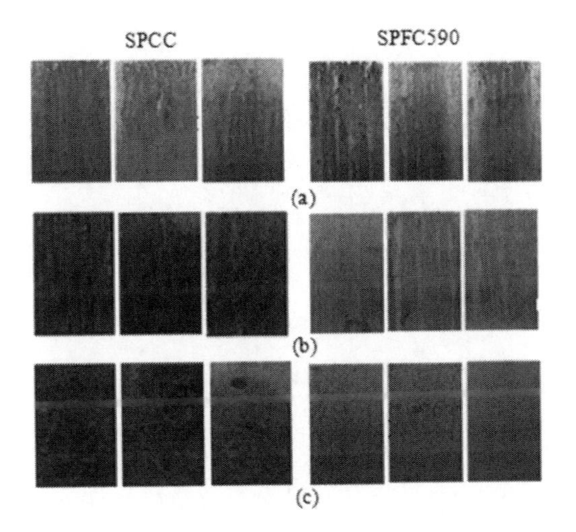

Figure 3. Typical appearance after salt spray test in SPCC and SPFC590. (a) U, (b) US_7, (c) US_7M_3.

Table 4. Corrosion area rate (%) in salt spray time of 7 hours.

Speci.	Solution \ Ht	1 min	2 min	3 min
SPCC	U	71.2	68.5	57.1
	US_7	93.8	89.5	78.7
	US_3M_3	94.3	73.3	39.5
	US_7M_3	93.7	76.1	7.7
	$US_{11}M_3$	90.9	87.5	36.6
	US_7M_1	36.7	35.2	33.2
	US_7M_5	46.6	44.8	13.3
SPFC 590	U	78.8	72	66.6
	US_7	68.3	64.2	63.2
	US_3M_3	83.7	60.6	24.4
	US_7M_3	21.0	18.7	1.5
	$US_{11}M_3$	80.7	76.6	24
	US_7M_1	35.6	31.0	22.7
	US_7M_5	42.3	31.6	14.8

Figure 3(b) shows the US_7 solution, and Figure 3(c) shows the US_7M_3 solution. Corrosion appeared to reduce with the increase of hardening time, regardless of the type of specimen and solution.

Table 4 shows the corrosion area rate in salt spray after 7 hours according to the hardening time of each solution. Here, Ht refers to the hardening time.

Figures 4(a) and 4(b) show the results of Table 4. It can be seen that the corrosion area rate decreases according to the increase in the hardening time. In Figure 4(a) and 4(b), US7M_3 shows the superior corrosion resistance at 3 minutes of hardening time. In this case, the corrosion area rate is 7.7% and 1.5%, for the SPCC and SPFC590 specimens, respectively. Therefore, US_7M_3 is the best solution.

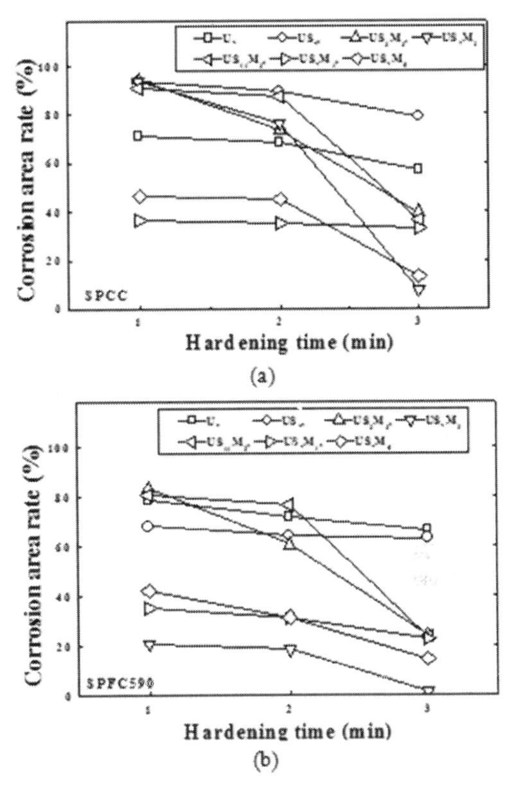

Figure 4. Results of salt spray test of (a) SPCC, (b) SPFC590.

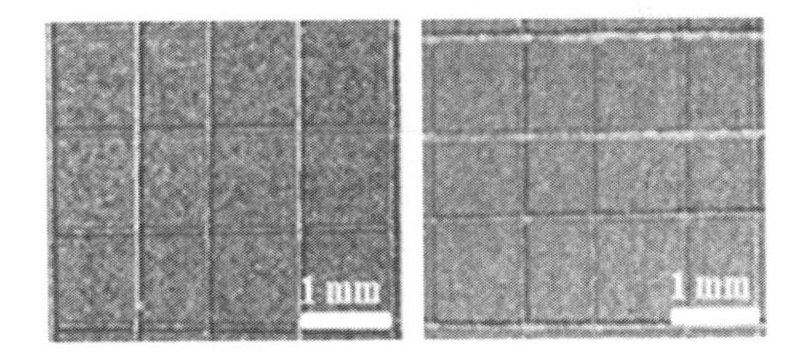

Figure 5. Results of adhesion test from specimen with coating of US_7M_3 solution. (a) SPCC, (b) SPFC590.

Figure 5 shows the typical surface after the cross-cutting test of the specimens with coating of US_7M_3 solution. The specimens were no peeling, and showed superior adhesion.

At boiling water temperature, the stability of the coating layer is very important. Figure 6 shows the typical results of the stability test and rubbing test of the specimens. The surfaces of both specimens did not change.

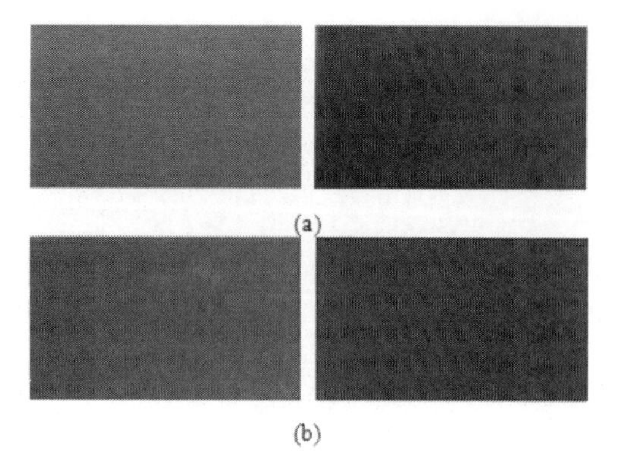

Figure 6. Results of boiling and rubbing test from specimen with coating of US_7M_3 solution. (a) SPCC, (b) SPFC590.

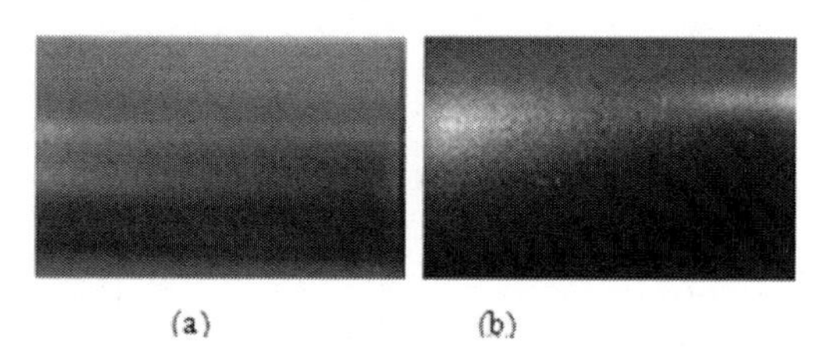

Figure 7. Results of 180° bending test from specimen with coating of US_7M_3 solution. (a) SPCC, (b) SPFC590.

The typical results of the bending test at 180 degrees are shown in Figure 7. In the seven types of solution, the occurrence of cracks and peeling on the coating surface was not observed.

Figure 8 show SEM photography of coating surface. Figure 8(a) shows the U solution, Figure 8(b) shows the US_7 solution, and Figure 8(c) shows the US_7M_3 solution. Figure 8(a) was made of a large crack by the shrinkage. Figure 8(b) was formed a lot of fine crack by the addition of SiO_2 polysilicate. But in the Figure 8(c), a melamine curing agent played a cross-linking role in the synthesis of SiO_2 polysilicate and urethane resin. So cracks did not occur at all. US_7M_3 is the optimal composition ratio in the range of solutions in this study.

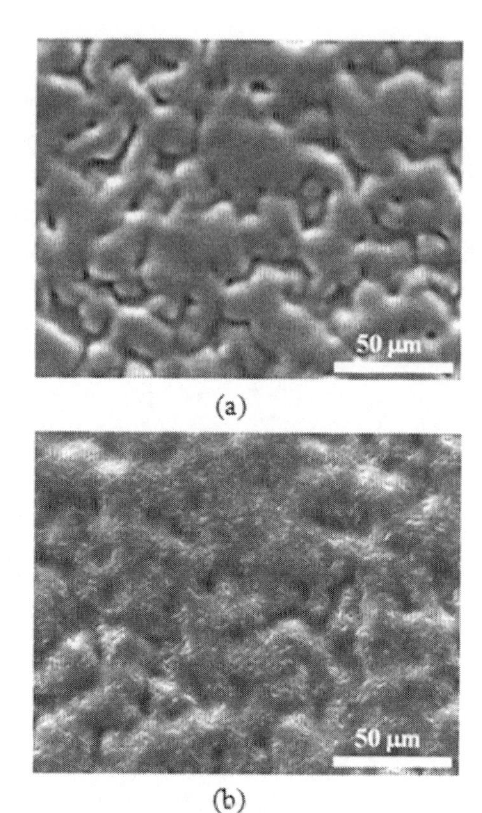

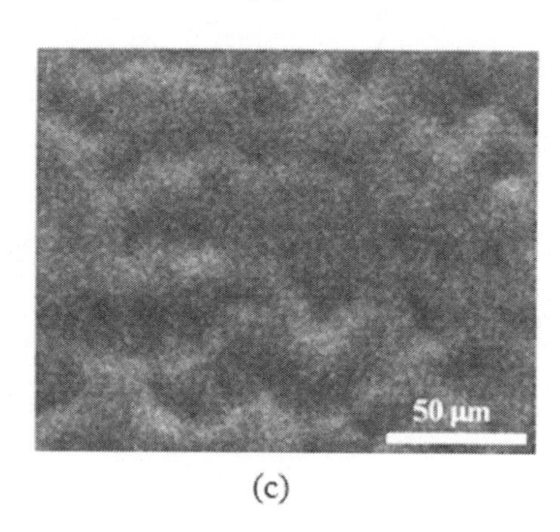

Figure 8. SEM photography of coating surface with hardening time of 3 minutes. (a) U solution, (b) US_7 solution, (c) US_7M_3 solution.

4 CONCLUSIONS

Seven types of organic/inorganic hybrid solutions were coated on two specimens of cold rolled steel used for automobiles, which were named the SPCC and SPFC590 specimens. The characteristics of the corrosion resistance and coating solution were evaluated. The organic/inorganic hybrid coating solution containing the Si component showed excellent corrosion resistance for the steel sheet for automobiles. The seven types of solutions also showed excellent cracking and peeling resistance after cross-cutting, and excellent boiling water, rubbing, and bending resistance.

ACKNOWLEDGEMENT

This work was supported by the Small and Medium Business Administration for 2015 Installation Business of Research Institute (C0268714).

REFERENCES

Bajat, J.B., Mišković-Stanković, V.B., Bibić, N. & Dražić, D.M. 2007. The influence of zinc surface pretreatment on the adhesion of epoxy coating electrodeposited on hot-dip galvanized steel. *Progress in Organic Coating*, 58: 323–330.

Nam, K.W. Development of Cr-free hybrid solution according to composition of organic/inorganic. (unpubl.)

Seo, H.S., Moon, H.J., Kim, J.S., Ahn, S.H., Moon, C.K. & Nam, K.W. 2010a. Corrosion resistance according to the heat treatment temperature of Cr-free coating solution on zinc coated steel. *Journal of Ocean Engineering and Technology*, 24(5): 60–66.

Seo, H.S., Moon, H.J., Kim, J.S., Ahn, S.H., Moon, C.K. & Nam, K.W. 2010b. Corrosion resistance of zinc coating steel coated Cr-free coating solution according to the heat treatment time. *Journal of Ocean Engineering and Technology*, 24(5): 67–74.

Seo, H.S., Moon, H.J., Kim, J.S., Ahn, S.H., Moon, C.K. & Nam, K.W. 2010c. Corrosion resistance of galvanized iron by treating modified Si organic/inorganic hybrid coating solution. *Journal of Ocean Engineering and Technology*, 25(1): 32–38.

Stven, J.H., Lowe, C., James, T.M. & John, F.W. 2005. Migration and segregation phenomena of a silicone additive in a multiyear organic coating. *Progress in Organic Coatings*, 54: 104–112.

Wang, D. & Gordon. P.B. 2009. Sol–gel coatings on metals for corrosion. *Progress in Organic Coatings*, 64: 327–338.

Zheludkevich, M.L., Miranda, S.I. & Ferreira, M.G. 2005. Sol-gel coatings for corrosion protection of metals. *Journal of Maters Chemistry*, 15: 5099–5111.

Advanced Materials, Mechanical and Structural Engineering – Hong, Seo & Moon (Eds)
 ISBN: 978-1-138-02908-8

Development of a robot machining program using tool center point-based transformation

C.H. She & J.J. Huang
Department of Mechanical and Computer Aided Engineering, National Formosa University, Huwei, Yunlin, Taiwan

ABSTRACT: Robot is the science and technology in a variety of equipment consisting of mechanical, electronic, control, computers, sensors, and artificial intelligence, and is one of the elements for the automation of production. The required robot numerical control program can be converted by the commercial software. However, such software is expensive. This study develops a robot machining program algorithm using tool center point-based transformation. The Staubli industrial six-axis robotic configuration is investigated in this research. The robotic manipulator's orientation can be obtained by the coordinate transformation matrix. A window-based transformation software dedicated to the Staubli industrial robot is developed to generate the corresponding Euler angle. The results indicate that the proposed algorithm can be applied to robot offline programming and practical machining.

1 INTRODUCTION

Robot consists of many technologies including mechanical, electronic, control, computers, sensors, and artificial intelligence, and is one of the important elements for the automation of production. According to the definition in ISO 8373 (2012), industrial robot is an automatically controlled, reprogrammable, multipurpose manipulator programmable in three or more axes, which may be either fixed in place or mobile for use in industrial automation applications. To control the end effector of the robot accurately, motion programming of the robot is necessary.

Traditionally, teach and repeat mode is sufficient for industrial robot programming in a wide variety of tasks, especially in which a robot will perform a simple task such as spot welding. However, this mode is only useful for the point-to-point path in which the robot moves to these points from a pose to a pose. Programming a robot by teaching mode is time-consuming, particularly with increasing complexity of the task.

In the case of the continuous path, the robot must follow an exact path such as arc welding, flame cutting, and deburring (Angeles et al. 1988, Valente & Oliveira 2004). There are many commercial robot software systems such as SprutCAM (2015), Robotmaster (2008), which can provide the off-line programming. With the aid of software systems, the robot set-up time can be reduced so that the robot can still be in production while its next task is being programmed. However, these dedicated software systems are expensive. Many manufacturers use the traditional CAD/CAM software and CNC machine tools for product design and manufacturing of parts, and their related technical ability is well developed. The required CNC program should be converted by the expensive commercial software while importing the robotic manipulator. It will not only save considerable cost of inputs, but also integrate with the existing CAD/CAM software if they can apply the existing CAD/CAM modules and theoretical basis in the field of robotics to carry out the path programming. This study develops a Tool Center Point (TCP)-based robotic NC program converter by the coordinate transformation matrix and inverse kinematics. It can integrate the traditional CAD/CAM module and is suitable for a commercial six-axis robotic manipulator.

2 ROBOT COORDINATE TRANSFORMATION

2.1 *Coordinate definition*

In order to specify the pose of the robot's end effector, the coordinate system should be established. Joint and Cartesian coordinate systems are often used for robot programming. Figure 1 shows the Staubli robot coordinate transformation system. In the joint coordinate system, the rotational axes are specified for each robot axis. In this study, the six-axis robotic manipulator for Staubli (2010) is used as an example and the six robot joint angles (J1, J2, J3, J4, J5, J6) should be entered to define the end effector's position and orientation

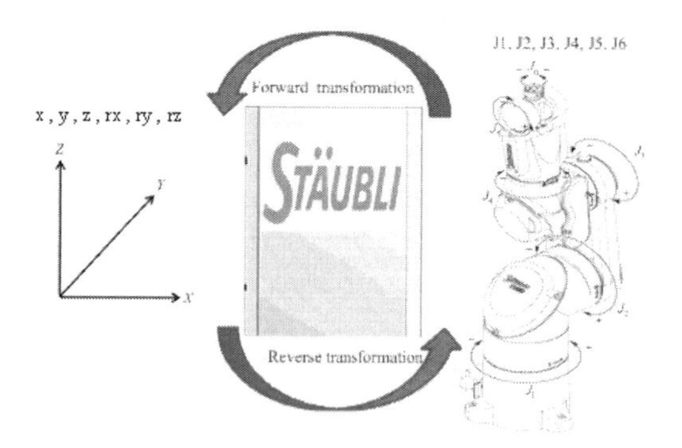

Figure 1. Staubli robot coordinate transformation system.

correctly. The joint coordinate system is used in the point-to-point path. The Cartesian coordinate system should be employed in the continuous path. Typically, the Tool Center Point (TCP) data including the position (x, y, and z) and orientation (rx, ry, and rz) of the end effector are used when referring to the robots. The TCP will be jogged or moved to the programmed target position. The tool center point also constitutes the origin of the tool coordinate system. In this study, the TCP-based robotic NC program is developed.

2.2 *Cutter location matrix determination*

Most of the current CAD/CAM systems can produce the cutter location data including the cutter tip location p and tool the axis a, as shown in Figure 2. To represent the robot pose using the cutter location data, two mutually orthogonal unit vectors o and n should be determined. The following equations are obtained by the definition of orthogonal unit vector:

$$a_x n_x + a_y n_y + a_z n_z = 0 \tag{1}$$

$$n_x^2 + n_y^2 + n_z^2 = 1 \tag{2}$$

Since there are only two equations to determine three unknown parameters (n_x, n_y, n_z), infinite possible solutions can be obtained. For convenience, we can assume that $n_z = a_x$. Therefore, n_x and n_y can be determined as follows:

$$n_x = \frac{-a_y n_y - a_z a_x}{a_x} \tag{3}$$

$$n_y = \frac{-a_x a_y a_z \pm \sqrt{a_x^4 - a_x^6 + a_x^2 a_y^2 - a_x^4 a_y^2 - a_x^4 a_z^2}}{a_x^2 + a_y^2} \tag{4}$$

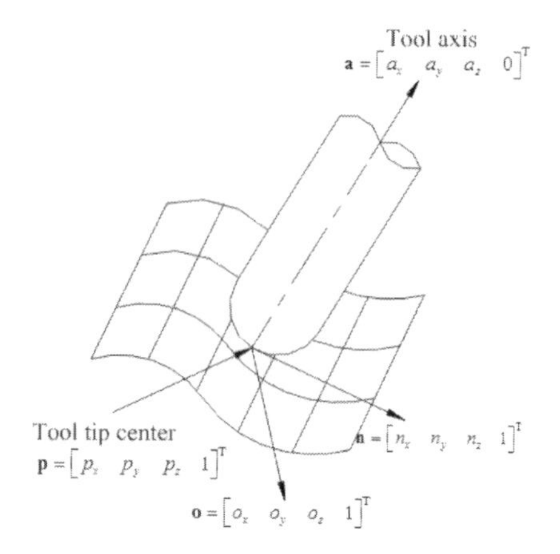

Figure 2. Geometric definition of cutter location data.

Once the two vectors a and n are obtained, the vector o can be derived by the right hand rule and expressed as follows:

$$o_x = a_y n_z - a_z n_y \tag{5}$$

$$o_y = a_z n_x - a_x n_z \tag{6}$$

$$o_z = a_x n_y - a_y n_x \tag{7}$$

The cutter location matrix represented by [n o a p] will be used to determine the pose of the robot, where [n o a] is the orientation and p is the TCP position.

2.3 *Euler angle transformation*

Manipulating objects in space by the robot requires the specification of the mutual position and orientation between the robot's end effector and an object. The position of the end effector is described by the Cartesian coordinate of the tool center point, while its orientation can be expressed in various ways. Euler angles are one of the several ways to specify the relative orientation of two coordinate systems. There are twelve possible conventions of Euler angles, in which different definitions are used by various robot manufacturers. For example, Z-Y'-X" is for KUKA (2010) while Z-Y'-Z" is adopted by Kawasaki (2001). In this study, without loss of generality, the X-Y'-Z" definition of Staubli is used, as shown in Figure 3.

When using Euler angles, it is assumed that coordinate transformation is performed by rotation around the origin of the base coordinate system. The X-Y'-Z" convention means that the transformation first involves a rotation by the angle ϕ about the X axis, then by the angle θ about the Y

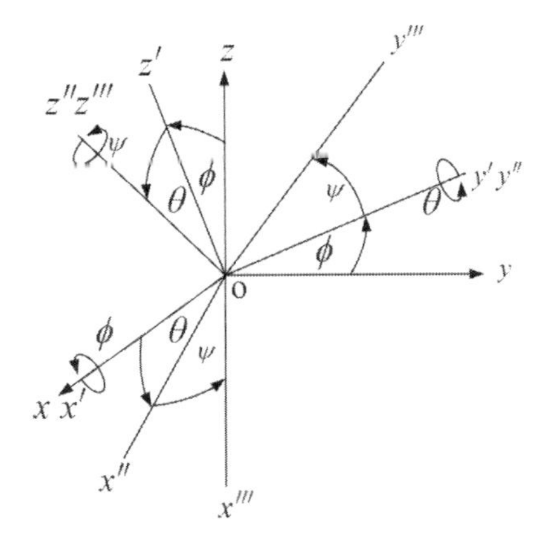

Figure 3. Euler angle coordinate system transformation.

axis, and finally by the angle ψ about the Z axis. Using a 4 × 4 homogeneous coordinate transformation matrix, the spatial transformation between the tool coordinate system and the robot base coordinate system can be expressed as follows:

$$Euler(\phi, \theta, \psi) = \mathrm{R}ot(x, \phi)\,\mathrm{R}ot(y, \theta)\,\mathrm{R}ot(z, \psi)$$
$$= \begin{bmatrix} C\theta C\psi & -C\theta S\psi & S\theta & 0 \\ C\phi S\psi + S\phi S\theta C\psi & C\phi C\psi - S\phi S\theta S\psi & -S\phi C\theta & 0 \\ S\phi S\psi - C\phi S\theta C\psi & S\phi C\psi + C\phi S\theta S\psi & C\phi C\theta & 0 \\ 0 & 0 & 0 & 1 \end{bmatrix} \quad (8)$$

where "C" and "S" refer to cosine and sine functions, respectively.

It is worth mentioning that determination of Euler angles from the transformation matrix is necessary in this transformation. Let us assume that a cutter orientation matrix has the following form:

$$\mathbf{T} = \begin{bmatrix} n_x & o_x & a_x & 0 \\ n_y & o_y & a_y & 0 \\ n_z & o_z & a_z & 0 \\ 0 & 0 & 0 & 1 \end{bmatrix} \quad (9)$$

Equating Equation (8) and Equation (9), and taking the corresponding elements of the two matrices, the following equations can be obtained:

$$n_x = \mathrm{C}\theta\mathrm{C}\psi \quad (10)$$

$$n_y = \mathrm{C}\phi\mathrm{S}\psi + \mathrm{S}\phi\mathrm{S}\theta\mathrm{C}\psi \quad (11)$$

$$n_z = \mathrm{S}\phi\mathrm{S}\psi - \mathrm{C}\phi\mathrm{S}\theta\mathrm{C}\psi \quad (12)$$

$$o_x = -\mathrm{C}\theta\mathrm{S}\psi \quad (13)$$

$$o_y = \mathrm{C}\phi\mathrm{C}\psi - \mathrm{S}\phi\mathrm{S}\theta\mathrm{S}\psi \quad (14)$$

$$o_z = \mathrm{S}\phi\mathrm{C}\psi + \mathrm{C}\phi S\theta S\psi \quad (15)$$

$$a_x = \mathrm{S}\theta \quad (16)$$

$$a_y = -\mathrm{S}\phi\mathrm{C}\theta \quad (17)$$

$$a_z = \mathrm{C}\phi\mathrm{C}\theta \quad (18)$$

Three Euler angles can be obtained by inverse kinematics and two possible solutions will result in the same orientation of transformation. The solutions are determined as follows:

$$\begin{bmatrix} \phi_1 \\ \theta_1 \\ \psi_1 \end{bmatrix} = \begin{bmatrix} \mathrm{atan2}\left(\dfrac{-a_y}{\mathrm{C}\theta_1}, \dfrac{a_z}{\mathrm{C}\theta_1}\right) \\ \mathrm{asin}(a_x) \\ \mathrm{atan2}\left(\dfrac{-o_x}{\mathrm{C}\theta_1}, \dfrac{n_x}{\mathrm{C}\theta_1}\right) \end{bmatrix} \quad (19)$$

$$\begin{bmatrix} \phi_2 \\ \theta_2 \\ \psi_2 \end{bmatrix} = \begin{bmatrix} a\tan 2\left(\dfrac{-a_y}{C\theta_2}, \dfrac{a_z}{C\theta_2}\right) \\ \pi - a\sin(a_x) \\ a\tan 2\left(\dfrac{-o_x}{C\theta_2}, \dfrac{n_x}{C\theta_2}\right) \end{bmatrix} \quad (20)$$

where atan2(y, x) is the function that returns angles in the range of $-\pi \leq \theta < \pi$ by examining the signs of both y and x (Paul 1981).

3 RESULTS AND DISCUSSION

3.1 *System implementation*

To validate the effectiveness of the developed algorithm, a window-based transformation software is developed using the Borland C++ Builder program. Figure 4 shows the snapshot of the system initiation dialogue where two transformation functions (Euler angle and cutter location) can be employed.

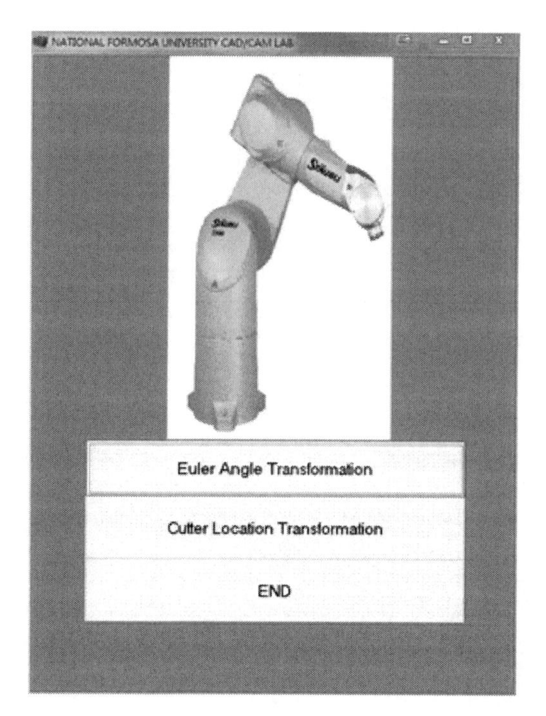

Figure 4. Initiation dialog of the developed software.

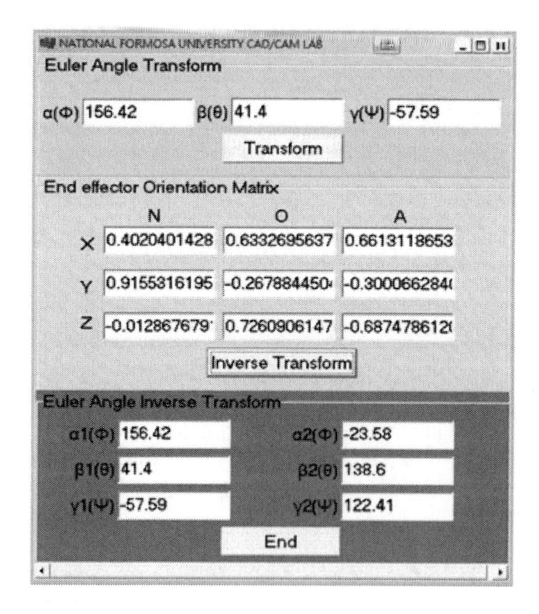

Figure 5. End effector orientation matrix obtained by Euler angle transformation.

First, the Euler angle transformation is verified. As shown in Figure 5, three Euler angles (156.42°, 41.4°, and −57.59°) are entered and the corresponding end effector orientation matrix is gener-

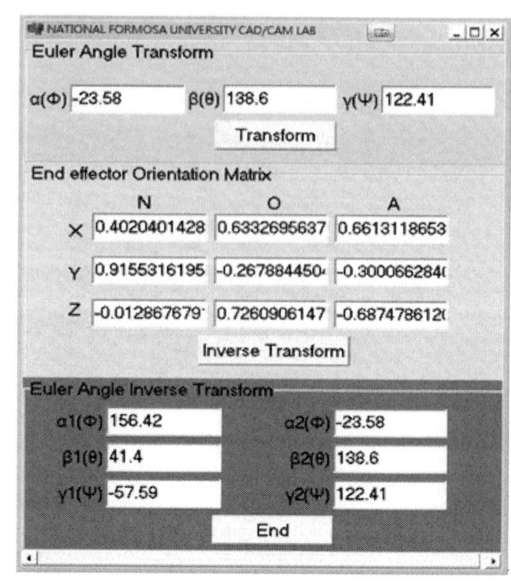

Figure 6. End effector orientation matrix verified by the other possible Euler angle transformation.

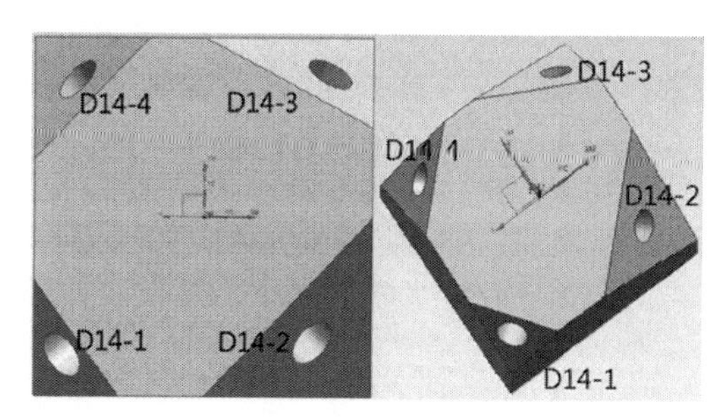

Figure 7. Example of a geometric model for drilling operation.

ated. Two possible Euler angles can be obtained by clicking the "Inverse Transform" button according to Equation (19) and Equation (20). It should be noted that Euler angles (−23.58°, 138.6°, and 122.41°) are the other possible solution and can be entered again in the software, as shown in Figure 6, in which the same end effector orientation matrix is obtained.

3.2 *Machining example*

A drilling operation is taken as an example where four holes on the oblique plane are drilled with a 14 mm drill, as shown in Figure 7. The cutter location file for each hole is generated by the NX CAM software. Figure 8 shows the cutter location trans-

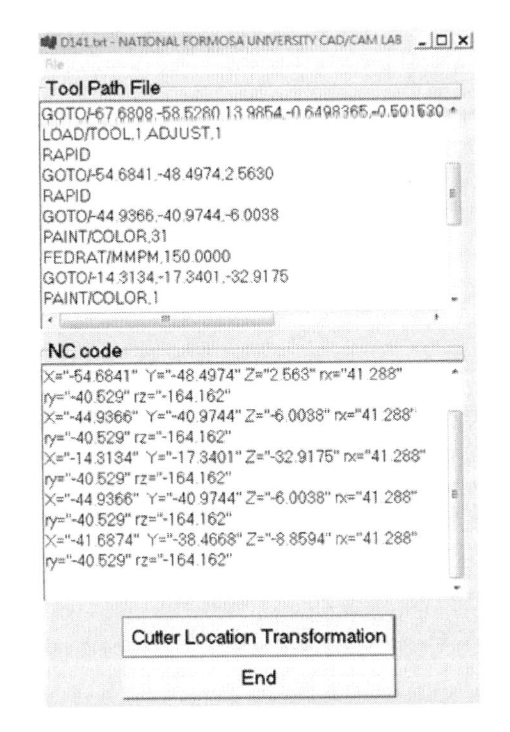

Figure 8. Robot NC code generated by the developed software.

formation process and the robot program of the D14-1 hole with TCP position (x, y, and z), and Euler angles (rx, ry, and rz) can be obtained for further machining application.

4 CONCLUSION

With the continuous development of industrial robot technology and expanding applications, robotics plays an important role in automated production. This paper developed a robot offline programming method instead of using the expensive commercial dedicated robotic software. The cutter location file generated by the traditional CAD/CAM software can be converted to the robotic program with TCP position and Euler angles. The result of this study can effectively enhance the added value of the existing CAD/CAM software modules and break through the bottleneck of the NC programming core technology while importing the robotic manipulator.

ACKNOWLEDGMENT

The authors gratefully acknowledge the financial support of the National Science Council of Taiwan under grant NSC 99-2221-E-150-040.

REFERENCES

Angeles, J., Rojas, A. & Lopez-Cajun, C.S. 1988. Trajectory planning in robotics-path applications. *IEEE Journal of Robotics and Automation,* 4(4): 380–385.

ISO 8373: 2012. *Robots and robotic devices—Vocabulary.*

Kawasaki, 2001. Robot C Series Controller Operation Manual.

Kuka. 2010. *System Software: Operating and Programming Instructions for System Integrators.*

Paul, R.P. 1981. Robot *Manipulators: Mathematics, Programming and Control.* Cambridge: MIT press.

Robotmaster 2008. *Tutorial Series Version 3.0.1200.1.*

SprutCAM 2015. http://www.sprutcam.com.

Staubli 2010. VAL3 reference Manual.

Valente, C. & Oliveira, J. 2004. A new approach for tool path control in robotic deburring operations. *ABCM Symposium Series in Mechatronics,* 1: 124–133.

Advanced Materials, Mechanical and Structural Engineering – Hong, Seo & Moon (Eds)
 ISBN: 978-1-138-02908-8

The noise prediction of structures with an adaptive acoustical boundary element integral method

Y.Y. Miao, T.Y. Li, X. Zhu & W.J. Guo
School of Naval Architecture and Ocean Engineering, Huazhong University of Science and Technology, Wuhan, Hubei, China

ABSTRACT: The acoustical boundary element method is a widespread tool for noise prediction, and the near-field sound pressure and pressure gradient around the vibrating structures are necessary in some domains of science and engineering. However, the nearly singularity will arise when the field point is very close to the structure boundary, which leads to poor precision with the standard gauss integral. In this paper, an adaptive boundary element integral method and a criterion are presented for this problem. The adaptive integral method is implemented by dividing the element into subelements according to the criterion. The convergence analysis and example show that the adaptive boundary element integral method is accurate and efficient.

1 INTRODUCTION

Over the past few decades noise pollution has emerged as an important issue in modern society and provides much of the impetus for the development of noise prediction and reduction techniques. After decades of development, the acoustical boundary element method has become an important method for problems on structural vibration and sound radiation. There are still some problems in the boundary element method worthy of further research, including the nearly singular integral, and so on.

A general regularization algorithm of evaluating the physical quantities at points very near the boundary was proposed (Zhou et al. 2003), in which the integral factors leading to singularities were transformed out of integral symbols with integral by parts. Hence, the nearly strongly singular and nearly hyper-singular singular can be semi-analytically evaluated. Furthermore, an analytical integral algorithm for two-dimensional elasticity problems was proposed (Niu et al. 2007), and this algorithm was extended to treat 2D orthotropic potential problems (Zhou et al. 2007) and 2D anisotropic potential problems (Zhou et al. 2008). Wei et al. (2012) combined the finite element method and the boundary element method to analyze the noise of underwater vehicles, in which the singular integral was calculated by dividing the element into subelements averagely. Croaker et al. (2015) predicted the scattering of flow-induced noise when a body immersed in the flow, and the singular volume and surface integrals were evaluated by the adaptive subelements method. Gao et al. (2008) combined the adaptive subelements method and the analytical method to calculate kinds of 2D singular integrals, and precise results were achieved. The subdivision process was based on the empirical formula, and the subelements sizes were determined by iteration (Gao & Davies. 2000). This adaptive subelements method and empirical formula were also adopted by Li (Li & Ye 2010), Wang (Wang & Gao 2010) and Qu (Qu et al. 2014).

In this paper, an adaptive boundary element integral method is presented. The element is subdivided hierarchically, and the number of subdivision is decided based on a criterion.

2 THE ACOUSTICAL BOUNDARY INTEGRAL EQUATION

The Helmholtz boundary integral equation for acoustical radiation problems in free space is:

$$C(z)p(z)=\int_S\left[G(z,y)\frac{\partial p(y)}{\partial n}-p(y)\frac{\partial G(z,y)}{\partial n}\right]\mathrm{d}S \tag{1}$$

where, $G(z,y)=e^{-ikr}/(4\pi r)$ is the fundamental solution of the boundary integral equation, S is the smooth structure boundary, k is the wave number, z is the field point, y is the source point, $r=|z-y|$ denotes the distance between z and y, $p(z)$ denotes the field point sound pressure, $p(y)$ denotes the source point sound pressure, $C(z)$ denotes the coefficient of $p(z)$. When z is in S, $C(z) = 0$; z is out of S, $C(z) = 1$; z is on S, $C(z) = 0.5$. $\partial p(y)/\partial n = -i\rho\omega v$, n is the inward normal unit vector at y, v is the element normal velocity, ρ is the medium density, ω is the circular frequency.

When the field point is close to but not on the structural boundary, r is not perpendicular to n. Hence G and its derivative are not zero, making the weakly singularity and strong singularity coexisting. In fact, the weakly singularity is much easier to be removed than the strong singularity with the adaptive boundary element integral method, so the focus of this paper is on the strong singularity.

3 THE ADAPTIVE SUBDIVISION METHOD

In this paper, the adaptive subdivision scheme and convergence analysis are based on the standard element (the vertex coordinates are [−1,−1], [1,−1], [1,1] and [−1,1]), for elements are mapped to the standard element by isoparametric transformation in the 2D gauss integral. The length unit is meter which is omitted in this article. As subpoints of the near field points may locate inside, at the vertexes and edges of the element respectively, various adaptive subdivision schemes are shown in Figure 1.

The black points in Figure 1 are subpoints of the near field points. The row (1) is the average subdivision scheme stemming from the work of Wei et al. (2012), and the rows (2)~(5) are the hierarchical subdivision schemes will be presented as the topic.

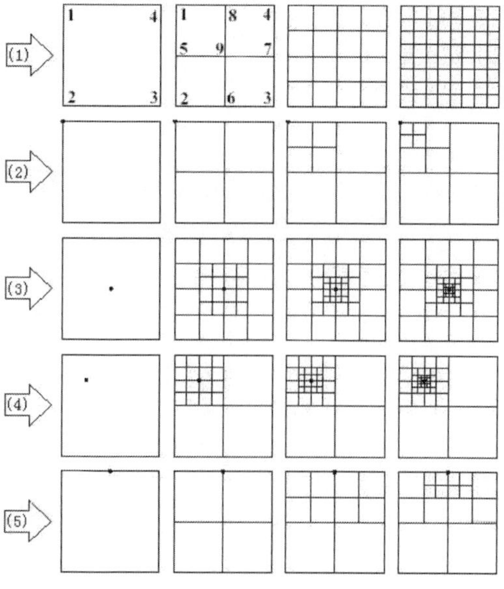

Figure 1. The adaptive subdivision schemes.

The initial singular boundary element is divided into temporary subelements, so the integral on the initial element is transformed to Gauss integral on subelements. The subelements can further be divided into smaller elements until integral converges at an allowable tolerance. The average subdivision scheme is carried out by dividing the initial element into four subelements, and every subelement is divided into four subelements again until the integral is converged. The hierarchical subdivision scheme is carried out by dividing the initial element into four subelements, and only the subelements in which the subpoint of the near field point is located needs subdivision, while other subelements remain unchanged.

After subdivision, the strong singular integral on the initial element becomes the summation of the integrals on subelements:

$$\begin{aligned} IR &= \int_s \frac{\partial G}{\partial n}\mathrm{d}s = \sum_{j=1}^{N}\int_{s_j}\frac{\partial G}{\partial n}\mathrm{d}s \\ &= \sum_{j=1}^{N}\int_{-1}^{1}\int_{-1}^{1}\frac{\partial G}{\partial n}J_j\mathrm{d}s_1\mathrm{d}s_2 \\ &= \sum_{j=1}^{N}\sum_{u=1}^{Ng}\sum_{v=1}^{Ng} w_u w_v\frac{\partial G}{\partial n}J_j \end{aligned} \tag{2}$$

where, s is the initial element, s_j is the subelement, N is the number of subelements, Ng is the number of gauss points, J_j is the Jacobian of subelements, wu and wv are weight coefficients.

The numbers of subelements corresponding to Figure 1 are:

$$\begin{cases} N_1 = 4^{num} \\ N_2 = 3num+1 \\ N_3 = \begin{cases} 4^{num}, num=1,2 \\ 12num-8, num \ge 3 \end{cases} \\ N_4 = \begin{cases} 4, num=1 \\ 12num-17, num \ge 2 \end{cases} \\ N_5 = 6num-2 \end{cases} \tag{3}$$

where, num is the number of subdivisions.

Take the x-coordinate as the example to explain the adaptive boundary element integral method. For the average subdivision scheme, the coordinate relations between the initial element and the subelements are:

$$\begin{cases}[x_1 \quad x_5 \quad x_9 \quad x_8]^{\mathrm{T}} = T_1 X_0 \\ [x_5 \quad x_2 \quad x_6 \quad x_9]^{\mathrm{T}} = T_2 X_0 \\ [x_9 \quad x_6 \quad x_3 \quad x_7]^{\mathrm{T}} = T_3 X_0 \\ [x_5 \quad x_2 \quad x_6 \quad x_9]^{\mathrm{T}} = T_4 X_0 \end{cases} \tag{4}$$

where, $T_1 \sim T_4$ are coordinate transfer matrices, and X_0 is the x-coordinate of the initial element:

$$\begin{cases} T_1 = \begin{bmatrix} 1 & 0 & 0 & 0 \\ 1/2 & 1/2 & 0 & 0 \\ 1/4 & 1/4 & 1/4 & 1/4 \\ 1/2 & 0 & 0 & 1/2 \end{bmatrix} \\ T_2 = \begin{bmatrix} 1/2 & 1/2 & 0 & 0 \\ 0 & 1 & 0 & 0 \\ 0 & 1/2 & 1/2 & 0 \\ 1/4 & 1/4 & 1/4 & 1/4 \end{bmatrix} \\ T_3 = \begin{bmatrix} 1/4 & 1/4 & 1/4 & 1/4 \\ 0 & 1/2 & 1/2 & 0 \\ 0 & 0 & 1 & 0 \\ 0 & 0 & 1/2 & 1/2 \end{bmatrix} \\ T_4 = \begin{bmatrix} 1/2 & 0 & 0 & 1/2 \\ 1/4 & 1/4 & 1/4 & 1/4 \\ 0 & 0 & 1/2 & 1/2 \\ 0 & 0 & 0 & 1 \end{bmatrix} \\ X_0 = [x_1 \quad x_2 \quad x_3 \quad x_4]^T \end{cases} \tag{5}$$

As the adaptive subdivision process goes on, the coordinates of subelements for *num* subdivision are:

$$\prod_{t=1}^{num} T_{mt} X_0 \tag{6}$$

where, $Tmt = Tm$ and $m = 1 \sim 4$.

For the hierarchical subdivision scheme, only the subelements in which the subpoint of the near field point is located needs subdivision, hence the coordinates of subelements according to the row (2) in Figure 1 are:

$$T_1^{num} X_0, T_2 T_1^{num-1} X_0, T_3 T_1^{num-1} X_0, T_4 T_1^{num-1} X_0 \tag{7}$$

The adaptive subdivision processes for rows (3)~(5) are similar to the row (2), the coordinate relations between the initial element and the subelements are referred to the Equation 7.

4 THE CRITERION AND CONVERGENCE ANALYSIS

As the adaptive subdivision is around the subpoint, actually the location relationship between the subpoint and the minimum subelements is referred to the row (2) in Figure 1. Therefore, *num* can be determined in a numerical way.

Assume that the quadric strong singular integral is:

$$Ir = \int_s \frac{1}{r^2} \frac{\partial r}{\partial n} \mathrm{d}s \tag{8}$$

Set the near field point exactly above the vertex of the standard element, making the vertex as the subpoint. As the near field point gets close to the subpoint, the convergence of Ir is investigated with the adaptive subdivision method. If Ir is convergent within three subdivisions and the error of the convergent integral versus the integral without subdivision is no more than 1%, then the distance between the near field point and the element is called as the critical distance U. When the distance is longer than U, the integral on the element is not singular and the field point is not near field point, either. $E = U/l$ is the critical proximity, where l is the average edge length of the element. For the standard element, U is 0.08 and E is 0.04 with 10 points gauss integral, accordingly the criterion for subdivision is:

$$H / (l / 2^{num}) \geq 0.04 \tag{9}$$

where, H is the distance between the arbitrary near field point and the element.

For the standard element, $k = 1$ rad/m, the coordinates of four near field points are [0, 0, 0.01], [1, 1, 0.01], [0, 0, 0.001] and [1, 1, 0.001], i.e. H is 0.01 and 0.001; the strong singular integrals are *IR1*, *IR2*, *IR3* and *IR4*; the errors of integrals for each subdivision versus the convergent integrals are *er1*, *er2*, *er3* and *er4*, respectively. The convergence curves with 10 points gauss integral are shown in Figure 2. In the same way, for two rectangle elements (the width is 2 and the lengths are 4 and 8), H is set as 0.01, the subpoints of four near field points are

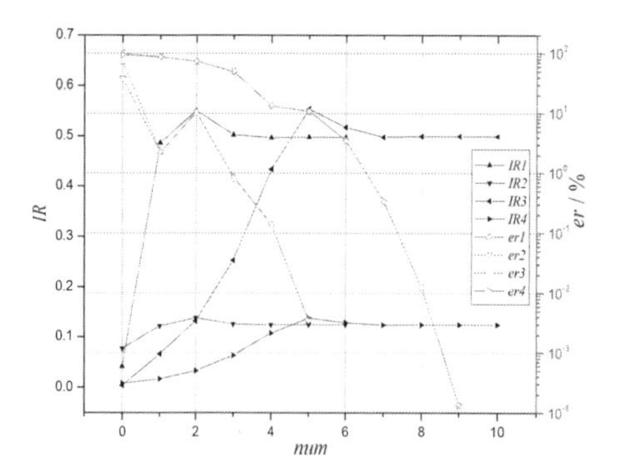

Figure 2. The convergence curves of the standard element.

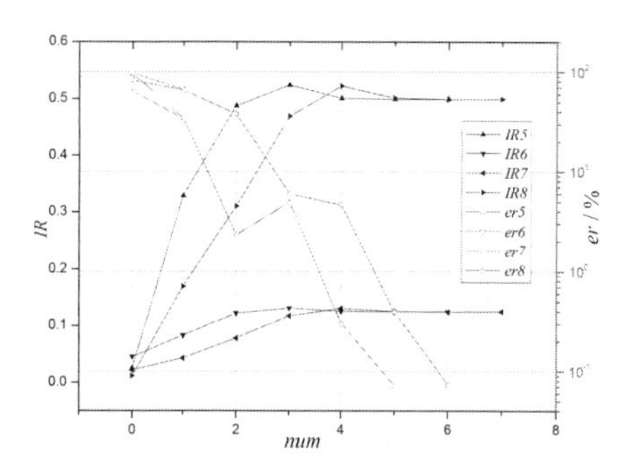

Figure 3. The convergence curves of the rectangle elements.

also at the centre and the vertex; the strong singular integrals are *IR5*, *IR6*, *IR7* and *IR8*; the errors are *er5*, *er6*, *er7* and *er8*, respectively. The convergence curves are shown in Figure 3.

As shown in Figures 2 and 3, whatever standard elements or rectangle elements, the real parts of *IR1~IR8* converge as soon as the numbers of subdivision meet the criterion of the Equation 9. The imaginary parts of *IR1~IR8* are not demonstrated, for them are convergent at the beginning, which also declares that the strong singularity is caused by the real part. Furthermore, the results obtained with the hierarchical subdivision scheme are the same as that with the average subdivision scheme. It can be found that regions far from the singular point in the element influence the singular integral weakly and do not need subdivision, hence the hierarchical subdivision scheme is better than the average subdivision scheme.

5 THE NUMERICAL EXAMPLE

The acoustical radiation problem of a pulsating sphere in water is taken for example to illustrate validity and usage of the adaptive boundary ele-

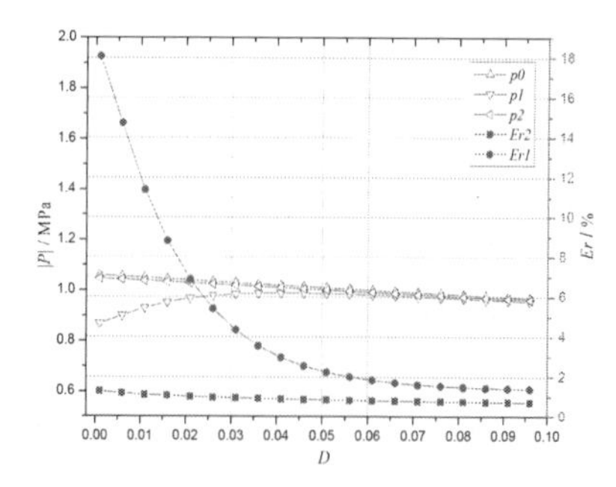

Figure 4. The results obtained by using two methods.

ment integral method. The sphere of radius $a = 1$ is divided into 242 random quadrangles. There is no rectangular element in the mesh, meaning that this method does not be confined to rectangular elements. The sound velocity $c = 1500$ m/s, $k = 1$ rad/m, $\rho = 1000$ kg/m^3, $v = 1$ m/s, the analytical solution of the field point sound pressure is:

$$p = \frac{i\rho\omega a^2 v}{R(1+ika)} e^{-ik(R-a)} \tag{10}$$

where, R is the distance between the sphere centre and the field point.

A number of field points are distributed in the radial direction with the distance to the sphere surface $D = 0.001\sim0.1$, and the nearest field point is close to an element vertex. The element size is $0.2\sim0.39$, so $l = (0.39 + 0.20)/2 = 0.295$ and $num = 4$ ($H = 0.001$). *p0* is the analytical field point sound pressure, *p1* is the numerical field point sound pressure with the standard gauss integral, *p2* is the numerical field point sound pressure with the adaptive integral, $|P|$ is the sound pressure amplitude, *Er* is the error of the numerical result versus the analytical result, $Er1 = ||p1| - |p0|| / |p0| \times 100\%$, $Er2 = ||p2| - |p0|| / |p0| \times 100\%$, results obtained by using two methods are compared in Figure 4.

It is obvious that *p2* agrees with *p0* very well, while *p1* does not. With sharp variation, *Er2* is always less than *Er1*. The maximum of *Er1* is more than 18%, while *Er2* is about 1% all the way. The total computation time for two numerical methods is equal, for subdivision of few elements hardly changes the total amount of calculation. Hence, the adaptive boundary element integral method is much better than the conventional boundary element method for the prediction of near-field noise.

6 CONCLUSIONS

For near-field noise prediction of vibrating structures, an adaptive boundary element integral method is presented. The element is divided hierarchically instead of averagely, for the regions far from the singular point in the element influence the singular integral weakly and do not need treatment. Compared with the conventional boundary element method, the adaptive boundary element integral method provides more accurate results.

In order to determine the number of subdivisions, a criterion is proposed by changing the proximity. Convergence analyses of strong singular integrals for different elements and different near-field points show that this criterion is precise and very practical for the adaptive boundary element integral method.

REFERENCES

Croaker, P. 2015. Strongly singular and hypersingular integrals for aeroacoustic incident fields. *International Journal for Numerical Methods in Fluids,* 77(5): 274–318.

Gao, X.W. 2000. Adaptive integration in elasto-plastic boundary element analysis. *Journal of the Chinese institute of engineers,* 23(3): 349–356.

Gao, X.W. 2008. An adaptive element subdivision technique for evaluation of various 2D singular boundary integrals. *Engineering analysis with boundary elements,* 32(8): 692–696.

Li, Y. 2010. The interaction between membrane structure and wind based on the discontinuous boundary element. *Science China Technological Sciences,* 53(2): 486–501.

Niu, Z. 2007. Analytic formulations for calculating nearly singular integrals in two-dimensional BEM. *Engineering Analysis with Boundary Elements,* 31(12): 949–964.

Qu, S. 2014. Adaptive Integration Method Based on Sub-Division Technique for Nearly Singular Integrals in Near-Field Acoustics Boundary Element Analysis. *Low Frequency Noise, Vibration and Active Control,* 33(1): 27–46.

Wang, J. 2010. Structural multi-scale boundary element method based on element subdivision technique. *Chinese Journal of Computational Mechanics,* 27(2): 258–263.

Wei, Y. 2012. Numerical prediction of propeller excited acoustic response of submarine structure based on CFD, FEM and BEM. *Journal of Hydrodynamics, Ser. B,* 24(2): 207–216.

Zhou, H.L. 2003. Regularization of nearly singular integrals in the boundary element method of potential problems. *Applied Mathematics and Mechanics-English Edition,* 24(10): 1208–1214.

Zhou, H.L. 2007. Analytical integral algorithm in the BEM for orthotropic potential problems of thin bodies. *Engineering Analysis with Boundary Elements,* 31(9): 739–748.

Zhou, H.L. 2008. Analytical integral algorithm applied to boundary layer effect and thin body effect in BEM for anisotropic potential problems. *Computers & Structures,* 86(15–16): 1656–1671.

Advanced Materials, Mechanical and Structural Engineering – Hong, Seo & Moon (Eds)
© 2016 Taylor & Francis Group, London, ISBN: 978-1-138-02908-8

Experimental study on the determination of High-Performance Concrete "Standard Table" mix design applicable to nuclear power plants

D.G. Kim, H.J. Lee, E.A. Seo & J.H. Lee
Korea Institute of Civil Engineering and Building Technology, Goyang, South Korea

M.S. Cho
Korea Electric Power Research Institute, Daejeon, South Korea

ABSTRACT: To optimize tendon designs used in post-tensioning systems in containment buildings of a nuclear power plant (APR1400), it is necessary to evaluate the applicability of High-Performance Concrete (HPC) to nuclear power plants instead of currently used concretes. Therefore, this study conducted a step-by-step replacement by considering fly ash and blast furnace slag maximum replacement rates proposed in the regulations of ACI 318, 349, ASME, KEPIC, and CP-C2 to evaluate the effect of mix proportion that applied mixed cement of a two-component system, in which only fly ash or GGBS is mixed with cement. In addition, three representative mixes such as fly ash, blast furnace slag, and silica fume were selected with reference to the existing literature via experiments using three-component systems in which GGBS and silica fume were mixed with cement. The strengths and weaknesses of each material in terms of characteristics such as binder, mutual complementary contribution, and applicability of HPC mix to nuclear power plants were evaluated. Based on this result, "Standard Table" mix design that is appropriate to the compression strength of 6,000/8,000 psi and 10,000 psi was determined. This study result is expected to contribute to performance improvements in the construction of Korean-style nuclear power plant structures.

1 INTRODUCTION

Structures related to the safety shutdown function in power plants in Korea include reactor containment building, auxiliary building, spent nuclear fuel building, and building for essential service water intake and drainage. Non-safety-related structures include turbine building and auxiliary boiler building (Concrete Structure Design Criteria Commentary 2007, Conllins & Mitchel 1991).

ACI 301, 304R, 305R, 306R, 308R, and 309R in the Construction Package Specification were used as a reference for concrete manufacturing and construction-related criteria and ASME Subsection NE, ASTM, KSF for Materials of Concrete was used as a reference to calculate the ratio of water/binder and admixture amounts required to produce standard mixing. For the classification due to the use environment and target strength, standard mixing is produced by dividing environments into local, general, and severe cold in Finland, and scorching heat in Saudi Arabia.

It is necessary to verify the mixing performance using various admixtures in terms of rapid response to nuclear power plant exportation and improvement direction for concrete quality of local nuclear power plants with respect to 6,000 psi and 8,000/10,000 psi strength, which are the strengths used in the current local nuclear power plant (APR1400) containment building. To optimize tendon designs used in post-tensioning systems in containment buildings of the nuclear power plant (APR1400), it is necessary to evaluate the applicability of High-Performance Concrete (HPC) to nuclear power plants (CEB-FIP, 1995).

Therefore, this study conducted a step-by-step replacement by considering fly ash and blast furnace slag maximum replacement rates proposed in the regulations of ACI 318, 349, ASME, KEPIC, and CP-C2 to evaluate the effect of mix proportion that applied mixed cement of a two-component system, in which only fly ash or GGBS is mixed with cement (ACI 318-09, 2009). In addition, three representative mixes such as fly ash, blast furnace slag, and silica fume were selected with reference to the existing literature via experiments using three-component systems, in which GGBS and silica fume were mixed with cement. The strengths and

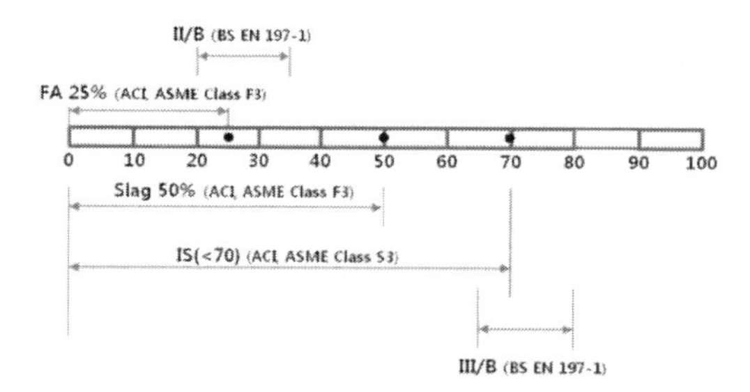

Figure 1. Comparison of admixture replacement rate in local and overseas technologies according to the use environment.

weaknesses of each material in terms of characteristics such as binder, mutual complementary contribution, and applicability of HPC mix to nuclear power plants were evaluated (Figure 1).

2 EXPERIMENT TO PRODUCE THE "STANDARD TABLE" MIX DESIGN

As a binder to manufacture concrete, normal Portland cement of local A company, with a specific gravity of 3.15, and fly ash of local H company, with a specific surface area of 3,375 cm^2/g, were used. River sand was used as fine aggregates whose specific weight, assembly rate, and fineness modulus were 2.56, 2.85, and 1.52%, respectively. In general, coarse aggregates used in HPC manufacturing consist of a smaller maximum size of aggregates than normal performance concrete in terms of flowability, so crushed stone of granite whose maximum size is 20 mm is cleaned and used for HPC.

Mix proportions used in this experiment were a normal concrete mix (OPC): a mix in which 10, 20, and 30% of cement weight were replaced with Fly Ash (FA); a mix in which 10, 30, 50, and 70% of cement weight were replaced with blast furnace slag, a mix in which 50% of cement weight was replaced with fly ash and blast furnace slag, a mix in which 50% of cement weight was replaced with blast furnace slag and silica fume, and a mix in which 25% of cement weight was replaced with fly ash and silica fume. Table 1 details the mix design of HPC.

For the existing design criteria strength of 6,000 psi mix, the actual strength exceeded 7,000 psi and 40% and 35% were applied as the W/B ratio of 8,000 psi HPC and 10,000 psi HPC, respectively.

To evaluate the flow characteristics of HPC using flowability evaluation, slump flow was measured according to the ASTM C 143 method, and air content, which was one of the important factors for HPC strength and durability, was measured according to the ASTM C 231 test method.

Table 1. Determination of mixing design.

Mix	Replacement rate (%)	W/B	Water	Cement	FA	BS	SF
OPC	0			515	0.0	–	–
FA10	10			463	52	–	–
FA20	20			412	103	–	–
FA30	30			360	154	–	–
BS10	10	35		463	–	52	–
BS30	30		175	360	–	154	–
BS50	50	40		257	–	257	–
BS70	70			154	–	360	–
TB1	50			257	103	154	–
TB2	50			257	–	231	26
TB3	25			386	103	–	26

*FA: fly ash, BS: blast furnace slag, SF: silica fume. Notation example: BS30FA25SF5 (blast furnace slag 30%, fly ash 25%, and silica fume 5% in a binder).

For testing the setting time of cement, KS F 2436 and ASTM C 191 were used as references, which were directly related to the concrete pouring completion time specified in CP-C2.

The results found that the total slump flow value was 580–610 mm, the fly ash volume content ratio increased by 10%, 20%, and 30%, and the slump flow value increased. In addition, the FA replacement ratio increased and the slump flow value increased by approximately more than 3% compared with that of OPC. This was due to the increase in slump flow caused by the ball bearing effect of spherical fly ash particles. The slump flow test result on fine powder of blast furnace slag by replacement ratio indicated that as the replacement ratio of the furnace slag fine powder increased, there was an overall increase in the slump flow (Table 2).

Most mixes replaced with performance-enhanced materials met the air content requirement of 4.5 ± 1.5% in the mix design, but most of them showed lower values compared with that of OPC. This result was obtained because air content was decreased due to the void filling effect of fine particulates caused by high fineness of mixed materials. However, since this result can be improved by adjusting the amount of the Air Entraining (AE) agent to reach the target value, it was not considered as an application factor in construction sites.

A trend of increase in air content was also found as the mixing amount of fly ash increased, which will increase the amount of fly ash input replaced along with increases in fine aggregates at the mix proportion increase, thereby causing increases in the percentage of void between high fineness particles and fine aggregates (Lankant 1984, Fibersteel 1984, Naman & Gopalaratmam 1983).

Table 2. Physical characteristic measurement results.

Mix	Replacement ratio (%) FA	GS	SF	Slump(mm)	Slump flow (mm)	Air content (%)	Setting time of concrete tA (hr:min)	tB (hr:min)	tB–tA (hr:min)
OPC	–	–		65	590	4.5	5:16	8:52	3:36
FA10	10	–		–	600	4.0	7:38	11:28	3:50
FA20	20	–		–	605	3.8	7:55	12:35	4:40
FA30	30	–		–	610	3.9	8:16	13:10	4:54
BS10	–	10		–	580	3.5	7:25	11:06	3:41
BS30	–	30		–	595	3.5	7:42	11:50	4:08
BS50	–	50		–	610	3.4	8:10	12:48	4:38
BS70	–	70		–	600	4.0	9:08	13:52	4:44
TB1	20	30	–	–	595	3.2	8:55	13:30	4:35
TB2	–	45	5	–	590	3.1	8:15	12:55	4:40
TB3	20	–	5	–	600	3.8	7:36	11:25	3:49

Table 3. Determination of mix design of 6,000 psi grade HPC.

6,000 psi	W/B (%)	Water	Cement	FA	BS	SF	Coarse Agg.	Fine Agg.	Use environment
FA20	40	162	324	81	0	0	1027	750	Current (APR1400)
SF5	40	155	368	0	0	19	1057	772	Severe cold
FA25	40	155	291	97	0	0	1044	762	General
BS25FA25	40	155	194	97	97	0	1041	760	
BS30FA30	40	155	155	117	116	0	1037	757	
BS50	40	155	194	0	194	0	1054	769	
BS65SF5	40	155	116	0	252	20	1048	765	Scorching heat

Table 4. Determination of mix design of 8,000 psi grade HPC.

8,000 psi	W/B (%)	Water	Cement	FA	BS	SF	Coarse Agg.	Fine Agg.	Use environment
SF5	34	155	433	0	0	23	1094	677	Severe cold
FA25	34	155	342	114	0	0	1078	667	General
BS25FA25	34	155	228	114	114	0	1073	664	
BS30FA30	34	155	182	137	137	0	1069	661	
BS50	34	155	228	0	228	0	1090	674	
BS65SF5	34	155	137	0	296	23	1083	670	Scorching heat

3 DETERMINATION OF "STANDARD TABLE" MIX DESIGN FOR HPC ACCORDING TO THE STRENGTH LEVEL

3.1 *Determination of mix design at 6,000 and 8,000 psi strength*

In this study, the Water/Binder ratio (W/B) was set to 40% for 6,000 psi and 34% for 8,000 psi according to the related technical standards and preliminary test results. The replacement ratio according to the admixture was set to 20–25% for fly ash, 25–65% for slag, and 55% for silica fume in terms of the heat of hydration and DEF (delayed ettringite formation) control according to mass concrete characteristics (Tables 3 and 4).

Flowability as per strength was increased from 4 ± 1 inch to 7.0 ± 1.5 inch, which is the current local nuclear power plant slump criteria, for 6,000 psi

Table 5. Determination of mix design of 10,000 psi grade HPC.

10,000 psi	W/B (%)	Water	Cement	FA	GS	SF	Coarse Agg.	Fine Agg.	Use environment
SF5	28	155	526	0	0	28	1041	644	Severe cold
FA25SF5	28	155	388	138	0	28	1016	628	General
BS25FA20SF5	28	155	277	111	138	28	1016	628	
BS30FA25SF5	28	155	222	138	166	28	1009	624	
BS45 SF5	28	155	277	0	249	28	1032	638	
BS65 SF5	28	155	166	0	360	28	1027	635	Scorching heat

and 8,000 psi. (ACI 363 1992, Baalbaki 1997, Aïctin 1998).

The replacement ratio according to the admixture was set to 20–25% for fly ash, 25–65% for slag, and 55 for silica fume in terms of heat of hydration and DEF.

3.2 *Determination of mix design at 10,000 psi strength*

In this study, the Water/Binder ratio (W/B) was set to 28% for 10,000 psi according to the related technical standards and preliminary test results.

The replacement ratio according to the admixture was set to 20–25% for fly ash, 25–65% for slag, and 55% for silica fume in terms of heat of hydration and DEF control according to mass concrete characteristics (Table 5).

4 EVALUATION OF MECHANICAL PROPERTIES

4.1 *Compressive strength*

The compressive strength test method was used in accordance with the test methods specified in ASTM C39. The evaluation result indicated that compressive strength at 91 days of aging met the target design criteria strength in all mixes, and a mix with only silica fume had excellent strength development at an early stage (Figure 2).

4.2 *Tensile strength and Poisson's ratio*

The tensile strength evaluation result indicated that all mixes had 4–6 MPa (571–857 psi) at 91 days of aging (Figure 3). The Poisson's ratio was in accordance with the ASTM C496 evaluation method, and the result indicated that a Poisson's ratio at 91 days of aging had a range of 0.2–0.25% in all mixes and a mix using only fly ash had a low early Poisson's ratio (Table 6).

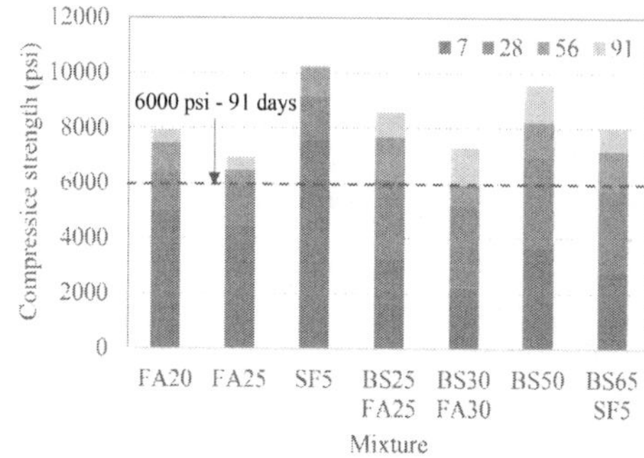

(a) Compressive strength (6,000 psi)

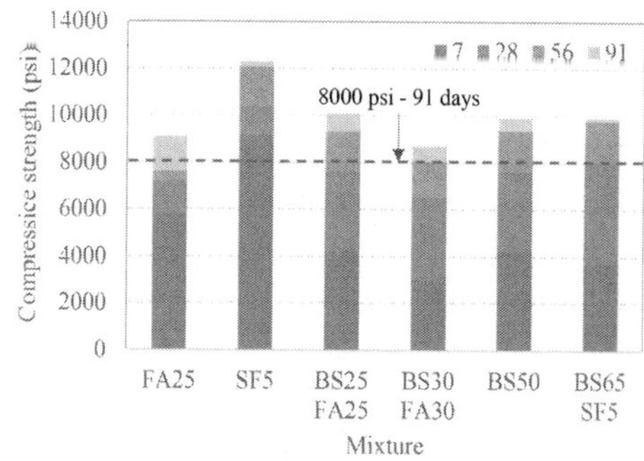

(b) 8,000 psi strength

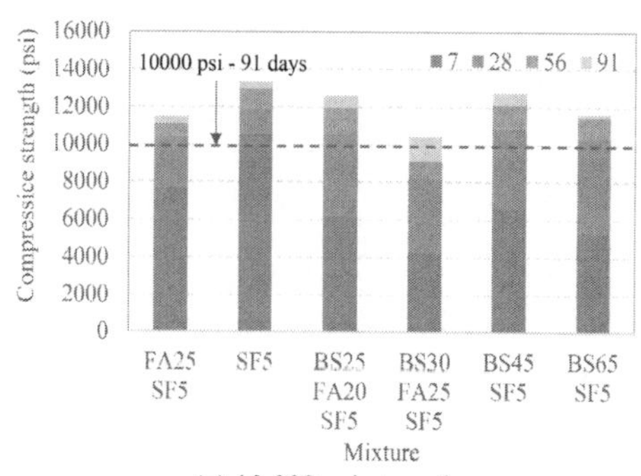

(c) 10,000 psi strength

Figure 2. Test result of compressive strength.

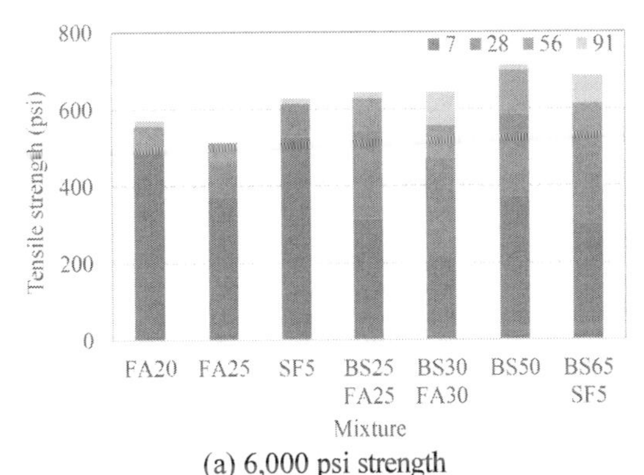

(a) 6,000 psi strength

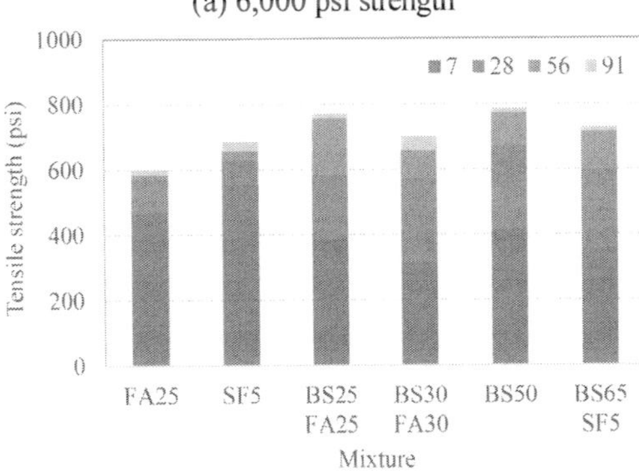

(b) 8,000 psi strength

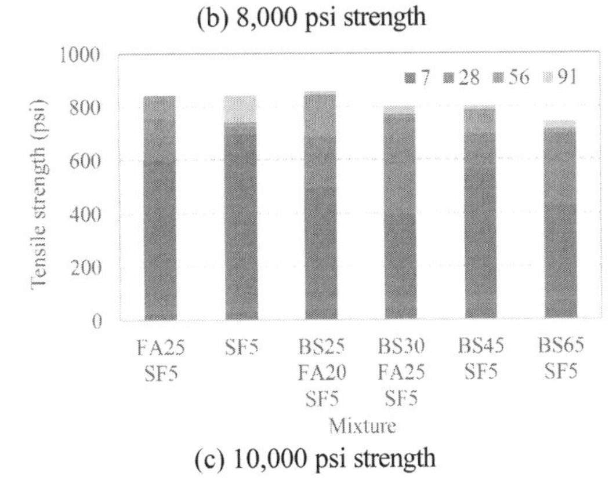

(c) 10,000 psi strength

Figure 3. Test result of tensile strength.

5 CONCLUSION

In this study, compressive strength in all mixes met the goal design strength and tensile strength was 4–6 MPa (571–857 psi). A relationship between compressive strength and Young's modulus was also studied with reference to the ACI, Euro Code 2, CEB-FIP, and Finland technical standards, and the result indicated that compressive strength and Young's modulus had a similar trend to the ACI and CEB-FIP estimation equations.

Table 6. Poisson's ratio of HPC.

		Poisson's ratio (%)			
Compressive strength	Mix type	7 days	28 days	56 days	91 days
6,000 psi	FA20	0.20	0.21	0.21	0.21
	FA25	0.15	0.18	0.19	0.21
	SF5	0.21	0.23	0.23	0.24
	BS25 FA25	0.19	0.20	0.20	0.21
	BS30 FA30	0.15	0.18	0.18	0.21
	BS50	0.19	0.20	0.21	0.21
	BS65 SF5	0.16	0.18	0.20	0.20
8,000 psi	FA25	0.16	0.18	0.21	0.22
	SF5	0.22	0.22	0.23	0.23
	BS25 FA25	0.20	0.20	0.21	0.21
	BS30 FA30	0.18	0.20	0.20	0.20
	BS50	0.19	0.20	0.20	0.21
	BS65 SF5	0.18	0.20	0.20	0.20
10,000 psi	FA25 SF5	0.20	0.20	0.21	0.22
	SF5	0.23	0.23	0.23	0.23
	BS25 FA20 SF5	0.20	0.22	0.22	0.23
	BS30 FA25 SF5	0.19	0.20	0.21	0.21
	BS45 SF5	0.19	0.22	0.22	0.22
	BS65 SF5	0.18	0.21	0.21	0.21

Furthermore, a relationship between compressive strength and Young's modulus was also studied with reference to the CEB-FIP and Finland technical standards. The test results of compressive strength indicated that compressive strength and tensile strength conformed to the CEB-FIP estimation equation.

Based on this result, "Standard Table" mix design that is appropriate to the compression strength of 6,000/8,000 psi, and 10,000 psi was determined. This study result is expected to contribute to performance improvements in the construction of Korean-style nuclear power plant structures.

ACKNOWLEDGMENT

This work was supported by the Nuclear Power R&D Program of the Korea Institute of Energy Technology Evaluation and Planning (KETEP) grant funded by the Korea Government Ministry of Knowledge Economy (No. 20141510101 69A).

REFERENCES

ACI Committee 363. 1992. State-of-the-art report on high-strength concrete.

Aïctin, P.C. 1998. *High-Performance Concrete*, E&FN SPON, London and New York, ISBN 0-419-19270-0.

American Concrete Institute (ACI) Committee 318–09, 2009. *Building Code Requirements for Structural Concrete and Commentary (ACI 318-09)*, Farmington Hills: MI.

Baalbaki, W. 1997. *Analyse experimentele et previsionnelle du module d'elasticite des betons*. PhD. Thesis. Universite de Sherbrooke: 158pp. Quebec: Canada.

CEB. 1995. *High Performance Concrete Recommended Extensions to the Model Code 90-Research Needs*. CEB Bulletin No. 228.

Conllins & Mitchell. 1991. *Prestressed concrete structures.* Prentice Hall, Englewood Clitls: NJ.

Fibresteel, Fibercrete Properties and Pavement Design. 1981. Technical Manual, No. CI/SfB, Australian Wire Industry: 46. Five Dock: Australia.

Korea Concrete Institute. 2007. *Concrete Structure Design Criteria Commentary*, Kimoondang Publishing: 82–154.

Lankard, D.R. & Newell, J.K. 1984. *Preparation of Highly Reinforced Steel Fiber Reinforced Concrete Composites.* International Symposium, American Concrete Institute: 287–304. Detroit: USA.

Naaman, A.E. & Gopalaratnam, V.S. 1983. Impact properties of steel fiber reinforced concrete in bending. International *Journal of Cement Composites and Lightweight Concrete (Harlow)*, 5(4): 225–233.

Advanced Materials, Mechanical and Structural Engineering – Hong, Seo & Moon (Eds)
© 2016 Taylor & Francis Group, London, ISBN: 978-1-138-02908-8

A study on the effect of a ball shooting direction in a ball swaging process on the roll static attitude parameter using finite element analysis

K. Panupich & T. Kiatfa
Department of Mechanical Engineering, Khon Kaen University, Khon Kaen, Thailand

ABSTRACT: The purposes of this research are to explore the base plate and actuator arm deformation that affect the roll static attitude occurring in the ball swaging process, which is the main component determining the quality of assembly and the head stack assembly with the actuator arm. By shooting a ball though the base plate, the component located on the head stack assembly, the base plate plastic deformation takes place and expands in the radial direction. The base plate then adjoins with the actuator arm. Using the finite element method to reproduce the ball swaging process, we repeated to study the effect of the ball shooting direction. The study was done by creating the three-dimensional finite element model to analyze and explain the characteristics of the base plate and actuator arm deformation that have an effect on the roll static attitude taking place in the ball swaging process.

1 INTRODUCTION

Ball swaging process is a method of joining the Head Gimbal Assembly (HGA) with the actuator arm by firing the ball through the base plate. The radius of the base plate is expanded and attached to the actuator arm. There is a contact pressure between the base plate and the actuator arm. The contact pressure directly affects its retention torque. After measuring the position and the torque of the gimbals in the ball swaging process, it has been found that the suspension had the Roll Stack Attitude (RSA) and the Pitch Stack Attitude (PSA). Consequently, the distance between the gimbals and the platters and the power to press the gimbals or the roll static attitude were found to be inconsistent with the determined values. Previously, there were several remedies such as changing the diameter of the ball, coating the ball surface to reduce the friction between the ball and the base plate, changing the ball direction, and modifying the surface of the spacer. However, the problems could not be resolved completely because there has been no study related to the stress analysis, roll static attitude, roll static attitude and pitch static attitude that occurred when the ball was fired through the base plate.

Recently, the hard disk drive has been developed rapidly with respect to the capacity, the speed and the access time; moreover, it is versatile. Apart from being a device for storing the data in the computer, it has also been used in the car, mobile phone, or even the food storage and other materials. Therefore, there is a tendency that the technology development around the hard disk drive industry would grow continually. This leads to the constant research and the development around the structural design or the hard disk drive assembly. Pielr (1992) initially studied the application of the finite element method to analyze the ball swaging process. He found that the finite element method could compare the stress level and the structure of the prototype. He suggested the manufacturer to develop the prototype that fit the ball swaging process in order to reduce the damage to the prototype. In 1992, Seagate Technology Incorporation patented the process to predict the retention torque after the ball swaging process by using the relationship between the retention torque of the gimbals and the driving pin used in firing the ball in a direct line. Such relationship was applied to the computer system to control the quality of the ball swaging process. In 1999, International Business Machines Corporation patented the application of a lubricant on the ball with several substances. It was found that the reduction in the friction generated when the ball was fired through the base plate reduced the distortion on the suspension, position and the placement of the gimbal plane. Kamnerdtong et al. 2005 used the finite element method to predict the stress behavior and the changes after the ball swaging process. There were four parameters used in this research that affected the quality of the ball swaging process: the size of the ball, the firing speed, the ball direction and the friction between the ball and the base plate. Aoki & Aruga (2007) analyzed the ball swaging process using the finite element method and compared the analysis of the

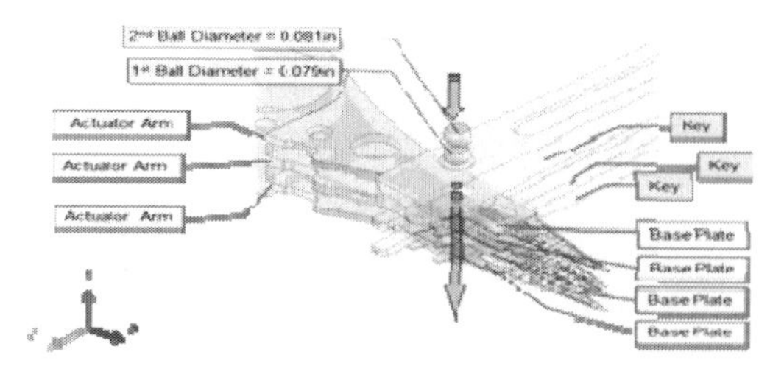

Figure 1. The ball swaging process.

ball swaging process with the test result. The interval that the gimbals lifted up or down was from the distortion of the two components supporting each other, namely the base plate and the actuator arm. The distortion of the base plate curved up, similar to an opened umbrella. Surachet et al. (1979) studied the head gimbal assembly using the ball swaging process. The finite element method was used to study the flow of the component, which helped to understand clearly the distortion and stress that occurred on the component. The factor being studied was the speed used to fire the ball, the size of the ball and the direction of the shooting.

2 FINITE ELEMENT ANALYSIS

This research studied the variable that affected the quality of the ball swaging process. The analysis considered the roll static attitude that occurred in the ball swaging process, and the change in the ball shooting direction. The analysis of the ball swaging process using the finite element method consisted of the following three main steps:

1. Finite element method
2. Processor
3. Result analysis

Each step is elaborated below.

2.1 *Finite element method*

Finite element prototype consisted of creating the finite element method for each component, dividing the elements, determining the specification of the materials, and determining the criteria and conditions to align with the ball swaging process.

2.1.1 *Creating the finite element method for each component*

The steps to create the finite element method for each component in the general ready-made finite element program could be done by two steps. The first step was to create the finite element method from the ready-made CAD program and then to put in the data file. This research created the prototype from the ready-made Solid Works program, then modified and reduced the complexity of the prototype. The second step was to create the prototype from the finite element program.

2.1.2 *Dividing the elements*

Choosing the element type to suit the finite element method was required for such particular task in order to yield the accurate and closest result to the reality. For this research, every element used in the analysis was divided into a cube without a node in the middle. This was because we used the Explicit Algorithm, which would not allow the node in the middle. The analysis of the ball swaging process chose to use Solid 164 element type with 8 nodes that had 9 degrees of freedom.

2.1.3 *Defining material properties*

The quality of the materials could be determined in two ways. The first way was to determine the quality of rigid body materials, which consisted of the ball and the key, without causing any distortion. And the second way was to determine the quality of deformed body materials, which consisted of the base plate, the actuator arm, the gimbals and the hinge, by using the finite element method with specified quality of bilinear kinematic hardening elastic-plastic material. The material properties are listed in Table 1.

2.1.4 *Boundary condition*

The steps to determine the boundary condition for the finite element method involved the determination of the boundary condition in accordance with the type of the ball swaging process as close as possible. As shown in Figure 2, after determining the material qualities and the boundary condition, the next step involved the determination of the pairs between the

Table 1. Materials properties.

Components	Materials	Modulus of slasticity (GPa)	Poisson's ration	Density (kg/m^3)	Yield stress (MPa)
Arm	Aluminum 6061	71.02	0.33	2,724.00	275
Baseplate	Stainless steel 304	190.00	0.32	8,017.80	262
Ball	Stainless steel 304	190.00	0.32	8,017.80	262
Key	Stainless steel 304	190.00	0.32	8,017.80	262
Hinge	Stainless steel 304	190.00	0.32	8,017.80	262

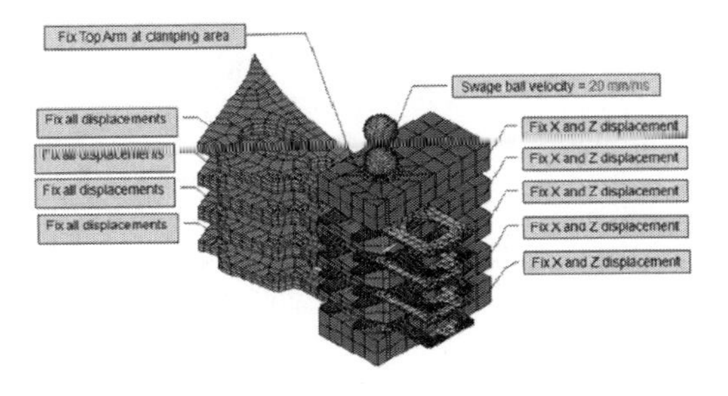

Figure 2. The finite element model.

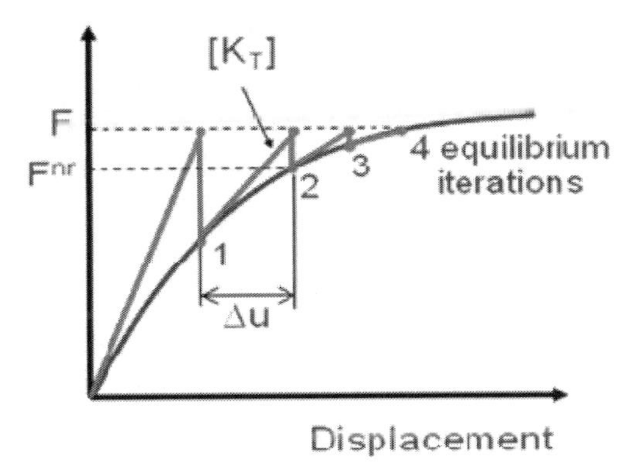

Figure 3. Newton-Raphson method.

components to identify the surface of the pair components and to determine the boundary condition. The type of problem being analyzed used the surface-to-surface contact. The coefficient of the friction between the ball and the base plate was 0.05 and that between the actuator arm and the base plate was 0.5.

2.2 *Processor*

An explicit analysis was used with the Newton-Raphson method to calculate the average results of the finite element model. More details are given below.

The load was applied with gradual increments. Also, equilibrium iterations were performed at each load increment to drive the incremental solution to equilibrium. Iterations continued until $\{F\} - \{Fnr\}$ (the difference between the external and internal loads) was within a tolerance, as shown in Figure 3.

$$[KT]\{Du\} = \{F\} - \{Fnr\} \tag{1}$$

where

$[KT]$ = tangent stiffness matrix
$\{Du\}$ = displacement increment
$\{F\}$ = external load vector
$\{Fnr\}$ = internal force vector

2.3 *Result of the analysis*

The result from the analysis using the finite element method was not able to show the roll static attitude that occurred in the ball swaging process. Therefore, the result from the deformation of the base plate and the actuator arm was used to calculate the tip roll and tip height position. Generally, the roll static attitude was determined as the energy against the gimbals. That roll static attitude pressed the gimbals to reach an equilibrium while functioning. Roll static attitude could be written as follows:

$$RSA\ Head\ Up = HGA\ RSA - Tip\ Roll$$
$$RSA\ Head\ Down = HGA\ RSA + Tip\ Roll \tag{2}$$

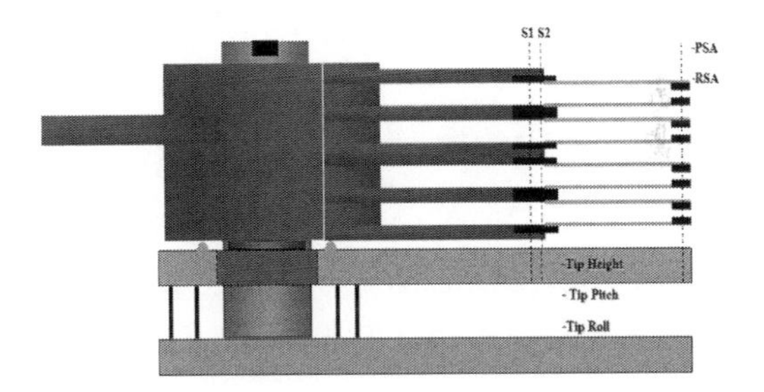

Figure 4. Tip height, tip roll and tip roll.

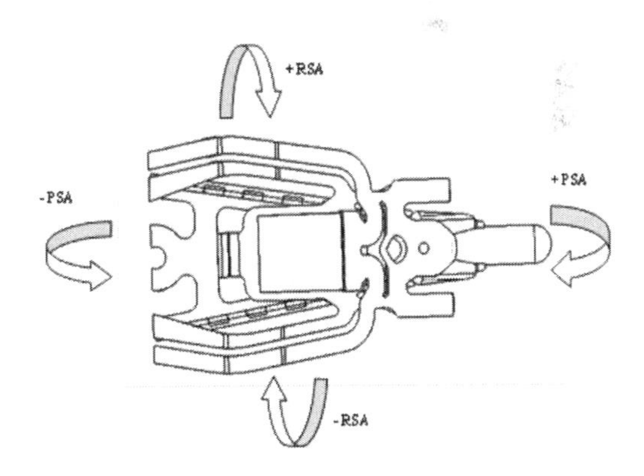

Figure 5. Roll static attitude.

3 RESULT

The analysis of the ball swaging process by the FEM showed that the stress, retention torque and roll static attitude of other components during swaging the ball through the base plate leads to the following two situations:

1. Situation 1. Shooting the ball through the base plate causes the base plate to expand in the

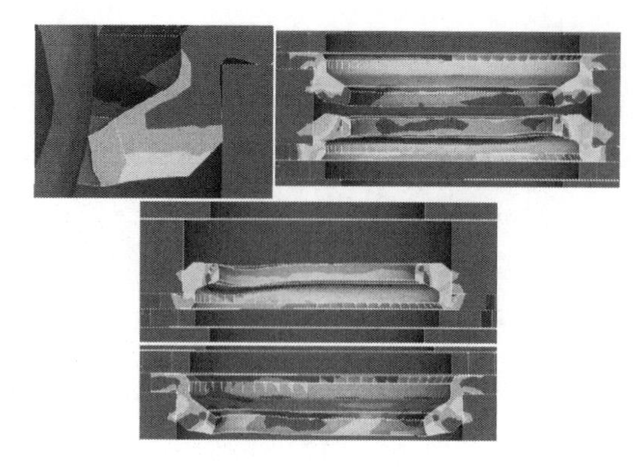

Figure 6. The von Mises stress and plastic deformation.

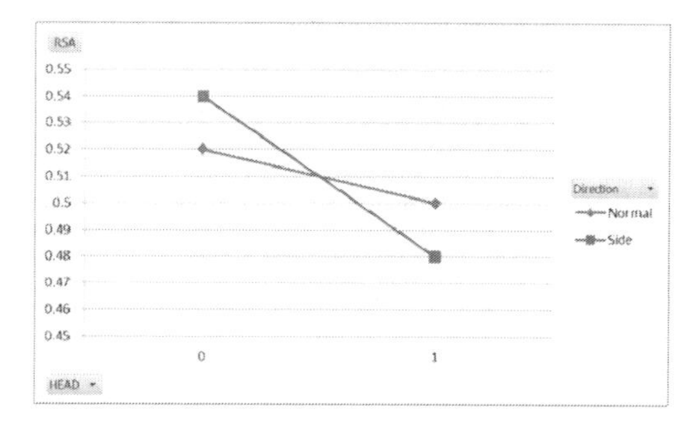

Figure 7. The roll static attitude.

radial direction and to compress against the actuator arm to produce the contact pressure.

2. Situation 2. Friction between the ball and the base plate causes the base plate mass to move in the ball movement direction.

These situations lead to retention torque and plastic deformation of the base plate, as shown in Figure 6. A characteristic resultant from the ball swaging process, which is required in the HSA, is the strength to support the retention torque and the roll static attitude after the ball swaging process.

In addition, the ball shooting direction changes the behavior of the HSA plane. This can be attributed to the plastic deformation of the base plate and the actuator arm. The result indicated that the side shooting direction induces the base plate and actuator arm deformation and leads to the change in the roll static attitude more than the normal shooting direction, as shown in Figure 7.

4 SUMMARY

The present work studies the behaviors and characteristics of the base plate and actuator arm deformation after a ball swaging process. The results from the FEM analysis indicated that the contact pressure occurred at the surface area of the base plate and the actuator arm. Based on Figure 7, the roll static attitude after the ball swaging process resulted in the following two main variables.

1. Contact pressure occurring at the surface areas of the base plate and the actuator arm.
2. Friction between the surface areas of the base plate and the actuator arm.

The ball shooting direction induces the base plate and actuator arm deformation and leads to the change in the roll static attitude.

ACKNOWLEDGMENT

This study was supported by Cooperate Project between National Electronics and Computer Technology Center (NECTEC) and Industry/University Cooperative Research Center (I/UCRC) in HDD Component, Khon Kaen University.

REFERENCES

Aoki, K. & Aruga, K. 2007. Numerical Ball Swaging Analysis of Head Arm for Hard Disk Drives. *Microsystem Technologies*, 13: 943–949.

Fossum, R.E. 2006. *Suspension base plate with boss tower having tapered swage ball- engaging surface*. US Patent, US 7,130,156 B1.

International Business Machines Corporation. Armonk. N.Y. 1999. *Etched/Lubricated Swage Ball for used in DASD Suspension-Arm Attachment*. United States Patent. US 5, 879,578.

Kamnerdtong, T., Chutima, S. & Ekintumas, K. 2005. *Effect of Swaging Process Parameters on Specimen Deformation*. The 8th Asian Symposium on Visualization. 50.1–50.7.

Pattaramon, J., Rattharong, R. & Chatchapol, S. 2009. Optimizad baseplate geometry for ball swaging process by using finite element analysis. *Songklanakarin Journal of Science and Technology*, 31(5): 533–540.

Piela. A. 1992. Study on the applicability of the finite element method to the analysis of swaging process. *Archives of Metallurgy*, 37: 425–443.

Seagate Technology Incorporation, Scotts Valley, Calif. 1992. *Apparatus for Predicting Twist out Torque in a Swage Mount*. United States Patent. US 5, 142,700.

The PC Guide. 2004. Construction and operation of the hard disk. Retrieved February 28, 2008, from http://www.pcguide. com/ref/hdd/op/index.htm.

Wadhwa. S.K. 1996. Material Compatibility and Some Understanding of the Ball Swaging Process. *IEEE Transactions on Magnetics*, 32(3): 1837–1842.

Yang, J., Lin, C.C. & Tabrizi, S. 2007. *Finite Element Simulation of Ball Swaging*, Process of Jointing HGA With Actuator Arm and Pitch static attitude Calculation. ASME Information Storage and Processing Systems Conference, Santa Clara, CA.

Advanced Materials, Mechanical and Structural Engineering – Hong, Seo & Moon (Eds)
© 2016 Taylor & Francis Group, London, ISBN: 978-1-138-02908-8

Determining the optimum elastic modulus and Poisson's ratio for an endodontic prefabricated post

W.M.K. Helal
College of Mechanical and Electrical Engineering., Harbin Engineering University, Harbin, P.R. China
Department of Mechanical Engineering, Faculty of Engineering, Kafrelsheikh University, Kafrelsheikh, Egypt

D.Y. Shi
College of Mechanical and Electrical Engineering, Harbin Engineering University, Harbin, P.R. China

ABSTRACT: In the dentistry field, many post materials are available in the market. The important question is which one is the best. Determining the optimum elastic modulus (E) and Poisson's ratio (υ) for an Endodontic Prefabricated Post (EPP) is the main important purpose of the present work. For analyzing the effect of the post material on the performance of the Endodontically Treated Teeth (ETT), six materials of EPP were selected. For this purpose, the Finite Element Analysis (FEA) was employed. The optimum values of E and υ of an EPP were determined through the optimization technique, and found to be equal to 193 GPa and 0.251, respectively.

1 INTRODUCTION

Nowadays, the treatment of our teeth has become a very important process in our life. At present, ETT is widely found in the treatment process of the tooth.

Two types of EEP can be found: parallel and tapered (Musikant & Deutsch 1984). According to the recommendations from a number of investigators, a parallel post is better than a tapered post (Standlee et al. 1978, Felton et al. 1991, Johnson & Sakamura 1978, Standlee et al. 1978, Qualtrough et al. 2003). Thus, a parallel post will be considered in this investigation.

For investigating and analyzing the structures, FEA is employed because it has been shown to be a useful tool (Asmussen et al. 2005, Chen & Xu 1994).

Recently, Helal and Shi (2015) determined the optimum material for an EPPP. However, the effects of the components of ETT on the performance of the post were neglected. The influences of the components of ETT will be taken into account in the present investigation.

Determining the optimum values of E and υ for an EPP was the main important purpose of the present work. For analyzing the effect of the post material on the performance of the ETT, the following six materials of EPP were selected in the present work: (1) Composite Resin (CR), (2) Fiber Reinforced Composite (FRFC), (3) Carbon (C), (4) Ni-Cr Alloy (NCA), (5) Grass Fiber (GF) and (6) Zirconia (ZC). The FEA was employed to investigate the aforementioned purpose. For determining the E and υ of an EPP used in the ETT, the optimization technique was used in the present investigation.

2 MATERIALS AND METHODS

2.1 *Materials*

The primary goal of the present investigation was to analyze the effect of the material post on the performance of the ETT. In order to investigate the previous goal, six different EPP materials were selected in the present work: (1) CR, (2) FRFC, (3) C, (4) NCA, (5) GF and (6) ZC. Mechanical properties, E (GPa) and υ (non-dimensional) of the materials used were taken from the literature and are listed in Table 1.

2.2 *Finite element analysis*

At present, for solving problems in biological structures, the FEA is widely used. For the present work, the ANSYS package was used to perform the FEA model. The model was built in ANSYS Rel.12.1 by means of the parametric design language (APDL). The structure of the present FE model includes dentin post, dentin, gutta-percha, periodontal ligament (PDL) and cortical bone, and it was taken from the literature, as shown in Figure 1.

In the present work, in order to generate the model in the x-y plane, plane elements (PLANE 82) were used. The element (PLANE 82) was used as a plane stress. The element was defined as having eight nodes with two degrees of freedom at each node.

It should be noted that all components of the FE model were considered as complete bonding. All exterior nodes on the surface of the cortical bone of the present FE model were considered as the fixed boundary condition. A compressive load of a magnitude of 200 N was applied on the center node at the top of the EPP surface, as shown in Figure 2. It should be noted that, in the present investigation, all materials used were considered to be homogeneous, linear elastic and isotropic.

2.3 *Optimization technique*

For solving the present problem, the optimization technique was developed. The model consisted of objective function, design variable and state variable. The design objective was to minimize the maximum von Mises stress in the EPP. For this problem, the design variables used were the E and υ of the EPP material.

For this problem, the objective function was to minimize the maximum von Mises stress in an EPP. In the present investigation, the constraints used are as follows:

1. To retain the value of the design variable "E" within the permissible limits used in the literature as 10 GPa ≤ E ≤ 210 GPa.
2. To retain the value of the design variable "υ" within the permissible limits used in the literature as 0.2 ≤ υ ≤ 0.5.

Table 1. Mechanical properties of the materials used in the present analysis.

Material	E (GPa)	υ	References
CR	16.6	0.24	(Song 2005)
FRFC	15	0.28	(Viguie et al. 1994)
C	118	0.28	(Yang et al. 2001, Lanza et al. 2005)
NCA	203.6	0.3	(Suansuwan & Swain 2001)
GF	85	0.22	(Li et al. 1996)
ZC	75	0.278	(Zhou et al. 2008)
Gutta-percha	0.96E-3	0.3	(Joshi et al. 2001)
Dentin	18.6	0.4	(Joshi et al. 2001)
PDL	0.0689	0.45	(Hong et al. 2001)
Cortical bone	13.7	0.3	(Hong et al. 2001)

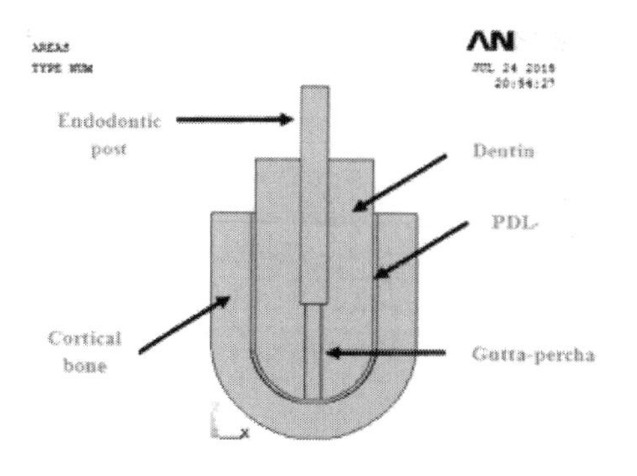

Figure 1. Structure of the present FE model.

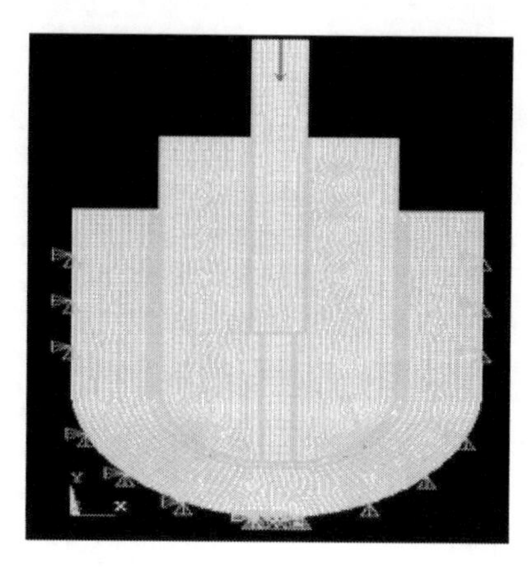

Figure 2. Boundary condition and external loads.

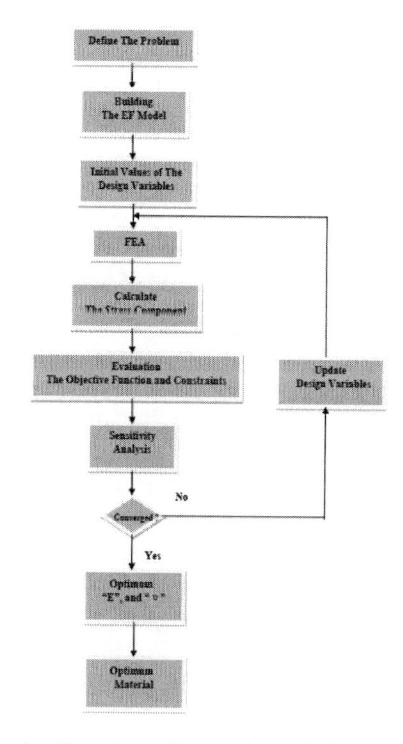

Figure 3. Flow chart of the optimum design procedure.

A flow chart of the optimum design procedure is shown in Figure 3.

3 RESULTS AND DISCUSSION

The primary goal of the present investigation was to analyze the effect of the material post on the performance of the ETT. Figure 4 shows the simulation cases and maximum von Mises stress on the dentin obtained from the six post materials.

According to the outcomes of this investigation, it was observed that the highest value of von Mises stress was found at the CR followed by FRFC, ZC, NCA and C, and the least value was found at GF, as shown in Figure 4.

Figure 5 shows the sample result of CR post (Svon) on the components of the ETT.

The analysis of the effect of the post material on the performance of the ETT using six EPP materials was carried out. The optimization technique was employed in order to determine the optimum E and υ for an EPP. After 9 iterations, the optimum values

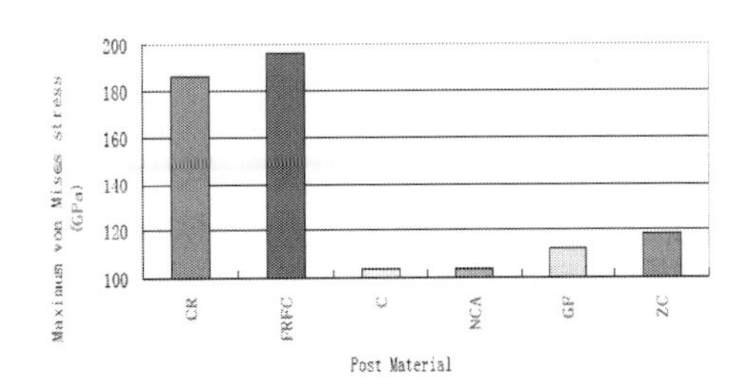

Figure 4. Maximum von Mises stress on the dentin obtained from using the six post materials.

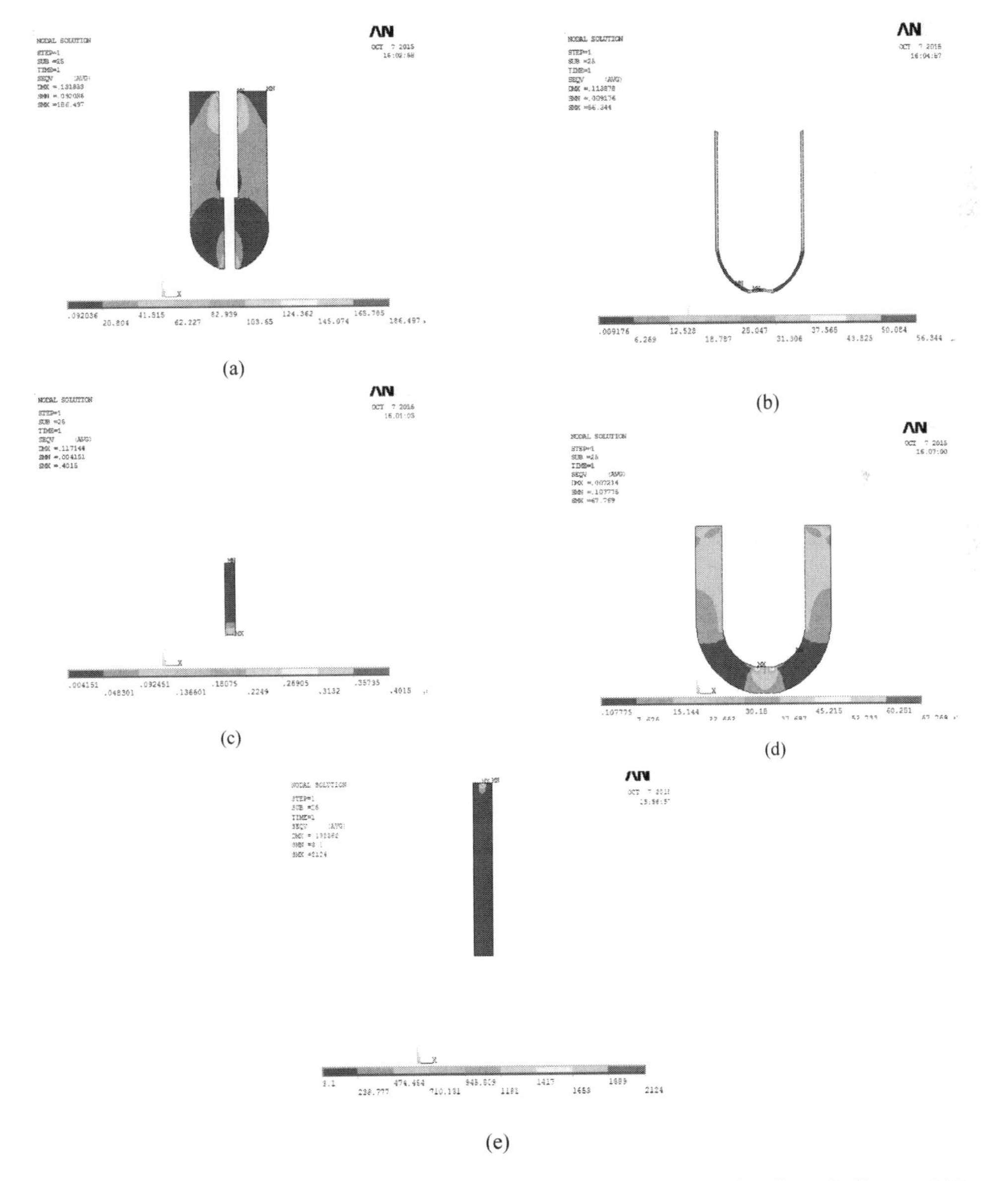

Figure 5. Sample result of CR post results (Svon) on (a) dentin, (b) PDL, (c) gutta-percha, (d) cortical bone and (e) post.

of E and υ of an EPP were found to be equal to 193 GPa and 0.251, respectively. These results indicate that the optimum material of an EPP used in the ETT is Stainless Steel AISI 302 (Fe/Cr18/Ni8).

4 CONCLUSIONS

The effect of the post material on the performance of the ETT was investigated. According to the outcomes of this investigation, it was observed that the highest value of von Mises stress was found at the CR followed by FRFC, ZC, NCA and C, and the least value was found at GF.

The optimization technique was employed in order to determine the optimum values of E and υ for an EPP. According to the results, the optimum values of E and υ of an EPP were found to be equal to 193 GPa and 0.251, respectively. These results indicate that the optimum material of an EPPP used in the ETT is Stainless Steel AISI 302 (Fe/Cr18/Ni8).

ACKNOWLEDGMENT

This paper was funded by the International Exchange Program of Harbin Engineering University for Innovation-oriented Talents Cultivation. The authors would like to thank Kafr El-Sheikh University, Egypt, for providing the facilities.

REFERENCES

Asmussen, E., Peutzfeldt, A. & Sahafi 2005. Finite element analysis of stresses in endodontically treated, dowel-restored teeth, *Journal of Prosthetic Dentistry*, 94(4): 321–329.

Chen, J. & Xu, L.J. 1994. A Finite Element Analysis of the Human Temporomandibular Joint, *Journal of Biomechanical Engineering*, 116(4): 401–407.

Felton, D.A., Webb, E.L., Kanoy, B.E. & Dugoni, J. 1991. Threaded endodontic dowels: Effect of post design on incidence of root fracture, *Journal of Prosthetic Dentistry*, 65(6): 179–187.

Hong, S.Y., Lisa, A.L., Anthony, M. & David, A.F. 2001. The effects of dowel design and load direction on dowel-and-core restorations, *Journal of Prosthetic Dentistry*, 85: 558–567.

Johnson, J.K. & Sakamura, J.S. 1978. Dowel form and tensile force, *Journal of Prosthetic Dentistry*, 40(6): 645–649.

Joshi, S., Mukherjee, A., Kheur, M. & Mehta, A. 2001, Mechanical performance of endodontically treated teeth, *Finite Elements in Analysis and Design, 37*(8): 587–601.

Lanza, A., Aversa, R., Rengo, S., Apicella, D. & Apicella, A. 2005. 3D FEA of cemented steel, glass and carbon posts in a maxillary incisor, *Dental Materials*, 21(8): 709–715.

Li, X., Bhushan, B. & McGinnis, P.B. 1996. Nanoscale mechanical characterization of glass fibers, *Materials Letters*, 29: 215–220.

Musikant, B.L. & Deutsch, A.S. 1984. A new prefabricated post and core system *Journal of Prosthetic Dentistry*, 52(5): 631–634.

Qualtrough, A.J., Chandler, N.P. & Purton, D.G. 2003. A comparison of the retention of tooth-colored posts, *Quintessence International*, 34(3): 199–201.

Song, G.Q. 2005. *Three-dimensional finite element stress analysis of post-core restored endodontically treated teeth,* master's thesis, Master of Science, University of Manitoba, Winnipeg, Manitoba (Canada).

Standlee, J.P., Caputo, A.A. & Hanson, E.C. 1978. Retention of endodontic dowels: Effects of cement, dowel length, diameter, and design, *Journal of Prosthetic Dentistry*, 39: 401–405.

Suansuwan, N. & Swain, M.V. 2001. Determination of elastic properties of metal alloys and dental porcelains, *Journal of Oral Rehabilitation*, 28(2): 133–139.

Viguie, G., Malquarti, G., Vincent, B. & Bourgeois, D. 1994. Epoxy/carbon composite resins in dentistry: Mechanical properties related to fiber reinforcements, *Journal of Prosthetic Dentistry*, 72(3): 245–249.

Yang, H.S., Lang, L.A., Guckes, A.D. & Felton, D.A. 2001. The effect of thermal change on various dowel-and-core restorative materials, *Journal of Prosthetic Dentistry*, 86(1): 74–80.

Wasim, M.K., Helal & Shi, D.Y. 2015. Stress Distribution Surrounding Endodontic Prefabricated Parallel Post, *Applied Mechanics and Materials*, 752: 39–43.

Zhou, G.H., Wang, S.W., Guo, J.K. & Zhang, Z. 2008. The preparation and mechanical properties of the unidirectional carbon fiber reinforced zirconia composite, *Journal of the European Ceramic Society*, 28: 787–792.

Advanced Materials, Mechanical and Structural Engineering – Hong, Seo & Moon (Eds)
© 2016 Taylor & Francis Group, London, ISBN: 978-1-138-02908-8

Compare of shell element and solid element in roll forming simulation

D.H. Yoon, D.H. Kim, Y. Zhang & D.W. Jung
Department of Mechanical Engineering, Jeju National University, Republic of Korea

ABSTRACT: Flexible roll forming process is an effective sheet metal forming process used to manufacture variable cross section profiles (Neugebauer 2006). To obtain a final part with desired shape, the flexible roll forming process should be systematically and accurately controlled. The major of this paper is to establish a flexible roll forming of steel SGARC440. Simulation is a very helpful and valuable work tool in the field of roll forming. The process and roll design for flexible roll forming of a sheet with varying cross-section was investigated. In order to obtain a constant bending width, rotational speeds, locations, orientations and dimensions of the rolls were calculated following the unfolding prediction of the sheet metal. The simulation is elaborated by using ABAQUS software linked to elastic plastic modules which we developed taking into account of interactions between these fields.

1 INTRODUCTION

Conventional roll forming process is one of the most common sheet metal forming processes where the strip is deformed by rotating rolls mounted on successive stands to produce a large variety of profiles continuously at high production rates. This process can only produce profiles with constant cross sections. However, a lot of components used in the automobile, railway cars, ship construction, and building industries have variable cross sections (Kleiner 2003). Therefore; flexible roll forming has been developed recently to produce variable cross section profiles.

Contrary to conventional roll forming process, in flexible roll forming process the forming rolls are not fixed in their position but are moved along a path which describes the desired bend line of the profile. The roll forming process is operated between roll forming multi-stages continuously. And roll forming a profile is formed in several forming steps from an unreformed strip to a finished profile. Nowadays there are more and more demands on such sheet metal, especially for the ones produced in single-piece or small-batch quantity, however, the widely used sheet metal forming processes such as stamping, stretching and hydroforming, etc. are only profitable for mass production, the large initial investments and long setup time make them unsuitable for manufacturing small-batch products, and the sheet metal forming process based on the principle of continuous forming and the usage of flexible forming tools will be an effective way to manufacture swept surface parts in the small-batch production (Kiuchi 1973).

The discontinuous flexible forming process, which is based on the usage of reconfigurable die, does not require expensive conventional die and time consuming setup operations, and thus is suitable and efficient for the small quantity production of doubly curved sheet metals (Kiuchi 1984).

In this study, flexible roll forming bending process is designed to predict the bending process and the sheet mechanical behavior after deformation.

2 OBJECTIVES AND APPROACH

The main objective of this study is to compare the differences in ABAQUS simulation with solid element and shell element. We develop two flexible roll forming bending processes with shell element solid element. The mechanical property was studied in this experiment. The CAD model of the desired shape was generated by the solid works and the simulation was carried out using a commercial finite simulation program ABAQUS (Rhodes, J. 1986).

The steps of this investigation were:

1. Model a given sheet which has a constant bending width and determine the rolls shape and dimension.
2. Establish two groups of experiments separately with shell element and solid element.
3. Define the roll forming process and conditions in ABAQUS such mesh and interactions.
4. Determine the rotation center of the roll and establish the boundary conditions for the roll make them moving forward as desired.
5. Evaluate effects of the process parameters based on the simulation results and determine of the optimal position and boundary conditions of the rolls and deformation steps.

3 SIMULATION AND ANALYSIS

The channel material is common structural sheet steel, SGACR 440 with a Young's modulus of 207 (GPa) and Poisson's ratio of 0.28 (Han 2002). The elastic-plastic properties, including the true stress and strain are given in Table 1 and the true stress-strain curve derived from these data is display in Figure 1.

With different design of rolls, sheet could be deformed successively into products with different section. Three experimental setups of flexible roll forming process are modeled. There three kinds of sheets and the dimensions are shown in Figure 2 and Figure 3. The width of the sheet is 900 mm and the length in longitudinal direction is 107.5 mm. The sheet has an edge with 3 degree. In these roll forming processes, the deformed sheet progressively passes through a series of rolls. The angles of the rolls are from left to right, 15°, 30° and 45°. The metal sheets with the thickness of 1 mm are used as the initial metal strip.

Table 1. Mechanical properties of the SGARC 440 sheet.

	Young's modulus	Tensile strength	Yield strength	Poisson's ratio
SGARC 440	207000 [MPa]	440 [MPa]	300 [MPa]	0.28

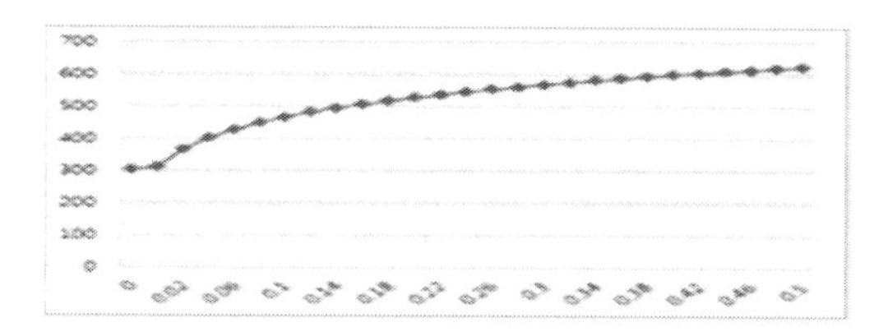

Figure 1. True plastic stress-strain diagram (mm/mm).

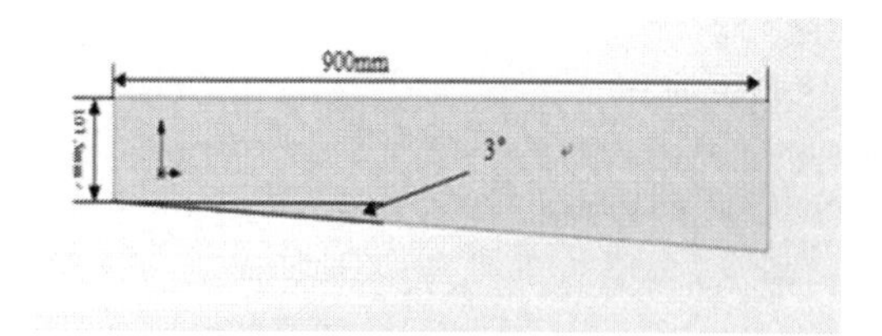

Figure 2. Dimension of sheet with 3 degree.

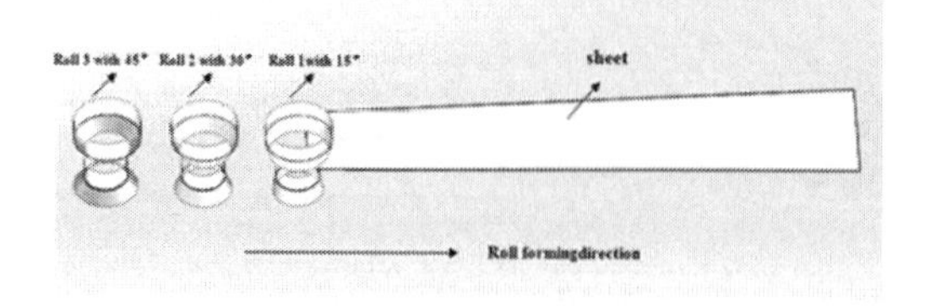

Figure 3. Roll forming process.

4 FINITE ELEMENT

ABAQUS has provided a wide range of elements available solving different problems. For this reason any combination of elements can be used to make up the model. Sometimes multi-point constraints are required for application of the necessary kinematic relations to form the model. All elements use numerical integration to allow complete generality in material. A good finite model in not only decides by the quality of the mesh also decide by the element type. In this paper we are mainly to compare the differences between the solid element and shell element in ABQUS/Explicit. Solid element used to model the widest variety of components. Solid elements names begin with "C" and next two letters indicate the dimensionality can be used in widely used different kinds of simulations. As the name implies, solid element can easily simulate a small piece of the parts. Solid element can build almost every shape and bare any load with the connection between the face and other element. The material description of three-dimensional solid elements may include several layers of different material, in different orientations, for analysis of laminated composite solids. Solid elements are provided with first-order (linear) and second-order (quadratic) interpolation, and we must decide which approach is more appropriate for the application (Wen 1994).

The ABAQUS shell element library provides elements that allow the modeling of curved, intersecting shells that can exhibit nonlinear material response and undergo large overall motions (translations and rotations). ABAQUS shell elements can also model the bending behavior of composites. Shell elements used to model structure with one dimension small enough to another two dimensions and the strain in the thick direction can be neglect. In ABQUS, shell element start with "S". There are two kinds of shell elements in ABAQUS. One is regular shell element and another is shell element based on continuum (E. 2006). In this paper, we just discuss about the regular shell element. Shell behavior that can be properly described with shear flexible shell theory and results in smooth displacement fields can be analyzed accurately with the second-order ABAQUS/standard thick shell element S8R.

5 DESIGN OF ROLL FORMING PROCESS

The sheet was meshed with shell element and solid element. Shell element types were evaluated and finally the quadrangular shell element (S4R) volume element with full integration was selected from the ABAQUS element library as the most stable element type for the Dynamic-implicit solver coupled with the roll forming design program. Solid element type is C3D3R. The mesh of the sheet is shown on Figure 4.

There is three possible ways to control the rigid contact bodies, by referencing the velocity, rotation and displacement or load. Each way can be chosen to execute the movement of the flattener rolls. For the sheet which has an angle with 3 degree is easier to use the velocity. By defining a new local coordinate system with the X-axis is parallel to the edge of the sheet. According to the Figure 5, the distance between every two group of rolls is 200 mm and the length of the sheet is 900 mm. The rolls were deigned to move forward in X-axis under the local coordinate system.

6 SIMULATION RESULT

The geometry of the strips after rolling deformation is shown in Figure 6 and Figure 7. The forming result is influenced by the thickness, the

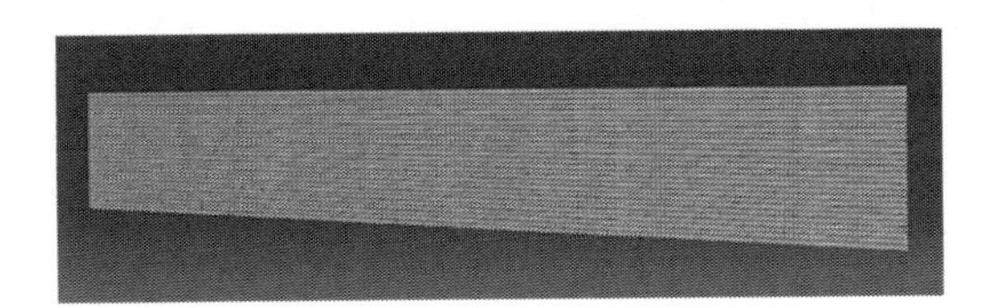

Figure 4. Mesh of sheet with 3 degree.

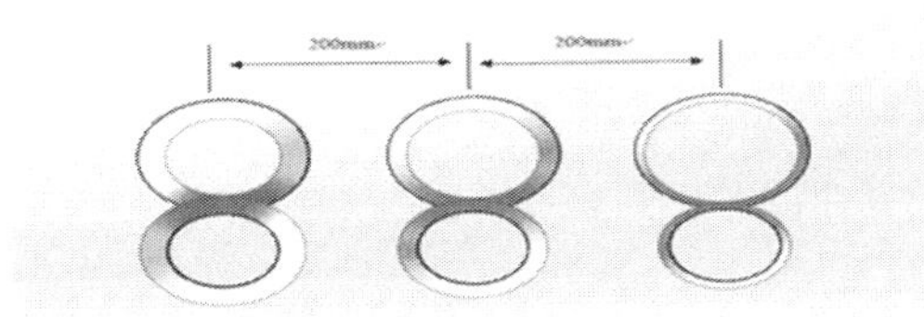

Figure 5. Distances between rolls.

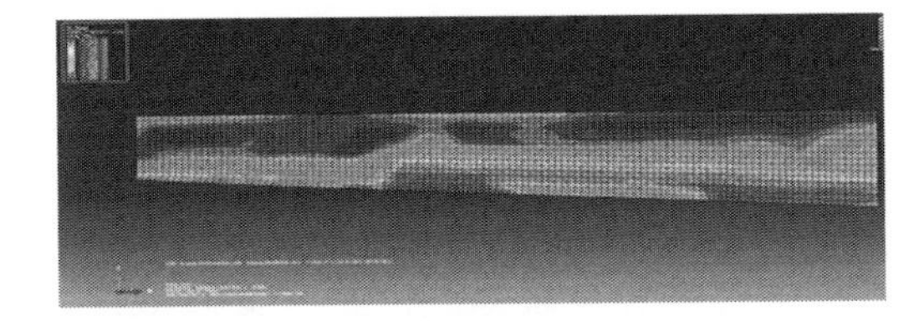

Figure 6. Shell element after deformation.

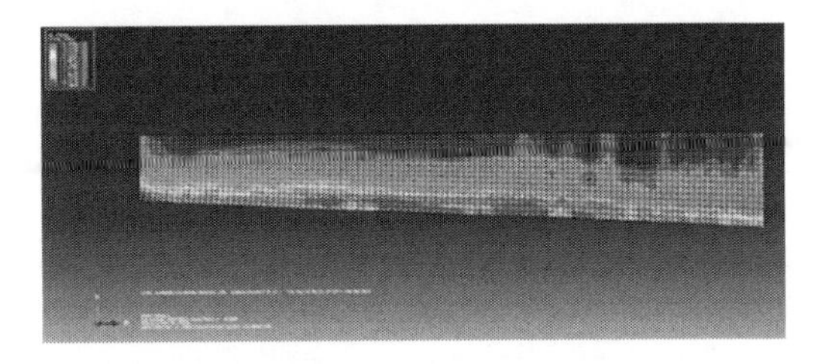

Figure 7. Solid element after deformation.

Figure 8. Flange wrinkling with solid element.

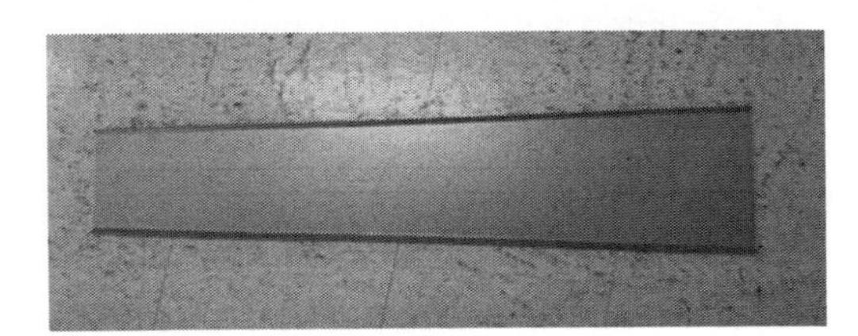

Figure 9. Experiment result.

material property, the roll diameter, the deformation rate of the strip, and the angular velocity of the roll. In simulation, only deformation areas are considered. The tensions before and behind of the deformation area are not considered.

The simulation with solid element results with flange wrinkling at the transition zone where the cross section changes is the major defect. As shown in Figure 8, the flange wrinkling at the transition zone is studied. The results showed that the strip deformation at the transition zone can be considered as a combination of two strip deformations observed in conventional roll forming process and the flanging process. According to shell element finite analysis results, the flange wrinkling only occurs in solid element simulation. Figure 9 shows the experiment result.

Compared with the two element types of deformation, the internal energy of the sheet has some differences. According to Figure 10 and Figure 11, the internal energy of solid element is much bigger than the shell element. That means the simulation is deviate from the actual Kinetic energy in deformation must be a small fraction (less than 1%) of

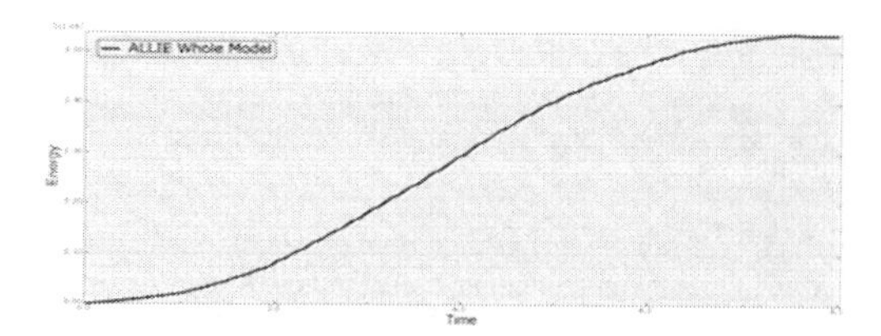

Figure 10. Internal energy with shell element.

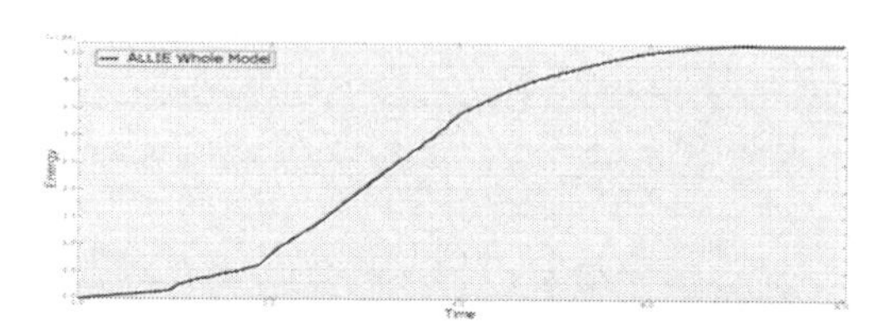

Figure 11. Internal energy with solid element.

the internal energy through all but the very beginning of the analysis. The additional requirements that the histories of kinetic energy and internal energy must be appropriate and reasonable are very useful and necessary, but they also increase the subjectivity of evaluating the results.

As mentioned in previous sections, the computer simulation results of solid elements have some differences with shell elements. The computer simulation time of solid element is much longer than shell element even solid element used one layer.

7 CONCLUSIONS

In this paper, a flexible rolls forming simulation with two kinds of finite elements. Different kinds of elements were analyzed and we can conclude that the model with shell element can more effectively predict the trends in the distribution of the strain-stress and energies. Through the simulation program, the important parameters are determined by taking account of roll stand distance and roll forming speed. With the study above, we can know that

1. Through the rigid plastic finite element method, the flexible roll forming process can be effectively analyzed.
2. The shell element is more suitable for this kind of roll forming and the result of shell element is better than solid element.
3. The solid element simulation consumes more time and easily results in flange wrinkling.
4. The above simulation results can be utilized in the real roll forming industry. It is either done by choosing suitable metal strips, or by modifying the process parameters, such as the roll design.

By comparing two kinds of elements in ABAQUS to find out the differences and through the finite element analysis, it will be useful to make an estimate of the reasonable roll forming process, formability, and the defects of the roll formed products.

ACKNOWLEDGEMENT

This research was supported by the Basic Science Research Program through the National Research Foundation of Korea (NRF) funded by the Ministry of Education (NRF- 2014R1 A1 A4 A01009199).

REFERENCES

E, D.X. 2006. The study on the law of plastic-deformation of tube bending without mandrel, *Journal of Plasticity Engineering,* 13: 5.

Han, Z.W., Liu, C., Lu, W.P. & Ren, LQ. 2002. Simulation of a multi stand roll forming process for thick channel section, *Journal of Material Process Technology*, 127: 382–387.

Kiuchi, M. & Koudobashi, T. 1984. Automated Design System of Optimum Roll Profiles for Cold Roll Forming, *Proc," 3rd Int Conf. on Rotary Metal Working Process*, 423–427.

Kiuchi, M. 1973. Analysis Study on Cold Roll Forming Process, *Report of the Institute of India Science*, 23: 1–23.

Kleiner. M., Geiger, M. & Klaus. A. 2003. Manufacturing of lightweight components by metal forming, *CIRP Annals*, 52(2): 521–542.

Neugebauer. R., Altan, T., Geiger, M., Kleiner, M. & Sterzing, A. 2006. Sheet metal forming at elevated temperatures, *CIRP Annals*, 55(2): 793–816.

Rhodes, J. 1986. A Semi-analytical Approach to Buckling Analysis for Composite Structures, *Composite Structures*, 35: 93–99.

Wen, B. & Pick, R.J. 1994. Modelling of skelp edge instabilities in the roll forming of ERW pipe, *Journal of Material Process Technology*, 41: 425–446.

Advanced Materials, Mechanical and Structural Engineering – Hong, Seo & Moon (Eds)
© 2016 Taylor & Francis Group, London, ISBN: 978-1-138-02908-8

Resonance vibrations of unbalanced two-bearing outboard rotor

Zh. Iskakov
The Institute of Mechanics and Machine Science, Almaty, Kazakhstan

ABSTRACT: The resonance vibrations of two-bearing outboard rotor with mass imbalance and disk skew are studied. External damping forces are introduced in all the motion equations. This allows to construct correctly the amplitude and phase-frequency characteristics and to describe the rotor behavior at the critical shaft speeds. Combined influence of the mass imbalance and disk skew on the rotor dynamics throughout the shaft whole rotation speed range is observed. Detection at the low speeds of the residual phase shift angles is of practical importance, it is extremely important in the rotor balancing, since to determine the mass eccentricity line the measurements at low speeds are used. To determine the unknown orientation of the disk skew line the values of phase shift angles at the high rotation speeds can be used. The resonance amplitudes and phase shift residual angles depending on the generalized imbalance and the shaft cantilevering value are studied.

1 INTRODUCTION

As is well known, high-speed rotary machines are widely used in many industries (electricity, electronics and radio, food, light, chemical, petroleum, medical, metallurgy, aerospace, nuclear, etc.). High efficiency, low weight and high power density, relatively low cost of production and operation, as well as a small environmental pollution lead to the expansion of applications of rotary machines. Consequently, it is not surprising that the rotary machines are studied for a long time. Despite this, many problems are unsolved, in particular related to the joint action of imbalanced mass and skewed disk to vibrations and stability, and subsequently stabilization of resonance vibrations of rotary machines.

Benson (1983) showed that the combined influence of skew and imbalance of disk mass, i.e. the action of centrifugal forces and gyroscopic active moments causes that processing rotor has the unusual phase characteristic, and besides the oscillations phase does not need to match the orientation of imbalance at low speeds. This may significantly change the methods of rotor balancing. Because frictional forces are only included in the equation of rotor translational displacement, the resonance amplitude of translational and angular displacements is indefinitely growing at the second critical speed, which gives a distorted picture of the amplitude- and phase-frequency characteristics of rotor. The mono-graph of Grobov (1961) studies the fluctuations of flexible rotor on elastic supports with non-linear characteristic, but does not concentrate interaction with generalized disk imbalance.

Considerable number of works is devoted to determination of position and orientation of imbalanced mass and skewed disk and, respectively, balancing methods of the rotor vibration control. Here, first of all the methods for determining the rotor mass imbalance and comparison of two methods of balancing are proposed (Gordeev & Maslov 2008) the first of them is based on the measurement of force, acting on the bearings of balancing machine and the other-the measurement of the corresponding deformation (Lingener 1991). Additional scales with a few pendulums are used for automatic balancing; motion of rotor is modeled on a computer (Artyunik 1992a, b). The problem, associated with the orientation sensor in the holospectrum technique, is still not resolved (Liao & Zhang 2010). Herewith the holospectrum technique works on the basis of the rotor balancing. Currently, to determine the imbalance, Initial Phase Vector (IPV) is used, but sometimes this method makes the equilibrium state is not defined. It contains the necessary modifications of this method. The principal compound of effects of a value of the major axis and the Initial Phase Angle (IPA) of precession orbit is to be replaced for IPV value. At that the magnitude and angular position of unbalanced mass can be determined exactly for the amount of the major axis and IPA value respectively. Estimation of value will not now depend on the sensor orientation, an optimal result will be provided. It was suggested an effective control scheme for transverse vibrations due to the imbalance of the rotor shaft, and theoretical study (Das et al. 2008). To do this, it's necessary to use an electromagnetic exciter, mounted on the stator in a place, convenient to operate transverse vibrations of the rotor through the air gap around the rotor. Suitable electromagnetic force of response is achieved by changing the control current, proportional to the displacement of the rotor section. This method provides control over the driving force in the air gap, freedom from hardship and loss of service and wear. The centrifuge with a system of automatic removal of centrifuge vibrations, generated by its imbalance is offered (Majewski et al. 2011). Centrifuge rotor rotates about a fixed axis point. Two or more balloons in the ring, which are attached to the rotor, can automatically eliminate its vibration. The balls, which are also called free elements, can change their position within the ring so as to compensate for dynamic forces. The equations that determine the system behavior, as well as graphics, describe the rotor vibration and behavior of a ball in the presence of imbalance. The article explains that the balls occupy the final position, when the rotor and balls are dynamically stable. Researches of hydroelastic vibrations of vertical gyro rotor, considering the energy source; it also offers practical methods for rotor control (Tolubaeva 2007).

The above review of the researches shows that combined influence of unbalanced mass and skewed disk to resonant vibrations and stability of rotary machines was explored a little, including limiting embodiments of the geometry rotor, it is very important to assess the magnitude and orientation of imbalances.

This article covers the external damping in all four motion levels for complete description of the dynamics of two-bearing console rotor with two generalized disc imbalances. Thanks to that, we can describe correctly the amplitude-frequency and phase-frequency rotor response curves, evaluate the influence of mass imbalance and disc skew on them, shaft overhanging and external damping, as well as to compare the oscillation amplitudes at critical speeds.

2 MOTION EQUATIONS

Figure 1 represents the rotor geometry. At the end of the shaft cantilever part with length a and with a bending stiffness EI the disk is fixed. The disk has a mass m, the polar moment of inertia Ip and equatorial moment of inertia I_T. The rotor rotates with the angular speed ω, its center of mass is displaced on the small distance e relative against disk mounting point S to the shaft and the disk itself is tilted to small angle τ with respect to a plane perpendicular to the shaft axis. The angle between the line of maximum disk skew and disk mass eccentricity vector is β. The distance between the supports is l, in support near the outrigger, the external damping χ_e arises. Shaft with a similar additional damping can be represented by the presence of additional rod, the dummy stiffness factor of which reproduces a viscous damping of support to rotary motion (Dimentberg 1959). Two Cartesian coordinate systems are used. The xyz system is fixed in space, where the axes x, y determine the position of point S, and the rotation axis z passes through the undeformed shaft axis. We will denote the shaft deviation angles in the planes x, z and y, z through θx and θy, correspondingly. All motions in the x and y directions are considered small and a motion in the z direction shall be neglected. The second system of coordinates XYZ rotates together with the disk, wherein Z is a polar axis and the X axis is drawn through the mass eccentricity vector.

The rotor motion equations are derived using the Lagrangian method (Jablonski 2007, Surin 2008). Using the expressions of the disk kinetic energy, the potential energy of the curved shaft (Grobov 1961), the dissipative Rayleigh function (Yablonskiy 2007), introducing the notation $u = x + iy$, $\theta = \theta_x + \theta_y$, $L = a + l$ and dimensionless quantities in accordance with the formulas.

$$\varepsilon = e/L, U = u/L, \bar{t} = t\left(2EI/mL^3\right)^{1/2}, \quad \Omega = \omega(mL^3/2EI)^{1/2}, \ \bar{I}_p = I_p/mL^2 \tag{1a}$$

$$\bar{I} = I_T/mL^2, \ \chi = \chi_e\,(2EI/mL^3)^{1/2}, \quad c_1 = \frac{1}{(a/L)^3}\frac{1+2(a/L)}{4-(a/L)}, c_2 = \frac{1}{(a/L)^2}\frac{2+(a/L)}{4-(a/L)}, \quad c_3 = \frac{1}{(a/L)}\frac{3}{4-(a/L)} \tag{1b}$$

We reduce the motion equations to dimensionless form (variables θ and τ are already dimensionless):

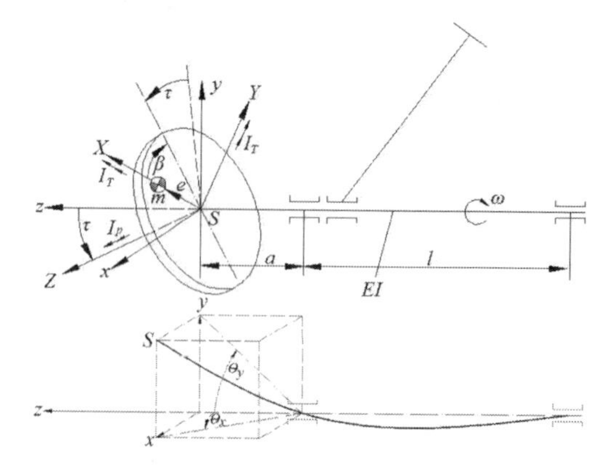

Figure 1. The geometry of the rotor (top figure) and the curved shaft (bottom figure).

$$U'' + \chi(6c_1U' - 3c_2\theta') + 6c_1U - 3c_2\theta = \varepsilon\Omega^2 e^{i\Omega\bar{t}},$$
$$\bar{I}_T\theta'' - i\Omega\bar{I}_p\theta' + \chi(-3c_2U' + 2c_3\theta') - 3c_2U$$
$$+2c_3\theta = (\bar{I}_p - \bar{I}_T)\tau\Omega^2 e^{i(\Omega\bar{t}+\beta)}. \tag{2}$$

Here the strokes denote derivatives with respect to dimensionless time $t\bar{\omega}$. Reducing the motion equations to the form (2) is convenient not only for studying the characteristics of vibrations under the influence of mass imbalance and disk skew, but also of their dependencies on the shaft cantilevering value and the coefficient of external damping.

3 STEADY-STATE VIBRATIONS

At a steady circular synchronous precession the solutions of motion equations is easy to seek in the polar coordinates as

$$U = |A|e^{i(\Omega\bar{t}-\gamma)}, \theta = |B|e^{i(\Omega\bar{t}-\delta)}, \tag{3}$$

where, |A| = amplitude of the disk linear motions; γ = angle of phase lag of this motion from the mass eccentricity vector; |B| and δ = similar characteristics of the angular motion of the disk (see opposite column).

Substituting (3) (2), discarding the time factor and entering the actual parameters:

$$H = \bar{I}_p - \bar{I}_T,$$
$$a_1 = (H\Omega^4 + 2c_3\Omega^2)\varepsilon - (3\chi c_2 H\Omega^3 \sin\beta)\tau + (3c_2H\Omega^2\cos\beta)\tau,$$
$$a_2 = (2\chi c_3\Omega^3)\varepsilon + (3\chi c_2 H\Omega^3 cos\beta)\tau + (3c_2H\Omega^2\sin\beta)\tau,$$
$$b_1 = (3c_2\Omega^2)\varepsilon + (-\Omega^2 + 6c_1)(H\Omega^2\cos\beta)\tau - (6\chi c_1 H\Omega^3\sin\beta)\tau,$$
$$b_2 = (3\chi c_2\Omega^3)\varepsilon + (-\Omega^2 + 6c_1)(H\Omega^2\sin\beta)\tau + (6\chi c_1 H\Omega^3\cos\beta)\tau,$$
$$\Delta_1 = -H\Omega^4 + (6c_1H - 2c_3)\Omega^2 + 12c_1c_3 - 9c_2^2,$$
$$\Delta_2 = \chi(6c_1H - 2c_3)\Omega^3 + 2\chi(12c_1c_3 - 9c_2^2)\Omega. \tag{4}$$

we will produce,

$$|A| = \left[\frac{a_1^2 + a_2^2}{\Delta_1^2 + \Delta_2^2}\right]^{1/2}, \ \gamma = \tan - 1\left[\frac{a_1\Delta_2 - a_2\Delta_1}{a_1\Delta_1 + a_2\Delta_2}\right] \tag{5}$$

$$|B| = \left[\frac{b_1^2 + b_2^2}{\Delta_1^2 + \Delta_2^2}\right]^{1/2}, \ \delta = \tan - 1\left[\frac{b_1\Delta_2 - b_2\Delta_1}{b_1\Delta_1 + b_2\Delta_2}\right] \tag{6}$$

We obtain the natural frequencies of undamped system from the equation

$$\Delta_1 = -H\Omega^4 + (6c_1H - 2c_3)\Omega^2 + 12c_1c_3 - 9c_2^2 = 0 \tag{7}$$

having solved it, we find the critical speeds for direct precession

$$\Omega_k = \left\{\left(3c_1 - \frac{c_3}{H}\right) \pm \left[\left(3c_1 - \frac{c_3}{H}\right)^2 + \left(\frac{12c_1c_3 - 9c_2^2}{H}\right)\right]^{1/2}\right\}^{1/2} \tag{8}$$

Formula (8) shows that the critical speeds depend on the design parameters H, c_1, c_2, c_3, varying by them considering the needs of technological procedures the optimal values of the critical speed required for efficient and safe operation of the rotary machine can be selected.

4 AMPLITUDE-FREQUENCY AND PHASE-FREQUENCY CHARACTERISTICS

In the numerical calculations for some typical versions of the rotor geometry, the following dimensionless parameters are used: the mass eccentricity and the disk skew angle assumed to be $\varepsilon = 0.01$ and $\tau = 0.02$, coefficient of external damping and the cantilevering degree are set to $\chi = 0.01$ and $a/L = 0{,}25$. Research results of dependencies $|A|(\Omega)$ and $\gamma(\Omega)$, $|B|(\Omega)$ and $\delta(\Omega)$ by the Formulas (5) and (6), respectively, indicate that the mass imbalance and the disk skew influence on the rotor dynamics throughout the shaft whole rotation speed range and the amplitude frequency and phase-frequency characteristics of the vibrations, similarly to results of Benson (1983), but in contrast, in the case of a thick disk (H = –0.1) at passing through the second critical speed the resonant amplitudes have the limited values, significantly smaller than in the first critical speed. This is illustrated obviously by the amplitude frequency characteristics of the linear motion of the rotor with thick disk, represented in the Figure 2.

In the Figure 3 and Figure 4 the $A|(a/l)$ graphs are given. From Figure 3 a moderate growth can be clearly seen in the first resonant amplitude with increasing of the cantilevering degree at the values of $\beta = \pm 90°$ and 180°. In $\beta = 0°$ the curve $|A|(a/l)$ has a minimum corresponding to $a/L = 0.25$.

The dependence conducts itself somewhat differently $|A|(a/l)$ near the second critical speed (Figure 4).

Until about the value of $a/L = 0.5$ the linear motions amplitudes at the values of $\beta = 0°$ and $\beta = \pm 90°$ increase sharply and then slowly decrease. At $\beta = 180°$ the resonance amplitude of |A| will have

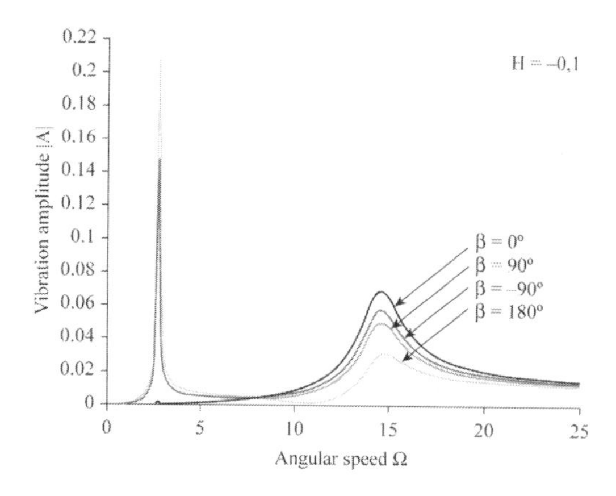

Figure 2. Amplitude-frequency characteristics. Thick disk at H = –0,1: a/L = 0,25; ε = 0,01; τ = 0,02; χ = 0,01.

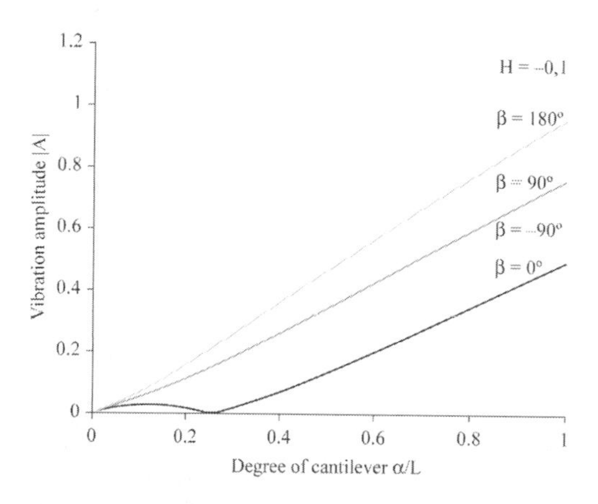

Figure 3. Dependence of the first resonant amplitude of the shaft cantilevering degree.

the maximum approximately near $a/L = 0.3$ peak, and the minimum near $a/L = 0.75$.

Dependence of the vibration amplitude near the first resonant speed from the damping coefficient is such hyperbolic law. Locations of curves depending on the value β are not changed throughout the range of the damping coefficient χ (Figures 5, 6). Around second resonant speed damping the coefficient greatly affects to values of the vibration amplitude.

By varying the values of the structural parameters ε, τ, β, H, c_1, c_2, c_3, a/L, χ, on the $|A(\Omega)|$ and $|B|(\Omega)$ dependencies the resonance amplitudes can be controlled, and the range of operating speeds desired for the technological procedure may be selected.

It is useful to derive a formula explicitly for residual phase angle of disk linear motion at the low rotation speeds. From (5) we will produce

$$\gamma_0 = \lim_{\Omega \to 0}(\gamma) = arctg\left[\frac{-\frac{3}{2}\left(\frac{H\tau}{\varepsilon}\right)\sin\beta}{\frac{C_3}{C_2} + \frac{3}{2}\left(\frac{H\tau}{\varepsilon}\right)\cos\beta}\right] \tag{9}$$

A detailed analysis of the Formula (9) shows that the residual phase angle γ_0 depends only on three parameters: the orientation angle of maximum skew line β, "the imbalances ratio" H τ/ε, which characterizes the relative importance of disk skew and mass eccentricity and structural parameters ratio c_3/c_2. In the Figure 7 the dependence diagrams (9) are shown. It is clearly seen that the residual phase angle can take any value within the range of $-180° \leq \gamma_0 \leq 180°$, especially if the ε→ 0 (single, disk skew only). Within a narrow

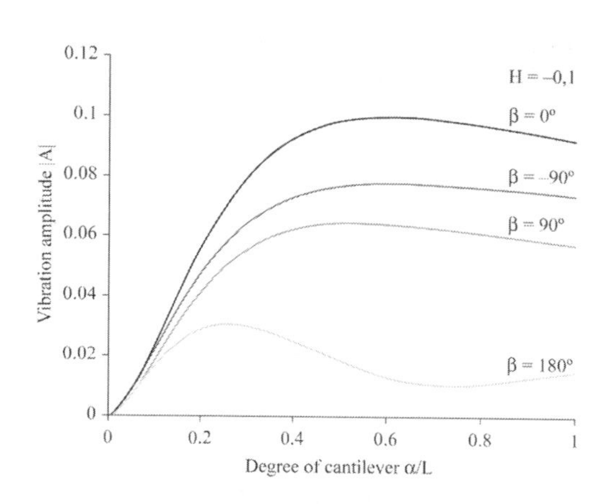

Figure 4. Dependence of the second resonant amplitude of the shaft cantilevering degree.

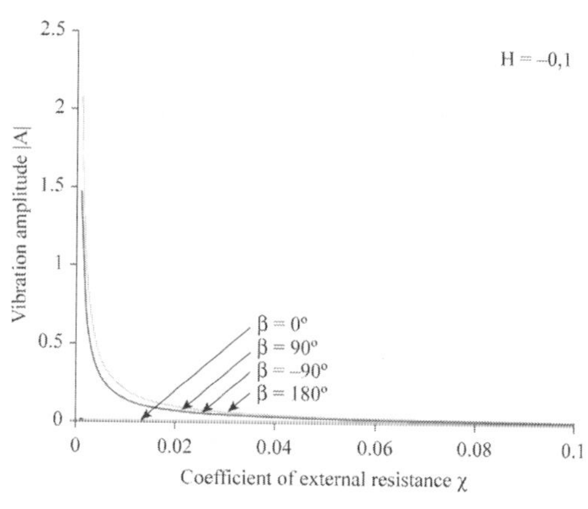

Figure 5. Effect of external resistance on the first resonant amplitude of linear motion. Case of a thick disk: H = –0.1; ε = 0.01; τ = 0.02; a/L = 0.25.

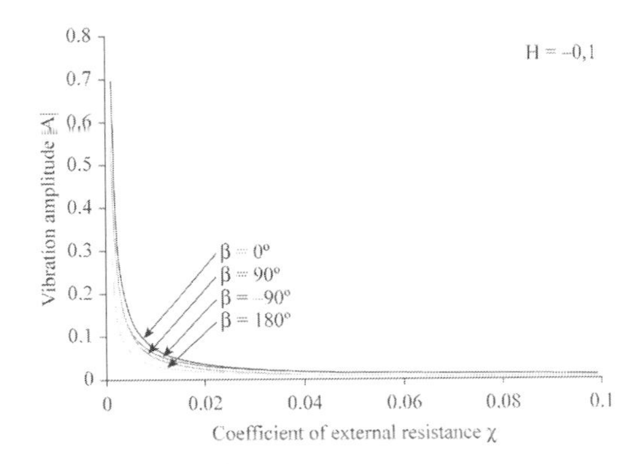

Figure 6. Effect of external resistance on the second resonant amplitude of linear motion. Case of a thick disk: H = –0.1; ε = 0.01; τ = 0.02; *a*/L = 0.25.

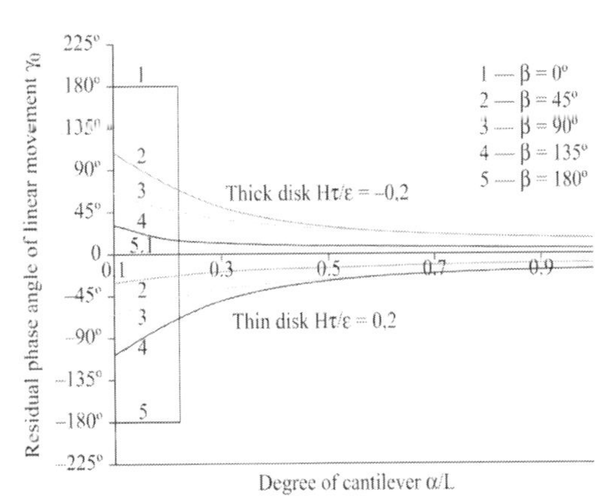

Figure 8. Dependence of the residual phase angle linear motion on the cantilevering value.

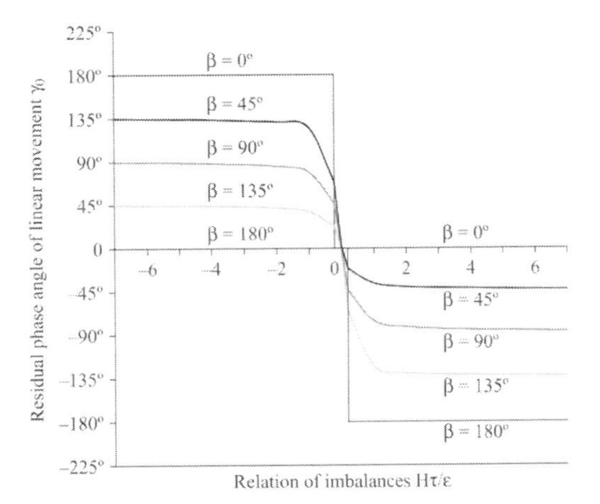

Figure 7. Dependence of a residual phase angle of linear motion of a disk on the relation of imbalances.

range of $|H\ \tau/\varepsilon| < 2c_3/(3c_2)$ the zero values of the residual phase angle γ_0 correspond to the values $\beta = 0°$ and 180°.

In the Figure 8 the dependency diagrams of the residual phase angle γ_0 on the value of cantilevering degree a/L at the constant values of the orientation angle β of disk maximum skew line. This indicates that with increase in cantilevering degree a/L the residual phase angle γ_0 decreasing tends to a definite limit value.

$$\tan-1\left[\frac{-\frac{3}{2}\left(\frac{H\tau}{\varepsilon}\right)\sin\beta}{1+\frac{3}{2}\left(\frac{H\tau}{\varepsilon}\right)\cos\beta}\right],$$

which is typical for the case of the single bearing outboard rotor.

Therefore, at low rotation speeds the processing rotor has a residual angle of phase shift linear motion, i.e. it unnecessarily is in the single phase with the line of the mass eccentricity. At the higher disk rotation speeds, all the curves of the thin disk linear motion tend asymptotically to the normal limit value equal to 180° and for the angular motion the asymptotes of phase shift angle are such that $\delta \to -\beta$. The obtained results are extremely important for the practice. For instance, in the rotor balancing to detect the line of mass eccentricity the measurement results of the linear motion phase shift angle at the low rotation speeds are often used. The designers should compare the results of phase measurements at the low and high speeds, with a view to judge the degree of influence of the disk skew based on the phase difference distinction degree from the value of 180°.

5 CONCLUSIONS

The resonance vibrations of two-bearing outboard rotor with mass imbalance and disk skew have been studied.

The differential equations of motion with allowance for the external damping are formulated, which allowed to describe correctly the amplitude and phase-frequency characteristics of the rotor, to assess the influence on them of joint generalized imbalances, the relative size of the cantilever portion of the shaft and the coefficient of external damping.

It was found that the rotor with a thick disk at the higher critical speed has the vibration amplitudes smaller rather than at a lower critical speed.

The found expressions for disk linear motion residual phase angle at the low speeds and results of studies may be needed to the engineers to estimate experimentally the value and orientation of the mass unbalance at the rotor balancing.

Calculations on the expressions of critical speeds and dependencies of the vibration amplitudes on the rotation speed allow to select the desired range of rotor operating speeds necessary for technological procedure and to control the resonance amplitudes.

This model of the rotor is more generalized as compared with the previously discussed ones and it can be easily extended to other types of rotors.

REFERENCES

Artyunik, A.I. 1992. *Investigation of resistance moments influence in supports of auto-balancer pendulums to rotor dynamics*. Deposit research papers. Ulan–Ude, Russia: East Siberian Institute of technology.

Artyunik, A.I. 1992. *Automatic balancing of rotors using the pendulum suspensions*. Deposit research papers. Ulan–Ude, Russia: East Siberian Institute of technology.

Benson, R.S. 1983. Steady oscillations of the console rotor with skew and imbalance of a disc. *Design and Manufacturing Engineering,* 105(4): 35–40.

Das, A.S et al. 2008. Vibration control and stability analysis of rotor shaft system with electromagnetic exciters. *Mechanism and Machine Theory,* 43(10): 1295–1316.

Dimentberg, F.M. 1959. *Bending oscillations of rotary shafts*. Moscow: Printing House AN USSR.

Gordeev, B.A. & Maslov, G.A. 2008. Detection of unbalance of rotor with acoustic methods. Balandin, D.V. & Erofeev, V.I. (eds), *Nonlinear vibrations of mechanical systems; Writings of VIII All-Russian Scientific Conference, Vol.* 2: 90–95. Nizhniy Novgorod: Publishing House Dialogue of Cultures.

Grobov, V.A. 1961. *Asymptotic methods for calculating the flexural vibrations of turbo machines' shafts*. Moscow: Academy of Sciences Press of the USSR.

Liao, Y.H. & Zhang, P. 2010. Unbalance related rotor precession behavior analysis and modification to the holobalancing method. *Mechanism and Machine Theory,* 45(4): 601–610.

Lingener, A. 1991. Balancing rotors store. *Mechanical Engineering,* 40(1): 21–24.

Majewski, T. et al. 2011. Automatic elimination of vibrations for a centrifuge. *Mechanism and Machine Theory,* 46(3): 344–357.

Surin, V.M. 2008. *Applied mechanics, tutorial*. Byelorussia, Minsk: Novoye Znaniye.

Tolubaeva, K.K. 2007. *Optimization of parameters and motion control for gyroscopic rotor system based on dynamic characteristics of the engine*. Dissertation of the candidate of technical sciences. Almaty: Mechanics and Engineering Institute.

Yablonskiy, A.A. 2007. *Oscillation theory course, tutorial*. St. Petersburg: BHV.

Advanced Materials, Mechanical and Structural Engineering – Hong, Seo & Moon (Eds)
© 2016 Taylor & Francis Group, London, ISBN: 978-1-138-02908-8

Production technology and performance of LY225 anti-seismic building steel

Z.Y. Chen, Z.G. Lin & H.Q. Liu
The Centre Iron and Steel Technology Research Institute of HBIS, Shijiazhuang, Hebei, China

X.S. Wang, M. Wei & D. Liu
Wuyang Iron and Steel Co. Ltd., Pingdingshan, Henan, China

ABSTRACT: The 225 MPa low-yield strength steel with 25 mm and 40 mm thicknesses for building aseismicity was developed through reasonable composition design and rolling technology. The microstructure of the LY225 steel plate was simple polygonal ferrite, and the grain sizes were grade 8–8.5 and grade 7.5–8, respectively. In addition, the yield strength is 225 MPa and 234 MPa, respectively, and the impact property below 0°C ranged between 285J and 294J. It had favorable high adaptability to the low-cycle fatigue property and welding property, and could meet the requirements of GB/T28905-2012 towards the LY225 low-yield strength steel mechanical property for building aseismicity.

1 INTRODUCTION

In recent years, severe earthquake disasters have occurred in many places around the world, which have bring enormous losses to people's life and property. People also have increasingly enhanced requirements in building aseismicity. In the aseismic design, the damages caused by earthquake can be reduced by an equitable distribution of earthquake inertia force and energy. Earthquake-resistant technique can be divided into two types: structural earthquake-resistant technique and energy dissipation earthquake-resistant technique.

Structural earthquake-resistant technique mainly absorbs seismic energy to achieve the purpose of earthquake resistance through the transformation process of the building pillar and beam, and resists destructive effects of earthquake depending on the property of structural components. Kobe earthquake shows that such design technique controls the earthquake-resistant property with difficulty, reflecting on the building's significance (Song et al. 2008), and such traditional structural earthquake-resistant technique presents certain defects. On the basis of such earthquake-resistant technique, it becomes necessary to make the size of the structural supporting member large enough to enhance the anti-seismic level of buildings, but this will increase construction costs and have effects on the appearance of buildings. Energy dissipation earthquake-resistant technique is a new earthquake-resistant technique, which mainly assimilates earthquake energy with help of energy dissipation damper. During an earthquake, these anti-seismic apparatuses will sustain seismic load in advance of other structural components and yield first, and assimilate earthquake energy depending on the alternate load lag to protect the safety of major structures and buildings. Energy dissipation earthquake-resistant technique has become a development tendency (Mitsuru et al. 1995, Zhou et al. 2001, Zhou et al. 1999).

Low-yield point steel for earthquake resistance is famous for its low inclusion, ultra-low yield point, narrow yield scope (±20 MPa), high ductility, high tenacity, repetitive endurance character and other features. It has favorable earthquake-resistant performance, and is the material for major components in the energy dissipation earthquake-resistant design. Furthermore, it is generally used in those countries and regions with frequent earthquakes and significant stadium construction (Tanemi et al. 1998, Song et al. 2010, Yamaguchi & Nagao 1998, Song et al. 2009). Due to the huge difficulty in production, at present, this type of steel mostly depends on import.

In this paper, a pilot production for 225 MPa earthquake-resistant steel was carried out in the pilot plant of HBIS with a reasonable composition design, rolling and after-treatment technology. In addition, this paper carried out the analysis of its structure property by a static CCT test of LY225, which laid a foundation for the commercial process of the LY225 low-yield strength earthquake-resistant steel plate.

2 TECHNICAL REQUIREMENTS AND COMPOSITION, PROCESS DESIGN

2.1 *Technical requirements*

According to the relevant provisions of GB/T 28905-2012 "low yield strength steel used in construction," the composition of LY225 low-yield point steel requirements is given in Table 1. The mechanical properties and process performance requirements are listed in Table 2.

The ground motion characteristic of earthquake wave is the direct factor of vibration behavior expressed under the geological process of specific buildings. The energy is transferred to the ground in the form of fluctuation. In addition, earthquake wave is not merely a transverse wave or longitudinal wave, but an irregular nonresonant wave. The duration of a strong shock at a time is 10 s~30 s (Hou et al. 2003). Relevant statistical data show that the duration of a strong shock usually is dozens of seconds. For example, Tangshan earthquake in 1976 was 23 s (T (0.70)), Mexico earthquake in 1985 was 20 s (T (0.70)), and Kobe earthquake 1995 was 20 s (T (0.70)). The frequency is mostly in the range of 1~3 Hz. The loading number of alternations giving rise to damages of buildings is generally below 200. These data show that an irregular low-cycle alternating load is applied to the building structural steel by earthquake, and the reciprocating transformation from bending (transverse wave) and pulling and pressing (longitudinal wave)

Table 1. The composition requirements of LY225 low-yield strength steel in GB/T28905-2012/%.

C	Si	Mn	S	P	N
≤0.10	≤0.10	≤0.60	≤0.015	≤0.025	≤0.006

Note: Nb, V, Ti, B or other elements should be added whenever appropriate.

Table 2. The mechanical property requirements of LY225 low-yield strength steel in GB/T28905-2012.

Yield strength/ MPa	Tensile strength/ MPa	Elongation/ %	Yield ratio	Longitudinal impact energy (0°C)/J
205~245	300~400	≥40	≤0.8	≥27

Table 3. Composition of LY225 low-yield strength steel/%.

C	Si	Mn	S	P	N
0.01~0.08	≤0.06	0.40~0.60	≤0.01	≤0.01	≤0.006

can give rise to plastic deformation and breakage of rolled steel. Thus, it can be seen that the form of producing anti-seismic building structural steel is high-strain, low-cycle fatigue (Kozaburo 1992). Steel for anti-seismic use will burden repeating alternating loads in earthquake. Therefore, in addition to the mechanical property that is internationally required, anti-seismic steel has a favorable low-cycle fatigue resistance.

2.2 *Composition and process design*

According to the stipulations of GB/T28905-2012, production characteristic and control ability of the field apparatus, designers should consider the requirements such as narrow yield strength scope (±20 MPa), high weldability, low-cycle fatigue property, delivery state of LY225 anti-seismic steel with a low yield point. The chemical component design should adopt the design thought of "low carbon, low silicon and low manganese", so that the matrix structure of steel can be simple polygonal ferrite. Table 3 presents the design component.

3 PILOT PRODUCTION

According to the requirements of the composition design presented in Table 3, the test adopts a 500 Kg vacuum induction melting furnace from a pilot plant of HBIS, to produce rectangle earthen bricks by vacuum casting. All the chemical components met the target of the design requirements after the test. A static CCT test was conducted on LY225 with a small patch of samples from the casting blank. The test results indicate that AC1 = 775°C and AC3 = 919°C for the tested steel, and the static CCT curve is shown in Figure 1.

We cut the casting blank into two rectangle castings with the size of 210 × 300 × 340 mm and 210 × 300 × 320 mm. The rolling testing scheme was made according to the static CCT curve of the LY225 steel. A two-stage rolling process was used

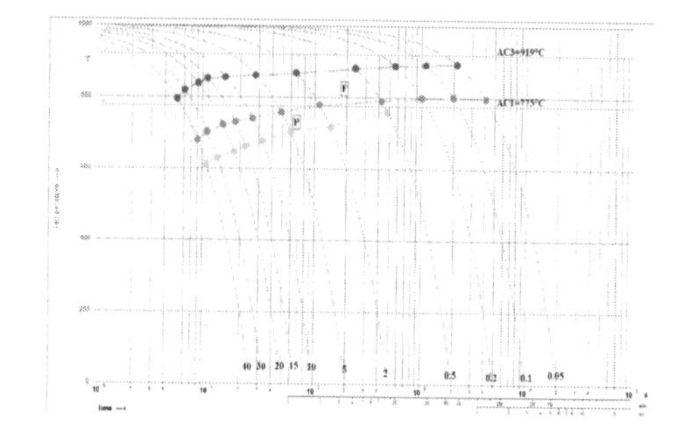

Figure 1. Static CCT curve of the tested steel

Table 4. The pilot rolling process parameters of LY225.

Serial number	Initial rolling temperature in stage I/°C	Temperature-holding thickness/mm	Initial rolling temperature in stage II/°C	Finish rolling temperature in stage II/°C	Self-tempering temperature/°C	Cooling way	Steel thickness/mm
LY225-1#	1100	90	856	790	710	Water cooling	25
LY225-2#	1090	90	850	760	670	Water cooling	40

for rolling in a 550 mm two-roll hot rolling mill in the pilot plant, and the reduction in pass ranged from 10 mm to 30 mm. The thicknesses of the finished steel plates were 25 mm and 40 mm. The rolling process parameters are listed in Table 4.

4 TESTING RESULTS AND ANALYSIS OF ANTI-SEISMIC STEEL LY225

According to the requirements of GB/T228.1-2010, we took two tensile samples of the steel plate along the rolling direction and used the CTM9100 microcomputer control electronic universal testing machine for tensile tests. We then took impact specimens, respectively, at the one-quarter thickness of the two steel plates for the impact test conducted below 0°C. The rolling state mechanical properties of trial-produced 1# and 2# steel plates are listed in Table 5. We cut the metallographic specimen and eroded it in 4% (volume fraction) Nital after polishing. We used the ZEISS Image A1 m type metallographic microscope to observe the microstructure. The metallographic structure of the test steel at the one-quarter thickness is shown in Figure 2, and the test result of metallography is summarized in Table 6.

It can be seen from the test data that the yield strengths of two trial-produced steel plates exceed the requirements of GB/T28905-2012, but the impact energies below 0°C are relatively low; their microstructures are F, and the grain size of the 1# steel, which has a high self-tempering temperature, is grade 7.5, while the grain size of the 2# steel, which has a low self-tempering temperature, is grade 8.5, and it has mixed crystals.

We conducted the performance test on the rolling state mechanical properties of the trial-produced steel plates after normalization of heat treatment. The performance is detailed in Table 7. It can be seen from the test data that all mechanical properties of the 1# and 2# test steel plates meet the requirements of GB/T28905-2012. We then conducted metallographic detection on the steel plates after normalization, and the metallographic structure at the one-quarter thickness is shown in Figure 3.

Table 5. The mechanical properties of LY225 before normalization.

Serial number	Location	Rel/ Mpa	Rm/ Mpa	A/ %	Longitudinal impact energy at the one-quarter thickness (0°C)/J		
LY225-1#	1/2	255	358	40.36	11.58	9.89	31.17
	1/4	268	375	36.8			
LY225-2#	1/2	244	347	50.34	18.63	22.1	18.47
	1/4	252	356	42.36			

(a) plate 1#

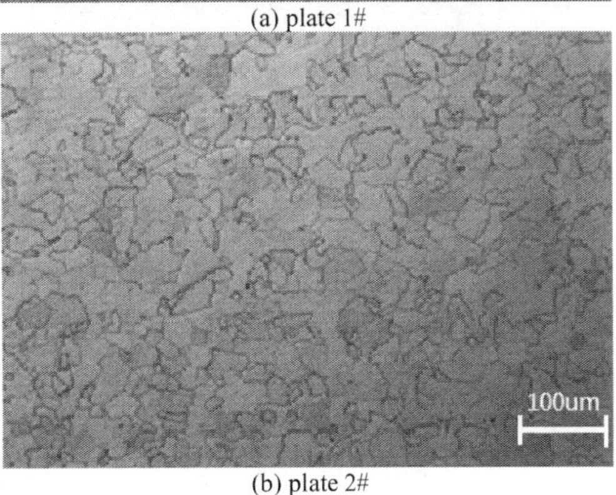

(b) plate 2#

Figure 2. Metallographic structure in the rolling state.

Table 6. Metallurgical test result of LY225 in the rolling state.

	A		B		C		D				
Serialnumber	Big	Small	Big	Small	Big	Small	Big	Small	Ds	Grain size	Microstructure
LY225-1#	0	0	0	0.5	0	0.5	0.5	1.5	0	8.5	F
LY225-2#	0	0	0	0.5	0	0.5	0.5	1.5	0	7.5	F

Table 7. The mechanical properties of LY225 after normalization.

Serial number	Normalizing process	Yield strength/MPa	Tensile strength/MPa	Yield ratio	Elongation/%	Longitudinal impact energy (0°C)/J		
LY225-1	900°C/38 min + water cooling	225	379	0.59	44.2	288	290	291
LY225-2	900°C/60 min + water cooling	234	393	0.60	49.5	285	294	287
Requirements in GB/T28905-2012		205~245	300~400	≤0.80	≥40		≥27	

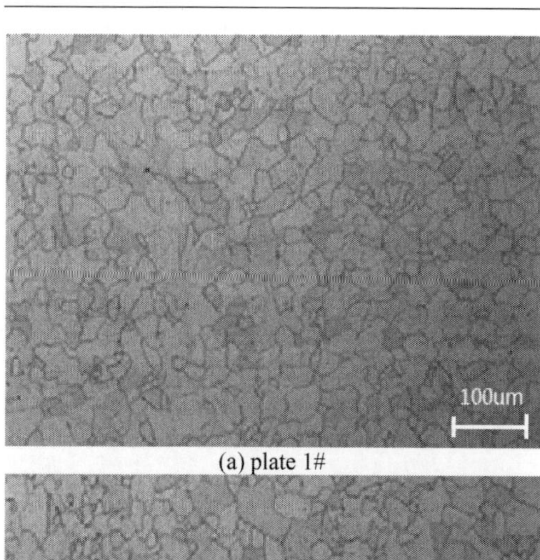

(a) plate 1#

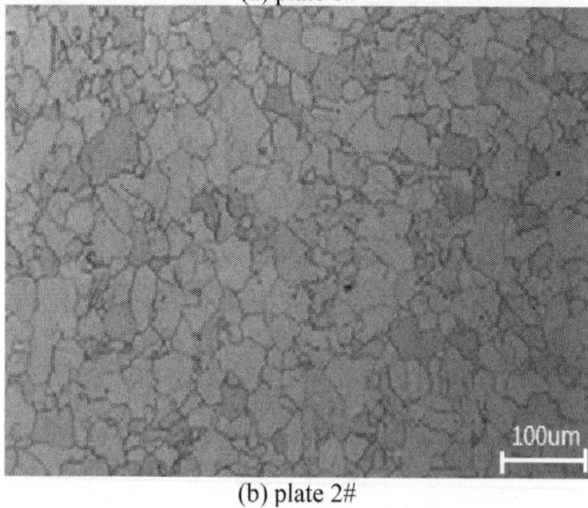

(b) plate 2#

Figure 3. Metallographic structure after normalization.

The result for metallographic detection is summarized in Table 8. It can be seen that the type of steel plate structure after normalization is pure ferrite, and the grain sizes of the two steel plates are

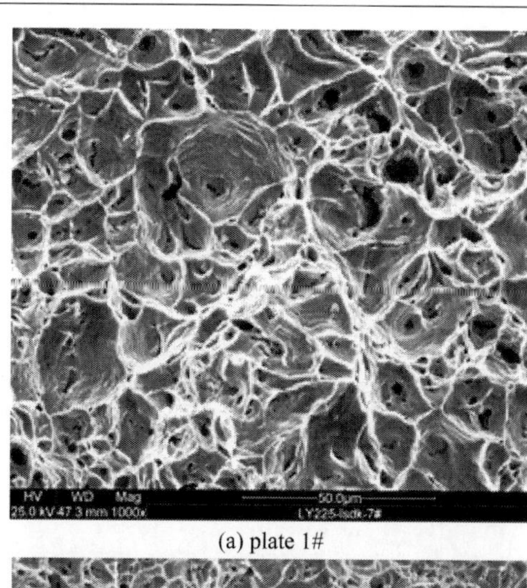

(a) plate 1#

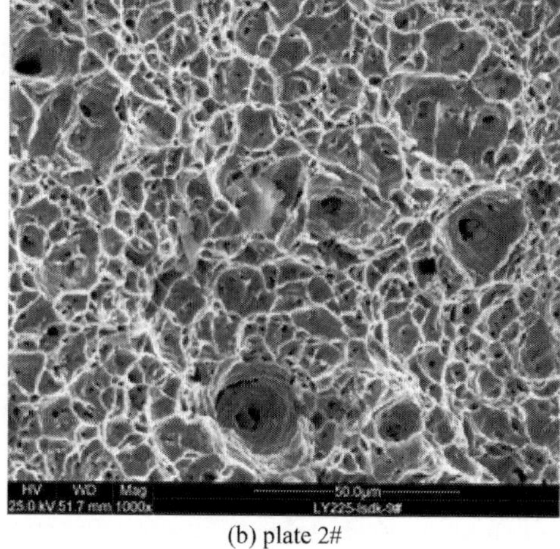

(b) plate 2#

Figure 4. Morphology of tensile fracture steel (SEM).

Table 8. Metallurgical test result of LY225 after normalization.

Serial number	Normalizing process	A		B		C		D			Grain size	Microstructure	Location
		Big	Small	Big	Small	Big	Small	Big	Small	Ds			
LY225–1	900°C/38 min +	0	0.5	0	0.5	0	0	0.5	1.5	0	8.0	F	1/2
	water cooling	0	0	0	0.5	0	0	0.5	1.5	0	8.5	F	1/4
LY225–2	900°C/38 min	0	0	0	0.5	0	0	0.5	1.5	0	7.5	F	1/2
	+ water cooling	0	0	0	0.5	0	0	0.5	1.5	0	8.0	F	1/4

Table 9. Low cycle properties of the 1# and 2# steel plates.

Strain amplitude ± Δε/%	Frequency/HZ	Strain rate/% . S^{-1}	Cycle times/times	Duration/s
0.5	1	2	1204	1204
0.5	2	4	1429	715
0.5	3	6	1006	335

grade 8–8.5 and grade 7.5–8 respectively. The liquid steel has a high degree of purity, and only contains Ti compounds (mixed with type D). Figure 4 shows the SEM morphology of the tensile sample fracture of the 1# and 2# steel plates after normalization. It can be seen that the two samples belong to ductile fractures with uneven sizes of dimples. The size of the dimple of the 1# steel plate ranges between 10 and 25 μm, while that of the 2# steel plate ranges between 5 and 15 μm. Generally, the dimple of the 2# steel plate is more even and precise than that of the 1# steel plate, thus the yield strength and strength of extension of the 2# steel plate are higher than that of the 1# steel plate.

Low-cycle fatigue test was conducted on the trial-produced LY225 steel plates according to the service environment of LY225 anti-seismic steel, the energy delivery and destruction characteristics in case of seismic wave. We prepared low-cycle fatigue testing samples for the 1# and 2# steel plates, and simulated the actual earthquake situation, as well as conducted the performance test on low-cycle fatigue in case of 1 HZ, 2 HZ and 3 HZ under the 0.5% strain range, respectively. The result is summarized in Table 9.

It can be seen from the test data that the duration time of strain cycle decreases with the elevation of strain frequency. The sample's failure cycles are more than 1000 times, and the duration time is more than 1.5 min. The duration time of a strong shock ranges from 10s to 30s, the frequency is in the range of 1–3 HZ, and the alternate loads are more than 200 times. Thus, the developed LY225 anti-seismic steel plate with 25 mm and 40 mm thicknesses can totally resist the destruction of a strong shock.

In addition, welding performance analysis was conducted on testing steel plates according to the carbon equivalent (Ceq) formula recommended by the International Institute of Welding:

$$Ceq = C + 1/6\ Mn + 1/5\ (Cr + Mo + V) + 1/15\ (Ni + Cu)$$

when Ceq < 0.35%, the steel plate has a good welding performance; when Ceq > 0.40% ~ 0.50%, it is difficult to weld the steel plate.

For the tested steel plate, Ceq = 0.035% + 0.43%* 1/6 = 0.11%, which is much less than 0.35%. It can be deduced that the tested steel plate has a good welding performance.

5 CONCLUSIONS

1. The composition design of "low carbon, low silicon, low manganese" and the water—cooling and normalizing processes after the two-stage rolling process were used to develop 225 MPa anti-seismic and low-yield point steel with 25 mm and 40 mm thicknesses. The microstructure of the steel plate was simple polygonal ferrite, and the grain sizes were grade 8–8.5 and grade 7.5–8, respectively. The yield strengths were 225 MPa and 234 MPa, respectively, and the impact performance below 0°C ranged between 285 and 294 J. All performances of the steel plate met the requirements of GB/T28905-2012 for the 225 MPa anti-seismic and low-yield point steel.
2. Low-cycle fatigue test in the case of 1–3 HZ and under the 0.5% strain range was carried out on the developed LY225 steel plate, and the failure cycles were more than 1000 times, and the duration time was more than 1.5 min. The tested steel plates exhibited a good low-cycle fatigue performance that can be used for the anti-seismic design of buildings.
3. The trial-produced LY225 steel plates with 25 mm and 40 mm thicknesses had an excellent welding performance.

REFERENCES

Hou, Z.Y., Liu, X.M. & Chen, R.M. et al. 2003. Microstructure and Composition De-sign of Anti-Seismic Building Structural Steels, *Metal Heat Treatment,* 28(4): 21–24.

Kozaburo, O. 1992. Recent Trend of Technology for Steel Plates Used in Building Construction, *Nippon Steel Technical Report,* (54): 27–36.

Mitsuru, S. & Hideji, N. et al. 1995. Development of Earthquake-resistant Vibration Control and Base Isolation Technology for Building Structure. *Nippon Steel Technical Report,* (66): 37–46.

Song, F.M., Wen, D.H. & Li, Z.G. et al. 2008. Application and Development of Low Yield Point Steel, *Hot Working Technology,* 37(6): 85–88.

Song, F.M., Wen, D.H. & Li, Z.G. 2009. Development of 225 MPa Low Yield Point Steel Used For Earthquake Resistant, *Material & Heat Treatment,* 38(12): 62–63, 69.

Song, F.M., Wen, D.H. & Li, C. et al. 2010. Low Cycle Fatigue Characteristic of Ultra-Low Yield Point Steel. *Journal of Iron and Steel Research,* 22(5): 37–40.

Tanemi, Y., Toru, T. & Toshimichi, N. et al. 1998. Seismic Control Devices Using Low-yield-point steel, *Nippon Steel Technical Report,* 77: 65–72.

Yamguchi, T. & Nagao T. 1998. Seismic Control Devices Using Low-Yield-point Steel. *Nippon Steel Technical Report,* 77: 65–72.

Zhou, Y., Xu, T. & Yu, G.H. et al. 1999. Recent Advances in Research, Development and Applications of Seismic Energy Dissipation, *Earthquake Engineering and Engineering Vibration,* 19(2): 122–131.

Zhou, H.T., Xu, Z.D. & Zhang, X.H. 2001. Study, Application and Development of Energy Dissipation and Damping Control, *Journal of Xi'an University Of Architecture And Technology,* 33(1): 1–5.

Advanced Materials, Mechanical and Structural Engineering – Hong, Seo & Moon (Eds)
© 2016 Taylor & Francis Group, London, ISBN: 978-1-138-02908-8

A study of wind load and snowdrifts on a wide greenhouse

K.P. You, Y.M. Kim, S.Y. Paek & B.H. Nam
Long-Span Steel Frame System Research Center, Department of Architecture Engineering, Chonbuk National University, Jeonju, Korea

J.Y. You
Department of Architecture Engineering, Songwon University, Gwangju, Korea

ABSTRACT: Damage is caused to single-span-type greenhouses by heavy snowfall in Korea, and the M shape collapses when excessive snow load on the roof occurs frequently regardless of the area damaging all the roof rafters. Therefore, this paper examined the effect of wind and snowdrifts on the wide greenhouse shape using wind tunnel simulation. The test results indicated that the snowfall on the roof of the arched greenhouse differed depending on the shape and location. The wind tunnel test results indicated that the distribution of mean wind pressure coefficients of the wide greenhouse and the magnitude of the coefficient that applies to the roof surface differed according to the wind direction angle. More snow was accumulated on the lower part of the roof surface than on the upper part of the roof surface.

1 INTRODUCTION

Greenhouses have a structure that is very vulnerable to wind and snowdrifts due to their composition of vinyl and pipes. Damage to vinyl greenhouses constitutes more than 90% of total farm facility damage, requiring urgent evaluation of their structural safety. The accurate evaluation of wind and snowdrifts, which exert major loads on vinyl greenhouses, is especially important. The concept of the wind tunnel test for assessing the phenomenon of rooftop snow-fall was first proposed by Taylor (1980). He proposed a wind tunnel test using artificial snow to review the phenomenon of rooftop snowdrifts. However, this type of test was first conducted by Isyumov (1989). Hoxey (1983) examined the surface pressure of the earth, transformed the spectrum by analyzing the pressure distribution of wind on vinyl greenhouses, and suggested design guidelines. Tahouri (1990) conducted a wind tunnel test of the average pressure distribution on the surface of the earth in turbulent conditions to investigate the impact of flow factor on building load. Okada et al. (1999) compared the gust effect coefficient and wind pressure coefficient according to the codes and standards of each country. Blackmore et al. (2006) suggested a new alternative to the guidelines suggested in EN1991-1-4 by comparing the wind tunnel test on curved ceilings and the external pressure co-efficient according to the span length division of each country. This study attempts to evaluate the effect of wind load and snowdrift types on the ceilings of wide greenhouses.

2 WIND TUNNEL TEST

For the wind pressure and snowdrift test performed, experiments using two wind tunnels were conducted in this paper. The wind pressure test was conducted in an Eiffel-type wind tunnel with a test section size of 1.7 m (height) × 2.0 m (width). A test on the phenomenon of snowdrifts was conducted in a closed-type wind tunnel with a test section size of 0.6 m (height) × 0.9 m (width). The boundary layers formed within the wind tunnel were in a suburban area.

2.1 *Wind pressure*

Regarding the wind pressure test, the wind pressure on the greenhouse was measured using a multi-point wind pressure anemometer that was installed on each side of the model. As the wind pressure test used in the greenhouse model satisfied the geometric similarity condition, a test model was fabricated in 1/40, and the wind tunnel test was performed on the ceiling surface, both sides, and the front. Table 1 presents the test model data for the greenhouse, and Figure 1 shows the wind tunnel test model that was installed inside the measuring part of the wind tunnel.

Regarding the location of the pressure measurer, a wind pressure tap was installed in constant distance, with 252-point pressure measurers installed on the ceiling surface, 164-point measurers on the side, and 39-point measurers on the front. Figure 2 shows the map plan of the greenhouse pressure measurers.

Table 1. Test model data for greenhouse.

Proto type		Model			
Width (mm)	Roof Height (mm)	Width (mm)	Roof height (mm)	Tap	Model Scale
14000	3700	350	92.5	489	1/40

Figure 1. Model installed inside the wind tunnel.

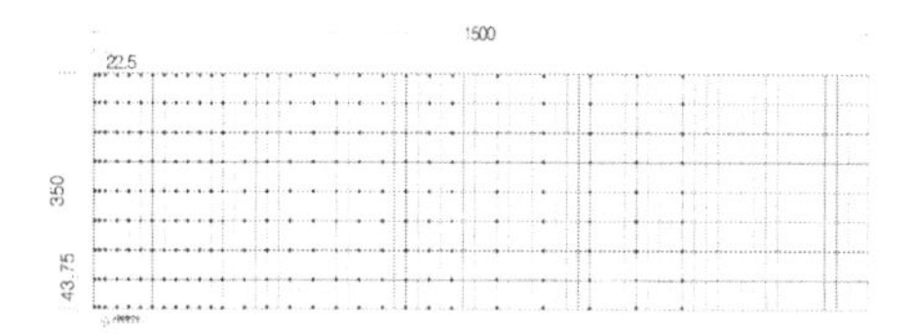

Figure 2. Map plan of greenhouse pressure measurers.

This paper used a wind velocity of 6 m/s. For the test wind direction, a wind pressure test was conducted using four wind angles (0°, 45°, 90°, and 135°). The test wind direction was used, as shown in Figure 3.

The mean wind pressure coefficient was obtained using Equation (1). Figure 4 shows the distribution of the mean wind pressure coefficients of the wide greenhouse according to the change of wind angle. The distribution of the mean wind pressure coefficients of the greenhouse ceiling changed as the wind angle changed. When the wind angle was 0°, a large mean wind pressure coefficient was measured at the entrance door ceiling. However, as the distance between the entrance door and ceiling increased, a distribution of smaller mean wind pressure coefficients was observed. The wind angle of 45° means the wind affected the corner of the wide greenhouse. The largest mean wind power coefficient was measured in the corner ceiling, and the measured value was measured to be as high as 80% larger than the value in the case of 0°. Smaller mean wind pressure coefficients were also observed with increasing distance from the corner in the case of a 45° direction. While the wind angle of 0° yielded a smaller mean wind pressure coefficient distribution on the overall ceiling surface, the wind angle of 45° showed results with a smaller distribution on the sides than the upper part of the ceiling surface. A more than 76% difference in mean wind pressure coefficients was observed between the side of the ceiling surface and the upper part in the case of a wind angle of 45°. A mean wind pressure coefficient of more than −1.17 was distributed on the upper part of the ceiling surface overall in the case of a wind angle of 90°. This can be attributed to the impact on the longer side of the vinyl greenhouse in the case of a wind angle of 90°, which was different from the local impact on the vinyl greenhouse in the cases of other wind angles. In the case of a wind angle of 90°, the maximum mean wind pressure coefficient was 20% larger than the case of a wind angle of 0° and 40% smaller than the case of a wind angle of 45°. However, upon examining the distribution range of the maximum mean wind pressure coefficient, it was found that the distribution was broad in the upper part of the ceiling surface in the case of a wind angle of 90°, while the distribution was focused on the corner of the ceiling in the cases of wind angles of 0° and 45°. Moreover, in the case of a wind angle of 90°, more than 88% of the difference in mean wind pressure coefficients was observed between the upper and lower parts of the ceiling surface.

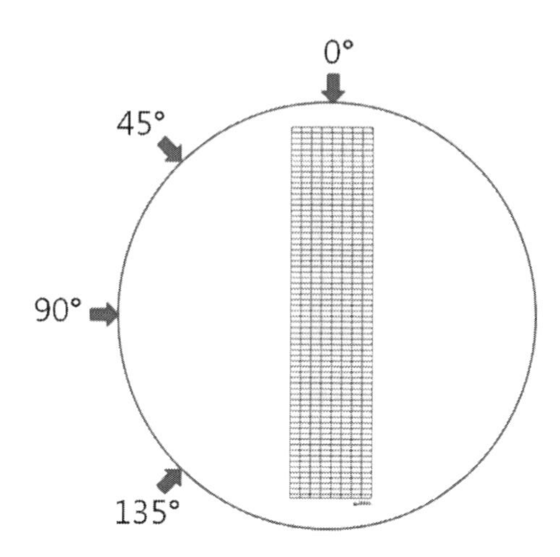

Figure 3. Test wind direction.

$$C_p = \frac{P}{P_{ref.}} \tag{1}$$

where, C_p = Pressure Coefficient. P = Pressure at pressure tap location. P_{ref} = Measure mean wind pressure at the reference height in the wind tunnel.

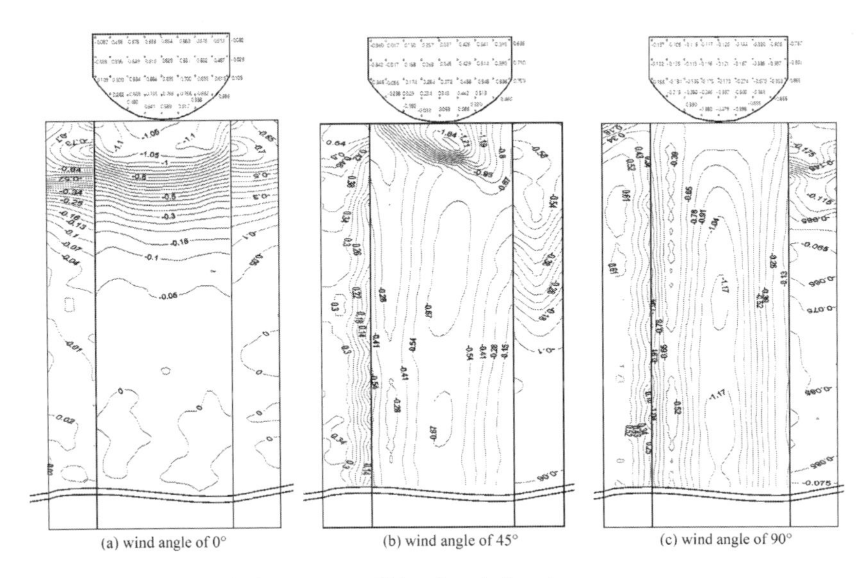

(a) wind angle of 0° (b) wind angle of 45° (c) wind angle of 90°

Figure 4. Distribution of mean wind pressure coefficient by wind angle.

3 SNOWDRIFT TEST

For testing the snowdrift patterns on structures using simulated snow, a dust generator and nozzle that could distribute snow inside the wind tunnel were used. A dust generator is a device that supplies sodium bicarbonate at a constant rate during a snowdrift test. Since the snow-blowing nozzle creates diverse snowdrift patterns depending on its location, determining the location was difficult. In this paper, snow was supplied at a rate of 2 cc per second through the discharge test. Figure 5 shows the nozzle installed in the dust generator and wind tunnel. Table 2 presents the snowdrift test model data for the wide greenhouse. Figure 6 shows the appearance of the snowdrift test model.

3.1 *Accumulation coefficient*

The experiment was performed by setting the speed of the experimental wind on the constantly supplied simulated snow to 1 m/s. The height of the greenhouse was set as a baseline for the experimental wind speed. For the baseline accumulation coefficient, a wind tunnel floor that had no impact on the greenhouse was chosen. The accumulation coefficient (R), which is obtained by dividing the snowdrift depth measured on a vinyl greenhouse roof by the snowdrift depth on the wind tunnel floor, can be written as Equation (2).

$$Accumulation\ coefficient(R) = \frac{The\ snowdrift\ depth\ on\ roof}{The\ snowdrift\ depth\ on\ the\ ground\ plate} \quad (2)$$

(a) Dust generator (b) nozzle

Figure 5. Dust generator and nozzle shape installed inside the wind tunnel.

Table 2. Snowdrift model data of wide greenhouse.

Proto type		Model			
Width (m)	Roof Height (m)	Width (mm)	Roof height (mm)	Scale	Velocity
15	5.2	214	74	1/70	1m/s

Figure 6. Snowdrift test model.

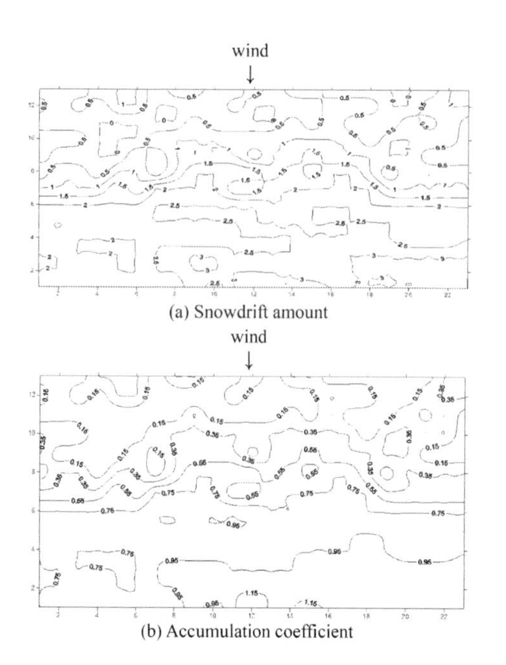

(a) Snowdrift amount

(b) Accumulation coefficient

Figure 7. Snowdrift depth in wide greenhouse roof and the distribution of accumulation coefficient.

Figure 7 shows the snowdrift depth on the wide greenhouse roof and the distribution of the accumulation coefficients. The experiment was performed using a 90° wind direction angle (the angle at which the average wind pressure coefficients are distributed on the broadest area of the roof surface in the wind pressure test). From the snowdrift test results, it was confirmed that snow accumulated more on the opposite side of the roof surface than on the roof surface of the wind-blowing direction. Similar to the snowdrift amount analysis, the accumulation coefficient distribution also showed higher snowdrift coefficients on the opposite side from the wind-blowing direction. In the wind pressure test, the distribution of negative pressure coefficients on the roof surface was highest on the upper part, with a lower density on the side corners. However, in the snowdrift test, the snowdrift amount in the lower part was found to be 33% larger than that in the upper part. This can be attributed to the movement of the snowdrift amount in the upper part to the lower part.

4 CONCLUSIONS

The test that investigated the distribution of mean wind pressure coefficients of a wide greenhouse and snowdrift patterns gave the following results.

In the distribution of mean wind pressure coefficients of the wide greenhouse, the magnitude of the coefficient that applied to the roof surface differed according to the wind direction angle. In the case of a wind direction angle in the short side (0°) and corner (45°) direction of the roof surface, a high negative wind pressure coefficient was measured on the local part of the roof. However, in the case of a wide-side (90°) wind direction angle, the wind pressure coefficients of negative pressure were largely distributed on the upper part of the roof surface. According to the analysis results of the snowdrift patterns of the roof surface, the snowdrift amount was broadly distributed on the upper part of the roof surface in the wind direction angle of the wide side. However, the snowdrift amount and accumulation coefficient were accumulated at least 33% more on the lower part of the roof surface than on the upper part in the snowdrift test, unlike in the wind pressure test. This can be attributed to the movement of the snowdrift amount in the upper part of the roof surface to the lower part.

ACKNOWLEDGEMENT

This research was supported by Basic Science Research Program through the National Research Foundation of Korea (NRF) funded by the Ministry of Education (No 2010-0024891).

REFERENCES

Blackmore, P.A. & Tsokri, E. 2006. Wind loads on curved roofs, *Journal of Wind Engineering and Industrial Aerodynamics,* 94: 833–844.

Hoxey, R.P. & Richardson, G.M. 1983. Wind loads on film plastic greenhouse, *Journal of Wind Engineering and Industrial Aerodynamics*, 11: 225–237.

Isyumov, N. & Mikitiuk, M. 1989. Wind tunnel model tests of snow drift on a two-level flat roof. *Proceedings of the 6th U.S. National Conference on Wind Engineering*, II: A6–47 to A6–58.

Okada, H., Tamura, Y., Uematsu, Y. & Kikitsu, H. 1999. Research project on propriety and reliability of designing wind load and wind pressure coefficients, *Proceedings of the Joint Meeting of the U.S.-Japan Cooperative Program in Natural Resources Panel on Wind and Seismic Effects*, 31: 79–85.

Tahouri, B., Toy, N. & Savory, E. 1990. Surface Pressure and Wake Flows Associated with Highly Curved Agricultural Buildings, *Journal of Wind Engineering and Industrial Aerodynamics*, 36: 319–327.

Taylor, D.A. & Schriver, W.R. 1980. Unbalanced snow distributions for the design of Arch-Shaped roofs in Canada. *Canadian Journal of Civil Engineering*, 7(4): 651–656.

Advanced Materials, Mechanical and Structural Engineering – Hong, Seo & Moon (Eds)
 ISBN: 978-1-138-02908-8

Shock wave speed of irregular honeycombs under dynamic compression

P. Wang, Z.J. Zheng & J.L. Yu
CAS Key Laboratory of Mechanical Behavior and Design of Materials, University of Science and Technology of China, Hefei, Anhui, P.R. China

S.F. Liao
National Key Laboratory of Shock Wave and Detonation Physics, Institute of Fluid Physics, CAEP, Mianyang, Sichuan, P.R. China

ABSTRACT: Strength enhancement and deformation localization are two typical features of cellular materials under dynamic crushing, which could be explained by the propagation of shock wave. A virtual 'test' of 2D Voronoi honeycombs under dynamic compression was carried out in this study. Cross-sectional stress method and strain field method were applied to estimate the shock wave speed in comparison with one-dimensional shock theory and several shock models. The relation is linear with a constant interval defined as the material impact parameter. The shock wave speeds predicted by local strain field method and cross-sectional stress method coincide but overestimated by the R-PP-L model. The R-PH model also overestimates the shock wave speed but better than the R-PP-L model. The shock wave speed predicted by the one-dimensional shock theory is very close to that predicted by the local strain field method and the cross-sectional stress method.

1 INTRODUCTION

Dynamic crushing behaviors of cellular materials have been extensively studied due to the excellent mechanical properties in energy absorption (Lu & Yu 2003). Deformation localization and strength enhancement have been observed in cellular materials under dynamic impact. The conception of "structural shock" was proposed by Reid & Peng (1997) to describe the shock-like propagation of deformation. The propagation of shock wave in irregular honeycombs under dynamic impact was studied by Zheng et al. (2005) and the shock wave speed was estimated from observing the change of the location of shock front by Liu et al. (2009).

Several shock models describing the dynamic behaviors of cellular materials have been proposed in the literature, such as the R-PP-L (rigid-perfectly plastic-locking) shock model (Reid & Peng 1997) the R-LHP-L (rigid-linearly hardening plastic-locking) shock model (Zheng et al. 2012) and the R-PH (rigid-plastic hardening) shock model (Zheng et al. 2013). Liao et al. (2013) proposed a strain field calculation method and obtained the shock wave speed through the strain distribution. Here, we propose a method to calculate the cross-sectional stress distribution from cell-based finite element models, and then give the shock wave speed. Finally, the relation between the impact velocity and shock wave speed is obtained and compared with those obtained from the shock models and the strain field calculation method.

2 NUMERICAL MODELS

The 2D Voronoi technique was employed to construct irregular honeycombs (Zheng et al. 2005). A sample of Voronoi honeycombs is illustrated in Figure 1. The cell irregularity k of the specimen is 0.3 and the relative density ρ_0/ρ_s is 0.1, where ρ_0 is the initial density of Voronoi honeycomb and ρ_s the density of matrix material. The length, width and thickness of the specimen are 50 mm, 50 mm and 1 mm, respectively. The average cell size, which is defined as the diameter of a circle whose area is equal to the average area of Voronoi cells, is about 2.5 mm.

The finite element code ABAQUS/Explicit was employed to perform this numerical test. The specimen shown in Figure 1 was modeled with S4R elements. The material of cell walls was assumed to be elastic, perfectly plastic with Young's modulus $E = 66$ GPa, Poisson's ratio $\nu = 0.3$, yield stress $\sigma_{ys} = 175$ MPa, and density $\rho_s = 2700$ kg/m^3. All nodes were constrained in the out-of-plane direction to simulate a plane strain situation. General contact was applied and the friction coefficient was assumed to be 0.02.

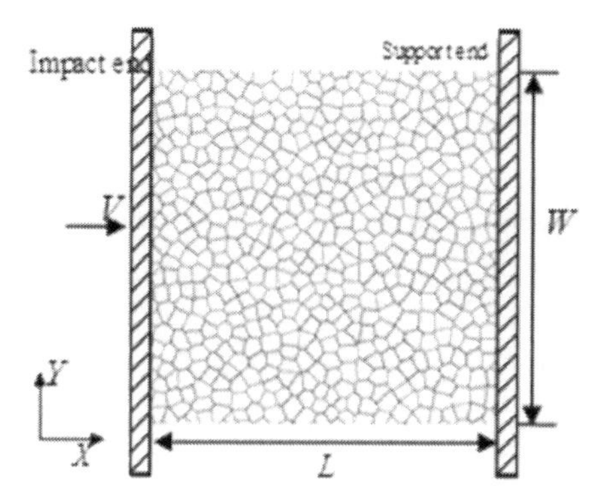

Figure 1. Finite element model of Voronoi structure.

The specimen was placed between two rigid plates: one is fixed and the other moves with a constant velocity, V, along the X direction, as schematically illustrated in Figure 1.

3 RESULTS AND DISCUSSION

3.1 *Shock wave speed based on cross-sectional stress method*

In this study, 100 cross-sections along the X direction were made to obtain the one-dimensional stress distribution. The nodal force generated by the element stress and the contact force caused by the inner contact of cell walls were extracted from the ABAQUS output data. Then, the cross-sectional internal force was obtained from the nodal force and the contact force. The corresponding cross-sectional stress was obtained as the cross-sectional force divided by the initial cross-sectional area of the specimen. The stress distribution of the Voronoi honeycomb at an impact velocity of 200 m/s for different nominal strains ε_N is shown in Figure 2, which shows the propagation of the shock front obviously.

Here, the location of shock front, Φ, is defined as the Lagrangian coordinate X where the absolute stress gradient reaches the maximum value, i.e,

$$\Phi = \left\{ X \mid \left| \partial\sigma / \partial X \right|_{\max} \right\}. \tag{1}$$

The stress distribution and the corresponding stress gradient at the nominal strain of 0.5 are shown in Figure 3. The location of shock front calculated by Equation 1 is marked with the red point. The shock wave speed could be obtained from the variation of the shock front location with the impact time, which is shown in Figure 4. It is found that the relation of the shock front location with impact time is linear which means that the shock wave speed is almost a constant during the response. Through linear fitting of the relation, the shock wave speed could be estimated by

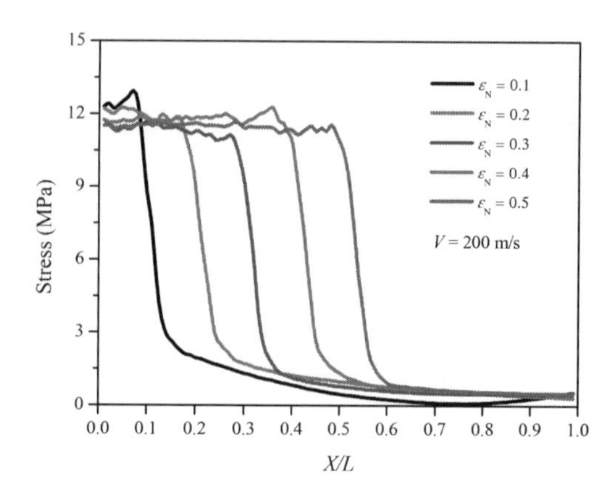

Figure 2. Stress distribution at an impact velocity of 200 m/s.

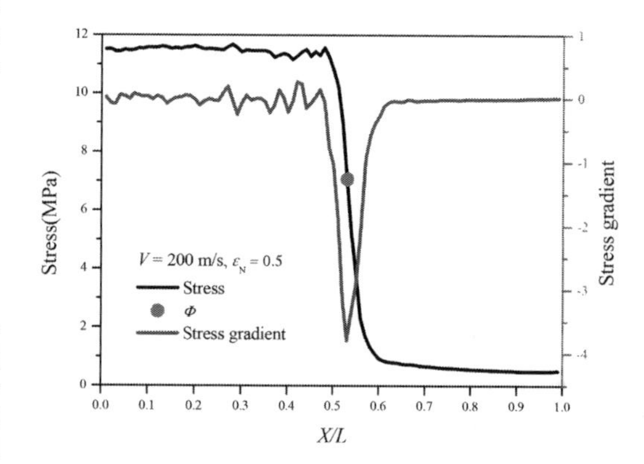

Figure 3. One-dimensional stress distribution and the corresponding stress gradient.

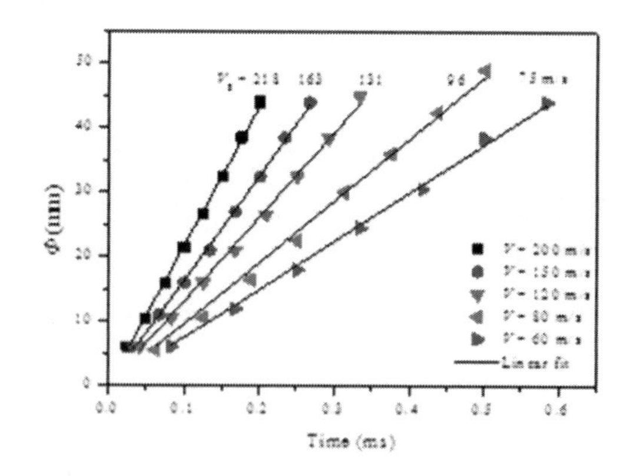

Figure 4. Variations of the shock front location with time for different impact velocities.

the slope of the line. Different shock wave speeds with respect to different impact velocities are obtained, as given in Figure 4, which shows that the shock wave speed increases with the increase of the impact velocity.

3.2 *Shock wave speed based on one-dimensional shock theory and cell-based numerical simulation*

The conservation relations of mass and momentum are formulated by

$$v_B(t)-v_A(t)=V_S(t)(\varepsilon_B(t)-\varepsilon_A(t)) \tag{2}$$

and

$$\sigma_B(t)-\sigma_A(t)=\rho_0 V_S(t)(v_B(t)-v_A(t)) \tag{3}$$

According to the stress wave theory (Wang 2007), where $V_S(t)$ is the shock wave speed, $\{\sigma_B, \varepsilon_B, v_B\}$ are the physical quantities behind the shock front and $\{\sigma_A, \varepsilon_A, v_A\}$ are those ahead of the shock front. Strain and particle velocity ahead of the shock front are zero in the idealization. R-PP-L model gives the shock wave speed as $V_S = V/\varepsilon_D$ (Liao et al. 2013), where ε_D is the densification (locking) strain and calculated to be 0.64 according to the efficiency of the energy absorption proposed by Tan et al. (2005) based on the quasi-static stress-strain relation shown in Figure 5.

For the R-PH model (Zheng et al. 2013) described as

$$\sigma=\sigma_0+\frac{C\varepsilon}{(1-\varepsilon)^2}, \tag{4}$$

the shock wave speed is formulated as

$$V_s=V+c, \tag{5}$$

where, c is a material impact parameter, defined by

$$c=\sqrt{\frac{C}{\rho_0}}. \tag{6}$$

Herein, c is calculated to be 26.4 m/s in this study.

Combing Equations 2 and 3 leads to

$$V_S(t)=\sqrt{\frac{\sigma_B(t)-\sigma_A(t)}{\rho_0(\varepsilon_B(t)-\varepsilon_A(t))}}, \tag{7}$$

which, means that the shock wave *speed* can be obtained if stress and strain ahead of and behind

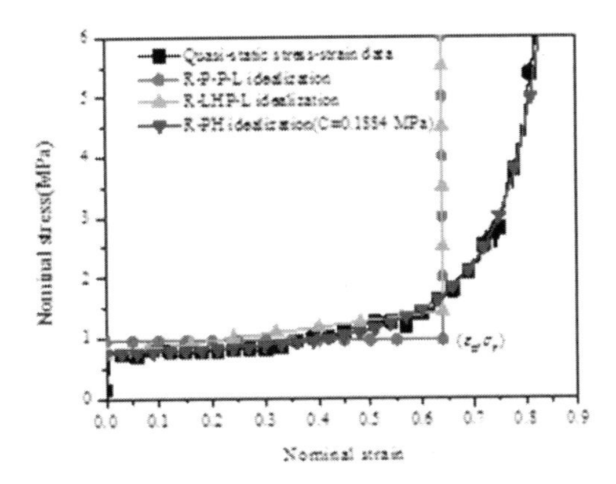

Figure 5. Quasi-static nominal stress-strain relation and its idealizations of Voronoi honeycomb.

the shock front are known. In this study, the stress distribution has been obtained as shown in Figure 2. The strain distribution can be obtained by the local strain field method proposed by Liao et al. (2013) in which the optimal cut-off radius of reference configuration is set as 1.5 times of the average cell radius R_c, as proved by Liao et al. (2014). The distributions of strain and stress at the nominal strain of 0.5 at the impact velocity of 200 m/s are presented in Figure 6.

To obtain the stress and strain across the shock front, the Lagrangian location range of $\Phi - 1.5R_c$ to $\Phi + 1.5R_c$ is excluded due to the averaging effect around the shock front. So the stress and strain ahead of the shock front are obtained by averaging over the Lagrangian location range on the right side of $\Phi + 1.5R_c$. The stress and strain behind the shock front are obtained by averaging over the Lagrangian location range on the left *side* of $\Phi - 1.5R_c$. Then, the shock wave speed based on Equation 7 at different nominal strains of the Voronoi honeycomb is obtained, as shown in Figure 7. The shock wave speeds corresponding to different impact velocities are obtained by averaging over the nominal strains. For simplification, this method will be called as one-dimensional shock theory method in the next content.

3.3 *Comparisons among the cross-sectional stress method, strain field method, one-dimensional shock theory and shock models*

Comparisons of the shock wave speed vs. the impact velocity relations predicted by the cross-sectional stress method, the local strain field method and one-dimensional shock models are shown in Figure 8. It is clear that the shock wave speed predicted by the R-PP-L model is overestimated than those of the strain field method and cross-sectional stress method, especially at higher

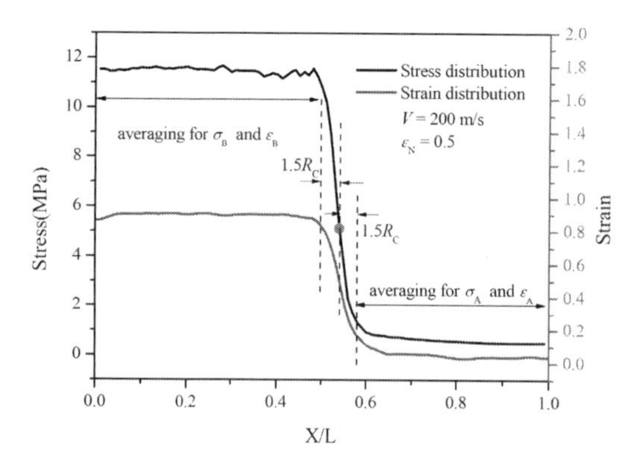

Figure 6. Distributions of stress and strain together with the idealizations.

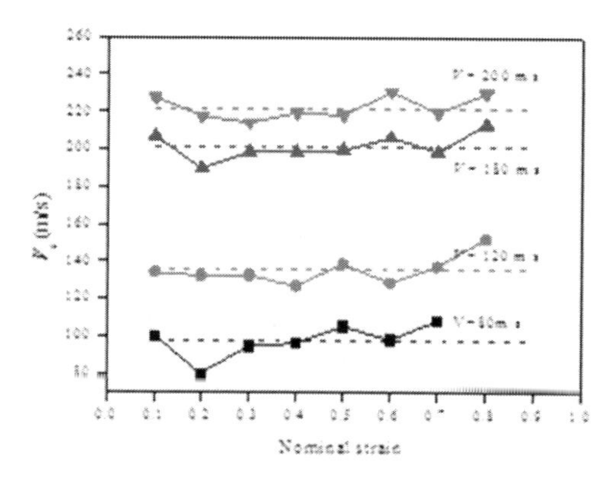

Figure 7. Variations of shock wave speed with nominal strain at different impact velocities.

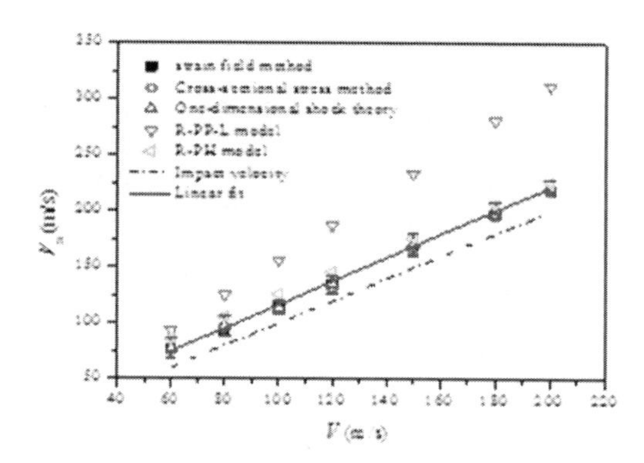

Figure 8. Comparisons of the shock wave speed between different methods.

impact velocities because the locking strain ε_D used by the shock models is a velocity-independent constant through the calculation while it should increase with the impact velocity as pointed by Zou et al. (2009).

According to the figure, a linear relation of the shock wave speed to the impact velocity is reasonable as pointed in Equation 5. Through linear fitting, the material impact parameters predicted by the local strain field calculation method and the cross-sectional stress method are respectively 16.0 m/s and 14.3 m/s with the standard derivations being 0.72 m/s and 0.88 m/s. It is shown that the two results almost coincide. Therefore, the material impact parameter c predicted by the R-PH model is also overestimated.

Linearly fitting the results of the one-dimensional shock theory leads the material impact parameter to be 18.4 m/s with the standard derivation being 1.2 m/s. It is found that the material impact parameter c is very close to that predicted by the local strain field calculation method and the cross-sectional stress method.

4 CONCLUSIONS

The propagation of shock front in Voronoi honeycombs under dynamic compression was investigated in this study. The shock wave speed vs. the impact velocity relation is linear with a constant interval defined as the material impact parameter. Comparisons of the relations predicted by the cross-sectional stress method, the local strain field calculation method, the one-dimensional shock theory and several shock models are made. The shock wave speeds predicted by local strain calculation field method and the cross-sectional stress method coincide. The shock wave speed is overestimated by the R-PP-L model and the R-PH model but the latter is better than the former. The shock wave speed predicted by the one-dimensional shock theory is very close to that predicted by the local strain calculation field method and the cross-sectional stress method.

ACKNOWLEDGEMENT

This work is supported by the National Natural Science Foundation of China (Project No. 11372308) and the Fundamental Research Funds for the Central Universities (Grant No. WK2480000001).

REFERENCES

Liao, S.F. et al. 2013. Dynamic crushing of 2D cellular structures: Local strain field and shock wave velocity. *International Journal of Impact Engineering*, 57: 7–16.

Liao, S.F. et al. 2014. On the local nature of the strain field calculation method for measuring heterogeneous deformation of cellular materials. *International Journal of Solids and Structures*, 51(2): 478–490.

Liu, Y.D. et al. 2009. A numerical study on the rate sensitivity of cellular metals. *International Journal of Solids and Structures,* 46(22–23): 3988–3998.
Lu, G.X. & Yu, T.X. 2003. *Energy absorption of structures and materials*. Cambridge, UK: Wood-head Publishing Ltd.
Reid, S.R. & Peng, C. 1997. Dynamic uniaxial crushing of wood. *International Journal of Impact Engineering,* 19(5–6): 531–570.
Tan, P.J. et al. 2005. Dynamic compressive strength properties of aluminium foams. Part I—experimental data and observations. *Journal of the Mechanics and Physics of Solids,* 53(10): 2174–2205.
Wang, L.L. 2007. *Foundations of Stress Waves*. Amsterdam, the Netherlands: Elsevier Science Ltd.
Zheng, Z.J. et al. 2005. Dynamic crushing of 2D cellular structures: A finite element study. *International Journal of Impact Engineering,* 32(1–4): 650–664.
Zheng, Z.J. et al. 2012. Dynamic crushing of cellular materials: Continuum-based wave models for the transitional and shock modes. *International Journal of Impact Engineering,* 42: 66–79.
Zheng, Z.J. et al. 2013. Dynamic crushing of cellular materials: A unified framework of plastic shock wave models. *International Journal of Impact Engineering,* 53: 29–43.
Zou, Z. et al. 2009. Dynamic crushing of honey-combs and features of shock fronts. *International Journal of Impact Engineering,* 36(1): 165–176.

Advanced Materials, Mechanical and Structural Engineering – Hong, Seo & Moon (Eds)
© 2016 Taylor & Francis Group, London, ISBN: 978-1-138-02908-8

Mechanical properties of steel fiber-reinforced HPFRCC according to compressive strength

J.J. Park, G.J. Park & S.W. Kim
Korea Institute of Civil Engineering and Building Technology, Goyang, Korea

ABSTRACT: Owing to the use of steel fiber and high fineness admixtures, HPFRCC (High Performance Fiber Reinforced Cementious Composites) exhibit outstanding compressive strength and ductility. These properties are attractive in protective structures since they offer the possibility to reduce significantly the cross-sectional dimensions. Accordingly, it is of interest to examine the mechanical properties of HPFRCC with regard to various strength levels for further applications to structures. Therefore, this study evaluates the mechanical performance of HPFRCC using steel fiber and developing compressive strength levels of 100, 140 and 180 MPa. The experimental results reveal that, under identical steel fiber ratio, the strength of HPFRCC improves its mechanical properties like the compressive strength, elastic modulus, flexural tensile strength and pullout strength.

1 INTRODUCTION

The economy of concrete made it historically one of the most preferred construction materials for the establishment of our modern infrastructural system (Aïtcin 2000). However, normal concrete presents deficiencies like the low strength developed in comparison to the weight and the brittleness. Among the achievements coping with such problems, High Performance Fiber Reinforced Cementious Composite (HPFRCC) achieves ductile behavior while developing significantly high compressive strength. HPFRCC exhibits remarkable strength development, ductility and durability to carbonation and alkali-reaction due to a very low Water-to-Binder ratio (W/B) compared to normal concrete, the absence of coarse aggregate, and the adoption of high fineness admixtures and steel fiber (Richard & Cheyrezy 1995).

In case of structural members supporting ordinary static loading, HPFRCC gains growing importance by enabling substantial reduction of the cross-sectional dimensions of the member (KICT 2006, Kim et al. 2006, Kamen 2009). Especially, the application of HPFRCC to civilian structures with protective functions presents the advantage of securing economic benefit by increasing the spatial exploitation as well as the stability to impact and blast. Accordingly, this study intends to investigate the mechanical properties of HPFRCC using steel fiber and with compressive strength levels of 100, 140 and 180 MPa for the mix design of HPFRCC providing adaptability to various strengths together with economic efficiency.

2 TEST METHOD AND PURPOSE

2.1 *Materials and fabrication of specimens*

The test in this study considers HPFRCC made of type-1 Portland cement, fine aggregates with grain size smaller than 0.5 mm, filler including 98% of SiO_2, Silica Fume (SF) produced in Norway, class-F Fly Ash (FA) from Korea, pulverized Blast-furnace Slag (BS) with fineness of approximately 6000, and polycarboxylate superplasticizer with density of 1.06 g/cm^3. The physical and chemical properties of the materials used in the mix composition are summarized in Table 1.

In order to evaluate the characteristics of HPRFCC with respect to the compressive strength, the specimens are fabricated by varying the W/B of the mixes so as to develop compressive strength levels of 100, 140 and 180 MPa, and by modifying the sand and filler. First, the 180 MPa-mix is manufac-

Table 1. Physical and chemical properties of materials.

Composition (%, mass)	Cement	SF	FA	BS
CaO	61.33	0.38	–	42.1
Al_2O_3	6.40	0.25	16.60	14.50
SiO_2	21.01	96.00	3.80	33.80
Fe_2O_3	3.12	0.12	5.58	0.01
MgO	3.02	0.10	0.82	5.80
SO_3	2.30	–	0.51	1.89
Specific surface (cm^2/g)	3,413	200,000	3,850	6,240
Loss ignition (%)	1.40	1.50	3.82	0.05
Density (g/cm^3)	3.15	2.10	2.13	2.91

tured with W/B of 20% and using sand from Australia (SiO_2 98%) and filler with average grain size of 4 μm. The 140 MPa-mix uses a larger W/B of 25%, Korean sand (SiO_2 90%) and filler with average grain size of 14 μm to lower the compressive strength. Similarly, the 100 MPa-mix adopts W/B of 30% and the sand and filler used for the 140 MPa-mix. The 140 MPa-mix exhibits relatively larger filler effect since the fineness of blast furnace slag is higher than that of fly ash. The mix is designed to increase the strength and the effect provided by the lower W/B. Moreover, steel fiber is introduced at volume fraction of 2% to HPFRCC to improve the toughness of HPFRCC. The steel fibers are admixed identically per considered level of strength for the purpose of examining the flexural strength with respect to the compressive strength. Table 2 arranges the physical properties of the steel fiber. The materials apart from the above mentioned ones are used identically in all the mixes. Table 3 lists the mix proportions adopted for the fabrication of the specimens.

HPFRCC was fabricated using a fan-type laboratory mixer with capacity of 120 liters and maximum speed of 100 rpm. Compaction was unnecessary since the mixed HPFRCC exhibited fluidity allowing self-leveling. The so-prepared specimens were subjected to 1 day of wet curing at temperature of 20°C and relative humidity of 65%. Then, after demolding, high temperature was conducted during 72 hours at temperature of 90 ± 2°C followed by 3 days of curing at 20°C and relative humidity of 65% prior to the test.

2.2 *Test method*

2.2.1 *Fluidity and compressive strength test*

The fluidity of fresh HPFRCC was measured by flow table test using the equipment suggested in ASTM C 1437. The compressive strength was measured as shown in Figure 1 using a UTM (Universal Testing Machine) with maximum capacity of 3,000 kN on ϕ100 × 200 mm cylinders in compliance with ASTM C 39. Sets of 3 LVDTs disposed at angle of 120° were installed to measure the displacement with respect to loading from which the elastic modulus of HPFRCC was calculated according to the formula proposed by ASTM C 469.

2.2.2 *Flexural tensile strength*

The flexural tensile strength was measured by 4-point bending test on 100 × 100 × 400 mm cubes in compliance with ASTM C 1609 (Figure 2). A UTM with maximum capacity of 200 kN was used for the test and the average value from 3 specimens is adopted.

2.2.3 *Pullout strength of fiber*

Pullout test was conducted on single fibers as shown in Figure 3 to evaluate the bond behavioral characteristics of the straight steel fiber according to the strength level of HPFRCC (Kim 2008).

The equivalent bond strength is calculated assuming only the bond strength developed through friction with identical size over the entire embedded length of the fiber. In this study, the embedded length is set as the half-length (6.5 mm) of the steel fiber (df = 0.2 mm, Lf = 13 mm).

In addition, the spacing between the grip system of the fiber and the top of the specimen was minimized so as to diminish the effect of the elongation of the fiber. The LVDT measuring the displacement was disposed to be exactly perpendicular to the direction of the load. Test was conducted at constant speed of 1 mm/min. The equivalent bond strength was derived from the pullout test using Equation 1 (Kim 2008).

$$\tau_{eq} = \frac{8 \times \text{Pullout work}}{\pi d_f L_f^2} \tag{1}$$

where, τeq = equivalent bond strength; Pullout work = total pullout energy required for pullout of fiber, defined as the area delimited by the load and slip; df = diameter of fiber; and, Lf = embedded length of fiber.

Table 2. Physical properties of steel fiber.

Type of fiber	Density (kg/cm³)	Tensile strength (MPa)	Length (l_f, mm)	l_f/d_f* (mm/mm)
Straight	7.8	2,500	13	65

*d_f = diameter of fiber (mm).

Table 3. Mix proportions of HPFRCC (in weight ratio).

Designation	W/B	Cement	SF	FA	BS	Filler	Sand	Superplasticizer	Steel fiber (volume fraction, V_f)	Flow (mm)
HPF 180	0.2	1	0.25	0	0	0.30 (4 μm)	1.10	0.02	2%	220
HPF 140	0.25	1	0.1	0	0.2	0.30 (14 μm)	1.10	0.0075	2%	240
HPF 100	0.3	1	0.1	0.2	0	0.30 (14 μm)	1.20	0.005	2%	245

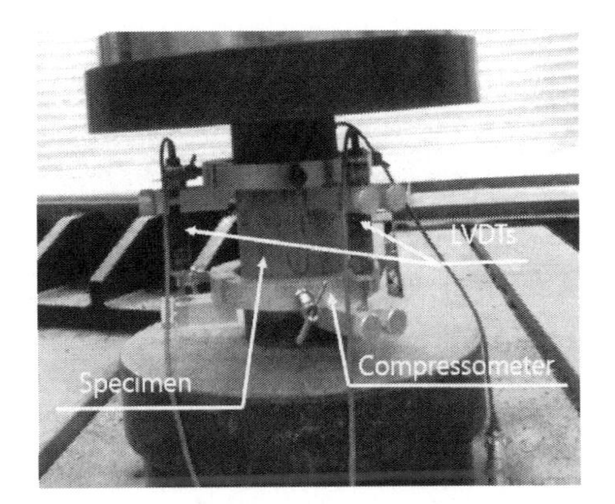

Figure 1. Test method for compressive strength.

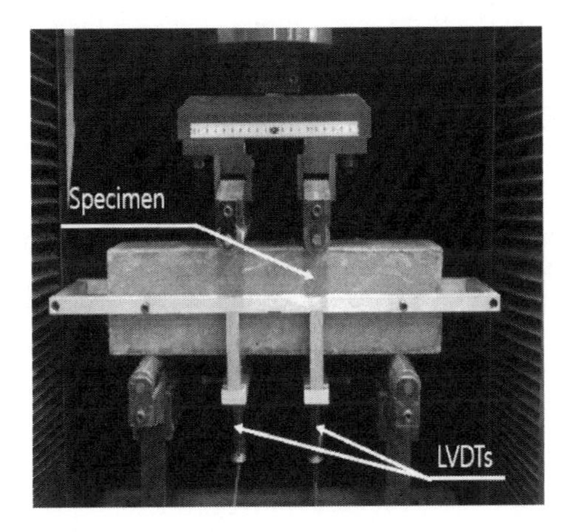

Figure 2. Test method for flexural tensile strength.

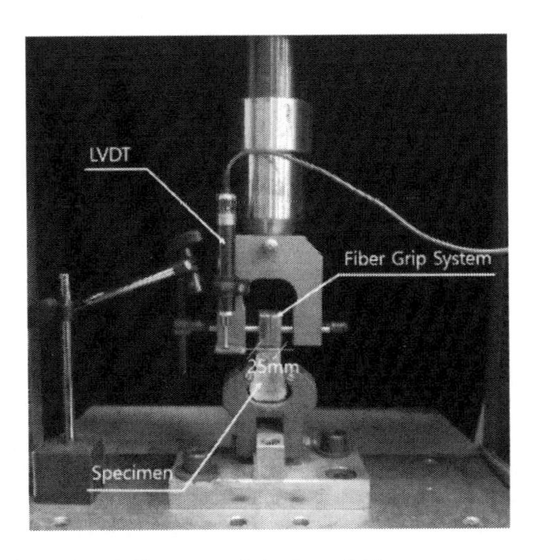

Figure 3. Pullout test of single steel fiber.

3 TEST RESULTS

3.1 *Compressive characteristics*

The flow measured in each mix runs around 210 ± 20 mm as listed in Table 3. Figure 4 plots the compressive strength and the elastic modulus according to the strain. The compressive strength appears to be 112 MPa for HPF 100, 150 MPa for HPF 140 and 202 MPa for HPF 180, which indicate that the intended design strength is satisfied. The elastic modulus is seen to increase by about 32% from 35.9 GPa at compressive strength of 112 MPa to 47.1 GPa at 202 MPa. The strain tends to increase with higher compressive strength.

3.2 *Flexural tensile characteristics*

Owing to the adoption of steel fiber, HPFRCC exhibits outstanding strain hardening behavior compared to normal concrete and shows improved toughness (Richard 1995). As shown in Figure 5,

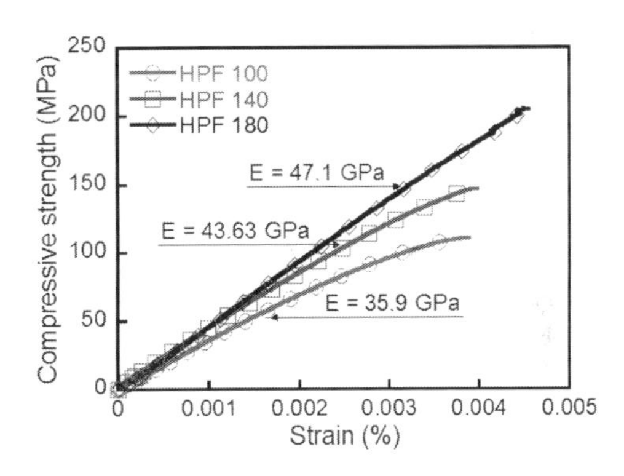

Figure 4. Compressive strength and elastic modulus of HPFRCC.

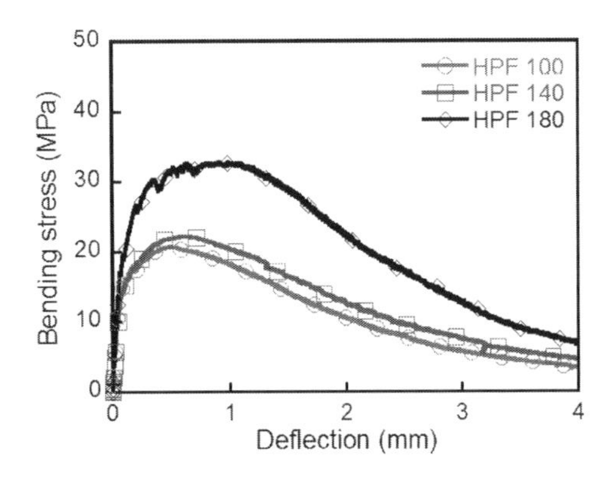

Figure 5. Flexural tensile strength of HPFRCC.

the flexural tensile strength is seen to be 19.1 MPa for HPF 180, 23.3 MPa for HPF 140, and 32.2 MPa for HPF 100. The flexural tensile strength increased by maximum 69% with respect to the increase of the compressive strength. The different flexural tensile strengths developed by the 3 mixes despite of their identical contents in steel fiber can be explained by the lower initial cracking strength of the matrix observed during the bending test for the mixes HPF 140 and HPF 100 with lower compressive strength than HPF 180. This lower initial cracking strength resulted in lower final flexural strength even in occurrence of strain hardening by the fiber after cracking.

3.3 *Pullout characteristics*

Figure 6 plots the pullout behavior of the straight steel fiber embedded in the matrix of HPFRCC according to the compressive strength. The indicated values correspond to the average values of 10 tested specimens for each mix.

In view of Table 4 summarizing the pullout test results, the maximum pullout load of HPF 100 is 28.9 N and the corresponding equivalent bond strength (τ_{eq}) runs around 6.7 MPa. Besides, the maximum pullout loads of HPF 140 and HPF 180 are respectively 34.6 N and 44.9 N, which represent an increase of 29.7% and 55% compared to HPF 100. The corresponding equivalent bond strengths are respectively 8.5 MPa and 9.7 MPa, representing an increase of about 29.8% and 45%. This means that the pullout resistance of the fiber in the matrix augments according to the increase of the compressive strength of the HPFRCC matrix.

In addition, even if the initial pullout load in all the mix compositions shows large difference, this difference reduces as much as the pullout length of the fiber lengthens to exhibit gradually similar behavior at the end. This confirms the results of a previous research (Kim 2008) which, in case of smooth surface of the straight steel fibers, observed the absence of mechanical resistance brought by the shape of the fibers during the pullout test and the existence of only frictional resistance acting on the surface of the fiber.

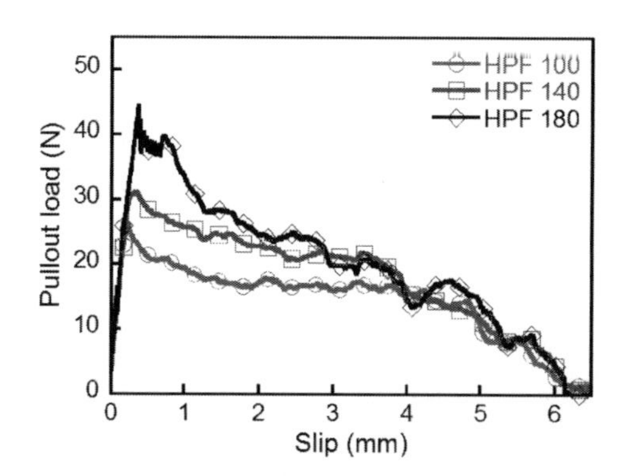

Figure 6. Pullout behavior of straight steel fiber embedded in HPFRCC matrix.

Table 4. Pullout behavior test results of single fiber in HPFRCC.

Designation	Slip at max. pull-out load (mm)	Max. pullout load (N)	Max. pullout stress (MPa)	Pullout energy (N-mm)	Equivalent bond strength (τ_{eq}, MPa)
HPF 180	0.29	44.9	1429.8	129.2	9.7
HPF 140	0.23	34.6	1100.4	113.0	8.5
HPF 100	0.14	28.9	918.8	88.9	6.7

4 CONCLUSIONS

This paper analyzed experimentally the mechanical characteristics of HPFRCC with respect to the compressive strength. The following conclusions could be derived from the results.

1. The elastic modulus of HPFRCC tended to decrease with lower compressive strength. For the flexural tensile strength also, the final flexural strength appeared to be smaller for HPFRCC with lower compressive strength despite of the occurrence of strain hardening by the fibers after cracking according to the difference in the initial cracking strength.
2. In view of the pullout behavior of HPFRCC, the pullout resistance increased with respect to higher compressive strength level of the HPFRCC matrix.
3. Consequently, in case of identical volume fraction of steel fiber, the strength of HPFRCC appeared to improve the mechanical performance like the compressive strength, elastic modulus, flexural tensile strength and pullout strength.

ACKNOWLEDGEMENT

This research was supported by a Construction Technology Research Project 13SCIP502 (Development of impact/blast resistant HPFRCC and evaluation technique thereof) funded by the Ministry of Land, Infrastructure and Transport.

REFERENCES

Aïtcin, P.C. 2000. Cements of yesterday and today—Concrete of tomorrow. *Cement and Concrete Research,* 30:1349–1359.

Chindaprasirt, P., Jaturapitakkul, C. & Sinsiri, T. 2005. Effect of fly ash fineness on compressive strength and pore size of blended cement paste. *Cement and Concrete Composites,* 27(4): 425–428.

Kamen, A., Denarié, E., Sadouki, H. & Brühwiler, E. 2009. UHPFRCC ARC tensile creep at early age. *Materials and Structures,* 42(1): 113–122.

KICT. 2006. *Development of technology to improve the durability of concrete bridges*, Report No. KICT 2006–89. Korea Institute of Civil Engineering and Building Technology: Korea.

Kim, D.J., Tawil, S.E. & Naaman, A.E. 2008. Loading rate effect on pullout behavior of deformed steel fibers. *ACI Material Journal,* 105(6): 576–584.

Kim, S.W., Kang, S.T. & Han, S.M. 2006. Characteristics and application of ultra high performance cementitious composite. *Magazine of the Korea Concrete Institute,* 18(1): 16–21.

Richard, P. & Cheyrezy, M.H. 1995. Composition of reactive powder concrete. *Cement and Concrete research,* 25(7): 1501–1511.

Wan, H., Shui, Z. & Lin, Z. 2004. Analysis of geometric characteristics of GGBS particles and their influences on cement properties. *Cement and Concrete Research,* 34(1): 133–137.

Advanced Materials, Mechanical and Structural Engineering – Hong, Seo & Moon (Eds)
© 2016 Taylor & Francis Group, London, ISBN: 978-1-138-02908-8

A study on design and visualization of enterprise knowledge on computer-aided innovation system

S.S. Chun
Korea Institute of S&T Evaluation and Planning, Seoul, South Korea

ABSTRACT: The enterprise knowledge map is created for compositing every context and networking generated semantic topic on business framework. And it is a knowledge space in progress, shaping system dynamics, business policy and events on the ground. In short, design and visualization of knowledge can be improved to understand by the analysis of a semantic networking and dynamic mapping of semantic topic. In this paper, we first analyze the semantic relationship and knowledge properties on organizing enterprise knowledge map and networks. We show that the way of visualization of knowledge content and designing of enterprise knowledge frameworks. For doing so, we let the user focus on some interested parts (called the semantic topic) rather than on a whole place space in a semantic-based knowledge map. Obviously, a set of places calculated in this way is a subset of every reachable place in knowledge map and networks. Therefore, generated semantic topic is used to optimize and translate between knowledge map and networks. And designed knowledge map and networks usually develop with new knowledge services and intelligent interface.

1 INTRODUCTION

1.1 *Knowledge map and networks*

Design and visualization of knowledge appears to be carried out differently to the way we are taught to understand the complex business process and system components, which is largely derived from the combination and structure of enterprise resource and knowledge relationship in domain contexts. The purpose of designing knowledge is to transform semantic topic, in such a way that the artifact being described is capable of producing those knowledge services (Gero 2006). In Figure 1, design and visualization of enterprise knowledge map on business domain analyzes business operations and produces information to help business users understand, improve and optimize business operations (Mario Roy 2003). And design and visualization of knowledge map in the business domain represents the semantic topic's elements and their relationships, which is labeled structure. In the visualization of the user interface design example, the topic's elements are the glazing and the frame, and their topology. In this paper, we first analyze the semantic relationship and knowledge properties on organizing enterprise knowledge map and networks. We show the way of visualization of knowledge content and designing of enterprise knowledge frameworks (Valverde & Sole 2007). For doing so, we let the user focus on some interested parts rather than on a whole place space in a semantic-based knowledge map.

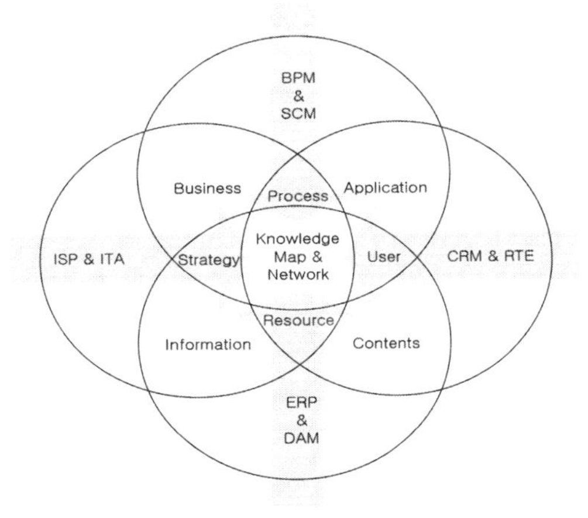

Figure 1. Resource and relationship of enterprise knowledge frameworks.

Obviously, a set of places calculated in this way is a subset of every reachable place in knowledge map and networks.

1.2 *Unified knowledge networks*

A prevalent and pervasive view of design and visualization is that it can be modeled using variables and decisions taken about what values should be taken by those variables (Devereausx and Ruecher 2006). The activity of designing of knowledge map and network is carried out with the expectation that the designed artifact will operate in the natural world and the social world (OCDE/GD102 1996). These impose constraints on the variables and their values. So, design could be described as a goal-oriented, constrained decision-making activity (Yamamoto 2005).

Design description of knowledge model represents the topic elements and their relationships. Computer-aided drafting systems have become the means by which the structure is transformed into a design description (Chun and Kwon 2002). Knowledge model is used to specify scopes in which one is interested, Computation Tree Logic (CTL) as Kripke structure; forward reachability analysis is used to find out a set of states inside it. Knowledge model is a tuple $<X, X_0, R, T, L>$, where

- X is a finite set of places.
- $X_0 \subseteq T$ is the set of initial places.
- $R \subseteq X \times X$ is the transition relation of places, which is assumed to be total: $\forall x \in X \bullet \exists s' \in X \bullet (x, x') \in R$.

That is, for every place $x \in X$, there exists a successor $x' \in X$ with $(x, x') \in R$.

- T is a finite set of atomic propositions,
- L: $T \rightarrow 2T$ assigns to each place the set of atomic propositions that are true in that place.

We show that the generation of a set of places calculated is a subset of every reachable place in knowledge map and networks. Therefore, generated place invariant is used to optimize and translate knowledge map and networks.

2 VISUALIZATION OF KNOWLEDGE

2.1 *Design prototypes*

A design prototype is a conceptual schema for representing a class of a generalized grouping of elements, derived from like design cases, which provides the basis for the commencement and continuation of a design (Filman 2003). Design prototypes do this by bringing together in one schema all the requisite knowledge appropriate to that design situation (Shah 2007). In addition, there is knowledge concerning the design prototype itself; this comprises the typology of the design prototype that identifies the broad class of the design prototype, as shown in Figure 2. In this case, $X_1 = \{x_1\}$ and $R=\{(x_2, x_1), (x_2, x_3), (x_3, x_1), (x_4, x_5), (x_6, x_5), (x_9, x_{10}), (x_7, x_{10}), (x_{12}, x_{13}), (x_{16}, x_{17}), (x_{14}, x_{15})\}$. $(x, x') \in R$ means that s moves to x' in a step. That is, x' is an image of x. Given the subset T of places, the set of images of T is image $(T) = \{x' \mid \exists x \bullet x \in T \wedge (x, x') \in R\}$. For example, image $(\{x_0\}) = \{x_1, x_2\}$;

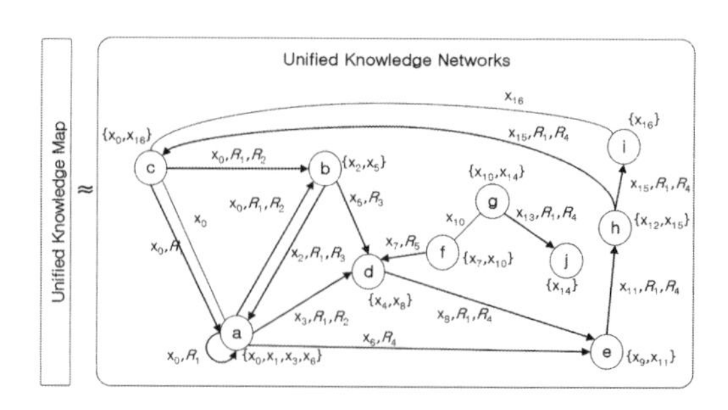

Figure 2. A translated knowledge networks from knowledge map.

that is, from x_0, we can move to both x_1 and x_2 in one step. A path is an infinite sequence of places in which each consecutive pair of places belongs to R. A place is reachable if it appears on some path starting from some initial place. In our example, x_{16} is reachable since there is a path $x_0, x_1, x_2, x_5, x_6, x_7, x_8, x_9, x_{11}, x_{12}, x_{15}, x_{16}$ that leads to that place, while x_{10} is unreachable since there is no path that leads to x_{10}, x_{13}, x_{14}. Every reachable place is computed by successively applying the function image from initial places X_0 until the fixed point is reached. The following procedure is to find out every reachable place. Thus, Pi is called the fixed point, which includes every reachable place. For convenience, let Q = Pi denote the set of reachable places. In Figure 2, $Q = \{x_0, x_1, x_2, x_5, x_6, x_7, x_8, x_9, x_{11}, x_{12}, x_{15}, x_{16}\}$. Only x_{10}, x_{13}, x_{14} does not belong to Q, since it is unreachable from the initial place x_0.

The length of place invariant generated in this naive way is n*m, provided $|Q| = n$ and $|T| = m$. Although it contains a tremendous amount of useful information, it is likely to be too complex for the user to understand.

Early models of organizational knowledge transfer looked at knowledge as if it was an object that could be passed on from the creator to a translator who would adapt it in order to transmit the information to the user (Tushman and Nadler 1978). This model implies a hierarchical top-down relationship between the generator of knowledge who holds the resource and the user who is locked in a dependency stance.

2.2 *Analysis of knowledge maps*

In general, the user is interested in some parts of a model rather than a whole place space. In this sense, scope is the extent of the place sequence over which the user has an interest (Ernst 2000). Dwyer classified basic kinds of scopes as follows. With the use of CTL, scopes can be formally specified as CTL formulas. To find out every place in a given scope, $\cdot\varphi$ is replaced with true (William Chan 2000). For example, 'AG φ' means φ is true globally. Thus 'AG true' implies every reachable place in a model. Moreover, 'A(true ∨ AG(¬α) W α)' implies a set of reachable places until r is reached (Chun Seung Su 2015).

2.3 *Backward reachability*

Given a scope in CTL, the function pre∀(P) is used to find out a set of places with backward reachability analysis:

$$pre\forall(P) = \{x \in X \mid \forall x' \bullet \text{if } (x, x') \in R \text{ then } x' \in P\}$$

It takes a subset P of places and returns the set of places that can make a transition only into P; that is, the set of predecessors of places in P. For any CTL formula φ, let ||[φ]|| denote the set of places, in which φ is true. Suppose we want to derive the global place invariant. Given a scope, the following shows how to find out the set of places in the scope: $\|[AG\ true]\| = \nu Z. (\|[true]\| \cap pre\ \forall\ (Z)) = X$.

Since the final result has meaningless places including $\{x_7, x_{10}, x_{13}\}$, it is clear that only $\{x_0, x_1, x_2, x_3, x_6\}$ is under the scope of 'place invariant before d'. Also, the union model as network composition shown in Figure 2 is only $\{\{x_0\}, \{x_0, x_1, x_3\}, \{x_2, x_5\}\}$. Except for them, the others are redundant, which makes place invariant lengthy. This is why another analysis technique is needed.

2.4 *Forward reachability*

The function post∃(P)

$$post\exists(P) = \{x' \in X \mid \exists x \bullet x \in P \wedge (x, x') \in R\}$$

takes a subset P of places and returns the set of places that can make a transition into P; that is, the set of successors of places in P. For any CTL formula φ, the set of Rφ(P) is defined as follows: $R\varphi(P) = \mu Z. ((P \cup post\exists(Z)) \cap \|[\varphi]\|)$

The μ is the least fixed-point operator. Rφ(P) is the set of places reachable from P going through only the places that satisfy φ. For example, in the union network of Figure 2, the set of places reachable from initial places going through only the places that satisfy ¬d is $R\neg t(\{x_0\}) = \mu Z.((\{x_0\} \cup post\ \exists(Z)) \cap \|[\neg d]\|) = \{\{x_0\}, \{x_2, x_5\}, \{x_0, x_1, x_3\}\}$.

The algorithm Find is defined below with the help of the functions Rφ (P) and post∃(P): it takes a CTL formula φ and a subset P of places, and returns the set of places reachable from P, in which φ is true. The following examples show how to find out a set of places against each scope. In this paper, compared with backward reachability analysis, forward reachability analysis gives a minimal set of places in a given scope. Given a model in Figure 2, for instance, the set Q of places in the scope 'before d' with forward reachability analysis is = $\{\{x_0, x_{16}\}, \{x_2, x_5\}, \{x_0, x_1, x_3\}\}$. Thus, optimized place invariant J in this scope is $J = (\neg a \wedge \neg b \wedge c \wedge \neg d \wedge \neg e \wedge \neg f \wedge \neg g \wedge \neg h \wedge \neg i \wedge \neg j) \vee (a \wedge \neg b \wedge \neg c \wedge \neg d \wedge \neg e \wedge \neg f \wedge \neg g \wedge \neg h \wedge \neg i \wedge \neg j) \vee (\neg a \wedge b \wedge \neg c \wedge \neg d \wedge \neg e \wedge \neg f \wedge \neg g \wedge \neg h \wedge \neg I \wedge \neg j)$.

Since $J \Rightarrow I$, J is a weaker, but comprehensible, place invariant than I that is the strongest place invariant.

3 APPLICATIONS OF KNOWLEDGE MAP

3.1 *Knowledge base for computer-aided innovation*

The participation of users at every phase of the knowledge development process has been identified as a key factor for its subsequent adoption (Bryant 1986). Informal communication networks are at the heart of the knowledge diffusion process. It is through them that peer stabile behavior and group norms are created, which will ultimately lead to the adoption of visualization and analyzing knowledge, as shown in Figure 3 and Figure 4. Openness to new knowledge is much easier when users need it (Dou 2014). Understanding users' need and providing information when the timing is appropriate of prime importance cannot be ignored by researchers. This study utilizes the notation of Kripke structure in designing resources, processes and services of the system. Kripke structure is the most optimal graphic notation for formal specification and could make verification of attitude of

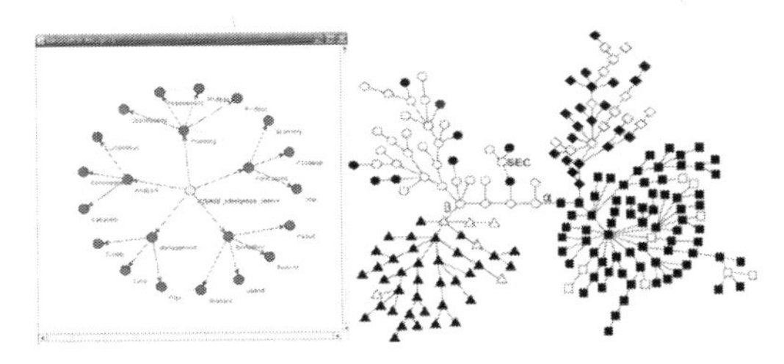

Figure 3. Visualization and analysis of knowledge networks.

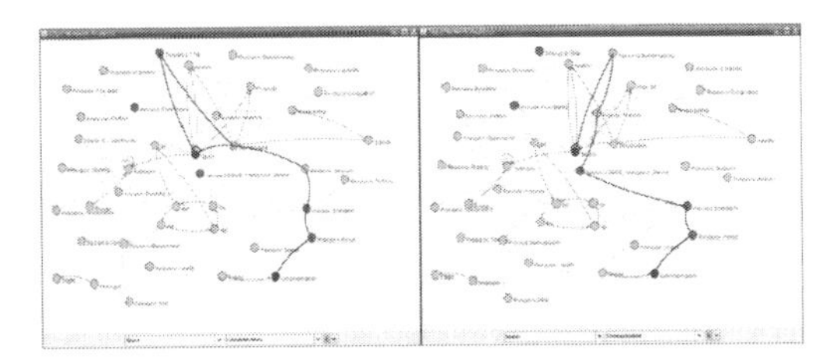

Figure 4. Trace and reachability analysis of knowledge networks.

the logic model (Liu 2006). And to build the service system of science of science policy, the definition should first be made on the processes of the upper level. Then, it is needed to classify the flow of work, activities, data and control according to component function and analyze specific model processes. For the services, the process model is extremely important because decision-making on R&D investment, establishment of strategies, and evaluation and adjustment of R&D programs are all sequentially processed. This is the reason that evidence-based knowledge is utilized in the activities of policy adoption, implementation and evaluation including decision-making. Therefore, integration of data, information, and knowledge should be designed in line with the flow of policy-making, considering semantic-based knowledge structuration. To classify information and knowledge on technology and innovation policy, the study understands in advance classical definition and typology of knowledge as well as criteria of classification, which was then abstracted and included in the processes of policy decision-making. Establishing knowledge database contributes to analyzing and structuring information so that more profound causes of specific phenomena could be captured and understood beyond simple understanding of disclosed superficial facts. It also refines the processes step by step by conducting an association analysis of stakeholders, information flow and models.

When designing a specific integration model, each Situation (S) is represented by unit of Topic (T), which equals Term (T) used in the natural language. The integration model conducts text-mining through processing all documents and texts within the system into the natural language, resulting in abstracting representative terms. It then generates a knowledge model through the analysis of relations between topics and layered structures of resources. Related to processes and tasks of the knowledge model, Address of document repository is defined as a process and Task is defined as a name of the specific document directory.

3.2 *Knowledge system*

For the purpose of establishing the service of computer-aided innovation policy, it is required to collect data on science and technology and systematically manage them. Next, tasks should be analyzed. Internal data and information connection as well as function are grasped as a result and then finally components are to be structured. Knowledge base integration of policy information has progressed through structuration of task-specific data and information. Component by process is defined as task flow and relation, and designed so as to combine an integrated information system and service.

With regard to collection of policy information, information has been collected on the basis of a component supporting tasks, as presented in Table 1. Unstructured information such as web documents, internal documents, thesis and patent specifications has been utilized for trend analysis based on text-mining and keyword abstraction, and presented through network visualization. Structured information such as program, project, performance and workforce could be analyzed statistically by exploiting on-line analysis.

The relation between components is established based on a relation between major task flow and information. For supporting service of innovation policy, components are configured by a process and organized by service, as summarized in Table 1.

Table 1. Specifications of the five test specimens.

Process	Components
Technology Policy	• Integrated information retrieval and management • Technology monitoring/Trend analysis on a real-time basis
Forecasting	• Community for future prediction • 2D matrix Delphi survey • Visualization of Roadmap
Planning	• Medium- and long-term plans, database on policy trends, other data query
Investment strategy	• DB for investment planning at a micro level and process supporting • Project analysis and management • Links to information on R&D programs
Validity analysis	• Benefit analysis of R&D • Instruments for ripple effect analysis • Instruments for cost/paper analysis
Investigation analysis	• Search function for statistical indices • DB of statistical tables and graphs
Program coordination	• Program track records/Budget planning • Budget requests, DB of mid-term project plans and deliberation materials
Program evaluation	• Query for in-depth evaluation results • Expert committee management

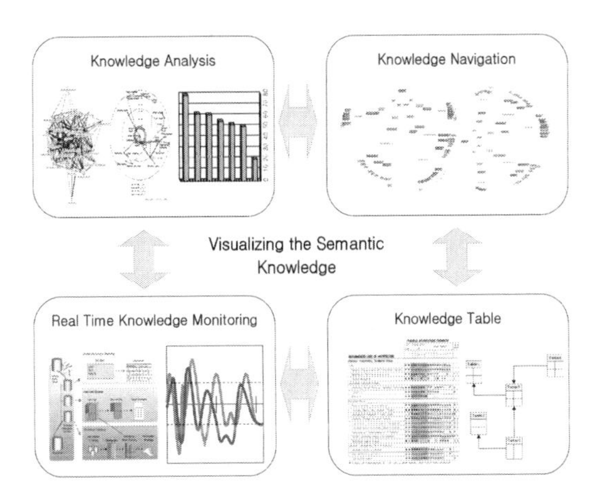

Figure 5. Visualization and analysis components of knowledge system.

In addition, effective component arrangement has required the realization of associated structure of policy data, management of document profiles, and cross-reference modeling. It has also led to the restructuring of general data and information into a process-based knowledge system.

3.3 *System utilization of knowledge models*

The adoption and utilization of new knowledge by group or a society often means the rejection of past practices, which may also have an impact on current political, economic or cultural equilibrium in the social system. The legitimacy of a new knowledge is then validated according to the values, the beliefs and the culture of potential users. All these semantic topics (called the knowledge factor) have to be taken into consideration if one wants to ease the process of generation, diffusion and utilization of knowledge within target groups. Knowledge does not exist without the context in which it is used. In other words, one should view knowledge less as a product or thing and more as a process used by a group of individuals to make sense of their world, as shown in Figure 5. The appeal and potential of knowledge visualization is increasingly recognized in a wide range of knowledge management (Campos 2013). In this paper, the visualization of enterprise knowledge networks, shown in Figure 5, aims to produce graphical representations of abstract knowledge structure for human users, and to rely on the enterprise descriptive framework of resources that can be utilized by the application agent. In this paper, we outline the origin of knowledge visualization and some of the latest advances in relation to the network analysis, as shown in Figure 5.

4 CONCLUSIONS

In previous work, every reachable place was to be considered to generate place invariant. In this paper, we first analyze the semantic relationship and knowledge properties on organizing enterprise knowledge map and networks. We show the way of visualization of knowledge content and designing of enterprise knowledge frameworks. Therefore, generated semantic topic is used to optimize and translate between knowledge map and networks. And designed knowledge map and networks usually develop with new knowledge services and intelligent interface. The complexity of place invariant is strongly dependent on the size of places to be considered. We hope that these results of the present study demonstrate the relevance of formal methods for a knowledge network-based business intelligence system in particular.

REFERENCES

Bryant, R.E. 1986. Graph-Based Algorithms for Boolean Function Manipulation, *IEEE Transactions on Computer*, 35(8): 677–691.

Campos, J.C. & Machado, J. 2013. A specification patterns system for discrete event systems analysis, *International Journal of Advanced Robotic Systems,* 10: 315–327.

Chan, W. 2000. *Temporal Logic Queries*, Proceedings of CAV 2000, LNCS 1855, Springer 1855: 450–463.

Chun, S.S. 2015. Effective Extraction of State Invariant for Software Verification, *Applied Mechanics and Materials,* 752–753: 1097–1104.

Chun, S.S. & Kwon, G.H. 2002. *Generation of State Invariants in Scopes*, Proceedings of International Conference on Computer and Information Science, 203, Aug.

Devereausx, Z.D. & Ruecher, S. 2006. Online issue mapping of International news and information design, *HUNANIT8*. 3: I-27.

Dou, W. 2014. *A Model-Driven Approach to Trace Checking of Temporal Properties with OCL*, Open Repository and Bibliography, S&T Centre, University of Luxembourg.

Ernst, M.D., Cockrell, J., Griswold, W.G. & Notkin, D. 2000. *Dynamically Discovering Likely Program Topics to Support Program Evolution*, Software Engineering, 1999. Proceedings of the 1999 International Conference on LA, CA, USA, 213–224.

Filman, R. 2003. *Semantic Services*, IEEE Internet computing, 4–6, July.

General Distribution OCDE/GD102, 1996. *The Knowledge based Economy, organization for economic cooperation and development*, OECD, Paris.

Gero, J.S. 2006. *Design Prototypes: A knowledge representation schema for design*, NSW 2006 Australia.

Liu, W.Y. 2006. *Using IDEF0/Petri net for Ontology based task knowledge analysis: the case of emergency*

response for debris flow, System Sciences, HICSS 4(04–07):76–76, Jan.
Mario, R., Robert, P. & Lise, D. 2003. Knowledge Networking: A strategy to Improve workplace Health & Safety Knowledge Transfer, *Electronic Journal on Knowledge Management,* 1(2): 159–166.
Shah, H. & Kourdi, M.E. 2007. *Frameworks for Enterprise Architecture*, IEEE IT Pro, Sept. 36–41.
Tushman, M.L. & Nadler, D.A. 1978. Information processing as an integrating concept in organization design, *Academy of Management Review*, 3(3): 613–624.
Valverde, S. & Sole, R.V. 2007. *Topology and evolution of technology innovation networks*, SPS2007, Physical Review E76.
Yamamoto, Y. 2005. Interaction design of tools for fostering creativity in the early stages of information design, *International Journal of Human Computer Studies,* 63: 513–535.

Advanced Materials, Mechanical and Structural Engineering – Hong, Seo & Moon (Eds)
© 2016 Taylor & Francis Group, London, ISBN: 978-1-138-02908-8

Nuclear containment structure mock-up test for performance evaluation of High-Performance Concrete

D.G. Kim & E.A. Seo
Korea Institute of Civil Engineering and Building Technology, Goyang, South Korea

ABSTRACT: Mock-up testing was performed to evaluate the workability and mechanical properties of High-Performance Concrete (HPC) applicable to nuclear containment structures by varying the admixture type, replacement ratio, and curing temperature (5°C, 20°C, 40°C). HPC test specimens were fabricated to simulate a nuclear containment structure. Optimal fluidity of HPC with high admixture contents was achieved by adding 0.5%–1% polycarboxylic-type superplasticizer. The slump and void content of the HPC satisfied the quality standards. The initial and final setting times were shortest at 40°C and increased with increasing admixture content and blast-furnace slag replacement ratio. The core compressive strength reached the target strength in the 5°C and 20°C specimens after 28-day curing. The 40°C specimen exhibited substantially elevated initial compressive strength, which decelerated at late-age stages. The mock-up test demonstrated that all the HPC mixes satisfied the quality standards of workability and target compressive strengths under their respective temperature conditions.

1 INTRODUCTION

Currently, approximately 430 Nuclear Power Plants (NPPs) are in operation worldwide because of their well-recognized resource efficiency, which is much higher than that of oil-fired power plants, and approximately 150 NPPs are planned for construction. After the 2011 Fukushima nuclear disaster, however, the structural safety of nuclear containment structures has become a global issue. Most safety-related NPP structures, including the nuclear containment structure, are concrete structures. In particular, the minimum wall thickness of a nuclear containment structure is 1.2 m (ACI 349 2006) to ensure protection against internal and external threats such as the release of radioactivity and airplane crashes or terrorist attacks. Concrete is a complex material, and its applicability should be tested based on its required performance and the environments to which it is exposed. There is a need for developing technology to prepare concrete that can ensure the structural safety of NPP structures and extend their service lives.

In this context, increasing the strength of High-Performance Concrete (HPC) for NPP structures from 6000 psi to 8000–10000 psi is under consideration, which involves various mixture techniques using pozzolanic admixtures such as Fly Ash (FA), Ground Granulated Blast-furnace Slag (GGBS), and Silica Fume (SF). However, the quality of HPC tends to deteriorate with increasing admixture replacement ratios. Therefore, quality improvement measures of HPC depending on the extent of admixture replacement are proposed (Ferraris et al. 2006, Nehdi et al. 1998).

The aim of this study was to determine the admixture type and applicability for HPC produced using various mixing techniques to introduce admixtures. To this end, we performed cement paste flow tests. The superplasticizers used for the test specimens were Lignin Superplasticizer (LS), melamine superplasticizer (PMS), naphthalene superplasticizer (PNS), and polycarboxylate superplasticizer (PC). As quality standards, we set the initial fluidity and slump loss at 150 mm or higher and 30%, respectively, 1 h after casting.

Additionally, we conducted a mock-up test on the HPC mixes at ambient temperature (20°C, Korea) and under extreme environments at hot (40°C) and cold (5°C) temperatures to evaluate their applicability under the respective temperature conditions (Kim & Han 2005). The test specimens had dimensions of 2.0 m (length) × 1.2 (width) × 1.0 (height), and the target compressive strengths were set at 6000, 8000, and 10000 psi.

2 H TYPE AND DOSAGE OF SUPERPLASTICIZER FOR QUALITY IMPROVEMENT OF HPC MIX

To determine the type and dosage of superplasticizer that can ensure the quality of HPC, we performed flow testing on the concrete mixes listed in Table 1. The four most commonly used superplasticizers in concrete mixes were used: LS, PNS, PMS, and PC. Table 2 presents the fundamental physical properties of these superplasticizers.

In a preliminary experiment, we tested the applicability of these four superplasticizers for HPC. As quality standards, we set the initial fluidity to 150 mm or higher and the slump loss at 30%, 1 h after casting. The test results identified PC as the most efficient superplasticizer in ensuring superior quality of all the HPC mixes irrespective of the target compressive strength (Figure 1).

Table 1. HPC mix and curing conditions.

No.	Mix	Compressive Strength (psi)	W/B (%)	Curing temperature
1	FA20	6000	40	20°C
2	BS50	8000	34	20°C
3	BS65SF5	8000	34	40°C
4	SF5	8000	34	5°C
5	SF5	10000	28	5°C

*FA: Fly Ash, BS: Blast furnace slag, SF: Silica fume. *Notation example: BS30FA25SF5 (30% blast-furnace slag, 25% fly ash, and 5% silica fume in a binder).

Table 2. Rheological properties of superplasticizers.

Type*	Specific gravity at 20°C	Viscosity (cp)	pH	Solid content (%)
LS	1.24	100–120	4.2–6.4	45
PNS	1.21	150–170	7.0–8.5	40
PMS	1.16	40	10–11	30
PC	1.05	120	6–8	20

LS: lignin superplasticizer, PNS: naphthalene superplasticizer, PMS: melamine superplasticizer, PC: polycarboxylate superplasticizer.

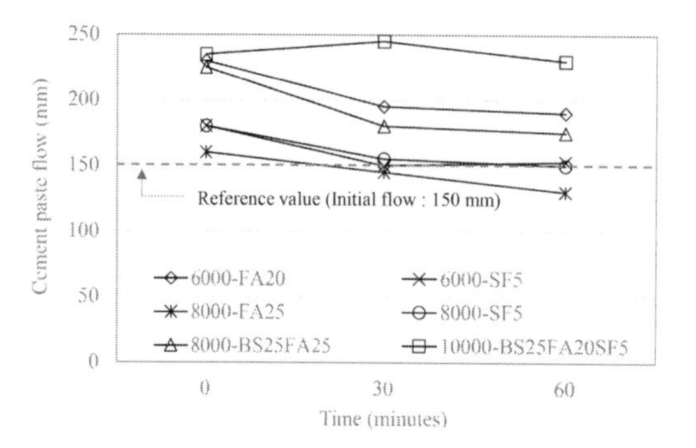

Figure 1. Result of application of PC to HPC mixes.

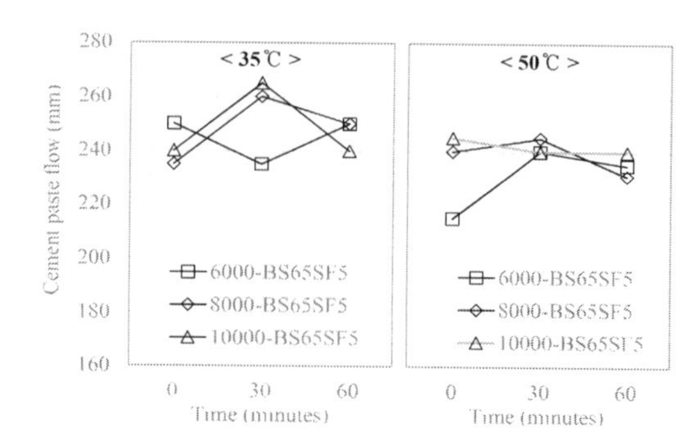

Figure 2. Results of mix flow testing on HPC with 0.5% polycarboxylate superplasticizer (PC) under hot temperature conditions.

Specifically, the appropriate mix dosage of PC for ensuring the HPC quality was observed to be 0.5%–1.0% of the total binder content.

Additionally, flow testing was performed on the 20°C, 35°C, and 50°C specimens of PC-added HPC to determine the superplasticizer dosage for different conditions of use (Figure 2). The optimal dosage was observed to be 0.5% at 35°C and 50°C. Because material segregation occurred at 20°C, however, the dosage should be reduced at this temperature.

3 MOCK-UP TEST

3.1 *Mix design of HPC & detail of specimens*

For performance evaluation of HPC applicable to NPP structures according to the target compressive strengths, we subjected the four HPC mixes (Table 3) to a mock-up test.

The variable for the flow test was the curing temperature of the mix. Normal temperature (20°C) was applied to FA20 and BS50, and hot (40°C) and cold (5°C) temperatures were applied to BS65SF5 and SF5, respectively.

In the mock-up test, the slump, void content, and concrete compressive strength (on concrete cores) were measured to evaluate the workability and mechanical properties of the HPC.

The test specimens (Figure 3) were designed to have a thickness of 1.2 m to simulate NPP structures and a length and height of 2 m and 1 m, respectively, to ensure adequate core collection. Cores were collected in the thickness direction of the test specimens to measure the compressive strength (Figure 4).

3.2 *Construction characteristics*

The target slump and void content of the HPC mixes investigated in this study were set to 140–216 mm and 3%–6%, respectively. All the

Table 3. Mixing design.

Mix	Compressive strength (psi)	W	C	FA	BS	SF	Coarse Agg.	Fine Agg.	AD (%)	AE (%)
FA20 (20°C)	6000	162	324	81	–	–	965	722	0.6	0.02
BS50 (20°C)		155	228	–	228	–	958	717	0.7	0.007
BS65SF5 (40°C)	8000	155	137	–	296	23	966	723	0.5	0.005
SF5 (5°C)		155	433	–	–	23	963	721	1.5	0.001
SF5 (5°C)	10000	155	526	–	–	28	915	685	2.0	–

– W: water, C: cement, FA: fly ash, BS: blast–furnace slag, SF: silica fume.
– Notation example: BS30FA25SF5 = 30% blast–furnace slag + 25% fly ash + 5% silica fume in a binder.

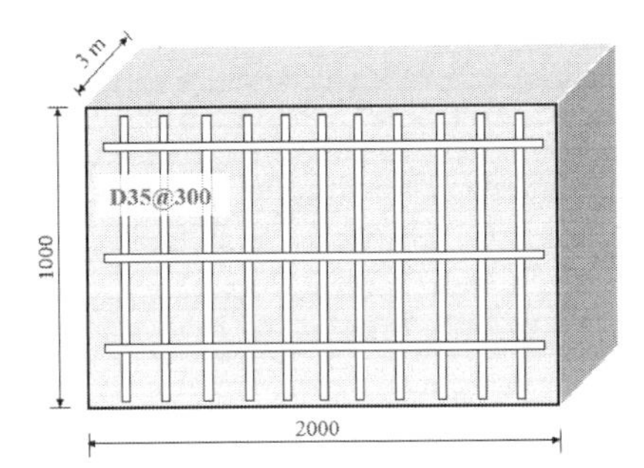

Figure 3. Details of test specimens.

Table 4. HPC mix and curing conditions.

Mix	Compressive strength (psi)	Target slump (mm)	Target void content (%)	Slump (mm)	Void content (%)
FA20	6000	120		120	5.3
BS50	8000			210	5.1
BS65SF5	8000	140–216		205	6
SF5	8000		3–6%	190	6
SF5	10000	550 (slump flow)		480	3

Figure 4. Core collection.

specimens of PC-added HPC satisfied the standards for void content and initial fluidity (Table 4).

3.3 *Compressive strength*

The compressive strengths of the core samples were measured in compliance with ASTM C 39, and the measurement results are presented in Figure 5 (ASTM C 39 2012). In most of the HPC mixes, the design strength requirements for 28-day curing were met. A BS50 mix exhibited the highest compressive strength of concrete, regardless of curing temperature.

The core samples mostly exhibited significant strength development at initial aging under the effect of the internal heat of hydration. This result is assumed to be in accordance with the general tendency that for high-temperature curing of concrete, significant compressive strength development usually occurs in the early curing ages and decreases or stagnates at the late-age stages.

3.4 *Setting property of HPC*

The setting properties of HPC were tested in accordance with the assessment method specified in ASTM C191 (Table 5). The initial/final setting times of HPC increased with increasing binder content (Nevil 1996). Irrespective of the mix type, the initial/final setting times tended to decrease as the curing temperature increased. The setting properties differed according to the admixture type, with BS65SF5 exhibiting the slowest initial/final setting times. The initial/final setting times were more greatly affected by the replacement

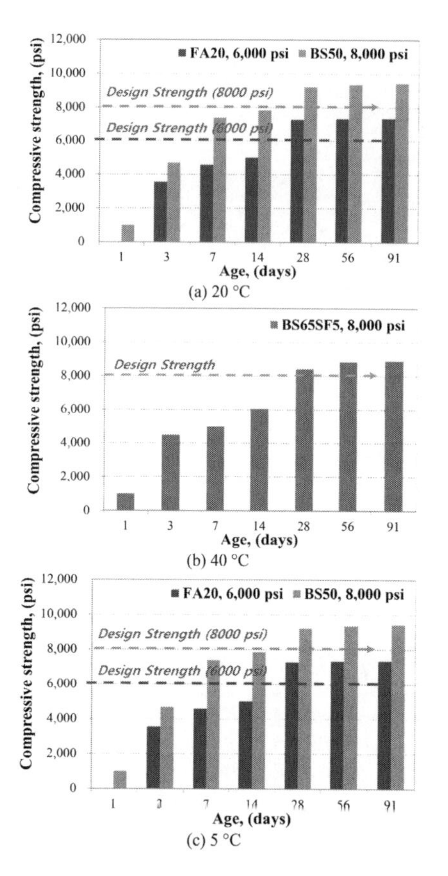

Figure 5. Compressive strength measurements of core samples under various temperature conditions.

Table 5. Setting time behavior results.

Mix	5°C		20°C		40°C	
	t_{ini}	t_{fin}	t_{ini}	t_{fin}	t_{ini}	t_{fin}
6000-FA20	17.42	25.92	10.92	14.92	4.92	5.67
8000-BS50	20.34	33.34	13.34	19.84	7.84	8.84
8000-BS65SF5	27.92	40.92	16.42	22.92	6.92	8.42
8000-SF5	24.92	33.92	15.67	19.92	8.92	9.42
10000-SF5	28.34	41.34	19.34	24.34	12.34	15.34

*t_{ini}: initial setting time (hour), t_{fin}: final setting time (hour).

ratio of the blast-furnace slag than by that of FA (Mehta 2004).

4 CONCLUSIONS

The mock-up test performed to validate the performance of HPC applicable to NPP containment structures yielded the following results and conclusions.

It is necessary to add superplasticizer to HPC containing replaced admixtures to satisfy the requirements for the workability and mechanical properties of HPC. The addition of 0.5%–1% PC is optimal for HPC. The superplasticizer dosage should be adjusted based on the conditions of use. Although 0.5% is the optimal dosage at 35°C and 50°C, this dosage should be reduced at 20°C.

For the mock-up test, we fabricated test specimens measuring 2 m × 1 m with a thickness of 1.2 m for the simulation of NPP structures. Temperatures of 5°C, 20°C, and 40°C were applied. All HPC test specimens achieved the target slump and void content values. The fastest initial and final setting times were measured at 40°C, and these setting times increased with increasing admixture content and blast-furnace slag replacement ratio.

For the 5°C and 20°C specimens, the core compressive strengths reached the target values mostly after 28-day curing. The 40°C specimen exhibited substantially elevated initial compressive strength development, which slowed down at late-age stages. The results of the performed mock-up test demonstrated that all the tested HPC mixes satisfied the workability and compressive strength standards under their respective temperature conditions.

ACKNOWLEDGEMENTS

This work was supported by the Nuclear Power R&D Program of the Korea Institute of Energy Technology Evaluation and Planning (KETEP) grant funded by the Korean government Ministry of Knowledge Economy (No. 2014151010169 A).

REFERENCES

ACI 349. 2006. Code requirements for nuclear safety-related concrete structures.

ASTM C 39. 2012. Compressive strength of cylindrical concrete specimens.

Ferraris, C.F., Obla, K.H. & Hill, R. 2006. The influence of mineral admixtures on the rheology of cement paste and concrete. *Cement and Concrete Research,* 31(2): 245–255.

Han, M.C. 2006. Prediction of setting time of concrete using fly ash and super retarding agent. *Journal of Korea Concrete Institute,* 18(6): 759–767.

Mehta, P.K. & Monteiro, P.J.M. 2004. *Concrete, microstructure, properties, and materials.* 3rd ed. New York: McGraw-Hill. 659.

Nehdi, M., Mindess, S. & Aitcin, P.C. 1998. Rheology of high-performance concrete: Effect of ultrafine particles. *Cement and Concrete Research,* 28(5): 687–697.

Nevil, A.M. 1996. *Properties of concrete.* 4th ed. New York: J. Wiley, 359–411.

Kim, J.M. & Han, C.G. 2005. Technical application for cold weather and hot weather concretes. *Journal of Korea Concrete Institute,* 17(1): 28–33.

Advanced Materials, Mechanical and Structural Engineering – Hong, Seo & Moon (Eds)
 ISBN: 978-1-138-02908-8

Simulation modeling to increase production capacity from prototype to production: A case study of a rocket motor parts manufacturing plant

T. Panyaphirawat, T. Pornyungyun & P. Sapsamarnwong
Defence Technology Institute (Public Organization), Ministry of Defence, Thailand

ABSTRACT: In this study, a simulation modeling by ARENA simulation software was used to investigate and analyze the required resources in order to increase production capacity from the prototyping phase to the production phase of a rocket motor parts manufacturing plant. From the simulation result, it was found that production capacity could be increased from 6 rocket motor sets/month of the current system to a required capacity of 18 motor sets/month or approximately 200 motor sets/year. The bottleneck in the production system was analyzed and resolved. The utilization rate at the system's bottleneck area could be reduced by more than 50% and queue time could be reduced by more than 7 hours. By using the line balancing principle, maximum production capacity could be achieved and more cost-effective.

1 INTRODUCTION

In the manufacturing industry, a newly developed product would go from a stage of research and development to preliminary design, detailed design and prototyping in order to reach a final goal of mass production. In general, during the designing and prototyping phase, the product is produced in a small amount only for testing and verification. Once the design is approved and all manufacturing processes are verified, only then will the product be produced in a large quantity to deliver to customers. For this reason, a production line for prototyping and mass manufacturing is normally set up differently with regard to the number of parts produced. Product development is the process of either designing, creating and marketing a newly designed product or improving an existing one to satisfy the customer requirements. In modern industries, manufacturers have to compete in an extremely demanding market and new products would be developed consistently to satisfy the needs of customers. Different techniques are used in order to analyze and optimize production capacity such as the use of design of experiments. However, with a complex production system, using the design of experiments alone can be costly and time-consuming. Thus, many engineers are turning towards system modeling and simulation software in order to find an optimal scenario suitable for their production requirements. Advantages of using process simulation software are (Pisuchpen 2010): 1) an ability to evaluate many potential alternatives and determine the best approach to the problem with less time and cost than the conventional method; 2) an ability to evaluate various system performance indices such as cost, production time and resource utilization simultaneously; 3) an ability to run a "What if" scenario to evaluate proposed changes; and, finally, 4) an ability to reduce the risk of inappropriate expenditure by running vigorous system simulations to verify all important managerial aspects before spending the company's capital. Many engineers successfully used simulation modeling to solve their production problems. Abed (2008) used a simulation study to increase the production capacity of a rusk production line. He thoroughly studied and analyzed several bottlenecks in his production system. After that, simulation experiments with seven different scenarios were developed in order to find the best approach. He found that by adding two new machines, replacing three old machines, modifying two other machines and decreasing the time in one process, the system performance could be improved by 50% in production with a decrease in total production time of 11.4%. Hasgul (Hasgul & Buyuksunetci 2006) used simulation models to evaluate the bottlenecks in a mixed-model production line in a refrigerator company. He found that by rearranging the vacuum station and changing AGV's cell selection rule, the cycle time of the system was reduced and the bottleneck problem at the vacuum station was solved. Ramis et al. (Ramis & Palma et al. 2001) used simulation modeling to evaluate different alternatives for process improvement of an ambulatory surgery center. They used statistical data taken from a clinical hospital where patients would enter and leave within the same day. From their study, it was found that by using two beds for patient preparation and five beds for post-operation recovery, the maximum throughput of 10 surgeries per day could be achieved. Line balancing is a method of leveling the workload across all processes in a value stream in order to remove the bottleneck or excess capacity and streamline the process for maximum output. Reducing production lead time, minimizing Work In Process (WIP) and maximizing resource utilization are the cornerstones of modern manufacturing strategies such as Lean, Quick Response and Just In Time (JIT) (Benjaafar 2002). Pisuchpen and Chansangar (2014) used line balancing to modify production line in order to increase the productivity of a vision lens factory. In their study, the productivity was improved by 8% and Work In Process (WIP) was decreased by 12%. In the present study, a possible bottleneck was found at S2: CNC Lathe. Therefore, a simulation model using ARENA simulation software was developed in this study to analyze and increase the production capacity of a rocket parts manufacturing plant in order to enhance the production line from prototyping to mass production using the line balancing principle.

2 MODELING THE PRODUCTION SYSTEM

2.1 *Parts, resources and production time*

The production system selected in this study was a 122 mm Multiple Launch Rocket System (MLRS) parts manufacturing plant. The objective of this study was to model a system and evaluate required resources needed in order to increase production capacity from the prototyping phase to the production phase. The expected production capacity of a production phase was forecasted to be 200 rocket motor sets/year. Each set of rocket motor consisted of a total number of 8 parts, as shown in Figure 1,

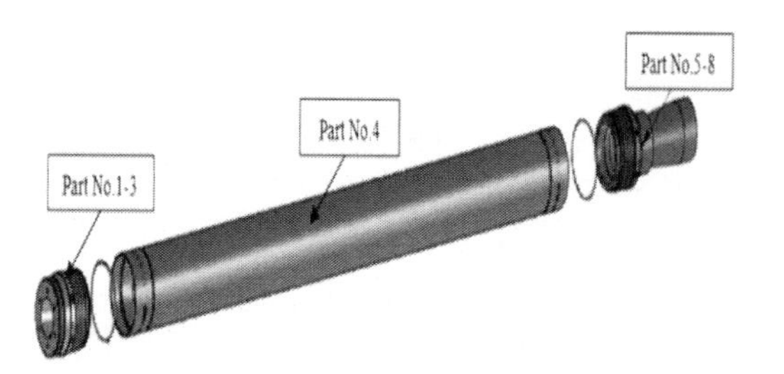

Figure 1. Rocket motor parts configuration.

Table 1. Production time and required resources for rocket motor parts production.

Part name	Resource used	Production Time (hr.)		
		Min.	Max.	Avg.
Head Plug (HP)	S2: CNC Lathe	5.5	7	6.5
	S7: Magnetic Test	0.5	1	0.7
	S10: QC Inspection	0.5	1	0.7
Igniter Case (IC)	S4: Vertical Machine Center	3.5	5	4
	S10: QC Inspection	0.5	1.5	1
Igniter Flange (IF)	S2: CNC Lathe	3	4	3.5
	S5: Radial Drill	0.2	0.7	0.5
	S10: QC Inspection	0.2	0.7	0.5
Motor Case (MC)	S2: CNC Lathe	1.5	2.5	2
	S6: Horning Machine	1.5	2.5	2
	S7: Magnetic Test	0.7	1.5	1
	S8: Flow Forming Machine	2.5	3.5	3
	S2: CNC Lathe	1.5	2.5	2
	S9: Hydrostatic Test	1	1.5	1.2
	S10: QC Inspection	0.3	1	0.5
Front Nozzle (FN)	S2: CNC Lathe	7.5	8.5	8
	S7: Magnetic Test	0.3	0.7	0.5
	S10: QC Inspection	0.5	1	0.7
Nozzle Graphite (NG)	S2: CNC Lathe	2	3	2.5
	S10: QC Inspection	0.3	0.7	0.5
Front Nozzle Assembly (FA)	S1: Heating Oven	3.5	4.5	4
	S2: CNC Lathe	1.5	2.5	2
	S10: QC Inspection	0.5	1	0.7
Rear Nozzle (RN)	S3: Manual Lathe	9	10	9.5
	S7: Magnetic Test	0.3	0.7	0.5
	S10: QC Inspection	0.5	1	0.7

which had to be produced before packing and delivering to the next manufacturing location. The production time and required resources for each part were recorded at different process stations and are listed in Table 1. Due to the small amount of parts produced during the prototyping phase, production time at all stations was regarded as a triangular distribution.

Raw material for each part had to undergo various manufacturing processes within 10 stations, as depicted in Figure 2. The queuing system in this simulation was regarded as a single-queue, multiple-server system for the purpose of evaluating a required number of servers for each station. First-In, First-Out (FIFO) concept was used for all stations except station 2 due to the fact that a high number of parts had to go through this station and some parts required longer processes to be finished. Therefore, at station 2, a workpiece with a longer process time was first served.

2.2 *Modeling the current system*

In order to analyze the current system to find production capacity, production time and utilization rates of required resources, the system simulation model was created using ARENA simulation software. The model can be separated into three sections: 1) Storage, 2) Production line and 3) Packing and shipping. The simulation was set up as 8 working hours per day and 20 days per month with 12 replications to simulate one year of production time.

2.2.1 *Storage section*

In this section, seven parts were created with a constant arrival rate. Each part was assigned a part number with a process sequence and time, as shown in Table 1 and Figure 2. Once all parts were created, they were sent to the production line via the sequence module. The model of the storage section is shown in Figure 3.

2.2.2 *Production line section*

In this section, 10 manufacturing stations were modeled and simulated. A single-queue, multiple-servers system was used with the FIFO concept for all stations except station 2. At station 2, workpiece with a longer process time had a higher priority in order to optimize the total production time. Working time at each station was according to Table 1. After the workpiece was completed, it was transferred to the next station via the sequence module. The model of the production line section is shown in Figure 4.

2.2.3 *Packing and shipping section*

This is considered the last section in the model. All finished parts will be transferred to this section prior to packing and shipping to the next manufacturing process at another location. A number of finished products will be counted at this section to evaluate the total number of rocket motors produced in 1-month period. The model of the packing and shipping section is shown in Figure 5.

2.2.4 *Simulation results of the current system*

The simulation results of the current system are summarized in Table 2. From the simulation result of the current system, it was found that the maximum production capacity of the rocket motor was 6 motor sets/month with the total production time of 126.55 hr. The resource with the highest utilization and longest queue time was S2: CNC lathe with 91.25% utilization and 8.81 hr queue time indicating a possible bottleneck. Therefore, in order to increase production capacity, the bottleneck at this area must be resolved.

The utilization rate of each resource in high to low order is shown in Figure 6.

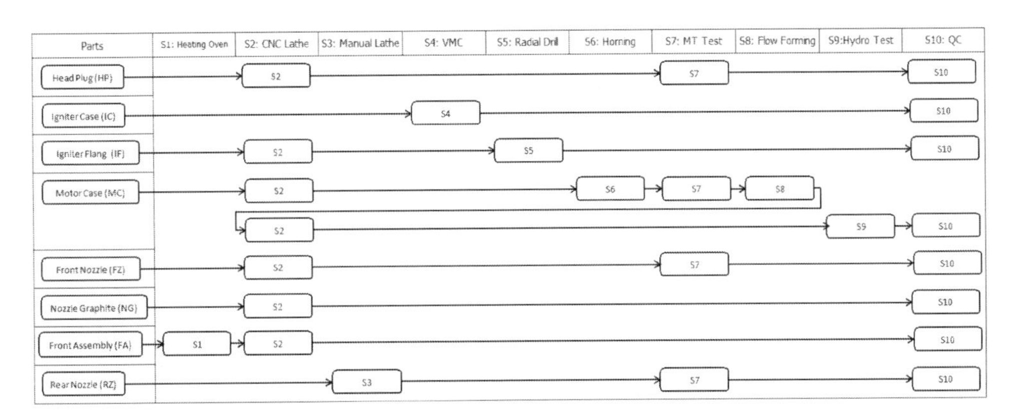

Figure 2. Rocket motor parts process and resource used.

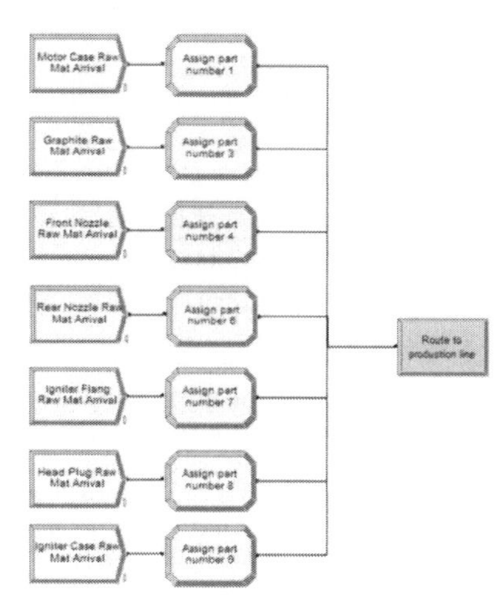

Figure 3. Model of the storage section.

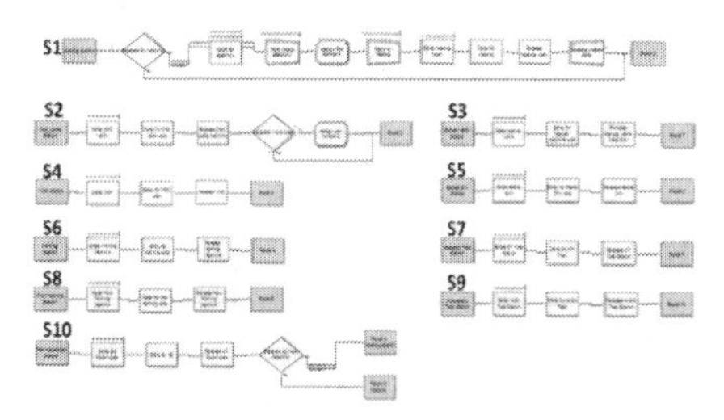

Figure 4. Model of the production line section.

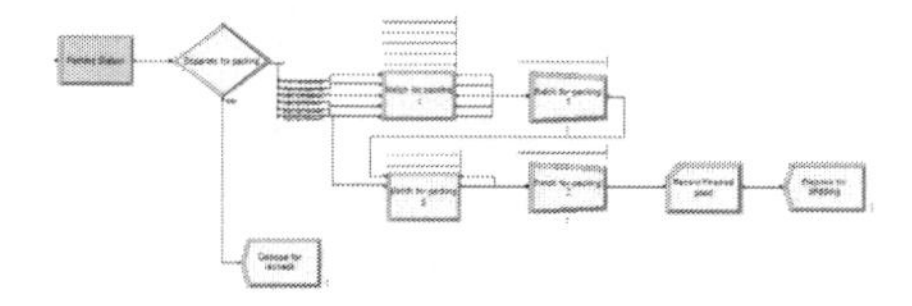

Figure 5. Model of the packing and shipping section.

Table 2. Simulation results of the current system.

Max Production Capacity (Set)	6			
Total Production Time (hr.)	126.55			
Production Time/Set (hr./Set)	21.1			
Resource Utilization Rate (%)				
S1	S2	S3	S4	S5
15.18	91.25	35.68	15.64	1.78
S6	S7	S8	S9	S10
7.45	10.67	11.23	4.59	20.54
Resource Queue Time (hr.)				
S1	S2	S3	S4	S5
0.0	8.81	0.05	0.02	0.0
S6	S7	S8	S9	S10
0.01	0.01	0.06	0.0	0.02

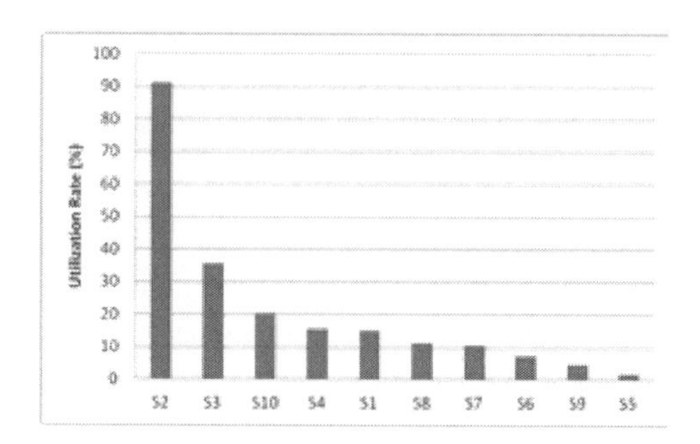

Figure 6. Resource utilization rate in a high to low order.

3 RESOLVING BOTTLENECK WITH LINE BALANCING CONCEPT

In this study, two simulation scenarios were proposed to compare the results between with and without the use of line balancing.

3.1 *Scenario 1: Increasing production capacity without line balancing*

In scenario 1, the system was modified by adding more resources without rearranging the manufacturing process. Two highest utilized resources namely S2: CNC lathe and S3: Manual lathe were chosen for controlled factors. The process was evaluated using the process analyzer tool in ARENA software. The simulation was evaluated in order to achieve a maximum number of 18 finished rocket motor sets per month or three times the current system. Therefore, a production time/set would be reduced from 21.1 hr./set of the current system to be less than 8.9 hr./set, as shown in the following equation:

$$production\ time = \frac{Maximum\ working\ hour}{Number\ of\ part\ produced} \quad (1)$$

The simulation was evaluated for 32 full factorial design using the process analyzer, and the result is shown in Figure 7 and Table 3. The minimum resources required to achieve 18 rocket motor sets/month were 3 servers at S2: CNC lathe and 1 server at S3: Manual lathe.

3.2 *Scenario 2: Increasing production capacity with line balancing*

In scenario 2, the production line was balanced by rearranging the workload at two stations, which were S2: CNC lathe and S3: Manual lathe, and combined a work at S1: Heating oven into 3 workpieces/batch. To rearrange the workload, the process of pre-machining at S3: Manual lathe was incorporated before sending to S2: CNC lathe for finishing. The rearranged process was used for the following parts: Head Plug (HP) and Front Nozzle (FN). After balancing the line, a simulation was run with 32 full factorial design using the process analyzer in order to evaluate the system performance. The simulation result of scenario 2 is shown in Figure 8 and Table 4. From the simulation result, it was found that a minimum number of resource required to achieve 18 rocket motor sets/month or production time less than 8.9 hr/set can be obtained by applying two servers at station S2: CNC lathe and S3: Manual lathe.

Table 3. Simulation results of scenario 1.

Max Production Capacity (Set)	6
Total Production Time (hr.)	39.85
Production Time/Set (hr./Set)	6.64

	Scenario Properties				Controls		Responses	
	S	Name	Program File	Reps	CNC Lathe Machine	Manual Lathe Machine	Record Finished	Entity 1.TotalTime
1		Scenario 1	17 : Sim D2 r	12	1	1	6	126.554
2		Scenario 1	17 : Sim D2 r	12	1	2	6	126.451
3		Scenario 1	17 : Sim D2 r	12	1	3	6	126.451
4		Scenario 1	17 : Sim D2 r	12	2	1	6	64.330
5		Scenario 1	17 : Sim D2 r	12	2	2	6	64.408
6		Scenario 1	17 : Sim D2 r	12	2	3	6	64.408
7		Scenario 1	17 : Sim D2 r	12	3	1	6	39.853
8		Scenario 1	17 : Sim D2 r	12	3	2	6	39.933
9		Scenario 1	17 : Sim D2 r	12	3	3	6	39.933

Figure 7. 32 Full factorial design evaluation result of scenario 1.

	Scenario Properties				Controls		Responses	
	S	Name	Program File	Reps	CNC Lathe Machine	Manual Lathe Machine	Record Finished	Entity 1.TotalTime
1		Scenario 2	3 : Sim D2 re	12	1	1	6	105.182
2		Scenario 2	3 : Sim D2 re	12	1	2	6	103.794
3		Scenario 2	3 : Sim D2 re	12	1	3	6	103.733
4		Scenario 2	3 : Sim D2 re	12	2	1	6	89.859
5		Scenario 2	3 : Sim D2 re	12	2	2	6	51.249
6		Scenario 2	3 : Sim D2 re	12	2	3	6	50.703
7		Scenario 2	3 : Sim D2 re	12	3	1	6	89.585
8		Scenario 2	3 : Sim D2 re	12	3	2	6	49.974
9		Scenario 2	3 : Sim D2 re	12	3	3	6	47.896

Figure 8. 32 Full factorial design evaluation result of scenario 2.

Table 4. Simulation results of scenario 2.

Max Production Capacity (Set)	6
Total Production Time (hr.)	51.25
Production Time/Set (hr./Set)	8.54

4 COMPARING THE RESULTS BETWEEN SIMULATIONS

From the simulation results, it was found that in order to increase the production capacity of rocket motor parts manufacturing from the prototyping phase to the production phase and reach a capacity of 18 motor sets/month or three times the current system, both scenarios can achieve a required objective. However, a production line using scenario 2 would require less capital investment and can reduce investment cost by more than 50% due to the fact that a CNC lathe machine is normally two to three times more expensive than a manual counterpart. A graphical comparison of the utilization rate and queue time between the current system compared with scenarios 1 and 2 is shown in Figures 9 and 10, respectively. From Figures 9 and 10, it was found that both simulation scenarios can resolve the bottleneck in a production system. The utilization rate at the bottleneck area of S2: CNC lathe can be reduced by more than 50% and the queue time can be reduced by more than 7 hours.

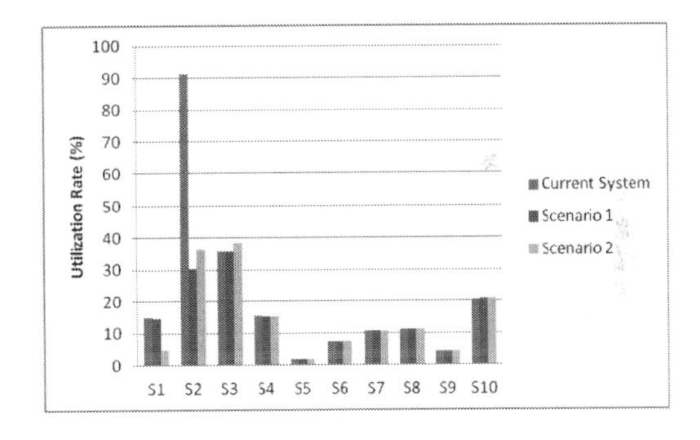

Figure 9. Comparing utilization rate between simulations.

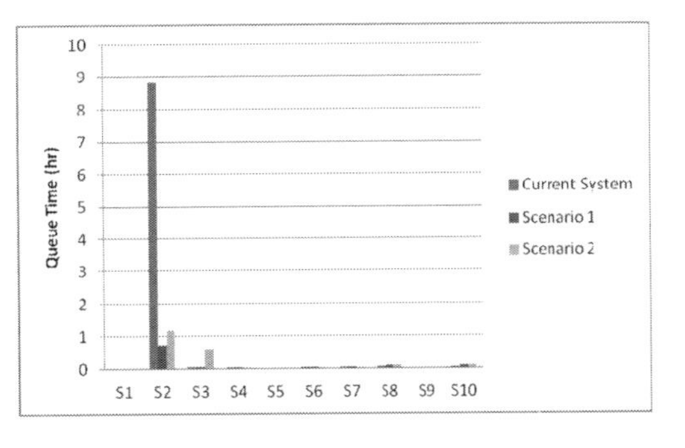

Figure 10. Comparing queue time between simulations.

5 CONCLUSION

Simulation modeling is a great tool for production planning. With a thorough study and evaluation of the production system, engineers can experiment with different setups using simulation software in order to improve the efficiency of the system. In this study, two scenarios were incorporated into simulation models to rearrange the workload between resources in production line and improve the production capacity of the system. Comparing between the current system and improved systems, it was found that both simulation scenarios can improve the production capacity from 6 rocket motor sets/months to 18 motor sets/moths or three times the current system. Also, both scenarios can reduce the utilization rate at the bottleneck area by more than 50% and reduce the queue time by more than 7 hours. However, with line balancing in scenario 2, capital investment can be cut by more than 50% due to the fact that some of the machining work was rearranged to manual lathe machine instead of CNC machine. Therefore, scenario 2 would be more cost-effective to be implemented in this case. For further study, engineers can incorporate other scenarios into a simulation, such as incorporating over time to increase the production capacity to increase products' variation into the system.

ACKNOWLEDGMENT

The authors would like to thank the Engineering Management Program, Department of Industrial Engineering, Faculty of Engineering, Kasetsart University for helping with simulation software.

REFERENCES

Abed, S.Y. 2008. A simulation study to increase the capacity of a rusk production line, *Science and Applications*, 5(9): 1395–1404.

Benjaafar, S. 2002. Modeling and analysis of congestion in the design of facility layouts, *Management Sciences*, 48(5): 679–704.

Hasgul, S. & Buyuksunetci, A.S. 2006. *Simulation modeling and analysis of a new mixed model production lines,* Proceedings of the 2005 Winter Simulation Conference, 1408–1412.

Pisuchpen, R. 2010. *Guide to arena simulation software*, Bangkok: Se-education.

Pisuchpen, R. & Chansangar, W. 2014. Modifying production line for productivity improvement: A case study of vision lens factory, *Songklanakar-in Journal of Science and Technology*, 36(3): 345–357.

Ramis, F.J., Palma, J.L. & Baesler, F.F. 2001. *The use of simulation for process improvement at an ambulatory surgery center,* Proceedings of the 2001 Winter Simulation Conference, 1401–1404.

Advanced Materials, Mechanical and Structural Engineering – Hong, Seo & Moon (Eds)
© 2016 Taylor & Francis Group, London, ISBN: 978-1-138-02908-8

Hydration heat of Ternary Blended Concrete applied to a nuclear containment Building

Y.H. Hwang
Department of Architectural Engineering, Kyonggi University Graduate School, Seoul, Korea

K.H. Yang & S.J. Kim
Department of Plant and Architectural Engineering, Kyonggi University, Suwon, Korea

ABSTRACT: This study estimated and evaluated the heat of hydration and the crack index based on thermal stress according to the casting height of a nuclear containment building using a finite element analysis program. As an important parameter of the analysis, Ternary Blended Concrete (TBC) based on low heat cement, to which ground fly ash and limestone powder were added, was used for the nuclear containment building to decrease the temperature, and its adiabatic temperature rise curve was determined. The heat of hydration and crack index were shown to be over 20% and 30% more effective than the control group (80% Ordinary Portland Cement + 20% fly ash), respectively. These results indicate that the use of TBC can minimize thermal cracks when the 1st-step casting height is set to 4 m.

1 INTRODUCTION

With the recent growth in the worldwide market for nuclear power plants, there has been an increasing interest in gaining a competitive edge in the export of nuclear power plants. In particular, as nuclear structures become more specialized and enlarged, and design criteria become stricter due to industrial demands, there has been a need for more detailed studies on mass concrete.

Mass concrete structures generally experience changes in volume because of the generation and dissipation of heat caused by the hydration reaction in cement and changes in the atmospheric environmental conditions. If there is a large change in volume or the confined force exceeds a certain limit, significant deformation or stress can be generated, causing damage such as cracks to the structure. Cracks weaken the stability, reduce service life, and degrade structural aesthetics, and so the change in the interior temperature of the concrete structure must be accounted for prior to construction; thus, there is a need for a measure to estimate and minimize such changes (Alhozaimy et al. 2015; Zhu 2014).

A nuclear containment building is a special structure, which is most sensitive to the heat of hydration and thermal stress of mass concrete. Therefore, pre-evaluation of the heat of hydration and thermal stress is essential to securing the structure's durability and safety. A general method to decrease the heat of hydration involves the use of a mineral admixture, fly ash, as a Supplementary Cementitious Material (SCMs) (Atis 2002) in cement to improve the concrete's flow, long-term strength and workability. However, the drawback of adding large quantities of FA is that it delays the early-age strength development of concrete. In particular, if formwork removal is delayed by the delay in the early-age strength development, the construction period may be extended, raising costs and possibly causing damage due to frost in the winter season from insufficient curing. Considering this disadvantage and the heat of hydration, regulations generally allow a maximum FA content of 25% in Ordinary Portland Cement (OPC) (ACI 2005). Numerous studies have attempted to improve the early-age compressive strength of concrete that has been supplemented with FA. Recently, Ternary Blended Concrete (TBC), based on Low Heat Cement (LHC), to which Ground Fly Ash (GFA) and Limestone Powder (LP) were added, was proposed and its mechanical properties and hydration characteristics were evaluated (Yang et al 2015; Kim et al. 2015). Although much research on low heat concrete has focused on reducing the heat generated by the concrete itself and increasing the concrete strength, there has been a lack of research on the performance of concrete with respect to heat of hydration and thermal cracks of nuclear containment buildings.

To account for the heat of hydration properties of concrete, the casting height is set to about 3 m in nuclear containment buildings. However, as nuclear facilities become larger, an increase in casting height is inevitable to shorten the period of construction for economic reasons and to reduce the number of construction joints. Although structural design and construction technology need to be taken into consideration when increasing the casting height in the construction of nuclear containment buildings, the cracking problem of concrete, in particular, due to heat of hydration and thermal stress needs to be examined.

Therefore, this study aims to use a finite element analysis program to predict and evaluate the heat of hydration and thermal crack index as a function of the casting height of the nuclear containment building. The mass concrete TBC, which has been studied earlier, was applied to the concrete mixture, and was compared with the concrete made by adding 20% FA to OPC, which is commonly used in nuclear containment buildings.

2 METHOD OF ANALYSIS

2.1 *Modeling*

For the analysis of the heat of hydration and thermal stress of a nuclear containment building, a finite element analysis software was used to model a cylindrical structure with a wall thickness of 1.2 m and a radius of 23 m (Figure 1). The wall heights were set to 3 m and 4 m to analyze the effect of 1-step casting. 3-D (solid) elements were applied to the model. Because a nuclear containment building has a symmetrical cross section, it was modeled as a 1/4 symmetrical building to reduce the time required for analysis. Continuity condition (condition of symmetrical section) was applied to the planes of symmetry. The basement of the structure was additionally modeled to account for any external constraints and the transfer of the heat of hydration generated from within the wall to the basement.

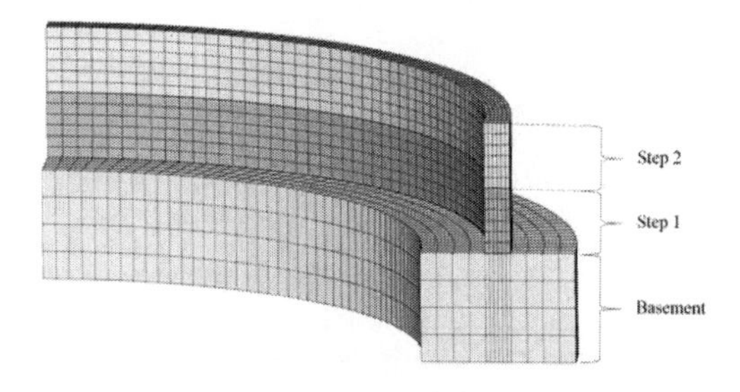

Figure 1. Modeling of a nuclear containment building.

2.2 *Concrete properties*

For the analysis of the containment building's heat of hydration, the mix test of the concrete commonly used in nuclear structures (control) and the Ternary Blended Concrete (TBC) was carried out to obtain data on their compressive strength and modulus of elasticity. The concrete's material properties and heat function, listed in Table 1, were used in the analysis. The Ternary Blended Concrete (TBC) was carried out to obtain data on its compressive strength and modulus of elasticity. The concrete's material properties and heat function, listed in Table 1, were used in the analysis. The 1-step casting interval was set to 14 days. We referred to the Korea Standard Specification for Concrete (KCI 2009) and earlier literature (Han et al. 2008; Kim et al. 2008) for other concrete material properties. The curing and initial casting temperature of the concrete was set to 20°C (standard), and the wall's convection coefficient value was assigned as 8 W/m^2 · °C for the wood form. The convection coefficient value of the top parts of the foundation and the wall was set to 13 W/m^2 · °C for open air.

2.3 *Adiabatic temperature rise curve*

Adiabatic temperature rise is an important property in the analysis of the heat of hydration of

Table 1. Input data for analysis.

		Control	TBC	Basement
f_{ck}, (MPa)		48.2	42.2	–
Constants factor for strength	A	2.683	3.262	–
	B	0.895	0.916	–
Elastic Modulus, (MPa)		32600	30500	23000
Poisson's Ratio		0.167	0.167	0.167
Expansion Coefficient		0.00001	0.00001	0.00001
Unit Weight, (kg/m^3)		2300	2300	2300
Conductivity (J/mhr°C)		10080	10080	10080
Specific Heat (Jg/N°C)		1050	1050	1050
Convection Coefficient (W/m^{2}°C)		Wood form (8) Open air (13)		

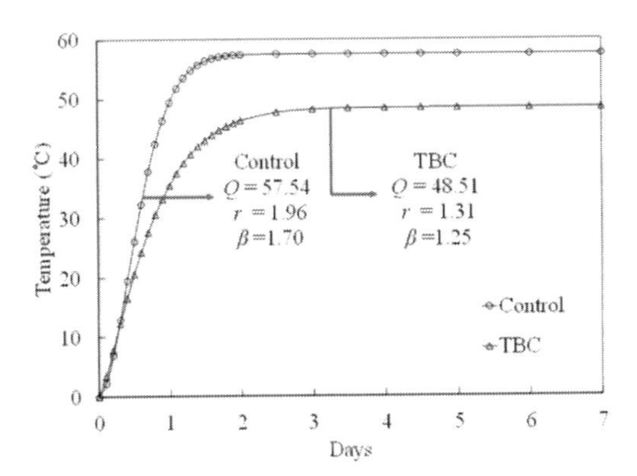

Figure 2. Adiabatic temperature rise curve.

concrete. The Korea Standard Specification for Concrete (KIC 2009) presents the maximum temperature rise Q_∞ and the rate of temperature rise r for each type of cement and different casting temperatures, but generally, they are obtained from the results of the adiabatic temperature rise experiments. Results of these experiments are standardized using the adiabatic temperature rise equation, accounting for the delaying effect of the heat of hydration on early-age strength development (KCI 2010), as follows:

$$Q(t) = Q_\infty(1 - e^{-rt^\beta}) \tag{1}$$

where Q_∞ is the maximum adiabatic temperature rise (°C); r is the rate of temperature rise; t is the age (days); $Q(t)$ is the adiabatic temperature rise (°C) on day t; and β is the constant obtained from the experiment.

The results from the adiabatic temperature rise experiments of the control and TBC were applied to Equation (1) and are shown in Figure 2. Q_∞ was determined to be 57.5°C for the control and 48.5°C for the TBC, which was lower by approximately 16%. Also, the rate of temperature rise was 1.96 for the control and 1.31 for the TBC, which was lower by about 33%.

3 ANALYSIS RESULTS

3.1 *Heat of hydration analysis results*

The heat of hydration of the nuclear containment building was evaluated at the point of maximum temperature at the center of the wall and at the point of maximum temperature at the horizontal surface. The temperatures due to heat of hydration at the 1- and 2-step casting height are given in Table 2. The maximum temperature at the center of the wall of the control at the 1-step casting was 70.5°C and that of TBC was 59.2°C, almost a 16% decrease from that of the control. The difference between the inner and outer temperatures was 20%. The maximum inner temperature and the difference between the internal and external temperature at the 2-step casting were the same as those of the 1-step casting. As the casting height

Table 2. Analysis results of hydration heat.

		Casting of 1 Step (°C)			Casting of 2 Steps (°C)		
		Center	Surface	ΔT	Center	Surface	ΔT
Control	3m	70.5	48.7	21.8	70.5	48.7	21.8
	4m	70.5	48.7	21.8	70.5	48.7	21.8
TBC	3m	59.2	41.8	17.4	59.2	41.8	17.4
	4m	59.2	41.8	17.4	59.2	41.8	17.4

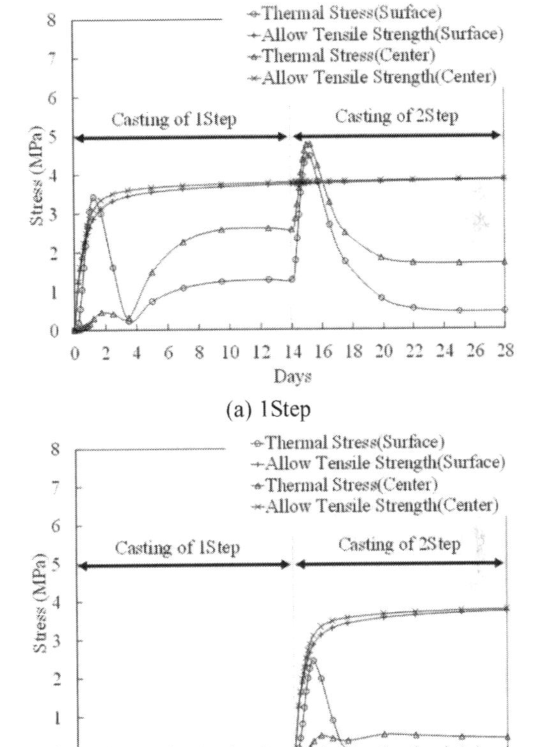

Figure 3. Thermal stress of control at the casting height of 3 m.

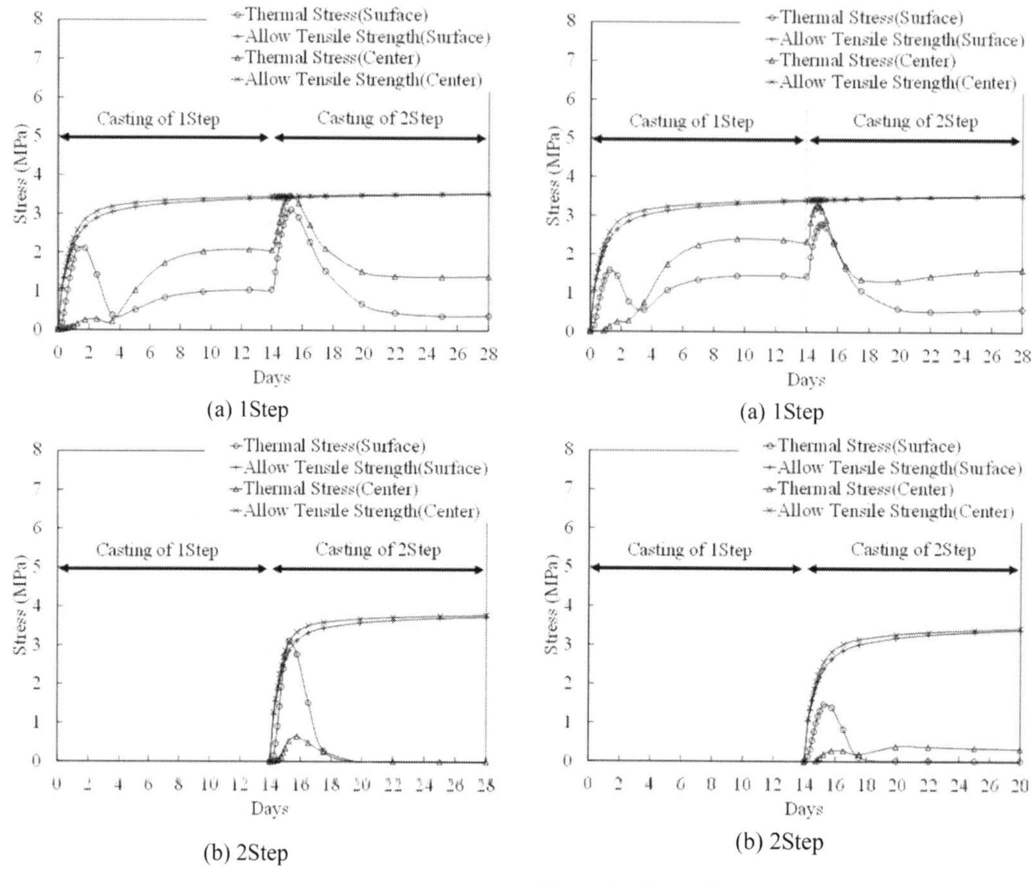

Figure 4. Thermal stress of control at the casting height of 4 m.

Figure 5. Thermal stress of TBC at the casting height of 3 m.

increased, there was a tendency for the distribution of temperatures at the center of the wall to expand along the height, but there were no changes in the maximum internal temperature and the difference between the internal and external temperatures.

3.2 *Thermal stress analysis results*

The thermal stress of the nuclear containment building wall was examined at a point of maximum thermal stress in the wall and at a point on the surface of the top part of the wall, which was vulnerable to thermal cracks. The thermal stress corresponding to the parameters at the 1-step and 2-step casting is shown in Figures 3–6.

In the case of the control, the thermal stresses at the center (3 m, 4 m) and surface (3 m) were lower than the allowable tensile strength. However, the thermal stress at the surface of 4 m was greater than the allowable tensile strength. In the case of the TBC, the thermal stress stayed under the allowable tensile strength even when the casting height was increased.

In addition, the thermal stress of the ordinary casted concrete tended to increase because of the later casted concrete as the casting steps increased. This result seemed to be due to the additional stress in the ordinary casted concrete generated by the external constraint of the later casted concrete.

Figures 3(a)–6(a) show the thermal stress variation occurring from the 1-step to 2-step casting, where the thermal stress in all cases exceeded the allowable tensile strength, regardless of the casting height. In the case of TBC, however, the thermal stress in the 1-step casting did not exceed the allowable tensile strength even after the 2-step casting. After the 2-step casting, all the thermal stress values were lower than the allowable tensile strength, the only exception being the value of the surface of the control at a casting height of 4 m, as shown in Figures 3(b)–6(b).

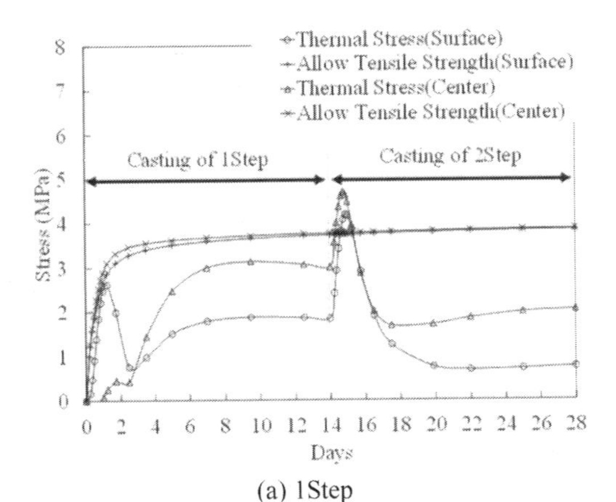

(a) 1Step

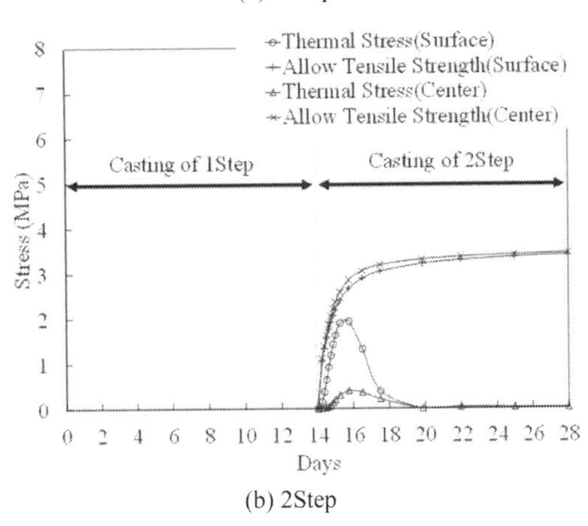

(b) 2Step

Figure 6. Thermal stress of TBC at the casting height of 4 m.

3.3 *Thermal crack index analysis results*

The evaluation of cracking in mass concrete can be made using the thermal crack index. The thermal crack index can be represented as a ratio of thermal stress to the allowable tensile strength of the concrete, as given in Equation (2). We reviewed the thermal crack index at the point where the thermal stress was evaluated. The probability of crack can be presented using the thermal crack index (Figure 7), as follows:

$$I_{cr}(t) = \frac{f_{sp}(t)}{f_t(t)} \tag{2}$$

where $f_{sp}(t)$ is the tensile strength of the concrete on t and $f_t(t)$ is the maximum thermal stress generated by the heat of hydration at t.

The crack index of the control and TBC is presented in Table 3 according to the casting height.

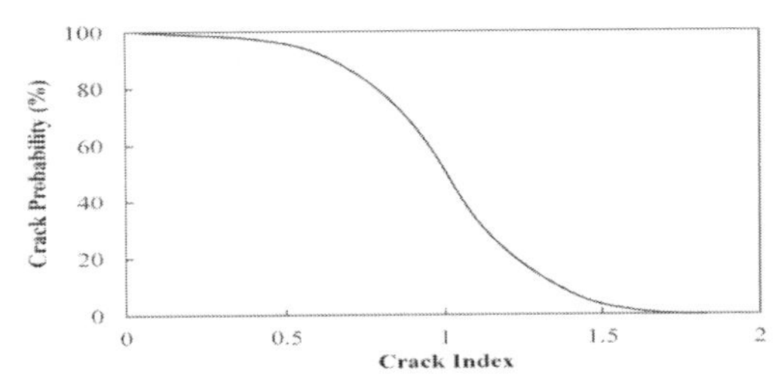

Figure 7. Crack probability.

Table 3. Analysis results of the crack index.

		Casting of 1 Step		Casting of 2 Steps	
		Center	Surface	Center	Surface
Control	3m	1.18	1.11	0.81	0.89
	4m	1.44	0.84	0.79	0.83
TBC	3m	1.40	1.57	1.05	1.22
	4m	1.65	1.15	1.00	1.10

In the case of the control, the crack index of the center and surface was 1.18 and 1.11 at the 1-step casting of 3 m, respectively, and the probability of crack was about 40%. The crack index of the center was 1.44 at the height of 4 m, with a probability of crack about 10%. And the crack index of the surface was 0.84, with a probability of crack about 70%. On the other hand, in the case of TBC, the crack index of the center and surface was 1.4 and 1.57 at the casting height of 3 m, respectively, with a probability of crack lower than 10%. And at the casting height of 4 m, they were 1.65 and 1.15, respectively, with a probability of crack about 50% on the center and 30% on the surface.

In the control after the 2-step casting, the crack index of the center and surface at the 1-step casting showed 0.81 and 0.89, respectively, with a probability of crack about 70%, at the casting height of 3 m, and they were 0.79 and 0.83 at the casting height of 4 m, respectively, with a probability of crack similar to that at the casting height of 3 m. In the TBC, the crack index of the center and surface at the 1-step casting was 1.05 and 1.22 at the casting height of 3 m, respectively. And at the casting height of 4 m, they were 1.00 and 1.10, respectively. And this was shown to have a crack probability of minimum 50%. At the casting height of 3 m and 4 m, in the control, the thermal crack index of the center and surface was under 1.00, with a probability of crack over 50%. On the other hand, in the TBC, the thermal crack index was over 1.00, with a crack probability of maximum 50%.

4 CONCLUSION

The results of hydration heat and thermal stress analyses of nuclear containment buildings with respect to the casting height of 1 step can be summarized as follows:

1. Results of hydration heat: the TBC showed a 16% reduction in maximum internal temperature compared with the control concrete, and the difference in internal and external temperatures was also lowered by 20%.
2. The thermal stress analysis results showed that for the control, after the 1-step casting, the thermal stress of the center and surface were lower than the allowable tensile strength, but after the 2-step casting, the thermal stress at 1 step exceeded the allowable tensile strength.
3. In the case of the TBC, all thermal stress values were under the allowable tensile strength, regardless of the changes in the casting height and increase in the number of casting steps.
4. In the case of the control, the crack index of the center had the lowest value of 0.79 at the casting height of 4 m, with a crack probability of about 70%. On the other hand, for the TBC, the lowest crack index was 1.0, with a crack probability of about 50%, which indicated a decrease in probability by 20% when compared with the control.

ACKNOWLEDGMENT

This work was supported by Kyonggi University's Graduate Research Assistantship 2015 and the Nuclear Power Core Technology Development Program of the Korea Institute of Energy Technology Evaluation and Planning (KETEP), and financial resource from the Ministry of Trade, Industry & Energy, Republic of Korea (No. 20131520100750).

REFERENCES

ACI. 2005. ACI 318–05: Building Code Requirements for Structural Concrete. ACI Committee 318: American Concrete Institute.

Alhozaimy, A., Fares, G., Alawad, O.A. & Al-Negheimish, A. 2015. Heat of Hydration of Concrete containing Powdered Scoria Rock as a Natural Pozzolanic Material. *Construction and Building Materials,* 81: 113–119.

Atis, C.D. 2002. Heat evolution of high-volume fly ash concrete. *Cement and Concrete Research,* 32(5): 751–756.

Kim, H.S., Han, M.C., Kim. J.Y. & Han, S.B. 2008. Hydration Heat Analysis of Mass Concrete Considering Low Heat Mixture and Block Placement. *Journal of the Architectural Institute of Korea Structure & Construction,* 24(1): 59–64.

Kim, H.S., Seo, K.Y. & Lee, D.U. 2008. Correction of Thermal Cracking Index by using Temperature Difference Inside and Outside of Internal Restricted Mass Concrete Structures. *Journal of the Regional Association of Architectural Institute of Korea,* 10(2): 283–289.

Kim, S.J., Yang, K.H. & Moon, G.D. 2015. Hydration Characteristics of Low-heat Cement Substituted by Fly Ash and Limestone Powder. *Materials,* 8: 5847–5861.

Kim, S.J., Yang, K.H. Lee, K.H. & Lee, S.T. 2015. *Mechanical Properties and Adiabatic Temperature Rise of Ternary Blended Concrete Based on Low Heat Cement.* The 2015 World Congress of Advances on Structural Engineering and Mechanics (ASEM15).

KCI. 2009. *Standard Concrete Specification.* KCI Committee: Korean Concrete Institute.

KCI. 2010. *MCP Thermal Crack Control of Mass Concrete. KCI Committee*: Korean Concrete Institute.

Zhu, B. 2014. *Thermal Stresses and Temperature Control of Mass Concrete.* Tsinghua University Press: Elsevier Incorporation.

Advanced Materials, Mechanical and Structural Engineering – Hong, Seo & Moon (Eds)
© 2016 Taylor & Francis Group, London, ISBN: 978-1-138-02908-8

Strain transfer error experiments and analysis on CFRP laminates using FBG sensors

C.S. Song, J.X. Zhang, M. Yang, J.G. Zhang & W. Yuan
School of Mechanical and Electronic Engineering, Wuhan University of Technology, Wuhan, Hubei, China

ABSTRACT: In order to obtain the strain transfer error of Fiber Bragg Grating (FBG) sensors on Carbon Fiber-Reinforced Polymer (CFRP) smart composite laminates, FBG sensors bonded on the surface of CFRP laminates and embedded in the middle layer as well as resistance strain gauges were used to monitor the strain of CFRP laminates. Compared with the finite element analysis, the experimental results proved that the FBG sensors are valid and reliable in the monitoring of CFRP laminates. During the monitoring of CFRP laminate stretching experiments, the strain transfer errors were caused by the transfer layers. The transfer errors of surface-bonded FBG sensors and embedded FBG sensors were calculated, and the experimental data were corrected by the calculated strain transfer errors. After the correction, the surface-bonded FBG strain was found to be close to the embedded FBG strain, and the measurement errors became smaller. This study is useful to monitor the strain of CFRP laminates in practical measurements.

1 INTRODUCTION

Carbon Fiber-Reinforced Polymer (CFRP) is frequently applied in various fields because of the high specific strength and specific modulus. Mechanical damage is generally generated around the stress concentration region in CFRP structural components due to static or cyclic loading (Spearing & Beaumont 1992). So, in order to make sure that CFRP laminates are valid and reliable, the stress and strain real-time monitoring of CFRP laminates is important.

Some sensing technologies such as soft X-ray radiography and ultrasonic C-scan are commonly used to detect damage. However, they cannot detect the microscopic damage in real time. To overcome these disadvantages, optical fiber sensor approaches have been studied. Among various sensors, Fiber Bragg Grating (FBG) sensors have excellent features such as small size, high sensitivity, no electric/magnetic induction, high accuracy and multiplexing capability. FBG sensors have already been adopted in the measurements of strain and temperature for various civil structures (Kersey et al. 1997).

Jeannot presented a method for the localization of an impact by using dynamic strain signals from FBG sensors (Frienden et al. 2012). Other studies (Leng & Asundi 2003, Wild & Hinckley 2009, Damien & Ptrice 2014, Chuanquan et al. 2014) have shown that FBG sensors are used in the structural health monitoring of smart composite materials. Many researchers have used FBG sensors that were embedded in composite laminates to detect the microscopic damage in composite laminates (Okabe & Yashiro 2012, Chanbers et al. 2007, Yashiro et al. 2005, Okabe et al. 2004). The use of metal-coated optical fiber sensors can determine the maximum strains of the composite structures (Kim et al. 2013). All these findings showed that FBG sensors have been widely used.

When using FBG sensors, we can find a tiny glued gap between the fiber grating sensors and the test substrate. And there is a big difference between the glue layer elasticity modulus and the naked fiber elastic modulus. This leads to a significant difference between the real strain and measured strain. The influence of the glued layer and the coating layer is always ignored, assuming that optical fiber sensors measured the strain on the attached surface. This assumption is acceptable for the long length of the bonding fiber, but for fiber Bragg grating that has a short fiber optic sensor, the strain of the matrix is not fully transmitted to the fiber Bragg grating. Considering that the shear layer absorbed energy that caused reduced strain reduce during transfer, Ansari made some corresponding amendments to the FBG strain transfer (Ansari & Libo 1998).

The fiber grating was pasted on the surface of a matrix, and the strain transfer coefficient was measured. Experiments proved that the different materials and packaging influence the strain transfer rate of fiber grating sensor (Pereira et al. 2013).

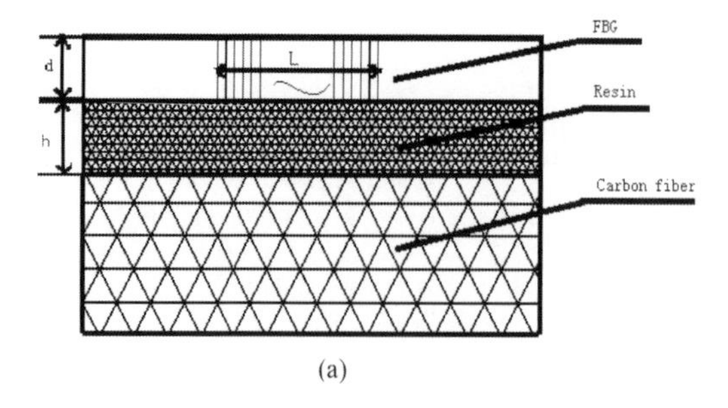

(a)

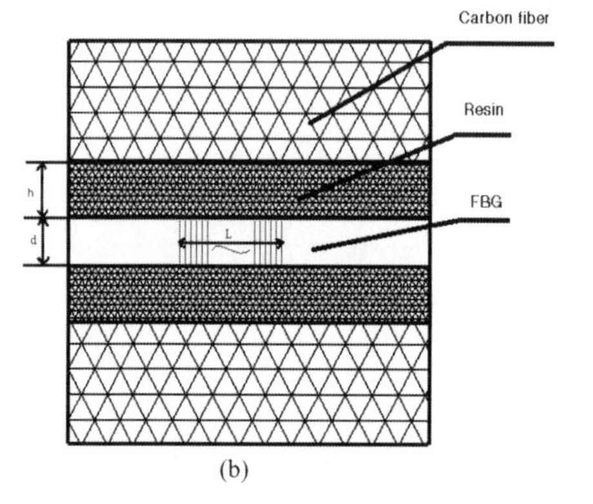

(b)

Figure 1. Model schematic of strain transfer error analysis; (a) FBG bonded on the surface, (b) FBG embedded in the middle layer.

So according to the actual condition, the optimum parameters of the paste can be calculated (Guo et al. 2011), and in order to get a low strain transfer error, we should get the optimum sensor and the localization of sensors.

So, in this paper, we monitored the strain of CFRP laminates by fiber Bragg grating sensor pasted on the surface and embedded in the carbon fiber composite material layers, and also monitored the strain of CFRP laminates by resistance strain gages pasted on the carbon fiber composite laminates. The results were compared with the strain results of Finite Element Analysis (FEA), and were corrected by the strain transfer error analysis formula.

2 THEORY AND METHOD

Figure 1(a) shows the FBG pasted on the surface and Figure 1(b) shows the FBG embedded in the middle layer. In the process of measurement, the fiber core was used for measuring the strain, but the fiber core was not in direct contact with the measured object. There was a strain transfer layer between them, including FBG sensor's inner protective layer and the external glue layer. The strain transfer layer affects the strain transfer efficiency of the fiber Bragg grating strain sensors. And the relevant corrections need to be considered. The fiber glued with the measured deformation substrate. Here h is the thickness of the strain transfer layer, d is the contact width of the fiber and strain transfer layer, L is the initial length of optical fiber working segments, ΔL is the deformation elongation of optical fiber working segments. $\Delta L'$ is the strain transfer layer error caused by shear deformation, $(\Delta L + \Delta L')$ is the true length of the measuring object, G is the shear elasticity strain transfer layer, E is Young's modulus of fiber, and A is the cross-sectional area of fiber. For the FBG pasted on the surface, the relationships between the strain transmission error and the true length of FBG sensor are shown as follows (Jiashang & Yu 2009):

$$\frac{\Delta L'}{\Delta L} = \frac{12EAh}{GdL^2} \tag{1}$$

For the FBG embedded in the middle layer of the specimen, the relationships between the strain transmission error and the true length of FBG sensor are shown as follows:

$$\frac{\Delta L'}{\Delta L} = \frac{12EAh}{G\pi dL^2} = \frac{3E\mathrm{d}h}{GL^2} \tag{2}$$

Strain transfer error analysis formula can be expressed as a formula of strain. According to Formula (1) and Formula (2), the strain transfer error analysis formula can be expressed as follows:

$$\varepsilon' = \frac{\varepsilon}{(1 + \frac{12EAh}{GdL^2})} \tag{3}$$

$$\varepsilon'' = \frac{\varepsilon}{(1 + \frac{3Edh}{GL^2})} \tag{4}$$

where ε is the true strain of the measuring object; ε' is the strain of the surface-bonded FGB sensor; and ε'' is the strain of the embedded FGB sensor.

According to Formula (3) and Formula (4), it is clear that the smaller the Young modulus of the fiber or the greater the shear modulus of the transfer layer is, the smaller the strain transfer error will be. This can provide some reference for choosing the material of the FBG measurement system. Also, from the geometric point of view, the smaller the fiber cross-sectional area and the strain transfer layer thickness are, the smaller the error strain transfer will be. These conclusions about the geometric sense can provide a clear improvement direction of the FBG production process and FBG adhesive technology in the process of using them.

3 FEA RESULTS

The finite element software ANSYS was used to calculate and obtain the strains of the specimens. The FEA model is shown in Figure 2. The specimen was 250 mm long and 25 mm wide. The 8-node three-dimensional laminates unit solid46 was optioned for meshing. The number of stacking piles was 26 and the layer thickness was 0.15 mm. The stacking sequence was [90°, 0°, 90°, 0°, 90°, 0°, 90°, 0°, 90°, 0°, 90°, 0°, 90°]s. While the CFRP specimens were treated as orthotropic material, so the material properties of three different directions can be defined according to Table 1. Loads were applied to the two ends of specimens, and the results are shown in Table 2 and Figure 3.

According to the Tsai-Wu criterion, the max failure index was 1 when the applied tension was 36.5 KN.

E_1: longitudinal modulus, E_2, E_3: transverse modulus; V_{21}: 21-direction Poisson's ratio, V_{31}: 31-direction Poisson's ratio, V_{32}: 32-direction Poisson's ratio; G_{12}, G_{13}: 12-direction, 13-direction shear modulus, G_{23}: 23-direction shear modulus; X_t: longitudinal tensile strength, X_c: longitudinal compressive strength; Y_t: transverse tensile strength, Y_c: transverse compressive strength; S: in-plane shear strength.

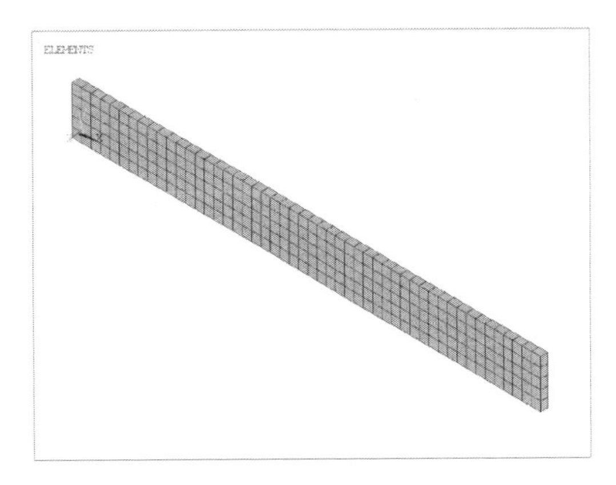

Figure 2. FEA model of the specimen.

Table 1. The mechanical properties of carbon fiber epoxy composites.

Parameter	Value Gpa	Parameter	Value Gpa
E_1	181	G_{23}	3.78
E_2	10.3	G_{13}	7.17
E_3	10.3	X_c	1500
V_{21}	0.28	X_t	1500
V_{32}	0.3	Y_t	40
V_{31}	0.28	Y_c	246
G_{12}	7.17	S	68

Table 2. The FEA results of the specimen's strain.

Tension KN	Strain uε	Tension KN	Strain uε
0	0	20	2132
5	533	25	2665
10	1066	30	3198
15	1599	35	3731

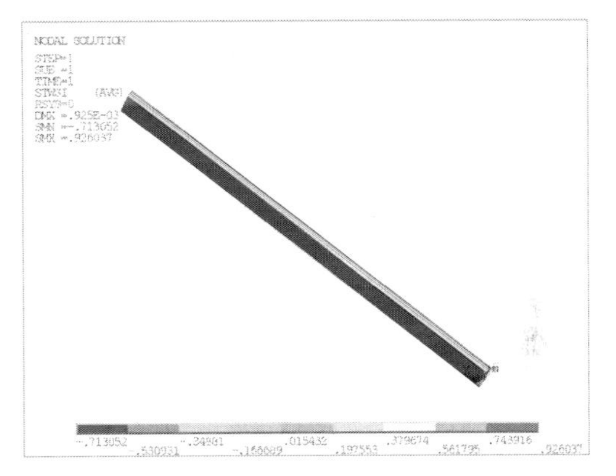

Figure 3. The max Tsai-Wu failure index.

4 EXPERIMENTS

In this research, CFRP T300/5280 laminates were used as specimens, and their stacking sequence was [90° 0°, 90°, 0°, 90°, 0°, 90°, 0°, 90°, 0°, 90°, 0°, 90°]s. Each specimen was 250 mm long and 25 mm wide (Figure 2). There were a total of 26 layers, and each layer was 0.15 mm thick.

The center wavelength of the fiber grating was 1300 nm, the diameter was 125 um, and fiber grating strain (uε) measurement ranged from 0 to 10000. A four-channel WUTOS optical fiber grating demodulator was used for FBG demodulation.

The actual measurement conditions are shown in Figure 1. Experimental specimens are shown in Figure 4. Resistance Strain Gauges (RSG) were bonded on the surface of the specimens, and FBG sensors were bonded on the surface and embedded in the specimens, respectively.

The schematic diagram of the experimental system is shown in Figure 5. INSTRON 5848 material testing machine was used in the tensile test, as shown in Figure 6. The load was applied until the specimen was damaged. During the tensile process,

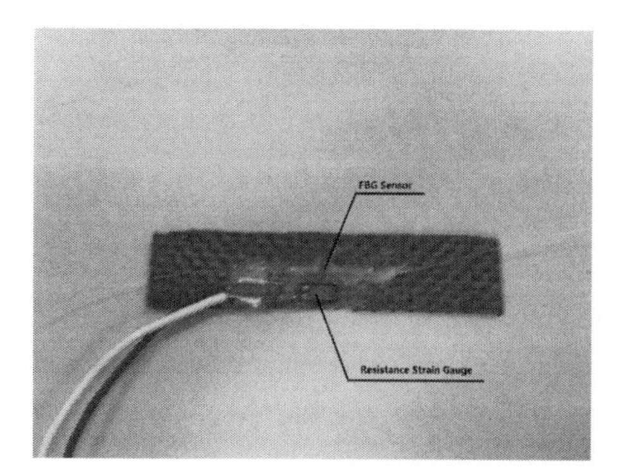

Figure 4. Experimental specimens.

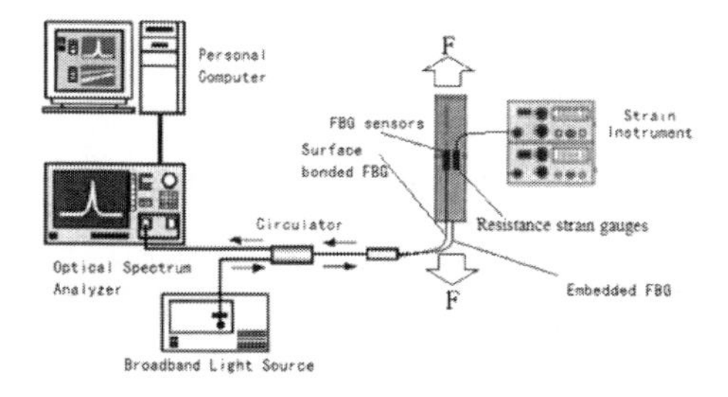

Figure 5. Schematic diagram of the experimental system.

Figure 6. INSTRON 5848 Material testing machine.

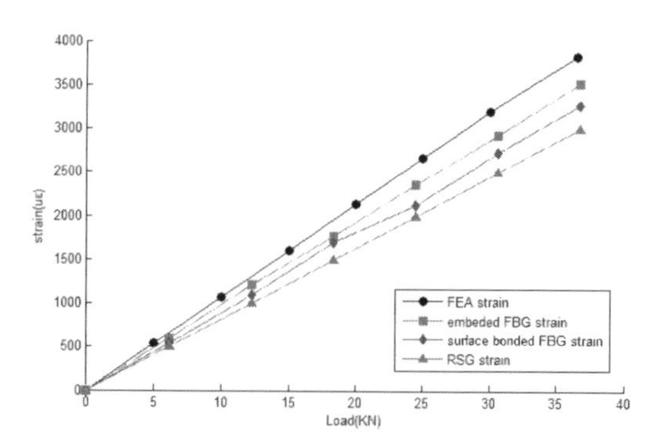

Figure 7. The FEA and experimental results.

the FBG wavelengths were collected by the FBG demodulation instrument. Carbon fiber specimen 1 stretched to 35.3 KN when the specimen was damaged, while the FEA analysis result was 36.5 KN. The theoretical and experimental results were similar.

The results of carbon fiber specimens are shown in Figure 7.

Figure 7 shows the relationship between the tension and strain of the specimens. It is clear that the specimen strain increases as the tension increases. According to the FEA results, the FBG results are closer than the RSG results, the error between the embedded FBG strain and FEA strain is 5.6%, and the error between the surface-bonded FBG strain and FEA strain is 12.4%. The embedded FBG results are closer than the surface-bonded FBG results.

5 ANALYSIS

A good linear relationship between the applied load and strain was observed, as monitored by FBG sensors and resistance strain gauges. In this experiment, the coating layer was not removed from the FBG sensors, leading to the strain transfer error between FBG strain and FEA strain. We can see that the error between the embedded FBG strain and FEA strain was lower than that between the surface-bonded FBG strain and FEA strain. This is mainly because the glued layer influenced the strain transfer rate of the embedded FBG sensors.

FBG sensors embedded in the composite laminates can more accurately measure the inner strain of composite laminates. In contrast, most sensors rely on the deformable object fixed by different glued layers to measure the strain of the object. So, to obtain a relatively accurate result of the bonded FBG strain, we should correct the result.

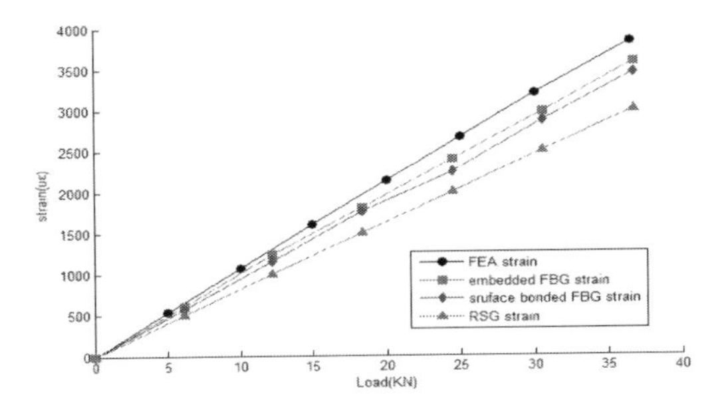

Figure 8. The FEA results and the corrected experimental results.

The relationship between ε' and ε'' is $\varepsilon' < \varepsilon'' < \varepsilon$, which is in conformity with the trend of the experimental results. While E is 74 GPa, A is 0.1252 π/4, h is 0.0625, G is 1 Gpa, d is 0.12 mm, and L is 10 mm, according to the Formula (3) and Formula (4), the results can be expressed as follows:

$$\varepsilon' = \frac{\varepsilon}{(1+\dfrac{12EAh}{GdL^2})} = 0.948\varepsilon \tag{5}$$

$$\varepsilon'' = \frac{\varepsilon}{(1+\dfrac{3Edh}{GL^2})} = 0.983\varepsilon \tag{6}$$

According to Formula (5) and Formula (6), the corrected strain can be expressed as follows:

$$\varepsilon_1' = \frac{\varepsilon'}{0.948} \tag{7}$$

$$\varepsilon_1'' = \frac{\varepsilon''}{0.983} \tag{8}$$

According to Formula (7) and Formula (8), the corrected strain of the embedded FBG and the strain of the surface-bonded FBG can be calculated. Figure 8 shows the FEA results and the corrected experimental results. After correction, we can see that the surface-bonded FBG strain is obviously close to the embedded FBG strain, and the measurement errors become smaller. Sometimes we can measure the strain of CFRP laminates by replacing FBG sensors bonded on the surface with FBG sensors embedded in the middle layer via correction. It is convenient to arrange the FBG as well as to obtain a relatively accurate result.

6 CONCLUSIONS

Based on the above experiments and analysis results, the following conclusions can be drawn:

1. FBG sensors and strain gauges are valid and reliable in monitoring the strain of CFRP laminates.
2. FBG sensors embedded in the composite laminates can more accurately measure the inner strain of composite laminates.
3. There is a strain transfer layer between FBG sensors and CFRP laminates, and the strain transfer error should be corrected in practical engineering applications. After correction, the surface-bonded FBG strain is found to be close to the embedded FBG strain.

ACKNOWLEDGMENT

This research was supported by the Independent Innovation of Foundation of Wuhan University of Technology (Project No. 155204009) and the Independent Innovation of Foundation of Wuhan University of Technology (Project No. 145204007).

REFERENCES

Ansari, F. & Libo, Y. 1998. Mechanics of bond and interface shear transfer in optical fiber sensors. *Journal of Engineering Mechanics,* 124(4): 385–394.

Chambers, A.R., Mowlem, M.C. & Dokos, L. 2007. Evaluating impact damage in CFRP using fiber optic sensors. *Composites Science and Technology,* 67(6): 1235–1242.

Chuanquan, L., Zhi, Z. & Jinping, O. 2014. Monitoring of structural prestress loss in RC beams by inner distributed Brillouin and fiber Bragg grating sensors on a single optical fiber. *Structural Control and Health Monitoring,* 21(3): 317–330.

Damien, K. & Ptrice, M. 2014. Fiber Bragg Grating Sensors toward Structural Health Monitoring in Composite Materials: Challenges and Solutions. *Sensors,* 14(4): 7394–7419.

Frieden, J., Cugnoni, J., Botsis, J. & Gmür, T. 2012. Low energy impact damage monitoring of composites using dynamic strain signals from FBG sensors Part II: Damage identification. *Composite Structures,* 94(2): 593–600.

Guo, W., Xinliang, L. & Hao, S. 2011. Analysis of Strain Transfer of Fiber Grating Sensors Adhered to the Structure Surface. *Metrology & Measurement Techniques,* 31(4): 1–4.

Jiashang, S. & Yu, W. 2009. Analysis and Correction of the Strain-Transfer Error in the Measurement with the Fiber Grating Sensor. *Measurement & Control Technology,* 28(3): 5–9.

Kersey, A.D., Davis, M.A. & Patrick, H.J. 1997. Fiber grating sensors. *Journal of Lightwave Technology,* 15(8): 1442–1463.

Kim, S.W., Jeong, M.S. & Lee, I. 2013. Determination of the maximum strains experienced by composite structures using metal coated optical fiber sensors. *Composites Science and Technology,* 78(1): 48–55.

Leng, J. & Asundi, A. 2003. Structural health monitoring of smart composite materials can by using EFPI and FBG sensor. *Sensors and Actuators A: Physical,* 103(3): 330–340.

Okabe, T. & Yashiro, S. 2012. Damage detection in holed composite laminates using an embedded FBG sensor. *Composites: Part A: Applied Science and Manufacturing,* 43(3): 388–397.

Okabe, Y., Tsuji, R. & Takeda, N. 2004. Application of chirped fiber Bragg grating sensors for identification of crack locations in composites. *Composites: Part A: Applied Science and Manufacturing,* 35(1): 59–65.

Pereira, G., Frias, C. & Faria, H. 2013. Study of strain-transfer of FBG sensors embedded in unidirectional composites. *Polymer Testing,* 32(6): 1006–1010.

Spearing, S.M. & Beaumont, P.W.R. 1992. Fatigue damage mechanics of composite materials. I: Experimental measurement of damage and post-fatigue properties. *Composites Science and Technology,* 44(2): 159–168.

Wild, G. & Hinckley, S. 2009. *Distributed optical fibre smart sensors for structural health monitoring: a smart transducer interface module.* In 2009 International Conference on Intelligent Sensors, Sensor Networks and Information Processing: 373–378.

Yashiro, S., Takeda, N. & Okabe, T. 2005. A new approach to predicting multiple damage states in composite laminates with embedded FBG sensors. *Composites Science and Technology,* 65(3): 659–667.

Advanced Materials, Mechanical and Structural Engineering – Hong, Seo & Moon (Eds)
© 2016 Taylor & Francis Group, London, ISBN: 978-1-138-02908-8

Liner wear in roll crushers: Microscopic investigations

R. Sinha & K. Mukhopadhyay
Department of Mining Machinery Engineering, Indian School of Mines, Dhanbad, India

ABSTRACT: Roll crushers are major size reduction equipment used in mining and allied industries. These industries involve a great deal of rock crushing. The main body of rolls in a roll crusher is protected against wear or damage by wear-resistant liners. The characterization of wear is essential for selecting a durable liner material. Mn-steel liner is mostly used in mines for crushing less abrasive feed materials. The composition of the Mn-steel liner is presented in this paper based on the data collected from a survey. Samples of Mn-steel liner material were collected from mines. Preparation of the liner samples were done to analyze wear behavior in crushing operation under the Field Emission Scanning Electron Microscope (FESEM). This paper reports on the mechanisms of wear in roll crusher Mn-steel liner through microscopic investigations.

1 INTRODUCTION

Wear in the roll crusher liner is a perennial problem. It is a complex phenomena involving service variables, environmental conditions, lubrication and corrosion. Wear resistance of a material is related to its microstructure and changes take place in the microstructure during the wear process. Microstructural investigation was performed in this paper to determine wear mechanisms. Scratching, surface deformation, micro-cracks and cavity formation have a major effect on liner material loss in a crusher (Lindqvist & Evertsson 2003). Investigations of liner wear will benefit in improving crusher performance, reduction in down time and uniform product size distribution by selecting a suitable liner material. At different zones of roll crusher, wear mechanism has been found to be different. The present study was performed to identify the wear zones and wear mechanisms with an aim to improve the quality of the liner material. The theoretical investigation was then confirmed by studying the worn-out liner material. In this study, the wear phenomena are examined for a smooth liner of a double roll crusher used for comminuting coal.

2 THE ROLL CRUSHER

A wide range of roll crushers are used in mining and allied industries; a double roll crusher is a common type of equipment used as a primary crusher for sizing coal. Roll crushers are used for sizing sintered coals, soft and medium hard rocks, fertilizers and salts. The crushing material is fed on the top of the whole width of the rolls with the help of feeders. Parameters such as roll diameter, roll rpm and gap between rolls determine the performance of the crusher. In a double roll crusher, one roll is fixed while the other is floating type for safeguarding the crusher from non-crushable material, and enables to set the gap between the two rolls. In addition, it compensates for the wear and tear of the roll surface. The rolls are individually driven through a suitable gearing arrangement. The crusher is equipped with a shear pin or a breaker bolt or spring loading attachment for protecting the crusher against any occasional non-crushable material entering into the crushing chamber. A double roll crusher is shown in Figure 1.

Figure 1. Image showing material loss/damage from the roller body surface of roll crushers.

Table 1. Liners and their composition.

	Material		
Compositions	NIHARD 4	Mn-steel	Mn-Cr steel
C %	2.80–3.00	1.05–1.35	0.55–0.70
Mn %	0.5	1	0.50–1.00
S %	≤0.05	11–14	0.05
Si %	1.9	0.05	0.75
P %	≤0.07	0.09	0.05
Cr %	8.5		1
Ni %	5.5		
Mo %			0.2

3 THE LINER MATERIALS IN THE ROLL CRUSHER

To replicate the problem of wear in roll crusher liners, some major steps are taken that are still in use for the roll crusher. One of them is to provide a better wear protection material for rolls to withstand good strength during crushing of feed materials. Hard phasing such as carbides, borides or nitrides is another step useful to agitate wear (Sesemann & Broeckmann 2013). In addition to hard phasing, Oberheuser (1996) remarked two types of wear protection mostly recognized in the majority of mining sectors as follows:

a. Wear-resistant tires of iron alloys provided on the rolls, e.g. NIHARD 4, Mn-Steel, manganese chromium cast iron.
b. On the roll body, sometimes hard-phasing layer is welded. This hard-phasing layer is renovated after abrasion from several runs.

The compositions of the liners are described in Table 1. The hardness of the liner decides its durability against wear performance. Hardness has an important effect in achieving the minimization of wear losses. Moshgbar et al. (1995) studied the fact that causes an uneven-wear profile of the liner and variables that are responsible for this effect. This uneven distribution of wear on liners is the cause for three-body abrasion.

4 THREE-BODY ABRASIVE WEAR

Abrasion is the removal of working surface from hard inclusion particles that results in cutting, scraping and surface deformation. In the three-body abrasive wear mechanism, the movement of hard particles is not constrained but is free to roll and slide down the surfaces. The common analogy is that wear takes place in roll crusher liners (Juri et al. 2011). Indentation plays a significant role in three-body abrasion, resulting in the formation of high-stress wear zones (Bingley et al. 2005). Such type of wear is found in high-pressure grinding rolls (Page et al. 2000).

4.1 *Literature survey on three-body abrasion*

A wide domain of wear rates have been varyingly reported for three-body wear, which depends on both the testing material and the apparatus for the test. It has been observed that the debris particle separates the two contacting surfaces that influence the wear rate. The in-between particle either rolls or slides over the surface. In most of the three-body wear analysis, scratching is observed along with surface deformation (Rabinowicz et al. 1961; Sassed et al. 1987). When the debris particles roll between the two contacting surfaces, reduced wear is produced during sliding. Other deformation modes may possibly arise. If the attack angle of the debris particle is low, then there is a possible cause of plowing wear. In addition to the plowing damage adhesion between the surface of the metal and the particle, large sub-surface shear strains are formed. This indicates the disjunctions between the contact surface of the metal and the debris particle after several runs of mass of debris particles. At higher attack angles, cutting also occurs (Hokkirigawa & Kato 1988). Another standpoint of three-body abrasive wear is to appraise the prospect of damage in the three-body wear condition, which might result in surface deformation due to the rolling and sliding of debris particles (Wang and Wang 1988). Significant work has been performed on dry three-body wear during indentation. Bingley and Roniotis (Bingley & Roniotis 2002) promote their work based on indentation for diverse metals under the three-body order.

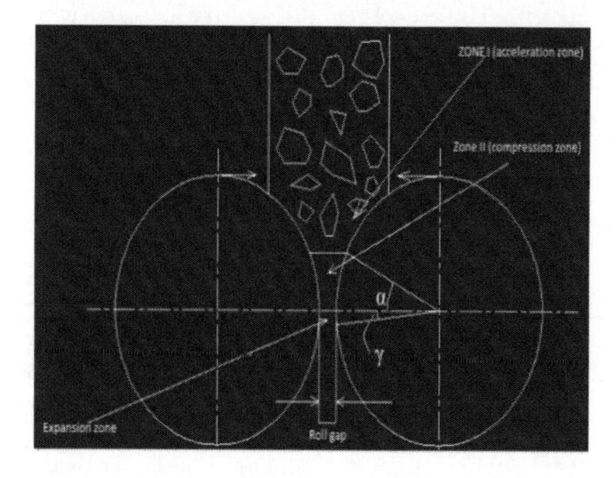

Figure 2. Wear zone division on the double roll crusher.

Liners are used in many applications of work. One of their major roles is to protect heavy mining crusher equipment against wear. P.P. Rosario et al. (2004) focused on liner wear of crusher by measuring the chamber profile of the crusher. They observed squeezing wear due to compression. Also, it has been found that feeding of ore particles from a certain height results in impact wear. This can be defined as repetitive exposure of contact body with ore particles (Ludema 1996). Osara & Tiainen (2001) observed that plowing, micro-cutting and micro-cracking are the wear phenomena that mostly govern the body of wear-resistant material under the impact condition. The microscopic photographs in their work show that plowing is the main wear mechanism with large crack exposed on surface layers of wear-resistant material. Under the three-body abrasive wear condition, Bingley & Schnee (2005) observed that the material undergoes surface deformation due to indenting of intermediate particles between the two contacting bodies, as observed in the microphotographs. They also identified that short-length surface scratches are the resulting behavior of wear mechanism under the three-body wear mechanism. In order to understand wear due to indentation mechanism, Bourithis & Papadimitriou (2005) performed their experimental work on the three-body abrasive wear mechanism to observe the effect of wear due to indentation. The cavity formation was identified with marks of networks of micro-cracks. As present, the work focuses on the roll crusher liner material; therefore, the study focuses on resolving the issue of wear from liners. In a roll crusher, the feed material acts as an intermediate body between the two rolls. The feed material accelerates or rotates between the two rolls before it is nipped for crushing. During the crushing process, it indents into the rolls' surface, leading to the removal of the metal. The feed consisting of rock materials acts as an intermediate body and the two rolls as two other bodies (Mishra & Finnie 1979). Theisen (1997) explained in his work about the three-body wear mechanism of the roll crusher. In his work, the roll surface was divided into two zones: sliding zone and compression zones.

4.2 *Microscopic observations and wear mechanisms*

By a thorough study of literature on the roll crusher, it has been found that less work has been done on the wear of roll crusher liners. Therefore, to understand the liner wear of the roll crusher, an initial step of microscopic investigations was performed on liner samples collected from mines. The sample was then washed with acetone to remove coal particles from its surface. The next step was the sample preparation for obtaining good microscopic results. To obtain a required size of the sample, the liner was cut by means of the Carbon Arc Cutting (CAC) process. The liner samples are shown in Figure 3.

The samples were tested under the Field Emission Scanning Electron Microscope (FESEM) to examine the class of damage. The EDS will help to determine the surface composition of the sample after crushing operation in the environmental condition. The microphotographs reveal the type of wear at different crushing zones. As already mentioned in this paper, there are two different wear zones in a roll crusher: i) sliding zone (I) and ii) compression zone (II). It was noticed that two different types of wear take place in these two zones. Wear starts in the sliding zone and continues up to the compression zone. Liner from the concerned mine in our study is Grade-I Mn-steel liner of the double roll crusher. Table 2 presents the hardness property of the liner with details of wear loss in mm. In the sliding zone, abrasive wear takes place when the feed material slides on the liner surface.

The liner material is damaged due to the continuous sliding action. The material from the damaged surface either leaves the surface or sticks on the surface. Such stuck material causes surface deformation. The cavity formation, abrasion and surface deformation are shown in Figure 4.

During the initial sliding of the feed material on the liner surface, shear force is developed. The shear force is non-uniformly distributed over the surface of the liner during the sliding motion of the feed material. On the other hand, the feed material also applies some load on the liner surface during sliding. The direction of the normal load

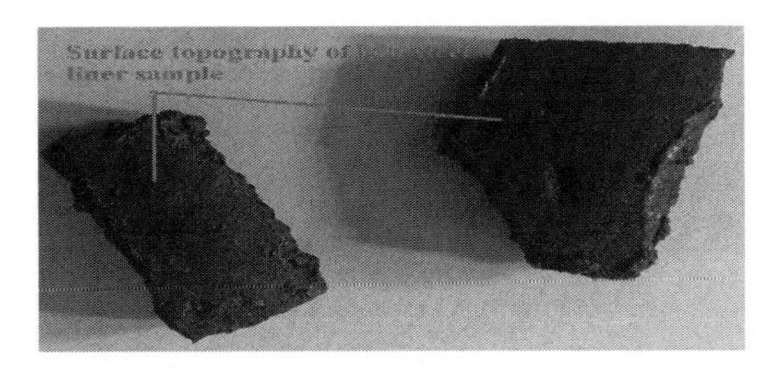

Figure 3. Prepared Mn-steel liner material samples for FESEM microphotographs.

Table 2. Data collected from the coal handling plant.

Mn-steel liner material	
Hardness	220–550 BHN
Roll diameter	1250 mm
Wear loss in mm/year	5 mm–6 mm

changes from angle 'α' to angle 'β' during feed material movement, as explained in Figure 5.

The overall feed material falling from the hopper induces some load on the roll surface, thereby increasing the overall load. The feed material, which is in contact with the roll surface continuously, changes its direction of the load from normal to tangential before it gets nipped in space between the rolls. The change in the direction of the load of the feed material from normal to tangential reaches a point where the load becomes critical, which is the minimum amount of load at which scratching or surface deformation begins to form in the unit area of the scar (Michael et al. 1992). This results in the increase of shear force on the liner surface. As the movement of the feed material reduces while moving towards the compression zone, the shear force attains a maximum value. At a point of time, the maximum shear force fails to resist the critical load of the feed material. When the critical load exceeds the maximum shear force, surface deformation is caused in the form of scratches on the liner surface. This surface deformation continues up to the end of the sliding zone (Gupta and Yan 2006). Thus, during the sliding motion of the feed material, surface deformation is observed with the presence of micro-cracks on the liner surface. The cavities are formed due to the impact caused by the falling of the feed material onto the moving rolls from a height. The formation of cavities is shown in Figure 6.

At the end of the sliding zone, the feed material stops sliding on the liner surface. This is the beginning of the nipping of the feed material. The compressive force exerted by the rolls on the feed material causes a deep cavity on the liner surface due to indentation. The indentation on the liner surface along the line of nipping throughout the width of the roll is uniformly distributed, as explained in Figure 7.

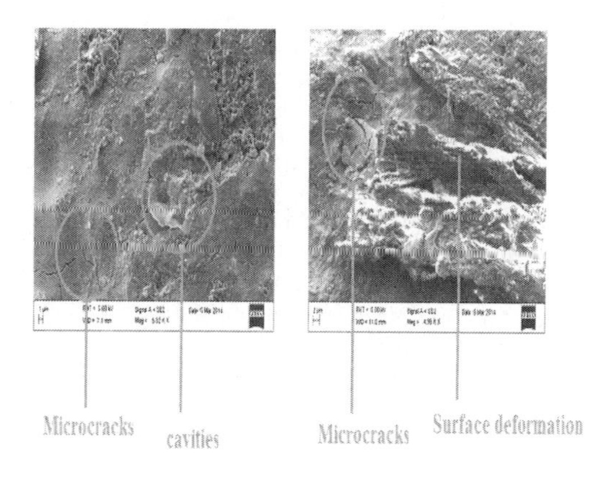

Figure 4. Groove formation with micro-crack marks identified as the first stage of material removal from the liner surface in the sliding zone.

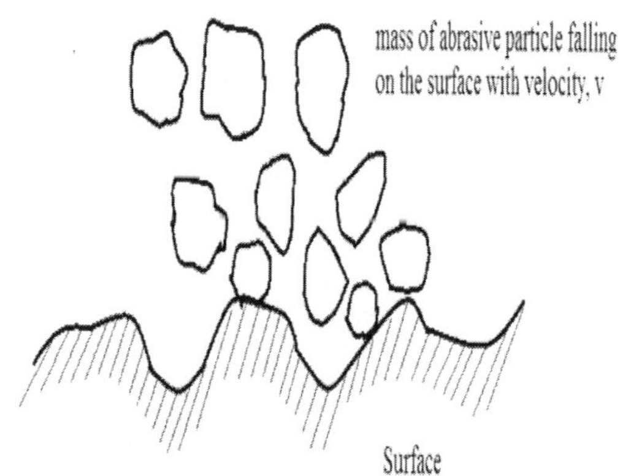

Figure 6. Explanation of cavity formation due to the impact of the feed material on the liner surface.

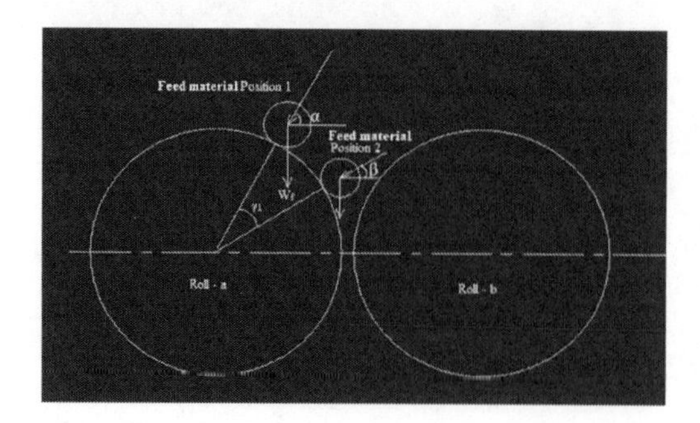

Figure 5. Load direction by feed material on the liner surface. (a) Description of different wear zones on the liner surface of the roll crusher. (b) Change in load direction at different positions of the feed material.

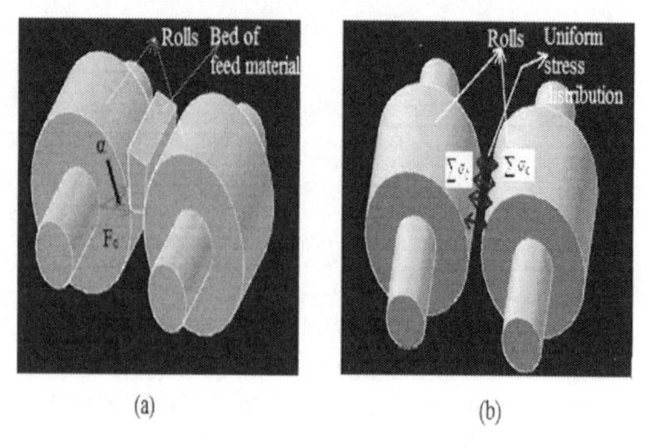

Figure 7. Uniform compressive stress distribution on the liner surface.

Flake-like structures are formed after the nucleation of micro-cracks from the liner surface, as shown in Figure 10. The detailed characterization of the liner provides us information about intermixing of coal with the liner surface. The respective change in surface chemical composition is shown in Figure 11.

(a) (b)

(c)

(i) & (ii) Deep cavity mark on liner surface and surface deformation respectively during indention in compression zone.
(iii) & (iv) Formation and growth of micro-cracks

Figure 8. Microphotographs showing images of (a) deep cavity formation, (b) micro-cracks on the liner surface and (c) micro-cracks formed due to stress concentration.

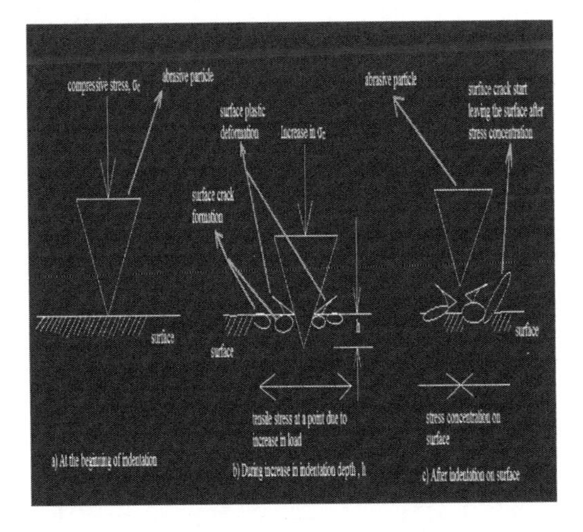

Figure 9. Explanation of micro-cracks formation and leaving of micro-cracks from the surface due to stress concentration, where 'h' is the increase in depth of indentation.

The content of Fe, Al and Na explains that the surface structure of the liner changes after handling a bulk of solid materials. Weight percentage of carbon content is changed due to intermixing of coal particles with surface grains of the Mn-steel liner. However, Al and Na are the impurities intermixed with the layers of the liner surface. This intermixing may develop some new bonding elements, which may result in the debonding of liner surface molecules. Due to the operation of

(i) Flake identified on liner surface

(ii) identification of cavities

Figure 10. Microphotograph of unit area of the liner surface with visible marks of flakes.

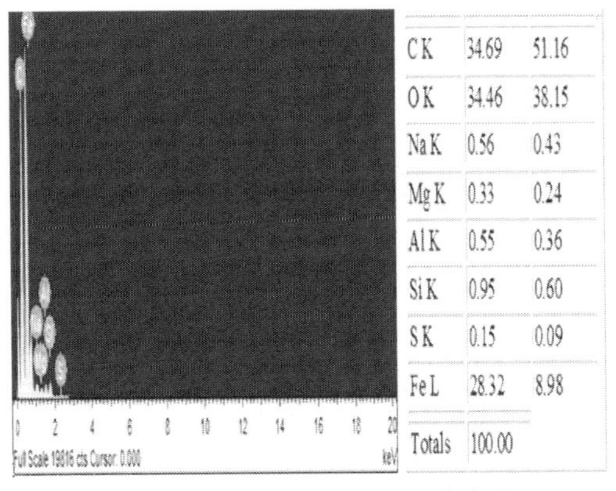

C K	34.69	51.16
O K	34.46	38.15
Na K	0.56	0.43
Mg K	0.33	0.24
Al K	0.55	0.36
Si K	0.95	0.60
S K	0.15	0.09
Fe L	28.32	8.98
Totals	100.00	

Where, C = carbon, O = oxygen, Na = Sodium, Mg = Manganese, Al = Aluminium, Si = silicon, S = sulphur, Fe = iron. K & L = quantum shells

Figure 11. EDS report analyzed for intermixing of coal with the liner surface.

the crusher in the environmental condition, some moisture content of coal has also been found from the EDS report. The amount of oxygen in weight percentage is also marked on the surface of the liner. Thus, it can be stated that the overall composition of the liner surface changes after dressing of coals to the desired size. The letters "K" and "L" represent the quantum shells. It is considered as the energy range used as a function of position.

5 CONCLUSIONS

The roll crusher used for coal mines is protected from wear damage by Mn-steel Grade-I liners. The liners have an ability of strain hardening that results in minimum damages in terms of wear and tear. It also provides the durability of performance in work. However, wear is an important issue that requires a close attention. Based on the literature survey, it has been found that wear in the double roll crusher is due to three-body abrasion. Wear in the roll crusher can be classified into two categories: wear in the sliding zone and wear in the compression zone. To identify wear mechanisms on the roll crusher, several studies were done. Based on the sample and data collection, it becomes important to perform a microscopic study using FESEM. Cavity formation, surface deformation and micro-cracks are the major types of damage observed under FESEM. The microphotographs provide a suitable explanation of wear in two different zones of the double roll crusher. This study also helps in identifying the valid conclusion of coal particles intermixing with the liner surface, which is another way of processing the cause of wear loss. This paper will help to take the next forward step to minimize the condition of wear of the roll crusher liner.

REFERENCES

Bingley, M.S. & Roniotis, D. 2002. *A comparison of two-body and threebody abrasive wear in various wet and dry environments*, Proceedings of the 6th International Tribology Conference, Austrib, Perth, Western Australia, (1) 181–188.

Bingley, M.S. & Schnee, S. 2005. A study of the mechanisms of abrasive wear for ductile metals under wet and dry three-body conditions, *Wear*, 258: 50–61.

Bourithis, L. & Papadimitriou, G. 2005. Three body abrasion wear of low carbon steel modified surfaces, *Wear*, 258: 1775–1786.

Gupta, A. & Yan, D.S. 2006. *Introduction to Mineral Processing Design and Operation*, Elsevier Science and Technology books, 142–146.

Hokkirigawa, K. & Kato, K. 1988. An experimental and theoretical investigation of ploughing, cutting and wedge formation during abrasive wear, *Tribology International*, 21: 51–57.

Juri, P., Mart V., Sergei, L., Kristjan, J. & Renee, J. 2011. Three-body abrasive wear of cermets, *Wear*, 271: 2868–2878.

Lindqvist, M. & Evertsson, C.M. 2003. Liner wear in jaw crushers, *Minerals Engineering*, 16: 1–12.

Ludema, K.C. 1996. *Friction, Lubrication*, Wear 1996, A text book in tribology, CRC Press.

Michael, N., Don, M.P. & Hertbert, G. 1992. *Mechanical Properties and Deformation Behavior of Materials Having Ultra-Fine Microstructures*, Proc. of NATO Advanced Study Institute, Porto-Novo, Portugal, June 28-July10: 309–310.

Mishra, A. & Finnie, I. 1979. *A Classification of Three-body Abrasive Wear and Design of a New Tester*, in; K.C. Ludema, W.A. Glaeser, S.K. Rhee (Eds.), Proc. Int. Conf. on Wear of Materials, Dearborn, MI, 313–318.

Moshgbar, M., Bearman, R.A. & Parkin, R. 1995. Optimum control of cone crushers utilizing an adaptive strategy for wear compensation, *Minerals Engineering*, 4/5(8): 367–376,.

Oberheuser, G. 1996. Wear protection of surfaces from high pressure grinding rolls - possibilities and limits, *International Journal of Mineral Processing*, 44–45: 561–568.

Osara, K. & Tiainen, T. 2001. Three-body impact wear study on conventional and new P/M + HIPed wear resistant materials, *Wear*, 250: 785–794.

Page, N.W., Yao, M., Keys, S. & McMillan, W. 2000. A High Pressure Shear Cell for Friction and Abrasion Experiments, *Wear*, 241(2): 186–192.

Rabinowicz, E., Dunn, L.A. & Russel, P.G. 1961. A study of abrasive wear under three-body conditions, *Wear*, 4: 345–355.

Rosario, P.P., Hall, R.A. & Maijer, D.M. 2004. Liner wear and performance investigation of primary gyratory crushers, *Minerals Engineering*, 17: 1241–1254.

Sassed, T., Oki, M. & Emory, N. 1987. *Effect of fine particles interposed between sliding surfaces on wear of materials*, in: Porch International Conference on Wear of Materials, ASME, New York, 185–190.

Sesemann, Y., Broeckmann, C. & Ofter, A.H 2013. *A new laboratory test for the estimation of wear in high pressure grinding rolls*, Wear, 302: 1088–1097.

Theisen, W. 1997. *Material aspects of machining wear resistant alloys*, Fortschr. Ber. VDI, Reihe, 2(428).

Wang, Y.L. & Wang, Z.S. 1988. An analysis of the influence of plastic deformation on three-body abrasive wear of metals, *Wear*, 122: 123–133.

Advanced Materials, Mechanical and Structural Engineering – Hong, Seo & Moon (Eds)
 ISBN: 978-1-138-02908-8

Study on tool electrode wearing and its compensation in electrical discharge milling

C.B. Guo, D.B. Wei & S.C. Di
School of Mechatronics Engineering, Harbin Institute of Technology, Harbin, China

ABSTRACT: In order to improve the compensation accuracy of Electrical Discharge milling (ED-milling) in titanium alloy machining, experiments were conducted to study the influence of electrical parameters on electrode wearing. Electrode shape wearing in the discharge process was analyzed. Brass and graphite were adopted as tool electrodes to study the influence of electrode materials. Wearing compensation strategy, which only considers wearing in the length direction, was proposed. Two kinds of predictive electrode length compensation strategies based on the process database and discharge current were designed, respectively, for simple shape machining. Two kinds of interval electrode length, namely detection and compensation strategies, based on the milling layer and time were designed, respectively, for ED-milling machining of complex cavity and curved surfaces. The compensation accuracy of different compensation strategies was discussed.

1 INTRODUCTION

A large number of high-strength, high-temperature-resistant materials such as titanium alloys and nickel-based high-temperature alloys are widely used in the aerospace industry. The machining of those difficult-to-machine materials by mechanical machining has low efficiency and high cost problems. Electrical Discharge Machining (EDM) is a non-contact machining method, which uses discharge energy supplied by pulse power to remove materials from the workpiece by melting and vaporization. There is no obvious force between the electrode and the workpiece in the discharge process. The machining efficient of the EDM process is independent of the strength and hardness of materials, and it is related to the physical property of materials such as electrical conductivity and thermal characteristic. The EDM process has a good performance in the machining of materials that has high electrical conductivity and low thermal conductivity, and it provides a good choice for machining of high-strength and high-temperature-resistance materials (Ho & Newman 2003).

Electrical Discharge milling (ED-milling) is similar to mechanical milling. It uses a simple tubular electrode that is controlled via the Computer Numerical Control (CNC) system to realize the desired geometric shape machining. The material is removed from the workpiece by discharge spark that is produced in the gap when the tool electrode is moved closely to the workpiece by the CNC system. This method needs no design and manufacture of a complex geometric shape electrode. The machining of a complex geometric shape workpiece can be achieved by controlling the moving path of the tool electrode using the CNC system. By using the ED-milling process, the time and cost for tool electrode machining are saved, which is helpful to shorten the processing cycle and reduce the production cost (Bayramoglu & Duffill 1994, Ding et al. 2006).

The tool electrode moving path of ED-milling is similar to that of mechanical milling, but the tool electrode has no contact with the workpiece in the ED-milling process, and the material is removed depending on the heat effect produced by the discharge process. Part of the tool electrode is also worn out in the discharge process; as a result, the tool electrode is shortened during the machining process, which will affect the machining accuracy (Mohd et al. 2007). The tool electrode wearing is influenced by the pulse width, discharge current, open gap voltage and polarity, and these parameters have a complex interaction effect with each other (Ho & Newman 2003, Saravanan et al. 2012). As a result, the accurate prediction of electrode wearing is difficult to realize. The tool electrode wearing compensation problem is one of the key technologies in ED-milling.

The tool electrode wearing can be reduced by designing a special geometric shape tool electrode (Mikesic et al. 2009) or by improving the power supply module (Han et al. 2009). Electrode wearing can be measured by the following methods: off-line prediction, on-line detection and tool electrode uniform wearing. The off-line prediction method is mainly based on previous experimental results and empirical operators (Zhou et al. 2008). The Tool electrode Wearing Rate (TWR) is predicted using the empirical formula or mathematical methods such as fuzzy algorithm and artificial neural network. The prediction accuracy of this strategy depends on the reliability of the experimental results, and the prediction error will be large when new processing parameters are applied. The on-line detection method (Bleys et al. 2002, Bleys et al. 2004) uses a probe or CCD image (Yan et al. 2009) to measure the electrode wearing. The accuracy of the measuring method for the CCD image is influenced by the refraction and reflection of the dielectric medium around the electrode. The tool electrode uniform wearing method (Yu et al. 1998) realizes only the axial direction wearing by keeping discharge occurring on the bottom of the electrode, but the cutting depth of each layer is limited.

The purpose of this research is to study the influence of process parameters on electrode wearing, analyzing the shape change of the electrode end. Based on analyzing the machining characteristic of different geometric shape workpieces, four types of electrode wearing compensation strategy are proposed.

2 EFFECT OF PROCESS PARAMETERS ON ELECTRODE WEARING

The influence of pulse width, discharge current and electrode polarity on electrode wearing is investigated by experiments. The machining conditions and parameters are listed in Table 1. A tubular graphite electrode with an outer diameter of 20 mm and an inner diameter of 5 mm is employed as the tool electrode. The titanium alloy TC4 is machined in 5% water in oil emulsion by ED-milling with high-pressure flushing. The TWR is calculated by using the following formula:

$$\eta = \frac{\Delta V_{electrode}}{\Delta V_{workpiece}} \tag{1}$$

$\Delta V_{electrode}$–the remove volume of the electrode;
$\Delta V_{workpiece}$–the remove volume of the workpiece.

The discharge lasting time is determined by pulse width. During the pulse width, the discharge voltage will maintain at around 20 V after the breakdown process, and materials will be melted or vaporized by high temperature that generated from the continuous discharge process. Part of the electrode material is also removed by high temperature in the workpiece removal process. Figure 1 shows the influence of the pulse width on electrode wearing when the discharge current is set at 30 A and the electrode is connected to the negative pole. It can be observed that the TWR reduced with the increasing pulse width. This is because the discharge times occurred at a certain length time that decreased when the pulse width was extended. It suggests that the decrease in discharge times is helpful to reduce electrode wearing. The standard deviation value was apparently high when the pulse width was set at 1000 μs. It is because the discharge process becomes unstable when the pulse width time was too long.

The discharge energy is mainly determined by discharge current, in order to decide how much of the material can be melted and vaporized in one discharge process. Figure 2 shows the influence of the discharge current on electrode wearing when pulse width is set at 500 us and the electrode is connected to the negative pole. It shows that the electrode wearing is increased with the improvement of discharge current. This is because the discharge energy of each discharge process is improved with increasing discharge current, resulting in the distribution of more energy on the electrode surface. As

Table 1. ED-milling machining conditions and parameters.

Machining conditions	Electrode material	Graphite
	Workpiece material	Titanium alloy TC4
	Dielectric medium	Emulsion (5%)
Machining parameters	Pulse width (μs)	250, 500, 750, 1000
	Pulse-off time (μs)	100
	Discharge current (A)	10, 30, 50, 80
	Electrode rotating speed (Rpm)	300
	Dielectric flushing pressure (MPa)	1.5
	Cutting depth (mm)	1.5
	Open gap voltage (V)	60
	Electrode size	Inner diameter 5 mm Outer diameter 20 mm

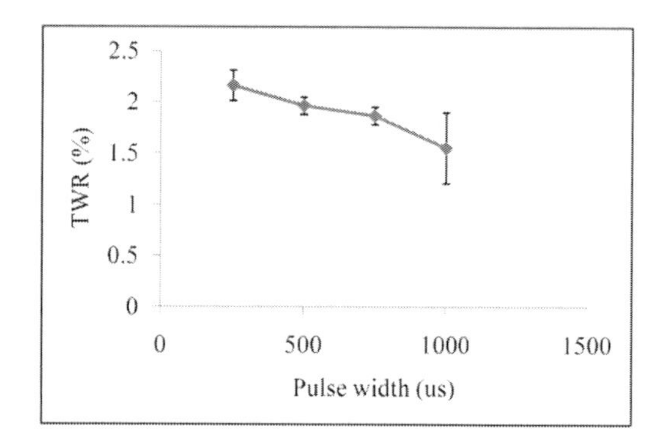

Figure 1. The influence of the pulse width on the TWR.

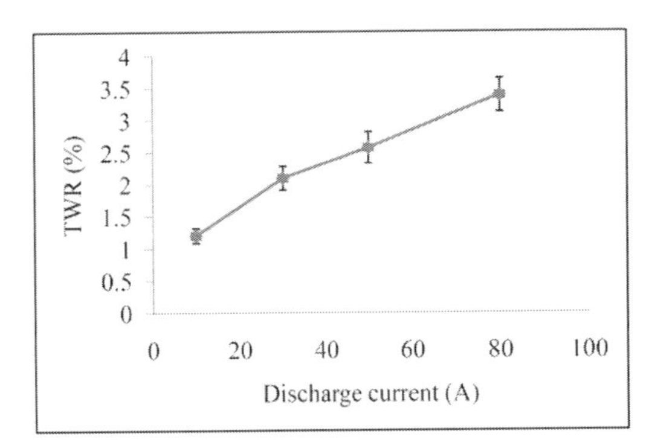

Figure 2. The influence of discharge current on the TWR.

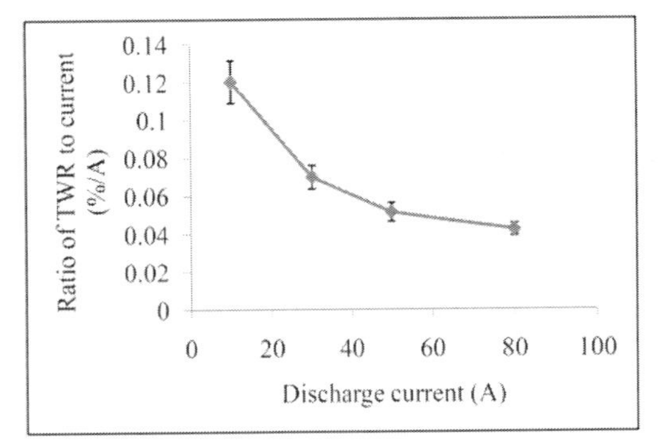

Figure 3. The influence of discharge current on the ratio of TWR to discharge current.

a result, more material is removed from the electrode by increasing the discharge current.

The ratio of TWR to discharge current reflects the tool electrode wearing by one unit discharge current. From Figure 3, it can be seen that with the increasing discharge current, the electrode wearing that is removed by 1 A current is decreased. It demonstrates that the electrode worn out by one unit of current decreases with the improvement of discharge current.

The polarity property is related to the bombarding effect of the electron and the ion on the electrode and the workpiece. This will influence the distribution of discharge energy on the electrode and the workpiece. The influence of electrode polarity on electrode wearing with the pulse width and the discharge current of 500 us and 30 A, respectively, is illustrated in Figure 4. It can be observed that a lower TWR is obtained when the electrode is connected to the negative pole, and the TWR increases by 2.5 times when the electrode is connected to the positive pole. This is because of the matching character between the electrode mate-

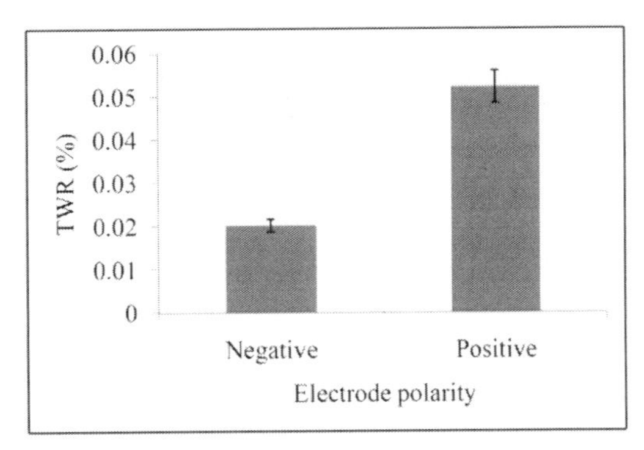

Figure 4. The influence of electrode polarity on the TWR.

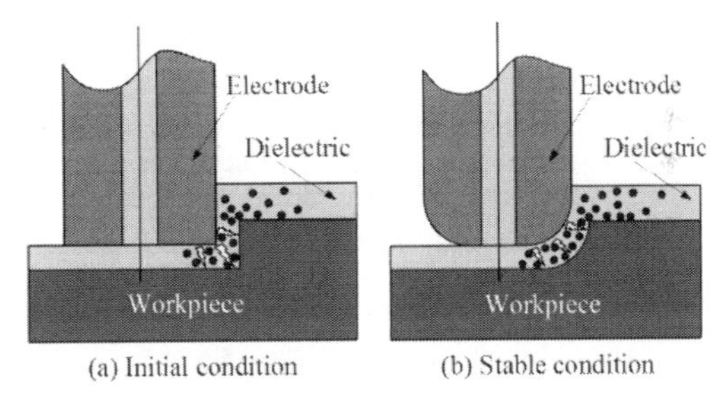

Figure 5. The wearing phenomenon at the bottom of the electrode.

rial and the workpiece material, and the TWR can be effectively reduced when matching polarity is applied in the discharge machining process. When the graphite electrode is employed to machining titanium alloys in the ED-milling process, the electrode wearing can be reduced by connecting the electrode to the negative pole.

3 TOOL ELECTRODE WEARING AND ITS COMPENSATION ANALYSIS

3.1 *The geometric shape wearing of the tool electrode*

In the ED-milling process, the tubular electrode that has a rectangle corner at the bottom is applied, as shown in Figure 5(a). As illustrated in Figure 5(a), most of the discharge will occur at the rectangle corner area of the electrode, and the corner area will be worn out prior to the other area. As a result, the rectangle corner at the electrode bottom is worn out and becomes a round corner during the discharge process. As shown in Figure 5(b), once a uniform round corner is formed at the bottom of the electrode, the continuous discharge sparks will be evenly distributed on the bottom surface of the electrode. As a result, the electrode will be worn out at the bottom surface uniformly, and the round corner formed at the bottom of the electrode will remain unchanged.

The copper electrode and graphite electrode are applied for ED-milling under the same machining conditions. In the machining process, after a certain time, a stable round corner is formed at the bottom of the electrode, as shown in Figure 6. From Figure 6(a), it can be observed that at the bottom of both copper and graphite electrodes, a round corner is formed, and the crater size on the copper electrode will be larger than that on the graphite electrode. This is because the copper has a melting point of 967°C, and the copper material is easily melted and vaporized by a high discharge energy and bombard effect, which result in large-sized craters on the copper electrode surface. When graphite material, which has a melting point of 3652°C, is adopted as the electrode, more discharge energy would be needed for melting and vaporizing graphite. Therefore, compared with the copper electrode, the graphite electrode that can be removed by one pulse discharge becomes smaller and its surface appears smooth with no obvious crater, as shown in Figure 6(b).

3.2 *Wearing compensation analysis*

In the ED-milling process, the tubular electrode is worn out in the axial direction and the radial direction by discharge spark. The wearing in the radial direction makes the electrode diameter become smaller, which influences the geometric shape accuracy of the workpiece in the plane direction. The wearing in the axial direction reduces the length of the electrode, which will influence the machining accuracy in the depth direction. The machined surface will not be straight if no compensation is made for electrode wearing, as shown in Figure 7.

In rough machining, in order to simplify electrode wearing compensation, compensation strategies that only consider the electrode wearing in the axial direction are proposed. The uniform wearing in the peripheral direction is achieved by the rotating electrode, which can guarantee the consistency of machining allowance.

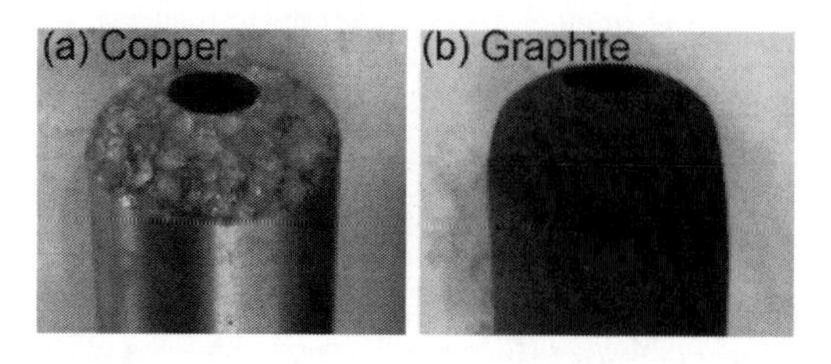

Figure 6. The shape of the bottom of the electrode after machining.

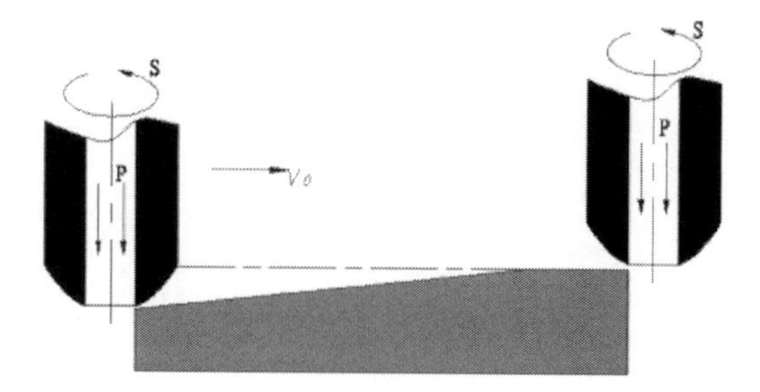

Figure 7. Machining result without considering wearing compensation.

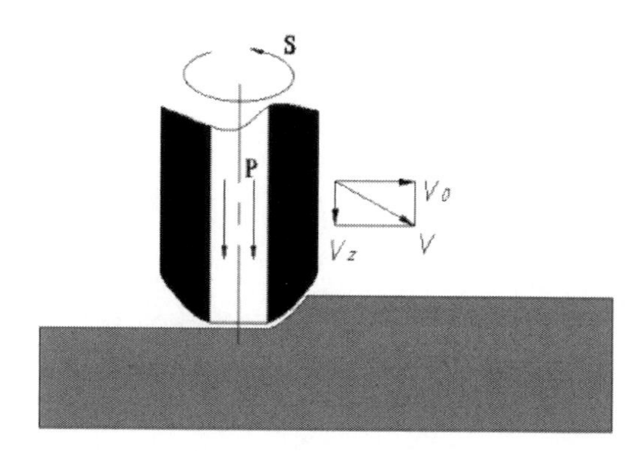

Figure 8. Machining with ideal wearing compensation.

Considering only the wearing in the axial direction of the electrode that was compensated in the proposed compensation strategy, for the convenient of wearing compensation, the calculation of the TWR can be simplified as follows:

$$\eta = \frac{\Delta Z}{L} \quad (2)$$

ΔZ–electrode wearing in the axial direction;

L–electrode moving length in the discharge machining process.

This method simplified the calculation of the TWR. It facilitated the realization of electrode wearing compensation in a convenient way, and its calculation result has a proportional relationship with that given in Equation (1).

Figure 8 shows the ideal compensation method, in which an axial feeding was added into the electrode moving path. The adjusted electrode moving path can combine the axial direction feeding and the plane direction feeding together, which can maintain the discharge occurring in the same depth, and the machined surface will be kept flat. This ideal compensation strategy has a high requirement for electrode wearing, and the electrode wearing process must be stable and uniform. However, in the real ED-milling process, the electrode wearing was influenced by the process parameters and the gap cooling effect. It is hard to keep the electrode wearing stable and uniform. As a result, this ideal compensation strategy will be difficult to realize in practical machining.

4 THE REALIZATION OF ELECTRODE WEARING COMPENSATION

4.1 *Electrode wearing compensation based on prediction*

4.1.1 *Compensation based on the process database*

The TWR of ED-milling under all kinds of process parameters was obtained by conducting experiments, and a processing database was established based on the experiment results. In ED-milling, the TWR was chosen from the TWR process database based on the processing parameters. The electrode wearing value was calculated using the TWR value multiplied by the machining length, as given in Equation 3. The wearing value was added into the discharge machining length evenly, as follows:

$$\Delta Z_1 = \eta_1 \times L_1 \tag{3}$$

For regular plane machining, the cutting depth per layer is constant, and only simple line and circle moving path are used. The length of the moving path can be calculated by the geometric method. The electrode wearing compensation procedure is illustrated in Figure 9. A 130 mm length grove was machined with a 0.05 mm machined depth deviations from the beginning point to the end point.

4.1.2 *Compensation based on discharge current*

In rough machining, discharge current is the key element in improving the machining efficient. Based on the character of the discharge current effect on electrode wearing, a compensation based on the discharge current was proposed. In this strategy, the discharge current changing process was detected in real time, and the relationship between the TWR and the discharge current was investigated. Its compensation accuracy depends on the detection accuracy of discharge current. In order to improve the reliability of TWR calculation, a large number of experiments were done to study the relationship between the TWR and the discharge current, as shown in Figure 2. It can be observed that the TWR increased with the improvement of discharge current. When the machining condition was fixed, the TWR empirical equation for current from 10 A to 80 A can be written as follows:

$$f(x) = p1*I_3 + p2*I_2 + p3*I + p4 \tag{4}$$

where $p1 = 1.591*10^{-8}$; $p2 = -2.47*10^{-6}$; $p3 = 0.0001441$; $p4 = -0.0003929$.

A 130 mm length grove was machined based on the TWR calculated by the above equation. The cutting depth error was about 0.03 mm from the beginning point to the end point.

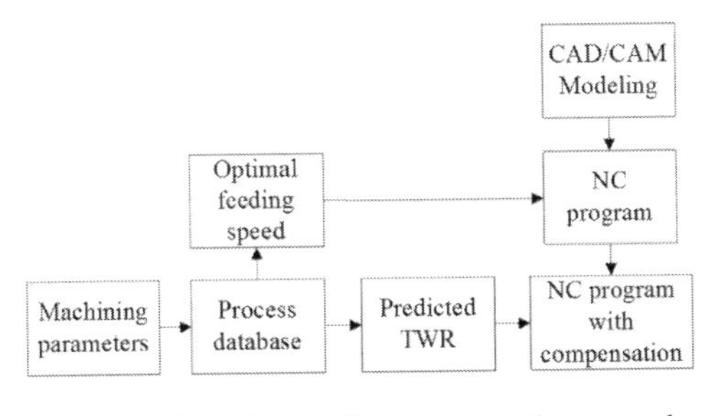

Figure 9. Electrode wearing compensation procedure based on the process database.

4.2 *Electrode wearing compensation based on interval detection*

For the machining of a complex geometric shape workpiece, the machining path will contain line, circle, arc and spiral motion, and the electrode moving length in the discharge process is difficult to calculate. Therefore, the above compensation strategy needs to be adjusted.

In the machining of a complex geometric shape workpiece, the geometric feature modeling and machining path planning are achieved by using the CAD/CAM software. In this paper, UG 6.0 was used to design the workpiece geometry, and the tool electrode moving path was planned through the milling module. The generated moving path was regular milling program for mechanical milling, and the electrode wearing was not taken into consideration. In order to guarantee the machining accuracy, compensation based on interval detection was proposed. The machining process will be interrupted after a preset machining time, and the electrode will be moved to a fixed position to measure its length. The wearing value can be obtained by comparing the measured electrode length with its original length. The compensation was realized by changing the length of the tool electrode. According to the interval pattern, the length of the electrode was classified into detection and compensation interval based on the milling layer and a preset machining time.

4.2.1 *Compensation based on milling layer interval detection*

The milling layer interval detection method is mainly used in the machining of the workpiece,

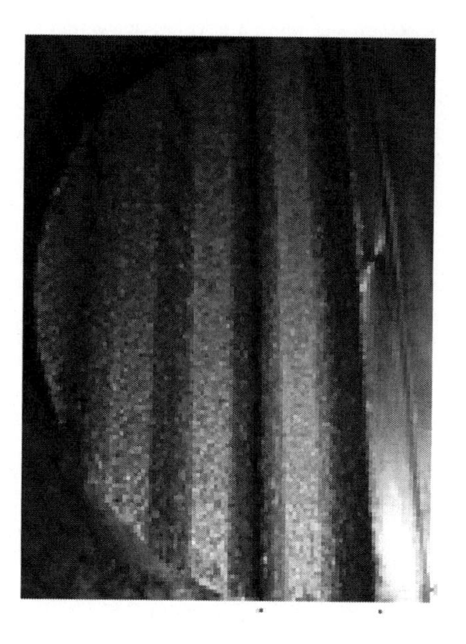

Figure 10. Grove machined with electrode wearing compensation.

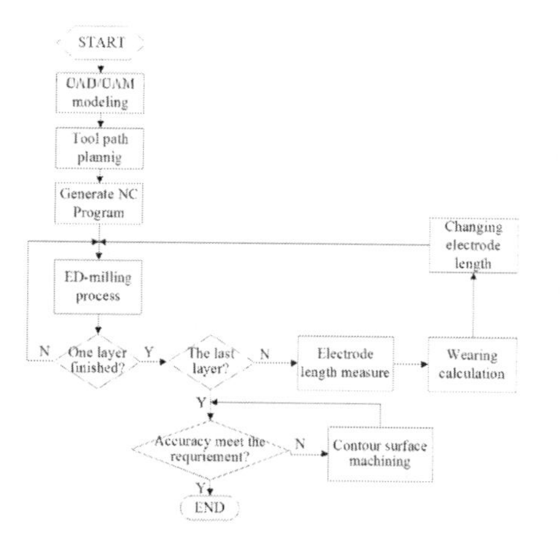

Figure 11. Electrode wearing compensation procedure based on milling layer interval wearing detection.

in which too much material is not removed by one layer. The electrode wearing was detected and compensated after one-layer machining was finished. The flow chart of the milling layer interval detection and compensation is illustrated in Figure 11. In this method, the electrode wearing in the fourth layer will be compensated in the next layer, and the machining accuracy depends on the electrode wearing value in one-layer machining.

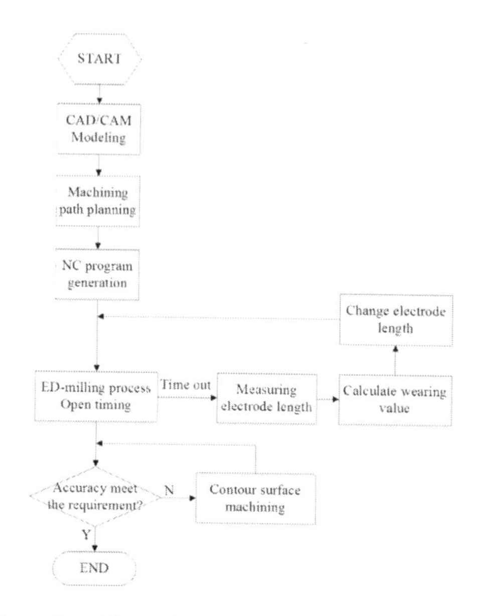

Figure 12. Electrode wearing compensation procedure based on time interval detection.

4.2.2 *Compensation based on time interval detection*

The time interval detection method was mainly used when too much electrode was worn out in one-layer machining or the tool path layer division is hard to recognize. When the machining time reaches the preset time, the machining process will be stopped for electrode wearing detection and compensation, and the program will return to the machining process after the wearing is compensated. The machining accuracy can be controlled by adjusting the interval time.

5 THE ACCURACY OF ELECTRODE WEARING COMPENSATION

The machining accuracy of ED-milling mainly depends on the electrode wearing compensation. The compensation value always deviates from the real value, regardless of the kind of compensation strategy applied. The aim of the compensation is to keep the accuracy deviation in an acceptable level. Figure 13 shows the machined surface property by different compensation strategies. It can be observed that the depth of the machined surface by the interval detection and compensation strategy changed orderly. This is because in the interval compensation method, the electrode wearing is always compensated after a preset interval time.

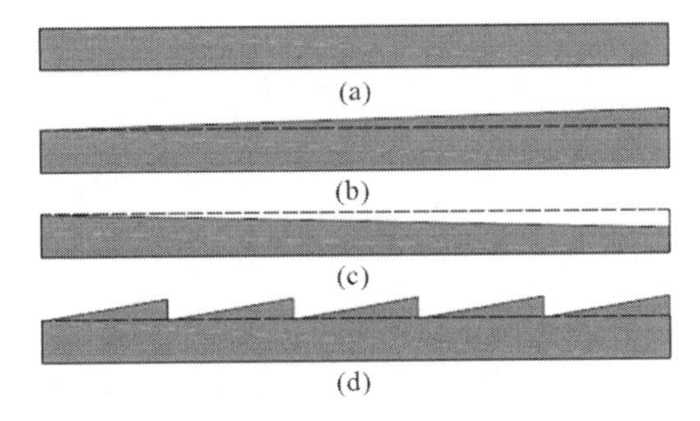

Figure 13. Surface property machined by different compensation strategies.

Figure 14. Machined titanium workpiece sample.

A complex geometric shape workpiece is machined by the time interval detection and compensation strategy, and its machining accuracy can be kept in 0.1 mm in all directions.

6 CONCLUSIONS

In the ED-milling process, the electrode wearing can be reduced by increasing the pulse width or decreasing the discharge current. The machining polarity has an obvious influence on electrode wearing in titanium alloy machining, and the electrode wearing can be reduced effectively by connecting the electrode to the negative pole. The electrode rectangle corner will be worn out prior to the other area, and a stable round corner will be formed after machining for a long period of time. The graphite electrode can result in a lower wearing compared with the copper electrode. For rough machining, the compensation process can be simplified by considering only the compensation in the axial direction. For the simple grove machining, the predicting compensation strategy based on the process database or discharge current can be applied. For the machining of a complex geometric shape workpiece, the interval detection and compensation strategy can be more effective. All compensation strategies have a deviation from the real wearing value, but the deviation can be controlled in an acceptable value by adjusting the compensation parameters.

REFERENCES

Bayramoglu, M. & Duffill, A.W. 1994. Systematic investigation on the use of cylindrical tools for the production of 3D complex shapes on CNC EDM machines. *International Journal of Machine Tools and Manufacture*, 34(3): 327–339.

Bleys, P., Kruth, J.P. & Lauwers, B. 2004. Sensing and compensation of tool wear in milling EDM. *Journal of Materials Processing Technology*, 149(1–3): 139–146.

Bleys, P., Kruth, J.P., Lauwers, B., Zryd, A., Delpretti, R. & Tricarico, C. 2002. Real-time Tool Wear Compensation in Milling EDM. *CIRP Annals—Manufacturing Technology*, 51(1): 157–60.

Ding, S., Yuan, Z.R., Li, Z. & Wang, K. 2006. CNC electrical discharge rough machining of turbine blades. *Proceedings of the Institution of Mechanical Engineers, Part B: Journal of Engineering Manufacture*, 220(7):1027–1034.

Han, F., Wang, Y. & Zhou, M. 2009. High-speed EDM milling with moving electric arcs. *International Journal of Machine Tools and Manufacture*, 49(1): 20–24.

Ho, K.H. & Newman, S.T. 2003. State of the art electrical dis-charge machining (EDM). *International Journal of Machine Tools and Manufacture*, 43(13): 1287–1300.

Mikesic, I.J., Fleisig, R.V. & Koshy, P. 2009. Electrical discharge milling with oblong tools. *International Journal of Machine Tools and Manufacture*, 49(2): 149–155.

Mohd, A.N., Solomon, D.G. & Fuad, B.M. 2007. A review on current research trends in electrical discharge machining (EDM). *International Journal of Machine Tools and Manufacture*, 47(7–8): 1214–1228.

Saravanan, P.S., Antony, L.M., Satish, K.S., Varahamoorthy, R. & Dinakaran, D. 2012. Effects of electrical parameters, its interaction and tool geometry in electric discharge machining of titanium grade 5 alloys with graphite tool. *Proceedings of the Institution of Mechanical Engineers,Part B: Journal of Engineering Manufacture*, 227(1): 119–131.

Yan, M.T., Huang, K.Y. & Lo, C.Y. 2009. A study on electrode wear sensing and compensation in Micro-EDM using machine vision system. *The International Journal of Advanced Manufacturing Technology*, 42(11–12): 1065–1073.

Yu, Z.Y., Masuzawa, T. & Fujino 1998. Micro-EDM for three dimensional cavities—development of uniform wear method. *CIRP Annals—Manufacturing Technology*, 47(1): 169–172.

Zhou, M., Han, F. & Soichiro, I. 2008. A time-varied predictive model for EDM process. *International Journal of Machine Tools and Manufacture*, 48(15): 1668–1677.

Advanced Materials, Mechanical and Structural Engineering – Hong, Seo & Moon (Eds)
© 2016 Taylor & Francis Group, London, ISBN: 978-1-138-02908-8

Energy absorption characteristics of a new design of externally stiffened circular stepped tube based on free inversion

M.S. Zahran, P. Xue & M.S. Esa
School of Aeronautics, Northwestern Polytechnical University, Xi'an, China

ABSTRACT: A new structural configuration based on free inversion of stiffened circular stepped tubes is presented in the current paper to improve and enhance the energy absorption characteristics under axial impact loading. New configurations can be constructed by adding external longitudinal stiffeners distributed around the cross section of the small tube in circular stepped tube. Five configurations of stepped tubes are proposed, numerical analyses is used to simulate the inversion process of the proposed tubes by using the non-linear finite element code ANSYS-WORKBENCH/LS-DYNA. A comparative study is conducted to compare the energy absorption characteristics and deformation history between the newly proposed tubes and the conventional stepped tube. The results showed that addition of external longitudinal stiffeners on circular stepped tubes can greatly improve the specific energy absorption capability up to 51% compared with unstiffened circular stepped tubes. A newfound role of external longitudinal stiffeners added to the stepped tubes that controls the inversion and deformation mechanism is presented.

1 INTRODUCTION

Energy absorption systems are widely used in the engineering systems (such as car bumpers, subfloor of aircrafts, the bottom of lifts, train buffers, marine structures, machinery and so on) during impact or sudden accidents. The main goal of energy absorption systems is to protect the engineering systems from serious damage, prevent occupant fatalities, and minimize the severity of injuries in the event of crash impacts. The energy absorption characteristics considered include load–displacement history, total absorbed energy, specific absorbed energy, stroke crushing length, mean crushing force and crush force efficiency (Lu & Yu 2003, Olabi et al. 2007).

Thin-walled circular tubes are most frequently used as energy absorber in energy absorbing systems due to their high strength, high stiffness, low weight, inexpensive, easy in manufacturing, excellent energy absorption characteristics and load carrying capacity. The thin-walled circular tubes under axial impact loading can be classified into different types of deformation mechanisms like splitting (Niknejad et al. 2013), extrusion (Morasch et al. 2014), expansion (Salehghaffari et al. 2010), trigger (Soica & Radu 2014), stiffeners (Salehghaffari et al. 2011) and inversion (Zhang et al. 2009). This study concentrates on the free inversion mechanism by adding external longitudinal stiffeners. The inversion process has been extensively used as an energy absorption system due to its constant inversion load during the process, which obtains an ideal long stable crush load–stroke curve. The thin-walled tube can be inverted inside out or outside in. There are two ways to produce the inversion tube: the first one is free inversion and the second is inversion with a die (Al-Hassani et al. 1972).

The circular stepped tubes can be considered as a type of free inversion technique preferable for energy absorption systems, therefore it provides a desirable stable crush load–displacement curve response and can absorb a large amount of impact energy under axial loading (Higuchi et al. 2014). Furthermore, use of stiffeners (externally or internally- axial or lateral) considered as a way for improving the energy absorption characteristics on thin walled tubes have been studied by several researchers (Liu et al. 2015, Bich et al. 2013, Zhang & Suzuki 2007).

In the current paper, a new structural configuration based on free inversion of stiffened circular stepped tubes is presented to improve/enhance the energy absorption characteristics under axial impact loading. The new configuration of circular stepped tube can be constructed by adding external longitudinal stiffeners distributed around the cross section of small radius in circular stepped tube. Five configurations of Stepped tubes are proposed. Numerical analyses are conducted to simulate the inversion process of the proposed tubes using the non-linear finite element code ANSYS-WORKBENCH/LS-DYNA (Manual 2013). The numerical models are implemented under axial impact crushing scenario. A comparative study is conducted to compare the energy absorption characteristics of the newly proposed tubes and the conventional stepped tube. The study focuses on the variation of the external stiffener width, its influence on the energy absorption characteristics and the inversion mechanism scenario.

2 STRUCTURAL CONFIGURATION

Assuming an axial force F applies on a stepped tube with double cylinders with R1 and R2 as the small cylinder radius and large cylinder radius respectively, connected by flat connection with length CL as shown in Figure 1.

Adding external longitudinal stiffeners to the small tube shows the influence on decreasing the curling normal stress thereby increasing the resistance and the energy absorption of the tube. During the inversion process, the stiffeners will strike the flat connection length between upper tube and lower tube, which will effect on the curling process.

The ratio between the width of stiffeners and the length of step connection is the effective parameter during the inversion process, hence the effect of the ratio on the inversion deformation, stroke length, absorbed energy, the specific energy absorption and crush force efficiency are studied. The ratio between the stiffener width of small tube and the connection length has been assumed to be ∂ which means ∂S = SSW/CL where ∂S, SSW and CL are small tube stiffener width ratio, small tube stiffener width and connection length respectively as shown in Figure 1.

All the configurations have the same dimension of the circular stepped tube (R1 = 25 mm, R2 = 35 mm, CL = 10 mm, Thickness = 1.5 mm, L1 = 100 mm, L2 = 100 mm). Eight (N = 8) stiffeners are

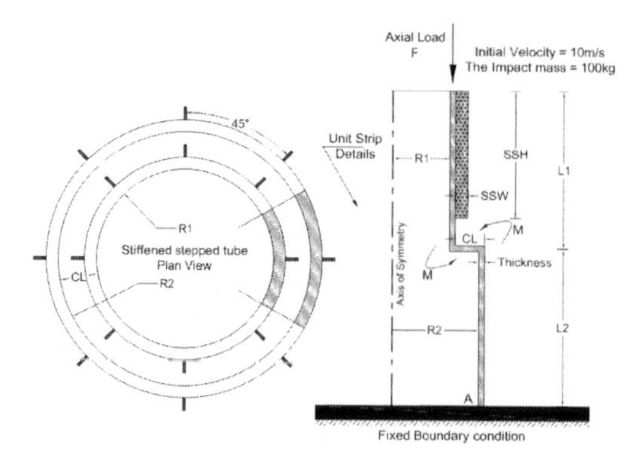

Figure 1. Schematic drawing shows the stiffened stepped tube parameters.

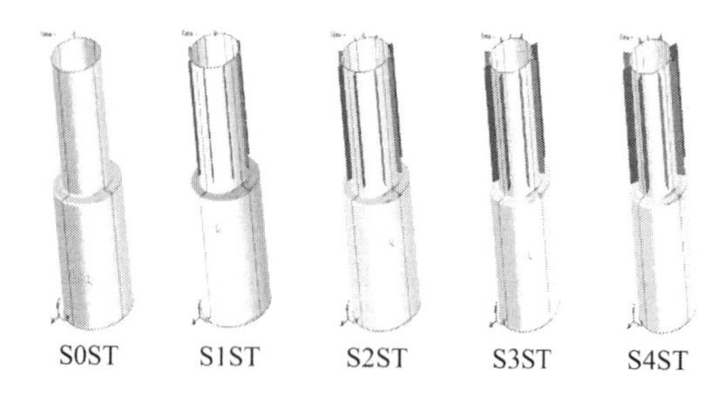

Figure 2. Three dimension numerical models (stiffened and unstiffened circular stepped tube).

distributed around the cross section of the small tube having rectangular geometric shape with height (SSH = 90 mm) and same thickness as of stepped tube but the widths are varied as shown in Figure 2.

3 FINITE ELEMENT MODELLING

3.1 *Numerical program*

The numerical program includes five numerical models (S0ST, S1ST, S2ST, and S3ST and S4ST) as shown in Figure 2. Furthermore, the differences between the models are the ratio (∂S) between the stiffener width of small tube (SSW) and the length of step connection (CL). A"0" in S0ST is used to identify the unstiffened stepped tube. A "1" in S1ST identifies a stiffened stepped tube in which ∂S = 0.25, in other words "S1ST" means SSW = 2.5 mm. Similarly "2" in S2ST identifies the stiffened stepped tube with ∂S = 0.5, in other words "S2ST" means SSW = 5 mm. A "3" in S3ST identifies the stiffened stepped tube in which ∂S = 0.75 in other words "S3ST" means SSW = 7.5 mm. A "4" in S4ST identifies the stiffened stepped tube in which ∂S = 1.00 in other words "S4ST" means SSW = 10 mm. All models are compared with unstiffened stepped tube (S0ST), as shown in Figure 2. This ratio is constrained by the length of connection (CL) hence it cannot exceed this length otherwise the inversion process is impossible.

3.2 *Description of finite element modeling*

The explicit non-linear finite element code LS-DYNA/WORKBENCH ANSYS (Manual 2013) is used to investigate the effect of external longitudinal stiffeners on energy absorption characteristics of aluminum circular stepped tube subjected to axial impact crushing. The sidewall of tubes is modeled with four-node shell elements with five integration points through the thickness and one integration point in the element plane

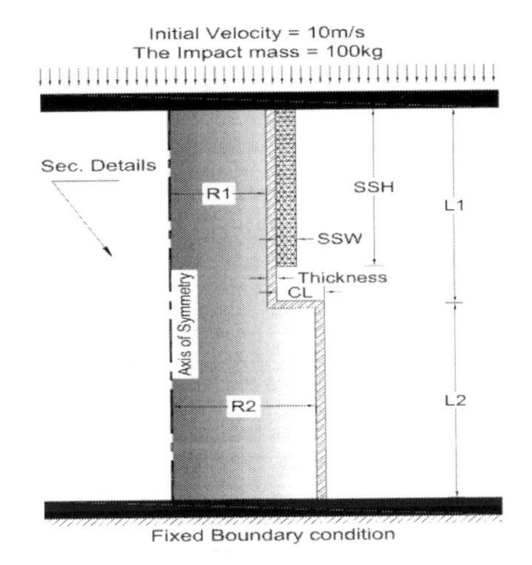

Figure 3. Section details and boundary conditions.

(Belytschko et al. 1984). The bottom-stepped tube is fixed to a stationary rigid plane, a rigid body with initial impact velocity of 10 m/s strikes the circular stepped tube axially and the mass of the rigid body is 100 kg as shown in Figure 3. Assume that the friction coefficient between all surfaces in contact is 0.25. All models are made of aluminum alloyAA6060-T4 with the following mechanical properties: Young's Modulus E = 68.2 GPa, initial yield stress σ_y = 80 MPa, the ultimate stress σ_u = 173 MPa, Poisson's ratio ν = 0.3, power law exponent n = 0.23 and the density ρ = 2770 kg/m^3 (Santosa et al. 2000). The material properties are set to be constant for all stepped tubes and all stiffeners.

4 PERFORMANCE INDEX FOR ENERGY ABSORBER

There are many parameters used to analyze the crushing response of structural members. Several authors to assess the performance and efficiency of the energy absorber (Lu & Yu 2003, Olabi et al. 2007) have presented these parameters. These parameters are defined below:

– Total Absorbed Energy (EA)
The total absorbed energy during axial crushing is calculated by the area under load–displacement curve from the following equation:

$$EA = \int F d\delta \tag{1}$$

where, F and δ are crushing load and crushing distance, respectively.

– Specific Energy Absorption (SEA)
SEA is defined as the ratio of the energy absorbed by a structure to its mass (m) and is given by

$$SEA = \frac{\int F d\delta}{m_t} \tag{2}$$

where, m_t is total mass of structure.

– Stroke Crushing Length (SL)
SL is the deformable distance of structure along the loading direction, which means the value of the displacement before starting the densification zone on the load- displacement curve.

– Crush Force Efficiency (CFE)
CFE is defined as the ratio of the mean crushing force (F_m) to the peak force (F_p) where the mean crushing force (F_m) is defined by dividing the total Energy Absorbed (EA) by the stroke crushing distance (SL). CFE can be obtained from the following equation:

$$CFE = \frac{F_m}{F_P} = \frac{EA}{F_P SL} \tag{3}$$

where. F_m and F_P are mean crushing force and peak crushing force, respectively.

5 NUMERICAL RESULTS AND DISCUSSION

5.1 *Interpretation of load–displacement behavior*

The load-displacement curves for models are shown in Figures 4–8. The response parameters are given in Table 1.

It is clear from Figures 4–8 and Table 1 that with increasing the width of stiffeners the absorbed energy increases in all models compared with the unstiffened circular stepped tube, consequently the specific energy absorption is improved by 51% (as in model S4ST for example). Actually increase in absorbed energy and specific energy absorption occurs due to two reasons:

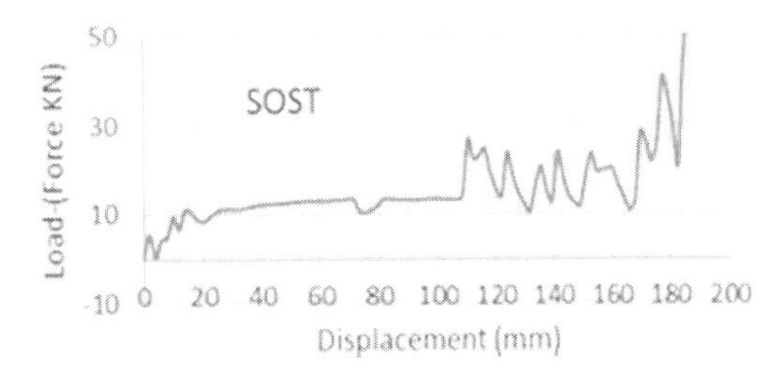

Figure 4. Load-displacement curve of model S0ST.

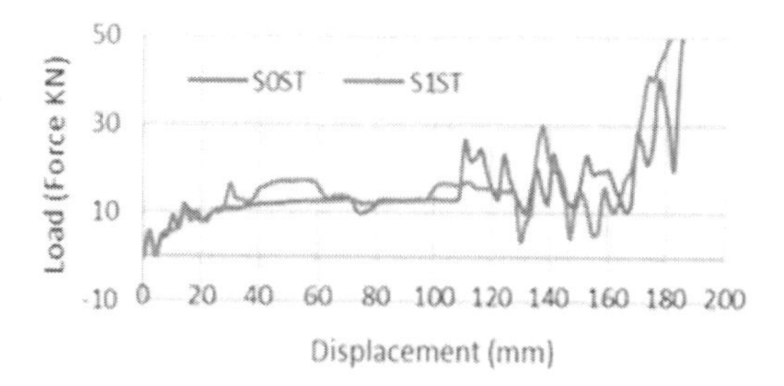

Figure 5. Load-displacement curve of model S0ST–S1ST.

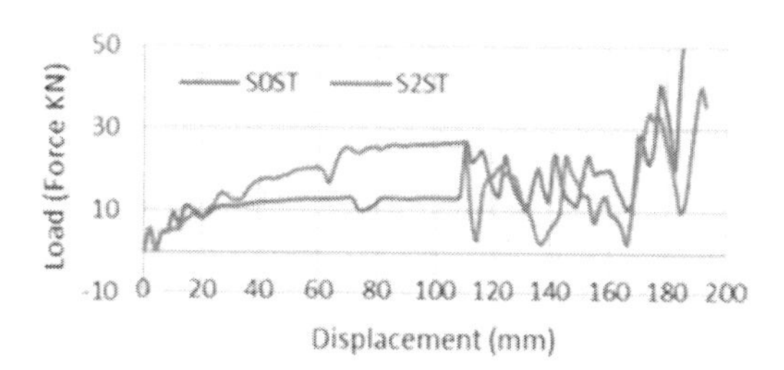

Figure 6. Load-displacement curve of model S0ST–S2ST.

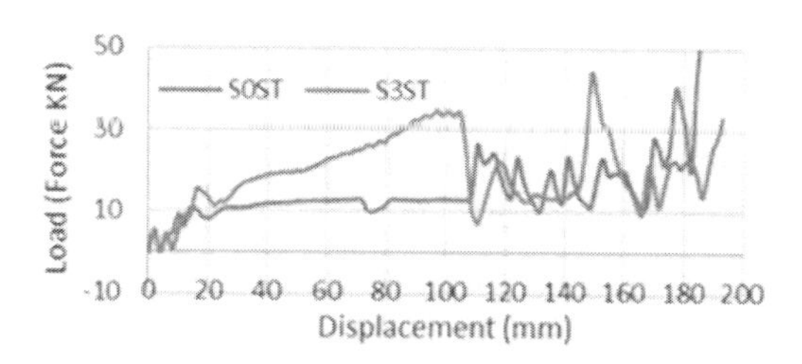

Figure 7. Load-displacement curve of model S0ST–S3ST.

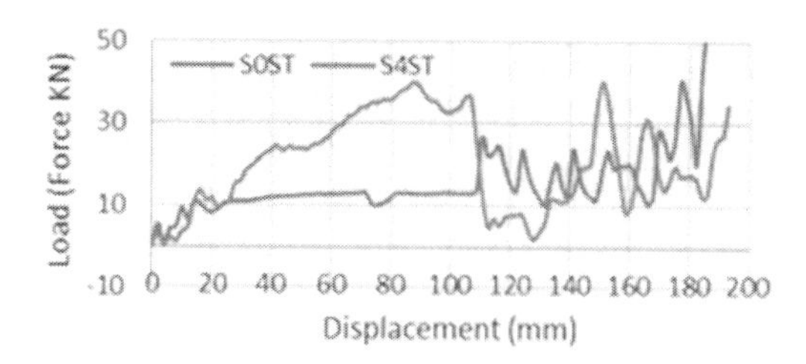

Figure 8. Load-displacement curve of model S0ST–S4ST.

1. Increase in the stiffness of stepped tube by adding the stiffeners in the small tube, which consequently make the inversion process hard to happened,
2. The friction between the stiffeners and wall of stepped tubes plays a significant role as high value of the friction can lead to high-absorbed energy, especially in model S4ST due to the fact that the stiffeners width (SSW) is the same as length of step connection.

Table 1. Summary of numerical results for energy absorption characteristics.

	Response parameters						
Models	Mass kg	SL mm	EA KJ	SEA KJ	F_p KN	F_m KN	CFE
S0ST	0.164	166	1.94	11.79	26.6	11.6	0.43
S1ST	0.171	162	2.21	12.85	30.1	13.5	0.45
S2ST	0.179	186	2.70	15.03	33.1	14.5	0.44
S3ST	0.186	186	3.29	17.59	43.2	17.7	0.42
S4ST	0.194	186	3.47	17.85	39.3	18.6	0.47

It is observed from Figures 4–8 and Table 1 that with increasing the width of stiffeners, the peak force and the mean crushing force are increased in all models compared with the unstiffened circular stepped tube. Consequently, the crush force efficiency is slightly increased (as in model S4ST). However, the crush force efficiency of unstiffened or stiffened stepped tube is higher than those of progressive buckling modes like plain tube (PC). It is evident from Figures 4–8 that the influence of external longitudinal stiffeners insignificantly effect the inversion stroke length.

5.2 *The influence of external longitudinal stiffeners on inversion mechanism*

The effect of external longitudinal stiffeners on the inversion mechanisms is shown in Figures 9–11, which illustrate progressive crushing deformation under axial impact.

Figure 9 shows the inversion process of unstiffened circular tube. It is observed that the small tube is inverted inside the large tube. This mechanism happens due to the difference in stiffness between the two tubes. Furthermore, it is observed from Figure 10 that the external longitudinal stiffeners lead to significant effect on the inversion mechanism. The large tube deforms with the small tube simultaneously .The simultaneous deformations occurs due to increase in the stiffness of the small tube by adding the stiffeners. Hence, the stiffness of the small tube is almost equivalent to the stiffness of the large tube. On the other hand, it is observed from Figure 11 that the models (S2ST-S3ST-S4ST) are deformed in a new inversion mechanism that only large tube is deformed during the inversion process. In this mechanism, the small tube works as a die to the large tube that has never happened before in the regular free inversion.

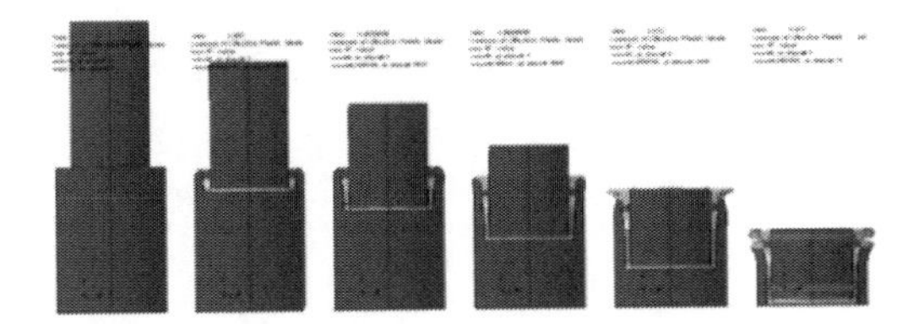

Figure 9. Progressive deformation of model S0ST.

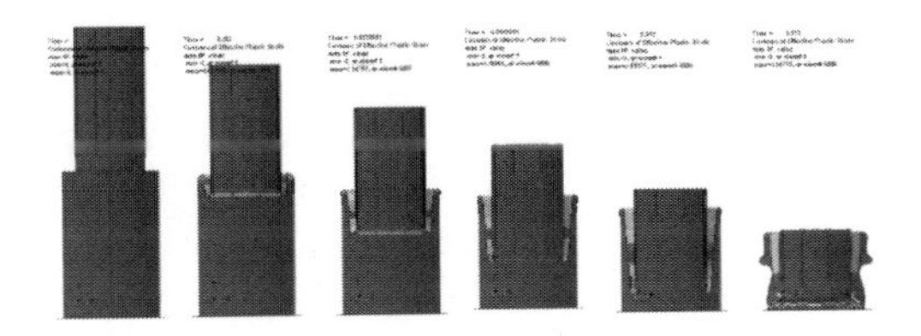

Figure 10. Progressive deformation of model S1ST.

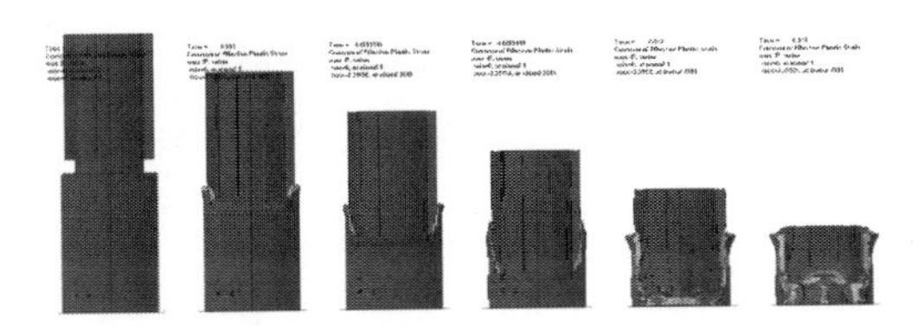

Figure 11. Progressive deformation of model S4ST.

6 CONCLUSIONS

In the current paper, a new structural configuration based on free inversion of stiffened circular stepped tubes is presented to improve and enhance the energy absorption characteristics under axial impact loading. Numerical analysis is carried out to simulate the inversion process of the proposed tubes using ANSYS-WORKBENCH/LS-DYNA. A comparative study is conducted to compare the energy absorption characteristics of the newly proposed tubes and the conventional stepped tube (S0ST). The main findings of this study can be drawn as follows:

1. Reinforcing the stepped tubes with external longitudinal stiffeners lead to improvement (up to 51%) of the specific energy absorption compared to the unreinforced section under axial impact load.
2. The stiffeners width is the most important parameter in the inversion and deformation mechanism after the end of curling. The friction between the stiffeners and wall of stepped tubes plays a significant role as high value of the friction can lead to high absorbed energy.
3. External longitudinal stiffeners can control the inversions mechanism, which means that stiffeners can force the inversion to happen to the small tube or the large tube or both which lead to several scenarios representing different cases and applications.

ACKNOWLEDGEMENT

The authors gratefully acknowledge the financial supports from National Natural Science Foundation of China under Grants 11472226 and 11072202.

REFERENCES

Al-Hassani, S., Johnson, W. & Lowe, W. 1972. Characteristics of inversion tubes under axial loading. *Journal of Mechanical Engineering Science*, 14: 370–381.

Belytschko, T., Lin, J.I. & Chen-Shyh, T. 1984. Explicit algorithms for the nonlinear dynamics of shells. *Computer Methods in Applied Mechanics and Engineering*, 42: 225–251.

Bich, D.H., Van Dung, D., Nam, V.H. & Phuong, N.T. 2013. Nonlinear static and dynamic buckling analysis of imperfect eccentrically stiffened functionally graded circular cylindrical thin shells under axial compression. *International Journal of Mechanical Sciences*, 74: 190–200.

Higuchi, M., Suzuki, S., Adachi, T. & Tachiya, H. 2014. Improvement of Energy Absorption of Circular Tubes Subjected to High Velocity Impact. *Applied Mechanics and Materials*, 566: 575–580.

Liu, S., Tong, Z., Tang, Z., Liu, Y. & Zhang, Z. 2015. Bionic design modification of non-convex multi-corner thin-walled columns for improving energy absorption through adding bulkheads. *Thin-Walled Structures*, 88: 70–81.

Lu, G. & Yu, T. 2003. *Energy absorption of structures and materials*, Elsevier.

Manual, F. 2013. *ANSYS, Release 15.0 ANSYS Documentation*. ANSYS Inc., Canonsburg, PA.

Morasch, A., Matias, D. & Baier, H. 2014. Material modelling for crash simulation of thin extruded aluminium sections. *International Journal of Crashworthiness*, 19: 500–513.

Niknejad, A., Rezaei, B. & Liaghat, G.H. 2013. Empty circular metal tubes in the splitting process–theoretical and experimental studies. *Thin-Walled Structures*, 72: 48–60.

Olabi, A.G., Morris, E. & Hashmi, M. 2007. Metallic tube type energy absorbers: a synopsis. *Thin-Walled Structures*, 45: 706–726.

Ramakrishna, S. & Hamada, H. 1997. Energy absorption characteristics of crash worthy structural composite materials. *Key Engineering Materials*, 141–143: 585–622.

Salehghaffari, S., Rais-Rohani, M. & Najafi, A. 2011. Analysis and optimization of externally stiffened crush tubes. *Thin-Walled Structures*, 49: 397–408.

Salehghaffari, S., Tajdari, M., Panahi, M. & Mokhtarnezhad, F. 2010. Attempts to improve energy absorption characteristics of circular metal tubes subjected to axial loading. *Thin-Walled Structures*, 48: 379–390.

Santosa, S.P., Wierzbicki, T., Hanssen, A.G. & Langseth, M. 2000. Experimental and numerical studies of foam-filled sections. *International Journal of Impact Engineering*, 24: 509–534.

Soica, A. & Radu, G. 2014. The influence of triggers geometry upon the stiffness of cylindrical thin walled tubes. *Open Engineering*, 4: 101–109.

Zhang, A. & Suzuki, K. 2007. A study on the effect of stiffeners on quasi-static crushing of stiffened square tube with non-linear finite element method. *International Journal of Impact Engineering*, 34: 544–555.

Zhang, X., Cheng, G. & Zhang, H. 2009. Numerical investigations on a new type of energy-absorbing structure based on free inversion of tubes. *International Journal of Mechanical Sciences*, 51: 64–76.

Advanced Materials, Mechanical and Structural Engineering – Hong, Seo & Moon (Eds)
© 2016 Taylor & Francis Group, London, ISBN: 978-1-138-02908-8

Well-defined pH-responsive triblock glycopolymer architectures for controlled loading and release doxorubicin

I.A. Altoom, S.Y. Zhu, Q.Y. Yu, X.Z. Jiang & M.F. Zhu
State Key Laboratory for Modification of Chemical Fibers and Polymer Materials, College of Material Science and Engineering, Donghua University, Shanghai, China

ABSTRACT: To explore drug loading and release behavior of pH-responsive block glycopolymers used as an anticancer drug delivery carriers, poly(ethylene glycol)-*b*-poly(2-(diethylamino) ethyl methacrylate)-*b*-poly(2-gluconamidoethyl methacrylate) (PEG_{113}-*b*-$PDEA_{41}$-*b*-$PGAMA_{34}$), was synthesized via Atom Transfer Radical Polymerization (ATRP) by polymerized poly (2-gluconamidoethyl methacrylate) (GAMA) and 2-(diethylamino) ethyl methacrylate (DEA) monomers using a poly(ethylene glycol)-based (PEG-based) macroinitiator without protecting group chemistry. proton nuclear magnetic resonance (^{1}H NMR) and Gel Permeation Chromatography (GPC) were used to characterize their structures, their self-assembly behaviors and drug loading and release behavior of Doxorubicin (DOX) as the model of an anticancer drug were further investigated in detailed by Transmission Electron Microscopy (TEM), zeta-potential, and ultraviolet-visible spectrophotometer (UV-Vis). The results show glycopolymers self-assembly into micelles at basic conditions and dissociate to be unimers at acidic conditions, this pH-responsiveness endows block glycopolymer to possess the ability to load the DOX at physiological condition (pH 7.4) and smart release at mimicking tumor acidic environments (pH 5.5), however, the glycopolymer at pH 5.5 Phosphate Buffered Saline (PBS) released medium possesses faster drug release profile compared with the released medium at pH 7.4 PBS. This study provides a promising light for designing novel drug delivery carrier for potential applications in cancer chemotherapy.

1 INTRODUCTION

Recently, self-assembly of block copolymers into core-shell micelles and mesophases of various morphologies is a subject of much concern for their potential applications in biomedicines as nano-sized carriers for anticancer drug and gene delivery (Rapoprt 2007). To improve water solubility of micelles prepared, poly (ethylene glycol) (PEG) chain was generally chosen by incorporating onto the surface of micelles, which also prevents further aggregation via steric hindrance of outer PEG shell (Yuan et al. 2011). Different block glycopolymers have been prepared by controlling radical polymerization with or without protection chemistry and explored their potential applications in details, especially as special drug or insulin delivery systems focusing on pH induced dynamic interaction with phenyl bronic acid groups (Guo et al. 2014, Sun et al. 2013). The responsiveness of pH-responsive block glycopolymer have been widely investigated, and this work was generally focused on design of block, preparation process, and pH induced micellization (Jiang et al. 2006). To study the effect of drug-loaded micelle as in interactions on loading and release from the copolymer micelle nanoparticles, we used drug loading and release behavior of micelles with a hydrophobic drug (Guzman et al. 2007). DOX is dissolved, entrapped, encapsulated or attached to a nanoparticles micelle depending on the method of preparation.

2 EXPERIMENTS

2.1 *Materials*

CuBr, 2,2-bipyridine (bpy), D-gluconolactone, 2-aminoethyl methacrylate hydrochloride, hydroquinone, 2-bromoisobutyryl bromide, and were purchased from Sigma-Aldrich without further purification; monohydroxy-capped poly(ethylene glycol) (PEG_{113}-OH) with a mean Degree of Polymerization (DP) of 113 (Mn = 5000 g/mol, M_w / M_n = 1.10) was purchased from Sigma-Aldrich and dried at the 40°C under vacuum overnight prior to use. 2-(Diethylamino) ethyl methacrylate (DEA) was purchased from J&K and purified by vacuum distillation; *N*-Methyl Pyrrolidone (NMP) was purchased from Sigma-Aldrich, dried over CaH_2 and obtained by distillation at reduced pressure. Chloroform, Triethylamine (TEA), Dimethyl Sulfoxide (DMSO), and Phosphate Buffered Saline (PBS) were purchased from Sinopharm Chemical Reagent Co. Ltd. and purified according to standard procedures. PEG_{113}-Br and GAMA monomer were prepared according to the reported procedures.

2.2 *Preparation of PEG-b-PDEA-b-PGAMA glyco-polymer with block architecture*

The triblock copolymer was prepared using one-pot synthesis method via successive ATRP using PEG-Br as the macroinitiator. A typical procedure was as follows: PEG-Br macroinitiator (0.250 g, 0.048 mmol), bpy (0.015 g, 0.096 mmol), and DEA monomer (0.539 g, 2.912 mmol, 0.496 mL) were dissolved in degassed NMP (2.0 mL) at room temperature, then the solution was degassed via nitrogen purge for 30 min. Copper (I) bromide (0.007 mg, 0.048 mmol) was added quickly, the mixture solution was stirred vigorously under nitrogen at 60°C. The DEA conversion was monitored at fixed time intervals by ^{1}H NMR, and after more than 90% conversion had been achieved in 24 hours, GAMA (0.223 g, 0.728 mmol) glycomonomer in degassed NMP was added. This reaction solution was further stirred at the 30°C for 20 hours. Then the reaction was terminated by liquid nitrogen, and exposed to the air. The mixed solution was purified by the dialysis method using a dialysis membrane with molecular weight cutoff of 3500 and then freeze-dried overnight, white powder (0.86 g) was obtained with yield 85%.

2.3 *Preparation of micelle from pH-responsive block glycopolymers*

The PEG-*b*-PDEA-*b*-PGAMA glycopolymers was molecularly dissolved in water at pH 3 overnight, and the solution was adjusted to pH 10 with concentrated NaOH aqueous solution to induce micelle formation at the final copolymer concentration of 1.0 g/L.

2.4 *Drug loading capacity and encapsulation efficiency*

To assess the drug concentration in the nanoparticle, UV-Vis absorption measurements were performed. The DOX concentration in the supernatant was analyzed by a UV-Vis spectrophotometer (TU-1901 Beijing Spectrum Analysis of General Instrument Co. Ltd) and calculated via a calibration curve with $R^2 = 0.99992$ prepared previously with different concentrations of DOX in distill water. Percentage of Encapsulation Efficiency (EE) of DOX was obtained using the Equation 1 below:

$$\textit{Efficiency}\ \% = \frac{\textit{mass of drug incapculation in micelle}}{\textit{mass of initial drug}} \times 100 \quad (1)$$

Percentage of Loading Capacity (LC) of nanoparticles was calculated by using the Equation 2 below:

$$\textit{Capacity}\ \% = \frac{\textit{mass of drug incapculation in micelle}}{\textit{mass of initial micelle}} \times 100 \quad (2)$$

2.5 *Preparation of DOX-loading nanoparticles*

Glycopolymer (40 mg) was dissolved in 8 mL of DMSO, and 8 mL of distilled water (pH 10) was added to stir polymer solution at a rate of 3 bulbs/minute via a syringe pump during stirring 4 hours. Then the micellar solutions were transferred into dialysis membrane and dialyzed against water for 48 hours to remove DMSO. Micellar solution was passed through 0.45 µm Nylon syringe filter, diluted the solution to a final concentration at 1.0 mg/mL and stored at 4°C. Doxorubicin Hydrochloride (DOX-HCl) (3 mg) and excess triethylamine (five mole ratios of DOX) were added into 1 mL chloroform and stirred overnight under dark condition. 15 mL of Micellar solution (1 mg/mL) was added into DOX solution and further stirred vigorously for 24 hours uncovered in a chemical fume hood to evaporate the chloroform. Drug-loaded micelles were then centrifuged at 5000 rpm for 10 minutes three times to remove remaining insoluble drug aggregates, and the final solution was filtered and measured by UV-Vis spectrophotometer at a wavelength of 480 nm to determine the loading capacity and loading efficiency.

2.6 *In-vitro release study*

In-vitro release of nanoparticle was studied in PBS buffered solution (0.01 M) at two different pH values; where one is acidic with a pH of 5.5, and the other is the physiological pH of 7.4. For this purpose, (5 mL, 1 mg/mL) drug-loaded micelle was transferred into a dialysis tube (MWCO: 14,000), and then the dialysis tube was dialyzed against 45 mL of PBS at 37°C and kept under shaking in an incubator at (100 revolutions per minute). At specific time intervals, 5 mL of the phosphate buffer solution was withdrawn from the release medium for UV-Vis absorbance measurements, while an equal amounts of fresh PBS solution were added into the media. The amount of releasing drug was calculated by using the calibration curve prepared previously, that was based on the absorbance intensity of DOX at 480 nm. In the assessment of drug release behavior, the accumulative amount of the released drug was calculated according to Equation 3 below, and the percentages of released drug from the nanocapsule were plotted versus time. All release experiments were carried out in triplicate, and the results were reported as mean ± standard deviation (n = 4).

$$Accumulative\ release\ \% = \frac{V_e \sum_{n=1}^{n} C_i + V_o C_n}{M_{DOX}} \times 100 \quad (3)$$

where, Ve = buffer solution replacement volume, 5 mL; Ci = concentration of the drug release medium, mg/mL; V_o = buffer solution volume, 50 mL; and M_{DOX} = micelles for the release of the drug content, mg.

2.7 *Characterization*

All ¹H NMR spectra were recorded using a Bruker Advance 400 NMR spectrometer with D_2O as the solvent at 25°C and tetramethylsilane as the internal standard. The polymer solution was adjusted by adding DCl and NaOD. The molecular weight and molecular weight distribution of polymer were measured on a waters breeze 1515 Gel Permeation Chromatography (GPC) equipped with two columns (PL mix-D from TOSOH Company). Dimethylformamide (DMF) containing 0.5 g/L LiBr was used as eluent at a flow rate of 1 mL/min at 80°C with PEG as calibration standards. The morphology of glycopolymeric micelles was observed with a Tecnai G220 TWIN (TEM) at an accelerating voltage of 200 KV; Samples in an aqueous solution at pH 10 at 2 g/L were prepared by dipping a formvar-coated copper grid, and followed by air-drying at ambient temperature. Dynamic Light Scattering (DLS) and electrophoresis measurements were observed with a nano zetasizer. Zeta-potentials were determined by electrophoretic measurements, carried out at 25°C by means of the nano zetasizer. The mean hydrodynamic diameter was determined from the obtained apparent diffusion coefficient through the stokes–einstein equation. UV-visible spectrophotometer (TU-1901 Beijing Spectrum Analysis of General Instrument Co. Ltd) used to determine the absorbance of the drug with time. (Shaker incubator TS-100C) machine used for shaker drug release to sample into release medium.

3 RESULTS AND DISCUSSION

3.1 *Synthesis of pH-responsive block glycopolymers*

pH-responsive block glycopolymers, PEG-*b*-PDEA-*b*-PGAMA, was prepared via ATRP using the PEG_{113}-Br as the macroinitiator and copper bromide and bpy as the catalysts. The polymerization solvent was N-Methyl Pyrrolidone (NMP) instead of methanol or isopropanol for better solubility of glycomonomer in NMP under mild condition. Triblock glycopolymer was obtained with block architecture of second and third block, PEG-*b*-PDEA-*b*-PGAMA, was prepared via successive polymerization of DEA with high conversion above 90% and then GAMA monomer using a PEG-Br macroinitiator. After polymerization and purification, block glycopolymer was characterized by ¹H NMR as shown in Figure 1. All signals of each proton of the block glycopolymer could be detected in D_2O/DCl at pH 3 as the characteristics at 3.64 ppm (c in PEG), 4.38 ppm (d in PDEA), 4.30–4.15 ppm and 3.9 ppm (d', i, j in GAMA) were observed, and then degree polymerizations of DEA and GAMA could be calculated to be 41 and 34 for PEG-*b*-PDEA-*b*-PGAMA by integrating their characteristic signal with that of PEG, thus block glycopolymer was denoted as PEG_{113}-*b*-$PDEA_{41}$-*b*-$PGAMA_{34}$. GPC traces showed each monomodal and symmetric peak with molecular weight M_n of 9248, and polydispersity M_w/M_n of 1.19 as shown in Figure 2.

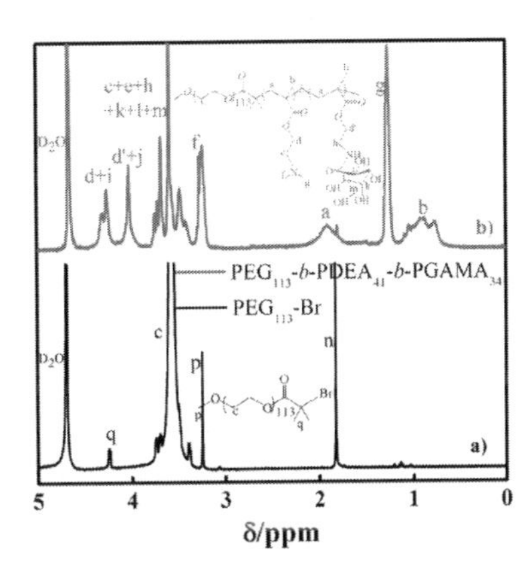

Figure 1. ¹H NMR spectra at different deuterated solvents of PEG_{113}-Br (a); and PEG_{113}-*b*-$PDEA_{41}$-*b*-$PGAMA_{34}$ glycopolymer at pH 3 (b).

3.2 *Micellar self-assembly of pH-responsive block glycopolymers*

In principle, PDEA block as a weak polybase with a pKa of 7.3 molecularly dissolved into acidic solution, and self-assembles into micelles at basic condition reversed (Zhu et al. 2007). PEG and PGAMA blocks dissolved in water over the entire pH range. Therefore, pH-responsive block glycopolymer would dissolve in water media as a weak cationic polyelectrolyte due to protonation of its tertiary amine groups pH below 6, and

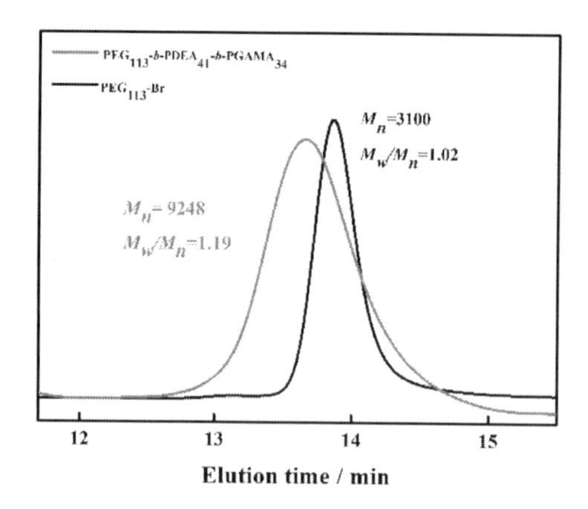

Figure 2. GPC traces of PEG_{113}-Br, PEG_{113}-b-$PDEA_{41}$-b-$PGAMA_{34}$ block glycopolymer.

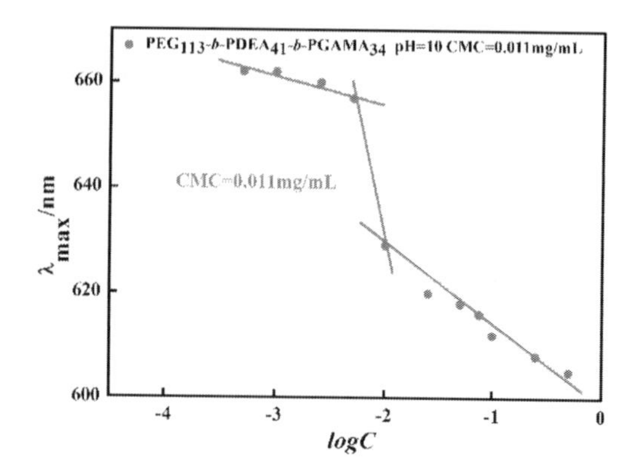

Figure 3. The maximum emission wavelengths of the Nile red in aqueous glycopolymer solutions at different concentrations.

Table 1. Molecular weight and distribution of the synthesized copolymers measured by GPC.

Polymer	M_n g/mol	M_n^b g/mol	M_w/M_n	EE %	LC %
PEG_{113}-*b*-$PDEA_{41}$-*b*-$PGAMA_{34}$	9248	22980	1.19	47	9.3

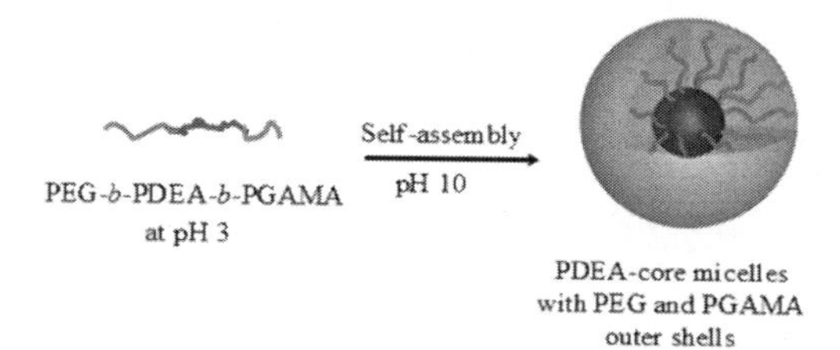

Figure 4. Schematic illustration of self-assembly of PEG_{113}-*b*-$PDEA_{41}$-*b*-$PGAMA_{34}$ pH-responsive block glycopolymer.

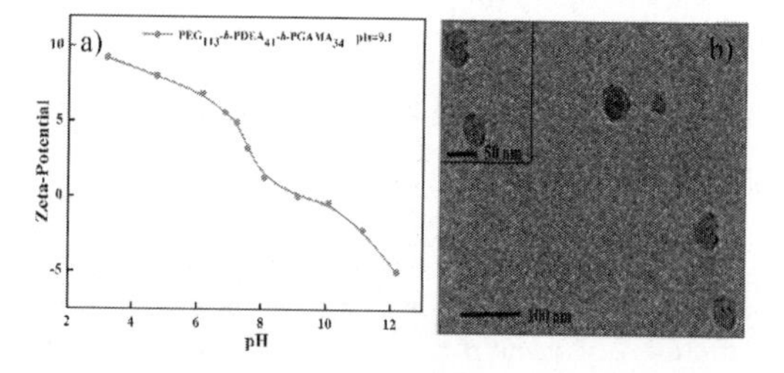

Figure 5. a) Zeta potential versus solution pH; b) TEM image of micelles at pH 10 prepared from the PEG_{113}-*b*-$PDEA_{41}$-*b*-$PGAMA_{34}$ pH-responsive glycopolymers at 2 g/L.

then converted to be micelles at pH 10. Before investigating their micellization behavior, Critical Micelle Concentrations (CMC) of this block glycopolymer at pH 10 was measured by fluorescent labeling method using Nile Red (NR) as the probe, and was calculated to be 0.011 mg/mL as shown in Figure 3. The self-assembly of block glycopolymer was confirmed by ^{1}H NMR as shown in Figure 1. The molecular components of glycopolymer could be proved by characteristics of the block at pH 3, but the signals of DEA at 4.38 and 1.4 ppm completely disappeared at pH 10. This indicates the formation of structurally PDEA-core micelles with mixed outer shells consisting of the solvated PGAMA and PEG blocks, as shown in Figure 4. The sizes of self-assembled micelles was characterized by DLS above pH 8, and presumably spherical micelles at pH 10 was observed with size of 40 nm in diameter, and their isoelectric points (pIs) were 9.1 measured by zeta potential as shown in Figure 5. The TEM, DLS and Zeta-potential studies indicated the successful pH-induced self-assembly of block glycopolymer at alkaline solution, two-layer micelles with comprising PDEA cores and mixed outer shell with PGAMA and PEG blocks was formed at pH 10. Due to the molecular architectures with block chain, glycopolymer possess CMC, micellar size, and pIs, which will probably affect the applications as a drug delivery system for hydrophobic anticancer drugs.

3.3 *Release DOX from the (PEG_{113}-b-$PDEA_{41}$-b-$PGAMA_{34}$) nanoparticles*

The sample presented a sustained release behavior of DOX within a burst release. As shown in Figure 6, about 39.4% and 90.4% by incorporating DOX

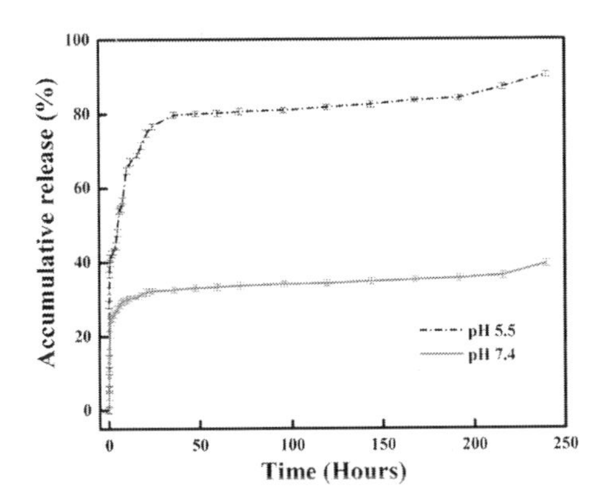

Figure 6. Release profiles of DOX from PEG_{113}-*b*-$PDEA_{41}$-*b*-$PGAMA_{34}$, The data are given as mean ± S.D. (n = 4).

was released at pH 7.4 and 5.5 during 240 hours. The sample possesses faster drug release profile in pH 5.5 PBS compared within pH 7.4 PBS. Micelles size, morphology, drug loading, and drug release were influenced by the molecular weight, chemical composition, and critical micellar concentration. The incorporation of DOX-loaded glycopolymer not only significantly improved the mechanical durability of nanoparticles, but also appreciably weakened the initial burst release of the DOX. This dissociation glycopolymer polymer displays surface erosion, leading to the release of encapsulated drug. Regarding to these results, the sample could be serving as low drug-release time, especially in physiological pH, provides a platform for the fabrication of targeted anticancer drug delivery system. This study discusses the pH induced self-assemble behavior of pH responsive block copolymers; and provides a discussion study of drug delivery systems for DOX as an anticancer drug.

4 CONCLUSIONS

In summary, well-defined PEG_{113}-*b*-$PDEA_{41}$-*b*-$PGAMA_{34}$ pH-responsive block glycopolymer was successfully prepared in this work by ATRP using a PEG-based macroinitiator. Moreover, it possesses faster drug release profile in pH 5.5 PBS compared within pH 7.4 PBS. This study provides a promising light for designing novel drug delivery carriers for their potential applications in cancer chemotherapy. Our results demonstrate the pH responsive block glycopolymer self-assembly into micelle at basic conditions and dissociate to be unimers at acidic conditions. This pH-responsiveness endows block glycopolymer to possess the ability to load the doxorubicin at physiological condition and smart release at mimicking tumor acidic environments.

ACKNOWLEDGEMENT

The authors gratefully acknowledged the financial support of the Natural Science Foundation of China (No. 21204010), the Fundamental Research Funds for the Central Universities (No. 2232014D3-35), the Research Program of Shanghai Science and Technology Commission (No. 13 NM1400102), National High-Tech Research and Development Program of China (No. 2012 AA030309).

REFERENCES

Guo, Q. et al. 2014. Phenylboronate-diol crosslinked glycopolymeric nanocarriers for insulin delivery at physiological pH. *Soft Matter*, 10(6): 911–920.

Guzman, M.L. et al. 2007. An orally bioavailable parthenolide analog selectively eradicates acute myelogenous leukemia stem and progenitor cells. *Blood*, 110(13): 4427–4435.

Jiang, X. et al. 2006. UV irradiation-induced shell cross-linked micelles with pH-responsive cores using ABC triblock copolymers. *Macromolecules*, 39(18): 5987–5994.

Rapoport, N. 2007. Physical stimuli-responsive polymeric micelles for anti-cancer drug delivery. *Progress in Polymer Science*, 32(8): 962–990.

Sun, L. et al. 2013. A pH Gated, Glucose-Sensitive Nanoparticle Based on Worm-Like Mesoporous Silica for Controlled Insulin Release. *The Journal of Physical Chemistry B*, 117(14): 3852–3860.

Yuan, W. et al. 2011. Synthesis and self-assembly of pH-responsive amphiphilic dendritic star-block terpolymer by the combination of ROP, ATRP and click chemistry. *European Polymer Journal*, 47(5): 949–958.

Zhu, Z. et al. 2007. Effect of salt on the micellization kinetics of pH-responsive ABC triblock copolymers. *Macromolecules*, 40(17): 6393–6400.

Advanced Materials, Mechanical and Structural Engineering – Hong, Seo & Moon (Eds)
© 2016 Taylor & Francis Group, London, ISBN: 978-1-138-02908-8

Development of long stroke length and recoverable strain using the double stepped tube under the axial impact

M.S. Esa, P. Xue & M.S. Zahran
School of Aeronautics, Northwestern Polytechnical University, Xi'an, China

ABSTRACT: New structural configurations of inversion retractable tubes are presented in the present work to improve the stroke length and the recoverable strain. The force–displacement curves and the deformation shapes of the inversion processes of the proposed tubes under the axial impact are simulated by using the non-linear finite element code LS-DYNA/WORKBENCH ANSYS, and then the characteristics of the proposed tubes are analyzed. A comparative study is conducted to compare the energy absorption characteristics of the new proposed double stepped tubes with the plain circular tubes and the normal single stepped tubes. The results indicate that the stroke lengths of the new proposed tubes are significantly longer; moreover, the plastic strain distributions and values are significantly lower. Furthermore, manufacturing the new design with shaped memory alloy material will be more beneficial than others.

1 INTRODUCTION

Thin-walled structural members are widely used for crashworthiness operation. Circular tubes have been known to be efficient energy absorbers per unit weight, and may be considered as the main part for energy-absorbing devices, even for frequent impact operations such as landing-gear or for sudden accidents such as other vehicles. A precise choice for the circular tubes' material, shape and mechanism may save a person's life or save a valuable equipment from damage. The ideal energy absorption can be achieved by maximum resistance and stroke efficiency, to improve the load-displacement curve characteristics, wherever long flat (small fluctuation) stroke occurs, in which the plateau force is kept lower than the failure force (Goel 2015).

Numerous researchers have succeeded to improve the characteristics of tubes subjected to axial load by using several techniques. These techniques can be classified based on the failure mechanism into two main groups, as shown in Figure 1. The first group is the most popular one in which the buckling deformation can be seen obviously, and a typical load-displacement curve shown in Figure 2 can describe these deformation characteristics.

The researchers used the trigger to minimize the peak force (Akita et al. 2013, Chiu & Jenq 2014, Siromani et al. 2014). Moreover, the stroke efficiency can be enhanced by using horizontal stiffeners (Salehghaffari et al. 2011, Liu et al. 2015) or vertical stiffeners, which can also be called multi-cell section (Chen & Wierzbicki 2001, Goel 2015). However, the energy absorption performance can be strongly improved by using filled materials including composite or cellular material (Rezaei et al. 2014, Zheng et al. 2014). The second group is the one that deforms with inversion of tube failure mechanism. A typical load-displacement curve

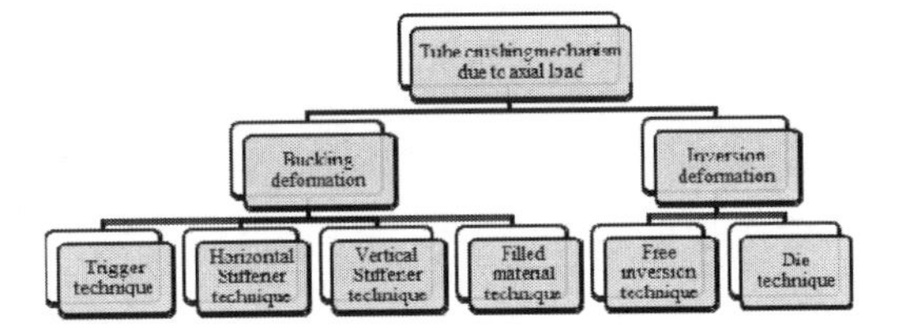

Figure 1. Classification of crashworthiness techniques with respect to the axial deformation shape.

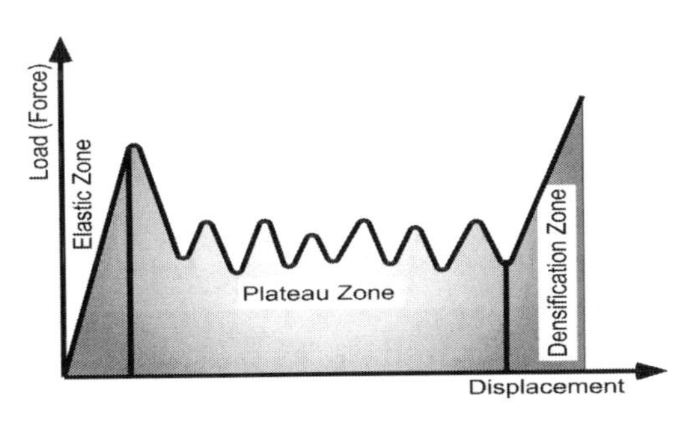

Figure 2. Typical load-displacement curve for an axial crushing circular tube.

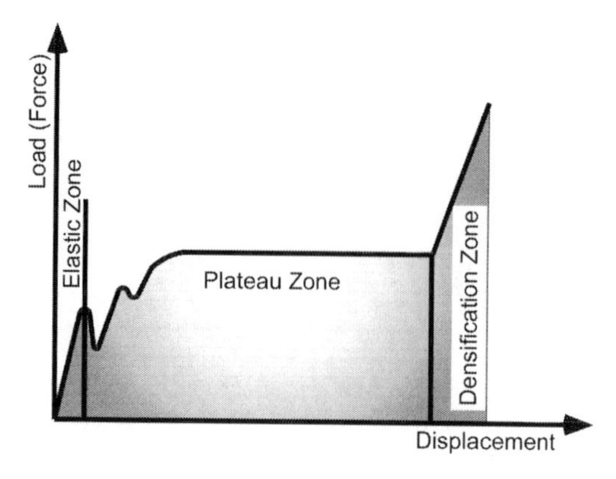

Figure 3. Typical load-displacement curve for an axial inversion of circular tube.

shown in Figure 3 can describe these deformation characteristics.

Free inversion (Qiu et al. 2013, Qiu et al. 2014) and die inversion (Gupta 2014, Luo et al. 2007, Niknejad et al. 2013, Rosa et al. 2003, Rosa et al. 2004) have almost the same behavior, which have some fluctuations during curling and then continue with the advantage of long, stable crushing stroke until the occurrence of densification. Generally, the free inversion needs a suitable fixture during axial compression, and is only feasible for ductile materials. Also inversions with a die are mainly achieved by axially compressing the tube onto a designed die (Al-Hassani et al. 1972).

The concerned section in this paper is the double stepped tube that can be considered as one of the free inversion techniques, as it dissipates the impact kinetic energy by the generated plastic deformation. The difficulty of the double stepped tube in manufacturing has been solved by Poor et al. (2014), and this solution occurs after passing through several phases to make a single stepped tube. However Asnafi & Skogsgårdh (2000) and Hwang & Chen 2005 contributed to these phases by using the same process of hydroforming.

The double stepped tube not only acts as an energy absorber but also has the ability to be a telescopic or retractable section if it is made of Shaped Memory Alloy (SMA) materials. Recently, SMAs such as NiTi have become important for impact applications due to their large recoverable strains and high capacity to dissipate energy (Cismasiu 2010). The main controller value in the recovering process is the occurred plastic strain during deformation. Moreover, in any stepped tube, the maximum plastic strain concentrated in the connection zones.

In the present paper, a program of comparative studies is conducted on the energy absorption characteristics of a plain circular tube with single and double stepped tubes. A new design of the double stepped tube has been developed that aimed to achieve the maximum length for a stable stroke; moreover, another new design in double identical connection shape has been developed that aimed to achieve the minimum of plastic strain values and distribution zones when subjected to the axial impact.

2 PROGRAM SETTING AND DETAILS

2.1 *New connection shape approach*

The flat connection with two sides of right angles has been used, as shown in Figure 4 (Zhang et al. 2009).

Figure 5 illustrates successive shots that are focused on the connection zone to explain the deformation process of the flat connection similar in the shape design to that used in the study by Zhang et al. (2009). During monitoring, the flat connection deformation, with a semi-half circle in both sides, is generated, starting from 0.002 s to 0.01 s. Therefore, an approach with a new design of s-shape connection is used to match the final deformation shape. The new s-connection can be described as two opposite half right circles, as shown in Figure 6. The reason for using this

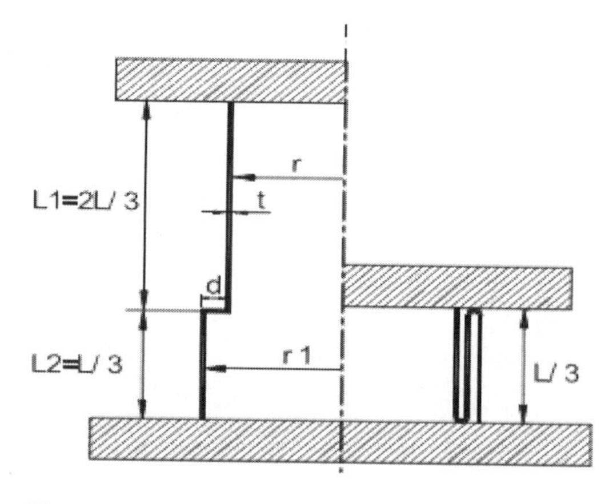

Figure 4. Section details of a stepped tube (Zhang et al. 2009).

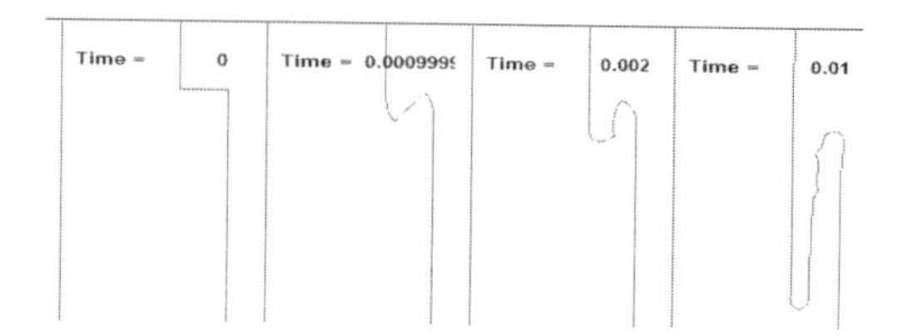

Figure 5. Successive shots show the flat connection deformation process.

s-shape is to prevent the aggressive plastic strain that occur in the flat connection, which is generated to make a straight line with the double right angle to be semi-half circles on both sides, starting from 0.002 as shown in Figures 5 and 6. This new design is aimed to minimize the plastic strain and then SMAs can be used to produce a reliable recoverable section.

The main contribution of this work is the new design of the double stepped tube and the new s-shape connection, as shown in Figure 7.

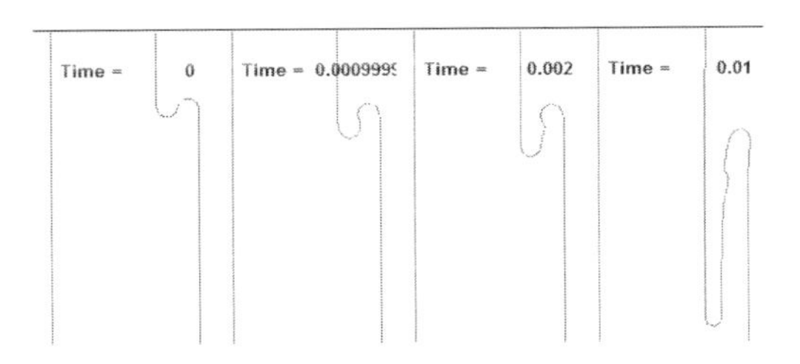

Figure 6. Successive shots show the s-shape connection deformation process.

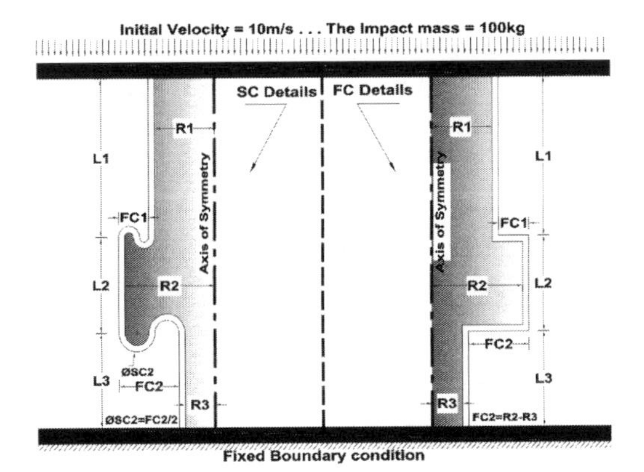

Figure 7. Details of section dimensions.

2.2 *Group details*

There are four groups of settings. The 1st stepped circular tube group (IST) is the normal shape version of single stepped circular tubes. The 2nd stepped circular tube group (IIST) is the normal shape version of double stepped circular tubes, and the objective of this group is to study the behavior of two different dimensions for connections with two different shapes (flat shape and S-shape). However, this can be done by using three different diameters of circular tubes R1, R2 and R3, as shown in Figure 7. The 3rd stepped circular tube group (IIIST) is also double stepped circular tube shape, but the objective of this group is to study the behavior of two identical connections with two different shapes (flat shape and S-shape). However, this can be done by using two equal diameters of circular tubes R1 and R3, as shown in Figure 7. The 4th group contains the popular shape used as the Reference CYLindrical (RCYL) shape to use the results as a comparative result. Moreover, the Flat Connection (FC) length is equal to the difference between two successive cylinders, also the S-shape Connection (SC) half circle Radius (RSC) is equal to the quarter of the difference between two successive cylinders. The model setup is shown in Figure 7. The initial velocity is 10 m/s, the impactor mass is 100 kg and the element type is shell element with a virtual thickness of 1.5 mm. All the group dimensions are listed in Table 1.

Their behavior and response analysis of the circular tubes under the axial impact load are obtained by using the explicit non-linear finite element code ANSYS-WORKBENCH/LS-DYNA (ANSYS, 2013). All the tubes are made of aluminum alloy AA6060-T4. The material has the following parameters: Young's modulus E = 68.2 GPa, initial yield stress $\sigma_y = 80$ MPa, ultimate stress $\sigma_u = 173$ MPa, Poisson's ratio $\nu = 0.3$, and density $\rho = 2770$ kg/m^3. The constants $C = 1.3 \times 10^6$ and $q = 4$ obtained from Symonds (1965) are used in the present study as strain rate constants

Table 1. Tubes dimensions (mm) and weight (kg).

		1ST TUBE			2ND TUBE			3RD TUBE			
Thickness	(t) = 1.5 mm	R1	L1	FC1	RSC1	R2	L2	FC2	RSC2	R3	L3
ISTFC	3.94E-02 kg	25	126	10	–	35	74	–	–	–	–
ISTSC	4.05E-02 kg	25	126	–	2.5	35	74	–	–	–	–
IISTFC	4.23E-02 kg	25	65	10	–	35	70	5	–	30	65
IISTSC	4.41E-02 kg	25	65	–	2.5	35	70	–	1.25	30	65
IIISTFC	4.24E-02 kg	25	55	10	–	35	90	10	–	25	55
IIISTSC	4.47E-02 kg	25	55	–	2.5	35	90	–	2.5	25	55
RCYL	3.26E-02 kg	25	200	–	–	–	–	–	–	–	–

FC (the length of the flat connection) and RSC (the radius of the half circle of S-Connection) ...{FC = 4 X RSC}, as shown in Figure 7.

Table 2. Strain hardening data.

Plastic strain (%)	0.0	2.4	4.9	7.4	9.9	12.4	14.9	17.4
Plastic stress (Mpa)	80	115	139	150	158	167	171	173

in the dynamic simulation for AA6060-T4. Strain hardening data for AA6060-T4 are listed in Table 2 (Langseth & Lademo 1994, Santosa et al. 2000).

3 RESULTS AND DISCUSSION

3.1 *Simulation results*

The successive section shots shown in Figure 8 describe the deformation scenario of ISTSC.

However, the force displacement curve of the 1st group, shown in Figure 9, exhibits the same characteristics of the free inversion tubes starting with the connections curling with a low peak force and smooth stroke distance until the section starts the densification. The mean crushing load (P_m) for RCY and IST are 12.2 KN and 8.5 KN, respectively. Furthermore, the densification for IST starts at a displacement of 132 mm.

In the ISTFC curve, an odd drop occurs as the large diameter cylinder is deformed to some extent, as shown in Figure 6, unlike ISTSC that has the same deformation but curls smoothly. The curvature of s-connection is shown in Figure 10.

The successive section shots, shown in Figure 11, describe the deformation scenario of IISTSC.

The force displacement curve of the 2nd group, shown in Figure 12, exhibits the same characteristics of the 1st group curve but with a longer stroke length. The stroke length is 129 mm in the 1st group, but it is 161 mm in the 2nd group. Furthermore, the change observed at the end of the plateau zone is due to the stroke generated by the third outer cylinder at the bottom fixed support. The mean crushing load (P_m) IIST is 8.75 KN.

Two drops occurred in both curves of IISTFC and IISTSC. However, these drops occurred due to the start of curling for the stronger (larger diameter) connection, as shown in Figure 13. Moreover, this means the end of inverting the smallest tube and the start of inverting the middle tube are stronger, wherever an improvement in the load-displacement curve appears obviously.

The successive section shots, shown in Figure 14, describe the deformation scenario of IIISTSC.

The force displacement curve of the 3rd group, shown in Figure 15, exhibits smooth curves; therefore, the double identical connection curling occurs in a simultaneous way that deforms smoothly by two identical elastic hinges. These deformation

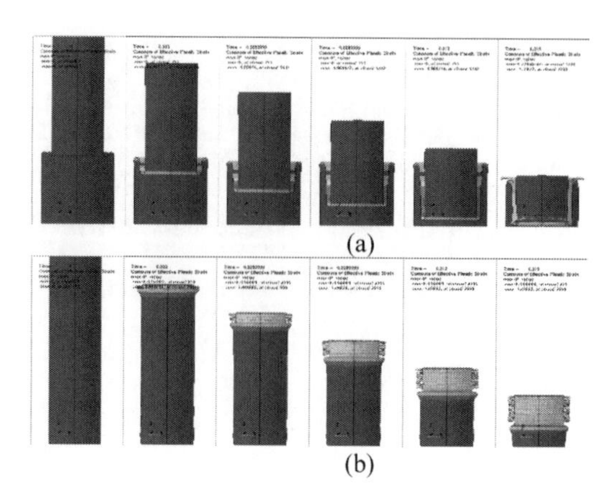

Figure 8. Successive shots describe the deformation scenario (A) ISTSC and (B) RCYL.

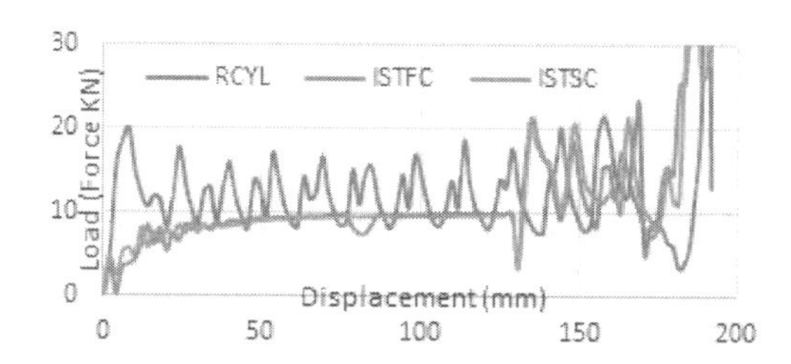

Figure 9. Force-displacement curve for RCYL, ISTFC and ISTSC.

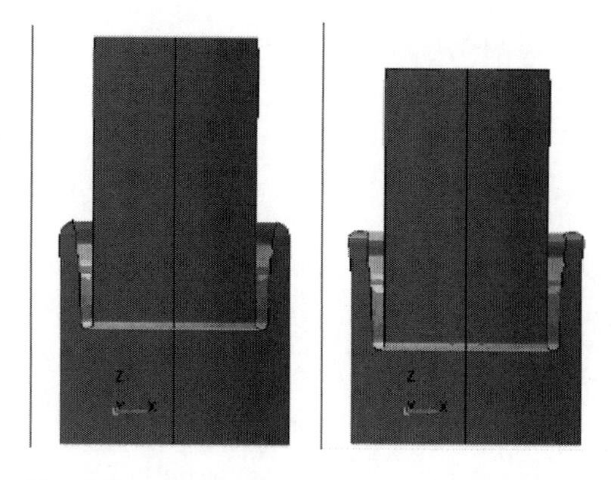

Figure 10. Two successive shots for displacements of 7.7 and 7.9 mm in ISTFC crush.

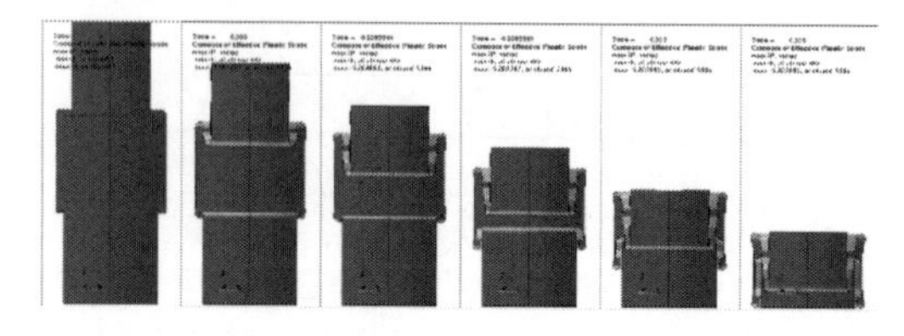

Figure 11. Successive shots describe the scenario of IISTSC deformation.

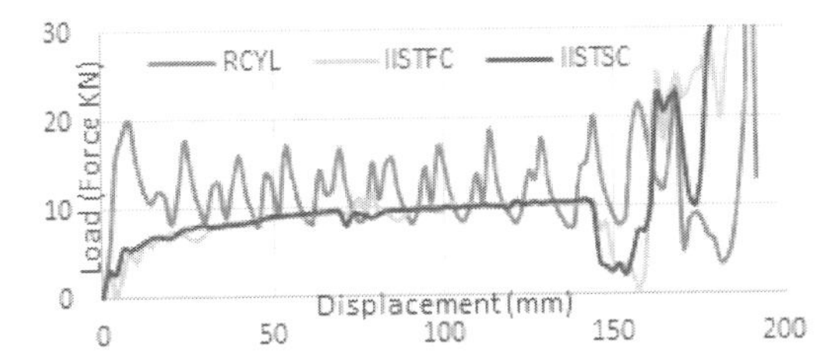

Figure 12. Force-displacement curve for RCYL, IISTFC and IISTSC.

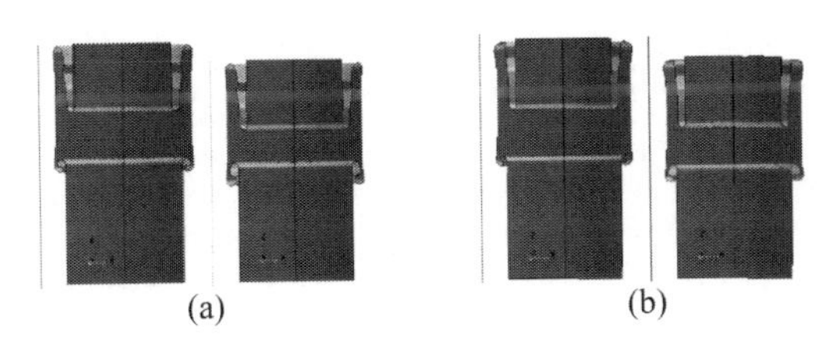

Figure 13. Two successive shots for displacements of 6.9 and 8.2 mm in (A) IISTFC and (B) IISTSC.

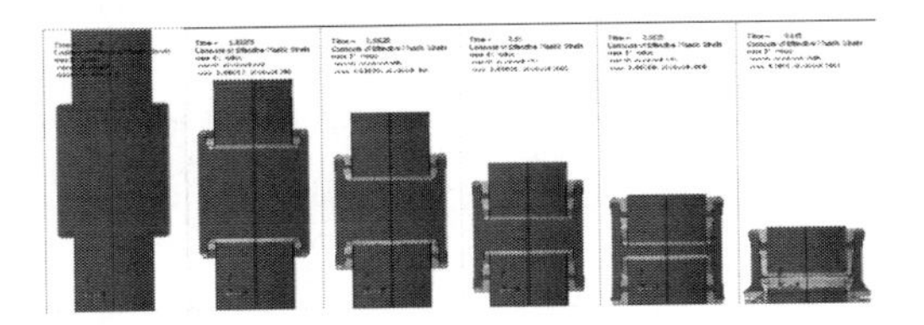

Figure 14. Successive shots describe the scenario of IIISTSC deformation.

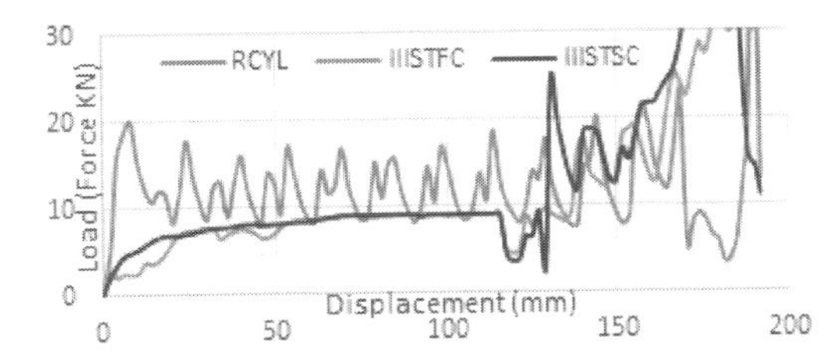

Figure 15. Force-displacement curve for RCYL, IIISTFC and IIISTSC.

mechanisms for the 3rd group are different from those for the 2nd group that start with small tubes curling and then middle radius tube curling, as shown in Figures 11 and 14. The mean crushing load (P_m) IIIST is 8.11 KN.

The most important action occurred by s-shape connection, as shown in Figure 16, which illustrates the effective plastic strain contour for IIISTFC and IIISTSC at the same values of displacement. The s-shape connection contributes not only to a lower max. value of plastic strain but also to the

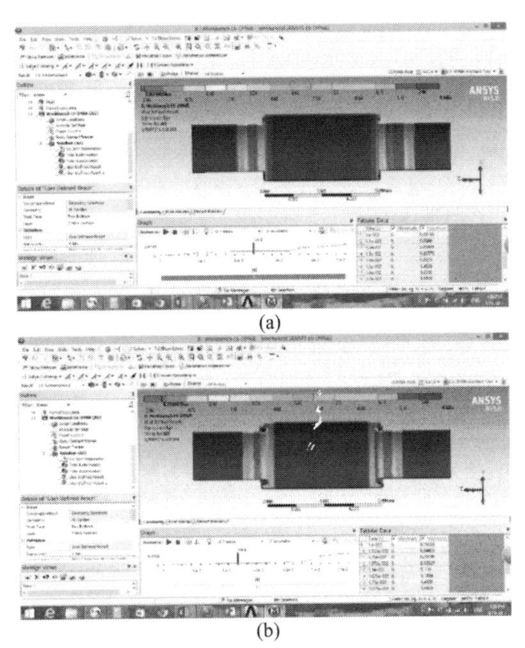

Figure 16. Effective plastic strain contour lines for (A) IIISTFC and (B) IIISTSC.

minimum distribution of the elements indicated by this value (the red color distribution). Therefore, s-connection is a better approach to be a recoverable section than flat connection, as it deforms smoothly and closer to the recoverable strain of the SMAs than the flat connection.

3.2 *Theoretical predictions and FEM result comparisons*

For RCYL mean force prediction, the equation of Abramowicz & Jones (1986), is used to calculate (Pm) for a cylinder axial crush with concertina mode of failure, which is given by

$$\frac{P_m}{M_o} = 25.230\sqrt{\left(\frac{2R}{t}\right)} + 15.09 \tag{1}$$

where $Mo = \sigma_0 t^2/4$ is the full plastic bending moment per unit length and σ_o is the plastic flow stress. However, by substituting in Equation 1, Pm for RCYL = 11.44 KN. Recalling the dynamic factor range proposed by Simhachalam et al. (2014) for aluminum alloys AA6060-T4 under different strain rates, the range for 10 m/s is from 1.1 to 1.13, so Pm for RCYL will be in the range of 12.58 KN to 12.93 KN, which agrees with an error minus 6% compared with the FEM analysis results of 12.2 KN, as given in Table 3.

Table 3. Theoretical predictions and FEM results (force unit in KN).

Groups	Simulation	Quasi-static theoretical prediction	Dynamic effect	Error (%)
RCYL	12.2	11.44	12.58–12.93	−6%
IST	8.5	7.3	8.03–8.25	+5.9%
IIST	8.75	7.65	8.41–8.64	+4%
IIIST	8.1	7.3	8.03–8.25	In range

Furthermore, for the inversion tube groups IST, IIST, and IIIST, the equation of Colokoglu & Reddy (1996) is used to calculate (Pm) for a free inversion tube, which is given by

$$P_s = 2\pi 6_o rt\left(\frac{b}{r} + \frac{t}{4b}\right) \quad (2)$$

where Ps is the predicted steady inversion load; r is the radius of the tube; and b is the knuckle radius with $b = \sqrt{rt}/2$. Then, for IST, and substituting IIIST in Equation 2, P_m calculations agree with accepted errors compared with the FEM analysis results, as given in Table 3.

But for substituting IIST in Equation 2 and taking the average from P_m that occurs through the smallest tube and then P_m that occurs through the middle tube, for the reason discussed before, P_m calculations agree with an accepted error compared with the FEM analysis results, as given in Table 3.

However, for the future work, another factors have to be added on the theoretical prediction formula to explain each type of stepped tube inversion individually.

4 CONCLUSIONS

A new structure configuration based on the free inversion of circular tubes is introduced in the present work to improve and enhance the energy absorption specs under axial impact loading. The new configuration consists of a double stepped tube, which can be done using flat connections or s-shape connections. Double stepped tubes with two different diameter connections are called IISTFC and IISTSC. Furthermore, double stepped tubes with two identical diameter connections are called IIISTFC and IIISTSC. The former groups plus the normal single stepped tubes ISTFC and ISTSC are mainly studied in this paper. Numerical analyses were conducted to simulate the inversion process of the proposed tubes by using the non-linear finite element code ANSYS-WORKBENCH/LS-DYNA. An impact with a 10 m/s initial velocity and 100 kg mass was investigated as an axial load, and the features of force–displacement curves and the deformation process of the mentioned groups were analyzed in the paper. A comparative study was conducted to compare the energy absorption characteristics of the new tube configurations with the normal straight cylinder RCYL tube that are most commonly used for energy-absorbing devices. The results indicate that the plateau zone and stroke length of IISTFC and IISTSC are significantly longer than those of ISTFC and ISTSC. Moreover, the plastic strain occurring in IIISTSC is significantly lower than the values and distributions in IIISTFC. This comparison result indicates that the proposed double stepped retractable tubes made of shaped memory alloys are possible and promising to be extensively adopted in future industrial applications.

Simple theoretical predictions of the mean forces for the straight tube and steady forces for inversion tubes were given by using the theories proposed by former researchers. The predicted mean and steady forces for all tubes compare very well with numerical results, and the error is acceptable. Finally, some discussion is made on the design of retractable tubes and the feasibility of SMAs.

ACKNOWLEDGMENT

The authors gratefully acknowledge the financial supports from the National Natural Science Foundation of China under Grants 11472226 and 11072202.

REFERENCES

Abramowicz, W. & Jones, N. 1986. Dynamic progressive buckling of circular and square tubes. *International Journal of Impact Engineering*, 4: 243–270.

Akita, R., Yokoyama, A., Koike, A., Kawamura, K., Sukegawa, Y. & Oohira, H. 2013. *Development of High Performance FRP Crush Box*. Proceedings of the FISITA 2012 World Automotive Congress, Springer, 869–878.

Al-Hassani, S., Johnson, W. & Lowe, W. 1972. Characteristics of inversion tubes under axial loading. *Journal of Mechanical Engineering Science*, 14: 370–381.

ANSYS 2013. *Release 15.0 ANSYS Documentation*. ANSYS Inc, Canonsburg, PA.

Asnafi, N. & Skogsgårdh, A. 2000. Theoretical and experimental analysis of stroke-controlled tube hydroforming. *Materials Science and Engineering: A*, 279: 95–110.

Chen, W. & Wierzbicki, T. 2001. Relative merits of single-cell, multi-cell and foam-filled thin-walled structures in energy absorption. *Thin-Walled Structures*, 39: 287–306.

Chiu, Y.S. & Jenq, S.T. 2014. Crushing behavior of metallic thin-wall tubes with triggering mechanisms due to quasi-static axial compression. *Journal of the Chinese Institute of Engineers*, 37: 469–478.

Cismasiu, C. 2010. *Shape Memory Alloys*. Janeza Trdine 9, 51000 Rijeka, Croatia www.sciyo.com.

Colokoglu, A. & Reddy, T. 1996. Strain rate and inertial effects in free external inversion of tubes. *International Journal of Crashworthiness*, 1: 93–106.

Goel, M.D. 2015. Deformation, energy absorption and crushing behavior of single-, double-and multi-wall foam filled square and circular tubes. *Thin-Walled Structures*, 90: 1–11.

Gupta, P. 2014. Numerical Investigation of Process Parameters on External Inversion of Thin-Walled Tubes. *Journal of Materials Engineering and Performance*, 23: 2905–2917.

Hwang, Y.M. & Chen, W.C. 2005. Analysis of tube hydroforming in a square cross-sectional die. *International Journal of Plasticity*, 21: 1815–1833.

Langseth, M. & Lademo, O. 1994. *Tensile and torsion testing of AA6060-T4 and T6 aluminium alloys at various strain rates*. Department of Structural Engineering, The Norwegian University of Science and Technology Trondheim.

Liu, S., Tong, Z., Tang, Z., Liu, Y. & Zhang, Z. 2015. Bionic design modification of non-convex multicorner thin-walled columns for improving energy absorption through adding bulkheads. *Thin-Walled Structures*, 88: 70–81.

Luo, Y., Huang, Z. & Zhang, X. 2007. FEM analysis of external inversion and energy absorbing characteristics of inverted tubes. *Journal of Materials Processing Technology*, 187: 279–282.

Niknejad, A., Rezaei, B. & Liaghat, G.H. 2013. Empty circular metal tubes in the splitting process–theoretical and experimental studies. *Thin-Walled Structures*, 72: 48–60.

Poor, H.Z., Menghari, H.G., De Sousa, R.J.A., Moosavi, H., Parastarfeizabadi, M., Farzin, M. & Sanei, H. 2014. A novel approach in manufacturing two-stepped tubes using a multi-stage die in tube hydroforming process. *International Journal of Precision Engineering and Manufacturing*, 15: 2343–2350.

Qiu, X., He, L., Gu, J. & Yu, X. 2013. A three-dimensional model of circular tube under quasi-static external free inversion. *International Journal of Mechanical Sciences*, 75: 87–93.

Qiu, X.M., He, L.H., Gu, J. & Yu, X.H. 2014. An improved theoretical model of a metal tube under free external inversion. *Thin-Walled Structures*, 80: 32–37.

Rezaei, B., Niknejad, A., Assaee, H. & Liaghat, G. 2014. *Axial splitting of empty and foam-filled circular composite tubes–An experimental study*. Archives of Civil and Mechanical Engineering.

Rosa, P., Rodrigues, J. & Martins, P. 2003. External inversion of thin-walled tubes using a die: experimental and theoretical investigation. *International Journal of Machine Tools and Manufacture*, 43: 787–796.

Rosa, P.A., Baptista, R.M., Rodrigues, J.M. & Martins, P.A. 2004. An investigation on the external inversion of thin-walled tubes using a die. *International Journal of Plasticity*, 20: 1931–1946.

Salehghaffari, S., Rais-Rohani, M. & Najafi, A. 2011. Analysis and optimization of externally stiffened crush tubes. *Thin-Walled Structures*, 49: 397–408.

Santosa, S.P., Wierzbicki, T., Hanssen, A.G. & Langseth M. 2000. Experimental and numerical studies of foam-filled sections. *International Journal of Impact Engineering*, 24: 509–534.

Simhachalam, B., Srinivas, K. & Rao, C.L. 2014. Energy absorption characteristics of aluminium alloy AA7XXX and AA6061 tubes subjected to static and dynamic axial load. *International Journal of Crashworthiness*, 19: 139–152.

Siromani, D., Henderson, G., Mikita, D., Mirarchi, K., Park, R., Smolko, J., Awerbuch, J. & Tan, T.M. 2014. An experimental study on the effect of failure trigger mechanisms on the energy absorption capability of CFRP tubes under axial compression. *Composites Part A: Applied Science and Manufacturing*, 64: 25–35.

Zhang, X., Cheng, G. & Zhang, H. 2009. Numerical investigations on a new type of energy-absorbing structure based on free inversion of tubes. *International Journal of Mechanical Sciences*, 51: 64–76.

Zheng, G., Wu, S., Sun, G., Li, G. & Li, Q. 2014. Crushing analysis of foam-filled single and bitubal polygonal thin-walled tubes. *International Journal of Mechanical Sciences*, 87: 226–240.

Advanced Materials, Mechanical and Structural Engineering – Hong, Seo & Moon (Eds)
© 2016 Taylor & Francis Group, London, ISBN: 978-1-138-02908-8

Analysis of urban spatio-temporal land use/cover change and expansion with their driving force through remote sensing and GIS in Samara city, Russia

K. Choudhary
Samara State Aerospace University, Samara, Russia

M.S. Boori
Samara State Aerospace University, Samara, Russia
American Sentinel University, Aurora, Colorado, USA

A. Kupriyanov & V. Kovelskiy
Samara State Aerospace University, Samara, Russia

ABSTRACT: A dynamic study was carried out on urban land use/cover change and its built-up area expansion in Samara city from 1985 to 2015. Landsat satellite imageries of four different time periods were acquired by the Global Land Cover Facility Site and the earth explorer site. Supervised classification methodology was employed using the maximum likelihood technique in ArcGIS 10.1 Software. The results indicate that urban expansion of Samara city is very fast with high speed of economic development. During the 30-year period, the built-up area of Samara city attained a net increase of 147.62 km^2, and reached 419.26 km^2 in 2015, which is nearly double the estimate in 1985. The annual average expansion was 14.76 km^2 and the annual growth rate was 5.43%. Here, the rate of urban expansion varied in different periods and was the most outstanding in various periods in the southeast region of the city. The built-up area of Samara city was highly correlated with population and socio-economic activities, which was the main driving factor for urban expansion. This information on urban expansion and land use/cover change is very useful to the local government and urban planners in the implementation of future plans for sustainable development of the city.

1 INTRODUCTION

Land use/cover studies have been a key area to know the changes in the balance of urban and natural resources. Urbanization is directly related to land use/cover in terms of land use pattern, urban space distribution, and social and economic pressure (Aljoufie et al. 2013). Remote sensing and GIS data are useful to study the urban dynamic change and expansion implementation of natural resources management (Boori et al. 2015). Remote sensing and Geographic Information System (GIS) technologies are used to achieve results from large landscape analysis (Shu et al. 2014), urban space structure and fractal shape (Dong et. al. 2007), analysis of urban expansion models and driving forces (Boori et al. 2014), land use survey (Aljoufie et al. 2013), land cover change (Ye et al. 2013), analysis of urban dynamic models (Henriquez et al. 2006), and analysis of farmland change and classification. As population is related to land use and built-up area, earlier studies have shown their correlation analysis. These studies performed by urban geoscience, human geoscience and economic geoscience scientists are based on subjects such as shape and direction of urban expansion, spatial evolution processes, dynamic mechanisms of urban expansion, transformation of farmland to urban land, and internal differentiation of urban land (Fan et al. 2008, Boori & Amaro 2010). However, no study has reported on the multi-temporal and long time-series expansion of the built-up area in Samara city.

Therefore, this paper presents a long time-series dynamic monitoring and analytical study on the processes of urban expansion and urbanization of Samara city in a period of nearly 30 years, based on multiple-sourced remote-sensing data. In addition, our study focuses on the regional rules governing urban expansion and aims to reveal the expansion processes of the built-up area in Samara city, together with their influencing factors and driving forces. The results would be meaningful to study the spatio-temporal processes and influencing factors for the expansion of the built-up urban area and for urban development and planning in the entire Volga River Delta region. This information can be used as a reference to study urbanization in the Volga River Delta region.

2 MATERIALS AND METHODS

2.1 *Samara city: the study area*

Samara region is situated in the southeast of the Eastern European Plain in the middle flow of the greatest European river, the Volga, which separates the region into two parts of different sizes, namely Privolzhye and Zavolzhye. The study area Samara, known as Kuybyshev from 1935 to 1991, is the sixth largest city in Russia and the administrative center of Samara Oblast. Its geographical coordinates are 53°12′10″N, 50°08′27″E. The region occupies an area of 53.6 km^2 (0.31% of the territory of Russia) and forms a part of the Volga Federal District.

2.2 *Data acquisition and processing*

Multi-temporal satellite data were used for the analysis of urban expansion and land use/cover change. All data were downloaded from the NASA and USGS websites. The specific satellite images used were Landsat TM (Thematic Mapper) from 1985–1995 and Landsat ETM+ (Enhanced Thematic Mapper plus) from 2005 and 2015, an image captured by a different type of sensor. In digital image process, geometric correction is used first, followed by band ratio, classification, and finally change detection. All four satellite images were classified through the maximum likelihood supervised classification in ArcGIS 10.1 software (Boori & Ferraro 2015). Also, secondary data such as field data and socio-economic data were used. In the past, the Samara region was covered by forests and grasslands (agriculture and open land); later, it was used for built-up and industrial purposes. After the classification, we found four major land cover classes: forest, built-up, water, and grassland. Here, grassland covers shrub-land, open space, and agriculture land.

After preprocessing and classification, land use/cover change detection and a post-classification detection method was employed. A pixel-based comparison was used to produce change information on the pixel basis and, thus, interpret the changes more efficiently, taking the advantage of "-from, -to" information. Classified image pairs of two different decade data were compared using cross-tabulation in order to determine the qualitative and quantitative aspects of the changes for the period of 1985 to 2015. A change matrix (Boori & Vozenilek 2014) was produced with the help of ArcGIS software. Quantitative areal data of the overall land use/cover changes as well as gains and losses in each category between 1985 and 2015 were then compiled.

Urban expansion rate and its dynamic change in the spatial structure of a city vary in a temporal sequence. The dynamism of land use class represents the change in the quantity of a certain land use class in a unit time (Hu et al. 2007), so this is a key index for evaluating the spatial change in urban expansion. By analyzing the dynamism of land use, the extent and rate of urban expansion can be compared quantitatively (Hu et al. 2007), according to the following formula:

$$\mathrm{LUDI} = \frac{Ua - Ub}{Ua} \times \frac{1}{T} \times 100\% \qquad (1)$$

where Ua and Ub denote areas of a certain land use class at time a and time b, respectively; T denotes the length of time from time a to time b. When T is in a unit of year, then the LUDI is the annual rate of change in area for this land use class.

3 RESULTS AND DISCUSSION

3.1 *Land use/cover change in Samara city*

The classified land use/cover change maps for 1985, 1995, 2005, and 2015 with an overall accuracy of 85%, show a substantial change in the landscape of the study area. These data reveal that in 2015, about 31.38% (616.14 km^2) area of the Samara block was under forest, 36.85% (723.49 km^2) under grassland, 10.42% (204.52 km^2) under water body, and 419.26% (21.35 km^2) under built-up land. During 1985, the area under these land categories was found to be about 34.32% (673.94 km^2) under forest, 43.68% (857.7 km^2) under grassland, 8.16% (160.13 km^2) under water body, and 13.83% (271.64 km^2) under built-up land (Table 1). Based on the statistics, about 147.62 km^2 (7.52%) area has been converted to built-up land in the last three decades. Consequently, 7.52% area from forest, grassland, and water body has been lost in the same period. First, the urban area was increased until 1995 and then reduced, but later it increased again due to increased population. Initially, the forest area was decreased and later increased due to government protection. Grassland class covers the highest area in the study area. It was increased and highest in 2005 but finally reduced. Water class is stable with small variation (Table 1).

Built-up urban area of Samara city was increased with fast economic development and deepening urbanization. The major built-up area of samara city was 271.64 km^2 at the end of 1985 but became 419.26 km^2 in 2015, which showed an increase by 147.62 km^2, an annual average expansion of 14.76 km^2, and an annual change rate of 5.43% in the 30-year period from 1985 to 2015 (Figure 1).

Table 1. Land cover classes of Samara city from 1985 to 2015.

	1985		1995	
Class	Area	%	Area	%
Built-up	271.64	13.83	375.04	19.1
Others	1691.79	86.17	1588.39	80.9
Total	1963.43	100	1963.4	100
	2005		2015	
Class	Area	%	Area	%
Built-up	254.69	12.97	419.26	21.35
Others	1708.74	87.03	1544.17	78.65
Total	1963.43	100	1963.43	100

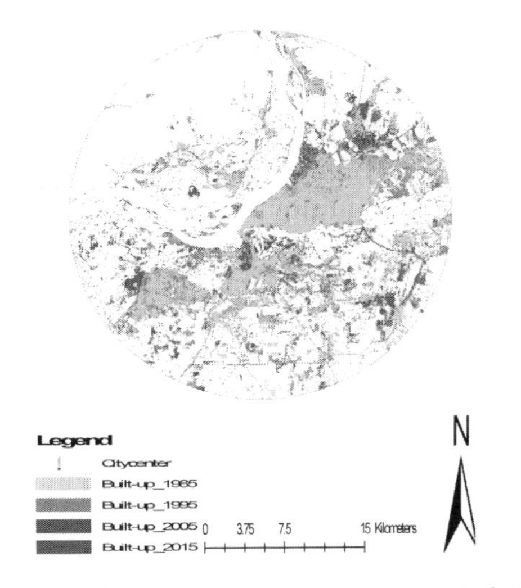

Figure 1. Samara city growth in different years from 1985 to 2015.

In 1985–1995, the built-up urban area of Samara city expanded by 103.4 km^2 at an annual average rate of 10.34 km^2, and mostly the expansion was in the north and south parts of the city and in the Volga River. In 1985–2005, the built-up urban area of Samara city reduced by −120 km^2 at an annual average rate of −12.03 km^2 and mainly in the form of radiation, especially near the city center and outer part of the city. In 2005–2015, the built-up urban area of Samara city expanded by 164.57 km^2 at an annual average rate of 16.45 km^2 and mainly in the form of radiation, especially in the east and south sides and first time in the west side of the city (Table 2). There was no city development in the north side of the Volga River and in the city due to government protection. Here, the government wants to preserve the north side of the Volga River as a natural heritage.

Table 2. Built-up area change of Samara city, Russia for the periods of 1985, 1995, 2005 and 2015.

Item	1985–1995	1995–2005	2005–2015	1985–2015
Expansion area (km^2)	103.4	−120.35	164.57	147.62
Expansion percentage	38.06	−32.08	64.61	54.34
Expansion rate (km^{2}a^{-1})	10.34	−12.03	16.45	14.76
Annual change rate (%)	3.8	−3.2	6.46	5.43

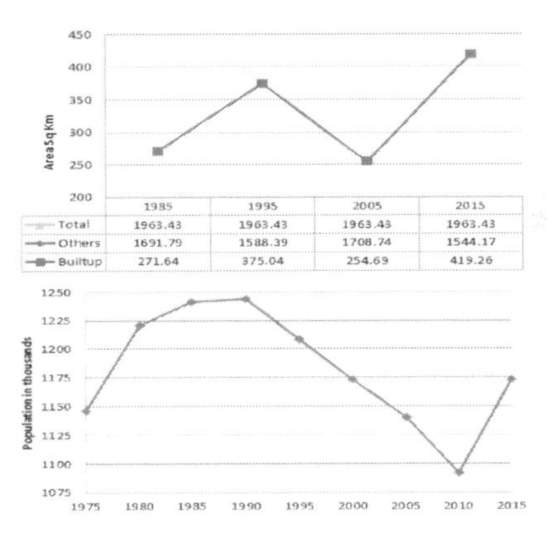

Figure 2. Population and built-up area change in Samara city, Russia.

Figure 2 shows the correlation between the built-up area and the population. The 1975 to 1990 census data show the increase in population from 1,150,000 to 1,244,000; meanwhile, the built-up area was also increased and was highest (375 km^2) until 1995. Since the breaking up of the Union of Soviet Socialist Republics (USSR), the population had been continually reducing until 2010, to about 1,164,000, so the built-up area was also reduced from 375.04 km^2 (1995) to 254.69 km^2 (2005). Now, in the last decade from 2005 to 2015, the population was increased. In 2015, the population was about 1,173,000, so the built-up area also began to increase with a very high speed due to economic development and infrastructure.

4 DRIVING FACTORS FOR URBAN GROWTH AND LAND USE CHANGE

Urban expansion and subsequent landscape changes are governed by geographical and socio-economic factors, such as population growth, policy, and economic development. The Samara region has highly developed industry and a diversified economy structure. Industry accounts for about 40% of the gross regional product. It includes production and processing and energy sectors. The development of the region's economy is based on high-tech processing industries with high added value: automobile manufacture, and aircraft and spacecraft manufacture, which accounts for up to 35% of the total volume of shipped production of processing industries; chemical and metallurgical enterprises have high degree of processing. The region manufactures 30% of new passenger cars made in Russia, 31% of polymer materials for floor, wall and ceiling coatings, 23% of anhydrous ammonia, 16% of sanitary products made from ceramics, 13% of ceramic floor tile, 7.7% of automobile gasoline and 9% of diesel fuel, 8.5% of plastics in primary forms, 7.3% of beer, 5.0% of confectionery products, and 4% of mineral fertilizers. Mining of minerals accounts for approximately 17% of industrial production. About 99% of them are fuel and energy raw materials. Production and distribution of energy resources accounts for about 11% of regional economy (Ministry of Economic Development 2015). That is why industry and infrastructure are the main cause of urban growth and land use change.

5 CONCLUSION

The results indicate that urban expansion with fast economic development of Samara city, Russia is accelerated by 147.62 km^2, which has increased the built-up area in the last three decades. The built-up urban area in 1985 was increased with an annual average expansion of 14.76 km^2 and an annual change rate of 5.43%. In the first decade from 1985 to 1995, the city grew in the east and south sides; however, in the next decade (1995–2005), it was reduced dramatically in the central and outer parts of the city. Finally, in the last decade from 2005 to 2015, the city again grew with a high speed of economic development due to government protection and foreign investments. Economic growth is the major driving force for the expansion of the built-up urban area. Population growth and residents' increased income are the two most direct driving forces for urban land expansion, while for the spatial development of a city, city infrastructure construction, city planning, and policy control contribute to a considerable extent, and the macroscopic pattern and development rate of urban expansion, natural geographic environment, and topography constrains contribute to a certain degree. This information on urban expansion and land use/cover change can provide scientific criteria for reasonable planning and decision making to the local government and urban planners for the implementation of future plans for sustainable development of the city.

ACKNOWLEDGMENT

This work was financially supported by the Russian Scientific Foundation (RSF), grant no. 14-31-00014 "Establishment of a Laboratory of Advanced Technology for Earth Remote Sensing."

REFERENCES

Aljoufie, M., Zuidgeest, M., Brussel, M. & Maar-seveen, M.V. 2013. Spatial-temporal analysis of urban growth and transportation in Jeddah city, Saudi Arabia. *Cities*, 31: 57–68.

Boori, M.S. & Amaro, V.E. 2010. Land use change detection for environmental management: using multi-temporal satellite data in Apodi Valley of northeastern Brazil. *Applied GIS*, 6(2): 01–15.

Boori, M.S. & Ferraro, R.R. 2015. Global land cover classification based on microwave polarization and gradient ratio (MPGR). *Geo-Informatics for Intelligent Transportation*, 71: 17–33.

Boori, M.S. & Vozenilek, V. 2014. Land-cover disturbances due to tourism in Jeseniky mountain region: A remote sensing and GIS based approach. *SPIE Remote Sensing*, 9245, 92450T: 01–11.

Boori, M.S., Vozenilek, V. & Burian, J. 2014. Land-cover disturbances due to tourism in Czech Republic. *Advances in Intelligent Systems and Computing*, 303: 63–72.

Boori, M.S., Vozenilek, V. & Choudhary, K. 2015. Land use/cover disturbances due to tourism in Jeseniky Mountain, Czech Republic: A remote sensing and GIS based approach. *The Egyptian Journal of Remote Sensing and Space Sciences*, 18(1): 17–26.

Dong, W., Zhang, X.L., Wang, B. & Duan, Z.L. 2007. Expansion of Urumqi urban area and its spatial differentiation. *Science in China Series D: Earth Sciences*, 50: 159–168.

Fan, F.L., Wang, Y.P. & Wang, Z.S. 2008. Temporal and spatial change detecting (1998–2003) and predicting of land use and land cover in core corridor of Pearl River Delta (China) by using TM and ETM+ images. *Environmental Monitoring and Assessment*, 137(1–3): 127–147.

Henriquez, C., Azocar, G. & Romero, H. 2006. Monitoring and modeling the urban growth of two mid-sized Chilean cities. *Habitat International*, 30(4): 945–964.

Hu, Z.L., Du, P.J. & Guo, D.Z. 2007. Analysis of urban expansion and driving forces in Xuzhou City based on remote sensing. *Journal of China University of Mining and Technology*, 17(2): 267–271.

Ministry of Economic Development. 2015. Investments and Trade of the Samara region, www.economy.samregion.ru.

Shu, B., Zhang, H., Li, Y., Qu, Y. & Chen, L. 2014. Spatiotemporal variation analysis of driving forces of urban land spatial expansion using logistic regression: A case study of port towns in Taicang city, China. *Habitat International*, 43: 181–190.

Ye, Y., Zhang, H., Liu, K. & Wu, Q. 2013. Research on the influence of site factors on the expansion of construction land in the Pearl River Delta, China: By using GIS and remote sensing. *International Journal of Applied Earth Observation and Geoinformation*, 21: 366–373.

Advanced Materials, Mechanical and Structural Engineering – Hong, Seo & Moon (Eds)
© 2016 Taylor & Francis Group, London, ISBN: 978-1-138-02908-8

The design and simulation of large range of eddy current displacement sensor circuit

Q. Qin & X.Q. Tian
National Engineering Research Center of Turbo-Generator Vibration, Southeast University, Nanjing, China

ABSTRACT: The eddy current displacement sensor is an analytical apparatus in the Turbine Supervisory Instrumentation system which can monitor rotating machinery features like vibration. It has abilities of good reliability, fast speed response and strong anti-interference. However, due to the limitations of measured dimensions and assembly space in the expansion error measuring procedures, a new kind of sensor, which has larger linear range but the same probe external diameter, need to be developed. In this paper, a new eddy current displacement sensor with large linear range is designed based on optimizations of measuring circuit. Theoretical analysis and experiments are also given in the paper.

1 INTRODUCTION

Eddy current sensor is widely used at the practical stage in various fields because of its many advantages, such as long-term reliability, wide measuring range, high sensitivity, high resolution, rapid response speed, anti-interference ability, simple structure. However, it also has its own corresponding disadvantage. It imposes restrictions on many aspects, including material selection, product development and the use in the development process. For example, the range limitation is one of its disadvantage. The domestic eddy sensor is only used in monitoring of radial and axial offset of rotating machinery with smaller displacement. Measurement range is small because it doesn't resolve the critical factors, including coil diameter, the signal processing circuit and the essence eddy current effect is nonlinear, it makes that measurement range is maintained at 1/5~1/3 the diameter of the probe.

In consideration of the shortcoming of the limitation in eddy current sensor, the paper used the way of the combination of model design and experiment. It used the technical solution that expands the eddy current sensor range by using nonlinear compensation circuit.

In this paper, expanding diameter 25 mm eddy current sensor linear range is the research target. Under the premise of maintaining the same outer diameter, it carried out research work in many aspects, in order to solve the limitation issues of eddy current sensor in the range of engineering applications.

2 CIRCUIT DESIGN AND SIMULATION OF EDDY CURRENT SENSOR

Measuring circuit of eddy current sensor makes the changeable equivalent inductance, equivalent impedance and quality factor convert into a standard output voltage or current (Yi & Zhu 1998). For three common ways of measuring conversion circuit, the performance of constant carrier amplitude modulation method of quartz crystal oscillator is best in stability. It can eliminate small changes of output variables caused by instability of frequency. Amplitude modulation and frequency modulation which their carrier frequencies are changeable are higher than constant carrier amplitude modulation method in sensitivity. The amplitude modulation method which carrier frequency is changeable is better than constant carrier amplitude modulation in the linear range of the measurement. Therefore, the paper used the amplitude modulation method because of its high sensitivity and linear range of the design requirements.

2.1 *Measuring circuit design*

Frequency modulation measurement circuit needs to make the sensor coil connect to capacitance connecting three point type oscillator, which constitutes the parallel resonant circuit. It is shown in Figure 1. When there is no measured conductor, the resonant frequency of the circuit is f_0 and the output voltage is U_0. When the measured conductor is closed to the sensor coil, the equivalent inductance of the coil will change. It will result in changes in the resonant frequency of the oscillator and resonance curve will move around and become flat. At this time, the output voltage waveform of oscillation circuit which is composed of the sensor circuit not only has some changes in frequency, but also has some changes in magnitude. This measurement method which directly takes the output voltage amplitude as the output is called amplitude modulation method which carrier frequency is changeable.

According to the principle of the amplitude modulation method, measuring circuit structure which realizes of the signal conversion is shown in Figure 2. It consists of the following parts.

2.2 *Oscillator*

The frequency amplitude measurement circuit which the paper designed is requested to let the exploring coil as inductance have access to oscillation circuit, and then it generates stable sine wave. Inductance parameter variation of the exploring coil is a direct result of changes in the output amplitude of the oscillation circuit. Therefore, the capacitor three point type oscillation circuits is best choice, because of its outstanding advantages of high sensitivity, good stability. The structure of the circuit is shown in Figure 3. The circuit consists of three parts. The first part is the transistor amplifier which plays the role of energy amplification. The second part is the positive feedback network which consists of three point type loop. The last part is frequency selective network which is used to determine the oscillation frequency. The circuit amplitude vibrating condition is AF>1. Phase vibration condition is $\varphi_A + \varphi_F = 2n\pi$.

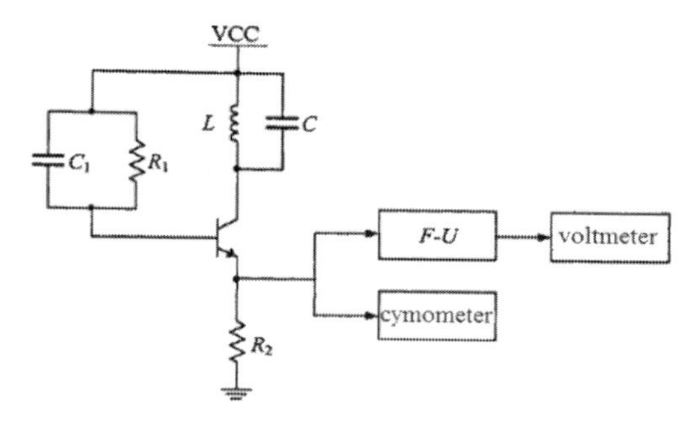

Figure 1. Frequency measurement circuit schematic.

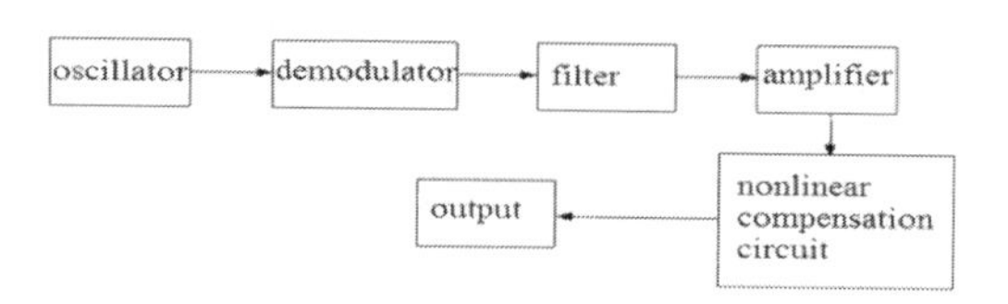

Figure 2. Measuring circuit structure.

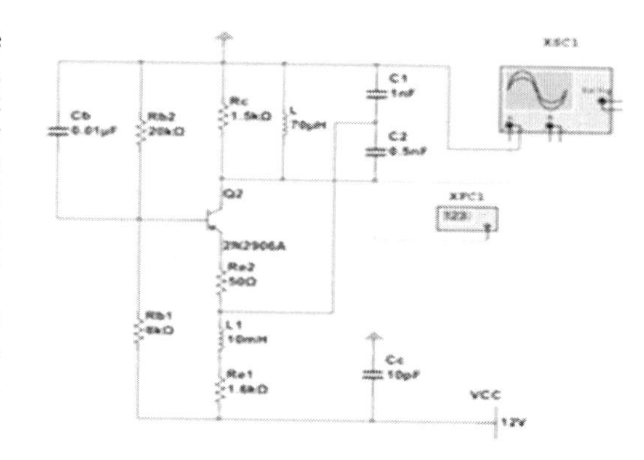

Figure 3. Three-point capacitance oscillator circuit.

The circuit was built and debugged in Multisim 12. After appropriate adjustments for each parameters, it obtained a stable sine wave. The output results of frequency counter and oscilloscope are shown in Figures 4 and 5.

2.3 *Demodulator and filter*

The output of eddy current sensor is direct current signal. Therefore, using the demodulator and the filter can complete this function (Wang & Zhang 2008). The function of the demodulator is abstract the low-frequency signal from high-frequency signal to achieve half-wave rectification according to the principle of unidirectional electric conductivity of diode. The filter is a device with frequency selective effect. It can make certain frequency signals through and the other frequency signals are blocked or decay. According to the selection of the excitation frequency, in order to get a direct current value of which the size is changing with the input signal frequency, it is needed to design a frequency of 10 KHz low-pass filter.

Commonly used detection circuit structure of charging and discharging is shown in Figure 6. In order to make sure the safety of the diode, the maximum average current and maximum reverse voltage should meet the following requirements, namely Formula (1):

$$\begin{cases} I_F > 1.1\dfrac{\sqrt{2}U_2}{\Pi R_L} \\ U_{\mathrm{RM}} > 1.1\sqrt{2}U_2 \end{cases} \tag{1}$$

Figure 7 is a schematic diagram of a low-pass filter. R_1 is a driver impedance. R_2 is a terminating impedance. The filter circuit is composed of three reactance elements L_2, C_2, C_3. While determining its parameters, it is needed to determine the values of fixed inductors. The values of each element are determined in accordance with the Formula (2).

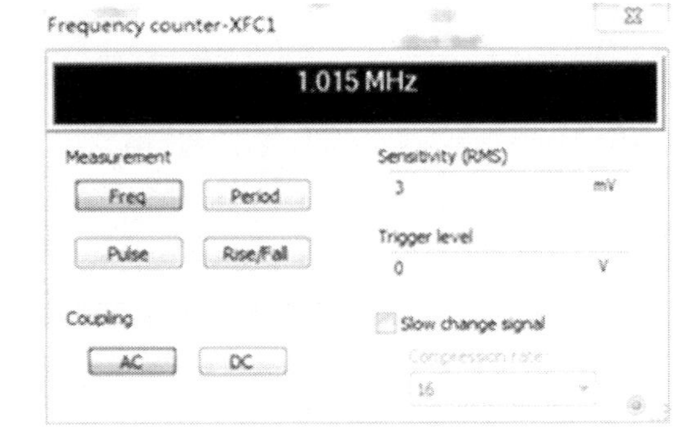

Figure 4. Frequency counter display results.

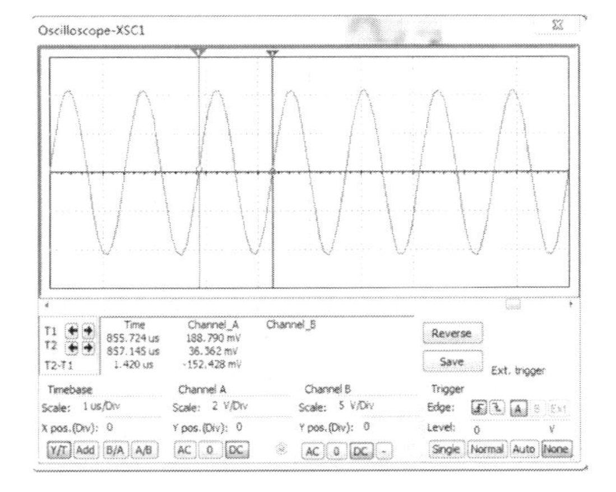

Figure 5. Oscilloscope display results.

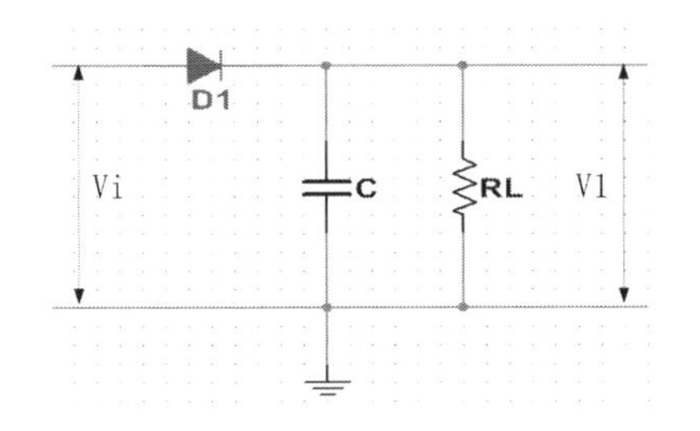

Figure 6. Diode detection circuit.

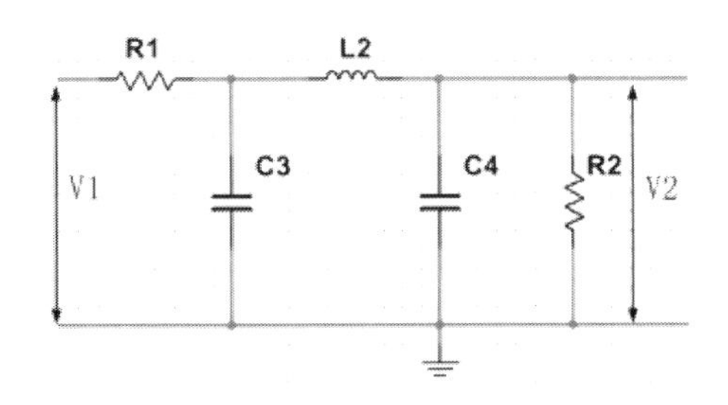

Figure 7. Low pass filter.

$$\begin{cases} 2\Pi f L_2 = R \\ R_1 = R_2 = R \\ C = \dfrac{1}{2\Pi f R} \\ C_3 = C_4 = C \end{cases} \tag{2}$$

The amplitude frequency characteristics are shown in Figure 8. According to the analysis from the figures, the longitudinal coordinates corresponding to the ordinate is –6.021 dB. Adding to the corresponding to the cut-off frequency. The corresponding ordinate is –9.021 dB. The cut-off frequency which is shown in Figure 8 is 10.085 KHz, it is consistent with the theoretical calculation value.

2.4 *Amplifier*

The eddy current sensor converts the displacement signal into a voltage signal, its output impedance signal is small. After detection and filtering, the voltage signal is still small, so there is a need for signal amplification processing. In order to avoid the signal saturation, difference proportion amplifier circuit is adopted which is composed of the integrated operational amplifier. The circuit schematic is shown in Figure 9.

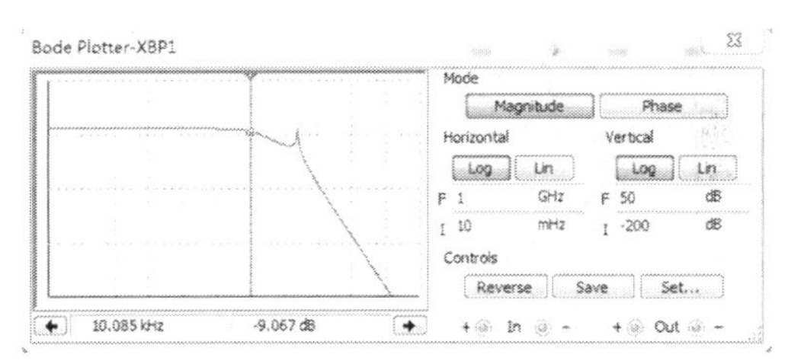

Figure 8. Filter amplitude-frequency characteristic.

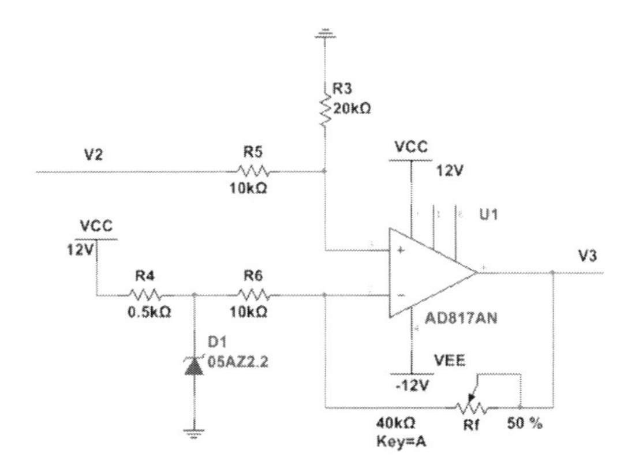

Figure 9. Amplifying circuit schematic.

The expression of input–output operation circuit is

$$V_3 = \left(1 + \frac{R_f}{R_6}\right)\frac{V_2 R_3}{R_5 + R_3} - \frac{R_f}{R_6} V_d \tag{3}$$

Selecting parameter symmetry, namely, $R_5 = R_6 = R$, $R_3 = R_f$ then

$$V_3 = \frac{R_f}{R_6}(V_1 - V_d) \tag{4}$$

By adjusting the feedback resistance value of R_f, amplification factor of the circuit is adjustable within 1–2 times.

2.5 *Nonlinear compensation circuit*

According to the eddy current sensor output impedance characteristics, the output has serious nonlinear. Therefore, it is needed to design a linear device to compensate (Yang & Peng 1999). It can make the output of the sensor have linear feature and expand its linear range and measurement range. In order to obtain good linear compensation and require real-time, simplicity and economy, the paper used the function compensation method as the foundation. The nonlinear compensation schematic is shown in Figure 10.

X is the measured displacement in the figure, v_1 is the output of the sensor, x has a nonlinear relationship with v_1, v_0 is the output after compensation. Using the correct link of nonlinear to eliminate nonlinear of eddy current sensor. Eventually, it makes the output of the measurement circuit and x have a linear relationship.

Assume that the sensor output is

$$v_1 = f(x) \tag{5}$$

The output of the expected measurement circuit is

$$v_0 = sx + b \tag{6}$$

Both s and b are constants in the formula.

Combing the Formula (5) with Formula (6), the relationship expression between input and output of linear is

$$v_0 = sf^{-1}(v_1) \tag{7}$$

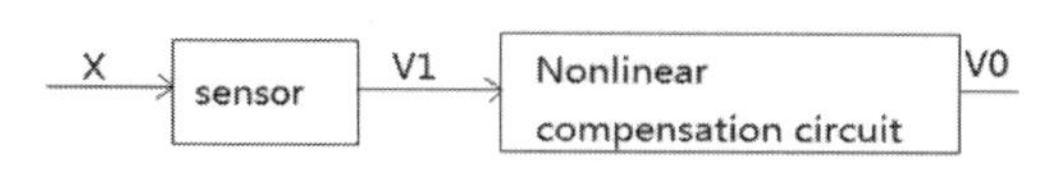

Figure 10. Nonlinear compensation circuit schematic.

According to the principle of nonlinear compensation, it is needed to calculate the inverse function of sensor output. Firstly, the paper used the circuit simulation of sensor measurement circuit to obtain the output characteristic curve, then regarded voltage as the input, regarded displacement as the output and conducted cubic polynomial function fitting (Wang & Zhang 2005). The fitting results from MATLABE are shown in Figure 11 and it gives the function about $x = f^{-1}(V_1)$.

$$x = 0.28v_1^3 - 1.56v_1^2 + 3.26v_1 + 2.05 \tag{8}$$

The output of the measurement circuit is

$$v_0 = 0.4x + 0.78 \tag{9}$$

Therefore the designed linearizer input-output relationship is

$$v_0 = 0.11v_1^3 - 0.62v_1^2 + 1.3v_1 + 1.60 \tag{10}$$

The paper designed the compensation circuit, according the Formula (10), it is shown in Figure 12. The circuit consists of operational amplifiers and a

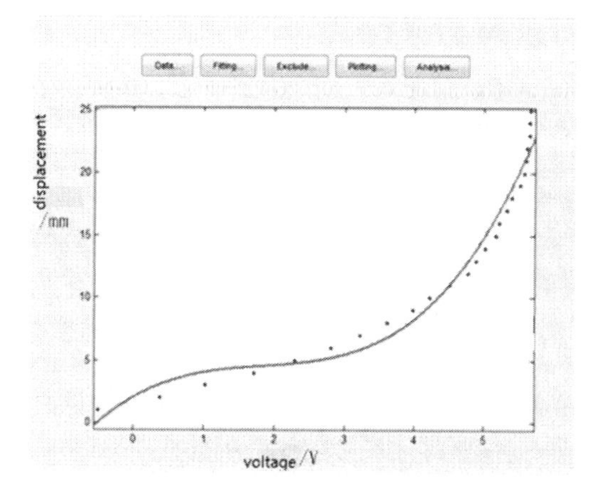

Figure 11. Voltage-displacement fitting curve.

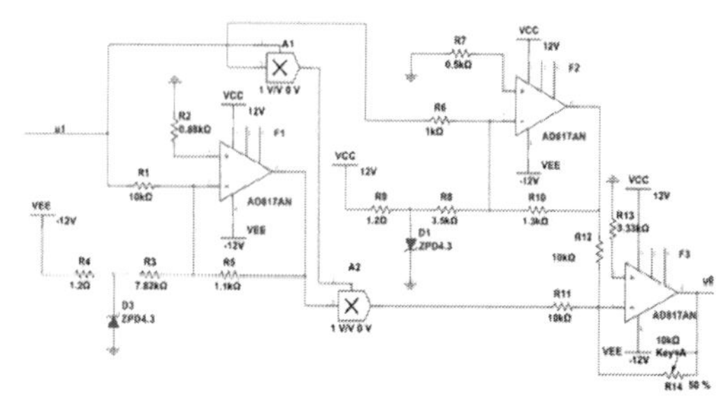

Figure 12. Nonlinear compensation circuit.

multiplier. The first amplifier F_1 implemented three times and quadratic term coefficient. The second amplifier F_2 implemented first term and constant coefficient. The last amplifier F_3 magnification time is one, it can inverse the output signal. The input-output experience of the Figure 12 is

$$u_0 = -R_{14}\left[-\frac{R_5}{R_{11}}\left(\frac{u_1}{R_1}+\frac{u_{D3}}{R_3}\right)u_1^2 - \frac{R_{10}}{R_{12}}\left(\frac{u_1}{R_6}+\frac{u_{D1}}{R_8}\right)\right] \tag{11}$$

2.6 *The eddy current sensor simulation experiment*

Hierarchical circuit design method divides a large complex project into several sub parts. It is advantageous for the design and management. Multisim12 simulation software (Xing et al. 2005) supports this level circuit design method, it can improve the system design of the modular and hierarchical. It can also increase the readability of the circuit design, improve the design efficiency, and reduce the cycle of the circuit design. After creating the module function circuit which designed in chapter 2.2 to 2.4 into the sub circuits. The paper set up the top circuit of eddy current sensor measurement circuit which is shown in Figure 13.

After conducting the test for equivalent resistance and equivalent inductance on the eddy current sensor impedance test experiment platform, the results of the experiment are shown in Table 1.

In the circuit testing, eddy current sensor probe can be equivalent to the concatenation of equivalent resistance and equivalent inductance. Connecting sensor probe equivalent components to the oscillator, according to values of equivalent resistance and equivalent inductance which the Table 1 recorded, using the top circuit of the eddy current sensor to simulate which Figure 13 designed, the displacement-voltage curve is shown in Figure 14. At the same time the output value after compensation needs a linear fitting, it is shown in Figure 15.

The linear equation after linear fitting is

$$u_0 = 0.417x + 0.526$$

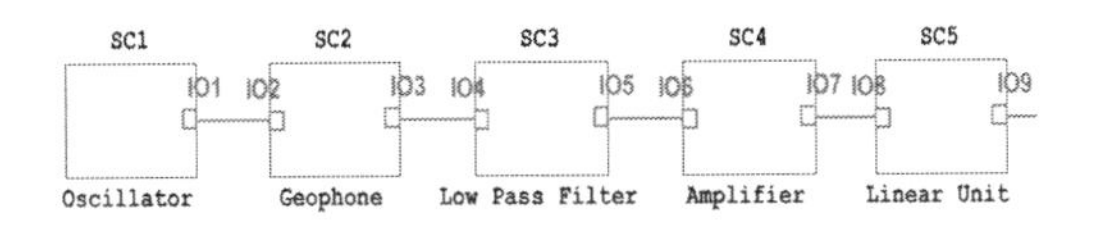

Figure 13. Sensor measure the top circuit.

Table 1. The coil impedance test data.

Displacement/mm	Equivalent resistance/Ω	Equivalent inductance/uh
3	32.10	68.17
4	28.84	70.33
5	26.60	71.94
6	24.97	73.15
7	23.80	74.10
8	22.97	74.81
9	22.34	75.38
10	21.87	75.83
11	21.52	76.18
12	21.25	76.47
13	21.05	76.70
14	20.88	76.89
15	20.75	77.05
16	20.66	77.18
17	20.58	77.29
18	20.51	77.38
19	20.46	77.46
20	20.43	77.53
21	20.39	77.58
22	20.36	77.63
23	20.34	77.67
24	20.32	77.71
25	20.31	77.74

Table 2. The eddy current sensor simulation results.

Displacement/mm	Output voltage before compensation/V	Output voltage after compensation/V
3	1.02	2.01
4	1.72	2.60
5	2.29	2.72
6	2.81	2.91
7	3.22	3.18
8	3.61	3.58
9	3.97	4.11
10	4.26	4.58
11	4.52	5.27
12	4.27	5.97
13	4.90	6.35
14	5.04	6.79
15	5.15	7.39
16	5.25	7.55
17	5.34	7.99
18	5.44	8.30
19	5.50	8.84
20	5.56	9.13
21	5.59	9.22
22	5.61	9.27
23	5.64	9.42
24	5.65	9.47
25	5.66	9.52

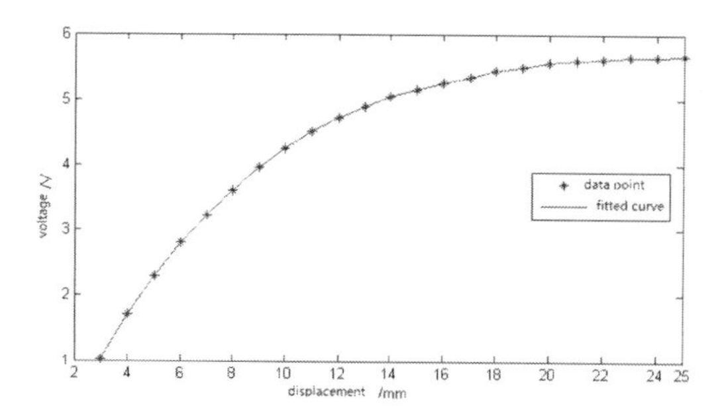

Figure 14. Eddy current sensor output before compensation.

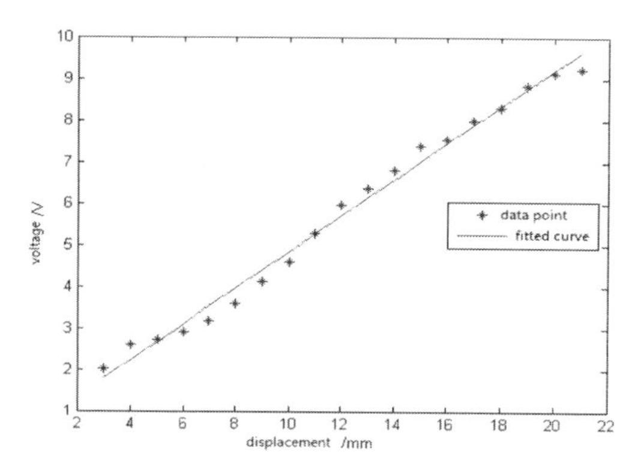

Figure 15. The linear output of the eddy current sensor.

The sensitivity of the eddy current sensor is

$$S = \frac{du_0}{d_x} = 0.417v/mm$$

And the relative error between the expected designed sensitivity and the sensitivity of the experiment is

$$\gamma = \frac{s - s_0}{s_0} \times 100\% = 4.25\%$$

The eddy current sensor nonlinear error is

$$\delta = \pm\frac{\delta_{max}}{U_N} = \pm 4.82\%$$

The linear range of the eddy current sensor is up to 19 mm through the nonlinear compensation, and it achieved the goal of expanding sensor linear range. At the same time, nonlinear error is less than 5% of the prescribed.

3 CONCLUSIONS

Measurement circuit is a key part of realization of displacement measurement based on eddy current sensor, it is also a key link in the process of expanding its linear measurement range. Based on the principle of variable frequency amplitude modulation measurement circuit, the paper used modular design method, designed and implemented the functional stable, simple structure of measurement circuit. In Multisim12 simulation software, through continuous debugging and modification, the expected results of measurement circuit are acquired. Using the optimized probe coil impedance parameters, the paper conducted comprehensive simulation of eddy current sensor measurement circuit and drawn conclusion that linear rang of the eddy current sensor is expanded from 7 mm to 19 mm, the nonlinear error of eddy current sensor is 4.82%, and the sensitivity is 0.4 v/mm. These datas conform the national standards.

REFERENCES

Wang, G. & Zhang, C. 2005. Eddy current sensor and its best characteristic curve fitting equation. *Instrument and application*, 3: 81–83.

Wang, H.X. & Zhang, S.Y. 2008. *Principles and Applications of Transducer*. Tianjin: Tianjin University Press, 87–92.

Xing, W. & Liang, Q. et al. 2005. *Multisim 7 circuit design and simulation applications*. Beijing: Tsinghua university press, 22–93.

Yang, F.Q. & Peng, G.M. 1999. Nonlinear compensation of sensors. *Mining research and development*, 19(4): 36–37.

Yi, J.Y. & Zhu, H.Z. 1998. Eddy current sensor linear circuit research in a wide range. *Instrument technology and instruments*, 6: 27–30.

Advanced Materials, Mechanical and Structural Engineering – Hong, Seo & Moon (Eds)
© 2016 Taylor & Francis Group, London, ISBN: 978-1-138-02908-8

A study on the 3D rebar placing automatize system development

H.S. Jang, S.Y. Kim & D.E. Kim
Department of Architectural Engineering, Hanyang University, Seoul, Korea

Y.S. Cho
Department of Architectural Engineering, Hanyang University, Ansan, Korea

ABSTRACT: This study was designed to develop a 3D rebar placing automatize system utilizing functions provided by BIM software for the continuous beam of RC structures that prioritizes the development among simple repetitive and labor intensive tasks in the field of architectural structures. Through the 3D rebar placing automatize system, it seeks to improve the functionality and productivity of BIM tools in the field of architectural structures, and thus to broaden the range of BIM applications. The development system is linked with the Open API on TS by converting the member properties, parameters and RC design values through the external module included in the development system based on the RC design DB and parameter DB constructed from the structural analysis and design results. Through this, it has the function to generate rebar placing models in which the structural models, splice and anchorage are reflected automatically.

1 INTRODUCTION

1.1 *Research background and purpose*

The field of architectural structures has still stuck to the conventional structural calculation method of calculating the stresses in members by constructing a 3D model (Lee 2010), and its application has been limited to the frame model due to a variety of reasons, such as structural characteristics, functional limitations of BIM tools and low productivity. In addition, the application of BIM has also been made only within the limited range of image support and clash detection through the conversion of 2D drawings into 3D ones (Eom et al. 2014). On the contrary, it is now being generalized in the plant structures that check for clashes with pipes or other equipment is conducted, or shop drawing for production is created through a 3D model (Lee 2010). In this regard, this study sought to develop a 3D rebar placing automatize system utilizing functions provided by BIM software for the continuous beam of RC structures that prioritizes the development among simple repetitive and labor intensive tasks in the field of architectural structures. Through the system, it attempts to improve the functionality and productivity of BIM tools in the field of architectural structures, and thus to broaden the range of BIM applications.

1.2 *Research trends*

BIM-based automatic rebar placement system enables the review of tasks related to the rebar placement that requires a lot of time and manpower to proceed by generating the rebar placement automatically before conducting the rebar work of building structures (MOLIT. 2013). If the automation technology is used, the structural performance can be secured by generating a 3D shape and location of the rebar automatically and calculating the proper splice and development length of the rebar.

Table 1. Development length-in the case of normal-weight aggregate concrete without being coated.

Rebar of D22 over (l_d)

Location	f차 (MPa)	f_y = 300	f_y = 350	f_y = 400	f_y = 500	f_y = 600
Bottom	21	$39.3d_b$	$45.8d_b$	$52.4d_b$	$65.5d_b$	$78.6d_b$
	24	$36.7d_b$	$42.9d_b$	$49.0d_b$	$61.2d_b$	$73.5d_b$
	27	$34.6d_b$	$40.4d_b$	$46.2d_b$	$57.7d_b$	$69.3d_b$
	30	$32.9d_b$	$38.3d_b$	$43.8d_b$	$54.8d_b$	$65.7d_b$
	35	$30.4d_b$	$35.3d_b$	$40.6d_b$	$50.7d_b$	$60.9d_b$
	40	$28.5d_b$	$33.2d_b$	$37.9d_b$	$47.4d_b$	$56.9d_b$
	45	$26.8d_b$	$31.3d_b$	$35.8d_b$	$44.7d_b$	$53.7d_b$
	50	$25.5d_b$	$29.7d_b$	$33.9d_b$	$42.4d_b$	$50.9d_b$
Top	21	$51.09d_b$	$59.54d_b$	$68.12d_b$	$85.15d_b$	$102.18d_b$
	24	$47.71d_b$	$55.77d_b$	$63.7d_b$	$79.56d_b$	$95.55d_b$
	27	$44.98d_b$	$52.52d_b$	$60.06d_b$	$75.01d_b$	$90.09d_b$
	30	$42.77d_b$	$49.79d_b$	$56.94d_b$	$71.24d_b$	$85.41d_b$
	35	$39.52d_b$	$45.89d_b$	$52.78d_b$	$65.91d_b$	$79.17d_b$
	40	$37.05d_b$	$43.16d_b$	$49.27d_b$	$61.62d_b$	$73.97d_b$
	45	$34.84d_b$	$40.69d_b$	$46.54d_b$	$58.11d_b$	$69.81d_b$
	50	$33.15d_b$	$38.61d_b$	$44.07d_b$	$55.12d_b$	$66.17d_b$

※ where, l_d: Development length; d_b: Nominal diameter (mm).

And since it is possible to perform detailed design on the rebar placement before construction, efficient construction management can be done as the rebar quantity is automatically updated even in the case of design changes due to the accurate calculation of the rebar quantity or change in site conditions. In addition, as the improvement of productivity due to the automation of design work leads to the reduction of design costs by more than 50%, and the minimization of reconstruction resulting from design errors brings about beneficial economic effects by shortening the construction period up to 5% (MOLIT. 2013), extensive research on the BIM-based automatic rebar placement is being conducted at home and abroad to reduce the cost of the rebar work.

2 THEORETICAL BACKGROUND

2.1 *Theory on the calculation of development length*

For calculation of development length, the DB for each development length of top reinforcement and bottom reinforcement was constructed by reflecting the table of Table 1 which is generally applied in practice.

2.2 *Theory on the calculation of anchorage of 90° standard hooks*

In the case of outer parts in which beams come in contact with lateral columns, if the tension steel is not extended as long as the required development length due to the limitation of the width of the column, hooks can be installed to have additional bond stress. According to KCI 8.2.5, if the horizontal length of the rebar in which a standard hook is installed is longer than the required development length l_{dh} of the hook proposed in the structural criteria, it is considered to be safely anchored. The calculation formula is shown in Equation (1).

$$l_{hb} = l_{hb} \times CorrectionFactor \geq (8d_b \ or \ 150mm) \quad (1)$$

where, $l_{hb} = \left(\frac{0.24\beta f_u}{\lambda\sqrt{f_{ck}}}\right)$; L_{hb}: development length; β: Coefficients of Reinforced coating (KCI 8.2.2(2)); Correction factor: (KCI 8.2.5(3)).

In addition, 90° standard hooks are calculated by extending more than 12db at the end of the semicircle bent as shown in Figure 1.

2.3 *Theory on the calculation of tension lap splice*

The continuous rebar was placed after calculating the class B tension lap splice length based on KCI 8.6.2. As for the splice position of the rebar, it is required that the number of splices should not exceed more than half in the same location even if the splice is not possible in the recommended splice section as shown in Figure 2 based on the KSEA Detailing Manual-2009.

2.4 *Double cut bar cutting point*

KSEA Detailing Manual-2009 specifies that as shown in Figure 3, the main rebar should be pushed as far as 0.25(1/4) of the span into the interior of the beam in the case of the top of the outer end, and as far as 0.3(3/10) of the span at the top of the inner end. It is also specified that if the lengths of the spans on both sides are different, centering on the column in the inner end, the lengths should be calculated based on the larger span value. Since the tension occurs into the lower part in the center of the beam, the rebar quantity at the bottom is generally increased in the central part rather than end part, and therefore it requires that the position in which the rebar in the central part is cut should be 0.125(1/8) times the span by reflecting the size of the bending moment.

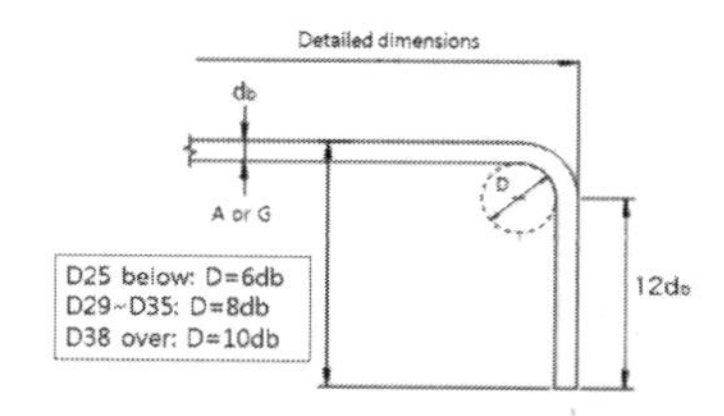

Figure 1. 90° Standard hooks (A: Start, G: End).

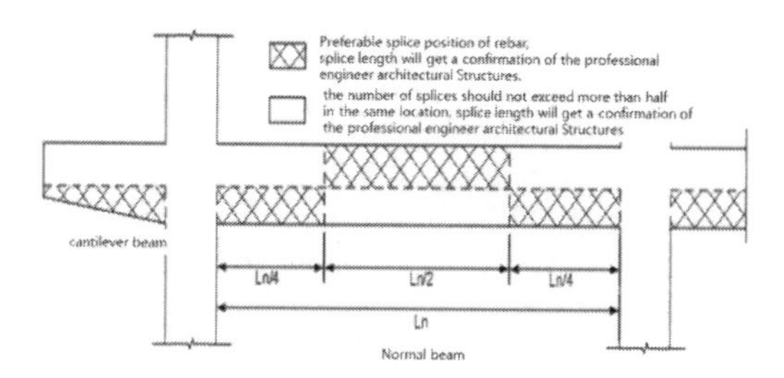

Figure 2. Splice position of beam reinforcement.

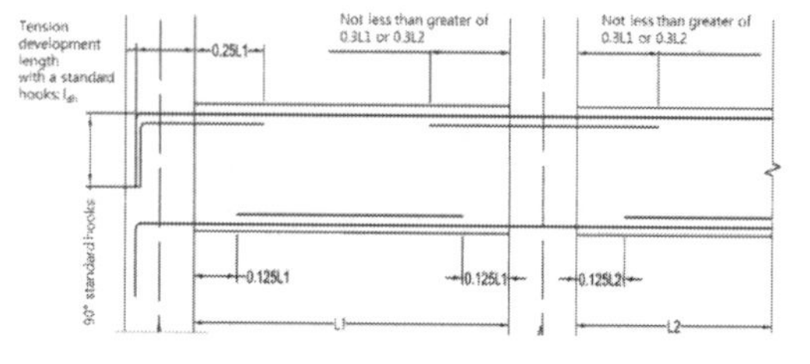

Figure 3. Rebar detailing of beam main bar.

3 DEVELOPMENT OF 3D REBAR PLACING AUTOMATIZE SYSTEM

3.1 *Information compatibility system*

Figure 4 shows the system for information compatibility between each software to construct a 3D rebar placing automatize system (hereafter referred to as 3DRPAS) through the utilization of structural BIM data. The 3DRPAS developed to conduct the structural detailed modeling performs the process of accepting and converting the data from anchorage and splice DB, mgt of Midas GEN (hereafter referred to as MG) and txt of BeST. Pro (hereafter referred to as BP). The converted data is implemented on TS in a direct connection method through the Open API of Tekla Structures (hereafter referred to as TS).

3.2 *Process model*

The process of 3DRPAS includes an external module developed to perform a series of processes from the structural analysis to the rebar modeling. The external module was developed in Visual C# (DotNET Framework) environment by using the Open API provided on TS.

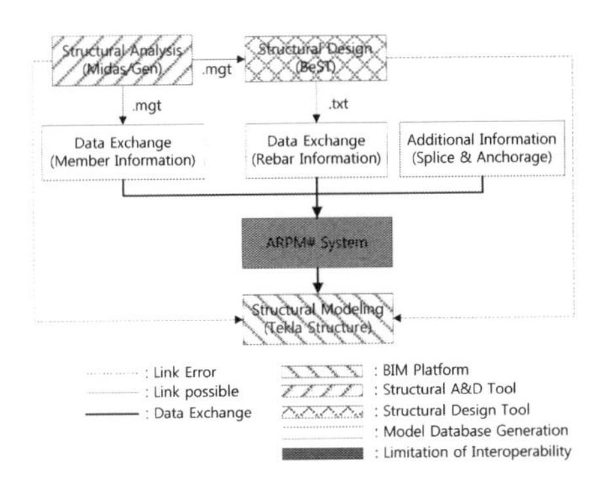

Figure 4. 3D RPAS information compatibility system.

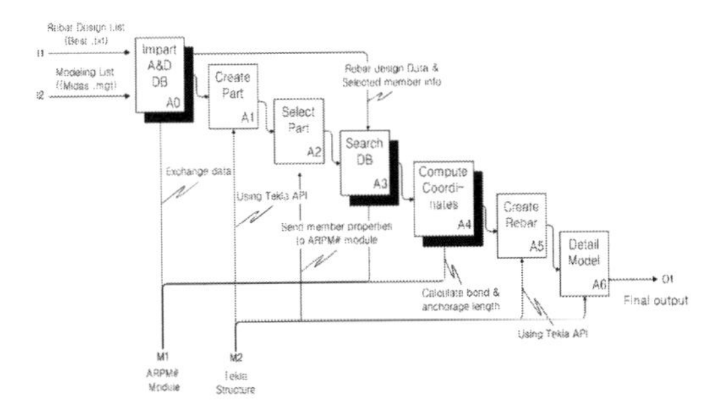

Figure 5. External module process.

Figure 5 shows the process model of the external module to which IDEFØ modeling technique is applied, which consists of seven activities (A0~A6), two system environments (M1~M2), two pieces of input information (I1~I2) and the final output (O1), and the contents by component are as follows.

4 CASE STUDY

4.1 *Selection of the target project*

In order to verify the applicability of the developed system, a dual-frame structure made of shear walls and reinforced concrete frame with twelve stories above ground as shown in Figure 6(a) was selected as the target project, and the design overview is shown in Table 2. The continuous beam was selected as the subject for application.

4.2 *Structural analysis phase*

A structural analysis model was produced in the first phase of the system application. After the production of the model, the structural analysis and member design suitable for the characteristics of the RC object are performed to review

Table 2. Structural overview.

Design overview	Contents
Location	Seoul
Basic frame	Reinforced Concrete frame
Design Code	KBC 2012 KCI-USD12 KSEA Detailing Manual-2009
Specified compressive strength	$f_{ck} = 24$ MPa
Rebar yield strength	KSD 3504 SD40 $f_y = 400$ MPa

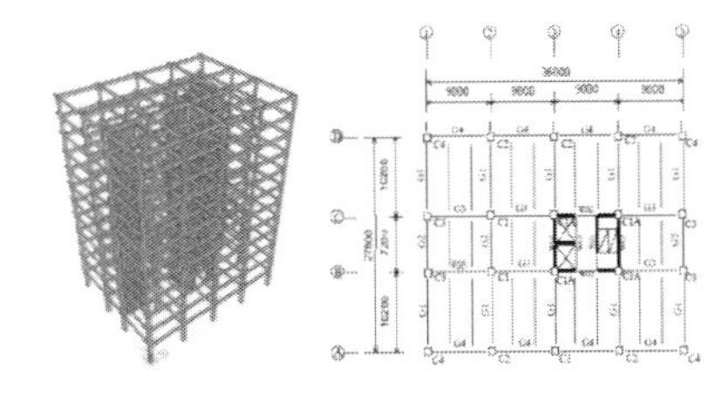

(a) Basic structural model (b) Structure floor plan

Figure 6. Structural analysis model for the target building and structure floor plan.

the structural stability and usability, and then the results of the member design, geometric information and information on member forces, including moment and shear force are generated. MG makes it possible to put out the values of analysis results and all the information generated in the process of the analysis and design of the target model in the form of a MGT file as shown in Figure 7. The MGT file is utilized as DB for automatic geometric modeling and RC design.

4.3 *RC design phase*

BP was applied in the RC design phase. As shown in Figure 8, an automatic collective design for beam placement by floor group can be made by importing a mgt file in which the structural analysis results of MG are included through the direct connection method on BP.

4.4 *3DRPAS implementation*

In the phase of the rebar detailing model construction, modeling is automatically conducted on TS by being linked with the design result DB of BP

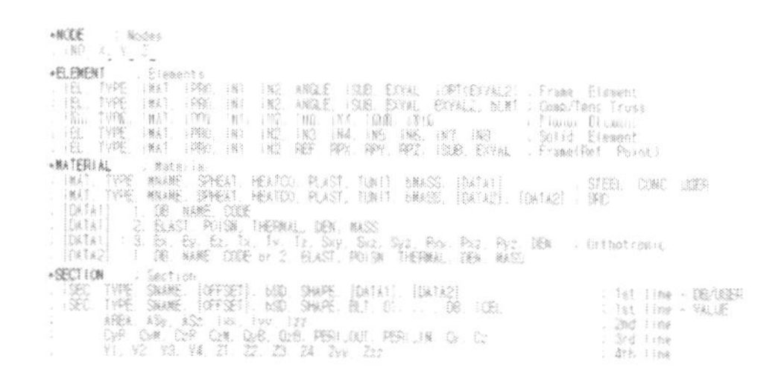

Figure 7. MGT data information.

```
batch Beam - RC-Beam Design      [ KCI-USD12 ]
================================================================
*fck = 24, fy = 400, fys = 400 N/mm2

*NAME = G1, Story = 3F
*B = 400,   H = 700
----------------------------------------------------------------
POS : Top-Bar Bot-Bar  S-Bar     |  Mu_neg  Ratio |  Mu_pos  Ratio |   Vu    Ratio
----------------------------------------------------------------
Int   7-D25,  3-D25, 2-D13@150 |  -595.1  0.894 |  157.3  0.490 |  260.1  0.529
Mid   3-D25,  3-D25, 2-D13@300 |  -127.1  0.411 |  266.9  0.864
Ext   7-D25,  3-D25, 2-D13@150 |  -585.8  0.880 |  173.2  0.540 |  257.3  0.523
----------------------------------------------------------------
```

Figure 8. BP reinforcement design information (txt).

(a) Imported DB (b) Implementation Rebar modeling

Figure 9. 3DRPAS implementation.

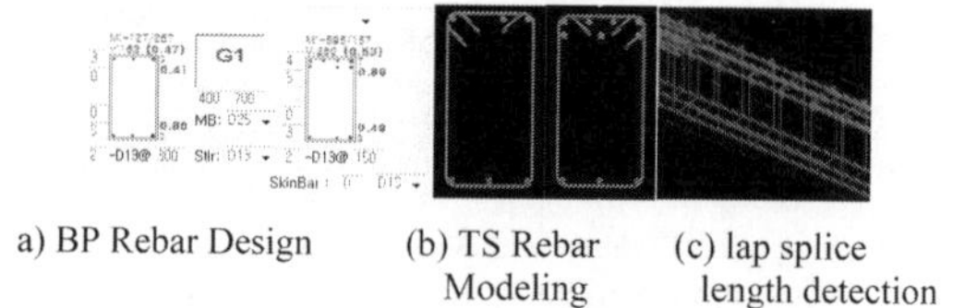

a) BP Rebar Design (b) TS Rebar Modeling (c) lap splice length detection

Figure 10. Evaluation on the reflection of RC design values.

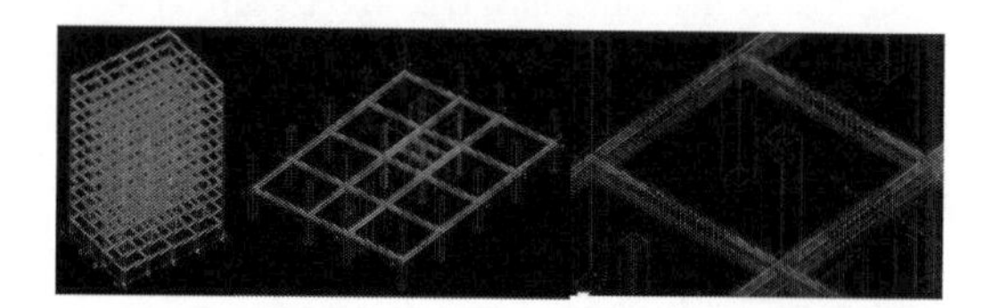

Figure 11. The final rebar details on TS model.

through 3DRPAS. As shown in Figure 9(a), DB of BP can be confirmed by importing BP design result values in txt format on 3DRPAS, and the modeling of the rebar detailing in consideration of anchorage and splice is automatically performed as shown in Figure 9(b).

Figure 10 compares the results of automatic rebar placement performed on TS through 3DRPAS and G1 designed on BP. Through 3DRPAS, the implemented RC design result was found to be the same as that of BP, and the placement of standard hooks, splice length and double cut bar could be confirmed through the rebar placement results on TS.

Figure 11 shows the results of 3DRPAS performed through 3DRPAS in which the rebar placement information considering anchorage and splice with respect to all the beam members of the target project on TS is automatically generated. As a result, this process can contribute to minimizing the waste of time and money as well as information errors caused by the repetitive tasks, improving the work efficiency and proceeding with the flow of the work smoothly.

5 CONCLUSIONS

This study was designed to develop a 3D rebar placing automatize system utilizing functions provided by BIM S/W, and the contents of the study can be summarized as follows.

1. The developed system is aimed at RC continuous beams and has functions of automatically

generating the model to which member properties, parameters and RC design values are applied on TS based on the bar list & schedule DB and member parameter DB constructed from the structural analysis and design results and performing the anchorage and splice suitable for the characteristics of the continuous beam.
2. 3DRPAS was equipped with an external module that can convert result values of each S/W to improve the interoperability between BIM softwares, and the converted result values are used to perform a 3D rebar placement by being linked with the Open API on TS.
3. An evaluation on the applicability was performed with the rebar placing model, and it was confirmed that the consideration of splice, anchorage and rebar spacing arrangement based on the initially planned algorithm was properly reflected in the automatic rebar placement.

ACKNOWLEDGEMENT

This work was supported by the National Research Foundation of Korea (NRF) grant funded by the Korea government (MEST) (NRF-2013R1 A1 A2012236).

REFERENCES

Lee, B.H. 2010. BIM Application for the Structural Engineering Field in Korea, *Journal of the Architectural Institute of Korea*, 54(1): 52–57.

Eom, J.U., Lee, J.H., Kim, J.H. & Choi, I.S. 2014. Introduce of Automated Modeling System for RC based on BIM, *Journal of the Computational Structural Engineering Institute of Korea*, 27(3): 18–24.

Ministry of Land. 2013. Rebar Detail Design and Construction Management Based on BIM.

Advanced Materials, Mechanical and Structural Engineering – Hong, Seo & Moon (Eds)
© 2016 Taylor & Francis Group, London, ISBN: 978-1-138-02908-8

An electrolyte with high thermal stability for the vanadium redox flow battery

D. Kim & J. Jeon
Department of Energy and Advanced Material Engineering, Dongguk University, Seoul, Korea

ABSTRACT: Vanadium Redox flow Battery (VRB) cannot be used in the widespread environment because the problem of generating V_2O_5 precipitation has restricted to its industrialization. To overcome this problem, this paper describes the thermally stable vanadium positive electrolyte using sodium oxalate as a supporting material for the VRB. To show the effectiveness of the proposed solution, thermal stability experiment at 60°C, viscosity measurement and electrochemical analysis (i.e., cyclic voltammetry, charge–discharge test, and scanning electron microscopy) were carried out, respectively. The experimental results indicate that the new electrolyte improves the charging and discharging capacity by about more than 18% for 30 cycles when compared with the pristine electrolyte, and can provide a higher thermal stability for the long-time operation of the VRB. Consequently, sodium oxalate is shown to provide an electrolyte–additive solution for high-temperature operations of the VRB.

1 INTRODUCTION

Vanadium Redox flow Battery (VRB) is a promising technology as an efficient Energy Storage System (ESS) for a wide range of applications such as large-scale, renewable and grid energy storage (Fabjan et al. 2001, Joerissen et al. 2004, Tokuda et al. 2000, Chakrabarti et al. 2007, Skyllas–Kazacos et al. 2011). By employing V (II)/V (III) and V (IV)/V (V) redox couples as negative and positive electrolytes, respectively, with sulfuric acid as the supporting electrolyte, the output power and capacity of the VRB are dependent on the volume and concentration of the electrolytes (Wu 2012). During the charging and discharging of the VRB, electrochemical reactions within the battery change the balance of the vanadium ions, respectively, (Faizur & Skyllas–Kazacos 2009) at the positive electrode:

$$VO_2^{+} + 2H^{+} + e \underset{\text{charge}}{\overset{\text{discharge}}{\rightleftharpoons}} VO^{2+} + H_2O E^{\circ} = 1.0V \quad (1)$$

at the negative electrode:

$$V^{3+} + e \underset{\text{discharge}}{\overset{\text{charge}}{\rightleftharpoons}} V^{2+} E^{\circ} = -0.26 \quad (2)$$

However, to date, the VRB cannot be used in the widespread environment because several problems have restricted its industrialization. For example, the positive electrolyte tends to generate V_2O_5 precipitation, which deposits on the carbon felt electrode and blocks the separator surface (Chang et al. 2012). Especially, the V (V) electrolyte suffers from thermal precipitation above 40°C, which leads to the limited specific energy density of the VRB (Rahman & Skyllas–Kazacos 1998). This situation causes the decline in the battery performance and may even cause battery malfunction (Lu 2001).

Several studies have reported on the effective methods of preventing and delaying the precipitate formation in the positive electrolyte of the VRB, such as increasing the concentration of aqueous sulfuric acid and stabilizing agent in the electrolyte (Kazacos 1990). Although a higher concentration of sulfuric acid can prevent V_2O_5 precipitation by forming a sulfate complex or achieving increased stability, it favors the precipitation of V (II, III, and IV) ions (Rahman & Skyllas–Kazacos 1998). However, the solubility of V (IV) sulfate decreases with increasing sulfuric acid concentration due to the common ion effect. Because of the reduced solubility of the V (IV) ions in the discharged positive half-cell electrolyte, it is not possible to increase the concentration of sulfuric acid above 6.0 M (Faizur & Skyllas Kazacos 2009).

This paper focuses on improving the thermal stability of the V (V) positive half-cell solution and the different factors that affect the precipitation of V (V) species. This supporting material, which is the sodium salt of oxalic acid with the formula $Na_2C_2O_4$, acts as a precipitation-preventing agent and can be added to the positive electrolyte of the VRB. The positive electrolyte is demonstrated through the thermal stability test, Cyclic Voltammetry (CV) measurement, charge–discharge test, and Scanning Electron Microscopy (SEM) analysis.

2 EXPERIMENTAL

2.1 *Preparation of the electrolyte*

V (IV) electrolyte solutions were prepared by dissolving $VOSO_4$ (Sigma-Aldrich, USA) in a sulfuric acid solution as H_2SO_4 (Samchun Chemical, Korea). The V (II, III and V) electrolyte solutions were prepared by electric charging the V (IV) solutions in an electrolytic cell (Li et al. 2011).

2.2 *Thermal stability experiment of V (V)*

The V (V) electrolyte as the thermal stability test solution was prepared by the electric oxidization of 1.8 M $VOSO_4$ + 3.0 M H_2SO_4 in the positive electrolyte of the electrolytic cell as the catholyte. During the thermal stability test, the test solutions (4 mL of 1.8 M V (V) + 3.0 M H_2SO_4) with different additives were placed in sealed vials and stored in a water bath at a constant temperature of 60°C. The test solutions were monitored regularly for the precipitate of V_2O_5 and the time at which a slight precipitation appeared was recorded. The precipitation reaction within the positive electrolyte changed the balance of the vanadium ions (Faizur & Skyllas–Kazacos 2009) as follows:

$$2VO_2^+ + H_2O \rightarrow V_2O_5 + 2H^+ \quad (3)$$

2.3 *Viscosity measurement*

The viscosity of V (IV) solutions (16 mL of 1.8 M $VOSO_4$ + 3.0 M H_2SO_4,) was measured with a DV-E viscometer (Brookfield, USA) using the UL Adapter (Brookfield, USA), and all viscosity measurements were carried out at room temperature.

2.4 *Electrochemical measurements*

A pristine electrolyte consisted of V (IV) solutions (1.8 M V (IV) + 3.0 M H_2SO_4) without any additive, and to this pristine solution, 0.05 wt% (weight percentage of the additive to the solution) sodium oxalate was added as the thermal stability agent. All electrochemical operations were performed at the WBCS3000 electrochemical workstation (WonA tech. Korea).

Cyclic Voltammetry (CV) measurement was carried out in a three-electrode electrochemical cell with a platinum wire as the counter electrode, a Saturated Calomel Electrode (SCE) as the reference electrode and a graphite plate (surface area 0.95 cm^2) as the working electrode. The potential scanning range was from 0.4 V to 1.4 V with the scan rate of 10 mV s^{-1} at room temperature.

The charge and discharge tests of the VRB were performed in a miniature flow cell, which consisted of a Nafion 115 membrane, two pieces of graphite foil (SGL, U.S.A.), and carbon felt (Toyobo, Japan) as the separator, positive and negative electrodes, respectively. The flow cell (active area about 9 cm^2) contained 30 mL anolyte and 30 mL catholyte, and a mixture solution of 1.8 M V (III) and (IV) + 3.0 M H_2SO_4, respectively. It cyclically pumped each electrolyte in the unit cell and controlled 9 mL min^{-1} by using a BT100-2J pump (Longer Precision Pump Co., China). The cell operation was carried out at 60°C for 30 cycles with a given galvanostatic charge and discharge between 0.7 V and 1.7 V under a current density of 40 mA cm^{-2}.

The Scanning Electron Microscopy (SEM) analyzed the precipitation of V (V) species (e.g., vanadium pentoxide) on the carbon felt electrode after the charge and discharge test of the miniature cell. The surface morphology of the felt electrode was observed by using JSM-7800F (JEOL, Japan) at an acceleration voltage of 15 kV.

3 RESULTS AND DISCUSSION

3.1 *Thermal stability test*

In order to investigate the effect of sodium oxalate on the thermal stability of V (V) at 60°C, sealed vials of each concentration were placed in the water bath. As shown in Figure 1, the precipitation reaction changed the balance of the vanadium

Figure 1. The test solution without the additive (1.8 M V (V) + 3.0 M H_2SO_4): (a) before and (b) after the thermal stability test at 60°C.

ions and appearance of the test solution (1.8 M V (V) + 3.0 M H_2SO_4). The precipitation time was periodically recorded and is presented in Table 1. As observed, the precipitation time was delayed by adding the additive in the vanadium positive electrolyte. Because the additive contained ⩦O (resonance structure) groups, it was adsorbed on the surface of the nuclei, which assisted in the dispersion of V (V) ions and inhibited precipitation (Wu 2012).

3.2 *Viscosity measurement*

The effect of sodium oxalate on the viscosity of 1.8 M $VOSO_4$ + 3.0 M H_2SO_4 was measured and is presented in Table 2. Viscosity, as an important property of electrochemical solutions, can affect the electrochemical performance. An increase in the additive dose led to a decrease in the diffusion coefficient of vanadium ions, which resulted in a decrease in the electrode reaction kinetics (Wu 2012). As shown in Table 2, the initial dose indicated lower viscosity than pristine (4.51 cps) but the higher dose of the additive increased, which means that it reduces the conductivity (Wu 2012) and electrochemical behavior of the solution (Chang et al. 2012). Thus, the 0.050 wt% concentration was selected as the suitable additive dose for the vanadium positive electrolyte to delay the precipitation at 60°C.

Table 1. Effect of sodium oxalate on the thermal stability of 1.8 M V (V) + 3.0 M H_2SO_4 electrolytes.

Additive dose	Precipitation time at 60°C
0 (pristine)	2.8 hr
0.01 wt%	8.4 hr
0.02 wt%	12.8 hr
0.05 wt%	18.4 hr
0.07 wt%	21.2 hr
0.10 wt%	24.3 hr
0.20 wt%	28.7 hr

Table 2. Viscosity of 1.8 M V (IV) + 3.0 M H_2SO_4 electrolyte and with the additive.

Additive dose	Viscosity
0 (pristine)	4.51 cps
0.01 wt%	4.47 cps
0.02 wt%	4.50 cps
0.05 wt%	4.52 cps
0.07 wt%	4.61 cps
0.10 wt%	4.76 cps
0.20 wt%	4.81 cps

3.3 *Cyclic voltammetry of the positive electrolyte*

In order to investigate the effect of sodium oxalate on the electrochemical property of the positive electrolyte, the CV of the test solution with sodium oxalate (0.050 wt%) was performed for 20 scans. Figure 2 shows the average cyclic voltammograms (from the 2nd cycle to the 20th cycle) of the electrolyte, 1.8 M $VOSO_4$ + 3.0 M H_2SO_4, and with the additive. As observed, first, the potential difference (between the oxidation and reduction peak potentials) of the solutions appeared to be similar, which means that the additive could not affect the reversibility of the electrode process (Chang et al. 2012). Second, sodium oxalate had a lower ratio of average anodic peak current ($\bar{I}_{PA}$) to average cathodic peak current ($\bar{I}_{PC}$), $\bar{I}_{PA}/\bar{I}_{PC}$, which means that the additive could improve the redox reversibility of V (IV)/V (V) (Wu 2012).

3.4 *Cell tests and SEM analysis*

Miniature cell operations were performed using the pristine electrolyte and with sodium oxalate. The electrolyte with the additive had a higher average coulombic efficiency (90.1%) than the pristine electrolyte (89.8%). Figure 3 shows the total charge–discharge curves using the electrolyte (Figure 3(a)) and with the additive (Figure 3(b)), which indicates

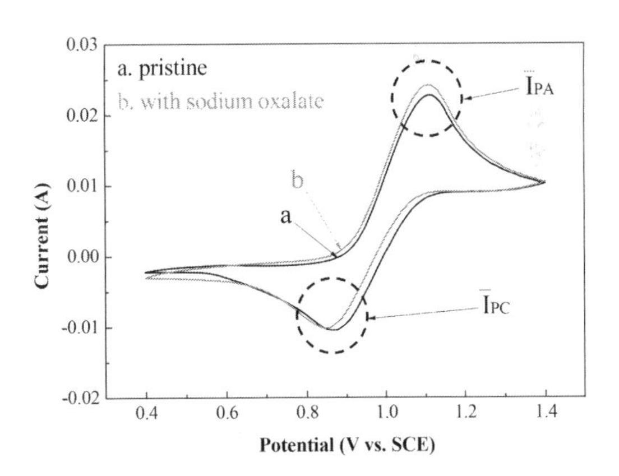

Figure 2. Cyclic voltammograms of the electrolyte (1.8 M VOSO4 + 3.0 M H_2SO_4): (a) pristine and (b) with sodium oxalate at a scan rate of 10 mV s^{-1}.

Table 3. Cyclic voltammogram data for the 1.8 M V (IV) + 3.0 M H_2SO_4 electrolyte and with the additive.

Additive	$\bar{I}_{PA}$ (mA)	$\bar{I}_{PC}$ (mA)	$\bar{I}_{PC}/\bar{I}_{PA}$
None (pristine)	24.0	19.1	1.26
Sodium oxide	23.8	19.2	1.24

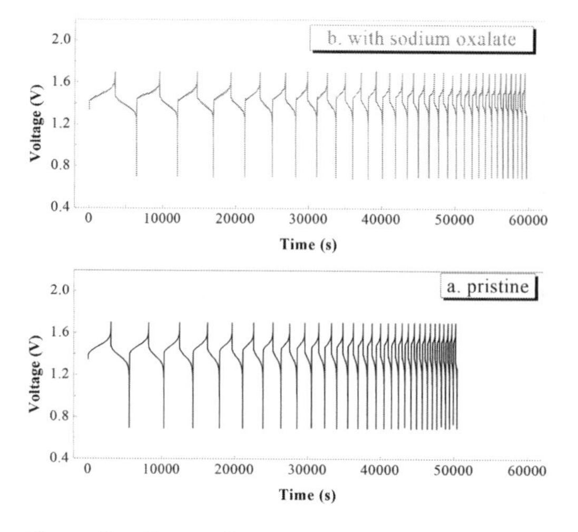

Figure 3. Charge–discharge curves for the 30-cycle operation: (a) pristine and (b) sodium oxalate at 60°C.

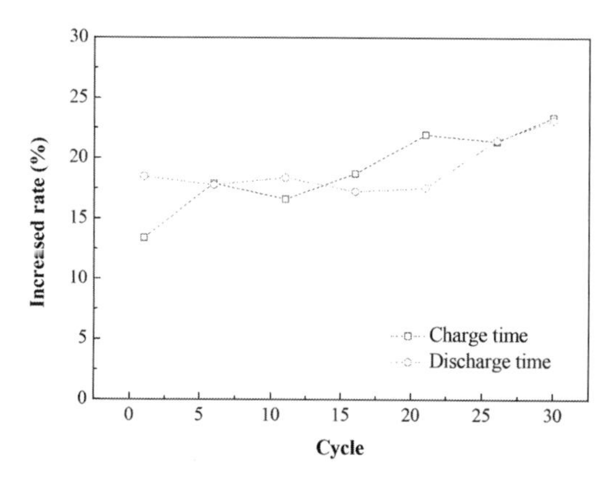

Figure 4. Increased rate of the charge and discharge capacity using sodium oxalate at the 1st, 6th, 11th, 16th, 21st, 26th, and 30th cycle.

the longer total charge capacity (18.05%) and the total discharge capacity (18.74%) than the pristine one. In addition, the charge–discharge capacity of the pristine electrolyte decreased more rapidly than using it with the additive as the cycles proceeded. This result indicates the effect on the reduction in the precipitation reaction at high temperatures.

Figure 4 shows the increased rates of the charging and discharging times using the electrolyte with sodium oxalate. As observed, the electrolyte with the additive had a longer charge (discharge) capacity than the pristine electrolyte as 13.39 (18.46)%, 17.89 (17.78)%, 16.59 (18.39)%, 18.72 (17.23)%, 21.95 (17.a55)%, 21.38 (21.48)%, and 23.42 (23.15)% at the 1st, 6th, 11th, 16th, 21st, 26th, and 30th cycle, respectively. This is due to

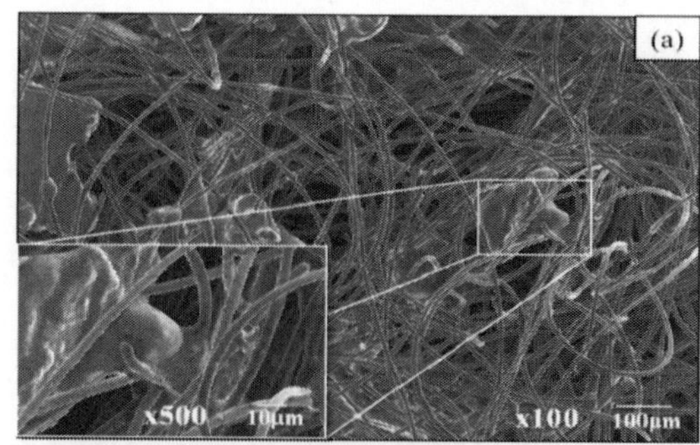

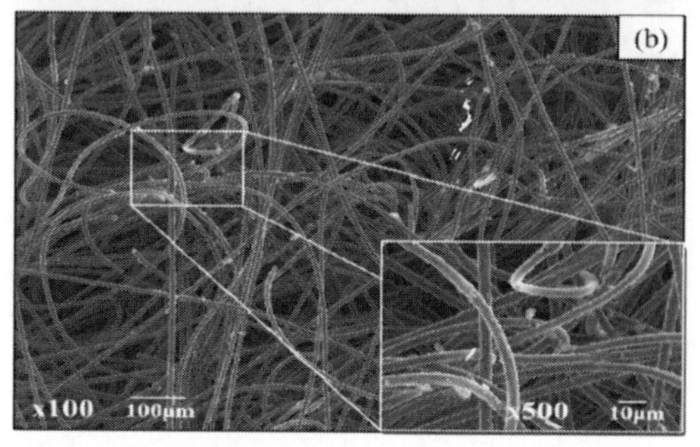

Figure 5. The SEM images of the carbon felt of the miniature cell after 30 charge–discharge cycles: (a) pristine and (b) sodium oxalate.

the fact that the additive could prevent the side reaction (i.e., 2 VO+ + $H_2O \rightarrow V_2O_5$ + 2 H^+) to generate the precipitate of V species at high temperatures. The search results are shown in Figure 5, in which SEM images show the surface of the carbon felt electrode in the positive half-cell after the 30th cycle of the cell operations. It can be seen from Figure 5(a) that few precipitates could be observed compared with Figure 5(b). Therefore, these results indicate that sodium oxalate enhances electrochemical reaction reversibility and the precipitation reaction could be delayed by the additive at high temperatures.

4 CONCLUSION

This paper focused on sodium oxalate as a thermal stability agent of the vanadium solution in the performance of the VRB. The experimental results found that thermal stability is enhanced by sodium oxalate (due to the precipitation reaction) in a vanadium positive electrolyte. In addition, the new electrolyte resulted in the improved charging and discharging capacity by about more than 18% for 30 cycles when compared with the pristine electrolyte. Consequently, the proposed electrolyte using sodium oxalate provides higher thermal stability and performance of the VRB at high temperatures.

ACKNOWLEDGMENT

This paper was supported by "Leaders in Industry-University Cooperation" Project, supported by the Ministry of Education.

REFERENCES

Chakrabarti, M.H. et al. 2007. Evaluation of electrolytes for redox flow battery applications. *Electrochimica Acta,* 52: 2189–2195.

Chang, F. et al. 2012. Coulter dispersant as positive electrolyte additive for the vanadium redox flow battery. *Electrochimica Acta,* 60: 334–338.

Fabjan, C. et al. 2001. The vanadium redox-battery: an efficient storage unit for photovoltaic systems. *Electrochimica Acta,* 47: 825–831.

Faizur, R. & Skyllas-Kazacos, M. 2009. Vanadium redox battery: positive half-cell electrolyte studies. *Journal of Power Sources,* 189: 1212–1219.

Joerissen, L. et al. 2004. Possible use of vanadium redox-flow batteries for energy storage in small grids and stand-alone photovoltaic systems. *Journal of Power sources,* 127: 98–104.

Kazacos, M. et al. 1990. Vanadium redox cell electrolyte optimization studies. *Journal of Applied Electrochemistry,* 20: 463–467.

Li, S. et al. 2011. Effect of organic additives on positive electrolyte for vanadium redox battery. *Electrochimica Acta,* 56: 5483–5487.

Lu, X. 2001. Spectroscopic study of vanadium (V) precipitation in the vanadium redox cell electrolyte. *Electrochimica Acta,* 46: 4281–4287.

Rahman, F. & Skyllas-Kazacos, M. 1998. Solubility of vanadyl sulfate in concentrated sulfuric acid solutions. *Journal of Power Sources,* 72: 105–110.

Skyllas-Kazacos, M. et al. 2011. Progress in flow battery research and development. *Journal of the Electrochemical Society,* 158: R55-R79.

Tokuda, N. et al. 2000. Development of a redox flow battery system. SEI. *Technical Review,* 50: 88–94.

Wu, X. et al. 2012. Influence of organic additives on electrochemical properties of the positive electrolyte for all-vanadium redox flow battery. *Electrochimica Acta,* 78: 475–482.

Advanced Materials, Mechanical and Structural Engineering – Hong, Seo & Moon (Eds)
© 2016 Taylor & Francis Group, London, ISBN: 978-1-138-02908-8

Cellular phone activity detection techniques

P.U. Okorie & M.I. Ogbile
Department of Electrical and Computer Engineering, Ahmadu Bello University, Zaria, Nigeria

ABSTRACT: This paper presents the techniques of a mobile phone activity detector. The device detects the signals radiated from a mobile phone due to incoming and outgoing calls, Short Message Service (SMS) and video transmission even when the mobile phone is kept in the "silent" mode. The device is a Radio Frequency (RF) detector circuit or sensor that detects the presence of a mobile phone in the active mode used in making incoming and outgoing calls as well as sending SMS messages. The device senses the presence of any mobile phone activities within a distance of 3 m, i.e., when the signal is detected, the buzzer sounds, alerting the operator. The first stage is the power supply and the second stage, a design using a down converter, a Voltage Controlled Oscillator (VCO), and a band-pass filter was investigated for cellular phone detection. The performance of this technique using hardware and computer modeling is discussed and the results are presented. The new system is an accurate and practical solution for detecting cellular phones in a secure facility.

1 INTRODUCTION

Imagine a scenario in which you are an IT security consultant in Immigrations or a staff of the National Agency for Food Drug Administration and Control (NAFDAC) recently hired to investigate a drug. The company holds sensitive information stored on computers that are closely monitored by cameras and surveillance devices. The building is heavily guarded by security men at all entrances and all staff and visitors are mandated to put on security tags equipped with RFID chips. Visitors must be escorted to ensure that no drug information leaks out. Evidently, some of the computers come with Bluetooth and are not disabled by IT security. Using a Bluetooth connection does not look conspicuous on the camera because there are no wires that are plugged into the computer.

In another scenario, you are an Examination Officer at a University in charge of ensuring that all students abide by the rules and regulations, especially the restriction on the use of mobile phones in an examination. A student is found to be assessing a website on his phone for solutions to the questions being asked.

Therefore, the organization needs a way to detect cellular phones in these situations. There are few existing cellular phone detectors for this purpose in the market today that can catch the employee prior to the information leaks out.

This paper seeks to address the need to prevent the use of mobile phones in areas where they are not allowed to be used, such as in examination halls, banks, meetings, confidential rooms, and security checkpoints especially with the spate of insecurity in the country.

The mobile phone is also called the "Global System for Mobile Communication" (GSM), which allows multiple users to share a single radio channel through a technique called "Time Division-Multiplexing" (TDM). A caller is assigned a specific time slot for transmission; the slots are in groups of six, forming a channel. This allows multiple users to share a single channel simultaneously without interference. The first generation of technology in GSM uses the Advanced Mobile Phone Service before the introduction of the second and third generation (www.cellbusters.com 2008). Another technique, known as frequency hopping, was introduced into the GSM in order to reduce interference.

Cell phone signal detector is an alternative to more expensive measures against denial-of-service by service providers. This work concentrates on designing a system that will detect the presence of GSM signals from an unauthorized user in restricted areas. The cell phone signal detector project is an advanced device that finds various applications in the modern fields of communication and surveillances. This work is very useful for private meetings, examination hall, defense establishments, military camp, hospitals, filling station, and homes.

A mobile phone detector is an electronic device that is capable of detecting signals of interest. It can sense signals radiated from an activated mobile phone from a distance of about 3 m. The circuit can detect incoming and outgoing calls, Short Message

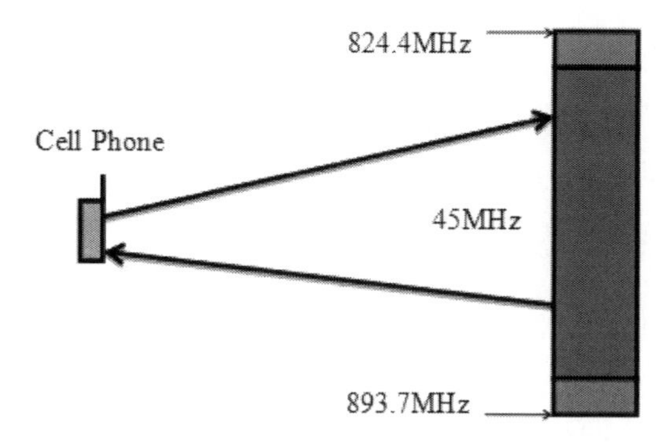

Figure 1. Frequency division multiple access.

Service, and video transmission. When a signal from an activated mobile phone is detected, the device produces both an audible and visible alert. The alarm continues until the transmission signal is no longer detected. The device also incorporates a counter that records the number of Radio Frequency (RF) transmission signals detected by the device during the operation.

2 CELLULAR PHONE COMMUNICATION STANDARDS

Currently, the three main technologies used by cellular phone providers are 2G, 3G, and 4G. Each generation of technology uses a different transmission protocol. Transmission protocols dictate how a cellular phone communicates with the tower. Some examples shown in Figure 1 are as follows: Frequency Division Multiple Access (FDMA), Time Division Multiple Access (TDMA), Code Division Multiple Access (CDMA), GSM, CDMA2000, Wideband Code Division Multiple Access (WCDMA), and Time-Division Synchronous Code-Division Multiple Access (TD-SCDMA). All of these protocols typically operate in the 824–894 MHz band. Some protocols, such as GSM (depending on the provider), will use the 1800–2000 MHz band (www. howstuffworks.com 2009).

Figure 1 shows a good example of how a cellular phone transmission (FDMA) works. Each phone call uses a different frequency within the 45 MHz bandwidth. FDMA is normally used for analog transmissions and capable of digital transmissions (www. howstuffworks.com 2009).

3 FEATURES OF CELLULAR PHONE DETECTOR AT A GLANCE

Bluetooth is a secure wireless protocol that operates at 2.4 GHz. The protocol uses a master–slave structure and is very similar to having a wireless USB port on your cellular phone. Devices such as a printer, keyboard, mouse, audio device, and storage device can be connected wirelessly. This feature is mainly used for hands-free devices, but can also be used for the file transfer of pictures, music, and other data such as:

- Deploys as a standalone or network device depending on your environment,
- Detects all cell phones in all countries (standby, texts, data, and calls),
- Detects all protocols including 2G, 3G, 4G, GSM, CDMA, WIFI, and Bluetooth,
- Audio alert announcement, visual light-emitting diode (LED) alert, or silent logging modes,
- Easily adjustable detection radius for small rooms or larger areas,
- Works out of the box for quick and easy deployment,
- Sensitivity adjustable without a connected computer (ideal for restricted areas),
- Intuitive web interface for advanced setup and administration,
- Future proof design: simply add new channels as cellular networks evolve,
- Includes audio-out jack and SMA connector for an optional external antenna,
- Controls third party devices using the included relay interface,
- Easily adds or disables detection channels via the browser (e.g., Wifi detection off).

4 STATE OF THE ART

Since cellular phone detection is a more recent problem, there are only a few articles that have already researched on this area. Two articles were published in 2007 and have provided a good analysis: the first article, "Detecting and Locating Cell Phones in Correctional Facilities," was written by EVI Technology, LLC; the second article, "Cell Phone Detection Techniques," was written by a contractor hired by the U.S. Department of Energy (DOE).

This article details the growing problem with cellular phones in correctional facilities and points out the constraints used to develop their solution. According to the research, cellular phones in a correctional facility are used to operate criminal enterprises, threaten witnesses, harass victims, plan uprisings, and undermine security. Their problem is monitoring, controlling, and locating cellular phones in a correctional facility (http://iiw.itt.com 2010). EVI's possible solutions include physical search, non-linear junction detectors, signal jamming, shielding, network provider location-based screening, RF detection, and their custom proprietary solution. They rule out all solutions except their own custom solution that uses a system of networked sensors that are controlled by a central computer. EVI uses proprietary software that determines the cell phone's location and detects any RF emissions. The location of detected cellular devices is displayed on a 14 facility map (http://iiw.itt.com 2010). This solution was developed for detecting cellular phones in a prison. It relies on the condition that the cellular phone remain stationary, which makes sense in the case of prison facilities since the movement is limited. EVI's detection system finds cellular phones after they have already been in the facility for at least 30 min or if someone is making a cell phone call. EVI does not provide any details on the signal detection technique used since it is all proprietary. Also, there have been no reviews or articles stating that this system works as advertised.

This paper examines the detection of cellular phones when a person is entering a secure facility or cellular phone-restricted area. The detection technique studied requires measuring a cell phone's electromagnetic properties and determining an identifiable signature. It shows the most potential for measuring the RF spectrum in the range about 240–400 MHz (outside the cellular phone band, http://inspire.ornl.gov). The DOE contractor recommends the development of a cellular phone detector by measuring the RF spectrum. Spurious emissions from cellular phones are monitored and recorded when the phone is in the standby or transmitting mode (http://inspire.ornl.gov). This method has some advantages (http://iiw.itt.com 2010):

- No external signal is required for detecting the phone.
- The band of frequencies is limited by the FCC and is likely to be used by most manufacturers.
- The system could potentially detect more than 15 cellular phones.
- This method could work on future generations of cellular phones.
- The system could potentially detect cellular phones even when they are switched off.

We put forward a proposed path for determining whether this technique is possible, and perform some preliminary testing. We also provide an alternative path that would detect cellular phones based on their RF reflecting material in the cellular phone filters on all phones. Cellular phones could then be detected whether they are switched on or off (http://inspire.ornl.gov 2010). We also mention an evaluation that was conducted in 2003 on the available commercial cellular phone detectors. Bechtel-Nevada and Sandia National Labs found that none of the detectors at that time were effective when the phone was switched off or in the standby mode (http://inspire.ornl.gov).

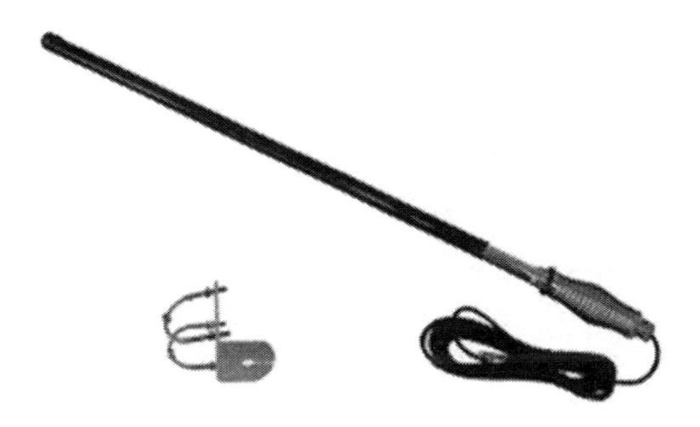

Figure 2. Antenna.

5 SYSTEM DETECTION

The cellular phone detector provides maximum protection for your environment by continuously scanning for cell phone and user-selectable RF transmissions.

Once the cellular or appropriate RF activity is detected, the device can be set to activate a choice of alerting options or to silently log all the detected activities.

An antenna (or aerial) is an electrical device that converts electric power into radio waves, and vice versa. It is usually used with a radio transmitter or radio receiver. During transmission, a radio transmitter supplies an electric current oscillating at radio frequency to the antenna's terminals, and the antenna radiates the energy from the current as electromagnetic waves (radio waves). During reception, an antenna intercepts some of the power of an electromagnetic wave in order to produce a tiny voltage at its terminals, which is applied to a receiver in order to be amplified.

When the device is active, a visual indicator is enabled as well an audible indicator. The indicator stage consists of components such as transistors, capacitors, resistors, and LED.

6 PROPOSED METHODS

The cell phone activity detector consists of various modules coupled together to form a single circuit. This section discusses the procedures and techniques to be used in the design, construction, and implementation. The system block structure and circuit employed in the accomplishment of this work are shown in Figures 3 and 4, respectively.

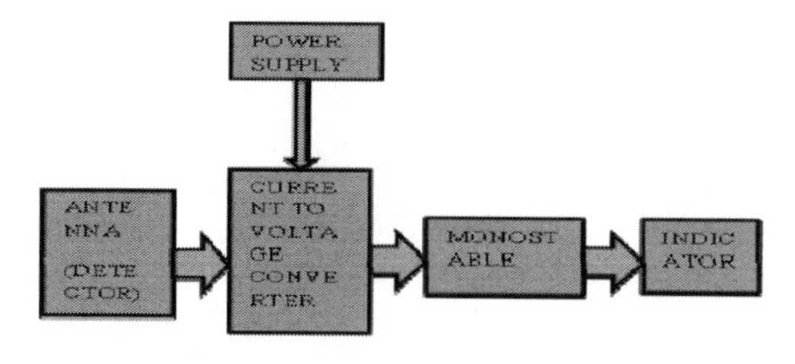

Figure 3. Schematic block diagram.

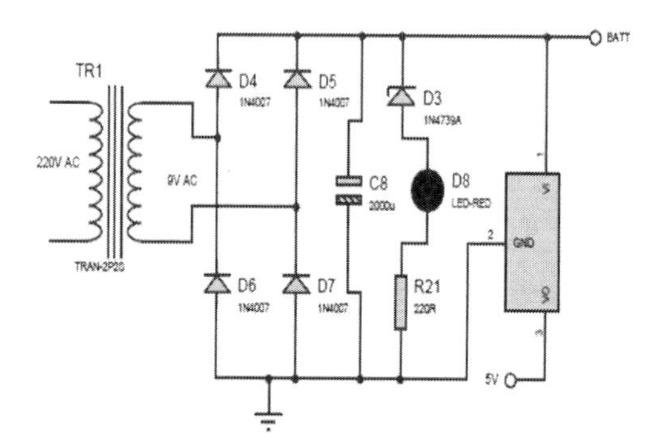

Figure 4. Power circuit.

This design can detect GSM Communication signals at 900 MHz (http://iiw.itt.com 2010). The design consists of two signal detectors, each with their own dipole antenna, inductor, and diode. Each dipole antenna is tuned to 900 MHz. When the antennas resonate at 900 MHz, a charge is induced in the inductor. A diode then demodulates the signal, which is amplified by an op amp and passed along to a 3.5 mm headphone jack. The design does not describe what sound you will hear when a cellular phone is being used. A schematic and parts list is provided (http://iiw.itt.com 2010).

7 SYSTEM OPERATION AND DESIGN

The power requirements for most of the components that make up this device are within the range of 3–15 V. Considering size, portability, and cost, a rechargeable 9 V DC cell is required, but for scarcity of the rechargeable 9 V cell, a primary cell is used instead. An AC source is also incorporated in order to supply power when AC is available and to charge the cell. In order not to blow the primary cell, the switch makes and breaks the connection of the battery to the circuit when required, so that AC power is not allowed into the non-rechargeable cell.

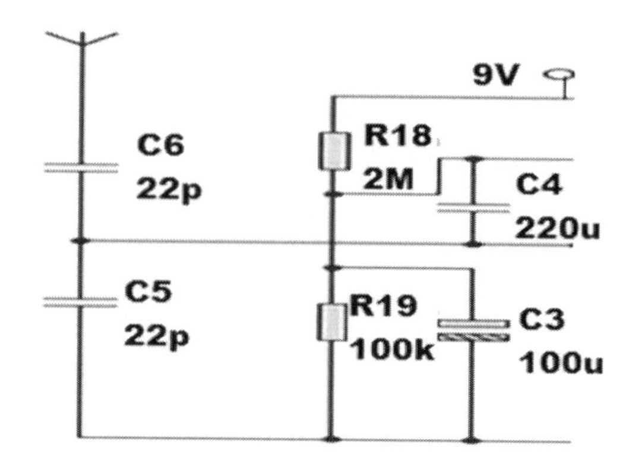

Figure 5. Antenna circuit.

The transformer (TR1) receives a 220 V AC supply and steps it down to a 9 V AC. Diodes D4, D5, D6, and D7 together act as a bridge rectifier to convert the Alternating Current (AC) supply into the Direct Current (DC). Capacitor (C8) filters the ripples present in the rectified waveform of the supply. The LED D8 indicates that there is an external AC power supply to power the device and charge the battery.

An ordinary RF detector using tuned LC circuits is not suitable for detecting signals in the GHz frequency band used in mobile phones. The transmission frequency of mobile phones ranges from 0.9 to 3 GHz with a wavelength of 3.3 to 10 cm. So, a circuit detecting gigahertz signals is required for a mobile bug. In this circuit, a 0.22 μF disk capacitor is used to capture the RF signals from the mobile phone. The lead length of the capacitor is fixed as 18 mm with a spacing of 8 mm between the leads to achieve the desired frequency. The disk capacitor along with the leads acts as a small gigahertz loop antenna to collect the RF signals from the mobile phone, as shown in Figure 5.

The capacitor, gain of the amplifiers, and the length of the antenna is critical in this circuit. The length of the antenna is calculated as follows:

$$\lambda = (300{,}000 \text{ km/h})/900 \text{ MHz} = 33.3 \text{ cm}$$

Then, antenna length = $\lambda/2 = 33.3/2 = 16.6$ cm. Thus, the length of the antenna is 8.3 cm.

8 DISCUSSION

Cellular phone technology is gaining new data capabilities very rapidly. New features such as Bluetooth, high resolution cameras, memory cards, and Internet make them ideal for getting data in and out of secure facilities. A cellular phone uses many different transmission protocols such as FDMA or CDMA. These protocols dictate how a cellular phone communicates with the tower. Typically, cellular phones in the United States operate in the range of 824–894 MHz.

The results obtained agree with the experiments conducted, which are useful for the application in cell phone detection.

9 CONCLUSION

Many businesses depend on keeping information protected. Currently, the only way to ensure that no one brings a cellular phone into a secure facility is to search everyone entering and exiting which requires a lot of manpower and money to implement. The implementation of this device can be extensively used for the detection of mobile phones in areas where a detector can be placed at strategic places in such a manner that the detection array (approximately 1 m) of the antenna overlaps. In so doing, this would make it possible to cover all areas of the facilities in question and prevent unauthorized access to confidential information. There is little or no need for personnel except when routine maintenance is to be carried out as the device incorporates an external AC power source as well as a rechargeable DC cell. Also, the device gives out an audible sound as well as a visual indication whenever a device radiates within the area of coverage of the device.

REFERENCES

http://www.cellbusters.com, 2008. Cellbuster Cell Phone Detector, Cellbusters INC. Accessed September.

http://www.howstuffworks.com/cellphone.htm/printable, 2009. How Cell Phones Work, How Stuff Works A Discovery Company. Accessed April.

http://iiw.itt.com/files/cellHound_wpCellPhonesIn-Prison.pdf, 2010. Detecting and Locating Cell Phones in Correctional Facilities, EVI Technology, LLC. June 2007. Accessed February.

http://inspire.ornl.gov, 2010. Cell Phone Detection Techniques, U.S. Department of Energy. October 2007. Accessed January.

Advanced Materials, Mechanical and Structural Engineering – Hong, Seo & Moon (Eds)
 ISBN: 978-1-138-02908-8

Transient analysis for a non-effective grounded network with distributed generations

L.S. Li, S.D. Zhang & H.J. Liu
State Grid Shandong Electric Power Company, Electric Power Research Institute, Jinan, China

R.R. Fu & X.Q. Ji
Shandong University of Science and Technology, Qingdao, China

ABSTRACT: This paper presents a simulation model of a non-effective grounded network with a distributed generation in order to analyze the transient characteristics. The influence of the different types and different positions of the distributed generations is analyzed. Simulation results show that when the distributed generation is accessed into the fault branch, it has obvious impacts on transient characteristics of the fault branch. Due to the presence of the inverter of the distributed generation, the value of zero-mode transient component becomes smaller. Although, the SFB has changed when the distributed generation is connected, the main energy of the zero-mode current is still concentrated in the SFB range.

1 INTRODUCTION

In recent years, many countries have gradually developed a Distributed Generation (DG) in the distribution network. In the event of a fault, it should immediately be able to locate the fault section in the distribution network and isolate it; then restore the power supply, so as to reduce the user's power outage time, and to ensure power supply reliability and power quality. The distribution network fault location is one of the basic functions of the distribution automation. Many research works have been carried out in this area (Ji et al. 2005, Ma et al. 2009, Sun et al. 2009). However, when DG is connected to the distribution network, the power flow and short-circuit current will be changed too, which leads to the traditional distribution network fault location method being no longer in application. LIU et al. 2013 studied the fault location technique for the distribution system considering DG based on short-circuit current, but a threshold needs to be found to reliably identify the short-circuit current supplied by the power source with the short-circuit supplied by the DG, and only then the fault location can be realized. Li et al. (2005) has suggested the influence of distributed generations on the distribution of network loss, but voltage and power quality is also analyzed.

In this paper, the model of a non-effectively grounded network is formed, and a single phase to ground fault is analyzed by, respectively, connecting the ideal power source and the distributed generation with the fault line and a normal line.

2 MODELS AND DYNAMIC EQUATIONS

2.1 *Simulation model*

The simulation model is shown in Figure 1. The system consists of an 110 kV substation and 6 overhead lines. The main transformer T0 of substation is Y_0/Y connection mode, and the transformers on each feeder are Δ/Y connection mode. The neutral point of the low voltage side of T0 is connected with an arc suppression coil through switch *S*.

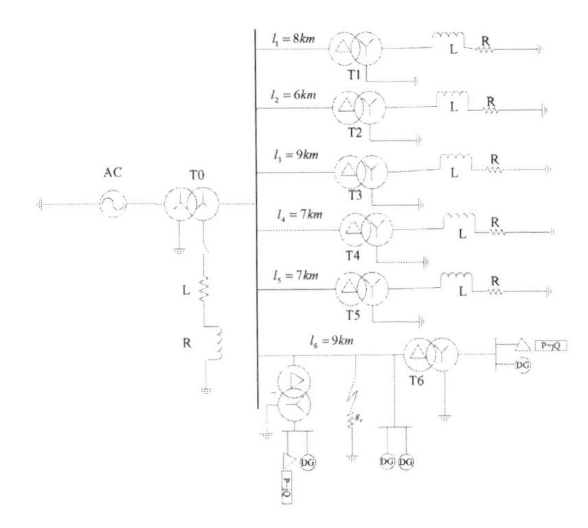

Figure 1. Simulation model.

2.2 *Parameters*

In the simulation model, the power source is an ideal voltage source, which has a zero internal resistance. The rated voltages of the main transformer T0 are 110 kV/10.5 kV, and the rated voltages of feeder transformers are 10 kV/0.4 kV. The positive and zero sequence impedance of each line are $Z_1 = (0.17 + j0.38)\ \Omega/km$ and $Z_0 = (0.23 + j1.72)\ \Omega/km$ respectively. The positive and zero sequence admittance to the ground are $b_1 = j0.3045\ \mu S/km$ and $b_0 = j1.884\ \mu S/km$.

The type of accessed DGs includes 380 V ideal power source, solar photovoltaic source using P/Q control strategy, and micro-gas turbine.

2.3 *Coordinate transformation*

For a three-phase system, due to the coupled inductance and capacitance among conductors or the earth, the transient model of the three phase line is much more complex than a single phase line model. In order to facilitate the analysis of the transient process, the Karrenbauer transform method is applied, which can transform a coupling three-phase system into a non-coupling of the 0 mode, 1 mode, and 2 mode system (Zhang & Xu 2008, Zhang & Zhao 2008).

2.4 *Conditions of simulation*

1. Simulation step. In order to improve the accuracy of transient simulation, the calculating step is set to 50 μs.
2. Sampling frequency. Sampling frequency is 20 k Hz.
3. Initial phase angle. The fault initial phase angle is set to 90°, 45°, 15°, and 0° respectively.
4. Ground resistance. The ground resistance is set to 10 Ω, 100 Ω, 1000 Ω, 10 kΩ, and 100 kΩ respectively.
5. Distributed power and its position. Different types and locations of DG are selected to study the effect of the transient fault characteristics.

3 ZERO MODE TRANSIENT ANALYSIS

3.1 *Case 1: Without DG*

In the above PSCAD model, a single-phase ground fault is set in line 6. The initial phase angle of faulty line is set to 90°, and the ground transition resistance is 10 Ω. The zero mode voltage U0, zero mode current im1, im2 of normal line 1 and line 2, and zero mode current I0 of fault line are measured. Figure 2 shows the corresponding results.

It can be seen from Figure 2:

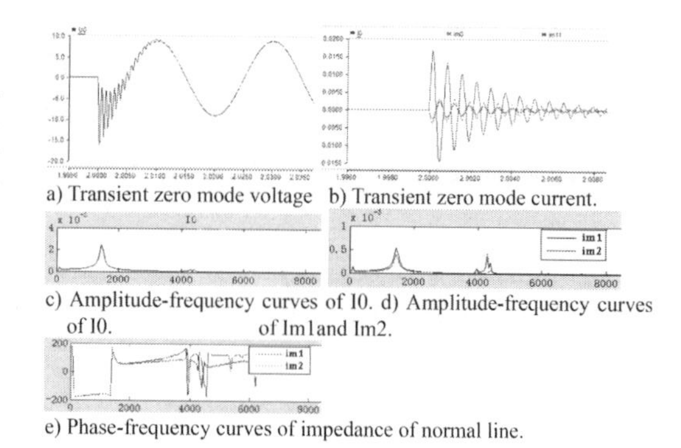

a) Transient zero mode voltage b) Transient zero mode current.

c) Amplitude-frequency curves of I0. d) Amplitude-frequency curves of Im1and Im2.

e) Phase-frequency curves of impedance of normal line.

Figure 2. Simulation result of case 1.

1. The zero mode voltage of fault line changed little before and after short circuit;
2. The zero current of fault line is larger than that of normal line, and they have opposite polarity. The maximum value of the transient zero mode current is far greater than its steady-state value;
3. There exist multiple resonant points in transient processes, the main resonance amplitude is larger when the ground transition resistance is small;
4. Main resonance frequency is within the Selected Frequency Band (SFB), which contains most transient zero mode energy. In SFB, transient zero mode energy is concentrated near the main resonance frequency.
5. Within the SFB, the amplitude of zero mode current of fault line is larger than that of normal lines. However, outside the SFB, zero current of the fault line may be smaller than that of normal lines.

3.2 *Case 2: DG connected to fault line*

The fault line is, respectively, connected with an ideal power source, a photovoltaic cell, and a micro-gas turbine by comparing the different effect on the zero mode transient voltage and current, the influence of DG supply access faulty line on non-effective grounding fault was analyzed. When the single-phase grounding fault occurs, the measured transient curves are as follows:

From the results illustrated in Figure 3, it can be seen that compared to the ideal voltage source, when the PV cell and gas turbine generator accesses to the fault line, the zero mode current amplitude is smaller. The zero mode current of the fault line is larger than that of the normal line. The polarity of zero mode current of the normal line and fault line is opposite. For each line, the maximum value of the transient zero mode current is far greater than its steady state value.

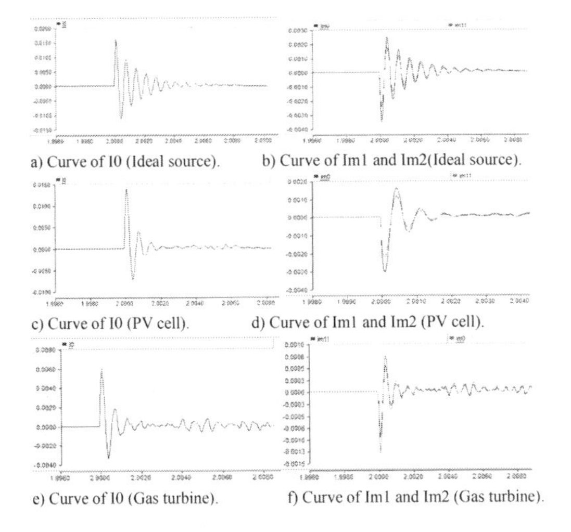

a) Curve of I0 (Ideal source). b) Curve of Im1 and Im2(Ideal source).
c) Curve of I0 (PV cell). d) Curve of Im1 and Im2 (PV cell).
e) Curve of I0 (Gas turbine). f) Curve of Im1 and Im2 (Gas turbine).

Figure 3. Curve of transient zero modal current.

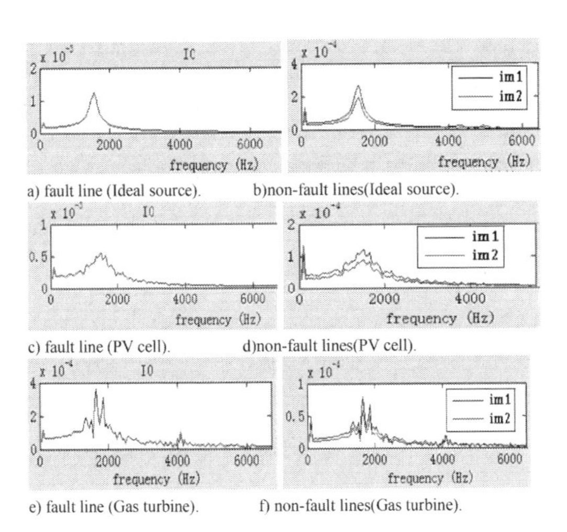

a) fault line (Ideal source). b)non-fault lines(Ideal source).
c) fault line (PV cell). d)non-fault lines(PV cell).
e) fault line (Gas turbine). f) non-fault lines(Gas turbine).

Figure 4. Amplitude-frequency characteristics.

Figure 4 illustrates the amplitude-frequency characteristics of the corresponding zero modal currents.

It can be seen that there exist multiple resonance processes in the transient process. Compared to the case of ideal power source, there are more harmonic and smaller amplitudes when PV cell and micro gas turbine access the network. The main energy of the zero mode current is still focused on the range of SFB characteristic frequency band.

In the range of SFB, the zero mode current amplitude of the faulty line is larger than that of the correct line; outside the range of SFB, the zero mode current of the faulty line and the correct line cannot be determined.

3.3 *Case 3: DG connected to non-fault line*

In this case, the ideal power source, the photovoltaic cell and the micro-gas turbine are connected to a non-fault line. When a single-phase grounding fault occurs, the transient data are recorded, which were shown in Figure 5 and Figure 6.

As can be seen from Figure 5, no matter what is the ideal source, PV cell or gas turbine accessing the grid, the amplitude and fluctuations of the zero mode current of the fault line is the same.

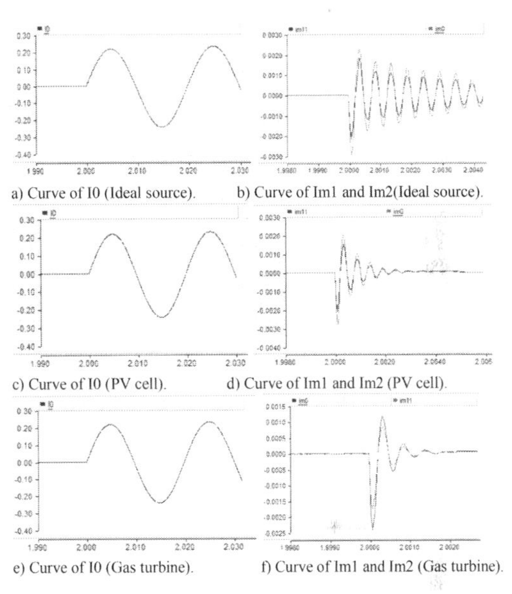

a) Curve of I0 (Ideal source). b) Curve of Im1 and Im2(Ideal source).
c) Curve of I0 (PV cell). d) Curve of Im1 and Im2 (PV cell).
e) Curve of I0 (Gas turbine). f) Curve of Im1 and Im2 (Gas turbine).

Figure 5. Transient zero mode current of case 3.

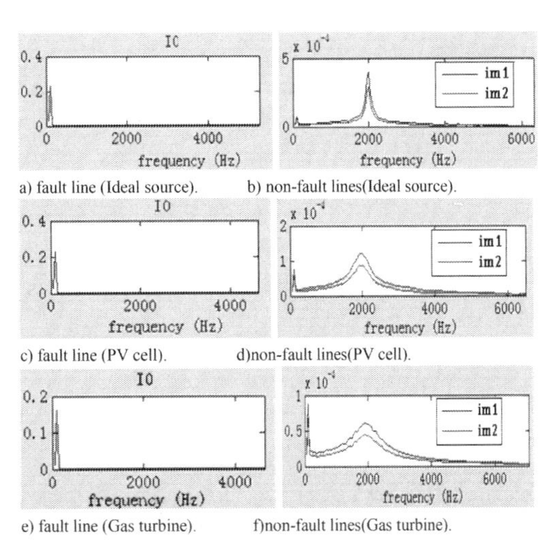

a) fault line (Ideal source). b) non-fault lines(Ideal source).
c) fault line (PV cell). d)non-fault lines(PV cell).
e) fault line (Gas turbine). f)non-fault lines(Gas turbine).

Figure 6. Amplitude-frequency characteristics.

However, for non-fault lines, the zero mode current amplitude is greater and the transient process duration is longer when PV cell and gas turbine generator is connected.

As depicted in Figure 6, multiple resonance processes only exist in the non-fault lines. The main energy of zero mode current is still focused on the range of SFB characteristic frequency band.

In the range of SFB, the zero mode current amplitude of the fault line is larger than that of the non-fault line.

4 CONCLUSION

The single phase to ground fault in non-effective grounded network is simulated in detail. Some important conclusions can be drawn from the corresponding results. When DG is connected to the fault line, the inverter of DG has great influence on transient characteristics of the fault line. Due to the difference in the impedance characteristics when DG gets accessed, the range of SFB has been changed. However, the energy of zero mode current is still mainly concentrated in the range of SFB. When distributed generation accesses the fault line, the impact of DG on the fault line and the non-fault line is not obvious, but the SFB has changed.

REFERENCES

Ji, T., Sun, T.J. & Xue, Y.D. 2005. Current status and development of fault location technique for distribution network. *Power System Protection and Control,* 33(24): 32–37.

Li, B. & Li, X.Y. 2005. Distributed generation sources and their effects on distribution networks. *Power System Technology,* 9(3): 45–49.

Liu, J., Zhang, X.Q. & Tong, X.Q. 2013. Fault location containing distributed power distribution grid. *Automation of Electric Power Systems,* 37(2): 36–43.

Ma, S.C., Gao, H.L. & Xu, B.Y. 2009. A survey of fault location methods in distribution network. *Power System Protection and Control,* 37(11): 119–124.

Sun, J.L., Li Y.L. & Li S.W. 2009. A protection scheme for distribution system with distributed generation. *Automation of Electric Power Systems,* 33(1): 81–84.

Zhang, B.H., Zhao, H.M. & Zhang, W.H. 2008. Faulty line selection by comparing the amplitudes of transient zero sequence current in the special frequency band for power distribution networks. *Power System Protection and Control,* 36(13): 5–10.

Zhang, X.H., Xu, B.Y. & Pan, Z.C. 2008. *Study on Single-Phase Earthed Faulty Feeder Selection Methods in Non-Solidly Grounded Systems.* The Third International Conference on Electric Utility Deregulation and Restructuring and Power Technologies, 06–09 April 2008, WeiHai: China.

Advanced Materials, Mechanical and Structural Engineering – Hong, Seo & Moon (Eds)
 ISBN: 978-1-138-02908-8

A study on the integration of topic-based knowledge model for computer aided innovation

S.S. Chun & S.R. Lee
Korea Institute of S&T Evaluation and Planning, Seoul, South Korea

ABSTRACT: Knowledge resources and management innovation is essential in knowledge-based economic environment and diverse knowledge management. Knowledge management systems are very important in creating new values. However, structuring existing information, technology for connecting knowledge, and integrated operation system has been a very difficult problem. In this research, integrated knowledge management systems have been designed between planning and assessment, which are the core processes of corporate management system.

1 INTRODUCTION

1.1 *Knowledge management and CAI*

The establishing method of knowledge-based through the connection between information processing and semantic based information has been suggested for knowledge management. In particular, the applications of knowledge resource and the management innovation in the knowledge-based economic environment are essential. And they are being developed as diverse knowledge management systems for creating new values. In the early creative phases and in the business activities of the NPD (New Product Development) process, standard office tools are often used as the sole software (CAD, CAE, CAM etc.) support (Yezersky 2007).

The gap in the ICT support is the area for application in the growing field of Computer Aided Innovation (Leon 2008). The purpose of knowledge management is to establish a knowledge base by organizing meanings in given information and to analyze or apply them as simulation methods in complex problems or social network analysis. Furthermore, it is possible to do strategic analysis on the meaningful evidence, logic, trend, transition, and forecast through the behavior knowledge model as well as static indicators and regression analysis. In this paper, we will first analyze the semantic relationship and the knowledge properties on organizing the enterprise knowledge map and networks.

1.2 *Enterprise knowledge system*

However, in structuring existing information, technology for connecting knowledge, and integrated operation system are very difficult problems.

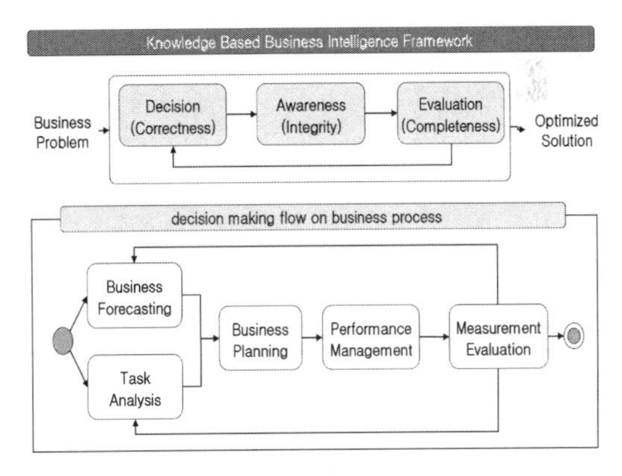

Figure 1. Unified framework for integration of enterprise knowledge model.

Regarding these problems, there are active researches on RTE (Real Time Enterprise), ERP (Enterprise Resource Planning), BPM (Business Process Management), and CRM (Customer Relationship Management). On the other hand, there are few researches on mutual connection and analysis of knowledge and structuring of integrated operation.

This research designs the integrated knowledge system between planning and evaluation, which are the core processes in intelligent corporate management and through the improved knowledge technology. Knowledge-based and real-time knowledge operation system was established to explain real-time knowledge analysis and application cases as shown in Figure 1.

2 GETTING KNOWLEDGE MODEL

2.1 *Using knowledge system*

The development of knowledge-based industrial economy and the change to a mutually connected social structure creates diverse complexity problems in the organizations, companies, and the country, which requires fast and strategic decision-making. Moreover, the rapid changes of external environment and increase of complexity system require more analytical information processing and knowledge utilization system for the investment strategy and decision-making. Therefore, rapid recognition, prompt response strategy on the changes of external environment, specific implementation, and evaluation system are the core values in the corporate competition. Also, the supporting intelligent integrated knowledge system is the overall frame for connecting a company's vision with its organizational structure, strategy, and external environment. In particular, the structured knowledge-based model with meanings is useful for an integrated society's value chain from decision-making to expression of opinions. These may be actively used for establishing policy and establishing future strategies (System Dynamic, Info Matrix, Simulation, etc.). Meanwhile, analysis on knowledge map and social network in the technology policy (SRI-BI, RaDUS, NTIS, euro-CRIS, NSF) and management consulting (Arthur Andersen, B. Allen & Hamilton, IBM, McKinsey, Cap Genini, Ernst & Young, KPMG, Gartner) are used as core devices for the original value creation competition.

Generally, knowledge management means systematically managing accumulated information that is considered valuable that is based on experience and assumption. Intelligent knowledge management means information or data being processed through the user's cognitive awareness, interpretation, and analysis and being established as a system, which continuously create additional value and the knowledge with long life cycle. This accompanies a growth model for knowledge management activities, maturity management through process evaluation, and the supporting platform with the knowledge management method of SRI-BI (Stanford Research Institute-Business Intelligence), which can define core elements by stages for knowledge management intelligence.

2.2 *Definition and integration of knowledge model*

The integrated knowledge base which is defined in the previous chapter has a structure to build every resource as recognizable semantic resource and

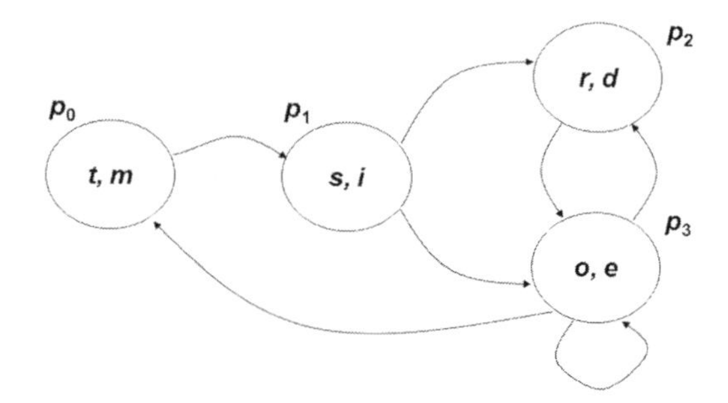

Figure 2. Abstract knowledge model of Kripke structure derived from process model and activities.

to take customized approach depending on situations. Therefore, every resource in the business has hierarchy as service, process and resource, and also has mutual relevance depending on the process. Thus, the given topics are extracted through hierarchy from the knowledge base and the causal relationship, and may create a knowledge map as illustrated in Figure 2.

Figure 2 shows that the given topics as 'Topic = {t, m, s, i, r, d, o, e}' use hierarchy of the process and the relationships between topics are deduced as hierarchy and consecutive relationships between resource and process that the information process of Kripke structure in Figure 2. The network model created through this process was made to analyze meaningful relationships and the approach among topics to help understand the finite elements and dynamic connection of knowledge and support integration between the knowledge model and selective analysis and combination (Chan 2000). In this paper, as shown in Figure 2, business activities in technology innovation were defined as abstracted process model. Here we denoted variables of the process—model (M), process (P), relations (R), and label (L). The relationship between the processes (p) on abstracted process model (M) is denoted by transition, interaction, reference, etc. A process (p) has element (E) and element (E) has proposition (X). Each process (P) in the model is defined as follows: each transition (p, p') has relation (R). Kripke structure's process model is defined as follows.

$\forall s \in P \bullet \exists p' \in P \bullet (p, p') \in R$
M: Process Model = $<P, P_0, R, X, L>$
P: Process = Set of finite process (p)
P_0: $P_0 \subseteq P$: Set of initial process
R:(p, p')$\in R \subseteq P \times P$ = Process transition ($\forall p \in P$, $\exists p \in P$)
X: True|False = Set of AP (Atomic Proposition)
L: P $\rightarrow$ 2X = assigns to each process the set of atomic propositions that are true in that process.
P_0: Business Foresight = {Trend t, Road Map m}
P_1: Business Planning = {Strategy s, Invest i}

P_2: Performance Management = {Research r, Development d}
P_3: Measurement Evaluation = {Outcome o, Evaluation e}

In this study, heterogeneous model analysis is easy to use for conceptual learning and analysis of knowledge model. And the topics to satisfy in a given model, the path of the property violated the state are all presented. In this way, analysis of knowledge model can be continuously refined. The model properties are expressed in CTL (Computation Tree Logic). According to CTL, the passage of time representation of the property it is easy to define a specific scope in this model. CTL's syntax and semantics of temporal quarter is defined by the BNF (Backus Normal Form) as follow.

$\Phi ::= true|x|\neg\Phi|\Phi_1 \vee \Phi_2|AG\,\Phi|A(\Phi_1\,W\,\Phi_2)$

As existing in CTL, 'A' is universal operator and 'G' is global operator. And 'W' is until operator. Intuitively 'AG Φ' is always true in all paths (true) and '$A(\Phi_1\,W\,\Phi_2)$' means that Φ_1 is forever true until Φ_2 be true for all paths in model. Assuming a finite model M and CTL formula are true in the process, P_0 is written as $p_0 \models \Phi$. A transition (p, p') in the model (M) indicated reachable from current process (P_n) to next process (P_n+_1). Abstract process model (M) has the set of the transition (R). Transition relation $(p_0, p_1) \in R$ and by a single transition from p_0 to p_1 is reachable. Topic based knowledge map M is equivalent to knowledge network N.

Knowledge Map $M \rightarrow$ Knowledge Network N, $M \rightarrow N = \{X \rightarrow S, X_0 \rightarrow S_0, R, T \rightarrow \text{true}, L \rightarrow E\}$

Unified Knowledge Network: $\bigcup_{n=1}^{m} Nn = N_1 \cup N_2 \cup \ldots \cup N_m$

if Knowledge Network N_1
$= \{S_1, s_0 \in S_1, R_1, T_1: \forall|E=\{\text{true}\}, E_1: \cup\{\}\}$

if Knowledge Network $N_m = \{S_m, s_0 \in S_m, R_m, T_m: \forall|E = \{\text{true}\}, E_m: \cup\{\}\}$

then $\bigcup_{n=1}^{m} Nn = \{S_1 \cup S_m, s_0 \in S_1 \cup s_0 \in S_m, R_1 \times R_m, T_1 \cap T_m: \forall E=\{\text{true}\}, \cup E_1:\{\} \vee E_m:\{\}\}$

2.3 *Topic detection and model analysis*

In the understanding of knowledge model, the structural traits are available on knowledge map and the relations between topics are available on the approach analysis of network model. However, in order to understand the implied meaning of knowledge, the interaction and effects among topics should be analyzed in-depth as they are the semantic unit that comprises the knowledge.

In this chapter, Selton index was referred for explaining characteristics of connection degree, closeness, and betweenness on certain topics.

Table 1. Large policy data from text mining to analyze the value of a keyword relationship.

Topic	Energy	City	Bio	Nano	Content
Count	92,069	27,591	27,302	4,334	4,649
Linkage	46,711	7,833	5,206	974	1,138
Distance	3,582	1,640	1,234	111	68
Citation	6,576	1,970	1,950	309	332
Distribution	0.03	0.03	0.05	0.05	0.07

Selton index, the network-based topic model, was used for efficiency analysis of the topic, and this research analyzed the relations through figures in Table 1 in the higher model which includes the accessibility analysis and process. At first, the intensity of relations may be quantification based on the connection intensity among certain topics. *Pi* is the number of *i* in certain knowledge, and *Pj* is the number of *j* in certain knowledge, and *Pij* is the number of *i* and *j* in certain knowledge.

If the intensity of relations is r, $r = \frac{P_{ij}}{\sqrt{P_i P_j}}$ the connection relations are divided into degree, closeness and medication, and they have absolute value and relative value. The Degree Centrality (C) has higher connection intensity when certain topics reveal in other major knowledge, and it has central role in the knowledge. The Closeness Centrality has central role when certain topics are close to other topics, and distance (D) D_{ij} is the shortest course which connects between *i* and *j* topics, and *g* is the number of overall node. Meanwhile, the Between Centrality is interpreted as central role when certain topics have more mediator roles in the relations between topics. g_{jk} is the shortest course number between two topics (*j* and *k*), and $g_{jk}(i)$ is the frequency of passing topic *i* between two topics *j* and $k(j \neq k)$. And *g* is the number of overall node. Through centrality analysis between topics, we can interpret meaningful characteristics of the knowledge model, and expand the range with the difference and connection between the knowledge models.

Linkage centrality:

$$\sum_{i=0}^{n} P_{ig} \quad \textit{Absolute value} \tag{1}$$

$$r = \frac{rA}{\sum_{n=0}^{n} P - 1} \quad \textit{Relative value} \tag{2}$$

Distance centrality:

$$C_i = \left[\sum_{j=1}^{n} d_{ij} \right] - 1 \quad \textit{Absolute value} \tag{3}$$

$$C_i = (g-1)\left[\sum_{i=1}^{n} d_{ij} \right] - 1 \quad \textit{Relative value} \tag{4}$$

Citation centrality:

$$C_B(i) = \frac{\sum_{j \prec k} g_{jk}(i)}{g_{jk}} \quad \textit{Absolute value} \tag{5}$$

$$C'B(i) = \frac{\sum_{j \prec k} \frac{g_{jk}(i)}{gjk}}{\frac{(g-1)(g-2)}{2}} \quad \textit{Relative value} \tag{6}$$

3 SYSTEM INTEGRATION

3.1 *Knowledge integration*

The previous chapters explained the designs of process improvement and integration structure. This chapter focuses on the enterprise knowledge integration method in the organization and creation and the connected analysis on the knowledge relevant to the models between processes (Chun 2015). This analyzes hierarchies and the relations between major topics in the interesting knowledge, structuring them as a model to understand the implicational meanings of the knowledge (Shah & Kourdi 2007). New knowledge is created in the relationships of a very complex and diverse elements including interaction, learning, exchange and ideation of existing knowledge and it creates new values (Janssen & Kuk 2006). Regarding this, recent Semantic Web service technology supports such complex knowledge structure to integrate effectively. This Semantic Web technology includes a web service around dispersion process and semantic resource operation and designs the knowledge base through W3C standard technology which is useful for meaningful interoperability. The general resource is designed with W3C RDF/X as standard and the service knowledge took SWSI (Semantic Web Services Initiative) of W3C as reference. The resource and service technology in the abstraction model stage are applied with restriction (Wang & Vitvar 2007). The detailing technology of web service profile process for discovery, contract, and policy is needed and this is defined with W3C SWSL (Semantic Web Services Language)-Rules and SWSL FOL (First Order Logic language) standard as reference model.

As such, every semantic resource is structured as a recognizable situation awareness system, and may be defined as the following. Meanwhile, the semantic-based service integrate in different dispersion environment is available in the system of W3C SWSA (Semantic Web Services Architecture) framework, WSMF (Web Services Modeling Framework) of EU-DERI (Digital Enterprise Research Institute) and TSC (Triple Space Computing) (W3C 2007). Semantic Web editor is used for the modeling of knowledge resource and the code creation as shown in Figure 3, and the compatibility with the topic map is available with conversion standard syntax. Knowledge model is being mapped from the collected knowledge resource as a meaningful structure in the design of internal knowledge of an organization, and it consists of meaningful relations between core topics created through resource analysis and concepts. The method of designing existing knowledge model builds ontology through formal language and integrated resource. However, the detailed rule is difficult, the significant cost and expertise are also required in the assessment and analysis of this model. So, this chapter creates a new abstraction knowledge map which satisfies interesting topics given from integrated knowledge based on the intelligent process and has relation analysis on this to change a knowledge network model. The knowledge map is a classification and expression on the relations and concepts among knowledge, in the knowledge management, and it may have the meanings of knowledge such as type, characteristic, or pattern. The design of knowledge

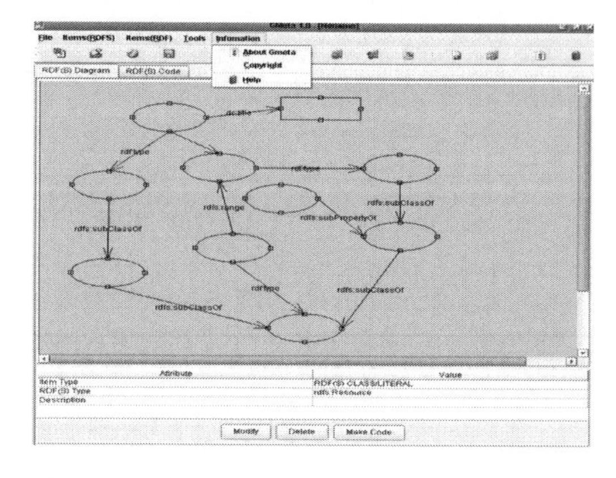

Figure 3. Utilizing the extracted keywords and metadata generated Web Service Modeling.

model has elements such as situation, relations, characteristic, calculation or restriction. Therefore, it is available to have semantic definition and assumption in the knowledge model. And through these understanding and application of tacit knowledge, it was possible (Dou 2014). To deduce implicated meaning in the information is a process of understanding the meaningful structure and relations of the knowledge. Therefore, to know how the given topics are located and have relations in the knowledge structure is very useful in the interpretation and evaluation. The existing knowledge management and knowledge analysis researches have limited all the knowledge as semantic search and classification or static relations between the data. This research structures a knowledge map around the user's topics and changes it to a knowledge network through frequency analysis and elimination of abundance. In particular, the knowledge analysis is available through course analysis among topics using approaches through the transition relations between knowledge by changing knowledge around topics to network model and visualizes them as groups.

3.2 *System integration*

In securing a competitive advantage, the advancement of knowledge management system is generally realized through real-time RTE system. The RTE deduces rapid response and innovation for environmental changes are based on the Sense & Response infrastructure, and it minimizes the responding time and restructuring time based on the evaluation reference index as shown in Figure 3. It is the infrastructure of organizational nervous network especially with activity automation and cost reduction, self-service, customer satisfaction, system cost reduction, fast detection, practice and innovation and restructuring. Such advanced integrated knowledge management system needs systematic saving and management of the organizational knowledge that requires safety, approach, security and diversity of media. In order to secure the safety, we established data base management system in the work activities unit for reliability of knowledge base, consistency of save knowledge, and elimination of abundance.

Moreover, in order to guarantee the accessibility, role models for all the users of knowledge were defined and a SSO (Single Sign On) was built, which allows the approaching to knowledge based on the situation model depending on the time and space. Also, the recommended model was defined to secure strategic value maintenance of the knowledge and quality preservation, and a Condition-Based Approach Control (CBAC) system was developed that the rights are dynamically controlled.

The diversity of media includes contents of management system which is available with the web editing and publishing to express in a diverse form including the test or image of the knowledge (Yasunaga & Yoon 2004). In consideration of the knowledge service classification in Figure 3, they are integrated as service combination methods based on SOA. The real-time knowledge analysis on the National Science Technology 577 strategy was conducted in relations to the real time service. Knowledge is collected in real time through crawling, and it is registered on the integrated knowledge base, then analyzed on certain topics and created the analysis for value 4. For the collection of knowledge, internal institutes, journals and press were classified, and they were saved as URL (40) and electronic documents (text series: 77, 873, 382). Total of 557 strategies underwent text analysis on 40 topics to create the analysis value to establish the strategy. Such values were further reflected on the definition of knowledge network model and the weight for similarity and relevance.

4 CONCLUSION

This research designs integrated knowledge management system between planning, which is the core process in the aspect of corporate management intelligence and its evaluation. And it builds a knowledge base and real-time knowledge operation system through the improved knowledge technology to explain the cases of analysis and its application. We have suggested extraction algorithm of semantic keyword from complex technical document. To do this, we have defined the process model of technology innovation, extracted semantic keyword and relationship property from technical document of each process. The extracted keywords and property are utilized to create and analyze knowledge model. Particularly, for the integration and analysis of knowledge model, the high abstraction level of topic-based knowledge models became hierarchy in this research, and it was automatically analyzed with the model available as network analysis, and the improved network analysis method was suggested.

REFERENCES

Chan, W. 2000. Temporal Logic Queries, *Proceedings of CAV 2000*, LNCS 1855, Springer.

Chun, S.S. 2015. Effective Extraction of State Invariant for Software Verification, *Applied Mechanics and Materials*, 752–753: 1097–1104.

Dou, W. 2014. *A Model-Driven Approach to Trace Checking of Temporal Properties with OCL*, Open Repository and Bibliography, S&T Centre, University of Luxembourg.

Heitmeyer, C. et al. 1998. Using abstraction and model checking to detect safety violations in requirements specifications, *IEEE Transactions on Software Engineering*, 24(11): 927–948.

Janssen, M. & Kuk, G. 2006. *A Complex adaptive system perspective of enterprise architecture in Electronic Government,* IEEE HICSS'06, 04: 1–10.

Leon, N.R. 2008. The future of computer aided innovation, *IFIP International Federation for Information Processing,* 277: 3–4, Springer.

Shah, H. & Kourdi, M. 2007. *Frameworks for Enterprise Architecture*, IEEE IT Pro, Sept. 36–41.

W3C, 2007, *Web Service Description Language (WSDL)*, Ver. 2.0 recommendation.

Wang, X. & Vitvar, T. 2007. *WSMO-PA: Formal Specification of Public administration service model on Semantic Web Service Ontology,* IEEE ICSS'07.

Yasunaga, Y. & Yoon, T. 2004. *Technology Roadmapping with Structuring Knowledge and Its Advantages in R&D Management*, IEEE IEMC'04, 581–585.

Yezersky, G. 2007. General Theory of Innovation, *IFIP International Federation for Information Processing*, 250: 45–55, Springer.

Advanced Materials, Mechanical and Structural Engineering – Hong, Seo & Moon (Eds)
© 2016 Taylor & Francis Group, London, ISBN: 978-1-138-02908-8

Electrochemical synthesis of lanthanum hexaboride ultrafine particles

L.V. Razumova, O.V. Tolochko & T.V. Larionova
Peter the Great St. Petersburg Polytechnic University, St. Petersburg, Russia

J.M. Ahn
Chungnam National University, Yuseong-Gu, Daejeon, Republic of Korea

ABSTRACT: Ultrafine powders of pure LaB_6 were prepared by the electrochemical synthesis using lantanium- and boron-containing compounds in halide melts. It is shown, that the following interdependent parameters: a total concentration of La and B bearing components, a ratio of La/B and a current density are determinant for the reaction product phase composition. The parameters led to pure LaB_6 synthesis were determined as following: (i) temperature is 700°C, duration is 60 min; (ii) total (La + B) concentration is 10^{-3} mol/cm^3 and La/B = 1/4; (iii) current density is above 0,6 A/cm^2. The synthesized particles predominantly had round shape with the mean size less than 100 nm. The particles size can be controlled by the current density and total concentration of La and B bearing components.

1 INTRODUCTION

Metals borides form a very large group of compounds that in general possess high electric and thermal conductivities, high hardness and heat resistance, as well as various functional properties (Samsonov et al. 1975, Takeda et al. 2004). Although lanthanum hexaboride has been known for many years as an effective electron and ion emitter because of low work function (Kerley et al. 1990, Boustani et al. 2001), new applications exploiting its unusual electron emission properties are still being investigated (Delley & Monnier 2004, Schelm et al. 2003). Nanostructured lanthanum hexaboride has been reported (Bao et al. 2010) to exhibit enhanced thermionic emission compared to that of coarse-grained LaB_6. Besides, nanocrystalline LaB_6 particles have been predicted to provide important advantages for infrared absorbers.

Synthesizing of LaB_6 has always been difficult and requires high temperature processing and often post-synthesis cleaning treatments. As mentioned in (Nishiyama et al. 2009) the preparation methods of LaB_6 powders can be roughly classified into the following five groups: direct reaction, borothermic reduction, carbothermic reduction, borocarbide reduction, and metallothermic reduction. These methods, however, are not suitable for nanocrystal production, as most of them require high temperatures—higher than 2300°C, and hence resulted in a formation of coarse particles.

Recently, a wide variety of technological routes has been employed to synthesize LaB_6 at lower temperatures (Wang et al. 2010, Selvan et al. 2008, Zhang et al. 2008). The main drawback for most of these methods is that they require considerable post-synthesis treatment to obtain pure LaB_6.

In (Shapoval & Malyshev 1999) an electrolysis from melted salts was suggested as a perspective way for production of the metal borides particles having nanometer dimensions. This method has been successfully applied in (Malyshev et al. 2010) for the preparation of molibdenum, tungsten and chromium borides.

In the present work we explore the possibility to utilize an electrochemical process for production of submicron LaB_6 particles at a comparatively low temperature without any post-synthesis treatments.

2 EXPERIMENTAL

Electrosynthesis of the LaB_6 compound was carried out from fused mixtures of NaCl–KCl–$LaCl_3$–$NaBF_4$, where $LaCl_3$ and $NaBF_4$ were added as sources of La and B, and mixture of NaCl and KCl—as a solvent. Volt-Ampere Characteristic (VAC) was measured on W electrode. Electrochemical synthesis of LaB_6 was performed under potentiostatic and galvanostatic conditions in a graphite crucible with W cathode. Experiments were fulfilled under argon atmosphere in order to protect the graphite crucible and tungsten electrode.

Scanning electron microscopic observation were fulfilled with SUPRA 55VP-25-78 microscope. Chemical composition was assessed by means of

Energy Dispersive X-ray Spectroscopy (EDS). The phases present in the samples were analyzed by X-ray diffractometer (DRON-2) using CuK-α monochromatic radiation. Phase identification utilized the ASTM database. The crystalline domain size of the samples was calculated using the Scherrer equation:

$$D = 0,9\ \lambda/\cos\theta\ (\beta - \beta_i), \quad (1)$$

where, D is the mean crystallite diameter, λ–X-ray wave length; β–Full Width at Half Maximum (FWHM) of XRD peak; β_i–instrumental FWHM; θ–diffraction angle.

3 RESULTS AND DISCUSSION

The optimal technological parameters were ascertained by varying voltage and current density, total concentration of La and B in solution, La/B atomic ratio, synthesis duration and temperature that ranged from 700°C to 1000°C.

As varying the voltage on the bath and current density has the greatest influence on the course of the experiment, first, the Volt-Ampere characteristics of the melt was studied. Figure 1 shows the three Volt-Ampere characteristic of i) NaCl-KCl itself, ii) with additions of $LaCl_3$, iii) and with simultaneous additions of $LaCl_3$ and $NaBF_4$.

Volt-Ampere curve of NaCl-KCl melt has a typical for ionic melt view. As $LaCl_3$ had been added a reduction wave appeared at potential of about −3.0 V. When $LaCl_3$ and $NaBF_4$ were added simultaneously there was distinctive increase in current within the range of potentials of (1.6–2.0) V observed, what is about 1.0 more positive than the reduction wave of NaCl-KCl-LaB_3 melt. So the reaction proceeds at potentials those are more positive than the individual reduction potentials.

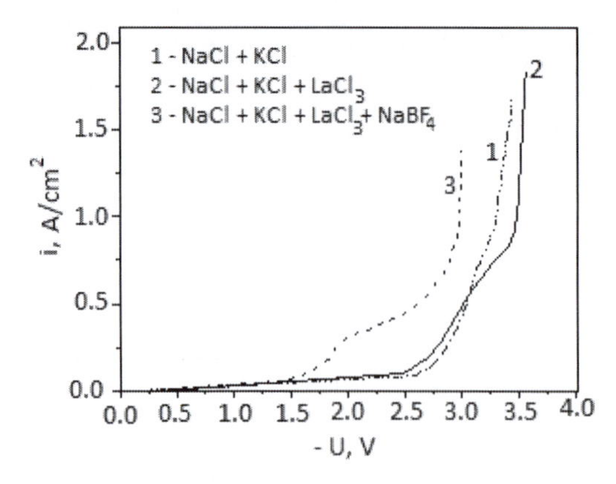

Figure 1. VAC of NaCl-KCl (1), NaCl-KCl-LaB_3 (2), NaCl-KCl-LaB_3-$NaBF_4$ (3) melts at 700°C.

After electrolysis we have got some dark powder, which deposited on the cathode as a mixture of salts, so called "cathode salt pear". The admixture salts contained in the "pear" are easily dissolved in the water. To remove the admixture the cathode salt pear was thoroughly washed in the distilled water and centrifuged. Figure 2 demonstrates the total scheme of the process.

Chemical analysis of the final product shows La, B and insignificant admixtures of oxygen and chlorine. Chemical analysis of the electrolyte after electrolysis showed that boron and lanthanum were extracted from the melt almost completely.

Identification of the reaction products by XRD revealed the presence of LaB_6, LaB_4, La_2O_3, LaOCl and $LaBO_3$ phase depending on the process conditions, except them there were KCl and NaCl occasionally observed. It was found that LaB_6 phase forms in the deposit product when current density exceeds 0.6 A/cm³ and there were only insignificant traces of other phases observed if current density exceeds 0.9 A/cm³. Some typical XRD patterns depending on current density are presented in Figure 3.

Investigations of the influence of source materials showed strong dependence of the main reaction products from La/B atomic ratio and total concentration of these components in the melt. The atomic ratio La/B varied as 1/1, 1/2, 1/4 and 1/6 by changing the concentration of $LaCl_3$ and

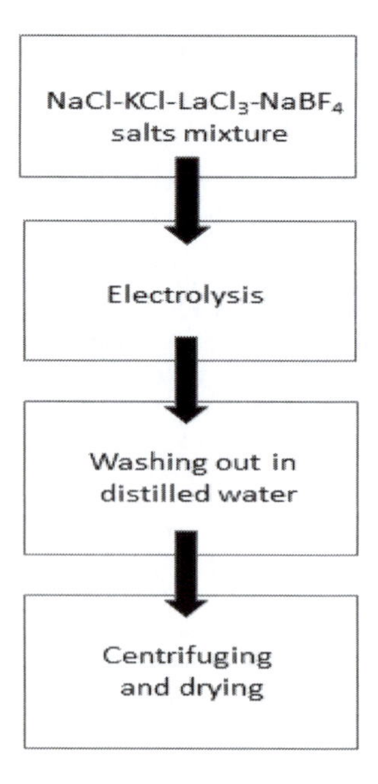

Figure 2. The process scheme utilized to produce lanthanum borides particles through electrochemical method.

$NaBF_4$ with (La+B) total concentration in the melt equaled to 10^{-3} mol/cm^3. Experiments at La/B = 1/1 showed the appearance of LaB_4.

Increasing of $NaBF_4$ concentration in the melt up to La/B = 1/2, leads to appearance of lanthanum hexaboride phase, however considerable amount of LaB_4 and La_2O_3 was also observed. Maximal quantity of LaB_6 phase appeared when La/B ratio was set as 1/4. At La/B = 1/6 some admixtures phases besides LaB_6 have been observed, identification of those did not succeed on the base of ASTM data and was not in the focus of this work. However, when the total (La + B) concentration had been decreased down to 10^{-4} mol/cm^3 the maximum of pure LaB_6 in the cathode product forms at La/B = 1/6. The phase composition depending on the La/B ratio at current density of 0.9 A/cm^3 presented in the Table 1.

Optimal duration for the synthesis was chosen as 60 minutes. Longer synthesis leads to the delamination of the cathode deposit in the salt melt, making difficult to collect the reaction product.

Investigations of the temperature influence on the reaction showed, that temperature of 700°C is optimal, because it is sufficiently high for synthesis to proceed, but rising the temperature leads to a decrease in the bath stability, as well as coarsening product particles.

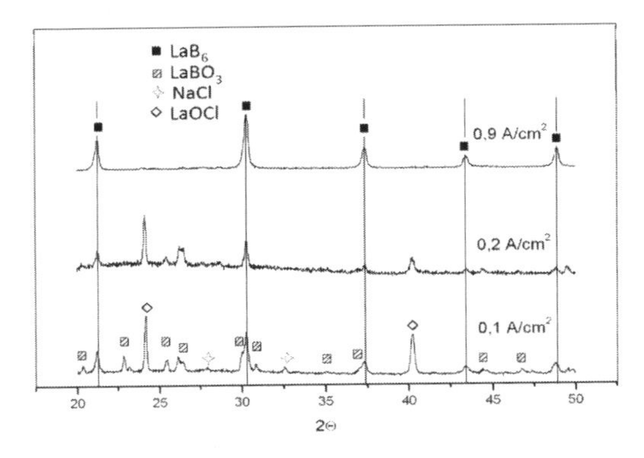

Figure 3. XRD patterns of cathode products synthesized at different current densities at melt temperature of 700°C.

Table 1. Phase composition in depending on La/B ratio and total (La+B) Concentration (C) at current density of 0.9 A/cm^3.

La:B	Phase composition	
	C = 10^{-3} mol/cm^3	C = 10^{-4} mol/cm^3
1:1	LaB_4, La_2O_3	LaB_4, La_2O_3
1:2	LaB_4, LaB_6, La_2O_3, $LaBO_3$	LaB_4, LaB_6, La_2O_3 $LaBO_3$
1:4	LaB_6	LaB_4, LaB_6
1:6	LaB_6, other borides	LaB_6

Figure 4 shows two different magnifications SEM microphotographs of the LaB_6 particles synthesized at total (La + B) concentration of 10^{-4} mol/cm^3 and current density of 0.6 A/cm^2. As revealed by SEM (Figure 4a) the product powder consists of fine agglomerates of size less than 1 mkm. The micrographs made at higher magnification (Figure 4b) reveals that the particles themselves have round shape and very narrow size distribution with average value of about 100 nm. At lower current density cubic shape particles have been also found.

With increasing the current density the particles size decreased. Measurements made using SEM microphotographs shows bigger particles size values compared with those calculated from the XRD results according to (1). Apparently, the particle has a crystalline core with amorphous shell.

(a)

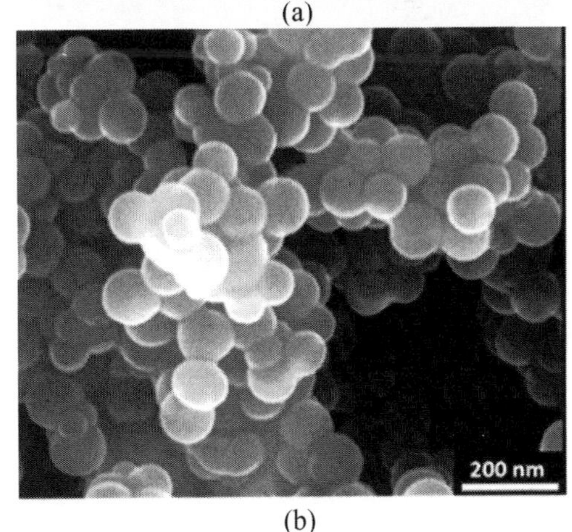

(b)

Figure 4. Selected micrograph of the product. Total La+B concentration is 10^{-4} mol/cm^3, current density is 0.6 A/cm^2 at different magnification: (a) X3000, (b) X60000.

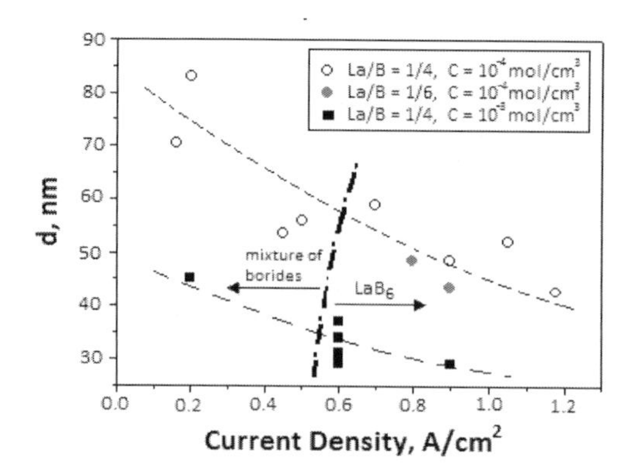

Figure 5. Dependence of the coherent dissipation area size on the current density at the different total (La+B) Concentration (C) and La/B ratio; the lines are provided as guides for eyes.

Thus, it was found that density above 0.6 A/cm^3 at La/B = 1/4 if total (La + B) = 10^{-3} mol/cm^3 or at La/B = 1/6 if total (La + B) = 10^{-4} mol/cm^3 leads to pure LaB_6 formation as the cathode deposit. The area of pure LaB_6 and a dependence of crystalline domain size on the current density and concentration are shown in Figure 5. The reaction running at higher current density results in synthesis of finer particles. An increase of total (La + B) concentration results in refinement of the powder product. Obviously, the nucleation process at the higher concentration prevailed over growth that leads to smaller particles formation.

4 CONCLUSIONS

Ultrafine LaB_6 powders were prepared by the electrochemical synthesis method using lantanium- and boron-containing compounds in halide melts.

Electrochemical synthesis parameters were determined as following: (i) temperature is 700°C, duration is 60 min; (ii) total (La + B) concentration is 10^{-4} mol/cm^3 at La/B = 1/6; or (La + B) concentration is 10^{-3} mol/cm^3 at La/B = 1/4 (iii) current density is above 0,6 A/cm^2. Powder particles had round or cubic shape with the mean size less than 100 nm. The particles size can be controlled by the current density and total concentration of the La and B bearing components.

REFERENCES

Bao, L.H., Zhang, J.X., Zhou, S.L. & Wei, Y.F. 2010. Preparation and characterization of grain size controlled LaB_6 polycrystalline cathode material. *Chinese Physics Letters*, 27: (107901) 1–4.

Boustani, I., Buenker, R., Gurin, V.N., Korsukova, M.M., Loginov, M.V. & Shrednik, V.N. 2001. Formation and stability of free charged lanthanum hexaboride clusters at field evaporation. *Journal of Chemical Physics*, 115(7): 3297–3307.

Delley, B. & Monnier, R. 2004. Properties of LaB_6 elucidated by density functional theory. *Physical Review B*, 70(19): 193403.

Kerley, E.L., Hanson, C.D. & Russell, D.H. 1990. Lanthanum hexaboride electron emitter for electron impact and electron-induced dissociation Fourier transform ion cyclotron resonance spectrometry. *Analytical Chemistry*, 62(4): 409–411.

Malyshev, V.V., Gab, A., Shakhnin, D., Popescu, A.M., Constantin, V. & Olteanu, M. 2010. High Temperature Electrochemical Synthesis of Molibdenum, Tungsten and Chromium Borides from Halide-Oxide Melts. *Romanian Journal of Chemistry,* 55(4): 233–238.

Nishiyama, K., Nakamur, T., Utsumi, S., Sakai, H. & Abe, M. 2009. Preparation of ultrafine boride powders by metallothermic reduction method. *Journal of Physics Conference Series,* 176(1): 137–145.

Samsonov, G.V., Serebryakova, T.I. & Neronov, V.A., 1975. *Borides*, Moscow: Metallurgiya.

Schelm, S. & Smith, G.B. 2003. Dilute LaB_6 nanoparticles in polymer as optimized clear solar control glazing. *Applied Physics Letters,* 82(24): 4346–4348.

Selvan, R.K., Genish, I., Perelshtein, I., Moreno, J.M.C. & Gedanken, A. 2008. Single step, low-temperature synthesis of submicron-sized rare earth hexaborides. *The Journal of Physical Chemistry C,* 112: 1795–1802.

Shapoval, V.I. & Malyshev, V.V. 1999. Electrochemical synthesis of titanium diboride in the dispersed state from halide melts. *Powder Metallurgy and Metal Ceramics,* 38(l): 1–12.

Takeda, M., Fukuda, T., Domingo, F. & Miura, T. 2004. Thermoelectric properties of some metal borides. *Journal of Solid State Chemistry*, 177(2): 471–475.

Wang, L.C., Xu, L.Q., Ju, Z.C. & Qian, Y.T. 2010. A versatile route for the convenient synthesis of rare-earth and alkaline-earth hexaborides at mild temperatures. *Crystengcomm*, 12: 3923–3928.

Zhang, M.F., Yuan, L., Wang, X.Q., Fan, H., Wang, X.Y., Wu, X.Y., Wang, H.Z. & Qian, Y.T. 2008. A low-temperature route for the synthesis of nanocrystalline LaB6. *Journal of Solid State Chemistry*, 181: 294–297.

Author index